C0-BRC-173

QUANTITATIVE MEASUREMENTS AND CHEMICAL EQUILIBRIA

Ernest H. Swift
CALIFORNIA INSTITUTE OF TECHNOLOGY

Eliot A. Butler
BRIGHAM YOUNG UNIVERSITY

W. H. FREEMAN AND COMPANY
San Francisco

A Series of Books in Chemistry

Linus Pauling and Harden M. McConnell, Editors

Printed in the United States of America

Library of Congress Catalog Number: 78–161009

International Standard Book Number: 0–7167–0170–7

1 2 3 4 5 6 7 8 9

Preface

Although use has been made in this text of much of the material in the senior author's *Introductory Quantitative Analysis*, this book represents more than a routine revision. It has been written in response to three trends that seem apparent in the current teaching of chemistry during the first two collegiate years. These are: an increasing emphasis on theory at the expense of experimental work, the disappearance of the qualitative analysis course, and an increasing use of elementary work in quantitative analysis in the freshman year. A brief review and assessment of these trends seems in order.

Thirty years ago the freshman course was largely descriptive in nature, with more minutiae than could possibly be correlated or absorbed, and there was need of a more theoretical approach. Linus Pauling's texts presented a much-needed change in emphasis. Today one may question if this change has not been carried to the point that the experimental method is falling into disrepute. Thirty years ago the qualitative course was usually given in the latter part of the freshman year and was subject to the criticism that pressure to cover too much ground in the laboratory left the student with an impression of a large number of sloppily done experiments accompanied by a mass of unrelated factual material. The authors realize the limitations and pressures responsible for the virtual disappearance of the qualitative course but view its demise with concern; and this concern is intensified by the lack of factual knowledge concerning inorganic chemistry shown by some graduate students.[1] Qualitative analysis, properly taught, can provide a systematized background of descriptive inorganic chemistry, and can show that simple, qualitative experiments can be designed to provide valuable information. The quantitative analysis course of thirty years ago was usually given in the sophomore year to students majoring in chemistry or related disciplines, but even for such students there was undue emphasis on methods for specific applications and the laboratory work was often a repetitive drill in a relatively few techniques.

Even earlier the senior author had been impressed by what now seems a pioneering project in freshman instruction initiated at Caltech by A. A. Noyes. This effort resulted in a small booklet first published in 1928 for use in the freshman laboratory entitled "An Introduction to the Chemistry of Solutions." The following comments are taken from the preface of that booklet:

> The preparation of this text has been made necessary by the fact that the idea underlying the course is quite different from those embodied in nearly all college text books of freshman chemistry. This idea is to substitute an intensive quantitative training in a few of the most important chemical topics for the usual informational survey of the subject, which commonly includes a larger mass of detailed phenomena and a greater variety of principles than can be assimilated and which is supplemented by a confusing multitude of minor, isolated laboratory experiments . . . in the laboratory work there is given at the start practice in the exact manipulation of volumetric analysis . . .

Use of the booklet was abandoned after Noyes' death, but the experiment bore fruit in 1956 when a major part of the quantitative analysis course that had been given in the sophomore year was moved into the freshman year, but with an emphasis on the objective stated by Noyes.

This text represents an effort by the authors to conform to Noyes' objectives by providing an intensive training in certain of the fundamental techniques of quantitative chemical measurements and in the application to these methods of the more important principles of chemical equilibria. In addition, by the use of simple qualitative or semiquantitative experiments and a strong emphasis on periodic relationships and on the properties and reactions of related elements, an effort has been made to minimize some of the tedium of the usual quantitative laboratory work and to regain some of the more constructive aspects of a qualitative course.

In most cases these simple experiments supplement the conventional quantitative determinations and can be done with such simple equipment as calibrated droppers or measuring pipets. They serve in some cases to

[1] See Derek Davenport. *J. Chem. Ed.*, **47**, 271 (1970).

provide preliminary training which expedites carrying out the quantitative procedure, in others to demonstrate the effect of various conditions on the equilibria involved, and in others to illustrate periodic relationships or the application of a reaction to constituents other than the one being determined.

In the earlier experiments in this text an effort has been made to state the manipulative directions in considerable detail. It is hoped, first, that this will help the overworked instructor, and, second, that it will enable the student to attain more quickly results that will provide him with some confidence in his ability. This detailed presentation is minimized in later experiments and the use of special problems or variants of the standard procedure is encouraged. Classical gravimetric and titrimetric methods have been used in these earlier experiments. They have been selected in part because of their historical interest and references have been given which in certain cases show the gradual development of such methods by successive workers. These methods also present thoroughly developed procedures for which standards of performance have been established by which both student and instructor can judge performance.

We have introduced in this text the use of gravimetric titrations made with small plastic wash bottles as burets. Based on experience in developing the method and on its use with students at Brigham Young University, the authors believe that this technique, when used with top-loading or single-pan balances, has distinct advantages over volumetric titrations for both student and professional use.

This text has been designed for use in the laboratory work of the freshman year with more professionally oriented courses, or in the sophomore year when an adequate training in quantitative measurements has not been obtained in the first year. It has been used in preliminary form in such courses at Brigham Young University.

The authors wish to acknowledge and express their appreciation for the cooperation and constructive criticism received from Professors K. P. Anderson, C. L. Cluff, N. K. Dalley, and J. R. Goates during the use of portions of this text in preliminary form at Brigham Young University. In addition, many valuable suggestions and corrections have been received from teaching assistants and students. Mr. Paul Steed prepared the index and gave helpful criticism during the preparation of the manuscript.

Liberal use has been made of material from *Introductory Quantitative Analysis* by the senior author and from *Qualitative Elemental Analysis* by the senior author and W. P. Schaefer. We are indebted to Dr. Schaefer and to many colleagues and students for their assistance in the development of this material.

Contents

List of Procedures and Experiments xi
Introduction xv

SECTION I GENERAL PRINCIPLES 1
CHAPTER 1 Quantitative Measurements and the History of Chemistry 3
2 Definition of Terms and Units 11
3 Evaluation of Experimental Data 22
4 Chemical Reactions. Types and Principles 38
5 Chemical Reactions. Precipitation 58
6 Chemical Reactions. Oxidation-Reduction 69
7 Chemical Reactions. Ionization 86
8 Chemical Separations. Periodic Properties of the Elements and Chemical Separations 116
9 Chemical Separations. Separations Involving Heterogeneous Systems 133

SECTION II EQUIPMENT AND TECHNIQUES 159
CHAPTER 10 Instructions Regarding Laboratory Equipment and Operations 161

SECTION III GRAVIMETRIC MEASUREMENTS AND METHODS 211
CHAPTER 11 Measurement of Mass 213
12 General Principles of Gravimetric Measurements 247
13 Operations of Gravimetric Measurements 274
14 Gravimetric Methods 284

SECTION IV TITRIMETRIC MEASUREMENTS AND METHODS 307

CHAPTER 15 Titrimetric Methods 309

16 Titrimetric Methods. Operations 317

17 Precipitation Methods 342

18 Oxidation-Reduction Reactions. Permanganate Methods 391

19 Oxidation-Reduction Reactions. Cerimetric Methods 432

20 Oxidation-Reduction Reactions. Iodometric Methods (I) 447

21 Oxidation-Reduction Reactions. Iodometric Methods (II) 466

22 Oxidation-Reduction Reactions. Elements Frequently Involved in Titrimetric Methods 489

23 Ionization Reactions. Acid-Base Equilibria 515

24 Ionization Reactions. Titrations with Monoprotic Acids and Monohydroxide Bases 537

25 Ionization Reactions. Titrations of Polyprotic Acids 566

26 Ionization Reactions. Total Cations by an Ion-Exchange Process 593

27 Ionization Reactions. Titrations Involving Complex Compounds 604

SECTION V ELECTRICAL AND OPTICAL METHODS 615

CHAPTER 28 Electrolytic Methods. Gravimetric 617

29 Electrolytic Methods. Coulometric 640

30 Optical Methods 656

APPENDIXES 681

APPENDIX 1 Ionization Constants of Selected Acids and Bases (at 25°C) 683

2 Dissociation Constants of Complex Ions 685

3 The Solubility-Product Values of Certain Slightly Soluble Compounds (at Room Temperature) 687

4 Selected Standard and Formal Half-Cell Potentials (at 25°C) 689

5 Species Formed by Certain Elements in Various Solutions 696

6 Densities of Water and Air at Laboratory Temperatures 701

7 Reagents and Chemicals 702

INDEX 708

Procedures and Experiments

A Procedure usually contains an Outline, a Discussion, Instructions, Notes—and occasionally contains one or more Experiments. (See page xvii regarding these terms.) Procedures are numbered serially within each chapter.
The titles of Experiments are *italicized* in the following table; an Experiment that is included in a Procedure is given the number of that Procedure.

CHAPTER 11 MEASUREMENT OF MASS 213

11-1 Determination of the Point of Rest of an Equal-Arm Balance 229
A. Method of Long Swings 231
B. Method of Short Swings 233

11-2 Determination of the Sensitivity of an Equal-Arm Balance 234

11-3 Weighing an Object on an Equal-Arm Balance 236
A. Method of Restoring the Original Point of Rest 237
B. Method of Using the Balance Sensitivity 238

11-4 Weighing an Object on a Single-Pan Balance 243

CHAPTER 14 GRAVIMETRIC METHODS 284

14-1 Gravimetric Determination of Chloride as Silver Chloride 285

14-2 Gravimetric Determination of Sulfate as Barium Sulfate 291

CHAPTER 16 TITRIMETRIC METHODS 317

16-1 Calibration of a Flask 321

16-2 Calibration of a Pipet 324

16-3 Calibration of a Volumetric Buret 328

16-4 Use of a Gravimetric Buret 335

CHAPTER 17 PRECIPITATION METHODS 342

17-1 Preparation of a Standard Solution of Silver Nitrate 351

17-2 Titration of Chloride with Silver Ion. Chromate as an Internal Indicator 356

17-2 *Reactions of Silver Compounds* 367
— Determination of the Silver in an Alloy— General Discussion (17-3, 17-4, 17-5) 371
17-3 Preparation of a Thiocyanate Solution 372
17-4 Standardization of a Thiocyanate Solution 374
A. Volumetric 380
B. Gravimetric 382
17-5 Determination of the Silver in an Alloy 383
17-5 *Titration of Mercuric Ion with Thiocyanate* 385

CHAPTER 18 OXIDATION-REDUCTION REACTIONS. PERMANGANATE METHODS 391
— *Stability of the Permanganate End Point* 398
— *Effect of Mn(II) on the Permanganate-Oxalate Reaction* 400
18-1 Preparation of a Permanganate Solution 402
18-2 Standardization of a Permanganate Solution. Arsenious Oxide as Primary Standard 406
18-2 *Reactions of Permanganate with Arsenious Acid* 413
18-3 Determination of the Iron in an Ore. Reduction by Stannous Chloride and Titration with Permanganate 417
18-3 *Titration of Fe(II) with Permanganate* 421

CHAPTER 19 OXIDATION-REDUCTION REACTIONS. CERIMETRIC METHODS 432
19-1 Preparation of Ceric Sulfate Solutions 436
19-2 Standardization of Ceric Sulfate Solutions Against Arsenious Oxide 439
19-3 Determination of the Iron in an Ore. Reduction with Stannous Chloride and Titration with Ceric Sulfate 442

CHAPTER 20 OXIDATION-REDUCTION REACTIONS. IODOMETRIC METHODS(I) 447
— *Reversibility of the Iodine-Arsenious Acid Reaction* 456

CHAPTER 21 OXIDATION-REDUCTION REACTIONS. IODOMETRIC METHODS(II) 466
21-1 Preparation of a Thiosulfate Solution 467
— Standardization of Thiosulfate Solutions (21-2, 21-3) 471
21-2 Standardization of Thiosulfate Solutions against Potassium Dichromate 472
21-3 Standardization of Thiosulfate Solutions against Potassium Iodate 475
21-4 Iodometric Determination of Copper in an Alloy 477

CHAPTER 22 OXIDATION-REDUCTION REACTIONS. ELEMENTS FREQUENTLY INVOLVED IN TITRIMETRIC METHODS 489

— *Reactions of Peroxide with Various Oxidation States of Manganese* 504

CHAPTER 24 IONIZATION REACTIONS. TITRATIONS WITH MONOPROTIC ACIDS AND MONOHYDROXIDE BASES 537

24-1 Preparation of Carbonate-Free Solutions of Sodium Hydroxide 539

24-2 Preparation of Hydrochloric Acid Solutions 541

24-3 Reaction of Strong Acids and Strong Bases. Determination of Titrimetric Ratios 542
A. The Use of Methyl Orange as Indicator 543

24-3A I. Titration of a Weak Base with a Strong Acid—Methyl Orange as Indicator 544
II. Titration of a Weak Acid with a Strong Base—Methyl Orange as Indicator 544

24-3 B. The Use of Phenolphthalein as Indicator 545

24-3B I. Titration of a Weak Acid with a Strong Base—Phenolphthalein as Indicator 547
II. Titration of a Weak Base with a Strong Acid—Phenolphthalein as Indicator 547

24-4 Standardization of a Sodium Hydroxide Solution against Potassium Hydrogen Phthalate 551

— *Use of KIO_3 as an Acidimetric Standard* 554

24-5 Standardization of a Hydrochloric Acid Solution against Sodium Carbonate 556

24-6 Determination of the Total Alkalinity of Carbonate-Hydroxide Materials 560

CHAPTER 25 IONIZATION REACTIONS. TITRATIONS OF POLYPROTIC ACIDS 566

25-1 Analysis of Phosphoric Acid Solutions 572

25-2 Analysis of Strong Acid-Phosphate-Hydroxide Solutions 575

25-3 Use of the Glass Electrode *p*H Meter. Determination of Titration Curves and Indicator Transition Ranges During Titrations of Polyprotic Acids 578

25-4 Titration of Amino Acids 581

CHAPTER 26 IONIZATION REACTIONS. TOTAL CATIONS BY AN ION-EXCHANGE PROCESS 593

26-1 Determination of Total Cations by Ion-Exchange 599

CHAPTER 27 IONIZATION REACTIONS. TITRATIONS INVOLVING COMPLEX COMPOUNDS 604

27-1 Preparation and Standardization of a Solution of EDTA 611

27-2 Determination of Nickel by Titration with EDTA 613

CHAPTER 28 ELECTROLYTIC METHODS. GRAVIMETRIC 617

28-1 Electrolytic Determination of Copper 635

CHAPTER 29 ELECTROLYTIC METHODS. COULOMETRIC 640

29-1 Titration of Arsenic with Electrolytically Generated Bromine 649

CHAPTER 30 OPTICAL METHODS 656

30-1 Spectrophotometric Determination of Manganese in Steel 673

Introduction

To the Student: The material in the following Introduction is intended to explain the objectives of this book and to aid you in organizing your approach to this course. The topics briefly summarized here are discussed at length later in the book. Read the explanations and carry out the instructions in this introduction, then proceed with your laboratory and class work as directed.

OBJECTIVES

This book has several objectives. The first is to provide you with a competence in applying quantitative measurements to chemical systems. The systems selected for use here are essentially those employed in quantitative chemical analysis, but the principles and techniques involved in the measurements are applicable to situations in other fields of science and engineering. A competence in applying these measurements implies that when presented with a new problem you can decide what measurements are required, apply them in an efficient manner, and then critically evaluate the reliability of your data—this last topic is treated in Chapter 3.

Secondly, these systems have been selected to illustrate the major types of chemical reactions and the principles applicable to them. The objective is to provide you with a competence in the use of these principles in the control of chemical reactions. A general treatment of these principles is given in Chapters 4 through 8.

Finally, in the selection of these systems an effort has been made to extend your background of descriptive inorganic chemistry and your appreciation of the periodic relationships determining the behavior of the elements and their compounds.

ORGANIZATION OF THIS BOOK

This book contains five sections. *Section I* contains a discussion of the application of physical measurements to chemical systems, definitions of terms and units used in this book, a discussion of the evaluation of experimental data, and, finally, a review of the principles that are the foundations of any chemical system. These principles are presented in general chemistry texts; where it seems appropriate, you are advised to review them. Consequently, although there will be some review, the treatment here will be based upon the assumption that you have some familiarity with these principles, and only topics directly related to your laboratory work will be covered.

Section II deals with the techniques of quantitative chemistry. It includes instructions regarding the preparation and use of equipment you will need in the laboratory and advises you on how best to keep your notebook. This section also provides you with detailed directions for your preliminary laboratory work.

Methods of constructing certain pieces of useful equipment of glass are described first. Then directions and suggestions are given for some of the more common operations of quantitative chemistry. The laboratory work of analytical chemistry, and especially of quantitative methods, requires not only an appreciation and use of scientific principles and factual material, but also a high degree of artistic skill. Some people are so fortunate as to easily acquire a facility for this type of work, while other have to acquire it by painstaking and thoughtful practice. *It is a pleasure to see the clean, orderly desk and the precise and efficient movements of one who has acquired this laboratory skill. One cringes to see the sloppy and ineffectual efforts of a careless or indifferent worker.*

Section III is concerned with gravimetric methods, *Section IV* with titrimetric methods, and *Section V* with methods involving electrical and optical measurements.

At the end of most chapters you will find a collection of Questions and Problems that are related to the preceding material.

PRELIMINARY INSTRUCTIONS REGARDING STUDY METHODS

The process of performing a quantitative chemical analysis for a single constituent has become known as a *determination*. In this book the material relating to a determination is called a Procedure and consists of four sections. The first section is a brief outline paragraph, which is written to assist and guide you through the subsequent sections, and to serve as a help in review. Second, there is usually a Discussion that treats the general principles and essential reactions involved. The third section consists of the Instructions, which contain the detailed working directions for performing the experimental work. And finally, there are Notes; these contain helpful suggestions as to techniques or interpretation of experimental observations. Occasionally, a Procedure will also contain one or more Experiments; these are intended to illustrate experimentally certain features of the determination.

Before coming to the laboratory to carry out any procedure, study the preceding Discussion until you understand clearly the general principles and the fundamental chemical reactions involved. Then read the Instructions until the purpose of each operation and the technique involved is clearly comprehended. *The laboratory work can be performed much more rapidly and effectively if the purpose of each operation is understood and the work carefully planned before entering the laboratory.* Before beginning the laboratory work of a procedure, prepare a list of the required apparatus or chemicals and obtain them. While carrying out the Instructions, read the Notes to which reference is made. Use time required for the cooling of precipitates or the evaporation of solutions to carry out duplicate determinations or to prepare for the next experiment.

After completing the procedure, study the material related to it contained under "Questions and Problems." Write out the answers to the questions or be prepared to present them in the classroom. Work such problems as your instructor may direct.

PRELIMINARY INSTRUCTIONS FOR LABORATORY WORK

Before beginning your laboratory work be sure that you have the proper equipment and that you can perform certain essential operations properly. Begin your first laboratory session as follows:

(*a*) Obtain a notebook. Read pages 162–163 regarding its use.

(*b*) Obtain a desk assignment, and check your equipment as directed on page 161.

(*c*) Have available or construct four stirring rods as directed on pages 165–

167. Note the suggestions regarding proper adjustment and use of gas burners on pages 163–165.

(*d*) Have available or construct at least two, preferably four, calibrated droppers as directed on pages 167–170.

(*e*) Have available or construct a wash bottle as directed on pages 171–175.

SECTION I

GENERAL PRINCIPLES

CHAPTER 1

Quantitative Measurements and the History of Chemistry

This book is concerned with the application of quantitative measurements of various kinds to chemical systems. It therefore seems appropriate to begin with a chapter reviewing the history of chemistry up to about the beginning of this century in order to note how closely the development of such quantitative measurements has paralleled the development of chemistry as a *science*.

Chemistry as a *technology* is of much more ancient origin, for there are pre-Hellenic records showing that people used natural products to color fabrics, to treat skins, and to make ceramic pottery and molten-silicate vessels; that they worked with various metals, including gold, silver, lead, mercury, and iron; that they performed distillations and evaporations—the evidence shows that they even performed filtrations.

It seems probable that the very first quantitative measurements were made of the volume of solutions and of the volume of particulate solids, such as grains, although we have evidence that quantitative measurements were also made of weight in earliest recorded history. Weights dating from 2600 B.C. have been found, and the principle of the equal-arm balance was known to early civilized man. Such measurements of volume and weight were made to serve commercial and technical needs.

The quality of articles of commerce was already subject to some control in this early period, and the modern fire assay for gold and silver comes to us almost unchanged in principle from the pre-Hellenic era. This technique required some means of weighing the metals, and it may have been the first application of a quantitative measurement to a chemical system. At the present time it seems unexpected to find that this analytical process, depending upon reactions at high temperatures (analysis in the "dry way"), should have preceded the now prevalent methods based upon reactions in solutions (analysis in the "wet way").

In this early period various philosophical systems were developed in attempts to explain the nature of things, but experimental science was to develop later.

Let us turn now to the age of the alchemists and their search for a means of converting so-called base metals into gold. The beginnings of alchemy date from around 300 A.D., but alchemy's most active period in Western civilization extended from the eleventh into the fifteenth century. Although alchemy is often looked back upon with condescension, we should remember that, based upon the knowledge available to the alchemists, there is no reason to consider their quest for gold as absurd. Why shouldn't they have expected to obtain gold from lead when they could obtain lead from litharge (PbO)?

Although the alchemists' objective was never attained, their efforts resulted in the discovery of many important reactions and the preparation of many important compounds. Viewing the historical record, we can conjecture that the period would have been much more fruitful except for the great secrecy with which much of the work was done. Among the better known alchemists were Albertus Magnus (1206–1280), Roger Bacon (1214–1294), and Raymund Lullus (1235–1315).

The predominance of alchemy ended in the iatro-chemical (*iatro*, Greek for physician) period, so called because emphasis shifted from the search for gold to a search for agents that would alleviate man's physical infirmities. One of the first and perhaps the best known of the iatro-chemists was Paracelsus (1493–1541).[1] He broke from the alchemists with the statement "the object of chemistry is not to make gold but to prepare medicines," and he attacked the physicians of the time by writing "You want to prescribe medicines and yet cannot make them: Chemistry solves for us the secrets of therapy, physiology, and pathology; without chemistry we are trudging in darkness."

Qualitative analysis in the wet way—that is, the *detection* of constituents by reactions taking place in solutions—had its early development in this period with the devising of tests that depended upon the formation of characteristic precipitates or colors. This development was the beginning of the interest in reactions taking place in solutions, that is, of modern solution chemistry. Near the end of the period Robert Boyle (1627–1691)

[1] There is some uncertainty about his name. He is said to have sometimes called himself Philippus Aureolus Theophrastus Paracelsus Bombast, ab Hohenheim.

devised qualitative tests for a number of substances, and was one of the first, if not the first, to use hydrogen sulfide as a precipitant; this reagent was later to become one of the most important precipitants for separating groups of elements in systems of qualitative analysis. Without taking note of it, Boyle also demonstrated the principles underlying quantitative *gravimetric* analysis; that is, methods by which the quantity of a constituent is determined by a weight measurement. He dissolved metallic silver in *aqua fortis* (HNO_3), added "sea salt" to the solution, washed and weighed the precipitate, and showed that the precipitate was heavier than the original metal; the stoichiometric relationship between the weight of his metal and that of his precipitate escaped him. Boyle was also one of the first to make quantitative measurements on gases and as a result, formulated the law that carries his name. Boyle's law states that the volume of a gas is inversely proportional to the confining pressure.

The next era in chemical history is known, because of the theory that dominated it, as the *phlogiston* period. The phlogiston theory was based on the belief that all combustible substances possessed a common component, phlogiston, which escaped in the process of combustion. For example, a piece of wood was composed of phlogiston and ash, and when burned the phlogiston escaped, leaving the ash. When metals "were calcined in the air" the ash that was obtained was evidence that the original metal was composed of phlogiston and the ash! In spite of what now seem to be obvious absurdities, the theory did explain many known facts at the time—perhaps there is a lesson here. Also, in spite of its obvious faults, the phlogiston theory did represent an advance in that it was the first effort to explain chemical phenomena from a unifying theoretical basis. It was during the phlogiston period, which roughly coincided with the seventeenth and eighteenth centuries, that the application of quantitative measurements to chemical systems began and that chemistry began to become an experimental science.

One of the first to make quantitative measurements was Joseph Black (1728–1799), who was able to use the apparatus devised by Boyle to show that carbon dioxide was different from air. Black thus opened the way for the discovery of oxygen, hydrogen, nitrogen, and other gases. It was the ability to make quantitative measurements of gas volumes that led to the realization that not all gases were air and to the eventual abandonment of the phlogiston theory. Much credit for this feat must be given to Lavoisier (1743–1794) who by quantitative measurements was able to show that there was no net change in weight during a chemical reaction. Although several others, among them Priestley (1733–1804) and Scheele (1742–1786), both ardent phlogistonists, had produced oxygen and realized that it had different properties from ordinary air, it was Lavoisier who recognized that it was a pure substance, an element, and that oxygen was involved in combustion. Lavoisier laid the foundations for the work of the next period, which was to firmly establish the difference between elements and compounds and to formulate the law of combining weights.

The century following Lavoisier's death, in 1794, has been called the Age of Quantitative Chemistry. It was the period when the equal-arm balance was developed into the remarkable laboratory instrument that it is today. It was the period in which Proust (1754–1826) stated his law of constant proportions, namely, "that different samples of a substance contain its elementary constituents in the same proportions". It was also the period in which Dalton (1766–1844) formulated his atomic theory and, following that, his law of multiple proportions, which stated that when two elements form more than one compound the ratios of the weights are small integers. It was also within a century of Lavoisier that Berzelius (1779–1848) realized that Dalton's atomic weights were in many cases not sufficiently accurate to prove the validity of the theory and embarked on a series of quantitative analyses that enabled him in 1814 to publish a more extensive and accurate table of atomic weights. Berzelius was an ingenious and able experimenter; the results of his efforts to automate the filtration process are shown in Figure 1-1. This

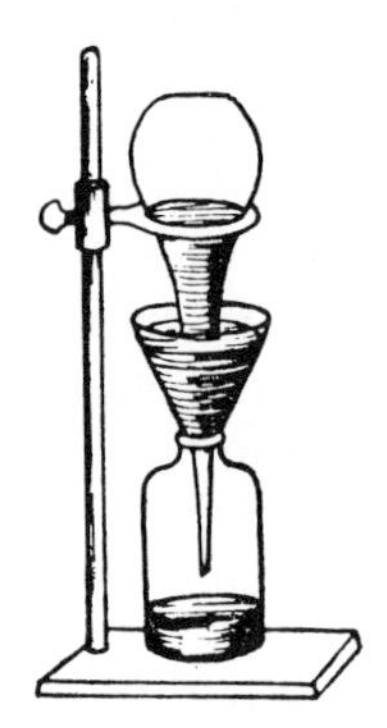
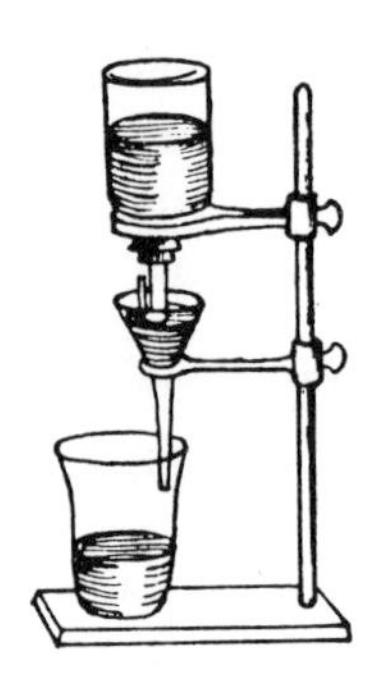
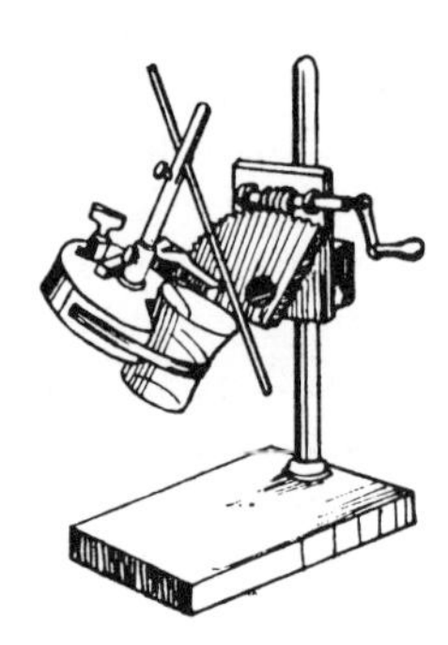

FIGURE 1-1
Automation of filtration by Berzelius (as shown in his *Lehrbuch der Chemie*, published in 1836).

was the century in which many of the present methods of both gravimetric and volumetric methods were developed, including, as you will see later, a surprising number of the procedures that are given in this book. When we consider their accuracy and realize that these procedures were developed entirely from empirical experimentation, without any knowledge of chemical equilibria or of the ionization theory, we develop an intense admiration for the chemists of the nineteenth century. Improvements in the accuracy of analysis of minerals led to the discovery of new elements and in improved accuracy in atomic-weight determinations. Finally, with a larger number of elements and more accurate atomic weights, relationships between the elements began to be apparent and eventually resulted in the periodic table.

It was during this period that titrimetric[2] methods of analysis were developed. Such methods are indirect in nature, since instead of converting the constituent of interest into some suitable form and weighing it directly, the quantity of a solution of a reagent required to react with the constituent is measured; this process is known as a *titration.* When the volume of the solution is measured the process is a *volumetric titration*; when the weight of the solution is measured the process is a *gravimetric titration.* Volumetric titrations have been much more extensively used because of the time required in making repeated weighings on the conventional equal-arm balance; it is probable that the recent development of more rapidly weighing balances may alter this situation. Gay-Lussac (1778–1850) is generally credited with initiating titrimetric methods when in 1835 he determined the silver in a solution by adding a standard sodium chloride solution until further addition caused no observable precipitate formation. This method, with some modification for observing precipitate formation, is one of the most accurate of titrimetric methods, has been used for atomic-weight determinations, and is still used in the mints of some governments.

The application of electrolytic processes to quantitative determinations serves to illustrate the lag that frequently occurs between the discovery of scientific phenomena and their application. Davy (1778–1829) had electrolyzed salts of sodium and potassium in 1808; by 1832, Faraday (1791–1867) had formulated his laws relating to the electrolysis of solutions. However, these results remained essentially unused analytically until 1864 when Wolcott Gibbs (1822–1908) made an *electrogravimetric* determination of the copper in a solution by electrolytically reducing and plating the copper on a weighed cathode consisting of a platinum crucible; after the electrolysis was completed he dried and reweighed the crucible. Electrogravimetry subsequently became a recognized analytical tool; examples of electrogravimetric methods are given in Chapter 28. Such methods were refined until early in the present century they were used to determine more accurately the value of the faraday—that is, the quantity of electricity required to produce one equivalent of chemical change. Again, in spite of the fact that this work demonstrated that one could use an electrolytic cell as a *coulometer*—that is, an instrument to measure the coulombs of electricity involved in an electrolysis—it was not until about 30 years later that this process was reversed, and the quantity of a substance was determined by measuring the

---

[2] The word is probably derived from the French word *titre*, which was used to indicate the purity of noble metals.

quantity of electricity involved in the electrolytic process. In 1938 two Hungarian chemists, instead of using a standard bromine solution as a titrant, electrolytically generated bromine in a solution and measured the quantity of bromine produced by means of a coulometer. They called the process a *coulometric* titration. An example of one type of a coulometric process is given in Chapter 29.

Quantitative measurements of other electrical properties of substances in solution were to lead to the theory of the ionization of electrolytes and the beginning of the modern period of physical chemistry. Clausius (1822–1888), a professor of physics, concluded as a result of measurements on the conduction of electricity through a solution that the positively and negatively charged parts of molecules were not fixed with regard to each other; Hittorf (1824–1914) and Kohlrausch (1840–1910) showed that these charged particles moved through the solution with differing mobilities; Kohlrausch later developed the alternating-current method of measuring the conductivity of solutions. Then Arrhenius (1859–1927) used such conductivity measurements in his doctoral thesis to obtain data that enabled him to conclude that the formation of the charged particles, the ions, occurred when an electrolyte was dissolved in the solution and not as the results of the passage of an electrical current. The publication of Arrhenius' doctoral thesis in 1883 was followed by the work of Ostwald (1853–1932), who applied the theory of mass action proposed in 1867 by Guldberg and Waage to Arrhenius' ionization theory and thus began the modern era of solution chemistry.

Quantitative measurements of the potentials at the electrodes of cells enabled Nernst (1864–1941) to test the equation that he had developed in 1889 from theoretical considerations. This equation established the relation between the potential at an electrode and the concentrations of the substances reacting at that electrode. The Nernst equation is discussed in Chapter 6 and is fundamental to any quantitative treatment of oxidation–reduction reactions. Conversely it is the basic principle involved in instruments for determining the concentrations of constituents in solution by potentiometric measurements. The modern glass-electrode pH meter is an example of such an instrument and is described in Chapter 23.

Spectral methods of analysis appear to have originated during the phlogiston period with the observation by Marggraf (1709–1782) in 1762 of the colors caused by sodium and potassium compounds in a flame. In 1882 John Herschel, the son of the astronomer, studied the spectra of the colors given to a flame by various compounds and stated "It is very probable that these colors originate from the molecules of the colored substances which after being transformed to the vapor state are in vigorous motion." He seems to have been misled by traces of sodium in his compounds with its intense yellow color into concluding that at certain temperatures all flames became yellow. In 1826 Talbot (1800–1877) studied flame spectra by passing the light through a slit and a prism and then on to a screen. He was able to distinguish between strontium and lithium and stated "I do not hesitate to state that by optical analysis the smallest amounts of these two substances can be distinguished at least as well, if not better, than by any other method." This was a prescient statement indeed for that time. However, it is doubtful that he attributed the spectra to the metal rather than its compounds. This step seems to have been made possible by the experiments of Ångström (1814–1874) and of Alter (1807–1881). The latter stated in 1854, "The spectrum emitted by an element differs from all others in its number of bands, intensity, and position, so that the element can be identified simply by observation. . ."

In 1859 Kirchhoff (1824–1887), a physicist, and Bunsen (1811–1899), a chemist, published the results of their measurements with an improved spectroscope and initiated modern developments in analytical spectroscopy. Their work was also an early demonstration of the value of interdisciplinary cooperation; their first report noted the discovery of a new element!

The process used by Kirchhoff and Bunsen is known as *emission spectroscopy* because it consists of providing enough energy to a material to volatilize its elements and excite their electrons from their normal states to higher energy states; when the electrons return to their normal states, they emit characteristic radiation. This process and the equipment it requires has been refined and developed until its availability is indispensible to a modern analytical laboratory. An emission spectroscopic analysis is especially valuable when dealing with an unknown material, since the presence or absence of most of the known elements can be ascertained in a very short time, and semiquantitative information regarding those present can be obtained.

To a certain extent the development of emission spectroscopy was paralleled by the development of absorption spectroscopy. In this process the light absorbed is measured rather than that emitted. The earliest developments in absorption spectroscopy were known as *colorimetry*, because the concentration of a colored substance in a solution was estimated by the intensity of the color of the solution, that is, by the amount of absorption of those wavelengths of visible light that were responsible for the color. In these earlier methods, begun around 1840, the intensity measurement was made by visual comparison with known solutions. The major developments in this process have taken place in this century with improvements in the optical systems which have extended the spectral measurements into the ultraviolet and infrared regions, and with more quantitative methods for measuring the light absorbed due to the substitution for the human eye of photoelectric measurements. Absorption measurements are capable of providing much more detailed information than emission methods regarding the arrangement of atoms in a molecule and have made significant contributions to the advances of both inorganic and organic chemistry during the past few decades. There is given in Chapter 30 a more detailed discussion of the principles and applications of absorption spectroscopy and an example of its use for the determination of the manganese in a steel.

Other spectral measurements, such as the rotation of polarized light by certain types of compounds, and changes in the refractive index of solutions, were developed during the last century but space is not available to trace their development and influence on the advance of chemistry. The same limitation applies to techniques developed in this century, such as mass spectrometry, x-ray diffraction techniques, and nuclear magnetic resonance, to mention only a few.

This brief historical sketch has shown a rather consistent pattern regarding the application of physical measurements to chemical systems. In general as the accuracy of a measurement has increased a better and more detailed understanding of the chemistry involved has been made possible. Frequently preliminary theories have had to be revised or discarded as the result of better or different data. In this text we will study the application of certain physical measurements to chemical systems. The apparent objective will be to make an accurate determination of the quantity of some constituent. We hope, however, that it will be obvious that these measurements have contributed to our present understanding of these systems and in all probability will contribute in the future to better and more complete understandings of such systems.

The History of Analytical Chemistry by Ferenc Szabadráry (New York: Pergamon, 1966) is an excellent source of detailed information on this topic.

CHAPTER 2

Definition of Terms and Units

Accuracy and brevity in the communication of scientific ideas and facts require that terms be rigorously defined and clearly understood. Unfortunately, there has not been general agreement as to the definition of all of the terms used in the chemical literature. For this reason, terms used in this book for expressing the quantities and the concentrations of chemical substances are defined below, beginning with certain metric units.

UNITS OF THE METRIC SYSTEM

Length

The metric unit of length is the *meter*, which is defined as 1,650,763.73 times the wavelength of the orange-red light of krypton 86. Formerly the meter was defined as the distance between two scratches on the standard platinum–iridium meter bar kept at the International Bureau of Weights and Measures in Paris. This bar, now a secondary standard, is still used as a convenient reference for comparison purposes. For all metric units, the prefixes *centi* and *milli* are used to designate 1/100 and 1/1000, respectively, of the unit to which they are applied. Thus the centimeter is 10^{-2} meter and the millimeter is 10^{-3} meter or 10^{-1} centimeter.

Mass

The metric unit of mass is the *gram*; it is one-thousandth the mass of the standard kilogram. The standard kilogram is a platinum–iridium weight, which also is kept in Paris. When this standard kilogram was prepared the intention was that it have the same mass as 1000 cm^3 of water at its maximum density, 3.98°C. Later measurements indicate, however, that the standard kilogram actually weighs about 28 mg more than was intended.

Volume and Capacity

The metric unit of volume, which can be derived from the metric unit of length, is the *cubic centimeter*, abbreviated cc or cm^3. Until 1964 the liter was the unit of capacity and was defined as the volume occupied by a kilogram of water at its maximum density; the liter differed, therefore, from 1000 cm^3 by 28 parts in a million. This discrepancy was removed in 1964 by the Twelfth General Conference on Weights and Measurements in Paris, France, which defined the liter as 1000 cm^3 and advised that the unit only be used in expressing volumes of gases and liquids.

UNITS OF CHEMICAL QUANTITY

Atomic-Weight Scale

The *atomic-weight scale* now in use is a relative system based on the assignment to the isotope ^{12}C of the weight 12.0000. Until 1961 two atomic-weight scales were in common use, one based upon naturally occurring oxygen (the chemical scale) and the other based upon the isotope ^{16}O (the physical scale). The unified scale of atomic weights now used removes problems of physical versus chemical atomic weights.

Moles

The *mole* is defined as Avogadro's number of specified units[1] and in this context is free from connotation of molecules. One can speak of a mole of hydrogen atoms (H), a mole of hydrogen ions (H^+), or a mole of hydrogen

[1] Historically *mole* was used as a shortened name for gram-molecular-weight, that is, the molecular weight expressed in grams. This older definition restricts the usefulness of the term and it forces the use of awkward names such as gram-formula-weight, gram-atomic-weight, and gram-ionic-weight, depending upon the nature of the substance being considered. Notice that the definition given above permits *mole* to serve equally well for atoms, ions, formula groups, and molecules. The above definition of the *mole* has been discussed by W. F. Kieffer (*The Mole Concept in Chemistry*. New York: Reinhold, 1962).

molecules (H_2), and in each case clearly mean 6.023×10^{23} units of the kind specified. It should be evident that a mole of H_2 molecules weighs 2.01594 g and a mole of H atoms weighs only half as much.

Atomic Weights

The *atomic weight* is the weight of a mole of the atoms of an element. The atomic weight of aluminum is 26.9815 g. This number written correctly with dimensions is 26.9815 g/mole. (The atomic weight has another set of dimensions—atomic-mass units/atom—but this set has its usefulness elsewhere.)

Molecular Weights

The term *molecular weight* is used for a substance that exists as molecules, and is defined as the weight in grams of one mole of the substance. The molecular weight of I_2 is 253.8 g; the number with dimensions is 253.8 g/mole.

Formula Weights

When a substance does not exist as discrete molecules or when there is doubt concerning its specific extent of aggregation, the term *formula weight* is used. NaCl is ionic in the solid state as well as in solution; the formula weight of NaCl is 58.443 g. Again the number with dimensions is 58.443 g/mole.

Equivalents

Those weights of different substances that react with each other are said to be equivalent to one another. Different approaches must be made in determining this equivalent relationship depending upon whether or not there is oxidation and reduction. Reactions in which there is no oxidation and reduction are called *metathetical* reactions (from the Greek word *metathesis*—to place differently). In such reactions one mole of hydrogen atoms (1.00797 g) is adopted as the standard of reference, so one *equivalent* of a substance is defined as the weight of it which reacts with one mole of hydrogen atoms in any of its compounds. In the reaction

$$Ag_2SO_4 + 2HCl = 2AgCl(s) + H_2SO_4$$

one mole of Ag_2SO_4 consists of two equivalents, since one mole reacts with two moles of HCl. The equivalent weight (weight of an equivalent) of Ag_2SO_4 is

$$311.8 \frac{\text{g}}{\text{mole}} \times \frac{1 \text{ mole}}{2 \text{ equivalents}} = 155.9 \frac{\text{g}}{\text{equivalent}}$$

One mole of $FeCl_3$ is three equivalents in the reaction

$$FeCl_3 + 3AgNO_3 = 3AgCl(s) + Fe(NO_3)_3$$

since one mole reacts with three moles of $AgNO_3$, and these would react with three moles of HCl. If the reaction does not include hydrogen, the extension is made to a hydrogen compound as was done here.

In acid–base reactions it is sometimes the case that not all of the available hydrogen atoms react. In such a case the number of equivalents per mole is determined by the number of reacting hydrogens. If one mole of sodium hydroxide is added to one mole of phosphoric acid, the reaction is

$$NaOH + H_3PO_4 = H_2O + NaH_2PO_4$$

and NaOH has one equivalent per mole because one mole reacts with one hydrogen of the H_3PO_4. Similarly H_3PO_4 has one equivalent per mole because one hydrogen per molecule reacts (or, additionally, because one mole reacts with one mole of NaOH, which is equivalent to one mole of hydrogen). However, if two moles of NaOH and one mole of H_3PO_4 are mixed, the reaction is

$$2NaOH + H_3PO_4 = 2H_2O + Na_2HPO_4$$

Here the H_3PO_4 is seen to have two equivalents per mole.

In *oxidation–reduction* (*redox*) reactions a different approach must be made in order to determine the number of equivalents per mole. In the reaction

$$Zn(s) + 2HCl = ZnCl_2 + H_2(g)$$

it is apparent that two moles of hydrogen atoms have reacted with one mole of zinc. Here the hydrogen atoms have been changed from the $+1$ to the 0 oxidation state. In redox reactions a unit change in oxidation state of one mole of a substance is defined as one equivalent. Therefore the quantity of a substance that reacts with one mole of hydrogen atoms in either a metathetical or redox reaction is one equivalent. Again, in the reaction

$$FeCl_3 + HI = FeCl_2 + \tfrac{1}{2}I_2 + HCl$$

the $FeCl_3$ has one equivalent per mole. Here the ferric chloride is reduced to ferrous chloride; the oxidation-state change of iron is from $+3$ to $+2$. It should be apparent that the number of equivalents per mole and the number of grams per equivalent can be determined only if the reaction involved is definitely stated. If there is more than one type of atom of the element being oxidized or reduced in the compound this must be taken into account. In the oxidation of thiosulfate ion by iodine,

$$2S_2O_3^{2-} + I_2 = S_4O_6^{2-} + 2I^-$$

the sulfur changes oxidation state from +2 to + $2\frac{1}{2}$. Since there are two sulfur atoms in the ion, there is one equivalent per mole of thiosulfate ion.

We emphasize again that one cannot specify the equivalent weight of a substance unless the reaction involved is likewise specified.

UNITS OF CONCENTRATION

Concentrations can be and are expressed in many different units. We will use only a few of these units and will define explicitly the concentrations to be used.

Volume Concentrations

For many years in analytical chemistry interest has been centered on units of volume concentration, which state the quantities of a solute per liter of solution. Until the recent development of high quality single-pan balances, weight measurements were so much more time consuming than volume measurements that volume concentrations were almost universally used in analytical work. *Volume-molar* and *volume-formal* (commonly abbreviated to *molar* and *formal*) concentrations both state the number of moles of solute per liter of solution. The formal concentration specifies the substance added to prepare the solution without reference to the molecular or ionic species in which it may exist in the solution. The molar concentration refers to the concentration of the particular molecule or ion being considered. For example, a solution that contains 12 g of sodium hydrogen sulfate, $NaHSO_4$, per liter is 0.1 formal (0.1 F) in $NaHSO_4$. This salt is completely ionized to sodium and hydrogen sulfate ions so the sodium-ion concentration is 0.1 molar (0.1 M). However, the hydrogen sulfate ion, HSO_4^-, is partly ionized to hydrogen and sulfate ions, and from measurements which have been made the solution is 0.034 M in SO_4^{2-} and H^+ and 0.066 M in HSO_4^-.

Square brackets are used to denote molar concentrations. The molar concentrations of the 0.1 F $NaHSO_4$ solution just considered are

$$[Na^+] = 0.1\ M$$

$$[H^+] = 0.034\ M$$

$$[SO_4^{2-}] = 0.034\ M$$

$$[HSO_4^-] = 0.066\ M$$

Volume-normal (commonly abbreviated *normal*) concentrations state the number of equivalents of solute per liter of solution. As has been pointed out, the equivalent weight that is assigned to a substance depends upon the reaction in which it is involved. Therefore confusion may result unless the reaction for which the solution is to be used is clearly understood. Thus objections have been raised to the use of this unit; however, in stoichiometric calculations the use of equivalents and of normal concentrations is very convenient.

Normal concentrations will be used frequently in the text where stoichiometric considerations are foremost. Molar concentrations will be used in equilibrium considerations where it is desired to express the concentration of some particular species of ion or molecule as it exists in the solution.

Weight Concentrations

Concentrations based upon the weights (masses) of solute and solvent are of considerable usefulness in certain measurements of physical chemistry. For gravimetric titrations we will use a system of weight concentrations based upon the quantity of solute per kilogram of solution.[2] The special advantages of weight concentrations are (1) the concentration is not a function of temperature, since only masses are involved, and (2) the inherent accuracy of weighings makes it possible to determine the concentrations very accurately. The *weight-molarity* and the *weight-formality* state the number of moles of solute per kilogram of solution. Again the distinction is that the weight-molarity refers to the concentration of the particular molecule or ion being considered whereas the weight-formality refers to the substance added without reference to its state of ionization or aggregation. *Weight-normality* states the number of equivalents of solute per kilogram of solution.

If 10.00 g of KIO_3 are dissolved in water and the solution so prepared weighs 952.3 g, the weight-formality is 0.0490_7 *WF*. The potassium-ion and iodate-ion concentrations are 0.0490_7 *WM*, and if the solution is to be used in a reaction in which the iodate is reduced to iodide, I^-, the solution is 0.294_4 *WN*.

As mentioned earlier, to be completely consistent, the volume concentrations should be expressed as *volume-formal*, *volume-molar*, and *volume-normal*. We will yield, however, to tradition; the terms *formal*, *molar*, and *normal*, without prefix mean the volume concentrations. Where the weight concentration is used the prefix *weight* will be given explicitly.

Table 2-1 contains a list of these concentration units, their symbols, and their definitions.

---

[2] Note that this is different from the *molal* concentration, which is defined as the number of moles of solute per kilogram of *solvent*. The molal concentration is especially convenient in physico-chemical measurement of the colligative properties of substances; in freezing-point depression; in boiling-point elevation; in vapor-pressure lowering; and in the measurement of osmotic pressure.

TABLE 2-1
Volume and Weight Concentrations

Designation	Abbreviation or Symbol	Definition and Dimensions
	VOLUME CONCENTRATIONS	
Volume-formal or formal; formality	*VF* or *F*[1]	Moles of solute per liter of solution (moles/l)[2]
Volume-molar or molar; molarity	*VM* or *M*[1]	Moles of solute per liter of solution (moles/l)
Volume-normal or normal; normality	*VN* or *N*[1]	Equivalents of solute per liter of solution (eq/l)
	WEIGHT CONCENTRATIONS	
Weight-formal; weight-formality	*WF*	Moles of solute per kg of solution (moles/kg)[2]
Weight-molar; weight-molarity	*WM*	Moles of solute per kg of solution (moles/kg)
Weight-normal; weight-normality	*WN*	Equivalents of solute per kg of solution (eq/kg)
Molal; molality	*m*	Moles of solute per kg of solvent

1. The shorter form (*F*, *M*, *N*) is used in this text for volume concentrations except where confusion with the weight concentration might result.
2. Formal concentrations are used to express total concentration of a substance without reference to its state of association or dissociation. Molar concentrations give the concentration of the stated species.

Standard Solutions

In titrimetric analysis much use is made of standard solutions. These are solutions that have accurately known concentrations. If such a standard solution has a known volume concentration, it is called a *volumetric standard solution*, since it can be used as the titrant in a volumetric titration. If a standard solution has a known weight concentration it is called a *gravimetric standard solution*, since it can be used as the titrant in a gravimetric titration. The terminology of titrations is defined in more detail in Chapter 17.

DIMENSIONS AND THEIR USE

Calculations in the physical sciences are much more easily made if one is consistent in the use of the dimensions or units. We will show all sample calculations in this text by a dimensional approach. The next short sections give a list of the more common abbreviations used in the book and examples of the use of the dimensional approach.

Abbreviations

Shown below are many of the units commonly used in this book along with their abbreviations.

meter, **m**; centimeter, **cm**; millimeter, **mm**
gram, **g**; kilogram, **kg**; milligram, **mg**
liter, **l**; milliliter, **ml**; cubic centimeter, **cc** or **cm**^{3}
mole, **mole**; millimole, **mmole**; (note that *m* is *not* an acceptable abbreviation for mole—it serves for meter and milli)
equivalent, **eq**; milliequivalent, **meq**

A Note on Calculations

The magnitudes of the weights and volumes used in analytical chemistry are such that milligrams, millimoles, and milliliters are often more convenient to work with than are grams, moles, and liters. Thus a titration may require 0.02723 liter or 27.23 ml of a 0.1045 N solution of $AgNO_3$. One can calculate the number of equivalents:

$$0.02723\ \text{l} \times 0.1045\ \text{eq/l} = 0.002846\ \text{eq}$$

or the number of milliequivalents:

$$27.23\ \text{ml} \times 0.1045\ \text{meq/ml} = 2.846\ \text{meq}$$

Notice that the number of equivalents per liter is the same as the number of milliequivalents per milliliter, since numerator and denominator are multiplied by the same factor. Similarly the molecular weight has the dimensions grams per mole or milligrams per millimole. For I_2 the molecular weight is

$$253.8\ \text{g/mole} \qquad \text{or} \qquad 253.8\ \text{mg/mmole}$$

The careful and consistent use of dimensions in calculations will keep you from many errors.

QUESTIONS AND PROBLEMS

1. Write a single expression complete with dimensions for the formal concentration of a sodium chloride solution that consists of A g of NaCl dissolved in water and diluted to B ml.

2. In the preceding problem, if A is 0.207 g and B is 996 ml, calculate the formal concentration of the sodium chloride solution using dimensions and keeping the appropriate accuracy in the answer. *Ans.* 0.00356 *F*.

3. One g ($\pm$0.01 g) of each of the following compounds is taken to prepare a liter of solution:

$AgNO_3$	$KSCN$	$BaCl_2$
$BaCl_2 \cdot 2H_2O$	Br_2	$K_4Fe(CN)_6 \cdot 3H_2O$
$Na_3AsO_4 \cdot 12H_2O$	$KMnO_4$	$K_2Cr_2O_7$
$Na_2S_2O_3 \cdot 5H_2O$		

(*a*) Calculate the formal concentration of each substance in the resulting solutions.
(*b*) Assuming that each of the salts is completely ionized, calculate the molar concentration of the anion and cation in each case. (Neglect the hydrolysis of certain of these ions.)

4. Calculate the formal concentration of H_2O in pure water at 20°C. *Ans.* 55.4 *F*.

5. One-tenth formal solutions of the overlined compounds shown below are prepared. State the normality of each of the solutions when used in the reaction indicated.

(*a*) $\mathrm{AgNO_3 + \overline{KIO_3} = AgIO_3(s) + KNO_3}$
(*b*) $\mathrm{3AgNO_3 + \overline{FeCl_3} = 3AgCl(s) + Fe(NO_3)_3}$
(*c*) $\mathrm{AgNO_3 + \overline{KSCN} = AgSCN(s) + KNO_3}$
(*d*) $\mathrm{BaCl_2 + \overline{K_2CrO_4} = BaCrO_4(s) + 2KCl}$
(*e*) $\mathrm{\overline{K_2Cr_2O_7} + 2BaCl_2 + H_2O = 2BaCrO_4(s) + 2HCl + 2KCl}$
(*f*) $\mathrm{\overline{K_2CrO_4} + \overline{3FeCl_2} + 8HCl = CrCl_3 + 3FeCl_3 + 2KCl + 4H_2O}$
(*g*) $\mathrm{\overline{2FeCl_3} + \overline{SnCl_2} = 2FeCl_2 + SnCl_4}$
(*h*) $\mathrm{\overline{10FeSO_4} + \overline{2KMnO_4} + 8H_2SO_4 = 5Fe_2(SO_4)_3 + K_2SO_4 + 2MnSO_4 + 8H_2O}$
(*i*) $\mathrm{\overline{2KMnO_4} + \overline{3MnSO_4} + 2H_2O = 5MnO_2(s) + K_2SO_4 + 2H_2SO_4}$
(*j*) $\mathrm{\overline{10KSCN} + \overline{12KMnO_4} + 13H_2SO_4 = 12MnSO_4 + 11K_2SO_4 + 10HCN + 8H_2O}$
(*k*) $\mathrm{\overline{KIO_3} + \overline{3SnCl_2} + 6HCl = KI + 3SnCl_4 + 3H_2O}$
(*l*) $\mathrm{\overline{KIO_3} + \overline{5KI} + 6HCl = 3I_2 + 3H_2O + 6KCl}$
(*m*) $\mathrm{\overline{KIO_3} + \overline{2KI} + 6HCl = 3ICl + 3KCl + 3H_2O}$
(*n*) $\mathrm{\overline{2Na_2S_2O_3} + \overline{I_2} = Na_2S_4O_6 + 2NaI}$
(*o*) $\mathrm{\overline{H_2SO_4} + \overline{Na_2CO_3} = Na_2SO_4 + H_2CO_3}$
(*p*) $\mathrm{H_2SO_4 + \overline{2Na_2CO_3} = 2NaHCO_3 + Na_2SO_4}$
(*q*) $\mathrm{\overline{H_3PO_4} + NaOH = NaH_2PO_4 + H_2O}$
(*r*) $\mathrm{\overline{H_3PO_4} + 2KOH = K_2HPO_4 + 2H_2O}$
(*s*) $\mathrm{\overline{H_2C_2O_4} + 2NaOH = Na_2C_2O_4 + 2H_2O}$

6. (*a*) What weight of anhydrous H_2SO_4 should be taken to prepare 1.0 liter of a 0.10 *F* solution? *Ans.* 9.8 g.
   (*b*) Can you state the normality of the solution?

7. Calculate the number of moles present in

   (*a*) 100 g of Fe metal
   (*b*) 100 g of Mg metal
   (*c*) 50 g of O_2 gas
   (*d*) 50 g of C_2H_4 gas
   (*e*) 50 g of $Na_2HPO_4 \cdot 12H_2O$ crystals

8. Calculate the equivalent weight of the overlined substance in each of the following stoichiometric equations:

   (*a*) $\overline{Mg(s)} + H_2SO_4 = MgSO_4 + H_2(g)$
   (*b*) $\overline{CaCl_2} + Na_2CO_3 = CaCO_3 + 2NaCl$
   (*c*) $2Fe(NO_3)_3 + \overline{3Na_2S} = Fe_2S_3(s) + 6NaNO_3$
   (*d*) $2NaOH + \overline{H_2C_2O_4} = Na_2C_2O_4 + 2H_2O$
   (*e*) $\overline{TiCl_3} + FeCl_3 = TiCl_4 + FeCl_2$
   (*f*) $\overline{K_2CrO_4} + BaCl_2 = BaCrO_4(s) + 2KCl$
   (*g*) $\overline{Na_2Cr_2O_7} + 2Pb(NO_3)_2 + H_2O = 2PbCrO_4(s) + 2NaNO_3 + 2HNO_3$
   (*h*) $\overline{K_2CrO_4} + 3FeCl_2 + 8HCl = CrCl_3 + 3FeCl_3 + 2KCl + 4H_2O$
   (*i*) $\overline{2K_2CrO_4} + 9KI + 16HCl = 2CrCl_3 + 3KI_3 + 8H_2O + 10KCl$
   (*j*) $\overline{K_2CrO_4} + 2Zn(s) + 8HCl = CrCl_2 + 2ZnCl_2 + 2KCl + 4H_2O$
   (*k*) $2FeCl_3 + \overline{2KI} = 2FeCl_2 + I_2 + 2KCl$

9. The concentration of a commercially available formic acid (HCOOH, formula weight 46.03) is 88% by weight. The solution has a specific gravity of 1.2. Calculate what volume of the acid should be taken to prepare 1 liter of 6 *F* reagent. *Ans.* 261 ml.

10. A phosphoric acid solution contains 85% by weight H_3PO_4 and has a specific gravity of 1.7. Calculate the normality of the solution when titrated with NaOH to Na_2HPO_4. *Ans.* 29.6 *N*.

11. Solutions were prepared by dissolving 5.00 g of each of the following compounds in water and diluting to 1 liter:

KIO_3	$Fe(NO_3)_3$	K_2CrO_4
$BaCl_2$	$K_2Cr_2O_7$	$Al_2(SO_4)_3$
$K_3Fe(CN)_6$	Na_3AsO_4	$NH_4Cl \cdot MgCl_2 \cdot 6H_2O$
KI_3		

   (*a*) Calculate the formal concentration of each compound.
   (*b*) Assume that each salt is completely ionized, neglect any hydrolysis reactions, and calculate the molar concentration of each resulting anion and cation.

12. The accompanying data relate to the weight-percentage composition and specific gravity of some of the concentrated acids of commerce.

Acid	Specific Gravity 20°/4°	Weight Percent
H_2SO_4	1.83	95
HNO_3	1.41	69
HCl	1.18	36
H_3PO_4	1.75	90

Calculate (*a*) the formal concentrations of these acids and (*b*) the normal concentration of each of these acids when it is used in a neutralization titration that results in the formation of the neutral salt.

13. Calculate the volume of each of these acids required to prepare 1 liter of a 6 *F* solution. What would be the normal concentration of each of these solutions? (Assume that the acid is converted to the neutral salt.)

14. Five g (± 0.01) of each of the overlined compounds shown in the over-all equations below were dissolved in 995 g of water. Calculate for each the weight formality of the resulting solution. State for each the weight normality of the solution when involved in the reaction shown. Explain your reasoning.

(*a*) $\overline{Pb(NO_3)_2} + Na_2SO_4 = PbSO_4(s) + 2NaNO_3$

(*b*) $Pb(NO_3)_2 + \overline{NaHSO_4} = PbSO_4(s) + HNO_3 + NaNO_3$

(*c*) $PbO_2 + \overline{Na_2SO_3} + H_2SO_4 = PbSO_4(s) + Na_2SO_4 + H_2O$

(*d*) $\overline{CuBr_2} + 2AgNO_3 = 2AgBr(s) + Cu(NO_3)_2$

(*e*) $\overline{2CuBr_2} + 4KI = 2CuI(s) + I_2 + 4KBr$

(*f*) $\overline{2CuBr_2} + 3KI + 2KSCN = 2CuSCN(s) + KI_3 + 4KBr$

(*g*) $\overline{K_2Cr_2O_7} + 2Pb(NO_3)_2 + H_2O = 2PbCrO_4(s) + 2KNO_3 + 2HNO_3$

(*h*) $\overline{K_2Cr_2O_7} + 6FeSO_4 + 7H_2SO_4 = Cr_2(SO_4)_3 + 3Fe_2(SO_4)_3 + K_2SO_4 + 7H_2O$

(*i*) $\overline{K_2Cr_2O_7} + 4Zn(s) + 14HCl = 2CrCl_2 + 4ZnCl_2 + 2KCl + 7H_2O$

CHAPTER 3

Evaluation of Experimental Data

In this chapter, a discussion is given of the validity of numbers when applied to measurements, of the accuracy and precision of measurements, types of errors, and criteria for the rejection of doubtful measurements.

In the laboratory an experimenter obtains data from experiments of many kinds, and from these data he attempts to draw conclusions. The data may come from his estimation of the position of a meniscus, his reading of a balance, or his measurement of time or temperature. There are many factors that affect every measurement that is finally recorded in his laboratory notebook. There may have been dust on the balance pan or fingerprints on the object being weighed. Parallax may have contributed an error in the reading of the buret. The experimenter may have transposed two digits in recording the number. Poor lighting may have made it difficult to recognize the start of a color change that was serving to indicate the completion of a reaction. The calibration marks on the buret may not correspond to the volume of the solution delivered. There may have been impurities in the reagents or there may have been side reactions.

It is evident that some of the errors listed could be avoided by the use of greater care and better technique, but that others would still remain. In this chapter we will consider problems of evaluating the data that one obtains in the laboratory.

THE RELIABILITY OF NUMBERS

When one is told that the population of a city is two million, one is not likely to suppose that there are 2,000,000 people, no more and no less, in that city. Casual conversation is not seriously affected by the uncertainty of what is meant by "two million" people (are there between 1.5 and 2.5 million? or are there between 1.9 and 2.1 million? or . . . ?). But when data are being given from physical measurements and these data are to be made the bases for decisions, numbers must be used in such a way that their reliability is clearly understood.

Scientific Notation

In order to show clearly the reliability of a number and to avoid the difficulties associated with writing and reading such numbers as 0.000017 and 273,000,000,000, the preferred method of writing a number is in scientific notation. The decimal point is placed after the first digit, only as many digits as carry significance are written to the right of the decimal, and the magnitude is indicated by a product of ten raised to the appropriate power. Thus the first number above becomes 1.7×10^{-5} and the second 2.73×10^{11}, if we assume that there are three meaningful figures in this second number. Already the problem of the reliability of the number as it was first written becomes evident; when one sees the isolated number 273,000,000,000, one has no way of knowing precisely what is meant. In the following sections we will discuss in greater detail techniques for indicating reliability. The scientific notation should be recognized as having been designed for convenience; when its use is inconvenient there seems little point in making a fetish of using it to be consistent. For example, when a chemist reads a buret, he estimates the volume reading to the hundredth of a milliliter. Hence, a reading might be 24.27 ml. Very few chemists would choose to write this as 2.427×10^1 ml; however, if the volume were expressed in liters, many would write 2.427×10^{-2} instead of 0.02427 liters.

From the above discussion it can be seen that a number should convey two kinds of information: (*a*) the magnitude of the quantity involved and (*b*) the accuracy to which this quantity is known. There are some numbers that are known to unlimited accuracy. There is, for example, exactly 1 atom of oxygen in a single molecule of H_2O. To take another example, the fraction $\frac{3}{7}$ can be expressed as a decimal fraction having as many meaningful (significant) digits as desired. However, numbers obtained in the laboratory are in general limited in accuracy by the nature of the experiment and the techniques used in obtaining the number. For this reason it is essential that the experimenter state a number in such a way that its limitations are clearly evident.

Significant Figures

The simplest means of indicating the reliability of a number is through the use of significant figures; according to this convention the digits composing the number are extended to, but not beyond, one digit whose value is uncertain. Thus if a length measurement is given as 1.02 cm the implication is that the length is not 1.01 cm and not 1.03 cm; the length is given to 3 significant figures. The same length could be recorded to the same accuracy (if we ignore the use of scientific notation) as 0.0102 m; there are still three significant figures since the zeros before the 1 only place the decimal point. In the preferred notation, 1.02×10^{-2} m, the number of significant figures is more clearly seen. If the same length were expressed in microns ($10^6\ \mu = 1$ m) the magnitude is given by 10,200 μ but the reliability is incorrectly stated, because the five significant figures imply that the length is between 10,199 and 10,201 μ. The proper statement of the length in these units is $1.02 \times 10^4\ \mu$.

If the uncertainty of the above measurement had been greater than ± 1, this fact should be indicated in giving the value. Thus, had the uncertainty been twice as great, this fact could have been indicated by reporting the measurement as 1.02 ± 2. A simpler, though less exact, method writes a number with an uncertainty greater than ± 1 as a subscript, thus 1.0_2.

Significant Figures in Mathematical Operations. When data are involved in mathematical operations, care must be used that the final value reflects properly the accuracy of the data; accuracy should not be lost, nor should the final value imply an unwarranted increase in the accuracy. Suppose that the following numbers, each properly expressed to its known reliability, are to be added: 1027.1, 1.33249, 0.2, and 7×10^{-3}. First write the numbers in a column for addition with the decimal points aligned:

$$
\begin{array}{lr}
 & 1027.1 \\
 & 1.33249 \\
 & 0.2 \\
 & 0.007 \\
\hline
\text{The sum is} & 1028.6
\end{array}
$$

Columns with unknown digits are not used. The number of significant figures in the sum is not limited to the least reliable number which was added. Thus 0.2 has only one significant figure, but it contributes to the final sum; 7×10^{-3} also has one significant figure but makes no contribution; 1.33249 has six significant figures, of which only two contribute.

The student should notice, however, that each number was used to the same *absolute* reliability. For convenience, assume that the numbers added above are lengths expressed in centimeters. Each number was added to the nearest 0.1 cm, even though the numbers of significant figures were very different.

The situation is different in operations involving multiplication. In the operation

$$\frac{3.60 \times 10^{-3} \times 2.9 \times 10^{2}}{0.351}$$

the result is correctly expressed as 3.0 because one number in the product is known only to two significant figures. The reliability of a result in a series of multiplications and divisions cannot possibly exceed that of the least reliable term.

Rounding Off Significant Figures. If the student is to maintain significant figures in arithmetical operations he must round off the number properly. If 27.3 and 19.76 are to be added, the answer is rounded to 47.1.[1] If the first digit to be dropped is greater than 5, the last digit retained is raised by 1; if the first digit to be dropped is less than 5, the last digit retained is unchanged. If the first digit to be dropped is 5 and is followed by zeros or by no digits, then the last digit retained is made the nearest even number. The following examples illustrate these rules. In each case the number is rounded to three significant figures.

1.237	becomes	1.24
24.92	becomes	24.9
29.65	becomes	29.6
29.651	becomes	29.7
4.315	becomes	4.32

There is an inherent ambiguity in the use of significant figures as a means of specifying the reliability of numbers. The numbers 1.2 and 9.7, for example, both have two significant figures. Yet as written here they mean between 1.1 and 1.3, and between 9.6 and 9.8, respectively. The uncertainty in 1.2 is ±1 part in 12 (approximately 8%) while that in 9.7 is only ±1 part in 97 (approximately 1%). Hence, if several numbers, each having the same number of significant figures, are to be multiplied and divided, the numbers may have quite different reliabilities and the problem may need careful examination

[1] A good practice commonly followed in making calculations is to carry one more significant figure in the arithmetical steps than is justified by the data. The answer is then rounded to the proper number of digits. This practice avoids the possibility of cumulative errors from rounding several numbers in a calculation.

in order to determine the reliability of the result. In the operation

$$\frac{29 \times 4.2 \times 1.1}{9.8 \times 31 \times 0.70} = 0.63 \tag{1}$$

each number has two significant figures. However, the least reliable number is 1.1, and this, indeed, is less reliable than the result as written. It would not be correct to round the number to 0.6, for it is known more reliably than that. As stated previously, a practice sometimes used is to write the value as 0.6_3. It is evident that a more critical method of expressing reliability is needed; this will now be discussed.

Explicit Statement of Reliability

While the use of significant figures is a helpful guide to the reliability of numbers, an approach that is usually to be preferred is an explicit statement of the reliability. The liquid level in a 50-ml buret can be read with certainty to 0.1 ml and can be estimated to ± 0.01 ml. Therefore, the volume of a solution used might be obtained from the data in the following way:

final reading	27.09 ± 0.01 ml
initial reading	1.00 ± 0.01 ml
volume used	26.09 ± 0.02 ml

The uncertainties are added in operations that involve addition and subtraction.

When only multiplication and division are involved, a good first approximation is to assign to the final value the fractional uncertainty of the least reliable number in the operation. For example, in the operation in Equation 1 the least reliable number has an uncertainty of 1 part in 11; the result then is written 6.3 ± 0.6. Actually each number in the multiplication operation contributes its own fractional uncertainty to the result. Consider the product of

$$(29.95 \pm 0.03)(2.004 \pm 0.006)$$

The percentage uncertainties of the numbers are 0.1% and 0.3%, so the uncertainty of the answer, 60.02 ± 0.24, is 0.4%.[2]

[2] The effect of the uncertainties can be seen readily if the binomial terms are multiplied in detail: $(29.95 \pm 0.03)(2.004 \pm 0.006) = 60.020 \pm 0.060 \pm 0.180 \pm 0.00018 = 60.02 \pm 0.24$.

ACCURACY AND PRECISION

Accuracy refers to the exactness with which a measurement agrees with the true value of that which is being measured;[3] *precision* refers to the close agreement of results with each other. Ordinarily these attributes are quite independent of one another and yet some correlations can be seen. For example, if several measurements that have poor precision are made, it is unlikely that the accuracy will be good. However, if in a determination two portions are weighed from a single sample by difference weighing (the sample is weighed, the first portion is removed and the sample is weighed again, the second portion is removed and again the sample is weighed; this requires three weighings for the two samples), it is possible that very poor precision but excellent accuracy can be obtained. An error in the intermediate weighing makes this unlikely sounding result possible.

Good precision and poor accuracy can occur readily in determinations in which all calculations depend upon a single measurement. An error in weighing during the preparation of a standard solution, for example, leads to a constant error in the use of that solution. No matter what care is exercised in subsequent steps of the determination, the result is affected by the initial error; therefore one can obtain high precision but low accuracy.

Relative Versus Absolute Errors

If a sample weighs 1.0273 g but its weight is recorded as 1.0275 g the error is +0.2 mg. This expression of the absolute error may provide all the information needed, but often the statement of the error is more useful if it is given in relative terms. The *relative* error in the above weighing is +0.02% or +0.2 part per thousand. This was calculated from the expression

$$\text{percentage error} = \frac{X - T}{T} 100$$

where T is the true and X is the measured value. If the sample had weighed 0.0102 g, the same +0.2 mg absolute error would have caused a relative error of +2%.

One must be especially careful of the terminology when working with data that are being reported in percentages. Consider a chloride sample, for example, that is 40.95% chloride. If the result from analysis is 40.86%, the absolute error is −0.09% and the relative error is −0.2%. The terms

[3] The *true value* is a relative term and can be established only within certain limits because it depends upon measurements, such as atomic weight determinations, or upon the care with which standard substances are prepared.

"absolute percentage error" or "absolute error expressed as percentages" are meaningless except in the case of a sample whose constituents are being reported in percentages.

EVALUATION OF DATA

In the quantitative determination of a substance, a chemist ordinarily works with very large and very small statistical *populations.* Consider the quantitative determination of the chloride present in a substance by the measurement of the quantity of silver ion that reacts with it. The number of chloride ions and silver ions is usually of the order of 10^{21}. On the other hand the number of separate determinations made is often only three or four. Statistical treatment of a very small number of determinations is difficult (imagine having two widely separated values for the composition of a sample). Nonetheless, a chemist must have some standard by which he can make decisions regarding his results. In the following sections we will classify experimental errors according to their origin and discuss methods for evaluating them.

Determinate Versus Indeterminate Errors

The errors that occur in a series of measurements can be classified into two general types. *Determinate errors* are those whose presence one can detect and for which, in principle, one can make corrections. *Indeterminate errors,* on the other hand, are random and unavoidable.[4] *Mistakes* do not fall into either of these classifications; the spilling of a sample is neither a determinate nor an indeterminate error.

Determinate Errors

Determinate errors, also called *systematic errors,* can in principle be assigned to a cause, although the relationship to that cause may be complex. The cause may be an instrument, the observer, or the method itself.

Most determinate errors can be assigned to four classifications: (1) errors of method, (2) errors of technique or observation, (3) instrumental errors, and (4) reagent errors.

---

[4] In addition to the above, the terms *constant* or *systematic* and *random* have been applied to these types of errors. Hurley (*J. Chem. Ed.*, **17**, 334, 1940) suggested the following definitions for indeterminate and determinate errors:

"Indeterminate errors are errors of measurement which have signs and magnitudes determined solely by chance, and therefore cannot be controlled by the observer.

"Determinate errors are errors of measurement which have signs and magnitudes determined by laws relating them to their causes, and therefore can be controlled or evaluated, provided the laws and causes can be discovered."

Errors of Method. Errors of method are caused by such factors as the solubility of the precipitate or the use of an indicator giving an error in the end-point determination. They can be recognized by systematically varying the steps of a procedure or by the use of a different method. Once recognized they often can be minimized by the use of appropriate correction factors.

Errors of Technique. Errors of technique are also called *personal errors* and are caused by the experimenter. One major contribution to this class is caused by improper observations, as for example consistently misreading the meniscus of a buret. A second major contribution arises from improper execution of required operations or neglect of certain details. Examples would be failure to mix a solution or failure to apply temperature corrections properly. Such errors are difficult for the experimenter to recognize because they are caused by his own carelessness or lack of skill or experience; the assistance of another experimenter or an instructor can be of value in correcting such errors.

Instrumental Errors. Sources of instrumental errors are the improper calibration, functioning, or use of an instrument. The calibrations of instruments should be checked when there is the least cause for doubt of their validity. An effective method of recognizing and correcting instrumental errors is the use of appropriate standards, such as Bureau of Standards certified samples, calibrated weights, or properly prepared standard solutions.

Reagent Errors. Reagent errors arise from impure or contaminated chemicals or reagents. They can be recognized by testing or analyzing the suspected source; they can be eliminated by replacing a contaminated reagent; they can often be minimized by making a *blank.* This involves carrying out the procedure without the sample to be analyzed and making a correction based upon the result obtained.

Indeterminate Errors

Indeterminate errors, also called *random errors*, are the result of unrecognized variables such as the inherent limitations of instruments and observer. If, for example, a meter stick is used for the measurement of length, the accuracy of the measurement is limited by the accuracy of the marks on the stick and by the observer's ability to estimate fractions of the smallest divisions. For each instrument–observer combination there is an inherent limit to the accuracy of the measurement. A large series of measurements of the same property by the same observer with the same instrument will result in a random scatter of values. This scatter can be eliminated, of course, if the measurement is made only to a coarse precision. For example, a meter stick can be used to measure the height of a mercury column to the nearest centimeter with small likelihood of any scatter. But if a much finer measurement is attempted—say to 0.1 mm—there will be a scatter of the results.

The Effects of Determinate Errors Versus the Effects of Indeterminate Errors

Determinate errors lower the accuracy of a method, but in many cases do not affect the precision. An impurity in a reagent will ordinarily cause a constant error (hence no lowering of precision); the error from the use of an uncalibrated buret may be different in sign and magnitude for different volumes measured. Statistical treatments do not help to eliminate determinate errors.

Indeterminate errors cause a lowering of precision. If these errors are truly random then the average of a large number of measurements will have high accuracy. Statistical methods can be properly applied in analysis of data affected by indeterminate errors.

Error Evaluation

An evaluation of the validity of a series of measurements that contain only indeterminate errors requires the use of methods of statistical analysis that are beyond the scope of this book. However, one can assume that indeterminate errors will conform to the normal frequency curve, the so-called Gaussian curve, when the frequency of occurrence of errors of varying magnitudes is plotted against the magnitude and sign of the errors as the abscissa. Such a curve is shown in Figure 3-1, and an inspection shows that

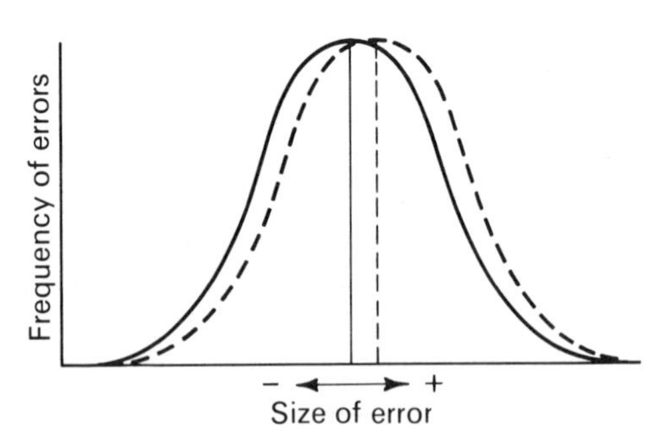

FIGURE 3-1

Normal distribution of errors. The effect of a determinate error is shown by the dotted line.

certain conclusions can be drawn. They are: (1) small errors occur most frequently, (2) very large errors occur very infrequently, and (3) positive and negative errors of a given size occur with the same frequency. From this last observation it follows that, statistically, the average value represents the true value. One should remember, however, that this curve is valid only for very large numbers of measurements and that deviations from it will occur more frequently with smaller numbers of measurements. This curve is applicable only to indeterminate errors; the effect of a constant error is shown by the dotted curve; the shape of the curve remains the same but it has been displaced by the value of the determinate error.

Some of the results of the application of the methods of statistical analysis and the terms commonly used are given below.

Mean. The mean is the arithmetic average of a set of results:

$$\bar{X} = \sum_i \frac{X_i}{n}$$

Here X_i $(X_1, X_2, X_3, \ldots, X_n)$ represents the individual results; $\bar{X}$ is the mean. If the five values from measurements are 73.27, 73.19, 73.24, 73.06, and 73.20, the mean is 73.19. As stated above, the probability of the mean being the true value increases with the number of measurements.

Median. The median of a series of measurements is that value that lies between an equal number of larger and smaller values. With an odd number of measurements, there is a unique median; with an even number of measurements the average of the central pair represents the value of the median. The median of the above set is 73.20.

Range. The range is the difference between the minimum and the maximum values in the set. It and the standard deviation are used to express the precision of the set. In the above set the range, r, is 0.21.

Deviation from the Mean. The deviations from the mean, $d = X - \bar{X}$, are 0.08, 0.00, 0.05, −0.13, 0.01.

Measures of Precision

Average Deviation. The average deviation is given by the expression

$$\bar{d} = \frac{\sum_i |X_i - \bar{X}|}{n}$$

For the five samples above this is 0.05. The average deviation was formerly used extensively as a measure of the precision of a set of measurements, but has been largely replaced by the standard deviation.

Standard Deviation. For small populations the standard deviation is

$$s = \left(\frac{\sum_i (X_i - \bar{X})^2}{n - 1} \right)^{\frac{1}{2}}$$

and for the five samples being considered, $s = 0.08$. It gives greater weight to large deviations and is generally considered to be a better measure of precision than the average deviation.

Confidence Limits

From the normal-distribution curve of Figure 3-1 and its mathematical derivation, one can state the fraction of results that fall within certain limits of accuracy, provided that a very large number of measurements has been made and that only random (indeterminate) errors occur. Under these circumstances, for example, 68% of the results fall between $-s$ and $+s$, and 99.7% of the results fall between $-3s$ and $+3s$. It is rare, however, that in measurements made on chemical systems the data can qualify for evaluation in terms of the normal-distribution curve, for not only is the number of measurements usually small but also the errors are seldom purely random. Therefore, a different treatment is used to permit one to state a probability that the true value[5] lies within some specified range. It must be emphasized that this treatment takes into account the small number of results or measurements being considered but assumes that the errors are truly random. Therefore, it is imperative that determinate errors be eliminated if the treatment is to be completely valid.

A certain **probability** can be calculated statistically that the true value lies within certain specific limits called *confidence limits*, and the interval between these limits is known as the *confidence interval* or *range*. The range is centered on the mean of the set of values and is specified in terms of that mean and the standard deviation.

Table 3-1 contains data from which the confidence limits for various probabilities can be calculated for data from small numbers of determina-

[5] The *true value* as used here for statistical purposes is defined as the mean of an infinite number of purely random measurements of the kind represented in the set of measurements being considered.

TABLE 3-1
Values of t for Calculating Confidence Limits for Various Probabilities

n (Number of Measurements)	Probability Levels		
	50%	95%	99%
2	1.00	12.70	63.66
3	0.82	4.30	9.93
4	0.77	3.18	5.84
5	0.74	2.78	4.60
6	0.73	2.57	4.03
7	0.72	2.45	3.71
8	0.71	2.37	3.50
9	0.71	2.31	3.35
10	0.70	2.26	3.25

tions. The confidence limits are given by the expression

$$\bar{X} \pm \frac{ts}{\sqrt{n}}$$

where n is the number of replicate determinations, t is obtained from the table, and s and $\bar{X}$ have their usual significance. For the set of five measurements on page 31 with the mean 73.19 and standard deviation of 0.08 the confidence limits for 95% probability are

$$73.19 \pm \frac{2.78 \times 0.08}{\sqrt{5}}$$

or 73.19 ± 0.10. Similarly, the confidence limits for 99% probability are

$$73.19 \pm \frac{4.60 \times 0.08}{\sqrt{5}} = 73.19 \pm 0.16$$

The precise meaning of these confidence limits is that if a very large number of such sets of measurements were made and confidence limits were calculated for each set, then 95% or, in the second example, 99% of these sets of confidence limits would include the true value.

Criteria for Rejection of a Measurement

One is not infrequently confronted with a series of measurements that includes one value that appears to deviate excessively from the others.

Various techniques have been suggested for deciding whether or not to discard such a value, and the criteria for rejection are easily established from statistical theory when there is a large number of measurements. But most often, especially in the chemical laboratory, the number of measurements is relatively small and the criteria for rejection is subject to uncertainties. Three rules that have been used in such situations are discussed below.

The 4$\bar{d}$ Rule. The 4$\bar{d}$ rule, at one time used quite extensively, requires that at least four measurements be contained in a series of measurements. If the deviation of a doubtful measurement from the mean is equal to or greater than four times the average deviation, the doubtful measurement is to be discarded and the mean and the average deviation are to be calculated from the remaining three measurements. Calculations based on the theory of probability with large populations indicate that the chances are about 100 to 1 that the doubtful measurement resulted from some mistake. In fact, the 4$\bar{d}$ rule was criticized because it appeared that doubtful measurements were being retained.

The 2.5$\bar{d}$ Rule. The 2.5$\bar{d}$ rule was proposed because of the above objections to the 4$\bar{d}$ rule. With this rule the chances were calculated to be roughly 20 to 1 that the doubtful value is invalid.

However, according to Blaedel, Meloche, and Ramsay,[6] the above chances are based on probability theory applied to large populations and are not valid with small numbers of measurements. They state that with four measurements the deviation of the doubtful measurement may exceed by 4$\bar{d}$ the average deviation of the other three almost 60% of the time. For this reason both the 4$\bar{d}$ and 2.5$\bar{d}$ rules have limits for rejection that are too low and lead to frequent rejection of valid measurements, with resultant loss of accuracy and estimates of precision that are too high. Rigid adherence to either of these rules with small numbers of measurements is not recommended.

The Q Test.[7] The Q test is applied by first obtaining the range, r, then obtaining the difference, d, between the values of the doubtful measurement and the one closest to it. Dividing d by r gives a quotient, Q, hence the name of the test. By reference to a prepared table, the value of Q is used to decide

[6] W. J. Blaedel, V. W. Meloche, and J. A. Ramsay, *J. Chem. Ed.*, **28**, 643 (1951). This article presents a discussion of the tests for rejection of a doubtful measurement from a relatively small number of measurements.

[7] A discussion of the criteria for rejection of data when dealing with small numbers of measurements and of the use of the Q test is given by R. B. Dean and W. J. Dixon (*Anal. Chem.*, **23**, 636, 1951).

TABLE 3-2
Number and Measurements and Q Values

Number of Measurements	Q for Rejection
3	0.94
4	0.76
5	0.64
6	0.56

as to the probability that the measurement is invalid and as to its rejection. Selected values from a more comprehensive table by Dean and Dixon[8] are given in Table 3-2; if Q exceeds the value given, the rejected value will be subject to unknown error 90% of the time.

The Q test has better statistical validity than the other two tests when a relatively small number of observations is in question. (Because of the time and effort required, it is usually practical for a chemist to make only two to four observations.) But this test tends to eliminate only very largely divergent values. It is also less convenient to apply because a table of Q values is required.

Thus it appears that there is no completely satisfactory rule for rejection that can be applied to experiments involving small numbers of measurements. As a result there is agreement that when faced with the problem of the rejection of a divergent value, any one of the above rules should be used only as a last resort and in conjunction with experience regarding the type of measurement being made.[9] The following considerations are suggested. First, carefully consider all possible sources of the divergence. Check and recheck all data, such as weight records, and calculations made therefrom. Review your notebook record of each operation to ascertain if there was any suggestion of errors, such as spillage of sample, precipitate, or solution, or leakage of a stopcock. As a last resort, if at all practical, make an additional measurement and then reconsider the situation.

Secondly, from your experience or that of your instructor, consider the precision to be expected from the method involved. Thus certain methods are more erratic than others because of known effects such as coprecipitation,

[8] Ibid.

[9] This problem is discussed in the following references: Dean and Dixon, op. cit., a presentation of the Q test; Blaedel, Meloche, and Ramsay, op. cit., a review of the three rules discussed above; H. A. Laitinen, *Chemical Analysis.* New York: McGraw-Hill, 1960, Chap. 28, an excellent introduction to the application of statistics in analytical measurements; I. M. Kolthoff, E. B. Sandell, E. J. Meehan, and S. Bruckenstein. *Quantitative Analysis.* (4th ed.) New York: Macmillan, 1969. Chapter 16 is a comprehensive discussion of errors in chemical methods and of criteria for the rejection of results.

induced reactions, and oxygen error. If the divergent result is within the expected spread caused by such effects it should be retained.

Thirdly, apply one of the above tests—the *Q* test is statistically the most valid for a small number of measurements. If the test indicates that the value should be retained, the median is likely to be a more reliable value than the average; it is less susceptible to the effect of a large deviation.

SUPPLEMENTARY REFERENCES

Bennett, C. A., and Franklin, N. L. *Statistical Analysis in Chemistry and the Chemical Industry*. New York: Wiley, 1954.

Gore, W. L. *Statistical Methods for Chemical Experimentation*. New York: Wiley-Interscience, 1952.

Kolthoff, I. M., and Elving, P. J. *Treatise on Analytical Chemistry*. Part 1. Vol. 1. New York: Wiley-Interscience, 1959.

QUESTIONS AND PROBLEMS

1. An object whose actual weight is 120.3142 g is weighed on various balances. Assuming that the weighing process is performed correctly to the limits of the instrument, how should the weight be recorded if the balance is sensitive (*a*) to only 1 mg; (*b*) to only 0.1 g? *Ans.* (*a*) 120.314 g; (*b*) 120.3 g.

   How would this last weight be expressed in milligrams? *Ans.* 120,300 $\pm$ 100 mg, or 120.3×10^3 mg.

2. A volumetric method for determining chloride was tested by making analyses of specially purified sodium chloride. After correction for all known *determinate* errors, the following results were obtained, calculated as the percentage by weight of chlorine in the salt: 60.56, 60.53, 60.60, 60.64, 60.62.

   Calculate (*a*) the arithmetical mean; (*b*) the absolute error (as a percentage) of this mean; and (*c*) the relative error (expressed as parts per thousand and as a percentage) of the above series of measurements. *Ans.* (*a*) 60.59; (*b*) -0.07%; (*c*) 1.2 parts per 1000, or 0.12%.

   NOTE: These answers are measures of the accuracy of the above measurements. However, they are calculated from (*a*) the experimentally determined atomic weights, and (*b*) the assumption that the sodium chloride is absolutely pure; such experimental uncertainties always enter into attempts to calculate the accuracy of measurements.

3. Calculate (*a*) the $\bar{d}$ (average deviation) of the series of measurements given in Problem 2, and (*b*) the relative $\bar{d}$, expressed as parts per thousand and as a percentage. *Ans.* (*a*) 0.04%; (*b*) 0.6 part per 1000, or 0.06%.

4. Calculate the standard deviation of the series of measurements given in Problem 2.

5. Since it can be shown that the precision of the mean is increased only by the square root of the number of observations in a series, it is obvious that it is not practical to increase the precision of a series greatly by increasing the number of measurements.

   A value is obtained from the mean of three measurements. How much more reliable would it be if (*a*) six and (*b*) nine measurements were made? *Ans.* (*a*) 1.4 times as reliable; (*b*) 1.7 times as reliable.

6. Frequently one measurement of a series will deviate widely from the others, and no explanation of the divergence can be found. Discuss the criteria for rejection of such a measurement. Apply the rejection rules to the measurements of Problem 2 and give your decision for each rule.

7. An analyst was asked to determine the percentage of potassium in a certain substance. He carried out six determinations and obtained the following results:

19.29	19.32	19.30
19.24	19.00	19.21

   (*a*) Should any determinations be discarded? (*b*) What value should the analyst report? (*c*) What are the $\bar{d}$ and the relative $\bar{d}$? (*d*) What is the standard deviation of the series? (*e*) The substance was potassium hydrophthalate, $KHC_8H_4O_4$. What are the analyst's absolute error and relative error? (*f*) Why is the accuracy of the determination so much less than the precision? What assumptions are made in calculating the accuracy? *Ans.* (*a*) Yes, No. 5; (*b*) 19.27% potassium; (*c*) 0.038%; 0.20%, or 2.0 parts per 1000; (*d*) 0.046; (*e*) +0.12%; 0.62%, or 6.2 parts per 1000. (*f*) Probably because of some unsuspected determinate error, such as, for example, coprecipitation in a gravimetric determination. Correctness of atomic weights and purity of compound.

8. The data shown in the following table were obtained by an analyst in carrying out the volumetric analysis of a low-grade iron ore.

Determination	Weight of Sample (g)	Volume of $KMnO_4$ Used in Titration (ml)
1	1.2463	20.80
2	1.4324	23.91
3	1.4892	19.82
4	1.2279	20.30

   The container of the permanganate solution was labeled as follows: "25.01 ml of this $KMnO_4$ solution are equivalent to 0.1262 g of iron."

   (*a*) Should any analysis be discarded?
   (*b*) What percentage of iron should be reported? How many significant figures are justified?
   (*c*) What are the $\bar{d}$ and the relative $\bar{d}$?
   (*d*) Were any of the above measurements carried to an unnecessary precision? Explain!

CHAPTER **4**

Chemical Reactions. Types and Principles

A large majority of the procedures of analytical chemistry are based upon reactions taking place in aqueous solutions. Your general chemistry text has discussed the unusual properties of water, particularly its effectiveness as an electrolytic solvent and its tendency to form hydrates, especially with dissolved positive ions. This discussion should be reviewed, since it will explain why aqueous solutions are used so extensively as a medium in which chemical reactions are carried out.

TYPES OF REACTIONS

Chemical reactions in aqueous solutions may be classified into three types:

1. Precipitation reactions,
2. Oxidation–reduction reactions,
3. Ionization reactions.

This classification is based upon the nature of the main changes that take place as reactants form products. There will be many reactions that are a combination of two or even all three of these types, but it is convenient to discuss the simpler cases first.

Reactions of the first type (precipitation) are characterized by the formation of a solid phase, called a *precipitate*, and the extent to which the reaction occurs is dependent upon the solubility of this solid phase.

Reactions of the second type (oxidation–reduction) are dependent upon the relative tendencies of certain atoms, ions, and molecules to lose or gain electrons.

Reactions of the third type (ionization) are dependent upon the relative tendencies of the reactants and the products to form *un-ionized* but soluble products. For example, the reactions between acids and bases in aqueous solutions are controlled to a great extent by the tendency of hydrogen ions (more precisely, the hydrated protons) to combine with hydroxide ions to form un-ionized water molecules.

Ionization and oxidation–reduction reactions are usually carried out in one-phase systems and for this reason are known as *homogeneous phase* or *homogeneous system* reactions. The formation of a precipitate (type-1 reaction) introduces a second phase and converts a homogeneous system into a *heterogeneous*, two-phase system.

The remainder of this chapter and Chapter 5 present a review of the law of chemical equilibrium, frequently called the mass-action law,[1] and its application to reactions taking place in heterogeneous systems. Homogeneous aqueous systems are treated in Chapters 6 and 7. It will be of help if you review the discussion and formulation of the law of chemical equilibrium given in your general chemistry text.

THE LAW OF CHEMICAL EQUILIBRIUM

The law of chemical equilibrium, or the mass-action law, states that for the simple case where A and B react to give C and D, as

$$A + B = C + D$$

equilibrium will be attained (at a given temperature) when the ratio $[C][D]/[A][B]$ has attained some constant value. The equilibrium condition is then

$$\frac{[C][D]}{[A][B]} = K.$$

Here the letters in brackets represent the *molarities*[2] of the reactants and products, and K is a number called the *equilibrium constant*. A consideration

[1] The term *mass-action law* is a misnomer, since, strictly speaking, the law is concerned with the effect of concentrations rather than masses upon chemical equilibria. The law was formulated by two Norwegians, C. M. Guldberg and P. Waage, in 1864. They considered the mass of the reactants within a "sphere of action," hence the concentration in mass per unit of volume.

[2] The concluding section of this chapter discusses limitations to the use of concentrations in equilibrium expressions. As explained in that section in a rigorous treatment corrected concentrations called activities would be used.

of this equilibrium expression will enable one to make a prediction as to the effect of changing the concentrations of any of the reactants after the system has reached equilibrium. Suppose at equilibrium an additional quantity of A is added. This will increase the concentration of A and the ratio $[C][D]/[A][B]$ will then be less than K; therefore the reaction will tend to proceed from left to right as written, increasing $[C]$ and $[D]$ and decreasing $[A]$ and $[B]$, until the equilibrium value of the ratio is restored.

Note that the effects that would be predicted from this law are in agreement with those that would be predicted from the more qualitative statement of the French chemist, Le Chatelier in 1888. This statement, generally known as *Le Chatelier's principle*, is discussed in your general chemistry text and can be stated as follows:

If any of the conditions that affect a system already at equilibrium are changed, the system will change in such a way as to tend to restore the initial equilibrium and to minimize the effect of the change.

From Le Chatelier's principle we predict that if an equilibrium exists between the reactants and products of the reaction formulated above, and if an attempt is made to increase the concentration of any one of these species, the equilibrium will shift in such a way as to minimize the change. That is, if we attempt to increase the concentration of either A or B the reaction will proceed from left to right and thus lessen the increase; if an additional quantity of either C or D is added, the reaction will proceed from right to left. As will be shown later, the law of chemical equilibrium enables us to calculate the concentrations of reactants and products existing after a system has again reached equilibrium.

The equation for a more generalized reaction can be written as

$$aA + bB + \cdots = fF + gG + \cdots$$

and the equilibrium expression formulated as

$$\frac{[F]^f[G]^g \cdots}{[A]^a[B]^b \cdots} = K.$$

The application of this generalized law to phase equilibria and precipitation reactions is considered in the following sections.

Phase Equilibria

Phase equilibria are not only of fundamental importance in analytical processes but also of fundamental importance in biological, geological, and industrial processes. The equilibrium between the oxygen in the air (the gas phase) and the oxygen dissolved in the blood (the liquid phase) is vital to many forms of life. Equilibria between minerals existing as solids and as liquids (in molten lava and magma) or as solids and dissolved salts (in brines) are important in the history of geological formations—from the remote past to the present. Pure chemicals are prepared industrially by crystallizing a compound as the solid phase from a solution with which it is in equilibrium.

Many of the operations of analytical chemistry are based upon reactions that convert the constituent or constituents of interest into compounds that can be separated as, or in, a second phase. Three types of phase separations are of predominant interest. They are:

First, solid–liquid separations, which involve the formation of an insoluble solid, or precipitate, in a solution. Many of the separations of systems of qualitative analysis are solid–liquid separations; in addition, they are the basis of gravimetric and titrimetric precipitation methods of quantitative analysis. The separation and determination of barium or of sulfate by precipitation as barium sulfate are examples.

Second, there are gas–liquid separations, in which a relatively volatile constituent is formed and separated from the liquid phase by passing a second gas through the solution.

Third, there are liquid–liquid separations in which a substance is extracted from one liquid phase into another, immiscible, liquid phase. The extraction of iodine from an aqueous solution by carbon tetrachloride is an example of such a separation.

A general discussion of phase equilibria is first given below. This discussion is followed by a more detailed consideration of each of the above types of phase separations.

The Phase-Distribution Law. This law can be stated as follows: *When a compound is in equilibrium (as the same molecular species) between two phases at a given temperature, the ratio of the concentrations (rigorously, the activities) of the compound in the two phases will have a constant value.* This statement can be formulated as follows for a substance in equilibrium between, for example, two liquid phases:

$$\frac{C_1}{C_2} = K_{\text{Distribution}} \tag{1}$$

Here C_1 is the concentration in one phase and C_2 that in the other. The units used for these concentrations may be any which are convenient; the value of $K_{\text{Distribution}}$ is independent of the units chosen, as long as both concentrations are expressed in the same units. However, it is customary to use partial pressures for gas-phase concentrations, and mole fractions for solid-phase concentrations. Therefore, in calculations using tabulated values of distribution ratios, the correct units must be used. Note that this law applies only when the same molecular species is present in each phase, and only to that particular species. For example, in hydrochloric acid solutions, iron(III) can exist in several forms, such as $FeCl^{2+}$, $FeCl_2^+$, $FeCl_3$, $FeCl_4^-$, or $HFeCl_4$. Of these species, only $HFeCl_4$ is extracted from the aqueous solution by ether; therefore, the distribution of Fe(III) between aqueous solution and ether is a distribution of only the species $HFeCl_4$.

Deviations from this law, similar to those from the mass-action law, will be found when it is applied to concentrated solutions, and activities should be used for rigorous calculations.

The *distribution law*, Equation 1, is a statement of the law of chemical equilibrium for the special case in which the same species is in equilibrium between two phases. These three facts should be noted concerning the distribution law: (1) The law can be derived from a consideration of the rates at which the molecules pass in each direction between the two phases. (2) The law agrees with the principle of Le Chatelier, since it implies that if a heterogeneous system is at equilibrium and the concentration or pressure of a substance is changed in one phase, the equilibrium will shift so as to minimize the change and restore the original equilibrium ratio. (3) The law indicates that the solubility of a pure liquid or solid compound in equilibrium with its solution in a given solvent should be constant, since the concentration (or more exactly, the activity) of the pure solid or liquid will have a constant value at a given temperature.

A further consequence of this law may be illustrated by the following example. A small amount of iodine (insufficient to saturate either layer) is added to water and carbon tetrachloride in a test tube. When equilibrium is reached, the ratio $[I_2]_{CCl_4}/[I_2]_{H_2O}$ will be equal to the distribution ratio, K. We may compare this with the ratio of the solubilities of iodine in CCl_4 and H_2O, and we find this ratio is the same as the distribution ratio. Thus, we may saturate separate aqueous and carbon tetrachloride solutions with iodine (that is, have two separate phases independently in equilibrium with a third phase, in this case solid iodine), and, if we then mix the carbon tetrachloride and water solutions, we find they are already in equilibrium. This of course holds true only if all phases are at the same temperature.

We may further illustrate this consequence of the distribution law by considering the system shown in Figure 4-1. Here we have a circular tube with solid iodine fixed in place at the bottom, water on one side, and carbon tetrachloride on the other. The water and CCl_4 are each saturated, that is, in equilibrium with the solid. If the tube is inverted and the water and CCl_4 brought in contact, the iodine concentration in the two phases will not change:

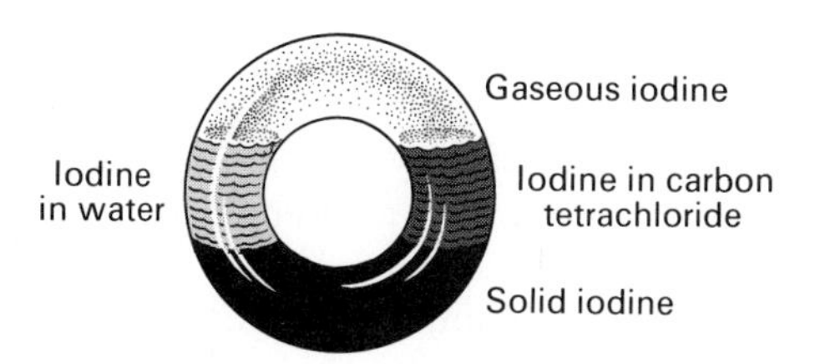

FIGURE 4-1
The distribution of iodine between solid, liquid, and gaseous phases.

they will already have a ratio equal to the distribution ratio of iodine between water and CCl_4.

If this were not true, the system could be arranged to produce perpetual motion. For example, assume that the iodine in the saturated CCl_4 phase were not in equilibrium with the iodine in the water. When the liquids then came in contact at the top of the tube, iodine would tend to pass from one into the other—let us say, from the CCl_4 into the water. But this would undersaturate the CCl_4 so that more iodine would dissolve on that side, and oversaturate the water, so that some iodine would precipitate. This process would continue indefinitely, and so a perpetual-motion system would result.

The phase-distribution law is also applicable to the distribution of a substance between gas and solid phases or gas and liquid phases. Since a concentration in a gas phase is proportional to the gas pressure in that phase, the equilibrium law for these cases can be written

$$\frac{P_1}{C_2} = K$$

This relation between the concentration of a gas in a solvent and the partial pressure of the gas above the liquid phase is known as *Henry's law* after the British chemist William Henry (1775–1836), who discovered it in 1803.

If we refer again to Figure 4-1, we can see that as a result of this law the concentration (or the more commonly used partial pressure) of iodine in a gas phase in equilibrium with any saturated iodine solution will be the same as that above solid iodine. This must be, for if the partial pressures of iodine above the two saturated solutions were not equal, there would be a continuous passage of iodine from one liquid to another, and again a perpetual-motion system would result.

Liquid–Liquid and Liquid–Gas Phase Equilibria

Procedures that involve the application of the phase-distribution law to the distribution of a substance between two liquids are used extensively both in analytical and industrial processes and are frequently called *phase-distribution* or *solvent-extraction* methods.

In Chapter 22 there is discussion of the iodine-monochloride endpoint. In this process iodine is extracted from aqueous solution into carbon tetrachloride, where it is not only much more soluble but also has a brilliant purple color and thus serves as the indicator in many oxidation-reduction titrations. In such titrations a small volume of carbon tetrachloride, which is immiscible with water, is shaken with the aqueous solution, which contains some iodine. As a result the iodine passes into the organic phase until the ratio of the iodine in that phase to the concentration in the water phase equals the distribution ratio. At equilibrium

$$\frac{[I_2]_{CCl_4}}{[I_2]_{H_2O}} = K = 87_{(25^\circ)}$$

Hence the iodine concentration in the carbon tetrachloride is 87 times that in the water; very low concentrations of iodine are easily detected by this technique.

The phase-distribution law applies similarly to gas-liquid phase separations. The equilibrium expression for carbon dioxide between the gas and aqueous phase is

$$CO_2(aq) = CO_2(g)$$

and the equilibrium constant is

$$K = \frac{P_{CO_2(g)}}{[CO_2](aq)}$$

As this equation shows, a solution that is saturated with CO_2 at a given pressure and temperature will have a fixed constant concentration of CO_2(aq). This equilibrium of CO_2 between the gas and aqueous phases is considered in Chapter 7 in connection with calculations of hydrogen-ion concentrations.

Solid–Liquid Phase Separations

A large proportion of the separations used in analytical chemistry involve the formation of relatively insoluble precipitates in aqueous solutions. In these separations we have the simultaneous application of the principles of both homogeneous and heterogeneous phase equilibria.

The Solubility-Product Principle. Consider the case of the precipitation of barium sulfate by the addition of a soluble sulfate to a neutral solution containing barium ion. (A neutral solution is specified so as to avoid considering the formation of HSO_4^-.) As soon as a precipitate is formed we have a solid phase and a liquid phase and therefore a heterogeneous system. When equilibrium is attained between the barium sulfate in the two phases (solid and liquid), the equilibrium relation is, according to the phase distribution law,

$$BaSO_4(s) = BaSO_4(aq) \tag{2}$$

and the equilibrium expression is

$$\frac{[BaSO_4](aq)}{[BaSO_4](s)} = K_D$$

As was indicated earlier in this chapter on page 42, the concentration of a solid is often expressed in terms of the mole fraction; since the mole fraction of a pure substance is 1, the equilibrium expression can be written[3]

$$\frac{[BaSO_4](aq)}{1} = K_D \tag{3}$$

In the solution we have a homogeneous equilibrium as follows:

$$BaSO_4(aq) = Ba^{2+} + SO_4^{2-}$$

The equilibrium expression is

$$\frac{[Ba^{2+}][SO_4^{2-}]}{[BaSO_4](aq)} = K_{\text{Ionization}}$$

---

[3] The identical result is obtained if one expresses the concentration of the solid $BaSO_4$ in some other units, such as moles/l. In such a case the expression

$$\frac{[BaSO_4](aq)}{[BaSO_4](s)} = K$$

is simplified by the fact that the concentration of barium sulfate in pure solid barium sulfate is constant. Therefore, when both sides of the equation are multiplied by $[BaSO_4](s)$ the expression becomes

$$[BaSO_4](aq) = [BaSO_4](s)K = K_D$$

and this is identical with Equation 3 above.

However, since $[BaSO_4](aq)$ has a constant value at a given temperature when in equilibrium with solid $BaSO_4$, we can write

$$[Ba^{2+}][SO_4^{2-}] = K_{sp}.$$

This constant, K_{sp}, is the ion-product constant or, as it is more commonly called, the *solubility product* for barium sulfate. One must remember that *this value is constant and is applicable only for solutions in equilibrium with the solid phase*, that is, for saturated solutions. Note also that the solubility-product expression is only a special case of the general law of chemical equilibrium.

Since the solid has constant activity and therefore the un-ionized compound has constant activity and concentration, the values for these terms are incorporated into the constant of the equilibrium expression.

Barium sulfate, like many salts, is predominantly an ionic compound, therefore the concentration of un-ionized $BaSO_4$ in an aqueous solution is very small; this amount can be neglected for all experimental calculations. Silver chloride and other compounds that are more covalent will be discussed later.

It will be apparent that for a saturated solution of a more complex salt, such as Ag_2CrO_4, the equilibria will be

$$Ag_2CrO_4(s) \rightleftarrows Ag_2CrO_4(aq) \rightleftarrows 2Ag^+ + CrO_4^{2-}$$

and the solubility product will have the form

$$[Ag^+]^2[CrO_4^{2-}] = K_{sp}$$

Thus the general statement can be made that in a saturated solution of a salt the product of the ion concentrations is a constant when each ion is raised to a power corresponding to the number of ions formed from a molecule of the salt.

Because of the uncertainty introduced in concentrated solutions by the electrical forces between ions, the so-called *interionic forces*—which are discussed later in this chapter—quantitative applications of the solubility-product principle are usually restricted to relatively insoluble substances and to solutions with total ionic concentrations less than approximately 0.01 M; useful qualitative predictions can be made with more concentrated solutions.

The Common-Ion Effect. Since the solubility-product principle states that for a salt such as $BaSO_4$ the product of the ion concentrations is a constant, it follows that as the concentration of one ion is increased, the concentration of the other must decrease. Therefore it is apparent that the quantity of $BaSO_4$ that will dissolve in a given volume of solution will be less if that solution already contains Ba^{2+} or SO_4^{2-} ions.

Let us attempt a more quantitative prediction in this regard by considering the solubility of $BaSO_4$ in a neutral solution that is initially 0.010 F in Na_2SO_4. First, we must determine the solubility product of $BaSO_4$. To do this we look up solubility data in a handbook or other reference book and find that one liter of pure water at room temperature dissolves 0.0023 g of $BaSO_4$. The molecular weight of $BaSO_4$ is 233.4; therefore, the saturated solution is 1.0×10^{-5} F. Since the concentration of un-ionized $BaSO_4$ is negligible, the concentrations of both barium and sulfate ion will be 1.0×10^{-5} M and the solubility product will be 1.0×10^{-10}, that is

$$K_{sp} = [Ba^{2+}][SO_4^{2-}] = (1.0 \times 10^{-5}) \times (1.0 \times 10^{-5}) = 1.0 \times 10^{-10}$$

With this value we may proceed with the problem.

If we assume that the Na_2SO_4 in a 0.010 F solution is essentially completely ionized, the concentration of sulfate ion in such a solution will be 0.010 M. When solid $BaSO_4$ is equilibrated with this solution we will have barium sulfate passing into solution as follows:

$$BaSO_4(s) = Ba^{2+} + SO_4^{2-}$$

Since there are sulfate ions already present in the solution, we must measure the quantity of $BaSO_4$ dissolving (its solubility) by the equilibrium concentration of barium ion. If the molar concentration of barium ion, $[Ba^{2+}]$, at equilibrium is set equal to x, then the molar concentration of sulfate ion, $[SO_4^{2-}]$, at equilibrium will be $0.010 + x$, since one mole of SO_4^{2-} must pass into solution with each mole of Ba^{2+}. Now we make use of the expression

$$[Ba^{2+}][SO_4^{2-}] = K_{sp}$$

and substituting we obtain

$$(x)(0.010 + x) = 1.0 \times 10^{-10}$$

Noting the small value of K_{sp} we first assume that x will be small compared to 0.010. Upon solving the equation we find that $x = 1.0 \times 10^{-8}$, and that our assumption was valid. We thus find that the $[Ba^{2+}]$ in solution will be 1.0×10^{-8}, and that the solubility of $BaSO_4$ in 0.010 F Na_2SO_4 is 1.0×10^{-8} formal.

This calculation leads to the conclusion that only one thousandth as much $BaSO_4$ will dissolve in a given volume of 0.010 F Na_2SO_4 as in water; if this were true, we could decrease the solubility of barium sulfate in a solution by a factor of 1000 if the solution is made 0.01 *M* in sulfate ion. Actually, the activities of SO_4^{2-} and Ba^{2+} in 0.010 *F* Na_2SO_4, are approximately 30% less than the molar concentrations because of the interionic effects mentioned below. Considering these effects the solubility will be approximately twice as great as previously calculated.

This common-ion effect is extensively used in analytical chemistry in order to obtain a more quantitative precipitation of a desired constituent. Specific applications will be pointed out in the discussions of the procedures in which it is used in this text.

Effect of the Hydrogen-ion Concentration on the Solubility of Salts. In the discussion above no consideration has been given to any effect of the hydrogen-ion concentration on the solubility of a salt. If a salt of a strong acid is present, there is essentially no effect. If the salt of a weak acid is present, this is no longer true. An increase in the hydrogen-ion concentration will decrease the anion concentration and therefore increase the solubility of the salt, because of the reaction

$$H^+ + A^- = HA$$

In the discussion of the solubility of $BaSO_4$ in water and in a sodium sulfate solution, care has been taken to state that the solutions were essentially neutral. When this is the case we are able to assume that in a sodium sulfate solution, the molar concentration of sulfate ion is equal to the formal concentration of sodium sulfate. This will not be true if the solution has a relatively high hydrogen-ion concentration.

Let us consider the effect obtained if the 0.010 *F* Na_2SO_4 solution considered above had also been 1.0 *F* in HCl. The proton of the HCl and the first proton of H_2SO_4 are essentially completely ionized; this is not true of the second proton of H_2SO_4. The ionization constant, K_A, for the dissociation reaction

$$HSO_4^- = H^+ + SO_4^{2-}$$

has a value of 1.2×10^{-2}, that is,

$$\frac{[H^+][SO_4^{2-}]}{[HSO_4^-]} = 1.2 \times 10^{-2}$$

As a result, in a solution 0.010 *F* in Na_2SO_4 and 1.0 *F* in HCl, there will be a tendency for the reaction

$$H^+ + SO_4^{2-} = HSO_4^-$$

to take place. If we let x equal the equilibrium $[SO_4^{2-}]$, the equilibrium $[H^+]$ will be $0.99 + x$ and the $[HSO_4^-]$ will be $0.010 - x$. Therefore we can write

$$\frac{[H^+][SO_4^{2-}]}{[HSO_4^-]} = \frac{(0.99 + x)(x)}{(0.010 - x)} = 1.2 \times 10^{-2}$$

Again we assume that x is small compared to 0.010 (and 0.99 also); we solve the expression and find that $x = 1.2 \times 10^{-4}$. Thus, because of the acid present in the Na_2SO_4 solution, the sulfate concentration has been decreased from 0.010 M to 1.2×10^{-4} M or almost 100-fold. This decrease in the sulfate-ion concentration means that the solubility of barium sulfate in such a solution, because of the acid present, is almost 100 times greater than in a neutral solution.

Limitations to the Use of Concentrations in Equilibrium Expressions. We have already noted the restriction that equilibrium calculations based upon concentrations are applicable with accuracy only in relatively dilute solutions. Although we might predict from the solubility product for silver sulfate that the solubility will decrease as the concentration of either silver ion or sulfate ion is increased, we would also predict that the solubility of silver sulfate would be unchanged by the addition of such salts as KNO_3 or $Mg(NO_3)_2$, which contain no common ions. Figure 4-2 shows the solubility of Ag_2SO_4 in solutions of various salts. We observe that in solutions containing common ions (Ag^+ or SO_4^{2-}) the solubility is decreased, but that instead of being unaffected in solutions of KNO_3 and $Mg(NO_3)_2$, the solubility is actually *increased*. The observed increase is caused by interionic forces in the solution and the explanation of these effects is commonly called the *interionic-attraction theory*.[4] The principal postulates of this theory are (1) most salts and the so-called strong acids and bases are practically completely ionized in aqueous solutions, and (2) because of the mutual attraction of the opposite charges of these ions, their *effective concentration* (activity) in solution is lower than the actual concentration.

The interionic-attraction theory, however, breaks down in the case of relatively concentrated solutions because other factors come into effect that may result in the "effective concentration" of a substance becoming much larger than its molar concentration. The magnitude of these various effects is dependent upon the total ionic concentration of the solution, the radius of

---

[4] This theory is frequently referred to as the *Debye–Hückel theory*, since these authors were the first to develop a quantitative treatment of the subject: P. Debye and H. Hückel. *Physik. Z.*, **24**, 185, 305 (1923). For a brief historical discussion of this topic and its application to the teaching of analytical chemistry, see B. Naiman. *J. Chem. Ed.*, **25**, 280 (1949).

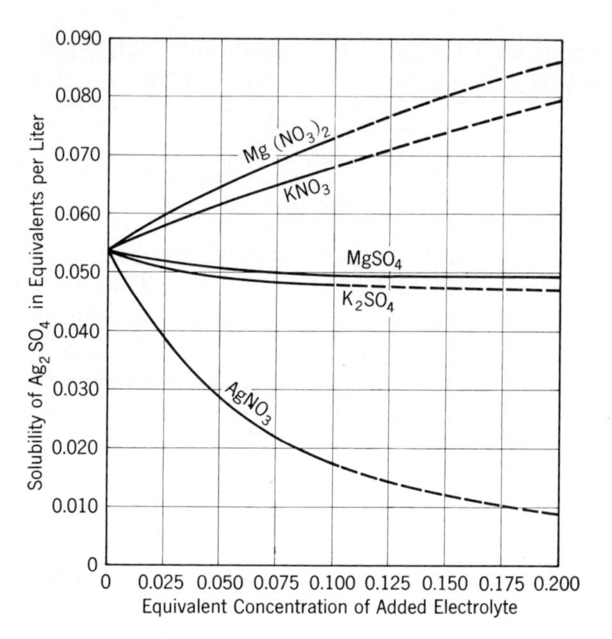

FIGURE 4-2

The solubility of silver sulfate in solutions of various salts. Broken lines are extrapolated values. [Data from article by A. A. Noyes. *J. Am. Chem. Soc.*, **46**, 1107 (1924).]

and the number of unit charges on the individual ions, the dielectric constant of the medium, and the temperature. In order to obtain the effective concentration, or, as it is termed, *activity*, a, to be used in equilibrium expressions, it is necessary to correct the concentration, c, by an appropriate factor termed the *activity coefficient*, γ; thus, $a = c\gamma$.

The solubility product for silver chloride then becomes

$$K_{sp} = a_{Ag^+}a_{Cl^-} = [Ag^+][Cl^-]\gamma_+\gamma_- \approx [Ag^+][Cl^-]$$

where a_{Ag^+} and a_{Cl^-} are the activities of the species. In very dilute solutions the activity coefficients approach unity and the solubility product is given accurately by the product of the ionic concentrations. Experimentally one cannot isolate the positive from the negative ions to measure their individual activity coefficients, so a mean activity coefficient is used:

$$\gamma_\pm = \sqrt{\gamma_+\gamma_-}$$

It is found that in all dilute solutions of the same total "ionic strength" the activity of any particular ionic species is the same. The ionic strength is defined by the expression

$$\mu = \tfrac{1}{2}\Sigma c_i z_i^2$$

where c_i is the concentration of the ith species and z_i is the charge on that species. Thus for a 0.10 F solution of KNO_3 the ionic strength is

$$\mu = \tfrac{1}{2}([K^+](1)^2 + [NO_3^-](-1)^2) = 0.10$$

In a solution which is 0.2 F in HNO_3 and 0.5 F in $BaCl_2$ the ionic strength is

$$\mu = \tfrac{1}{2}(0.2 \times 1^2 + 0.2 \times (-1)^2 + 0.5 \times 2^2 + 1.0 \times (-1)^2) = 1.7$$

One sees in Figure 4-2 that at equivalent concentrations $Mg(NO_3)_2$ has a greater effect than has KNO_3 upon the solubility of Ag_2SO_4. The ionic strengths, then, of the two solutions should be different in order to explain this observation, since the activity coefficient depends upon the ionic strength. We will calculate the ionic strength of 0.05 F KNO_3 and of 0.025 F $Mg(NO_3)_2$. For the 0.05 F (0.05 N) KNO_3 the ionic strength is numerically equal to the concentration

$$\mu = \tfrac{1}{2}(0.05 \times 1^2 + 0.05 \times (-1)^2) = 0.05$$

However, for the 0.025 F (0.05 N) $Mg(NO_3)_2$ the ionic strength is

$$\mu = \tfrac{1}{2}(0.025 \times 2^2 + 0.05 \times (-1)^2) = 0.075$$

which is 50% higher than that for the KNO_3 and which should, therefore, be expected to have a greater effect upon the activity coefficients and hence the solubility of silver sulfate.

The activity coefficients of electrolytes in quite dilute solutions can be calculated with considerable accuracy from theoretical considerations that are beyond the scope of this text; in addition, the activity coefficients of many substances have been experimentally derived from measurements of vapor pressure, freezing point, and electromotive force. In order to illustrate the magnitude of this effect as well as the agreement between theory and experiment, the calculated and observed values for a number of common electrolytes are shown in Table 4-1. It is to be emphasized that these are the activity coefficients of the electrolytes in pure solutions of the concentrations given, and that in the presence of other electrolytes the activity coefficients will have other values dependent upon the total ionic concentrations of the solution and the charges on the individual ions.

Because of the interionic attraction effects discussed above, it is unsafe to make quantitative use of equilibrium relationships unless the total ionic concentrations are below approximately 0.01 F or unless the activities, and

TABLE 4-1
Activity Coefficients of Typical Electrolytes (25°C)

Molality	0.005	0.01	0.05	0.10	0.50	1.00	3.00
Calculated for B^+A^-	0.93	0.90	0.81	0.76	0.62	0.56	0.49
Observed:							
For HCl	0.93	0.90	0.83	0.79	0.76	0.81	1.32
For NaCl	0.93	0.90	0.82	0.79	0.68	0.66	0.71
For KOH	0.93	0.90	0.81	0.76	0.67	0.68	0.90
For $AgNO_3$ (0°)	0.92	0.90	0.79	0.72	0.50	0.39	—
Calculated for $B^{2+}A_2^-$ or $B_2^+A^{2-}$	0.78	0.71	0.52	0.44	0.27	0.23	—
Observed:							
For $BaCl_2$	0.78	0.72	0.56	0.50	0.44	0.40	—
For $Pb(NO_3)_2$ (0°)	0.76	0.69	0.46	0.37	0.17	0.11	—
For K_2SO_4	0.78	0.71	0.53	0.44	0.26	0.21	—
Calculated for $B^{2+}A^{2-}$	0.56	0.46	0.24	0.16	0.066	0.045	—
Observed:							
For $MgSO_4$	0.57	0.47	0.26	0.19	0.091	0.067	—
For $CuSO_4$	0.56	0.44	0.23	0.16	0.066	0.044	—

not the concentrations, are used. In Figure 4-3 this effect is shown by plotting the solubility of thallous chloride against the equivalent concentration of various added salts. In this same figure the dotted line shows the calculated solubility (assuming complete ionization) in the presence of salts having a common ion. The increased solubility in the presence of the salts with no common ion is to be noted, and especially the effect of the binegative sulfate ion, which, having a molar concentration only half that of the nitrate, causes a greater solubility increase.

To summarize, it is seen that the deviations to be expected may be relatively large unless the total ionic concentration of the solution is small (0.01 F or less) and unless only singly charged ions are present. Because the calculation of the activity coefficients of individual ions in complex solutions is too involved to be treated here, concentrations will be used in mass-action expressions, but it will be understood that except in extremely dilute solutions of simple electrolytes, the results so obtained are of value only as an approximation of the effect to be expected.[5]

Effect of Complex and Un-ionized Compound Formation on Solubility. In deriving the solubility product of $BaSO_4$, the assumption was made that the concentration of un-ionized aqueous $BaSO_4$ is negligible for analytica

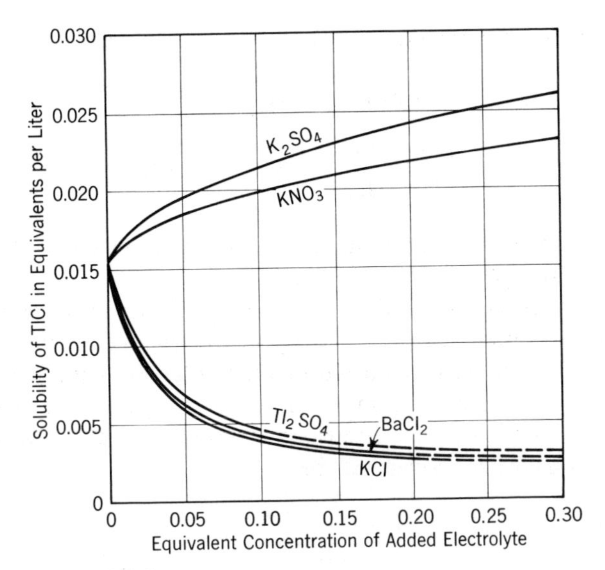

FIGURE 4-3

The solubility of thallous chloride in various salt solutions. Broken lines are extrapolated values. [Data from article by A. A. Noyes. *J. Am. Chem. Soc.*, **46**, 1107 (1924).]

considerations. With less ionic compounds this may not always be true; in addition, complex-ion formation can be a significant analytical factor with such compounds. These effects are illustrated by silver chloride.

Measurements by Jonte and Martin[6] on saturated solutions of silver chloride indicate that because of the reaction

$$AgCl(s) = AgCl(aq)$$

the concentration of aqueous un-ionized AgCl in equilibrium with solid AgCl is 3.6×10^{-7} *M*. Furthermore, they found aqueous AgCl to be a weak

---

[5] We should point out that the source of a small error has been ignored in the above discussion, namely, the use of two different concentration scales. These are molality (moles of solute per kg of solvent) and molarity (moles of solute per liter of solution). Fortunately for our considerations here the difference in these units is small so long as dilute aqueous solutions are used. The physical chemist most commonly uses molality—for good reason—and for equally good reason the analytical chemist commonly uses molarity or weight molarity (see page 15). The error introduced by using the wrong concentration scale is much smaller than the usual error introduced in the estimation of the activity coefficient.

Some excellent reading on the subject of activity considerations is available: J. N. Butler, *Ionic Equilibrium.* Reading, Mass.: Addison-Wesley, 1964, Chap. 12; A. A. Noyes. *J. Am. Chem. Soc.*, **46**, 1098 (1924).

[6] J. H. Jonte and D. S. Martin. *J. Am. Chem. Soc.*, **74**, 2052 (1952).

electrolyte with an ionization constant of 5×10^{-4}; that is, for the reaction

$$AgCl(aq) = Ag^+ + Cl^-$$

$$\frac{[Ag^+][Cl^-]}{[AgCl](aq)} = K = 5 \times 10^{-4}$$

By making use of this information one obtains a value for the solubility product of AgCl, $[Ag^+][Cl^-]$ of 1.8×10^{-10}.

The first conclusion to be drawn from these data is that regardless of the excess of either chloride or silver ion provided in a saturated solution, the formal solubility of silver chloride in such solutions cannot be decreased below the equilibrium concentration of un-ionized AgCl, that is, 3.6×10^{-7} *M*. Thus, if the chloride-ion concentration is made 1.0 *M*, the equilibrium silver-ion concentration will be

$$[Ag^+] = \frac{1.8 \times 10^{-10}}{1.0} = 1.8 \times 10^{-10}$$

However, this is only the molar silver-ion concentration; the concentration of un-ionized silver chloride will be approximately 2000 times greater than this.

In addition to the un-ionized AgCl, complex ions, such as $AgCl_2^-$ and $AgCl_3^{2-}$, will also be formed in significant quantities in a solution 1 *F* in chloride ion, and the formal solubility of silver chloride will be the sum of all the various silver species present. The formation of the complex ion $AgCl_2^-$ can be represented as

$$AgCl + Cl^- = AgCl_2^-$$

and the formation constant for $AgCl_2^-$ has been calculated as 90, that is,

$$\frac{[AgCl_2^-]}{[AgCl][Cl^-]} = 90$$

Thus if $[Cl^-] = 1.0$, the $[AgCl_2^-]$ can be calculated as follows, for a solution in equilibrium with solid silver chloride:

$$[AgCl_2^-] = (90)[AgCl][Cl^-]$$
$$= (90)(3.6 \times 10^{-7})(1.0) = 3.2 \times 10^{-5}$$

The contribution of the $AgCl_2^-$ alone to the formal solubility of silver chloride in a 1.0 *M* chloride solution is greater than the solubility in pure water; the experimental solubility of silver chloride in 1 *M* chloride solutions is approximately 10^{-4} *F*, with $AgCl_2^-$ and $AgCl_3^{2-}$ being the predominant species.

The very important practical conclusion to be drawn from these calculations is that where there is the possibility of complex-ion formation, we must be especially careful about attempting to reduce the solubility of a compound by application of the common-ion effect.

SUPPLEMENTARY REFERENCES

Bard, A. J. *Chemical Equilibrium.* New York: Harper and Row, 1966.

Butler, J. N. Calculating Molar Solubilities from Equilibrium Constants. *J. Chem. Ed.*, **38**, 460 (1961).

Butler, J. N. *Ionic Equilibrium.* Reading, Mass.: Addison-Wesley, 1964.

Clifford, A. F. *Inorganic Chemistry of Qualitative Analysis.* Englewood Cliffs, N.J.: Prentice-Hall, 1961.

King, E. J. *Qualitative Analysis and Electrolytic Solutions.* New York: Harcourt, Brace, 1959.

Ramette, R. W. Meaningful Stability Studies in Elementary Quantitative Analysis. *J. Chem. Ed.*, **33**, 610 (1956).

Ramette, R. W. Solubility and Equilibrium of Silver Chloride. *J. Chem. Ed.*, **37**, 348 (1960).

QUESTIONS AND PROBLEMS

1. In the accompanying table are given the formal solubilities (formula weights per liter of saturated solution) of certain salts at room temperature. Calculate solubility products for these salts, proceeding on the basis that they are completely ionized and assuming, first, that no significant hydrolysis of the ions occurs, and, second, that the molar concentrations of the ions represent their activities. State in which cases these assumptions are least justified.

Salt	Formal Solubility
(*a*) $BaSO_4$	1.0×10^{-5}
(*b*) $BaCrO_4$	1.6×10^{-5}
(*c*) Ag_2CrO_4	8×10^{-5}
(*d*) $PbCl_2$	3.9×10^{-2}
(*e*) $ZnCO_3$	1.6×10^{-4}
(*f*) Ag_3PO_4	1.6×10^{-5}

Ans. (*d*) 2.4×10^{-4}.

2. The formal solubilities at room temperature of some typical salts are as follows:

(*a*) $AgIO_3$, 1×10^{-4}
(*b*) Cu_2S, 1×10^{-13}
(*c*) Tl_3PO_4, 7×10^{-3}
(*d*) $Ba(BrO_3)_2{\cdot}H_2O$, 1.9×10^{-2}
(*e*) $BaSeO_4$, 2.9×10^{-5}
(*f*) $Pb_3(PO_4)_2$, 1.7×10^{-7}
(*g*) $La(OH)_3$, 8×10^{-6}
(*h*) $La_2(C_2O_4)_3{\cdot}9H_2O$, 1.1×10^{-6}
(*i*) $Ce(IO_3)_4$, 1.8×10^{-4}
(*j*) $MgNH_4PO_4{\cdot}6H_2O$, 2×10^{-3}

Calculate the solubility products for these salts; assume (1) that these salts in their saturated solutions are completely ionized, (2) that the resulting ions are not significantly hydrolyzed, and (3) that the molar concentrations of the ions are approximately equal to their activities. State in which cases these assumptions introduce serious errors.

3. Calculate the molar concentrations of the silver and the chloride ions resulting after attainment of equilibrium when to 50.00 ml of 0.1000 *N* sodium chloride there are added (*a*) 49.90 ml, (*b*) 50.00 ml, and (*c*) 50.10 ml of 0.1000 *N* $AgNO_3$.

4. A solution containing 0.06005 g of CH_3COOH (acetic acid) is diluted to 1.500 liters. What is the formal concentration of the acetic acid? It has been calculated from conductance measurements that the acetic acid in this solution is 15% ionized. What is the molar concentration of the (*a*) CH_3COOH, (*b*) H^+, and (*c*) CH_3COO^- in the solution? Calculate the ionization constant of acetic acid from the data given.

5. (*a*) Chloroform (trichloromethane, $CHCl_3$) and water solutions saturated at 25°C with iodine are 0.169 *F* and 0.0013 *F* in I_2, respectively. Calculate a value for the distribution ratio for iodine between $CHCl_3$ and water. What assumptions are made regarding the formal solubilities? *Ans.* 130.
   (*b*) Calculate a distribution ratio for iodine between $CHCl_3$ and CCl_4 (assume the two liquids immiscible).

6. (*a*) An aqueous solution containing 2.0 mg of iodine in 10 ml of solution was shaken until equilibrium was established with 3 successive 5-ml portions of CCl_4. Calculate the mg of iodine remaining in the aqueous solution. *Ans.* 2.4×10^{-5} mg.
   (*b*) The process was repeated with a similar aqueous solution, but 5 successive 3-ml portions of CCl_4 were used. Calculate the mg of iodine remaining in the aqueous solution.

   What conclusion can be reached from a consideration of these values?

7. What would be the effect of the presence of KI in the aqueous solution upon the distribution of iodine between aqueous and CCl_4 solutions? (See page 447 regarding the tri-iodide ion.)

8. Make use of the data in the Appendix to calculate the $[Cu^+]$ in a saturated solution of CuCl in water; in 0.010 *F* NaCl. Are these the formal solubilities of CuCl in these solutions? What other factors would affect the formal solubilities? *Ans.* $[Cu^+] = 1 \times 10^{-4}$.

9. Proceed as directed in Problem 8 and calculate the $[Pb^{2+}]$ in a saturated solution of $PbCl_2$ in water.

10. Calculate the solubility of silver acetate in water; in 0.10 *F* HNO_3.

11. Calculate the *p*H at which ZnS would just begin to precipitate (assuming equilibrium) from a solution at room temperature that is 10^{-3} *M* in Zn^{2+} and saturated with H_2S at atmospheric pressure. (An aqueous solution in equilibrium with H_2S gas at 1 atmosphere pressure is 0.10 *M* in H_2S; see Appendix 1 for K_{sp} values.) *Ans.* *p*H = 0.5.

12. Calculate the formal solubility of silver thiocyanate in water; calculate also the formal solubility in 0.1 *F* ammonia; note the effect of the ammonia. (See Appendixes for necessary constants.)

13. Calculate the ionic strength of (*a*) 0.020 *F* $NaNO_3$ and (*b*) 0.020 *F* $Sr(NO_3)_2$. (*c*) In which of these solutions would you expect AgCl to have the greater solubility? Explain. *Ans.* (*b*) 0.060.

14. What formal concentration of Na_2SO_4 has the same ionic strength as a solution which is 0.010 *F* in $Fe(NO_3)_3$? *Ans.* 0.020 *F*.

15. According to an extended form of the Debye–Hückel relationship the activity coefficient, γ, in a solution of a salt, B_nA_m, is related to the ionic strength in dilute solutions by the equation

$$-\log \gamma = \frac{0.5 z_A z_B \sqrt{\mu}}{1 + \sqrt{\mu}}$$

In this expression, z_A and z_B are the charges on A and B, and 0.5 is the approximate value in water at 25°C of a constant that is dependent upon the temperature and dielectric constant of the solvent. Calculate the activity coefficient for 0.01 *F* $B^{2+}A^{2-}$. (The answer is given in Table 4-1.)

CHAPTER 5

Chemical Reactions. Precipitation

In Chapter 4 the law of chemical equilibrium was applied to heterogeneous systems, that is, to those systems containing more than one phase. In this chapter we will consider more detailed application of that law to those heterogeneous systems in which one of the phases is a solid.

A reaction that proceeds because a slightly soluble substance is formed is classed as a precipitation reaction. Use is made of such reactions in two different ways for the quantitative determination of substances: first, by gravimetric methods in which the substance is precipitated, dried, and weighed (Chapters 11 through 14); and second, by titrimetric methods in which the precipitating agent is added as a solution of known concentration until precipitation is essentially complete. From the measured volume or weight of solution used and its concentration, we can calculate the quantity of unknown present. These two applications are discussed in the following sections; discussions of titrimetric methods are given in Chapter 17.

Gravimetric Applications

A reaction such as that between silver ion and chloride ion,

$$Ag^+ + Cl^- = AgCl(s) \tag{1}$$

which results in the formation of a slightly soluble salt, can, in principle, be used for the quantitative estimation of either of the species involved in the reaction. To a solution of the species to be determined, say chloride, is added a solution of silver ion until further addition causes no more observable precipitate to form. A small excess of the precipitant is added, and the precipitate is filtered, dried, and weighed. From the weight of the precipitate of AgCl, the quantity of chloride that was present in the unknown sample can be calculated.

In broad outline, then, the gravimetric application of precipitation reactions is very simple. It is sufficient for our present purposes to state briefly the requirements that must be satisfied. They are: (1) a substance of very low solubility must be formed; (2) this precipitate must have such physical characteristics that it can be separated from the solution and washed free of interfering substances; and (3) it must be capable of being dried to a constant and known composition and being weighed. If these three requirements are not met, the reaction cannot be used successfully for a gravimetric method; if the requirements are met, there may yet remain considerable challenge in the effects of such factors as nucleation, crystal growth, coprecipitation, mixed-salt formation, and hydration. These factors will be discussed as they relate to specific procedures in Chapters 12 and 14.

Titrimetric Applications

The application of precipitation reactions to titrimetric methods also requires that certain qualifications must be satisfied. They are: (1) as for a gravimetric method, a precipitate of low solubility must be formed; (2) this precipitate must be of known and constant composition; (3) precipitation must take place with reasonable rapidity; and (4) a practical means must be available for determining the end point, that is, the point at which the titration is terminated. Let us consider again the reaction of silver ion and chloride ion to give the white precipitate of silver chloride, this time viewing the reaction as the basis of a titrimetric method for the determination of chloride. One of the oldest titrimetric methods, known as the Mohr method,[1] makes use of this reaction and employs the reaction between silver ion and chromate as the indicator reaction:

$$2Ag^{+} + CrO_4^{2-} = Ag_2CrO_4(s) \qquad \text{(red precipitate)} \qquad (2)$$

Thus, the reaction in Equation 1, *the titration reaction*, permits the stoichiometric calculation of the quantity of chloride present, and the reaction given

[1] F. Mohr. *Liebigs Ann. Chem.*, **97**, 335 (1856). It is of interest to note that many of the classical methods of analytical chemistry were developed experimentally before the formulation of the law of chemical equilibrium by Guldberg and Waage (1864) or the explanation of the ionic nature of electrolytes by Arrhenius (1887). The work of the chemists of that era provided a quantitative basis for the later developments in chemistry.

in Equation 2, *the indicator reaction*, shows when the titration is to be stopped.[2] As long as there is a significant amount of chloride ion in the titrated mixture, the concentration of the silver ion remains so small, because of the precipitation of silver chloride, that a permanent precipitate of the more soluble silver chromate does not form. When the amount of silver added becomes approximately equivalent to the chloride present, both ions are present in very small concentrations (approximately 10^{-5} *M*); then, further addition of the standard silver solution (usually approximately 10^{-1} *M*) causes the concentration of the silver ion in the solution to rise very rapidly. This results in the solubility product of silver chromate being exceeded and the precipitation of a perceptible amount of silver chromate, which is red-colored, as a result of the reaction in Equation 2. The perception of this precipitate is taken as the end point.

A clear distinction must be made between the *end point* and the *equivalence point.* The end point is the point at which the titration is stopped; the equivalence point is the point at which an amount of the titrant just equivalent to the substance being titrated has been added. Thus, if the end point and equivalence point coincide, there is no error; if those points do not coincide, the extent of the lack of agreement between them is the measure of the error in the titration.

The changes in the concentrations of the chloride and the silver ions during the course of the titration can be predicted from the equilibrium constant for the reaction

$$AgCl(s) = Ag^{+} + Cl^{-} \qquad (3)$$

As soon as enough titrant has been added to cause a precipitate of silver chloride to be formed, if equilibrium obtains, this equilibrium can be represented by the above equation.[3] Therefore, when equilibrium is attained

[2] One often reads such misleading statements as "after all the AgCl precipitates, Ag_2CrO_4 starts to precipitate." The principles of chemical equilibrium show the fallacy of such a statement. The addition of very large quantities of silver ion will decrease the chloride-ion concentration to some very low value but never to zero; the addition of more Ag^+ will only take the chloride ion to lower values. This is seen clearly in the solubility product for silver chloride:

$$K_{sp} = [Ag^+][Cl^-]$$

Any value chosen for $[Ag^+]$ still leaves a concentration greater than zero for $[Cl^-]$.

The $[Cl^-]$ will be lowered by addition of more Ag^+, but *the total chloride* in solution may be increased by added Ag^+ because of the formation of such complex species as Ag_2Cl^+. In addition, as explained in Chapter 4, AgCl is not a completely ionized salt and there will always be a small but fixed concentration of un-ionized AgCl in equilibrium with the solid AgCl; that is, AgCl(s) = AgCl(aq) and [AgCl(aq)] is 3.6×10^{-7} *M*.

[3] As has been explained on page 54, calculations of the solubility product for silver chloride from the formal solubility are somewhat in error because of the small but measurable concentration of un-ionized AgCl(aq) that exists in equilibrium with the solid AgCl. The concentration of this species is relatively small, however, and the shapes of the titration curves that we will now plot are not affected by the assumption of complete ionization of the AgCl *in solution.*

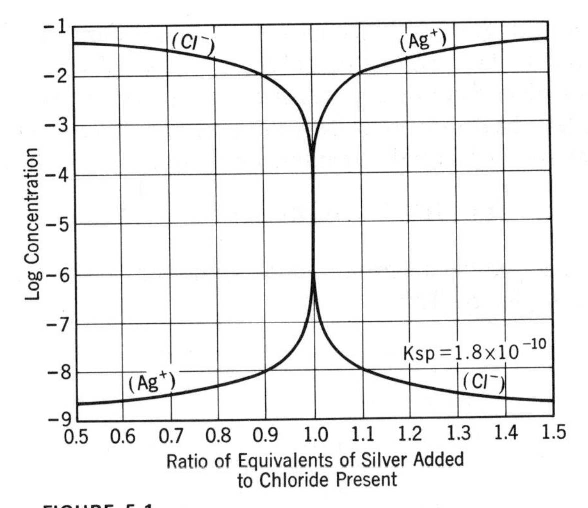

FIGURE 5-1

Changes in silver and chloride-ion concentrations when titrating a soluble chloride with silver nitrate.

(that is, the solution is saturated, but not supersaturated, with AgCl) the equilibrium expression can be formulated:

$$[Ag^+][Cl^-] = K_{sp} \tag{4}$$

This expression states that in dilute solutions in equilibrium with solid AgCl, the product of the concentrations of the two ions has a constant value at a given temperature. Figure 5-1 shows two curves obtained as the ratio of the equivalents of silver added to the equivalents of chloride initially present is increased. The curve labeled $[Ag^+]$ is obtained by plotting the log of the silver-ion concentration; the curve labeled $[Cl^-]$ is obtained by plotting the log of the chloride-ion concentration.

As an example of the method of calculating these values, consider the case in which the ratio of equivalents of silver added to chloride has the value 0.90. For simplicity assume that the volume remains constant throughout the titration. First, it is necessary to obtain the solubility product of silver chloride. This was calculated in Chapter 4 to be 1.8×10^{-10}. Therefore,

$$[Ag^+][Cl^-] = K_{sp} = 1.8 \times 10^{-10} \tag{5}$$

If a standard solution of silver nitrate is added to a 0.1 M Cl^- solution until the ratio of equivalents of silver ion added to equivalents of chloride ion initially present is 0.90, the concentration of the *excess* chloride ion will be 0.01 M (if the volume is assumed to be unchanged). This should be apparent when it is noted that only a tenth of the original chloride ion is in excess. Now, the *total* molar concentration of chloride ion will be 0.01 *plus* the molar solubility of silver chloride in a solution which was 0.01 M in chloride ion. Thus, if we let x be the silver-ion concentration then $(0.01 + x)$ is the chloride ion, and the solubility-product expression becomes

$$[Ag^+][Cl^-] = x(0.01 + x) = 1.8 \times 10^{-10}$$

It is easily seen that x must be very small compared with 0.01, so the additive term can be neglected. Thus

$$x(0.01) \approx 1.8 \times 10^{-10}$$

and

$$x \approx 1.8 \times 10^{-8}$$

The assumption that x is very small compared with 0.01 is seen to be valid; the silver-ion concentration is 1.8×10^{-8} M and the chloride-ion concentration is $(0.01 + 1.8 \times 10^{-8})$ or 0.01 M.[4]

The two curves of Figure 5-1 show the changes in the silver- and chloride-ion concentrations as one titrates a chloride solution with silver ion. The curves show that as the ratio of the equivalents of silver to chloride is increased, the chloride-ion concentration decreases and the silver-ion concentration increases until they are equal at the equivalence point. These concentration changes are in agreement with the requirement of the solubility product equation. Note that the concentrations of silver and chloride ions undergo great change in the immediate region of the equivalence point. Indeed, in this titration, when the ratio of equivalents of silver to chloride approaches unity, the change of the concentrations of these ions with a given change of the ratio of equivalents reaches its maximum value. This rapid change of concentration near the equivalence point is of fundamental importance, since by the addition of a small amount of the standard solution (in this case the silver) such a large change in the ionic concentrations of the solution is produced that some effect producing an end point will take place (in this case the formation of a visible amount of the red silver chromate precipitate). As will be observed later, this rapid change near the equivalence point is an essential characteristic of all "titration curves" and is one of the features that determine the precision of the particular method involved.

[4] In the solution of the example just given, a technique was used that is most useful in equilibrium problems. The expression was formulated as exactly as possible, and only after this explicit formulation were methods of simplification considered. A good rule is to write the exact expression, then to try approximations, and finally to check the approximations.

Consider in this respect the titration of chloride ion with silver ion in a solution that has been made 0.01 M in chromate ion. When silver chromate dissolves, the reaction can be represented as

$$Ag_2CrO_4(s) = 2Ag^+ + CrO_4^{2-}$$

and at equilibrium the solubility product for this reaction is

$$[Ag^+]^2[CrO_4^{2-}] = K_{sp}.$$

If pure water is saturated with silver chromate at 25°C the solubility of the salt is 8.4×10^{-5} F. If we assume that the dissolved salt is completely ionized and that there is no significant hydrolysis,[5] the concentrations of species are

$$[Ag^+] = 2 \times 8.4 \times 10^{-5} = 1.7 \times 10^{-4}\ M$$

and

$$[CrO_4^{2-}] = 8.4 \times 10^{-5}\ M$$

and the solubility product becomes

$$(1.7 \times 10^{-4})^2(8.4 \times 10^{-5}) = K_{sp} = 2.4 \times 10^{-12}$$

Therefore, if the chromate-ion concentration is 0.01 M, the silver ion must become

$$[Ag^+] = \sqrt{\frac{2.4 \times 10^{-12}}{0.01}} = 1.5 \times 10^{-5}\ M$$

in order to saturate the solution.

It is seen from Figure 5-1 that this concentration of silver ion will not be reached until after the ratio of silver ion to chloride has become greater than unity (that is, until after the equivalence point has been passed), but that it will be reached before the ratio has become as great as 1.001. Therefore, if the end point were taken at that concentration, the error in the titration would be less than 1 part in 1000, or 0.1%. In actual practice such solutions always become supersaturated to some extent; also, a finite amount of the silver chromate must be present before it can be perceived. Therefore the

---

[5] Chromate reacts with hydrogen ion to form $HCrO_4^-$, which polymerizes by the reaction $2HCrO_4^- = Cr_2O_7^{2-} + H_2O$ to give dichromate. If the hydrogen-ion concentration is about 10^{-7} or less, these reactions do not occur significantly. A quantitative treatment of this question is given in Chapter 17; in the calculation above, the $[H^+]$ is assumed to be below 10^{-7}.

actual end point will not be reached until the ratio of silver added to the chloride present is greater than the above value.

CALCULATION OF THE TITRATION ERROR

How can calculation of the theoretical error in the Mohr titration be made? First, if the titration is made without error—that is, if the end point and equivalence point coincide—then there must have been just as many equivalents of silver ion added as there were equivalents of chloride present. Thus

$$\Sigma \text{ eq Ag} = \Sigma \text{ eq Cl}$$

where the Greek letter Σ means summation or total, Σ eq Ag means "total equivalents of silver ion added" (whether precipitated or in solution), and Σ eq Cl means "total equivalents of chloride present". (The symbols Ag and Cl were purposely written without charges to emphasize that the summation is over all species that include silver or chloride.) The equality above can be restated as

$$\text{T.E.} = 0 = \Sigma \text{ eq Ag} - \Sigma \text{ eq Cl}$$

where T.E. is the titration error expressed in equivalents. If the end and equivalence points do not coincide, the T.E. is no longer zero but is given by the expression

$$\text{T.E.} = \Sigma \text{ eq Ag} - \Sigma \text{ eq Cl}$$

A word about the sign convention used here may be helpful. The titration is done with silver ion, and the *correct* result in the titration is, of course, the number of equivalents of chloride that were being titrated. The analyst determines how many equivalents of chloride were present from the number of equivalents of the known silver used. Therefore, if he uses too much silver ion, he will report too much chloride ion. This is a positive error; therefore the above expression is written with a positive sign on Σ eq Ag so that if the value of Σ eq Ag exceeds the value of Σ eq Cl, the T.E. is positive. Similarly a negative value of the T.E. signifies that less silver was added than the equivalent amount and that therefore a low result is obtained in the determination of chloride. This is called a negative error.

The titration error can easily be converted to a percentage titration error (% T.E.) by dividing the T.E. by the correct value for the equivalents of chloride and changing from decimal fractions:

$$\%\ \text{T.E.} = \frac{\Sigma \text{ eq Ag} - \Sigma \text{ eq Cl}}{\Sigma \text{ eq Cl}} \times 100$$

The silver that has been added is present at the endpoint as silver ion, as solid silver chloride, as un-ionized silver chloride, and as solid silver chromate; the chloride is present as chloride ion, as solid silver chloride, and as un-ionized silver chloride. Thus

$$\Sigma \text{ eq Ag} = \text{eq Ag}^+ + \text{eq AgCl(s)} + \text{eq AgCl(aq)} + \text{eq Ag}_2\text{CrO}_4\text{(s)}$$

and

$$\Sigma \text{ eq Cl} = \text{eq Cl}^- + \text{eq AgCl(s)} + \text{eq AgCl(aq)}$$

The percentage titration error is

$$\%\text{T.E.} = \left[\frac{(\text{eq Ag}^+ + \text{eq AgCl(s)} + \text{eq AgCl(aq)} + \text{eq Ag}_2\text{CrO}_4)}{\Sigma \text{ eq Cl}} - \frac{(\text{eq Cl}^- + \text{eq AgCl(s)} + \text{eq AgCl(aq)})}{\Sigma \text{ eq Cl}}\right] \times 100$$

$$= \frac{\text{eq Ag}^+ + \text{eq Ag}_2\text{CrO}_4 - \text{eq Cl}^-}{\Sigma \text{ eq Cl}} \times 100 \qquad (6)$$

The equivalents of silver chromate represent the amount of that precipitate that must be formed before it becomes visible. Experiments have shown that on the average 0.25 ml of 0.01 N $AgNO_3$ is required to produce a perceptible precipitate of silver chromate in 100 ml of a solution that was initially 6×10^{-3} F in K_2CrO_4; therefore, 2.5×10^{-6} equivalent of silver gives a perceptible precipitate. Since 6×10^{-3} mmole/ml $\times$ 100 ml $=$ 0.6 mmole of CrO_4^{2-} was present in the solution and only 2.5×10^{-3} meq (or mmole) of silver has been added, it is evident that the chromate-ion concentration is virtually unaffected by the addition. That is, if *all* the silver ion added were to precipitate as Ag_2CrO_4, only 2.5×10^{-3} meq or 1.3×10^{-3} mmoles of CrO_4^{2-} could be precipitated. This represents only about 0.2% of the chromate ion initially present, and the chromate-ion concentration is still 6×10^{-3} M.

The solubility product of Ag_2CrO_4 is 2.4×10^{-12}; from this and the known chromate-ion concentration we can calculate the silver-ion concentration:

$$[\text{Ag}^+]^2[\text{CrO}_4^{2-}] = 2.4 \times 10^{-12}$$

hence

$$[\text{Ag}^+] = \sqrt{\frac{2.4 \times 10^{-12}}{\text{CrO}_4^{2-}}} = \sqrt{\frac{2.4 \times 10^{-12}}{6 \times 10^{-3}}}$$

$$[\text{Ag}^+] = 2.0 \times 10^{-5}\ M$$

Since the solution volume is 100 ml there will be 2.0×10^{-6} mole or 2.0×10^{-6} eq of Ag^+ present in the solution; therefore 0.5×10^{-6} equivalent of silver has been precipitated as silver chromate. If we apply these conditions to a titration, taking the solubility product of AgCl as 1.8×10^{-10}, we see that in this solution with $[Ag^+] = 2.0 \times 10^{-5}$ *M* the chloride-ion concentration will be 9×10^{-6} *M*, and there will be (in the 100-ml solution) 0.9×10^{-6} equivalent of chloride. If we assume some reasonable initial quantity of chloride, say 25 ml of 0.1 *M* solution or 2.5×10^{-3} equivalent, all of the quantities are available for substitution in Equation 6 above and the percentage titration error can be evaluated:

$$\%\text{T.E.} = \frac{2.0 \times 10^{-6} + 0.5 \times 10^{-6} - 0.9 \times 10^{-6}}{2.5 \times 10^{-3}} \times 100$$

$$= 0.06\%, \text{ positive error}$$

From an inspection of the factors in the expression, it is seen that the magnitude of the percentage error is affected by the following: (1) the amount of Ag_2CrO_4 precipitated, (2) the chromate-ion concentration (which directly controls the concentration of silver ion and indirectly that of chloride ion), and (3) the equivalents of chloride in the sample. If more dilute solutions—say 0.01 *M*—are used, it should be apparent that the error becomes quite serious for a given volume of standard solution, since the term Σ eq Cl in the denominator of Equation 6 becomes much smaller. (Other examples of titration-error calculations are given in Chapter 17 in the discussion of the Volhard method for determination of silver in an alloy.)

Figure 5-2 is a composite plot of the region near the equivalence point of titration curves for various anions forming slightly soluble silver salts. It is seen that the position of the curve and the rate of change of the concentration of the ions involved near the equivalence point are characteristics of the solubility product of the salt. Thus, if a titration were made of the anion of a silver salt with a solubility product of 2×10^{-8} (corresponding to that for silver iodate), the concentration of the silver ion would reach 2×10^{-5} *M* when the ratio of the equivalents was only about 0.99, and thus a negative error of as much as 1% might result. In this regard it is to be remembered that one of the requirements set up for the use of a reaction as the basis of a precise volumetric method was that it should be quantitatively complete. Obviously, the more soluble the salt, the less complete its precipitation reaction. As will be observed later, it is a general feature of titration curves that the less complete the titration reaction, the less pronounced the inflection of the curve near the equivalence point. One should note that these curves are characteristic of other di-ionic salts; for example, the concentration curves for barium and sulfate ions in the precipitation of barium sulfate, $K_{sp} = 10^{-10}$,

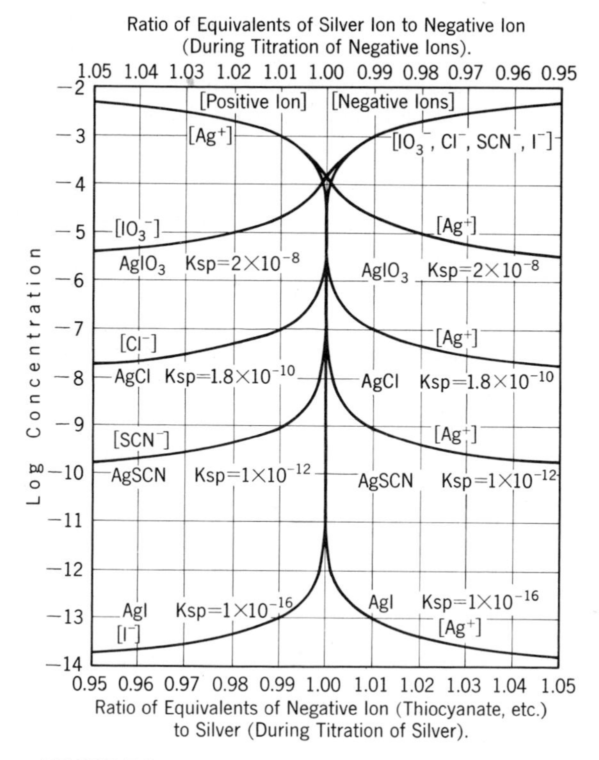

FIGURE 5-2

Concentration changes when titrating various anions with silver nitrate, or the reverse.

would approximately coincide with those above for AgCl. Unfortunately, barium sulfate does not meet the requirement for a titrimetric-precipitation process that precipitation occur rapidly; its solutions tend to supersaturate, that is, the precipitation equilibrium is attained slowly.

SUPPLEMENTARY REFERENCES

See the references given for Chapter 4, page 55.

QUESTIONS AND PROBLEMS

1. Calculate the volume of 0.10 N $AgNO_3$ that would be required to saturate with silver chromate 100 ml of a solution containing 0.30 g of K_2CrO_4. *Ans.* 0.012 ml.

2. A soluble chloride is to be titrated with silver nitrate, with a soluble chromate being used as indicator. Calculate what should be the chromate-ion concentration in order that the solution should become just saturated with silver chromate when an amount of silver nitrate just equivalent to the chloride present has been added. *Ans.* $[CrO_4^{2-}] = 1.1 \times 10^{-2}$.

3. Repeat the calculation requested in Problem 2 for the case in which a soluble iodide is to be titrated with silver nitrate, with a soluble chromate being used as indicator.

   What does the result of the calculation indicate in regard to the practicability of such a titration of iodide?

   NOTE: Regardless of the above consideration, the titration of iodide or thiocyanate by the Mohr method is not satisfactory because the absorptive tendency of the precipitate makes the determination of the end point uncertain.

4. Calculate the volume of 0.10 N $AgNO_3$ that would have to be added to 100 ml of a solution in order to raise the silver-ion concentration from 1.0×10^{-8} M to 1.0×10^{-5} M. *Ans.* 1.0×10^{-2} ml.

   In what way does this calculation apply to the practicability of the titration of Problem 3, without consideration of the effects mentioned in the note accompanying that problem?

5. If the formal solubility of AgCl in 3 F NH_3 is 1.0×10^3 times that of AgI, what is the ratio of the solubility products of the two halides? *Ans.* 10^6.

6. Calculate (with the aid of the solubility product and the complex-dissociation constant) the formal solubility of silver chloride in 1 F NH_3.

7. Calculate the formal solubility of silver bromide in 1 F NH_3. *Ans.* 2.8×10^{-3}.

8. How many milligrams of silver chloride could be dissolved by 50 ml of 6 F NH_3?

9. Pure water is saturated with both solid silver chloride and silver iodide. Calculate the ratio of the molar concentrations of chloride to iodide in the solution.

10. Solid AgCl and AgI were added to a solution that was initially 1 F in NH_3 until an excess of both solids was present. Calculate the ratio of the molar concentrations of chloride to iodide after attainment of equilibrium. Compare this value with that in Problem 9 above. (Note that this value sets a theoretical limit to the completeness of the separation of iodide and chloride as silver salts.)

11. Calculate the solubility of lead sulfate (expressed as mg of Pb per ml) in a solution 0.30 F in $(NH_4)_2SO_4$ and having (*a*) a pH of 3.0 and (*b*) a pH of 0.0. *Ans.* (*a*) 1.5×10^{-5} mg/ml.

CHAPTER 6

Chemical Reactions. Oxidation-Reduction

The subject of oxidation and reduction is discussed in detail in your general chemistry text; you should review this material. A brief review of certain aspects of the subject that are especially useful are presented below.

When an atom loses one or more electrons, it is said to be *oxidized*; when an atom gains one or more electrons, it is said to be *reduced*. Therefore, *oxidation* is a *loss of electrons* and *reduction* is a *gain of electrons*. In covalent compounds this loss is represented by a shift of electrons away from one element (oxidation) towards another (reduction). The term oxidation was originally used to describe the reaction of an element with oxygen to form one or more oxides; the removal of oxygen was a reduction.

An *oxidation-reduction* or *redox* reaction is one in which there is a transfer or shift of electrons from one reactant, the *reductant* (or reducing agent), to another, the *oxidant* (or oxidizing agent).

When chlorine reacts with iodide ion in an aqueous solution, the reaction can be formulated as the result of two changes in the number of electrons associated with these elements, thus

$$\begin{aligned} Cl_2 + 2e^- &= 2Cl^- \\ 2I^- - 2e^- &= I_2 \\ \hline Cl_2 + 2I^- &= 2Cl^- + I_2 \end{aligned}$$

The chlorine has acted as an electron acceptor (oxidant) and the iodide ion has acted as an electron donor (reductant). In the course of the reaction, the chlorine, having gained electrons, has been reduced; the iodide, having lost electrons, has been oxidized.

There are many oxidation-reduction reactions in which there is not a complete transfer of electrons. When the reaction

$$S + O_2 = SO_2$$

takes place, there is only a shift of the outer electrons of the sulfur atom toward the oxygen atoms and not a complete transfer. However, it is customary to state that the sulfur is oxidized and the oxygen reduced. Also, consider the electrolytic oxidation of manganous ion, Mn^{2+}, to MnO_2, in an aqueous solution. Two electrons are removed from the manganous ion by the anode, but the resultant increased positive charge on the manganese causes the formation of electron-pair bonds with the oxygen of the water molecules, and MnO_2 results. The anode reaction can be formulated as

$$Mn^{2+} + 2H_2O = MnO_2(s) + 4H^+ + 2e^-$$

Here again, even though the manganese is surrounded by more electrons, the outer electrons have been shifted toward the oxygen atoms.

There are also cases where a redox reaction can involve the shift of an atom, or atoms, from one molecule to another. As an example, studies have been made of the oxidation of sulfite by hypochlorite in which hypochlorite prepared with ^{18}O was used; after the reaction the ^{18}O was found with the sulfate. The reaction can be formulated as

$$Cl^{18}O^- + SO_3{}^{2-} \longrightarrow Cl^- + SO_3^{18}O^{2-}$$

In effect the oxygen has caused the shift of two electrons towards the chlorine and two away from the sulfur; as a result the chlorine atom is considered to have been reduced and the sulfur atom oxidized.

OXIDATION STATES AND OXIDATION NUMBERS

Chemists use a consistent system of assigning to atoms numbers that are related to the charge on the atoms. These numbers—called *oxidation numbers*—designate the *oxidation states* of the elements. For example, the oxidation numbers of several species are shown below.

Fe	0	Cl_2	0
Fe^{2+}	+2	Cl^-	−1
Fe^{3+}	+3	He	0

Oxidation states of the elements are commonly indicated by the symbol for the element followed by Roman numerals in parentheses; ferrous and ferric iron are thus Fe(II) and Fe(III). In this text, a minus sign preceding the number indicates a negative oxidation state; the chlorine in the chloride will thus be written Cl(−I). Two rules can be used to assign oxidation numbers in the above examples. They are:

First, each substance in its elementary state is assigned the oxidation number 0.

Second, a simple ion has an oxidation number of the same magnitude and sign as the charge on the ion.

A few additional rules enable one to assign oxidation numbers to elements in compounds and complex ions:

1. In most compounds hydrogen has the oxidation number +1. Exceptions are the metal hydrides (NaH, CaH_2, etc.).
2. In most compounds oxygen has the oxidation number −2. Exceptions are the peroxides (Na_2O_2, H_2O_2, etc.), the superoxides (KO_2, etc.), and the oxygen fluorides.
3. Metals in compounds have positive oxidation numbers.
4. In their stable compounds the alkali metals (Li, Na, K, Rb, Cs) have the oxidation number +1.
5. In their stable compounds the alkaline earth metals (Be, Mg, Ca, Sr, Ba) have the oxidation number +2.
6. The maximum positive oxidation number that an element can have is usually given by the group number at the top of the periodic table. (Cu, Ag, and Au are exceptions to this rule.)
7. In a covalent compound of known structure, the oxidation number of an atom is the charge it would have if each shared electron pair in the compound is assigned to the more electronegative of the two sharing atoms. Unless there is evidence to the contrary, when an electron pair is shared by two atoms of the same element, one electron is assigned to each atom.[1]

As a consequence of this rule, when fluorine, which is the most electronegative element, is combined with other elements, it is assigned an oxidation number of −1; oxygen, the next most electronegative element, when combined with an element other than fluorine is assigned the

[1] An example of an exception to this rule is the thiosulfate ion, which is discussed after Rule 8:

$$\begin{array}{c} \text{O} \\ \text{O}:\ddot{\underset{\cdot\cdot}{\text{S}}}:\text{S}^{2-} \\ \text{O} \end{array}$$

Here the electron pair is not equally shared by the two sulfur atoms.

oxidation number −2 (except as noted above); hydrogen, when bonded to nonmetals, has the oxidation number +1.

8. The algebraic sum of the oxidation numbers of the constituent atoms of a neutral molecule equals zero, and, in the case of a complex ion, the sum of the oxidation numbers must equal the ion charge. Application of these rules leads to an oxidation number of +2 for zinc in zinc oxide, of −3 for nitrogen in ammonia, and of +6 for sulfur in sulfuric acid or the sulfate ion.

When two or more atoms of the same element are present in a compound of unknown structure, application of the above rules gives the *average* oxidation state of these atoms. Thus, without reference to its structure, one obtains an average oxidation state of +2 for the sulfur atoms in sodium thiosulfate, $Na_2S_2O_3$, although structural studies have indicated that the oxidation states of the two atoms are +6 and −2, respectively. However, this average oxidation state is useful information, since it enables one to balance chemical equations.

OXIDATION AND REDUCTION REACTIONS

We can use the concept of oxidation numbers to define oxidation as a *change to a more positive oxidation number*. Each of the following steps is illustrative of an oxidation:

Mn^{2+}	to	MnO_4^-	Br^-	to	Br_2
+2		+7	−1		0
Cl^-	to	ClO_3^-	NH_3	to	NH_2OH
−1		+5	−3		−1

There is no requirement that the final oxidation number be positive; in the oxidation of ammonia (NH_3) to hydroxylamine (NH_2OH), the change is to a more positive (less negative) oxidation number, but both initial and final states are negative.

Reduction is the opposite of oxidation and therefore can be defined as a *change to a less positive oxidation number*. The reverse of each step shown above is a reduction.

As was stated above, the equation for an oxidation-reduction reaction can be separated into two reactions, each involving electrons, one of which involves an oxidation, the other a reduction. Reactions of this type are called *half-cell* reactions, because, as is shown below, they can be considered to be the reactions taking place at the two electrodes of an electrochemical cell. This separation of an over-all reaction into two half-cell reactions is illustrated below by the reduction of permanganate by ferrous iron.

Over-all reaction:

$$MnO_4^- + 5Fe^{2+} + 8H^+ = Mn^{2+} + 5Fe^{3+} + 4H_2O \quad (1)$$

Half-cell reaction; reduction of MnO_4^-:

$$MnO_4^- + 8H^+ + 5e^- = Mn^{2+} + 4H_2O \quad (2)$$

Half-cell reaction; oxidation of Fe^{2+}:

$$5Fe^{2+} = 5Fe^{3+} + 5e^- \quad (3)$$

The presence of the electrons in these half-cell reactions often makes it convenient to speak of reduction as a gain of electrons (electrons had to be added to the MnO_4^- side of the half-cell reaction to get the reduction to Mn^{2+}) and oxidation as a loss of electrons (Fe^{2+} lost electrons in becoming Fe^{3+}). In the case of the very simple half-cell reaction, Equation 3, it is clear that each ferrous ion did, indeed, lose an electron and become ferric ion. However, while five electrons are involved in Equation 2, it is much less clear just which atom gains them; the same is true of the nonionic reaction $S + O_2 = SO_2$. But in either Equation 2 or the SO_2 example, there is no problem assigning oxidation numbers regardless of which atoms are oxidized and which are reduced.[2]

Half-cell reactions such as those in Equations 2, 3 and

$$Cr^{2+} = Cr^{3+} + e^- \quad (4)$$

$$2Cl^- = Cl_2(g) + 2e^- \quad (5)$$

$$Ag(s) = Ag^+ + e^- \quad (6)$$

can be combined to give complete reactions. Thus, Equations 3 and 4 give

$$Cr^{2+} + Fe^{3+} = Cr^{3+} + Fe^{2+} \quad (7)$$

It is observed experimentally that, if one starts with equal concentrations of the above species, the reaction shown in Equation 7 tends to go as written rather than in the reverse direction. Therefore Fe^{3+} (the oxidizing agent or oxidant) is reduced by Cr^{2+} (the reducing agent or reductant); ferric ion oxidizes Cr(II) to Cr(III).

[2] The concept of oxidation numbers is extremely useful. The concept and rules were developed by chemists, and, although not infallible, they help systematize oxidation–reduction phenomena.

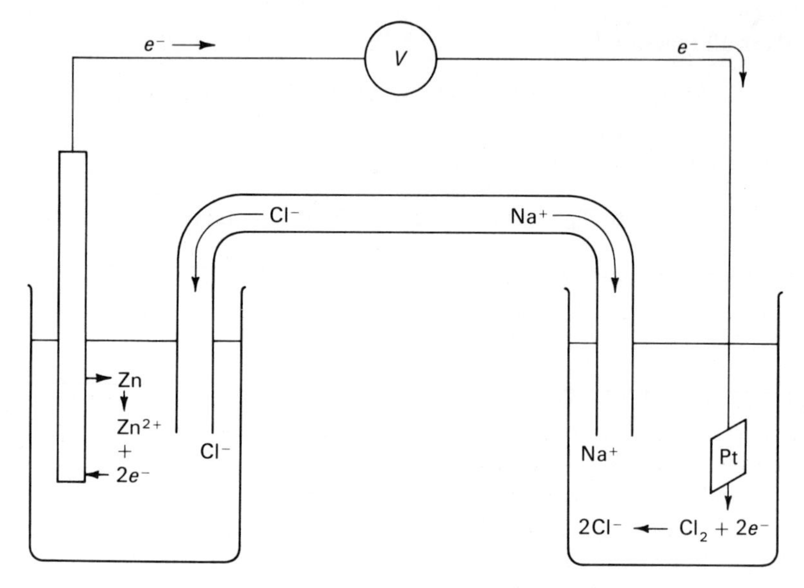

Half-cell reactions: $Zn = Zn^{2+} + 2e^-$

$2Cl^- = Cl_2 + 2e^-$

Cell reaction: $Zn + Cl_2 = Zn^{2+} + 2Cl^-$

FIGURE 6-1

An electrochemical cell.

ELECTRODE REACTIONS AND POTENTIALS

We will discuss in this section how the half-cell reactions introduced above can represent the reactions taking place at suitable electrodes in electrochemical cells. If a strip of zinc metal is placed in a beaker of water, there is a tendency for zinc ions to go into solution, leaving the metal negatively charged (Figure 6-1). The reaction cannot proceed appreciably, because the resulting separation of positive and negative charges creates an electrical potential that opposes any continuation of the reaction.[3] Similarly, if chlorine gas is passed over an inert but conducting metal strip (platinum is frequently used) in a beaker of water, there is a tendency for the chlorine molecules to draw electrons from the metal and to become chloride ions. Again the reaction is stopped by the potential resulting from the charge separation. Hence we have two electrodes with potentials of opposite sign;

[3] This opposition may be easier to recognize if it is observed that continued reaction would put more positive charges into the solution, which is already positively charged immediately around the electrode, and would put more negative charges onto the zinc strip, which is already negatively charged.

it would seem that if they were suitably connected in an electrical circuit, a flow of electrons would take place. There would thus be produced an *electrochemical cell*; each electrode and the solution of the substances involved in the production of its potential would constitute a *half-cell.*

If the two electrodes were connected, by means of any suitable conductor (usually a copper wire), a flow of electrons would take place momentarily. However, at the zinc electrode, positive ions (Zn^{2+}) would accumulate, and at the platinum electrode, negative ions (Cl^-) would accumulate; the potentials caused by these accumulated charges would check the further flow of electrons. Only if some means of transfer of these charges between the two solutions is provided can a continuous current flow. This is accomplished experimentally by connecting the two solutions by means of a *salt bridge*, which is usually an inverted U-tube dipping into each solution and filled with a solution of electrolyte (see Figure 6-1). This arrangement permits the migration of positive ions to the chloride–chlorine half-cell and of negative ions to the zinc–zinc-ion half-cell. If the two electrodes are now connected, a finite and continuous current will be obtained: electrons pass from the zinc electrode to the platinum electrode; positive ions are produced in the zinc–zinc-ion half-cell, negative ions in the chloride–chlorine half-cell; and a transfer of ions takes place through the salt bridge.

Our interest here is not in the flow of charge through the system, but is, rather, in the tendency for the over-all reaction to occur. Therefore, we must replace the ammeter by a potentiometer—an instrument designed to measure the electrical potential without permitting a current—so we can measure the potential of the cell without changing the cell conditions.[4] The potential measured across the electrodes is the algebraic sum of all the potentials in the system; in addition to the two electrode potentials, there are potentials at the junction of unlike liquids and at the junctions of unlike metals. The latter are canceled in the completed circuit, and the former can be kept small. The potential measured across the electrodes is, therefore, determined primarily by the values of the two electrode potentials. These are determined by the inherent tendency of the substances involved to take or give electrons:

$$Zn = Zn^{2+} + 2e^-$$

and

$$Cl_2 + 2e^- = 2Cl^-$$

[4] The potentiometer is discussed in Chapter 18. An analogy with a water system may help: at the bottom of a water tank, one could make a hole and measure the rate at which water comes out (analogous to the electrical current), or he could measure the total quantity of water that comes out (analogous to the total charge), or yet again he could measure the pressure without letting any water out (and this corresponds to the electrical potential).

All these equations involving electrons are essentially representative of the reactions taking place at suitable electrodes; they are termed *half-cell reactions.* If a value is arbitrarily assigned to the potential of one of these half-cells under specified conditions, then the potential for any other half-cell can, in principle, be measured by suitably combining it with this reference half-cell and measuring the potential of the complete cell. The reference half-cell that has been adopted is the *standard hydrogen electrode*,

$$H_2(\text{g, 1 atm}) = 2H^+ \text{ (unit activity)} + 2e^-$$

and it has been assigned the value zero volts. A list of the more important of these half-cell reactions, with their potentials in volts when all the reactants are present in standard conditions is presented in Appendix 4. Under standard conditions, the gases are at 1 atmosphere pressure, the ions or soluble compounds are at unit activity, and the solids and liquids are pure. The potential values under these conditions are called *standard half-cell potentials.*

According to the convention adopted here, half-cell reactions will be written with the electrons on the right side; a positive sign indicates that with respect to the reference hydrogen half-cell the tendency is for the half-cell reaction to proceed to the right, that is, to give up electrons; a negative sign indicates a tendency for the reaction to proceed to the left. For example, if the chloride-chlorine half-cell, which has a negative potential value, were combined with the hydrogen half-cell, with all the constituents at unit activities, the resultant cell reaction

$$Cl_2 + H_2 = 2H^+ + 2Cl^-$$

would tend to proceed from left to right; if under similar conditions the zinc–zinc-ion half-cell, which has a positive standard potential value, were combined with the hydrogen half-cell, the resultant cell reaction

$$Zn^{2+} + H_2 = 2H^+ + Zn$$

would tend to proceed from right to left.

EFFECT OF CONCENTRATIONS ON HALF-CELL POTENTIALS

The potential of a half-cell that is not at standard conditions can be calculated from the standard value by means of a thermodynamically derived equation, known as the *Nernst equation* (W. Nernst, 1864–1941); thi equation has the following form:

$$E = E^\circ - \frac{0.059}{n} \log Q. \quad (8)$$

Here E° is the standard potential (a *constant*); E is the potential of the half-cell when the various species have the activities specified in the term Q; Q, called the *activity quotient*, contains the activity of each species in the half-cell reaction raised to the appropriate power (see illustration below); n is the number of electrons involved in the half reaction; and 0.059 is a constant for half-cells at 25°C. Consider the half-cell reaction

$$Ag = Ag^+ + e^- \quad (9)$$

which has $E^\circ = -0.799$ v. Let us calculate the potential of the half-cell consisting of a silver metal electrode in a solution that is $1.0 \times 10^{-5}\,M$ in Ag^+. The activity quotient, Q, has the form of the equilibrium expression for the half-cell reaction, the electrons being excluded. That is, activities of products appear in the numerator, each having as its exponent the coefficient that appears in the half reaction. Activities of reactants similarly appear in the denominator. A simple way to ensure the correct application of the Nernst equation is to consistently write the half-cell reactions with the electrons and hence the oxidized species on the right. For Equation 9, the activity quotient is

$$Q = \frac{a_{Ag^+}}{a_{Ag(s)}} = \frac{a_{Ag^+}}{1} \approx \frac{1.0 \times 10^{-5}}{1}$$

Therefore, the potential is

$$E = -0.799 - \frac{0.059}{1} \log 1 \times 10^{-5} = -0.504 \text{ v}$$

Let us consider Equation 9 again. This moderately large negative value indicates that silver ion should show oxidizing tendencies and, in fact, should act as a better oxidizing agent than ferric ion, since the ferrous-ferric potential has the more positive value, namely,

$$Fe^{2+} = Fe^{3+} + e^- \qquad E^\circ = -0.782 \text{ v}$$

Experiments will show that if silver ion is added in excess to a ferrous-ion solution, there is a partial reduction of the silver ion to metallic silver, and an equivalent amount of ferric ion will be produced:

$$Ag^+ + Fe^{2+} = Ag(s) + Fe^{3+}$$

However, each tenfold decrease in the silver-ion concentration causes the potential of the silver half-cell to become more *positive* by approximately 0.06 v, and each tenfold increase in the *ratio* of the ferric ion to the ferrous ion, $[Fe^{3+}]/[Fe^{2+}] = Q$, causes the potential of the ferrous-ion–ferric-ion half-cell to become more *negative* by the same amount. It is apparent that, as the reaction proceeds, the values of the two half-cell potentials will very soon become equal, whereupon the reaction has reached an equilibrium, and no further change in the concentrations of the reactants will take place. When the half-cell potentials are equal, there is no longer a tendency for the reaction to proceed; this is the equilibrium condition.

If, however, thiocyanate is added to a solution of silver ion until the thiocyanate is 1 M, the concentration of the silver ion is decreased to 1×10^{-12} (K_{sp} AgSCN being 1×10^{-12}), and the silver–silver-ion potential becomes

$$E = -0.799 - \frac{0.059}{1} \log 1 \times 10^{-12} = -0.091 \text{ v.}^5$$

One concludes that metallic silver in the presence of thiocyanate ion should reduce ferric iron. In fact, a quantitative method based on this reaction has been proposed for the reduction of ferric solutions preliminary to titrating them with standard permanganate solutions.[6] In this method the thiocyanate-ion concentration is made approximately 0.01 M, so that the silver–silver-ion potential becomes -0.209 v. Since, *under equilibrium conditions, there can be only one potential value obtaining in a given solution*, other oxidizing or reducing agents present in this solution must have their concentrations so changed as to have this same potential value. Since the potential of the Fe(II)–Fe(III) couple must be -0.209 v, the ratio of ferric to ferrous iron can be obtained by substitution into the Nernst equation as follows:

$$-0.207 = -0.782 - \frac{0.059}{1} \log \frac{[Fe^{3+}]}{[Fe^{2+}]}$$

which gives to the ratio $[Fe^{3+}]/[Fe^{2+}]$ the value 1.7×10^{-10}. This would indicate that only 1.7×10^{-8}% of the ferric iron should remain unreduced.[7]

---

[5] It is to be noted that this value should represent the standard potential for the reaction

$$Ag + SCN^- = AgSCN(s) + e^- \qquad E^\circ = -0.091 \text{ v}$$

As the result of their measurements, Pearce and Smith (*J. Am. Chem. Soc.*, **59**, 2063, 1937) have calculated for this potential the value -0.095 v; the discrepancy probably arises from the use of concentrations rather than activities and the approximated value of the solubility product used above.

[6] Edgar and Kemp. *J. Am. Chem. Soc.*, **40**, 777 (1918).

[7] One should remember, however, that if the SCN^- is 1 M, the ratio of $[Fe(SCN)^{2+}]/[Fe^{3+}]$ is approximately 140 because of the reaction $Fe^{3+} + SCN^- = Fe(SCN)^{2+}$ (see page 108). Therefore the total Fe(III) remaining is about 2.4×10^{-6}%, still within the usual experimental errors.

Formal Potentials

The standard half-cell potential has been defined as that value at which the specific species involved in the half-cell reaction are at unit activities (or, for approximate calculations, concentrations and pressures). Unfortunately there are many half-cell reactions in which, especially under the conditions existing in analytical methods, the concentrations of the specific species involved are not known. As an example, although the standard potential of the half-cell reaction

$$Fe^{2+} = Fe^{3+} + e^-$$

is -0.78 v, one finds a value of -0.68 v upon measuring this half-cell in a solution that is 1 F in H_2SO_4 and 1 F in both Fe(II) and Fe(III). The shift to a less negative potential is due to a decrease in the Fe^{3+} concentration caused by the formation of one or more ferric sulfate complex ions. In many cases the formation constants of these ions are not known, therefore one cannot calculate the specific ion concentrations or the potential. This situation has led to the use of what are known as *formal half-cell potentials.* They are the values experimentally measured when (1) the reduced and oxidized substances are present at 1 F concentration and (2) the concentrations of any other species involved are explicitly stated. Thus the formal potential of the half-cell

$$\text{Fe(II)} = \text{Fe(III)} + e^- \qquad (1\ F\ H_2SO_4)$$

is -0.68 v and is applicable only to a 1 F sulfuric acid solution; if activity changes are neglected, the potential will remain constant as long as the ratio Fe(II)/Fe(III) remains one, but a tenfold change in this ratio will produce a potential change of approximately 0.06 v. Values for standard and formal potentials are shown in Appendix 4.

If adequate data were available for hydrolysis constants, for complex-ion-formation constants, and for activity coefficients of all species present, the formal potentials could be calculated from the standard potentials. Since such data are usually inadequate, there is an advantage in having available the experimentally measured formal potentials.

THE COMBINATION OF HALF-CELL REACTIONS TO GIVE COMPLETE REACTIONS

When two half-cell reactions are combined to obtain a cell reaction, the cell potential is the algebraic difference of the two half-cell potentials. One of the half-cell reactions will have to proceed from right to left in the whole reaction; it is the $E°$ of this half-cell reaction that is subtracted from the other.

As an example, consider the reaction in which HNO_2 is used to oxidize iodide to iodine. The half-cell reactions and half-cell standard potentials used to obtain the cell reaction and cell standard potential are as follows:

$$2I^- = I_2(s) + 2e^- \qquad E^\circ = -0.53 \text{ v}$$

$$NO + H_2O = HNO_2 + H^+ + e^- \qquad E^\circ = -0.98 \text{ v}$$

In order to obtain a balanced cell reaction, the electrons for each half-cell must be equal, therefore the NO-HNO_2 half-cell must be multiplied by two—this does not change the value of E°, only the number of electrons (the quantity of electricity) involved in the reaction. We now have

$$2I^- = I_2(s) + 2e^- \qquad E^\circ = -0.53 \text{ v}$$

$$2NO(g) + 2H_2O = 2HNO_2 + 2H^+ + 2e^- \qquad E^\circ = -0.98 \text{ v}$$

$$2I^- + 2HNO_2 + 2H^+ = I_2(s) + 2NO(g) + 2H_2O \qquad E^\circ_{cell} = +0.45 \text{ v}$$

Note that the value for the iodide–iodine standard half-cell potential and therefore for the cell potential applies only to a saturated iodine solution, that is, one in equilibrium with solid iodine.

It was stated above that a standard half-cell potential is assigned a positive value if the tendency of that half-cell is to proceed from left to right relative to the hydrogen half-cell; conversely, the standard half-cell potential is assigned a negative value if it tends to proceed from right to left. If this convention is used and two half-cell reactions and their potentials are combined to give a cell reaction and a cell potential, then a positive sign for the cell potential indicates a tendency for the cell reaction under standard conditions to proceed from left to right; a negative sign indicates the opposite tendency. The positive value of +0.45 v obtained above for the iodide–nitrous-acid cell reaction indicates a tendency for that reaction to proceed from left to right.

Conventions as to the Sign of Potentials

There is not general agreement as to the sign to be given half-cell potentials. In this book the potential of a half-cell reaction is given by writing the electrons on the right, with a positive sign when, with respect to the standard hydrogen half-cell, the reaction tends to proceed from left to right. This convention is more generally used in the United States. It has the advantages that a positive sign indicates that a half-cell reaction, or a cell reaction derived from two half-cell reactions, tends to proceed spontaneously as written. These are *half-cell potentials* and should not be termed "electrode potentials," since they do not represent the electrical potential on an external electrode as does European convention, which reverses the plus and minus signs.

Because of the uncertainty regarding the sign given potential values, one should be careful when using the literature to ascertain which convention is being used. It is useful to remember when using tables of potential values that under the convention used in this text, the potentials of the reactive metal half-cells have a positive sign, that is

$$\text{Li} = \text{Li}^+ + e^- \qquad E° = 3.05 \text{ v}$$

and the potentials of half-cells involving very strong oxidizing agents have a negative sign,

$$2\text{F}^- = \text{F}_2 + 2e^- \qquad E° = -2.65 \text{ v}$$

When the signs of these half-cell potentials are reversed, the European convention is being used.

CALCULATION OF EQUILIBRIUM CONSTANTS FROM STANDARD HALF-CELL POTENTIALS

The Nernst equation and half-cell potential values can be used to obtain additional quantitative information concerning the concentrations prevailing when a cell reaction is at equilibrium. When two half-cell reactions attain equilibrium, their half-cell potentials are equal. Then the two Nernst equations describing the two half-cell reactions can be equated and combined to give a very useful equation relating the cell potential under standard conditions, $E°_{\text{cell}}$, to the equilibrium constant, K, for the cell reaction. Thus for any two half-cells with the same value of n,[8]

$$E_1 = E_1^\circ - \frac{0.059}{n} \log Q_1$$

and

$$E_2 = E_2^\circ - \frac{0.059}{n} \log Q_2$$

(Q_1 or Q_2 may be adjusted to obtain the same value of n in both cases.) At equilibrium $E_1 = E_2$, therefore

$$E_1^\circ - \frac{0.059}{n} \log Q_1 = E_2^\circ - \frac{0.059}{n} \log Q_2$$

[8] If n is different in the half reactions, appropriate factors can be used to make the n's identical. Such a step was taken in the example of the I^-–HNO_2 reaction above.

and

$$E_1^\circ - E_2^\circ = \frac{0.059}{n} \log \frac{Q_1}{Q_2} \tag{10}$$

Now, $E_1^\circ - E_2^\circ$ is the standard cell potential, E_{cell}°, and Q_1/Q_2 is the equilibrium expression for the cell reaction. The ratio Q_1/Q_2 is constant even though neither Q_1 nor Q_2 is constant. This is best seen in Equation 10, in which the ratio is found to depend only upon E_1°, E_2°, and n—all constants. Therefore,

$$E_{\text{cell}}^\circ = \frac{0.059}{n} \log K \tag{11}$$[9]

Thus, for the iodide–nitrous-acid cell reaction discussed above (page 80),

$$E_1 = E_{(\text{I}^-,\text{I}_2)}^\circ - \frac{0.059}{2} \log \frac{1}{[\text{I}^-]^2}$$

and

$$E_2 = E_{(\text{NO},\text{HNO}_2)}^\circ - \frac{0.059}{1} \log \frac{[\text{HNO}_2][\text{H}^+]}{P_{\text{NO}}}$$

[9] While the Nernst equation relationship is readily observed experimentally, the equation itself can be rigorously derived from the first two laws of thermodynamics. The derived equation is

$$E = E^\circ - \frac{RT}{n\mathscr{F}} \ln Q$$

where E, $E°$, n, and Q have meaning as defined above, R is the molar gas constant in calorics per degree, T is the absolute temperature, and $\mathscr{F}$ is the faraday in calories per volt-equivalent. If we substitute numerical values for these terms and multiply by 2.3 to convert from natural to common logarithms, we obtain

$$E = E^\circ - \frac{1.99 \text{ cal/deg} \times 298 \text{ deg}}{n \text{ eq} \times 23{,}060 \text{ cal/v-eq}} \times 2.3 \log Q$$

or

$$E = E^\circ - \frac{0.059}{n} \log Q$$

The relationship between a whole cell potential and the equilibrium constant can also be derived thermodynamically by the observation that the standard free energy, ΔG°, is given by two expressions:

$$\Delta G^\circ = -RT \ln K$$

and

$$\Delta G^\circ = -E_{\text{cell}}^\circ n\mathscr{F}$$

Combination of these gives

$$E_{\text{cell}}^\circ = \frac{RT}{n\mathscr{F}} \ln K = \frac{0.059}{n} \log K$$

Adjusting the log term in E_2 to obtain the same value of n in both equations and equating E_1 to E_2 gives

$$E^\circ_{(\text{I}^-,\text{I}_2)} - \frac{0.059}{2}\log\frac{1}{[\text{I}^-]^2} = E^\circ_{(\text{NO},\text{HNO}_2)} - \frac{0.059}{2}\log\frac{[\text{HNO}_2]^2[\text{H}^+]^2}{(P_{\text{NO}})^2}$$

or

$$E^\circ_{\text{cell}} = -0.53 - (-0.98) = \frac{0.059}{2}\log\frac{(P_{\text{NO}})^2}{[\text{H}^+]^2[\text{HNO}_2]^2[\text{I}^-]^2} = \frac{0.059}{2}\log K$$

Since $n = 2$, we find that $\log_{10} K = 15.3$, and, therefore,

$$K = 2 \times 10^{15} = \frac{(P_{\text{NO}})^2}{[\text{H}^+]^2[\text{HNO}_2]^2[\text{I}^-]^2}$$

Hence, if standard potential values are available, predictions can be made as to the completeness of any redox reaction at equilibrium. It should be emphasized again that such predictions refer to *equilibrium conditions only* and that predictions cannot be made from cell potentials concerning the *rate* at which the equilibrium will be attained.

SUPPLEMENTARY REFERENCES

Latimer, W. M. *Oxidation States of the Elements and their Potentials in Aqueous Solutions.* (2nd ed.) Englewood Cliffs, N.J.: Prentice-Hall, 1952.

Lingane, J. J. *Electroanalytical Chemistry.* (2nd ed.) New York: Wiley-Interscience, 1958.

Sillen, L. G. Redox Diagrams. *J. Chem. Ed.*, **29**, 600 (1952).

QUESTIONS AND PROBLEMS

1. Apply the rules given in this chapter to calculate the oxidation number of:

   (*a*) Mn in $KMnO_4$
   (*b*) Si in Li_2SiF_6
   (*c*) Cr in $K_2Cr_2O_7$
   (*d*) N in $N_2H_5^+$
   (*e*) P in $H_2PO_2^-$
   (*f*) Mn in Mn_3O_4

   What can be stated regarding the oxidation states of the individual atoms of Cr, N, and Mn in (*c*), (*d*), and (*f*), respectively?

2. (*a*) Draw a diagram of the electrolytic cell resulting from passing an electrical current between two inert electrodes in a sulfuric acid solution. Show the constituents of the solution and all reactants and products.

(*b*) Combine the predominant anode and cathode half-cell reactions to give the cell reactions.
(*c*) What would be the predominant half-cell reactions if the sulfuric acid solution also contained copper sulfate?

3. (*a*) What is meant by half-cell potentials?
(*b*) By a standard half-cell potential?
(*c*) What convention will be used in this book regarding the sign assigned to half-cell potentials?

4. What would be the Mn(II)-Mn(VII) equilibrium half-cell potentials in solutions having the following concentrations?[10]

	Mn^{2+} (*M*)	MnO_4^- (*M*)	H^+ (*M*)
(*a*)	0.010	0.010	1.0
(*b*)	0.10	0.10	1.0
(*c*)	1.0×10^{-4}	0.10	1.0
(*d*)	0.10	1.0×10^{-4}	1.0
(*e*)	0.10	0.10	0.10
(*f*)	0.10	0.10	1.0×10^{-4}

(*g*) Formulate a general statement regarding the effect upon the Mn^{2+}-MnO_4^- half-cell potential (and the resultant oxidizing and reducing tendency) of changing the concentrations of oxidant and reductant but maintaining their concentration ratio constant.
(*h*) How would this statement have to be modified to apply to the half-cell

$$2Cr^{3+} + 7H_2O = Cr_2O_7^{2-} + 14H^+ + 6e^-$$

(*i*) (1) Compare the relative effects upon the Mn^{2+}-MnO_4^- half-cell potential of a 10-fold change in (*a*) [Mn^{2+}], (*b*) [MnO_4^-], and (*c*) [H^+]. (2) By what factor would the ratio [MnO_4^-]/[Mn^{2+}] have to change to equal a 10-fold change of the [H^+]? (Note that this [H^+] effect will be obtained in various degrees with the oxygen acid anions used as oxidants; examples are ClO_3^-, ClO_2^-, ClO^-, IO_3^-, IO_4^-, BrO_3^-, $Cr_2O_7^{2-}$, NO_3^-, and H_3AsO_4.)

5. Calculate the [Pb^{2+}] in a solution in which the Zn^{2+} is 0.10 *M* and which is in equilibrium with metallic zinc. *Ans.* 5×10^{-23}.

[10] Both the Mn^{2+}-MnO_4^- half-cell and the Cr^{3+}-$Cr_2O_7^{2-}$, mentioned in (*h*), have anomalous properties that make valid potential measurements difficult. Measurements of the Mn^{2+}-MnO_4^- half-cell cannot give a true equilibrium value, because the two oxidation states tend to react, although slowly, to give intermediate oxidation states such as Mn(III) and MnO_2. Measurements of the potential of the Cr^{3+}-$Cr_2O_7^{2-}$ at a platinum electrode give erratic values. However, the potential curves obtained in titrations of Fe(II) with either oxidant are those to be expected from equilibrium systems. It appears that the Fe(II)-Fe(III) half-cell rapidly reaches an equilibrium with both of the above half-cells and with the platinum electrode. Such half-cells have been called "potential mediators" and are used in experimental determinations of the standard potentials of other half-cells that cannot be directly measured at platinum or other inert electrodes. See H. A. Laitinen. *Chemical Analysis.* New York: McGraw-Hill, 1960, (pp. 293–296) for a discussion of such nonequilibrium electrode phenomena.

6. Calculate the formal solubility of silver bromide in a solution that is 0.100 F in NH_3. *Ans.* 3.3×10^{-4}.

   NOTE: The potential of a silver electrode immersed in a solution that was 0.100 F in silver nitrate and 1 F in NH_3 was found to be -0.319 v.

7. (*a*) Calculate the equilibrium constant for the reaction between permanganate and arsenious acid. (Assume that manganous ion and arsenic acid are the products.)
   (*b*) What does this value indicate regarding the possibility of quantitatively titrating H_3AsO_3 in an acid solution with a $KMnO_4$ solution?
   (*c*) What additional factors would have to be considered before making a valid prediction regarding experimental results?

CHAPTER 7

Chemical Reactions. Ionization

The two preceding chapters have given an introduction to the use of precipitation and oxidation-reduction reactions in chemical measurements. In this chapter we will consider ionization reactions, that is, reactions that are characterized by the formation or dissociation of compounds that are only slightly ionized.

These compounds are the result of positively charged elements forming covalent bonds with electron-pair donors, which can be simple or complex anions, or neutral compounds. The resulting species can be neutral molecules, cations, or anions. The following equations show examples of reactions that give such species:

$$H^+ + OH^- = HOH$$

$$H^+ + CN^- = HCN$$

$$Ag^+ + 2CN^- = Ag(CN)_2{}^-$$

$$Fe^{3+} + SCN^- = Fe(SCN)^{2+}$$

$$Cu^{2+} + 4NH_3 = Cu(NH_3)_4{}^{2+}$$

$$Hg^{2+} + 2SCN^- = Hg(SCN)_2(aq)$$

There are at present various definitions of acids and bases and of neutralization reactions; these definitions are discussed later. For our purposes the first two reactions shown above involve hydrogen ions and are classified as *acid–base* reactions: the first is the basis of *neutralization* reactions. The next three result in the formation of what are classified as *complex ions.* The last gives a neutral compound that can be called *a salt* or *a neutral complex.* It should be emphasized that regardless of the names applied, all of these reactions result from the tendency of the reactants to form covalent bonds; also, the extent to which these reactions proceed as written, or conversely the products ionize or dissociate, is dependent upon the strength of these bonds. The data needed for predicting the resulting equilibria are usually given as ionization constants for acids and bases, and as dissociation, or instability, constants for the complex species.

The discussion below is divided for convenience into two sections: Acid–Base Reactions, and Reactions Involving Un-Ionized Compounds.

ACID–BASE REACTIONS

The changing concepts of acids, bases, and of neutralization reactions present an interesting chapter in the history of chemistry. Ancient records contain observations that certain substances changed the color of some plant extracts, but the English scientist, Robert Boyle (1627–1691), appears to have been the first to attempt a comprehensive definition, as follows: acids act as solvents depending upon the nature of the substance being dissolved; they precipitate sulfur from solution in alkalies; they change the color of certain plant extracts to red; they restore the color of these extracts when changed by alkalies; they lose their corrosive properties by unification with alkalies. Lavoisier (1743–1794) thought that oxygen was a constituent of all acids,[1] but was forced to abandon this theory when the composition of the halogen acids was established. Liebig (1803–1873) appears to have been the first to attribute acidic properties to hydrogen, and he defined an acid as any substance with easily replaceable hydrogen atoms. Finally it was the work of the Swedish scientist Arrhenius (1859–1927) during the period from 1884 to 1890 that established the theory of the dissociation of electrolytes in aqueous solutions and led to the definitions of acids, bases, and neutralization reactions that, while overly restrictive, permit discussion of many simple acid–base reactions in aqueous solution. These three definitions follow:

An *acid* is a substance which produces hydrogen ions in water.

A *base* is a substance which produces hydroxide ions in water.

A *neutralization reaction* is the combination of hydrogen ions and hydroxide ions to produce water:

$$H^+ + OH^- = H_2O \qquad (1)$$

[1] He named the then newly discovered gas oxygen from the Greek words *oxys*, acid, and *genon*, to form.

It was assumed that common acids such as HNO_3 and HCl dissolve in water to produce hydrogen ions and that such bases as NaOH dissolve in water to produce hydroxide ions. It was further assumed that simple dissociation reactions such as

$$HCl \longrightarrow H^+ + Cl^-$$

and

$$NaOH \longrightarrow Na^+ + OH^-$$

yielded the H^+ or OH^- ions. Problems with this theory are seen when one observes that $NH_3(g)$, which contains no OH^- group, dissolves in water to produce hydroxide ions; or that HCl(g) is a covalent compound that must, therefore, require something more than simple dissociation to give acidic properties in water solution.

Moreover, the bare proton has such a large charge-to-size ratio that it bonds with water electron pairs and does not exist in appreciable concentrations in aqueous solutions, but it is hydrated:[2]

$$H^+ + H_2O = H_3O^+$$

Therefore, the properties associated with aqueous acid solutions are the result of the presence of this ion.[3] This hydration of the proton is in large part responsible for the ionization of predominantly un-ionized compounds such as HCl(g) in aqueous solution:

$$HCl(g) + H_2O = H_3O^+ + Cl^- \qquad (2)$$

In order to indicate that the proton is hydrated, the neutralization reaction of Equation 1 is more accurately written as

$$H_3O^+ + OH^- = 2H_2O \qquad (3)$$

Again, water is seen to be the product. The completeness of the acid–base reaction is fundamentally controlled by the extent to which the water molecule is ionized relative to the extent of the ionization reactions of the acid and the base.

The above theory not only fails to take into account the contribution that the solvent makes to the reactions of acids and bases, but also restricts acid–

[2] *Hydration* is the specific term for water combining with an ion or substance; *solvation* is the more general term for combination with the solvent.

[3] There is evidence that the hydronium ion forms a trihydrate, that is, $H_3O^+ \cdot 3H_2O$ or $H_9O_4^+$. Clever (*J. Chem. Ed.*, **12**, 637, 1963) reviews the evidence for these ions and proposes a structure for $H_9O_4^+$.

base considerations to reactions in which water is the product. However, there are reactions in nonaqueous media, such as liquid ammonia, that do not involve the formation of water but that are in many ways analogous to the above neutralization reaction. If hydrogen chloride gas is passed into liquid ammonia, the reaction taking place is

$$HCl(g) + NH_3 = NH_4^+ + Cl^- \tag{4}$$

The solvation of the proton again leads to the formation in the solution of a highly ionized compound, ammonium chloride. The NH_4^+ in Equation 4 and the H_3O^+ in Equation 2 are protonated-solvent molecules in their respective solutions. As the H_3O^+ reacts with the strong base, OH^-, in Equation 3 to produce the solvent, so NH_4^+ reacts with the strong base, NH_2^-, to produce the solvent:

$$NH_4^+ + NH_2^- = 2NH_3 \tag{5}$$

In the ammonia solvent, NH_4^+ displays acid-like properties in its effect upon indicators. If an equivalent quantity of sodium amide, $NaNH_2$, is added to a solution of ammonium chloride in ammonia, the acid properties of the solution are lost. Equation 5 is the strong-acid–strong-base reaction in ammonia just as Equation 3 is the strong-acid–strong-base reaction in water.

As a result of studies of many analogous reactions in nonaqueous solutions, the following more general definitions of acids, bases, and neutralization reactions are often used.

An *acid* is a compound containing one or more hydrogen atoms capable of producing hydrogen ions in solution.

A *base* is a compound capable of combining with hydrogen ions.

An acid is, then, a *proton donor* and a base is a *proton acceptor*. When an acid loses its proton it becomes a base called *the conjugate base* of that acid:

$$\text{acid} = \text{conjugate base} + \text{proton}$$

$$\text{proton donor} = \text{proton acceptor} + \text{proton}$$

The analogy between this formulation and that for the half-cell reactions of oxidizing and reducing agents is apparent. This analogy can be extended to the arrangement of acids and their conjugate bases according to their tendency to give up protons, just as half-cell reactions are arranged in standard potential tables according to their tendency to give up electrons.

To illustrate:

acid	= base	+ proton
H_3O^+	$= H_2O$	$+ H^+$
CH_3COOH	$= CH_3COO^-$	$+ H^+$
NH_4^+	$= NH_3$	$+ H^+$
H_2O	$= OH^-$	$+ H^+$
NH_3	$= NH_2^-$	$+ H^+$

In oxidation-reduction reactions the reductant of a half-cell reaction will tend to react with the oxidant of any half-cell reaction below it (see Appendix 4) when the concentrations involved are comparable. In the above series the acids will react with the bases below them with the formation of the conjugate bases and acids. Just as an oxidation-reduction reaction can be defined as

$$\text{reductant A} + \text{oxidant B} = \text{oxidant A} + \text{reductant B}$$

so, according to the above concept, a neutralization reaction can be defined as

$$\text{acid A} + \text{base B} = \text{base A} + \text{acid B}$$

In the classification of oxidation-reduction half-cell reactions, use is made of electron activity as measured by the resulting potential under standard conditions. In the classification of acid–base reactions, similar use could be made of the ionization constants of the acids, since the value of the constant indicates the hydrogen-ion activity when the activity of the acid and its conjugate base are equal.

The above extension of the concept of neutralization reactions was proposed independently in 1923 by Brønsted in Denmark and by Lowry in England. It has been of considerable value in the study of such reactions in nonaqueous media, and its application will be seen in certain of the discussions that follow.

It is interesting to observe that the generalized concept of acids and bases given by Brønsted and Lowry leads directly to the conclusion that the hydronium ion (H_3O^+) is the strongest acid that can exist in aqueous solutions and that the hydroxide ion is the strongest base in such solutions. A very strong acid such as $HClO_4$ reacts essentially completely with the *base* water to form the hydronium ion:

$$HClO_4 + H_2O = ClO_4^- + H_3O^+$$

Any acid stronger than H_3O^+ is converted to its conjugate base as it loses its proton to water.

Similarly a very strong base (proton acceptor) such as NH_2^- reacts with the hydrogen of the *acid* water to form hydroxide ion:

$$H_2O + NH_2^- = OH^- + NH_3$$

Any base stronger than OH^- is converted to its conjugate acid as it accepts a proton from water. This ability of water to limit the upper and lower hydrogen-ion concentrations attainable in aqueous solutions is termed *the leveling effect.* This effect causes all "strong" acids, such as HCl, HNO_3, and $HClO_4$, to appear to be equally ionized in water. The relative strengths of such acids can be determined in less basic solvents: perchloric acid is a much stronger acid than hydrochloric in acetic acid as a solvent.

The simple acids and bases that are discussed and used in experiments in this text qualify as such under the classical as well as under Brønsted's definitions. Therefore, since data in the literature are given in terms of the classical definitions, these definitions will be used throughout the text.[4] Also, as a concession to simplicity of presentation, the hydronium ion will be written as H^+ instead of H_3O^+.

Acid–Base Equilibria in Aqueous Solutions

The characteristic feature of neutralization methods in aqueous solution is the reaction of hydrated hydrogen ions with hydroxide ions to form neutral water molecules. This reaction will be written as

$$H^+ + OH^- = H_2O$$

and the equilibrium expression for the ionization (the reverse of the neutralization reaction) will be formulated as follows:

$$\frac{[H^+][OH^-]}{[H_2O]} = K_{\text{ioniz.}}$$

[4] Even more generalized concepts of acids and basis have been proposed. Thus, G. N. Lewis proposed that an acid be defined as an electron-pair acceptor and a base as an electron-pair donor. Thus, the proton on the hydronium ion is an electron-pair acceptor and the hydroxide ion is an electron-pair donor. The Lewis system is more general than either of those discussed above and can be extended to solvents and compounds that do not contain hydrogen. For example, when silicon dioxide is fused with calcium oxide, the acid (or electron-acceptor), SiO_2, reacts with the base (or electron donor), CaO, to produce the salt, Ca_2SiO_4. The booklets *Acids and Bases* and *More Acids and Bases* published in 1941 and 1944, respectively, by the *Journal of Chemical Education* contain interesting collections of articles on this subject.

In this expression the term H_2O can represent either the mole fraction or the molar concentration of the water and $K_{\text{ioniz.}}$ is the *ionization constant* of water. Since in pure water there are approximately 55.5 moles of water per liter,[5] it is obvious that in dilute solutions and for approximate calculations both the mole fraction and the concentration can be considered to be constant and therefore the expression can be simplified as follows:

$$[H^+][OH^-] = K_{\text{ioniz.}}[H_2O] = K_W$$

This constant, K_W, is the *ion product* of water although it is sometimes called the *ionization constant* of water. At 25°C its value is 1.0×10^{-14}, and for convenience this value will be used for all approximate calculations at room temperature. The ionization of water becomes greater at higher temperatures, and K_W is approximately 5×10^{-13} at 100°C. Thus, the temperature effect is unusually large because of decreased association of water molecules.

Reaction of a Strong Acid[6] with a Strong Base

The equation for the reaction of a "strong" acid, that is, one that is highly ionized (such as hydrochloric, nitric, perchloric, and the first hydrogen of sulfuric acid), with a strong base (such as sodium hydroxide or potassium hydroxide) can be written ionically as follows:

$$H^+A^- + B^+OH^- = HOH + B^+A^- \qquad (6)$$

[5] This value is the formal concentration; it states (page 15) the quantity of the substance present without specifying the state of association or dissociation. In the case of water the molar concentration is not known, for, because of hydrogen-bond formation, pure water is associated to an unknown extent into polymerized molecules, such as $(H_2O)_2$ and higher polymers, and the molar concentration of H_2O is unknown. However, in pure water at a given temperature these association equilibria remain constant; therefore, the molar concentration of H_2O remains constant and the above simple expressions can be employed.

[6] A strong acid can be defined as one that is so completely ionized that for practical mass-action calculations the concentration of the un-ionized acid can be neglected. The following rules by Pauling (*College Chemistry*. 3rd ed., San Francisco: W. H. Freeman and Company, 1964) are a useful aid in correlating strengths and ionization constants of acids.

1. The values of the *first* ionization constant of oxygen acids with a central positively charged atom can be approximated by the value of m in the formula $XO_m(OH)_n$ as follows: if m is zero, the acid is very weak, with $K_1 \leqq 10^{-7}$; for $m = 1$, the acid is weak, with $K_1 \approx 10^{-2}$; for $m = 2$, the acid is strong, with $K_1 \approx 10^3$; and for $m = 3$, the acid is very strong, with $K_1 \approx 10^8$. Examples:

$m = 0$: $HBrO$, $K = 10^{-9}$; H_3BO_3, $K_1 = 10^{-9}$; H_3SbO_3, $K_1 = 10^{-11}$

$m = 1$: $HClO_2$, $K = 10^{-2}$; H_2SO_3, $K_1 = 10^{-2}$; H_3PO_4, $K_1 = 10^{-2}$; H_5IO_6, $K_1 = 10^{-2}$

$m = 2$: $HClO_3$, $K \approx 10^3$

$m = 3$: $HClO_4$, $K \approx 10^8$

Footnote continued

This reaction reduces to

$$H^+ + OH^- = H_2O$$

and has the equilibrium expression

$$K_N = \frac{1}{[H^+][OH^-]} = \frac{1}{K_W} \tag{7}$$

Therefore the constant for such a neutralization reaction, K_N, has the value 10^{14}, which shows that the reaction is complete well within the usual measurements of quantitative analysis.

At the equivalence point of such titrations involving only strong acids and bases, the solution will be neutral; the hydrogen-ion concentration will be equal to the hydroxide-ion concentration and will be given by the expression

$$[H^+] = [OH^-] = \sqrt{K_W} = 10^{-7}$$

The Use of *p*H Values

The above expression and the value of K_W, the ion-product constant for water, indicate that for most operations in aqueous solutions, $[H^+]$ values will extend from around 1 to 10^{-14} (or less in strongly alkaline solutions). In using these values in plots (for example, titration curves), it is much more practical to use, rather than the concentrations, the logarithms of these concentrations. The same is also true in many of the calculations involved in the determination and use of hydrogen-ion concentrations. Also, in order to avoid negative numbers this practice has led to the convention of designating the hydrogen-ion concentration of a solution by a quantity called the *p*H and defined by the equation $pH = \log 1/[H^+] = -\log [H^+]$. (Rigorously, the hydrogen-ion activity rather than the concentration should be

---

2. The values of the successive ionization constants of the above type of oxygen acids, $K_1, K_2, K_3, \ldots$, decrease by a factor of approximately 10^{-5}. Examples:

H_3PO_4: $K_1 \approx 10^{-2}, K_2 \approx 10^{-7}, K_3 \approx 10^{-12}$

H_3AsO_4: $K_1 \approx 10^{-2}, K_2 \approx 10^{-7}, K_3 \approx 10^{-12}$

H_2SO_3: $K_1 \approx 10^{-2}, K_2 \approx 10^{-7}$

H_2SO_4: $K_1 \approx 10^{3}, K_2 \approx 10^{-2}$

The oxygen acids of the transition elements such as H_2MnO_4 and H_2CrO_4 deviate from this rule; an article by Tong and King (*J. Am. Chem. Soc.*, **75**, 6180, 1953) discusses these deviations.

3. With the exception of fluorine (HF, $K \approx 10^{-3}$), the hydrides of the halogens form strong acids in aqueous solutions. The hydrides of oxygen and its congeners form weak acids: H_2O, $K_1 \approx 10^{-14}$; H_2S, $K_1 \approx 10^{-7}$; H_2Se, $K_1 \approx 10^{-4}$; H_2Te, $K_1 \approx 10^{-3}$. The hydrides of nitrogen and its congeners tend to act as proton acceptors, and therefore act as bases in aqueous solutions.

specified.)[7] Thus it is seen that for solutions in which the hydrogen concentrations are 10^{-7} and 5×10^{-7}, respectively, the *p*H values are 7 and 6.3. This usage has been extended to other values such as $[OH^-]$ and equilibrium constants; for example $p\text{H} + p\text{OH} = pK_W$. Since the use of *p* values has become quite widespread, their meaning should be clearly comprehended.

Reaction of a Strong Acid with a Weak Base

The reaction of a strong acid with a weak base can be represented as follows:

$$H^+A^- + BOH = H_2O + B^+A^- \tag{8}$$

If the ionization of the salt B^+A^- can be assumed to be complete, the equation becomes

$$H^+ + BOH = H_2O + B^+ \tag{9}$$

and the equilibrium expression for this neutralization reaction has the form

$$\frac{[B^+]}{[H^+][BOH]} = K_N \tag{10}$$

The neutralization constant for this reaction is very different in appearance from that for the strong acid–base reaction (Equation 7) but the following steps show the relationship of the two constants:

$$\frac{1}{K_W} \cdot K_B = \frac{1}{[H^+][OH^-]} \cdot \frac{[B^+][OH^-]}{[BOH]} = \frac{[B^+]}{[H^+][BOH]}$$

But, from Equation 10,

$$\frac{[B^+]}{[H^+][BOH]} = K_N$$

---

[7] The term "*p*H" was originally defined as $p\text{H} = -\log[H^+]$, where $[H^+]$ represents the hydrogen-ion concentration; subsequently this definition was modified as follows:

$$p\text{H} = -\log a\text{H}^+$$

where $a\text{H}^+$ represents the activity of the hydrogen ion, and this modified definition is now commonly accepted. The uncertainties involved in the rigorous calculation of hydrogen-ion activities from measurements of electromotive force, and also the procedure used in the experimental standardization of the *p*H scale, are discussed by R. G. Bates (*Electrometric* pH *Determinations*. New York: Wiley, 1954). See also Bates. *Chem. Rev.*, **42**, 1 (1948). In addition to discussing the above problems, Letter Circular LC 933 of the National Bureau of Standards, entitled *Definition of* pH *and Standards for* pH *Measurements*, contains methods for the preparation of standard buffers and data for the calibration of *p*H meters at various temperatures.

Therefore,

$$\frac{K_B}{K_W} = K_N$$

Hence, the strong-acid–weak-base reaction is not as complete as the reaction of strong acids and bases; the completeness depends upon the strength of the base.

A simple expression for making an approximate calculation of the hydrogen-ion concentration at the equivalence point of a titration involving such a reaction can be obtained as follows. By definition, at the equivalence point, equivalent quantities of acid and base have been added. The resulting solution is identical with one of the same volume to which has been added an equivalent amount of the salt BA, indicated in Reaction 8. At equilibrium the same concentrations will obtain in such a solution, regardless of whether it is prepared by (*a*) adding equivalent quantities of H^+A^- and BOH or (*b*) by adding an equivalent quantity of the salt B^+A^-. Let C equal the total formal concentration of the salt BA, and assume that

$$[H^+] = [BOH]$$

Therefore,

$$[A^+] = C \qquad \text{and} \qquad [B^+] = C - [H^+]$$

Since

$$K_N = \frac{[B^+]}{[H^+][BOH]} = \frac{K_B}{K_W}$$

the expression becomes

$$\frac{C - [H^+]}{[H^+]^2} = \frac{K_B}{K_W}$$

When $[H^+]$ is small in comparison with C, the following simple approximate expression for calculating the hydrogen-ion concentration at the equivalence point is obtained:

$$[H^+] = \sqrt{\frac{K_W}{K_B}C}$$

Reaction of a Weak Acid with a Strong Base

When a weak acid reacts with a strong base, the reaction can be represented as follows:

$$HA + B^+OH^- = H_2O + B^+A^- \qquad (11)$$

Again assuming the ionization of the salt to be complete, we obtain for this equation[8]

$$HA + OH^- = H_2O + A^- \tag{12}$$

and the equilibrium expression

$$\frac{[A^-]}{[HA][OH^-]} = K_N$$

Again, not only the ionization constant for water $[H^+][OH^-] = K_W$, but also that of the acid, $[H^+][A^-]/[HA] = K_A$, has a determining effect on the equilibrium and on the value of K_N. By combining these last two expressions, thus eliminating $[H^+]$, we obtain the neutralization expression above,

$$\frac{[A^-]}{[HA][OH^-]} = \frac{K_A}{K_W} = K_N$$

The analogy between this expression and that for the reaction of a strong acid and a weak base is apparent. Such reactions will not be as quantitative as those between strong acids and bases, and the completeness of such reactions will vary with the strength of the acid.

Considerations similar to those used above lead to the expression

$$[OH^-] = \sqrt{\frac{K_W}{K_A}C}$$

by which the hydroxide-ion concentration at the equivalence point can be calculated. The hydrogen-ion concentration can then be obtained from the relation $[H^+][OH^-] = K_W$.

Reaction of a Weak Acid with a Weak Base

Reactions of this type are of minor importance analytically, because the experimenter will nearly always use a strong acid or base as the standard titrating agent.

For the reaction of a weak acid and a weak base, $HA + BOH = H_2O + B^+A^-$, the equilibrium expression is

$$\frac{[B^+][A^-]}{[HA][BOH]} = K_N$$

[8] It is of interest to note that on the basis of the more generalized concept of acids and bases, this equation represents the reaction of the acid HA with the base OH^- to give the acid H_2O and the base A^-; also, the neutralization constant is equal to the ratio of the ionization constants of the two acids.

Here the constants for water, K_W, for the acid, K_A, and for the base, K_B, are all involved. If the corresponding expressions are combined properly, the neutralization expression is obtained:

$$\frac{[B^+][A^-]}{[HA][BOH]} = \frac{K_A K_B}{K_W} = K_N$$

From this expression it is evident that the reaction of even moderately weak acids and bases will be far from complete.

The four examples of acid–base reactions just discussed include all types of neutralization reactions covered by the classical definition. Each of these will be approached in a more detailed and general treatment in Chapter 23; in that treatment the approximations used in the derivations are shown more clearly and explicitly. The partial neutralization of polyprotic acids by reactions such as

$$H_2A + OH^- = HA^- + HOH$$

requires special consideration and will be covered in Chapter 25. The reactions of strong acids with salts of weak acids, and of strong bases with salts of weak bases are discussed below.

Hydrolysis of Salts in Aqueous Solutions

In the discussion above, the reactions have been considered as taking place from left to right as written, have resulted in the formation of water, and have been termed *neutralization reactions.* It is apparent that for those cases in which the reaction is incomplete in this direction, if we start with the products on the right side of the equation, namely, a salt and water, then the reaction will proceed to the left until an equilibrium is reached that satisfies the equilibrium constant. Such reactions are termed *hydrolysis reactions*, and from the considerations given above it is clear why the hydrolysis of salts occurs. Moreover, it follows that, when the constants for the acids and bases involved are known, the equilibrium constants for such hydrolysis reactions can be calculated and that they will be the reciprocals of the neutralization constants. Once the hydrolysis constants have been obtained, the extent of such reactions can be predicted and the equilibrium conditions can be calculated.

Since the value of the ionization constant for water (K_W) changes much more rapidly with temperature than does that of most acids and bases, K_W being 0.11×10^{-14} at 0°C and 48×10^{-14} at 100°C, it is seen that neutralization reactions will be less complete and hydrolysis reactions correspondingly more complete as the temperature of the solution is increased.

Displacement Reactions

Reactions of this type involve the reaction of a strong acid with the salt of a weak acid or of a strong base with the salt of a weak base. If a strong acid is added to the salt of a monoprotic weak acid, the reaction that takes place can be represented as follows:

$$H^+A_s^- + B^+A_w^- = HA_w + B^+A_s^-$$

where A_s^- represents the anion of the strong acid, and A_w^- the anion of the weak acid. If the salts are assumed to be highly ionized, the equation can be simplified to

$$H^+ + A_w^- = HA_w$$

The equilibrium expression for this reaction is

$$\frac{[HA_w]}{[H^+][A_w^-]} = K_D = \frac{1}{K_A}$$

Such reactions are termed *displacement reactions* or, by some, *replacement reactions*, since it was formerly considered that the weak acid was displaced from its salt. The weaker the acid, the greater the value of K_D and the more complete the reaction. In terms of the Brønsted concept, a displacement reaction can be classified as a neutralization reaction as follows:

$$\text{acid A} + \text{base B} = \text{base A} + \text{acid B}$$

$$H_3O^+ + A_w^- = H_2O + HA_w$$

The equilibrium expression for the reaction is

$$\frac{[H_2O][HA_w]}{[H_3O^+][A_w^-]} = K$$

and in aqueous solutions the expression reduces to the one given above for a displacement reaction. In considering the change in the hydrogen ion concentration of the solution during such titrations, one observes (*a*) that the salt of a strong base and a weak acid will be alkaline by hydrolysis; (*b*) that during the titration the $[H^+]$ added will combine with the anion of the weak acid as long as an appreciable concentration of the salt remains; (*c*) that at the equivalence point there is essentially a solution of the weak acid; (*d*) that as an excess of strong acid is added, the $[H^+]$ will rise very rapidly.

Similarly, the equilibrium expression for the reaction of a strong base with the salt of a weak base is

$$\frac{[B_wOH]}{[B_w^+][OH^-]} = K_D = \frac{1}{K_B}$$

Changes in the hydroxide-ion concentration during the titration parallel the changes in the hydrogen-ion concentration during the titration with a strong acid of the salt of a weak acid, which was discussed above.

Buffer Systems

A solution that contains in significant concentrations (*a*) a weak acid and its salt or (*b*) a weak base and its salt is said to be buffered; each of these two constitutes a buffer system. When a buffer system is present in a solution, it decreases the change in hydrogen-ion concentration that otherwise would be caused by the addition of acid or base to that solution. We can illustrate the properties of such systems if we first calculate the *change* in hydrogen-ion concentration caused by adding 10 mmole quantities of a strong acid and of a strong base to separate 1 liter volumes of pure water and then compare this change with that obtained by adding the same quantities of acid and base to separate 1 liter volumes of a solution 0.10 F in CH_3COOH and 0.10 F in CH_3COONa.

We have noted earlier that in aqueous solutions $[H^+][OH^-] = 10^{-14}$. Because in pure water

$$[H^+] = [OH^-] = \sqrt{10^{-14}}$$

the water will initially have a hydrogen-ion concentration of $1.0 \times 10^{-7}\ M$. After 10 mmoles of strong acid have been added, the hydrogen-ion concentration becomes $1.0 \times 10^{-2}\ M$ (the hydrogen ion contributed by the water is negligible). Similarly, after 10 mmoles of base have been added to a liter of water, the hydroxide-ion concentration becomes $1.0 \times 10^{-2}\ M$, and, as seen from the ion product of water, the hydrogen-ion concentration is $10^{-12}\ M$. In both cases the hydrogen-ion concentration has changed by a factor of 100,000.

Now consider the effect of the addition of the same quantity of the strong acid (10 mmoles) to the acetic-acid–acetate-ion solution. Acetic acid is such a weak acid that we can assume as a first approximation that essentially all of the hydrogen ion added reacts with acetate ion to produce acetic acid as follows:

$$H^+ + CH_3COO^- = CH_3COOH$$

As a result, the quantity of acetate ion in the solution is decreased by 10 mmoles and that of the acetic acid is increased by the same amount; this changes the acetate-ion concentration from 0.10 to 0.09 *F* and the acetic acid concentration from 0.10 to 0.11 *F*. The concentration of hydrogen ion is given by the acid constant for acetic acid,

$$[H^+] = K_A \frac{[CH_3COOH]}{[CH_3COO^-]}$$

and the concentrations of other species can be written in terms of the $[H^+]$:

$$[CH_3COOH] = 0.11 - [H^+]$$

$$[CH_3COO^-] = 0.09 + [H^+]$$

Therefore

$$[H^+] = 1.8 \times 10^{-5} \frac{0.11 - [H^+]}{0.09 + [H^+]}$$

If the assumption is made that the $[H^+]$ is much less than 0.09 or 0.11, the expression becomes

$$[H^+] = 1.8 \times 10^{-5} \times \frac{0.11}{0.09}$$

$$= 2.2 \times 10^{-5}$$

The assumption concerning the $[H^+]$ is good, so the answer is acceptable.

The hydrogen-ion concentration has been changed in this case from 1.8×10^{-5} to 2.2×10^{-5} or by a factor of only 1.2. This is in striking contrast to the factor of 10^5 obtained in the absence of the buffer system. Also observe that this factor of 1.2 is the factor by which the ratio of acid to salt has changed.

If, instead of acid, 10 mmoles of a strong base is added to the acetic-acid–acetate-ion solution the following reaction can be assumed to be essentially complete:

$$CH_3COOH + OH^- = CH_3COO^- + HOH$$

As a result, the concentration of the acetic acid is decreased to 0.09 *F* and the concentration of the acetate ion is increased to 0.11 *F*. A calculation similar to that made above gives $[H^+] = 1.5 \times 10^{-5}$. The hydrogen-ion concentration has changed by the same relative amount, that is, by a factor of only 1/1.2 or 0.8.

Weak bases and their salts are also used as buffer systems for alkaline solutions. The analogy to the acid systems is obvious if the generalized ionization expression for the weak base is written as

$$BOH = B^+ + OH^-$$

Then the following is true:

$$[OH^-] = K_B \frac{[BOH]}{[B^+]}$$

$$[H^+] = \frac{K_W[B^+]}{K_B[BOH]}$$

One should realize that a buffer system is effective only when the components of the buffer system are present in relatively large amounts compared with the amount of added acid or base. For example, the buffer discussed above contained 100 mmoles each of acetic acid and acetate ion; if 150 mmoles of strong acid are added to such a solution, the capacity of the buffer will be far exceeded.

Another characteristic of a buffer is that it is most effective when the components of the buffer system are in nearly equal concentration. This is apparent if one observes that the $[H^+]$ changes just as the ratio of the components changes. The smallest change in this ratio caused by the addition of a given quantity of acid or base occurs when the concentrations of the components are equal. Moreover, to a first approximation the $[H^+]$ in a buffered solution does not change as the solution is diluted. That this is so follows from the dependence of $[H^+]$ upon the concentration ratio of acid to the salt; since both are in the same solution their concentrations change equally with dilution.

Chapter 20 gives an example of the use of a buffer system to control the *p*H in a redox titration; Chapter 23 contains a more detailed treatment of the theory of ionization reactions; and Chapters 23 through 26 give applications of ionization reactions to analytical methods.

Hydration of Anhydrides and Evaluation of Pseudo Acid and Base Constants

The anhydride of an acid is that substance which reacts with water to form the acid. For example, SO_2, SO_3, and CO_2 are the anhydrides of H_2SO_3, H_2SO_4, and H_2CO_3. In classical theory NH_3 was considered to be the anhydride of the base, NH_4OH, but in the Brønsted theory NH_3 itself is the base. The completeness of the hydration reaction of an acid or a base with water is dependent upon the bonding formed by the water electron pairs with the anhydride molecule. Sulfuric acid is formed when $SO_3(g)$ is passed into water, and in dilute solution the hydration reaction and subsequent ionization of the first proton is so complete that the concentration of $SO_3(aq)$ is quite negligible:

$$SO_3(g) + H_2O = H^+ + HSO_4^-$$

Sulfur dioxide, on the other hand, is much less completely hydrated; because of the decreased positive charge on the sulfur atom there is less tendency to form electron-pair bonds with the oxygen of the water. Consequently, when gaseous sulfur dioxide is passed into water there is an equilibrium established between the gas and liquid phases:

$$SO_2(g) = SO_2(aq) \qquad (13)$$

There is an additional equilibrium between the dissolved SO_2 and its hydrated species, H_2SO_3,

$$SO_2(aq) + H_2O = H_2SO_3 \qquad (14)$$

The situation is further complicated by the partial ionization of the sulfurous acid,

$$H_2SO_3 = H^+ + HSO_3^- \qquad (15)$$

The equilibrium expressions for Equations 13, 14, and 15 are

$$\frac{[SO_2(aq)]}{P_{SO_2}} = K_1$$

$$\frac{[H_2SO_3]}{[SO_2(aq)]} = K_2$$

and

$$\frac{[H^+][HSO_3^-]}{[H_2SO_3]} = K_3$$

If the pressure of the sulfur dioxide is kept constant both $[SO_2(aq)]$ and $[H_2SO_3]$ are kept constant. This can be seen readily if K_1 and K_2 are multiplied together:

$$\frac{[H_2SO_3]}{P_{SO_2}} = K_1K_2$$

Therefore, the sum of $[H_2SO_3]$ and $[SO_2(aq)]$, which will be designated as $[\Sigma\, SO_2]$, is constant (the pressure of SO_2 being kept at a fixed value), that is

$$[\Sigma\, SO_2] = [SO_2(aq)] + [H_2SO_3] \tag{16}$$

Now, by combining K_2 and Equation 16 we obtain

$$[H_2SO_3] = \frac{[\Sigma\, SO_2]}{1 + 1/K_2}$$

which reduces to

$$[H_2SO_3] = \frac{[\Sigma\, SO_2]}{K_4}$$

where $K_4 = 1 + 1/K_2$. If this is substituted into the expression for K_3 we obtain

$$\frac{[H^+][HSO_3^-]}{[\Sigma\, SO_2]} = \frac{K_3}{K_4} = K_{A_1}$$

This states that a relationship similar to that of K_3 is valid if $[\Sigma\, SO_2]$ appears in place of $[H_2SO_3]$. In this case K_{A_1} is in effect a pseudo acid constant but is the value conventionally to be found in tables. Thus, the usual practice in such a case is to redefine the concentration terms, $[SO_2(aq)]$ and $[H_2SO_3]$, so that either means $[\Sigma\, SO_2]$ and to use the combined constant, K_3/K_4, calling this in the case of polyprotic acids, K_1, the first acid constant. We will follow this convention. For example, in the appendix the first acid constant for H_2SO_3 is given as 1.3×10^{-2}. This is not the value for K_3 above, but is rather the value for K_3/K_4.

Other similar cases are encountered, and for each we will use the same convention. Thus, carbon dioxide dissolves in water to give $CO_2(aq)$ and H_2CO_3. Either $[CO_2(aq)]$ or $[H_2CO_3]$ will be used to mean $[\Sigma\, CO_2]$, that is $[CO_2(aq)] + [H_2CO_3]$.[9] As has been explained earlier, in regard to the hydration of the proton, $[H^+]$ or $[H_3O^+]$ is used to mean the sum of all the various hydrated protons in the solution. In addition for positive metal ions, $[M^{n+}]$ represents the sum of the concentrations of M^{n+} and all hydrated species.

[9] In the case of CO_2 the equilibrium constant for the hydration of CO_2, $CO_2(aq) + H_2O = H_2CO_3$, is known and has the value 2.6×10^{-3}, therefore the true first acid constant can be calculated and has the numerical value 1.7×10^{-4}. Thus carbonic acid is a much stronger acid than the pseudo constant indicates. D. M. Kern (*J. Chem. Ed.*, **37**, 14, 1960) gives an interesting account of the early evidence that the hydration reaction is slow in attaining an equilibrium and of the investigations leading to an equilibrium constant.

Ammonia presents an interesting case, in that the species in aqueous solutions has been conventionally written as NH_4OH and called ammonium hydroxide. Actually, according to a recent study, NH_3 is not hydrated to a detectable extent.[10] Therefore, NH_4OH should be recognized as only a convenient way of representing the classical base behavior of ammonia. Emerson *et al.*[10] provide data showing that other substituted ammonia compounds such as CH_3NH_2 and $(CH_3)_3N$ do hydrate measurably.

Acid–Base Reactions in Nonaqueous Solvents

As was explained above, the addition of a strong acid to water results in the reaction

$$H^+A^- + H_2O = H_3O^+ + A^-$$

In Brønsted–Lowry terminology the generalized reaction is

$$\text{Acid}_1 + \text{Base}_2 = \text{Acid}_2 + \text{Base}_1$$

In this reaction, water (the base) is the solvent and is present in excess; therefore, all highly ionized acids would be converted essentially completely to the acid H_3O^+. This ion is the strongest Brønsted–Lowry acid that can exist in significant concentrations in such solutions. As was stated earlier, this conversion of all stronger acids to the hydronium ion in aqueous solutions is known as the *leveling effect*. This leveling effect does not mean that all strong acids are completely or equally ionized in water, but only that with ionization constants ranging from 10^9 for $HClO_4$ to 10^3 for the first proton of H_2SO_4 (see Footnote 6, page 92), the fraction remaining un-ionized is so extremely small that it can be neglected for most experimental considerations.

The leveling effect restricts the number of Brønsted bases that can be quantitatively titrated by aqueous solutions of strong acids to constituents that are such strong bases that the generalized reaction

$$B^- + H_3O^+ = HB + H_2O$$

is complete within quantitative limits (examples of such bases are OH^-, $CO_3{}^{2-}$, and CN^-). Quantitative titrations of weak bases, such as the acetate ion and many other anions of organic acids, cannot be made.

In a more acidic solvent, such as anhydrous acetic acid, the addition of a strong acid results in the generalized reaction

$$H^+A^- + \text{Sol.} = \text{H·Sol.}^+ + A^-$$

---

[10] M. T. Emerson, E. Grunwald, M. Katlan, and R. Kromhout. *J. Am. Chem. Soc.*, **82**, 6307 (1960); see also J. Halpern. *J. Chem. Ed.*, **45**, 372 (1968).

When acetic acid is the solvent, $H{\cdot}CH_3COOH^+$ is the strongest acid attainable. The leveling effect toward acids in acetic acid is largely lost because $H{\cdot}CH_3COOH^+$ is so strong that so-called "strong acids," such as HCl and HNO_3, are no longer essentially completely ionized. In other words, it is harder to force CH_3COOH than H_2O to accept a proton; however, when CH_3COOH is protonated (accepts a proton) the result is $H{\cdot}CH_3COOH^+$, which is a much stronger acid than H_3O^+. The apparent order of ionization in acetic acid of the more common "strong acids" is $HClO_4 > HCl > H_2SO_4 > HNO_3$.

Because of the higher acid strength of $H{\cdot}CH_3COOH^+$, a solution of a strong acid can be used in acetic acid for the titration of much weaker bases than is possible in aqueous solutions. Perchloric acid is most frequently used as the titrant and a large number of methods, especially for the titration of the anions of organic acids (such as acetate, CH_3COO^-, and other carboxylate anions, $R{\cdot}COO^-$) have been developed. The general reaction for such titrations is

$$A^- + H{\cdot}CH_3COOH^+ = HA + CH_3COOH$$

The end points of such titrations can be obtained by *p*H meters or suitable acid–base indicators.

Space does not permit an extension of this discussion to basic and neutral solvents or to the more detailed theory of acid–base reactions in nonaqueous solvents. Such discussions, as well as numerous references to the original literature, will be found in the following sources:

Fritz, J. S. and Hammond, G. S. *Quantitative Organic Analysis.* New York: Wiley, 1957, Chap. 3.

Laitinen, H. A. *Chemical Analysis.* New York: McGraw-Hill, 1960, Chap. 4 and Chap. 5.

Latimer, G. W., Jr. The Nonaqueous Titration of Salts of Weak Acids. *J. Chem. Ed.*, **43**, 215 (1966).

Walton, H. F. *Principles and Methods of Chemical Analysis.* (2nd ed.) Englewood Cliffs, N.J.: Prentice-Hall, 1964, Chap. 15.

REACTIONS INVOLVING UN-IONIZED COMPOUNDS

The discussion above has been concerned primarily with reactions of the hydronium ion, a positively charged cation, with anions capable of acting as electron-pair donors to form various un-ionized complexes.

We will now consider reactions in which metal cations react with electron-pair donors to form un-ionized complexes of various types.

With the exception of the cations of the elements of columns Ia and IIa of the periodic table, which have the electronic structure of the nearest inert gas, there is a strong tendency for the cations of the remaining elements to form covalent bonds with other chemical species that can act as electron-pair donors. This tendency is so general and the effect so pronounced that it is an oversimplification to portray such cations in aqueous solutions as bare ions just as it was an oversimplification to portray the hydrogen ion as a bare proton. The typically metallic cations are hydrated in aqueous solutions and are surrounded by a sheath of water molecules. Thus, when almost colorless anhydrous cupric sulfate is dissolved in water the resulting solution is blue because of the formation of a hydrated ion or aquo complex, the reaction being

$$CuSO_4(s) + 4H_2O = Cu(H_2O)_4{}^{2+} + SO_4{}^{2-}$$

This same type of covalent bond between metallic cations and hydroxide oxygens is responsible for most metallic hydroxides being weak bases.

When another species capable of acting as an electron-pair donor and forming a covalent bond with a cation is added to a solution of a hydrated cation, there will be competition between water molecules and the added species for positions around the cation. The extent to which the added species replaces the water molecules will depend upon the relative strength of the bonds formed with the cation.[11] The term *ligand* (from the Latin word *ligatio*, to bind) is used for chemical species forming such bonds with a central atom; the resulting species is termed a complex and can be a cation, a neutral compound, or an anion.

Un-ionized Compounds Involving Inorganic Ligands

Because of the general hydration of cations in aqueous solutions the formation of nonaqueous complexes in such solutions is more properly represented as a substitution reaction. For example, when concentrated hydrochloric acid is added progressively to a blue solution of cupric sulfate a gradual color transition from blue to greenish results. This color change is caused by successive reactions such as

$$Cu(H_2O)_4{}^{2+} + Cl^- = Cu(H_2O)_3Cl^+ + H_2O$$

$$Cu(H_2O)_4{}^{2+} + 2Cl^- = CuCl_2(H_2O)_2 + 2H_2O$$

and finally over-all

$$Cu(H_2O)_4{}^{2+} + 4Cl^- = CuCl_4{}^{2-} + 4H_2O$$

[11] You should review your general chemistry text for a discussion of the factors, such as heat of hydration, ionic radii, and electronegativity, that affect the strength of such bonds.

Reactions such as those discussed above, which involve the formation of complex species other than water or acids and bases, are frequently used in volumetric methods. These methods may involve the formation of a neutral complex, usually a salt, or they may involve the formation of various types of complex ions. Brief discussions of examples of these methods are given below.

An example of a method involving the formation of an un-ionized salt is the titration of mercuric ion with thiocyanate ion to form un-ionized mercuric thiocyanate. With the exception of the elements of columns Ia and IIa most metals in their various oxidation states have cations that form complex ions with various inorganic anions. They also form stable neutral complexes, but these tend to be insoluble; examples are the hydroxides and oxides of most of them, the chlorides, bromides, and iodides of the unipositive oxidation states of column Ib elements, and of unipositive mercury ion, Hg_2^{2+}. The dipositive mercury ion is unusual in forming relatively un-ionized, but moderately soluble, salts with the above halides. The other elements of column IIb, zinc and cadmium, also have the $(n-1)d^{10}ns^2$ configuration and form similar (but much less stable) compounds with halides. Mercury is also unusual in that, although it coordinates additional halide ions to form complex ions such as $HgCl_3^-$ and $HgCl_4^{2-}$, these additional ions are much less closely held than are the first two. It is this property that makes it possible to develop volumetric methods based on the formation of a neutral salt. The constants given in Table 7-1 show this latter effect.

Thiocyanate, sometimes termed a pseudohalide, resembles a halide and forms compounds with similar properties with mercuric ion. For example, it forms moderately soluble but only slightly ionized $Hg(SCN)_2$. A discussion of the use of this reaction for the volumetric determination of mercury is given below. The titration and indicator reactions are

$$Hg^{2+} + 2SCN^- = Hg(SCN)_2(aq) \quad \text{(Titration reaction)}$$

$$Fe^{3+} + SCN^- = Fe(SCN)^{2+}_{red} \quad \text{(Indicator reaction)}^{12}$$

For convenience, in these equations and in the discussion that follows, the hydration of the ions has been omitted. The titration reaction is carried within quantitative limits by the formation of un-ionized $Hg(SCN)_2$. This is evident when you note that the dissociation constant for the reaction

$$Hg(SCN)_2 = Hg^{2+} + 2SCN^-$$

is 3×10^{-18}. (In addition a saturated solution of $Hg(SCN)_2$ at 25°C is only

[12] This titration is known as the Volhard method. See Footnotes 8, page 371, and 16, page 384, regarding the history of the development of the method.

TABLE 7-1

Dissociation Constants of Hg(II) Complexes with Halide Type Ligands

	Cl^-	Br^-	I^-	SCN^-
$HgL^+ = Hg^{2+} + L^-$	5.4×10^{-6}	1×10^{-10}	1×10^{-13}	—
$HgL_2 = HgL^+ + L^-$	3.1×10^{-8}	5×10^{-9}	1×10^{-11}	3×10^{-18}*
$HgL_3^- = HgL_2 + L^-$	7×10^{-2}	4×10^{-2}	2×10^{-4}	—
$HgL_4^{2-} = HgL_3^- + L^-$	1×10^{-1}	5×10^{-2}	6×10^{-3}	6×10^{-22}*

* These values are for $Hg(SCN)_2 = Hg^{2+} + 2SCN^-$ and $Hg(SCN)_4^{2-} = Hg^{2+} + 4SCN^-$; it appears that the intermediate species are unstable with respect to the reactions $2HgSCN^+ = Hg^{2+} + Hg(SCN)_2$ and $2Hg(SCN)_3^- = Hg(SCN)_2 + Hg(SCN)_4^{2-}$.

2×10^{-3} *F*,[13] therefore in a saturated solution the ion product constant is $[Hg^{2+}][SCN^-]^2 = 6 \times 10^{-21}$.) The appearance of the red color of the $Fe(SCN)^{2+}$ serves as the end point of the titration.

An experimental study of the end point has made it possible to calculate from equilibrium principles the end-point error of the titration. In this experimental study it was found that 0.10 ml of 0.010 *F* KSCN was required to produce a perceptible red color in 100 ml of a solution that was 0.013 *F* in $Fe(NO_3)_3$ and that was also 0.5 *F* in HNO_3. Frank and Oswalt[14] have calculated a value of 7.25×10^{-3} for the dissociation constant for the ferric thiocyanate complex, that is,

$$\frac{[Fe^{3+}][SCN^-]}{[FeSCN^{2+}]} = K_D = 7.25 \times 10^{-3}$$

Since the formal (or total) concentration of thiocyanate in the above solution was 1.0×10^{-5}, we can solve for the molar concentration of the $FeSCN^{2+}$ as follows:

$$\frac{[Fe^{3+}][SCN^-]}{[FeSCN^{2+}]} = \frac{(0.013 - x)(1.0 \times 10^{-5} - x)}{x} = 7.25 \times 10^{-3}$$

and obtain a value of 6.4×10^{-6} *M*. [The calculation is made easy if one observes that x must be smaller than 1.0×10^{-5} or the factor $(1.0 \times 10^{-5} - x)$ becomes negative and meaningless, since the factor represents the concentration of SCN^-. If x is smaller than 1.0×10^{-5}, then it is very much smaller than 0.013, so the x in the factor $(0.013 - x)$ can be neglected, and the quadratic equation can be solved as a linear equation.] Therefore, since the total

[13] In most titrations a white crystalline precipitate will be observed toward the latter part of the titration.

[14] Frank and Oswalt. *J. Am. Chem. Soc.*, **69**, 1321 (1947); they also substantiate other evidence that, contrary to statements found elsewhere, HSCN is a strong acid.

thiocyanate ion was 1.0×10^{-5} *F*, the concentration of thiocyanate ion is 3.6×10^{-6} *M*.

When near the end point of a titration of mercuric ion with thiocyanate, the species present in significant quantities are Hg^{2+}, $Hg(SCN)_2$, $Hg(SCN)_4{}^{2-}$, SCN^-, Fe^{3+}, and $Fe(SCN)^{2+}$. (The concentrations of $Fe(SCN)_2{}^+$, $Hg(SCN)^+$, and $Hg(SCN)_3{}^-$ are assumed to be negligible.) Usually near the end point the solutions become saturated, therefore, from the ion product of $Hg(SCN)_2$ and the concentration of thiocyanate ion, the mercuric ion is calculated to be 4.6×10^{-10} *M*. Once the $[Hg^{2+}]$ and $[SCN^-]$ are known, the concentration of $Hg(SCN)_4{}^{2-}$ can be calculated from the expression

$$\frac{[Hg^{2+}][SCN^-]^4}{[Hg(SCN)_4{}^{2-}]} = 6 \times 10^{-22} = \frac{(5 \times 10^{-10})(3.6 \times 10^{-6})^4}{[Hg(SCN)_4{}^{2-}]}$$

to be 1.4×10^{-10} *M*.

The titration error, expressed in equivalents (eq) is

$$\text{T.E.} = \Sigma \text{ eq SCN} - \Sigma \text{ eq Hg}$$

or

$$\text{T.E.} = [\text{eq } SCN^- + \text{eq } Fe(SCN)^{2+} + \text{eq } Hg(SCN)_4{}^{2-} + \text{eq } Hg(SCN)_2] - [\text{eq } Hg^{2+} + \text{eq } Hg(SCN)_4{}^{2-} + \text{eq } Hg(SCN)_2]$$

The $Hg(SCN)_2$ is present as $Hg(SCN)_2(aq)$ and, if a precipitate is present, $Hg(SCN)_2(s)$; however, the stoichiometry of the titration assumes equal equivalents of Hg and SCN^- in $Hg(SCN)_2$, therefore the equivalents of this compound can be dropped, and the equation written

$$\text{T.E. eq} = [\text{eq } SCN^- + \text{eq } FeSCN^{2+} + \text{eq } Hg(SCN)_4{}^{2-}] - [\text{eq } Hg^{2+} + \text{eq } Hg(SCN)_4{}^{2-}]$$

In 100 ml of the solution

$$\text{T.E. eq} = [3.6 \times 10^{-7} + 6.4 \times 10^{-7} + 1.4 \times 10^{-11} \times 4] - [4.6 \times 10^{-11} \times 2 + 1.4 \times 10^{-11} \times 2]$$

Since the last term is insignificant one finds 1×10^{-6} eq Hg (positive error).

If one were titrating 2.5×10^{-3} equivalents of mercury—equivalent to 25 ml of 0.1 *N* thiocyanate—the percentage error would be

$$\frac{1 \times 10^{-6}}{2.5 \times 10^{-3}} \times 100 = 4 \times 10^{-2}\% \text{ (+error)}$$

These calculations indicate that the titration should give highly accurate results and this conclusion has been verified by an experimental study.[15]

One of the best known and most extensively used volumetric methods involving the formation of a complex between a metal cation and an inorganic ligand is the so-called Liebig titration for determining cyanide. This method was developed over one hundred years ago[16] and is based on the following reactions:

$$Ag^+ + 2CN^- = Ag(CN)_2^- \qquad K = 1 \times 10^{21}$$

$$Ag^+ + Ag(CN)_2^- = Ag{\cdot}Ag(CN)_2(s) \qquad K = 2 \times 10^{12}$$

Upon addition of silver to an alkaline cyanide solution, $Ag(CN)_2^-$ is formed until the reaction is near the equivalence point, when the appearance of a white precipitate is taken as the end point. Although for convenience the solubility product of the precipitation is usually formulated as

$$[Ag^+][Ag(CN)_2^-] = K$$

structural studies have shown that the metal atoms are equivalent.

About forty years after Liebig developed the end-point method, Deniges[17] improved it by adding iodide as an indicator to the cyanide solution. When this is done, the less soluble yellowish silver iodide (rather than silver cyanide) precipitates; the colored precipitate tends to remain colloidally dispersed and is more readily perceived. The use of iodide also permits the titration to be done in the presence of ammonia, which can be used to prevent the precipitation as hydroxides of various metals that form ammine complexes.

As stated above and as shown in Table 7-1, mercuric ion is unusual in that there is a factor of 10^4 or more between the constants for the formation of the neutral halide-type complexes and those for the HgL_3^- types. This means that there will be a rapid change in concentration of the ligand titrant near the equivalence point. With most other cations there is relatively little difference in the stabilities of the various complexes, and the last ones are loosely held. This is illustrated in Table 7-2, where the values of the constants for the various cyanide and ammonia complexes with cadmium are shown. If the titration of cadmium ion with either of these ligands were attempted, one would find that the titrant concentration change at the various equivalence points would be too gradual to provide a practical end point.

Cases such as these are so common that volumetric methods based upon titrations of metal cations with inorganic ligands are of limited use. In recent

[15] I. M. Kolthoff and J. J. Lingane, *J. Am. Chem. Soc.*, **57**, 2377 (1935).
[16] J. Liebig. *Liebigs Ann. Chem.*, **77**, 102 (1851).
[17] G. Deniges. *Compt. Rend.*, **117**, 1078 (1893).

TABLE 7-2

Dissociation Constants of Cd(II) Complexes with Cyanide and Ammonia

	CN^-	NH_3
$Cd(II)L = Cd(II) + L$	2.9×10^{-6}	2.2×10^{-3}
$Cd(II)L_2 = Cd(II)L + L$	8.7×10^{-6}	7.9×10^{-3}
$Cd(II)L_3 = Cd(II)L_2 + L$	2.0×10^{-5}	3.6×10^{-2}
$Cd(II)L_4 = Cd(II)L_3 + L$	1.8×10^{-4}	1.2×10^{-1}

years titrations with organic ligands have been developed that avoid these limitations. They are discussed below.

Un-ionized Compounds Involving Organic Ligands

The ligands that have been discussed above have been capable of donating a single pair of electrons toward the formation of covalent bonds—such ligands are said to be monodentate. There are many compounds, more commonly organic in nature, that possess more than one pair of unshared electrons. When these pairs are at separated positions in a large molecule, each is capable of forming a bond with the same metal cation. Such *polydentate* ligands are classified as *bidentate, tridentate, quadridentate*, or *hexadentate*, depending on whether two, three, four, or six bonds are formed. An example of an organic bidentate ligand is the compound known as ethylenediamine[18] (abbreviation En), which has the structure

```
 H  H  H  H
 |  |  |  |
:N—C—C—N:
 |  |  |  |
 H  H  H  H
```

Each of the amine groups, like the ammonia molecule from which it is derived, has an available electron pair on the nitrogen atom. Both of these pairs are capable of forming a bond with a metallic cation; the resulting complex with the cadmium ion has the ring structure

```
    H  H
    |  |
 H—C—N—H
   |  |
   |  Cd2+
   |  |
 H—C—N—H
    |  |
    H  H
```

[18] The name *amine* is given to the group $-NH_2$. It is an ammonia molecule in which one of the hydrogens has been replaced by another element or organic group.

Complexes of this type, in which a polydentate ligand is attached to the central atom by more than one bond, are known as *chelate* (from the Greek word *chele*, a claw) compounds. One would expect intuitively that since the ethylenediamine is attached by two nitrogen bonds the complex would be more stable than that formed by a single ammonia molecule and this is true. The dissociation constant, K_1, for the $CdEn^{2+}$ complex is 3.4×10^{-6} compared with 2.2×10^{-3} for the monoammonia complex; two additional En molecules can be added and the corresponding constants are 2.8×10^{-5} and 8.5×10^{-3}. Although these ligands are held more tightly than the corresponding ammonia molecules (see Table 7-2), the difference in their constants is too small; they are not sufficiently stable for the compound to be a satisfactory titrant. In a search for polydentate ligands that could combine with cations in a 1:1 ratio and thus avoid the problem of successive and close dissociation constants, the compound triaminotriethylamine (tren) has been studied and found to be a satisfactory titrant for a number of metals.[19] The structure is

$$N \begin{cases} CH_2{-}CH_2{-}NH_2 \\ CH_2{-}CH_2{-}NH_2 \\ CH_2{-}CH_2{-}NH_2 \end{cases}$$

There are four nitrogen atoms able to form bonds. They form 1:1 complexes with many metals, and these complexes are more stable than those of ethylenediamine; the value of the dissociation constant for the cadmium complex is 5×10^{-13}. Other compounds having as many as six electron pairs for bond formation have been prepared; one of the most widely used is ethylenediaminetetraacetic acid, commonly called EDTA. This compound and its applications are discussed in Chapter 27.

SUPPLEMENTARY REFERENCES

Bates, R. G. *Electrometric* p*H Determinations.* New York: John Wiley, 1954.

Beck, M. T. *Chemistry of Complex Equilibria.* New York: Van Nostrand Reinhold, 1970.

Bell, R. P. *Acids and Bases.* New York: John Wiley, 1956.

Butler, J. N. *Ionic Equilibrium.* Reading, Mass.: Addison-Wesley, 1964.

Johnston, M. B., Barnard, A. J., and Flaschka, H. EDTA and Complex Formation. A. Demonstration Lecture. *J. Chem. Ed.*, **35**, 601 (1958).

Pokras, L. On the Species Present in Aqueous Solutions of Salts of Polyvalent Metals. *J. Chem. Ed.*, **33**, 152, 223, 282 (1956).

---

[19] Flashka and Soliman. *Z. Anal. Chem.*, **158**, 254 (1957); Reilley and Sheldon. *Chemist-Analyst*, **46**, 59 (1957).

Reilley, C. N., Schmidt, R. N., and Sadek, F. S. Chelan Approach to Analysis. I. Survey of Theory and Application. II. Illustrative Experiments. *J. Chem. Ed.*, **36**, 555, 619 (1959).

QUESTIONS AND PROBLEMS

1. Discuss the limitations to the so-called Arrhenius definition of acids, bases, and of a neutralization reaction.

2. Can the equation $3Na_2O + 2As_2O_5 = 2Na_3AsO_4$ be said to represent a neutralization reaction? Explain.

3. Predict the approximate value of K_1 for the following acids: H_4SiO_4, H_2SeO_4, HNO_2, H_5IO_6, H_3AsO_3, $HMnO_4$. Predict the approximate value of K_2 for the polyprotic acids. Explain.

4. What is meant by (*a*) a pseudo-acid constant, (*b*) the leveling effect?

5. Discuss the validity of the following equation, $Cu^{2+} + 4NH_3 = Cu(NH_3)_4{}^{2+}$, for describing the reaction when an excess of ammonia is added to a cupric sulfate solution.

6. Formic acid (HCOOH) in a 0.10 *F* solution has been found by conductance measurements to be 4.5% ionized at 25°C.

   (*a*) Calculate the acid ionization constant (K_A).
   (*b*) Calculate the molar $HCOO^-$ concentration in (1) 0.010 *F* and (2) 1.0 *F* acid solutions. *Ans.* (*a*) 2.1×10^{-4}.

7. (*a*) Show that in its aqueous solution the fractional ionization—that is, the fraction of the total acid ionized—of a very weak acid can be said to vary inversely as the square root of the concentration.
   (*b*) Show that in a 1 *F* solution the fractional ionization of such an acid is equal to the square root of the ionization constant.

8. Calculate the $[H^+]$ and the percentage ionization of the formic acid in

   (*a*) a solution 0.10 *F* in HCOOH and 0.020 *F* in HCOONa,
   (*b*) a solution 0.10 *F* in HCOOH and 0.10 *F* in HCOONa, and
   (*c*) 0.10 *F* in HCOOH and 0.020 *F* in HCl.
   (*d*) Derive a general equation relating the formal concentration of a weak acid, its acid dissociation constant, its fractional ionization, and the formal concentration of added strong acid or salt or the weak acid. *Ans.* (*b*) 0.21%.

9. Calculate the $[H^+]$ and the *p*H of a solution made by dissolving 0.10 mole of HCOOH and 0.10 mole of NaOH in water and diluting to exactly one liter. *Ans.* $pH = 8.3$.

10. Calculate the hydrogen-ion concentration of a solution 0.10 *F* in NH_3.

11. To 50 ml of 0.10 *F* acetic acid, 20 ml of 0.10 *F* NaOH were added; calculate the *p*H of the resulting solution (assume additivity of volumes). *Ans.* *p*H = 4.6.

12. To 200 ml of 7.5 *F* ammonia, 500 ml of 1.0 *F* HNO_3 were added and the resulting solution was diluted to 1 liter. Calculate the *p*H of the solution.

13. Compare the concentrations of CN^-, HCN, and OH^- in the following solutions:

 (*a*) 10 ml of 0.010 *F* HCN.
 (*b*) 10 ml of a solution containing 0.00010 mole of HCN.
 (*c*) 0.00010 mole of NaCN and 0.00010 mole of HCl in 10 ml of solution.
 (*d*) 0.00010 mole of NaCN and 0.00010 mole of HCN in 10 ml of solution. The ionization constant for HCN is 2×10^{-9}. *Ans.* (*b*) $[CN^-] = 4.5 \times 10^{-6}$.

14. The equilibrium constant for the reaction

$$Cu(H_2O)_4{}^{2+} + 4NH_3 = Cu(NH_3)_4{}^{2+} + 4H_2O$$

is 2×10^{13}. Calculate the hydrated copper ion concentration when ammonia gas is passed into a solution initially 10^{-3} *F* in Cu(II) in sufficient quantity to make the ammonia concentration 1 *M*. (Note how effective *an excess* of NH_3 is in reducing the concentration of the hydrated cupric ion; with lower NH_3 concentrations ions such as $Cu(NH_3)(H_2O)_3{}^{2+}$, $Cu(NH_3)_3(H_2O)^{2+}$, and so forth, are formed; in very concentrated ammonia solutions appreciable concentrations of $Cu(NH_3)_5{}^{2+}$ and $Cu(NH_3)_6{}^{2+}$ are present.)

15. *Buffer Solutions.*

 (*a*) Calculate the hydrogen-ion concentrations of solutions prepared by (1) mixing 100 ml of 1.0 *F* HCOOH and 100 ml of 1.0 *F* HCOOK; (2) diluting 10 ml of the solution from (1) to 100 ml; (3) adding 20 ml of 0.10 *F* HCl to 20 ml of the solution from (1); (4) adding 20 ml of 0.10 *F* HCl to 20 ml of water; (5) adding 20 ml of 1.0 *F* HCl to 20 ml of the solution from (1).
 (*b*) What characteristics of buffer solutions are illustrated by these experiments? *Ans.* (*a*-4) $[H^+] = 0.050$.

16. Calculate the *p*H of a 0.10 *F* solution of NH_4Cl. *Ans.* *p*H = 5.1.

17. Calculate the *p*H of a 0.010 *F* H_2SO_4 solution. *Ans.* *p*H = 1.82.

18. Calculate the *p*H of 50 ml of 0.010 *F* HCl after 0.10 g of Na_2SO_4 is added to the solution. Is this solution buffered effectively?

19. Calculate the $[H^+]$, the $[SO_4{}^{2-}]$, and the $[HSO_4{}^-]$ when to 100 ml of 0.10 *F* Na_2SO_4 one adds (*a*) 50 ml, (*b*) 100 ml, and (*c*) 150 ml of 0.10 *F* H_2SO_4. *Ans.* (*a*) $[H^+] = 0.015$.

20. (*a*) Calculate the $[Ba^{2+}]$ and the $[Ca^{2+}]$ that would be in equilibrium with each of the solutions of Problem 19.
 (*b*) Calculate the ratio of $[Ba^{2+}]$ to $[Ca^{2+}]$ for each solution. Derive an expression for calculating this ratio.

21. To 30 ml of a solution initially 1.0 *F* in HNO_3 were added 40 ml of 3 *F* $(NH_4)_2SO_4$. Calculate the resulting hydrogen-ion concentration. Is the solution buffered?

22. To 100 ml of a solution which contains 60 mmoles of HNO_3 there were added 40 ml of 6 *F* CH_3COONH_4. Calculate the hydrogen-ion concentration and the *p*H of the resulting solution. *Ans.* $[H^+] = 6 \times 10^{-6}$.

23. Calculate the weight of sodium acetate which must be added to 150 ml of a 0.10 *F* HNO_3 solution to give a buffer solution of *p*H 5. Assume no volume change upon addition of solid CH_3COONa.

24. Calculate the *p*H of a solution prepared by mixing 20 ml each of 3 *F* CH_3COONa, 1 *F* HNO_3, and 6 *F* CH_3COOH, and diluting to 100 ml. *Ans.* *p*H = 4.20.

25. A solution containing 0.060 g of CH_3COOH is diluted to 1.50 liters. What is the formal concentration of acetic acid? Conductance measurements have indicated that the acetic acid in this solution is 15% ionized. What is the molar concentration of the (*a*) CH_3COOH, (*b*) H^+, and (*c*) CH_3COO^- in the solution? Calculate the ionization constant of acetic acid from the data given.

CHAPTER 8

Chemical Separations. Periodic Properties of the Elements and Chemical Separations

PERIODIC PROPERTIES OF THE ELEMENTS

In connection with your study of this chapter you should review the following topics in your general chemistry text: the periodic system and the electronic structures of the elements responsible for this periodicity; the various types of chemical bonds; the electronegativity concept; and the effect of bond type upon the properties of compounds.

Your general chemistry text has pointed out that there is a gradual transition in the physical properties of the elements as one proceeds from left to right along a horizontal row (a *period*) of the periodic table. The elements of any vertical column are usually referred to as a *group* and have similar chemical and physical properties. There is generally an orderly change in these properties as you go from the top to the bottom of a column. This similarity of properties within the members of a group is especially pronounced in the columns of the periodic table labeled Ia and IIa, and VIa and VIIa (on the left and right sides, respectively, of the periodic table shown in the inside front cover). The elements of each of the shorter columns labeled Ib to VIIb, and VIII (these are the central elements of the long periods) likewise resemble each other in their properties. Also these "b" elements and the "a" elements of the same column number in the first two periods of eight have similar properties. However, these similarities are less pronounced than those that exist among the "a" elements of a given column.

The elements of the subcolumns from IIIb across to and including the subcolumns under VIII are called *transition elements.* These columns, which are headed by the elements from scandium (21) to nickel (28), contain elements in which an inner electronic shell is being filled. The elements of the subcolumns Ib and IIb, headed by copper (29) and zinc (30), are also frequently classified as transition elements because some of their properties are similar to those of the elements in columns IIIb through VIII.

Metals, Nonmetals, and Metalloids

The elements of columns Ia and IIa have relatively high electrical and thermal conductivities and their electrical conductivities decrease with increasing temperature. These elements have a pronounced luster and high reflectivity; they are ductile and malleable and flow under mechanical stress; they emit electrons relatively easily when subjected to radiation or thermal excitations. These properties are attributed to the presence of relatively loosely bound electrons in these elements.

Elements characterized by the properties listed above are called *metals*; elements lacking in these properties are called *nonmetals*; those that are intermediate in nature are called *metalloids.* There are no abrupt changes in elemental properties as one proceeds either horizontally across a row or vertically down a column of the periodic table; therefore, sharp dividing lines cannot be drawn to distinguish these three classes of elements.

The elements of columns Ia and IIa, respectively, are called the *alkali* and *alkaline earth elements.* The outermost electrons of these elements are easily emoved. After their removal these elements display the octet electronic configuration of the nearest noble gas. The alkali and alkaline earth elements are outstanding in the extent to which they exhibit the properties of metals.

On the other hand the elements of columns VIa and VIIa tend to *add* lectrons and assume the electronic configuration of the nearest noble gas. As elements they must share electrons to assume this configuration and thus they exist in the elemental state as molecules such as Cl_2 and S_8. These lements complete a stable outer octet of electrons by electron-pair bond ormation, leaving no free electrons available; therefore they are nonconducting. Their other properties are also different from the properties of metals, and they are consequently classified as nonmetals.

As you descend a given column of the periodic table the elements become nore metallic in their properties; the larger the atom the less firmly the uter electrons are held. As a result, the most metallic elements are found in the lower left-hand portion of the periodic table; the most nonmetallic are ound in the upper right-hand portion; and the metalloids divide them. The lements boron, silicon, germanium, arsenic, antimony, and bismuth are generally classified as metalloids. They extend diagonally between the metals and the nonmetals. The elements to the left of these metalloids in their respective rows are predominantly metallic in their properties and those to the right are predominantly nonmetallic. You will observe that a large proportion of the elements are classified as metals.

Thus we see that the properties of the elements leading to their classification as metals, metalloids, and nonmetals can be explained by the relative availability of loosely bound outer electrons. In the case of metals these electrons are readily available; in the case of nonmetals atoms tend to hold their electrons more strongly by sharing them with other atoms in order to achieve a stable octet configuration. This relation between the physical properties of the elements and their electronic structures is of interest here because the same structural factors are largely responsible for the types of compounds that the elements form. When an atom of one of the metallic elements of column Ia reacts with an atom of one of the nonmetallic elements of column VIIa the metallic element becomes an electron donor and is said to be *electropositive*; the nonmetallic element becomes an electron acceptor and is said to be *electronegative.* The resulting compound will be predominantly *ionic* in character, which means that it will consist mainly of *positive ions* (*cations*) of the metallic element and *negative ions* (*anions*) of the nonmetallic element. The crystals of such compounds are held together predominantly by electrostatic forces known as *ionic bonds*; such crystals tend to be extremely hard and to have high melting points.

It was stated above that the nonmetallic elements tended to share electrons as electron-pair bonds in order to attain a stable octet configuration and thus they exist as molecules. Such electron-pair bonds are called *covalent bonds.* When a predominantly metallic element reacts with a predominantly nonmetallic element there is essentially a transfer of electrons and an *ionic compound* (one held together by *electrostatic bonds*) is formed. When a nonmetallic element forms a compound by pairing one or more electrons with those of another nonmetallic element, the term *covalent bond* is again applied to the resultant electron pair. Since we find a gradual transition from metallic elements to nonmetallic elements, we find compounds in which the type of bonds range from predominantly ionic to predominantly covalent. The tendency of an element to attract the electrons of an electron-pair bond is called its *electronegativity.* By making use of the ionization potential and the electron affinity of an element, or by considering the reactions that the element undergoes and the properties of the resulting compounds, numbers can be assigned to the elements that give a relative measure of this tendency. Values of the relative electronegativities of the elements are shown in Figure 8-1. As you might expect, the pattern formed by the electronegativities of the elements conforms to the pattern of the periodic table.

H																
2.1																
Li	Be											B	C	N	O	F
1.0	1.5											2.0	2.5	3.0	3.5	4.0
Na	Mg											Al	Si	P	S	Cl
0.9	1.2											1.5	1.8	2.1	2.5	3.0
K	Ca	Sc	Ti	V	Cr	Mn	Fe	Co	Ni	Cu	Zn	Ga	Ge	As	Se	Br
0.8	1.0	1.3	1.5	1.6	1.6	1.5	1.8	1.9	1.9	1.9	1.6	1.6	1.8	2.0	2.4	2.8
Rb	Sr	Y	Zr	Nb	Mo	Tc	Ru	Rh	Pd	Ag	Cd	In	Sn	Sb	Te	I
0.8	1.0	1.2	1.4	1.6	1.8	1.9	2.2	2.2	2.2	1.9	1.7	1.7	1.8	1.9	2.1	2.5
Cs	Ba	La–Lu	Hf	Ta	W	Re	Os	Ir	Pt	Au	Hg	Tl	Pb	Bi	Po	At
0.7	0.9	1.0–1.2	1.3	1.5	1.7	1.9	2.2	2.2	2.2	2.4	1.9	1.8	1.9	1.9	2.0	2.2
Fr	Ra	Ac	Th	Pa	U	Np–No										
0.7	0.9	1.1	1.3	1.4	1.4	1.4–1.3										

FIGURE 8-1

Relative electronegativities of the elements. The values are taken from *General Chemistry* (3rd ed.) by L. Pauling. W. H. Freeman and Company.

A survey of this figure shows that the electronegativities of the elements of a given row of the periodic table increase from the typically metallic elements of column I to the nonmetallic elements of column VII. Note the series Na, 0.9; Mg, 1.2; Al, 1.5; Si, 1.8; P, 2.1; S, 2.5; and Cl, 3.0. Also, from the top of a given column downward, the electronegativities decrease, as with the halides: F, 4.0; Cl, 3.0; Br, 2.8; I, 2.5. The elements with the most metallic properties, found in the lower left-hand corner of the periodic table, have the lowest electronegativity values; the least metallic elements, found in the upper right-hand corner, have the highest electronegativity values. When a strongly metallic element reacts with a strongly nonmetallic element, there is a relatively large difference in their electronegativities; we can predict that the bond between them will be strong and predominantly ionic in character; we can also predict that the reaction will be accompanied by the evolution of a large quantity of heat. When the difference between the electronegativities of two elements is 1.9 units, the bond between them will show approximately equal amounts of ionic and covalent character. If this difference is less than 1.2, the bond will tend to be predominantly covalent in character; if the difference is greater than 2.6, the bond will be predominantly ionic. Differences between 1.2 and 2.6 indicate the bonds are intermediate in character and will range from predominantly covalent to predominantly ionic.

Oxygen Compounds of the Elements

Because of its abundance and reactivity oxygen plays a tremendously important part in the geological and biological processes taking place on the earth. For this reason it is appropriate to examine the types of compounds that the various elements form with oxygen, and then to extend this examination to the behavior of these compounds in water as a solvent. The discussions in your general chemistry text of oxygen, its occurrence and properties, and of water and its physical and chemical properties should be reviewed.

Oxygen is nonmetallic and with the exception of fluorine is the most electronegative element. As a result we expect that the bond between a strongly metallic element and oxygen will be predominantly ionic in character and this is indeed the case. Conversely, there are relatively small electronegativity differences between oxygen and the nonmetallic elements; the bonds in the oxides of these elements are predominantly covalent in nature. Thus the oxides of the metallic elements of columns Ia and IIa are nonvolatile solids; the oxides of the nonmetallic elements of columns VIa and VIIa are molecular compounds and exist as gases or relatively volatile solids. Typical examples of the metallic oxides are Na_2O and CaO; of nonmetallic oxides, Cl_2O, SO_2, and SeO_2. We can predict to what extent the bonds of an oxide will be ionic or covalent by making use of the electronegativity values given in Figure 8-1. The difference in the electronegativity values for oxygen and sodium is 2.6, and therefore we can predict that Na_2O is predominantly ionic; the difference in the values for oxygen and chlorine is 0.5, and therefore we can predict that the bonds in Cl_2O are predominantly covalent. It follows that the bonds in BcO and Al_2O_3 should be intermediate in character.

Behavior of Oxides with Water

Water is unique in the extent to which it occurs on earth, in its properties as a solvent, and in the extent to which it is used as a solvent medium in which to carry out chemical reactions. For this reason it is of interest to note the chemical behavior of the various types of oxides in aqueous solutions.

Ionic Oxides. The oxides of the alkali metals are predominantly ionic compounds and exist in the solid state as unipositive cations and binegative oxide anions. If sodium oxide, $(Na^+)_2O^{2-}$ is put into water (an ionizing solvent because of its high dielectric constant), there is a tendency for Na^+ and O^{2-} ions to pass into solution. However, since water is a diprotic acid with the first proton being much more ionized than the second, the following reaction takes place:

$$H_2O + O^{2-} = 2OH^-$$

This reaction proceeds so quantitatively to the right that the concentration of O^{2-} existing in aqueous solutions can be neglected for experimental purposes. This reaction is quite analogous to the reaction taking place when Na_2S is added to water and the aqueous solution is saturated with H_2S, that is

$$H_2S + S^{2-} = 2HS^-$$

It is apparent from the reaction between water and oxide ion given above that when an ionic oxide dissolves in water the hydroxide-ion concentration of the solution will increase, and the solution will become more alkaline.

Covalent Oxides. When the oxide of a predominantly nonmetallic element, such as Cl_2O, dissolves in water the reactions resulting are quite different from those obtained with an ionic oxide. Because of the covalent bonding (the electronegativity difference between chlorine and oxygen being only 0.5), there is no significant ionization into Cl^+ and O^{2-} when the oxide is dissolved in water. Instead, the following reaction takes place:

$$Cl_2O + H_2O = 2ClOH$$

Again, the oxide oxygen of the Cl_2O acquires a proton, but in this case, because of the covalent bonding between the chlorine and oxygen atoms, there is no significant ionization of the OH group as hydroxide. The difference between the behavior of Na_2O and Cl_2O with water is due to the fact that the chlorine in ClOH is more electronegative than the hydrogen; consequently, the electrons of the oxygen are shifted toward the chlorine and away from the hydrogen. As a result the bond between the oxygen and hydrogen is weakened and the hydrogen is more easily removed; in aqueous solution it can attach itself to a water molecule to form H_3O^+. When this occurs the solution becomes more acid. Sodium, being less electronegative than hydrogen, does not shift electrons away from the H–O bond; thus when NaOH dissolves in water the H–O bond is not broken but ionization gives Na^+ and OH^-. As stated above, ClOH does not ionize in aqueous solutions to give OH^- but does ionize to a small extent to give H^+ and ClO^-. Because of this the compound is conventionally written as HClO (or HOCl) and is called hypochlorous acid rather than chlorine hydroxide. Moreover, Cl_2O, chlorine monoxide, is considered to be an *acid anhydride*, that is, an oxide that upon hydration forms an acid.

Effect of Oxidation State on the Properties of Oxides

Consider the behavior of SO_2 (sulfur dioxide; oxidation number of sulfur, +4) and SO_3 (sulfur trioxide; oxidation number of sulfur, +6) in water.

Sulfur dioxide is one of a large number of molecules whose structure cannot be properly shown by the usual electronic representation, because the oxygen–sulfur bonds are neither single nor double bonds but have intermediate properties. The situation with sulfur dioxide is conventionally represented by writing the two electronic formulas

```
       ..              ..
      .O:             .O:
     /  ..           //
  :S                :S
    \\                \ ..
      O:               O:
      ..               ..
```

and saying that the actual structure is a *resonance hybrid* of the two. Since there is a difference of only one unit in the electronegativities of sulfur and oxygen, the bonds in sulfur dioxide will be predominantly covalent. Therefore there is little tendency for ionization into a sulfur cation and an oxygen anion when SO_2 is dissolved in water. Rather, the sulfur atom tends to complete a stable octet configuration and does this by attracting the electron pairs of a water molecule and hydrating to form H_2SO_3. The electronic formula of H_2SO_3 is

```
            ..
            S
       .. / | \ ..
    H—O  :O:  O—H
      ..  ..  ..
```

In this molecule, as in HClO, the electron-pair bonds formed between the oxygens and the sulfur cause a shift in electron density that results in a weakening of the hydrogen–oxygen bond. The hydrogen ions can then attach themselves to water molecules, forming H_3O^+, and the solution becomes more acid. With H_2SO_3, as with HClO, the weakening of the hydrogen–oxygen bond is of such a limited extent that only a small fraction of the protons is transferred to the solvent, predominantly as H_3O^+ ions. In other words, H_2SO_3 and HClO are weak acids.

When SO_3 is dissolved in water H_2SO_4 is formed. The increase in the oxidation state of the sulfur increases the effective positive charge on the sulfur atom. This gives it an increased attraction for available electron pairs; for this reason the tendency of SO_3 to hydrate is much greater than that of SO_2 and the energy of the hydration reaction, as shown by the heat evolved, is much greater.

In addition, because of the increase in effective positive charge, the sulfur atom in H_2SO_4 has a much greater attraction for electrons than the sulfur atom in H_2SO_3. This causes an even greater weakening of the hydrogen–oxygen bonds. As a result the protons of H_2SO_4 are much freer to attach themselves to water molecules and H_2SO_4 is a stronger acid (more highly ionized) than is H_2SO_3. This effect, caused by the charge on the positive element in oxides and oxygen acids, is general in character. One can state as a rule that as the oxidation state of an element increases, the properties of the corresponding oxides become more acidic in character; this means that the products of their reaction with water have less tendency to ionize as hydroxides and more tendency to ionize as acids, and the resulting acids become more highly ionized.

The acid ionization constants of the oxygen acids of chlorine illustrate this effect; thus

$$\begin{array}{ll} HClO & K_A = 10^{-8} \\ HClO_2 & K_A = 10^{-2} \\ HClO_3 & K_A = 10^{3} \\ HClO_4 & K_A = 10^{8} \end{array}$$

Another example is given by the oxides of manganese, which range from being predominantly basic to predominantly acidic in character. Thus MnO hydrates to give $Mn(OH)_2$, which, while only slightly soluble, ionizes to give Mn^{2+} and OH^- ions; it therefore becomes more soluble if the hydrogen-ion concentration of the solution is increased and less soluble as the hydroxide-ion concentration is increased. The bonds in MnO_2 have so little acidic or basic character that MnO_2 is not only extremely insoluble in water but also not appreciably dissolved by either concentrated acid or base. The oxides MnO_3 and Mn_2O_7 are both unstable and tend to decompose into oxygen and MnO_2, but their hydration products are more stable, and both H_2MnO_4 and $HMnO_4$ are strong acids.

Effect of Atomic Size on the Properties of Oxides

If the oxides and oxyacids of the elements of a given vertical column of the periodic table are examined, we find that for a given oxidation state the oxides become less acidic in nature and the acids become less highly ionized as we proceed down the column. This trend illustrates the effect of the atomic size of the element when the charge of its ion is constant. The smaller the atom the more concentrated is the charge and the greater is the attraction of the central atom for the electron pairs of the oxygen atoms. This leaves the oxygen atoms more positive and weakens the hydrogen–oxygen bond. Thus the ionization constant for the first proton of H_2SO_3 is approximately 10^{-2}, that for H_2SeO_3 is 10^{-3}, and that for H_2TeO_3 is 10^{-6}. In addition, whereas SO_2 and SeO_2 are quite soluble in water, TeO_2 is only slightly soluble; the saturated solution is only about 10^{-4} formal in TeO_2. Finally, H_2TeO_3 is not only a very weak acid, but the tellurium–oxygen bond is so weak that in a concentrated acid solution one of the hydroxide groups can be removed. Thus H_2TeO_3 not only behaves as an acid in neutral or alkaline solutions,

TABLE 8-1
Properties of the Alkaline Earth Elements

	Be	Mg	Ca	Sr	Ba
Atomic radii (Å)	0.89	1.36	1.74	1.91	1.98
Ionic radii M^{2+} (Å)	0.31	0.65	0.99	1.13	1.35
Heat evolved on hydration $MO + H_2O = M(OH)_2$ kcal mole		5.4	15.1	17.7	22.3
Ionization constant $\frac{[M^{2+}][OH^-]}{[M(OH)^+]}$	10^{-8}	10^{-2}	0.031		0.23
Solubility products: $[M^{2+}][OH^-]^2$	10^{-26}	10^{-11}	10^{-6}	10^{-4}	10^{-3}
$[M^{2+}][CO_3^{2-}]$		10^{-5}	5×10^{-9}	7×10^{-10}	2×10^{-9}
$[M^{2+}][SO_4^{2-}]$			6×10^{-5}	5×10^{-7}	1×10^{-10}

but it can behave as a base in acidic solutions. Oxides and their hydration products that act in this dual manner are said to be *amphoteric*, from the Greek word *amphoteros*, meaning *both*.

The effect of atomic size can also be illustrated with the metallic elements of column II, which are known as the alkaline earth metals, and certain of their properties are shown in Table 8-1. Of these elements beryllium is the least electropositive and has the highest electronegativity value, 1.5. Therefore its oxide has more covalent bonding than the oxides of the other elements. As a result beryllium hydroxide is not only a weak base but it is also relatively insoluble in water—although it is soluble in acids. In addition, because of its relatively high charge density, the beryllium atom of $Be(OH)_2$ can attract the electron pairs of an additional hydroxide group and form the $Be(OH)_3^-$ ion, and thus beryllium oxide becomes more soluble as the hydroxide-ion concentration is increased. Because of this dual behavior, beryllium oxide and hydroxide are said to be amphoteric.

As can be seen from Table 8-1 above, there is a relatively large increase in ionic radius, from 0.31 to 0.65 Å, in going from Be^{2+} to Mg^{2+}, and, in addition, the electronegativity of magnesium is 1.2 as compared with 1.5 for beryllium. Magnesium oxide is predominantly ionic in character and hydrates readily to $Mg(OH)_2$, which, while not highly ionized and only slightly soluble, does not show significant amphoteric properties. As seen from Table 8-1, calcium, strontium, and barium show successive increases in atomic radii. Their oxides are predominantly ionic; they show increasing tendencies to hydrate to form hydroxides; and their hydroxides become increasingly more soluble and more highly ionized.

We are now in a position to classify the oxides of the elements in accordance with their reactions with water or with aqueous solutions having various *p*H values. We will classify them as *basic oxides, acidic oxides*, and *amphoteric oxides.*

Basic Oxides

An oxide can be classified as basic: (1) if when treated with water it dissolves significantly and gives an alkaline solution (in other words, if it reacts with water to form a soluble ionized hydroxide), or (2) if the solubility of the oxide or hydroxide (*a*) *decreases* as the hydroxide-ion concentration of the saturated solution is increased, and (*b*) *increases* as the hydrogen-ion concentration is increased. Examples of the first type, which are the most basic, are the typically ionic oxides of the alkali metals and alkaline earth metals below beryllium. Examples of oxides of the second type are the oxides and hydroxides of the elements below aluminum in the third vertical column of the periodic table (the hydroxides of the rare earth elements are only slightly soluble but are highly ionized) and the oxides of the lower oxidation states of manganese, iron, cobalt, nickel, and copper. Thus MnO, Mn_2O_3, FeO, Fe_2O_3, CoO, NiO, and Cu_2O are all insoluble in concentrated hydroxide solutions. (MnO_2 is not included in this list; its classification is subject to question because of its extreme insolubility in both concentrated acids and bases.)

Acidic Oxides

An oxide can be classified as acidic: (1) if when treated with water it dissolves significantly and gives an acid solution (in other words, if it reacts with water to give a soluble ionized acid), or (2) if the solubility of the oxide or the corresponding acid (*a*) *decreases* as the hydrogen-ion concentration of the saturated solution is increased, and (*b*) *increases* as the hydroxide-ion concentration is increased. Typical oxides of the first type, which are the most acidic, are SO_2, SO_3, CrO_3, Mn_2O_7, and As_2O_5. Acidic oxides of the second type are SiO_2, SnO_2, and Sb_2O_5. These latter oxides are all relatively insoluble in concentrated, noncomplex forming acids, but are dissolved by concentrated hydroxide solutions. Anhydrous or ignited SiO_2 and SnO_2 are extremely resistant, but when freshly precipitated they behave as acidic oxides.

When discussing the solubility of these oxides in acids, reference is made to the hydrogen-ion effect only. Thus both hydrous silicon dioxide and hydrous tin dioxide (the so-called silicic and metastannic acids) are quantitatively precipitated by fuming perchloric acid, but both are dissolved by hydrofluoric acid (a much weaker acid) because of the formation of the complex ions SiF_6^{2-} and SnF_6^{2-}. Hydrous SiO_2 is insoluble in hydrochloric acid, but hydrous SnO_2 is dissolved because the hexachlorostannate ion, $SnCl_6^{2-}$, is formed.

Amphoteric Oxides

The oxide of an element, and the corresponding hydroxide or acid, can be classified as amphoteric if its solubility in water can be increased both by increasing the hydrogen-ion concentration and by increasing the hydroxide-ion concentration of the saturated solution. This effect has been mentioned above in connection with tellurium and beryllium oxides. Amphoteric oxides and their hydration products are usually relatively insoluble because of their intermediate nature. Their bonds are not sufficiently ionic to give rise to ionizable cations and hydroxide-ions, nor does the central atom attract water electron pairs sufficiently strongly to cause ionization of a proton. On the other hand, the bond has sufficient ionic character so that hydroxide ions can be removed by a high concentration of hydrogen ions, and in a high concentration of hydroxide ion the central atom will attract the electron pairs of an oxygen atom to form a hydroxide-complex anion.

Thus zinc oxide, which hydrates to give zinc hydroxide, behaves as an amphoteric oxide. This behavior can be illustrated by the following equations, which show that the addition of either acid or base to a saturated solution of the hydroxide will increase its solubility:

$$Zn(OH)_2(s) + H^+ = Zn(OH)^+ + H_2O$$

$$Zn(OH)_2(s) + OH^- = Zn(OH)_3^-$$

In many cases the energy of hydration of amphoteric oxides, that is, their tendency to form stable bonds with water molecules, is so small that it is difficult to assign a definite formula to the precipitates obtained from aqueous solutions. A precipitate that conforms to a formula M_pO_q can be designated an *oxide*; one that is predominantly basic and conforms to a formula $M_r(OH)_s$ can be designated a *hydroxide*. Precipitates such as the so-called aluminum and chromium hydroxides, which contain varying quantities of water when precipitated, and which lose their water on heating without giving evidence of a stable hydrate or hydroxide, are more properly called *hydrous oxides*. However, the term *hydroxide* is so generally applied to such precipitates that it will be used except when it is advantageous to indicate the uncertain composition of the product.

We may expect that there should be some correlation between the metalloid elements, which are intermediate in character between typical metals and typical nonmetals, and those elements forming amphoteric oxides. On looking for this relationship we must avoid being misled or confused by the effect of oxidation state on the acidic–basic properties of an oxide.

Typical amphoteric oxides are BeO, Al_2O_3, ZnO, Ga_2O_3, GeO_2, SnO, and PbO. The elements involved, like the metalloid elements, appear in a rough diagonal, dropping toward the right on the periodic table. The effect of oxidation state is illustrated by the oxides of chromium and manganese. Chromous oxide, CrO, is a basic oxide; chromic oxide, Cr_2O_3, is amphoteric, while chromium trioxide, CrO_3, hydrates to form chromic acid, H_2CrO_4; two H_2CrO_4 molecules dimerize to give $H_2Cr_2O_7$. As stated previously, MnO is a basic oxide; MnO_3 and Mn_2O_7, while unstable, can be considered to be the anhydrides of the strong manganic and permanganic acids, H_2MnO_4 and $HMnO_4$.

In summary, we have seen that the elements can be classified as metals, nonmetals, and metalloids. The properties characteristic of metals can be attributed to the presence of relatively loosely bound electrons, and the elements on the left side of the periodic table are most characteristically metallic because of the ease with which they lose electrons and assume a stable octet configuration. The elements to the right tend to acquire electrons in order to attain the same electronic configuration, and this leads to electron sharing by means of electron-pair bonds. The nonmetals thus tend to exist in the elementary state as relatively nonconducting molecules. The metalloids are intermediate in character and appear on the periodic table in a diagonal region extending from boron at the top of column IIIa down to bismuth in column Va.

The ease with which the elements in columns Ia and IIa lose electrons and the strong tendency of those in columns VIa and VIIa to gain electrons cause the compounds formed between elements of these columns to be predominantly ionic in character and to exist in the crystal primarily as cations and anions held together by electrostatic forces. The bonds in compounds formed between the elements of columns VIa and VIIa are predominantly covalent (electron-pair) bonds. The oxides of the predominantly metallic elements will be predominantly ionic in character and when treated with water will dissolve to give the metal cation and hydroxide ion; the resulting hydroxides are stable and highly ionized. The oxides of the predominantly nonmetallic elements contain covalent bonds and, in general, the nonmetallic elements will have such an attraction for electrons that when dissolved in water the hydrogen–oxygen bond will be weakened and an acid will be formed. The greater the oxidation number of the central element the greater this electronic attraction, the weaker the hydrogen–oxygen bond, and the more highly ionized the resulting acid. The oxides of the elements toward the center of the periodic table will tend to be intermediate in character and their energies of hydration will be low. The lower oxidation-state oxides of the more metallic of these elements will tend to be relatively insoluble, and the oxides of the metalloids will tend to be amphoteric or weakly acidic; the oxides of the highest oxidation states will tend to be acidic. These effects can be seen in Appendix 5, where the more common elements are listed in accord with their positions in the periodic table and the species formed by their various oxidation states in acid, neutral, and alkaline solutions, and in acid and alkaline sulfide solutions, are shown.

SEPARATIONS BASED ON PERIODIC PROPERTIES

The broad range of properties exhibited by the oxides of the elements suggests that it should be possible to take mixtures containing compounds of various elements, treat them with solutions having controlled *p*H values, and thus separate the basic oxides, the amphoteric oxides, and the acidic oxides from each other. Such separations would depend upon the different solubilities of the various types of oxides in these solutions. Such procedures are in fact used in quantitative methods for the separation of two or more elements and for the separation of groups of elements in systems or "schemes" of qualitative analysis.

Figure 8-2 shows the behavior of the more common elements upon treatment with approximately 1 *F* NaOH when in their commonly occurring oxidation states. (More comprehensive and detailed information regarding the behavior of these elements is given in Appendix 5.) The elements that would be precipitated are indicated by dark gray shading; the typically amphoteric elements are indicated by medium gray shading; thc typically acidic elements are indicated by light gray shading. The oxides of the strongly

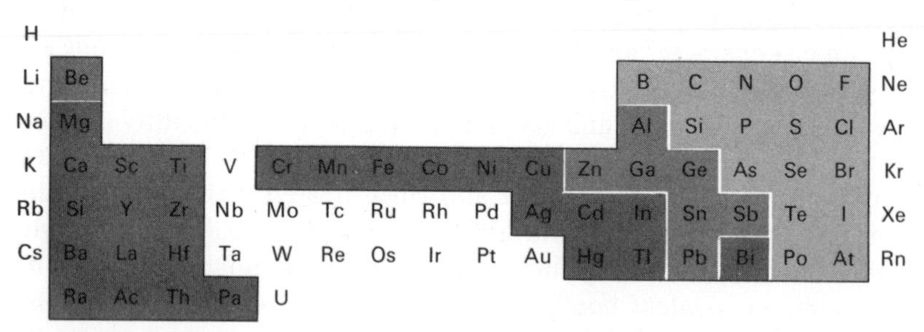

FIGURE 8-2

Behavior of the elements in concentrated hydroxide solutions. (They are assumed to be in their more stable oxidation states.) Elements forming insoluble basic hydrous oxides are indicated by dark gray shading. Elements forming amphoteric oxides and dissolving are indicated by medium gray shading. Elements forming acidic oxides and dissolving are indicated by light gray shading. Unshaded elements at the extreme left form basic oxides and soluble hydroxides. Unshaded elements in the central part tend to form amphoteric oxides in their more common oxidation states and are not precipitated.

metallic alkali metals from column Ia form highly ionized and soluble hydroxides and are not shaded. The properties of the oxides of the unshaded elements appearing in the central and lower portion of the figure are highly dependent upon their oxidation state; they normally occur in higher oxidation states and are not precipitated because of the formation of hydroxide or oxide anions. The oxides of the metalloid elements are in general sufficiently amphoteric to dissolve as hydroxide anions. The nonmetallic elements are present either in their negative oxidation states as simple anions or in their positive oxidation states when they form acidic oxides and are present as oxide anions; thus sulfur (unless an oxidant is used) can be present as sulfide, or as oxygen anions. In order to obtain more predictable behavior an oxidant, such as a peroxide, is usually provided when separations are attempted by the use of strongly alkaline solutions.

A system of analysis, qualitative and semiquantitative, based upon the above properties, was developed during World War II. At the beginning of that war our Chemical Warfare Service needed a means for the rapid identification of any new or unknown chemical agent that might be used by the enemy. Instrumental methods were not generally applicable; moreover, many analytical instruments are not portable, nor are they easily serviced under field conditions. To meet wartime needs systematic procedures were developed through several years' effort by a group of chemists at the California Institute of Technology. These procedures included a system of analysis that provided for the detection and estimation of 32 elements selected on the basis of their possible use in toxic agents and in other chemical munitions. A 20- to 40-mg sample was used; any element constituting 1% of the sample could be detected, and by means of volumetric and colorimetric methods the quantity of any element present could be estimated to within ± 0.3 mg.

In the first step of the system, the sample was fused with sucrose and an excess of sodium peroxide in a nickel bomb. The fusion mass was treated with water, giving an insoluble residue containing those elements whose oxides are so basic as to be essentially insoluble in this strongly alkaline solution. This residue was called the *basic element group*. The solution was divided into two portions. One portion was analyzed for the elements whose oxides are predominantly amphoteric (the *amphoteric element group*); the other portion was analyzed for the elements forming predominantly acidic oxides (the *acidic element group*).[1] A simplified tabular outline showing the result of the fusion procedure is shown in Table 8-2.

Although chemical warfare did not develop, this system was used for various analytical purposes by the Chemical Warfare Service throughout the

[1] These titles are not rigorously correct; it is not the elements, but their oxides that are basic, amphoteric, or acidic.

TABLE 8-2
Behavior of Certain Elements after Sodium Peroxide Fusion and Treatment with Water

Fusion Residue	Fusion Solution	
	Amphoteric Elements	Acidic Elements
Fe_2O_3, TiO_2, MnO_2 $Ni(OH)_2$, $Cd(OH)_2$, CuO	$Pb(OH)_4^{2-}$, $Zn(OH)_4^{2-}$ $Al(OH)_4^-$, CrO_4^{2-}	I^-, Br^-, Cl^-, F^- AsO_4^{3-}, PO_4^{3-}, SiO_3^{2-}
$Mg(OH)_2$, $CaCO_3$, $SrCO_3$, $BaCO_3$	$Sb(OH)_6^-$, $Sn(OH)_6^{2-}$	SO_4^{2-}, SeO_4^{2-}, $H_4TeO_6^{2-}$, $H_2BO_3^-$, CO_3^{2-}, NO_3^-

war, and was employed after the war in the identification of toxic agents found in enemy territory.[2]

A much more detailed discussion of the wide range of separations that can be made by control of the *p*H of solutions is given in the next chapter.

Sulfur is directly under oxygen in the periodic table; therefore one would expect to find similarities in the properties of the oxides and sulfides, and this is true. Also, similarities are found with adjacent elements in any column of the table. Just as the alkali metal oxides reacted with water to give the metal ion and OH^-, the sulfides react similarly but are more soluble than the corresponding hydroxides. Most of the elements in the central portion of the periodic table form insoluble sulfides that show little evidence of amphoteric properties. Farther to the right such properties become evident; the sulfides of arsenic, antimony, and tin are insoluble in acid solutions but because of their amphoteric properties dissolve in alkaline sulfide solutions as sulfo anions. Separations based upon sulfide-ion control are extensively used in both quantitative methods and in systems of qualitative analysis. These separations are also discussed in the next chapter.

QUESTIONS AND PROBLEMS

1. What are the predominant characteristics of metallic elements? Nonmetallic elements? Metalloid elements?

[2] A brief description of this system was published by Swift and Niemann in *Anal. Chem.*, **26**, 538 (1954); otherwise it has not been made generally available. Subsequently a similar but much simplified system of analysis based on periodic properties was developed for classroom use: E. H. Swift and W. P. Schaefer. *Qualitative Elemental Analysis.* San Francisco: W. H. Freeman and Company, 1962.

2. Write the electronic structures of lithium, magnesium, fluorine, sulfur, cesium, iodine, silicon, arsenic, and barium atoms. (Consult your general chemistry text if necessary.) Classify these elements as metallic, metalloid, nonmetallic. What features are common to the electronic structures of each class?

3. Classify the following as metallic, nonmetallic, or metalloid: Se, Ta, Rh, Ga, Po, Te, At, Hf, Eu, Tl. Is the decision always clear-cut? Why?

4. State whether the bonds in each of the following elements or compounds are predominantly covalent, predominantly ionic, or intermediate in character:

   (*a*) I_2 (*c*) MgI_2 (*e*) MgF_2
   (*b*) $SnCl_2$ (*d*) $MgCl_2$ (*f*) K_2S

5. Write equations for the reactions taking place when Rb_2O and BaO dissolve in water. (Note that predominantly ionic *soluble* oxides of this type are restricted to elements of columns I and II.) What is the effect on the *p*H of the solution?

6. Write equations for the reactions taking place when SO_2, P_2O_5 (properly P_4O_{10}), and Cl_2O_7 dissolve in water. What is the effect on the *p*H of the solution? To what region of the periodic table are elements whose oxides behave in this manner restricted?

7. Write equations for the reactions taking place when CrO_3 and Mn_2O_7 dissolve in water. Since chromium and manganese are classified as metals, explain why these oxides are acidic in character. What is the electronic structure of these elements? What is meant by the "transition elements"?

8. What differences or similarities exist among covalent and ionic oxides, metallic and nonmetallic oxides, and acidic and basic oxides? Give reasons for the existence of these relationships.

9. State clearly the effect of the oxidation state and of the atomic size of an element upon the properties of its oxides. Give examples.

10. Define a basic oxide; an acidic oxide; an amphoteric oxide.

11. Write the formulas of the oxides of each element of the second period of eight of the periodic table (Na to Cl); assume each element to have the oxidation state corresponding to the number of the vertical column in which it appears.

    Predict the products that would result upon treating each of these oxides with water. Write chemical equations where a significant reaction takes place.

    Where the oxide is insoluble, predict the effect of treating it with (*a*) 6 *F* HNO_3 and (*b*) 6 *F* NaOH. (Assume the oxide is reactive and not made inert by long standing or heating.)

    Classify the oxides as either basic, acidic, or amphoteric.

    How do the oxides change in classification as you proceed from left to right across the periodic table? Explain and give examples.

12. How does the oxidation state affect the basic, acidic, and amphoteric properties of the oxides of a given element? Explain and give examples.

13. Explain why ferric hydroxide can be precipitated at a lower *p*H than can ferrous hydroxide.

14. Note the precipitation *p*H of Fe(III) in Table 9-1 (page 135). Explain why a 0.1 *F* solution of ferric nitrate can be prepared by dissolving the salt in pure water.

15. Which of the cations of the following pairs of salts would be more extensively hydrolyzed in aqueous solutions if present in the same formal concentration and at the same temperature:

    (*a*) $FeCl_2$ or $FeCl_3$
    (*b*) $MnSO_4$ or $Mn_2(SO_4)_3$
    (*c*) $TiCl_3$ or $TiCl_4$
    (*d*) $SnCl_2$ or $SnCl_4$
    (*e*) $CrCl_2$ or $CrCl_3$
    (*f*) UCl_4 or UCl_5

16. Why do the alkaline earth group elements (with the exception of Be) show relatively little tendency to form complex ions? Why is Be the exception?

17. Compare the properties of Mn(II) with those *of* Mn(IV), Mn(VI), and Mn(VII), and account for the changes that accompany an increase in oxidation state.

18. Why would you predict that you would find a similarity between the analogous compounds of phosphorus and arsenic? Cite evidence to support the validity of this prediction.

CHAPTER 9

Chemical Separations. Separations Involving Heterogeneous Systems

In Chapter 8 we discussed separations that were based upon periodic properties of the elements. In this chapter we will consider separations involving multiple phase or heterogeneous systems. The emphasis in analytical chemistry upon precipitation methods involving aqueous solutions is the result of the historical development of this branch of chemistry. Because of the inherent liabilities in precipitation methods it seems unfortunate that more effort was not expended earlier on other methods, for example those based upon volatility and liquid-phase distributions. In recent years there has been much effort expended and great progress made in such non-precipitation methods of effecting separations. The three systems most used for chemical separations in both analytical and technical fields are solid–liquid, gas–liquid, and liquid–liquid. These systems will be discussed in that order in the following three sections; variants of these systems will be discussed in a fourth section.

SOLID–LIQUID SEPARATIONS. PRECIPITATION METHODS

Limitation of the time available for courses in quantitative analysis makes it necessary at the present time to resort to the use of idealized samples, which are purposely selected because they are free from constituents that would have to be separated before the desired method could be used to determine the constituent of interest. Under actual conditions this problem of the separation of elements becomes one of vital importance, requiring a background of judgment based on experience. Usually the only experience available as to the separation of the elements, singly or in groups, is that obtained in the course in qualitative analysis. Use of this experience is likely to result in disillusionment, since these courses are often so limited that they fail to impart an appreciation of the liabilities involved in the use of these separations.

Often a resort to more advanced texts or reference books on quantitative analysis is frustrating in that one can find only general statements, lacking in restrictions concerning the conditions under which proposed separations are valid or in verifying experimental data.[1]

A brief discussion of the more frequently used precipitation methods follows, especially of those involving the separation of certain groups of elements.

Methods Involving the Precipitation of Hydroxides and Hydrous Oxides

Methods involving *p*H control

Under this classification are included those procedures in which the separations are based upon control of the *p*H of the solution at a value that will cause the precipitation of certain elements as hydrous oxides or hydroxides. Such separations are extensively used in analytical work both for the separation of single elements and for the separation of groups of elements. There are two reasons for this extensive use. (1) There is a tremendous difference between the solubilities of the oxides and hydroxides of the elements. Compare, for example, the solubilities of ferric oxide and sodium hydroxide. Acidic oxides such as those of tetrapositive tin and silicon make an even more striking contrast with the soluble oxides. (2) The hydrogen-ion and hydroxide-ion concentrations of aqueous solutions can be varied and, by the use of buffer systems, controlled over a range exceeding 10^{16}-fold!

Table 9-1 shows approximate values of the *p*H at which precipitation of the hydrous oxides and hydroxides of certain elements occurs. As is stated in this table, these values and the solubility-product values that are given in

[1] Two exceptions to this statement are: W. F. Hillebrand, G. E. F. Lundell, H. A. Bright, and J. I. Hoffman. *Applied Inorganic Analysis.* (2nd ed.) New York: Wiley, 1953; and A. A. Noyes and W. C. Bray. *Qualitative Analysis for the Rare Elements.* New York: Macmillan, 1927. Both of these books contain a wealth of factual inorganic chemistry, supported by experimental data and conditions, as well as numerous references to the original literature. Unfortunately, the Noyes and Bray is out of print.

TABLE 9-1

Approximate Precipitation *p*H and Methods of Precipitation of the Oxides and Hydroxides of Certain Elements*
(Arranged in the Order of Their Precipitation as the Hydrogen-Ion Concentration is Decreased)

Element and Oxidation State	Precipitation *p*H	Methods of Precipitation and Remarks
W(VI), Si(IV)	<1	Conc. HCl, HNO_3, H_2SO_4, $HClO_4$
Sb(V), Sn(IV), Ta(V), Nb(V)	<1	Conc. HNO_3, $HClO_4$
Mn(IV)	<1	Conc. HNO_3 or $HClO_4$, with $KClO_3$
Pb(IV)	<1	Conc. HNO_3, anodic oxidation
Ti(IV), Ce(IV), Bi(III), Sb(III)	2–3	Bi and Sb precipitate as basic salts
Fe(III), Sn(II), Hg(II)	3–4	Hg(II) in presence Cl^-, pH 8–9
Hg(I)	4	
Al(III), Cr(III)	5–6	Fe(III), Al(III), Cr(III) frequently precipitated by following buffer systems: (*a*) $NH_3—NH_4^+$, (*b*) $CH_3COO^-—CH_3COOH$, (*c*) benzoate-benzoic acid, (*d*) $Ba^{2+}—BaCO_{3(s)}$
Fe(II), Cu(II), Pb(II)	6–7	
Cd(II), Zn(II), Ni(II), Co(II)	7–8	
Ce(III)		
Mn(II)	8–9	
Ag(I)	9	
Mg(II)	11	
Ca(II), Ba(II), Sr(II)	>11	
As(III), As(V)		Soluble, acidic oxides

NOTE: The *p*H values, which are taken largely from the data of Britton (*J. Chem. Soc.*, **127**, 2110–2159, 1925), are given only to show the approximate *p*H at which precipitation occurs in solutions 0.005–0.025 *F* in the various salts used: they do not indicate the *p*H at which quantitative precipitation can be obtained. In addition, this precipitation *p*H will be shifted significantly by temperature changes, by the concentration of the precipitating cation, by the presence of complex-forming anions or of cations which may cause the formation of stable colloidal systems, and, in certain cases, by a slow rate of hydrolysis and precipitation.

* See Appendix 3 for the solubility products of certain of these compounds.

Appendix O are subject to many uncertain factors and should be used with these limitations in mind.

As is evident from Table 9-1, if the elements were arranged successively according to their precipitation *p*H values, a series with no distinct breaks would result. Separation of adjacent elements in the series by *p*H-control methods would probably, therefore, be very difficult. With this limitation taken into account, *p*H-control separations can be placed into four general classes, as follows:

(*a*) Precipitations made in concentrated solutions of strong acids.
(*b*) Precipitations made in buffered solutions.
(*c*) Precipitations made in solutions of strong bases.
(*d*) Precipitations made as in (*a*), (*b*), or (*c*), but with additional use of complex-ion formation.

(*a*) *Precipitations Made in Solutions of Strong Acids.* The precipitates formed in solutions of strong acids are composed of certain elements that form insoluble acidic hydrous oxides in their highest oxidation state and of manganese and lead, which form similar oxides in the quadripositive state. Silicon is typical of these elements and silicic acid is frequently precipitated and removed from solutions of native materials by treatment with hot concentrated acid followed by filtration. It is found experimentally that the hydrous colloidal silicic acid that first forms must be dehydrated to approximately the composition $SiO_2 \cdot \frac{1}{2}H_2O$ before it can be coagulated sufficiently to be quantitatively retained by paper filters. This dehydration can be accomplished either (1) by evaporation with a volatile acid, such as hydrochloric acid, and by the heating of the dry residue, or (2) by fuming with a high-boiling-point acid, such as perchloric acid or sulfuric acid, which serves as a dehydrating agent. Hydrofluoric acid and fluorides must be absent because of the reactions $SiO_2(s) + 4HF = SiF_4(g) + 2H_2O$ and $SiF_4 + 2HF = H_2SiF_6(aq)$.

The elements tungsten, tantalum, niobium, antimony, and tin are precipitated as a group of acidic oxides in one system of analysis[2] by means of fuming perchloric acid.

(*b*) *Precipitations Made in Buffered Solutions.* Such solutions are used when it is necessary to control the *p*H within close limits. The hydrous oxides of tripositive iron, chromium, and aluminum, and of quadripositive titanium, are most commonly precipitated from buffered solutions. The purpose is usually to effect the separation of these elements, singly or as a group, from the dipositive elements (zinc, cobalt, nickel, manganese, and those of the alkaline-earth group). The buffer system most commonly used for this separation is ammonia and ammonium salts. A study of this method, showing the extensive coprecipitation that can occur when it is carried out under various conditions

---

[2] Noyes and Bray, op. cit. This reference contains experimental studies of the relative advantage of the use of nitric acid or perchloric acid for this precipitation.

with the first four dipositive elements mentioned above being present, has been made by Swift and Barton.[3]

Buffer systems composed of organic acids (such as acetic, benzoic, and succinic) and their salts, which give somewhat lower *p*H values, have been used extensively.

(*c*) *Precipitations Made in Solutions of Strong Bases.* Precipitations by the third class of methods are made by using relatively concentrated solutions of sodium or potassium hydroxide. Separations made from such solutions are based upon the oxides involved being basic, acidic, or amphoteric. Factors determining these properties are the periodic relationship of the elements and their oxidation state. Because of the latter, in many cases a strong oxidant, such as sodium peroxide, is used to produce a higher oxidation state and develop a more acidic, and soluble, oxide. Sodium hydroxide with sodium peroxide added is extensively used as a group reagent in qualitative systems and for certain quantitative separations.[4]

(*d*) *Precipitation Separations Involving Complex-ion Formation.* The use of the ammonia–ammonium-ion buffer system mentioned under (*b*) above usually involves the assumption that the tendency of ammonia to form complex ions is an additional important factor. This method can well be termed one of the classical analytical separations and, as is stated by Hillebrand and Lundell,[5] "One of the commonest operations the analyst has to perform . . ., with the object either of weighing the precipitated compound or of effecting a joint separation of two or more metals from others." That it may be inadequate even as a qualitative separation in certain cases is shown by experiments in which, with large amounts of aluminum or ferric iron (100 mg–200 mg) and with amounts of cobalt, zinc or nickel up to 20 mg, from 75% to 99% of the latter elements were found to be carried down in the precipitate. It has also been shown that a large quantity of zinc may be quantitatively precipitated by ammonia when a larger proportion of chromium is present and that manganese will in any case be partially precipitated by that reagent owing to its oxidation by the air. In a study of the conditions necessary for securing optimum separations Swift and Barton[6] found that better separations were obtained when an excess of ammonia was avoided.

---

[3] E. H. Swift and R. C. Barton. *J. Am. Chem. Soc.*, **54**, 2219 (1932).

[4] Data showing the extensive coprecipitation involved in this separation of iron, nickel, and cobalt from aluminum, chromium, and zinc are given by Swift and Barton, op. cit., p. 4155.

[5] See Footnote 1.

[6] See Footnote 3.

The statements made above show that methods involving the precipitation of hydrous oxides and hydroxides are likely to be subject to severe limitations because of coprecipitation effects and should be used with this fact in mind. In many cases, even under the most favorable conditions, one or even more reprecipitations will be necessary before coprecipitation can be reduced to acceptable limits.

Methods Involving Oxidation Reactions

Chemical Oxidation. Certain elements can be precipitated by being oxidized to a state in which the oxide of the element is insoluble. Manganese(II), for example, can be oxidized quantitatively in hot nitric acid solution by chlorate ion to MnO_2. Manganese dioxide is unique in being so devoid of either acidic or basic character that it can be quantitatively precipitated from concentrated solutions of either strong acids or strong bases.

In dilute alkaline solution, lead(II) can be oxidized by hypochlorite to give PbO_2. The lead(II) is in solution as plumbite, $Pb(OH)_4{}^{2-}$, and upon oxidation is precipitated as the dioxide.

Electrolytic Oxidation. Electrolytic precipitation methods in general are discussed in Chapter 28. Such methods most frequently involve the reduction and deposition of metals on the cathode. There are a few elements, however, that can be oxidized and electrolytically deposited as their oxides on the anode and hence separated from other elements. The best known example is Pb(II) which is anodically oxidized and deposited as PbO_2. Other elements that may be oxidized and at least partially deposited on the anode under various conditions are silver, cobalt, manganese, and bismuth.

Methods Involving the Precipitation of Sulfides

Separations involving the precipitation of sulfides under conditions that control the sulfide-ion concentration are probably even more extensively used than those involving the precipitation of oxides and hydroxides. The first reason for their widespread use is that the coprecipitation limitations mentioned above with hydrous oxides cause one to resort to the use of sulfides in spite of the unpleasant features associated with hydrogen sulfide. Second, like the hydrous oxides, there is a tremendous difference in the solubilities of the sulfides; for example, the solubility product of mercuric sulfide has the value 10^{-54}, whereas the sulfides of the alkaline-earth-group elements are relatively soluble. Third, the sulfide-ion concentration can be controlled, by means of pH control, through an even wider range than can the hydrogen-ion concentration. For example, one can calculate that a solution that is 1 F in a strong acid such as HCl and that is saturated with

TABLE 9-2
Behavior of Certain Elements in Various Sulfide Solutions

Element* and Oxidation State	Acid Solutions Saturated with H_2S			Alkaline Solutions	
	HCl 3 *F*	HCl 0.3 *F* or Thio-acetamide with 0.1 *F* HCl	$[H^+] = 10^{-6}$	$[H^+] = 10^{-9}$ $(NH_4)_2S$ NH_3	NaOH 1 *F* Na_2S 1 *F*
As(V)	As_2S_5[a]	As_2S_5[a]	As_2S_5[a]	AsS_4^{3-}[b]	AsS_4^{3-}
As(III)	As_2S_3	As_2S_3	As_2S_3	AsS_3^{3-}[c]	AsS_3^{3-}[c]
Hg(II)	HgS	HgS	HgS	HgS	HgS_2^{2-}
Cu(II)	CuS	CuS	CuS	CuS	CuS
Ag(I)	Ag_2S	Ag_2S	Ag_2S	Ag_2S	Ag_2S
Sb(V)	Sb_2S_5	Sb_2S_5	Sb_2S_5	SbS_4^{3-}	SbS_4^{3-}
Sb(III)	Sb_2S_3	Sb_2S_3	Sb_2S_3	SbS_3^{3-}[c]	SbS_3^{3-}[c]
Bi(III)	$BiCl_4^-$	Bi_2S_3	Bi_2S_3	Bi_2S_3	Bi_2S_3
Sn(IV)	$SnCl_6^{2-}$	SnS_2	SnS_2	SnS_3^{2-}	SnS_3^{2-}
Cd(II)	$CdCl_4^{2-}$	CdS	CdS	CdS	CdS
Pb(II)	$PbCl_4^{2-}$	PbS	PbS	PbS	PbS
Sn(II)	$SnCl_4^{2-}$	SnS	SnS	SnS[d]	SnS[d]
Zn(II)	$ZnCl_3^-$	$ZnCl_3^-$	ZnS	ZnS	ZnS
Co(II)	Co^{2+}	Co^{2+}	CoS	CoS	CoS
Ni(II)	Ni^{2+}	Ni^{2+}	NiS	NiS	NiS
Fe(II)[e]	Fe^{2+}	Fe^{2+}	Fe^{2+}	FeS	FeS
Mn(II)	Mn^{2+}	Mn^{2+}	Mn^{2+}	MnS	MnS

NOTE: The elements are arranged in approximately the order in which they precipitate as a concentrated HCl solution is progressively diluted and kept saturated with H_2S. Only the sulfides of arsenic are quantitatively precipitated from hot 9 *F* HCl.

* The sulfides of titanium, aluminum, and chromium are unstable in water because of hydrolysis to the hydrous oxides and H_2S.

Sulfides of V(IV) and V(V) are not quantitatively precipitated from acid solutions; thioanions are formed in alkaline sulfide solutions—acidification of such solutions results in only partial precipitation of VS_2 and V_2S_5.

[a] As_2S_3 and S are also formed. The rate of precipitation by H_2S is slow.

[b] The exact formulas of many of the sulfo anions have not been established.

[c] When S_2^{2-} is present the higher oxidation state is formed.

[d] SnS dissolves to a slight extent in concentrated monosulfide solutions; it is oxidized and dissolves in polysulfide solutions.

[e] Fe(III) is reduced to Fe(II) by H_2S in acid solutions; in alkaline sulfide solutions Fe_2S_3 is precipitated.

H_2S at room temperature and pressure has a sulfide-ion concentration of only 2×10^{-23} *M*, whereas solutions 1 *M* in sulfide ion can be attained by the use of an alkali metal sulfide and hydroxide. As with *p*H-control methods, buffer systems can be used to maintain the sulfide-ion concentration within quite close limits during precipitation reactions.

Table 9-2 shows certain of the frequently used separations of the common elements that are made by sulfide-ion control. Figure 9-1 shows the positions in the periodic table of those elements forming sulfides and sulfo-anions of

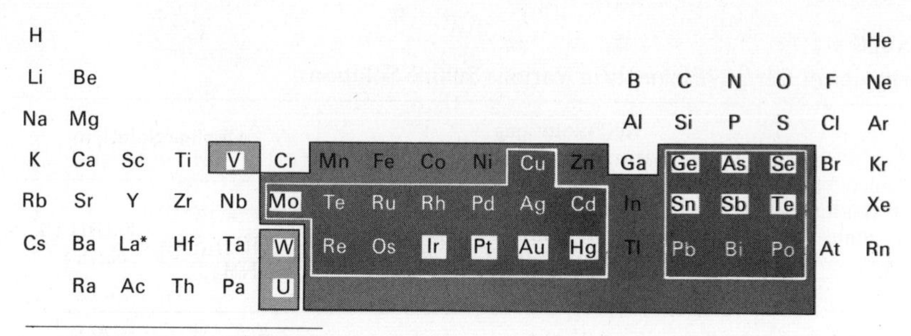

FIGURE 9-1

Behavior of sulfides in certain analytical separations. Elements forming sulfides in neutral aqueous solutions are in the major block with medium gray shading. Elements forming sulfides insoluble in approximately 0.3 *F* solutions of strong acids are in the two sub-blocks with darker shading. Elements forming sulfo anions in solutions containing sulfide and hydroxide are indicated by small white blocks. Elements forming sulfides insoluble in aqueous solutions only in certain oxidation states or under certain conditions are in the two small lightly shaded blocks. Certain elements in their higher oxidation states are reduced by sulfide, especially in acid solutions. Thus, Cr(VI) is reduced to Cr(III), the higher states of Mn to Mn(II), Fe(III) to Fe(II), V(V) to V(IV), Sb(V) to Sb(III), Tl(III) to Tl(I), and Au(III) to Au(I) and Au.

analytical importance; it is of interest to note that only those elements forming strongly basic and strongly acidic oxides are not included. It is seen that as the sulfide-ion concentration increases there are two effects: (1) additional sulfides of metals that are predominantly basic in character precipitate since their solubility products are exceeded, and (2) certain sulfides of metals that are more acidic in character and that precipitate at lower sulfide-ion concentrations dissolve through the formation of sulfo anions such as AsS_4^{3-} and SnS_3^{2-} as the sulfide-ion concentration is increased. These sulfo anions are analogous to the oxygen anions and hydroxide complexes formed by amphoteric and acidic oxides in alkaline solution: AsO_4^{3-}, $Al(OH)_4^-$, $Sn(OH)_6^{2-}$, and so forth.

In addition the table emphasizes another very important principle, namely, that the anions present have a very decided effect upon the solubility of the sulfides. Notice, for example, the sulfides precipitated from 9 *F* HCl and those from 4.5 *F* H_2SO_4. Studies have shown that the predominant factor preventing the precipitation of sulfides of Hg, Cu, Ag, Sb, and Bi from 9 *F* HCl is the formation of chloride complexes with the ions of these metals.

Two group separations often used in systems of qualitative analysis depend primarily upon the sulfides and their solubilities. The first group, called the hydrogen sulfide group, is precipitated by saturating a solution 0.3 *M* in H^+ and about 0.6 *M* in Cl^- with hydrogen sulfide. The third column in Table 9-2 shows the elements precipitated by this treatment. After this precipitate is removed, the solution is made just alkaline with NH_3 and again treated with H_2S. The second column from the right in Table 9-2 shows approximately the conditions of this precipitation; Zn, Co, Ni, Fe, and Mn precipitate as sulfides and Cr and Al as hydroxides. Thus, in two sulfide precipitations a group of nine elements and another group of seven are separated from each other and from alkaline earth and alkali-metal ions. Although space does not permit detailed discussion of these and subsequent separations in such a system, it should be mentioned that not only are the initial separations into groups accomplished by sulfide precipitations but so also are several subsequent separations into subgroups. Let us emphasize again that the extraordinary range of sulfide-ion concentrations that can be controlled in aqueous solution, along with the great spread in the solubilities of the sulfides of the elements, makes possible the unusual number of separations that can be effected by the use of the sulfides. We also see that since *p*H control can be used to control sulfide concentrations, sulfide separations can be grouped into the same four classifications as hydrous oxide separations (page 136).

Methods Involving Other Precipitants. Inorganic Precipitants

Chloride. Chloride, usually as ammonium chloride, has been used in quantitative analysis primarily as a specific precipitant for silver, but finds almost universal use in qualitative systems. Frequently the initial separation used in these systems is the precipitation of AgCl and Hg_2Cl_2, the silver group. When a large quantity of lead is present, the precipitate may also contain $PbCl_2$.

Carbonate. After removal of all other elements except alkaline earth and alkali metals, carbonate is used in qualitative systems for the precipitation of barium, strontium, calcium, and sometimes magnesium. One should remember that most metal carbonates are insoluble, so that carbonate ion is a nonspecific precipitant.

Sulfate. Some isolated attempts have been made in systems of qualitative analysis to use sulfate as a precipitant for a group composed of lead, barium, and strontium.

Oxalate. The precipitation of barium, strontium, and calcium (and, under certain conditions, magnesium) is sometimes made by an oxalate precipitation. Since a large number of cations form insoluble oxalates, the reagent is not specific. Oxalate is most commonly used for the precipitation of calcium and its separation from magnesium.

Phosphate and Arsenate. The compounds formed by these two anions are remarkably similar in type and solubility. Since most of the metals form insoluble phosphates and arsenates, these anions are nonspecific precipitants. Phosphate is most frequently used in gravimetric work for the precipitation of magnesium as $MgNH_4PO_4 \cdot 6H_2O$, which on ignition is converted to $Mg_2P_2O_7$. Arsenate precipitates an analogous compound. Other elements that form the type precipitate, MNH_4PO_4, are zinc, manganese, cadmium, nickel, and cobalt.

Methods Involving Other Precipitants. Organic Precipitants

Aside from oxalate, mentioned above, an increasing number of organic precipitants are being used for the separation of single elements or of groups.

Certain advantages and liabilities accompany the use of these reagents as precipitants. In many cases a specificity can be obtained that would be impossible with inorganic precipitants; this advantage is enhanced by the fact that many organic reagents are organic acids and thus an effective control of solubility can be attained by *p*H control. Many of the precipitates are not only exceedingly insoluble but also of such high molecular weight that they are exceptionally adapted for micro or semimicro procedures. Four liabilities may be cited: (1) the introduction of objectionable organic material into a systematic analysis, (2) in some cases these organic compounds are of such a low solubility that the precipitate is likely to be contaminated by precipitant, (3) organic reagents are frequently unstable on storage or in use, especially in the presence of oxidizing agents, (4) they are often expensive. Space does not permit a complete listing of the organic precipitants that have found extensive use. Only a few familiar examples are given below, followed by a reference to a more extensive treatment.

Dimethylglyoxime [diacetyldioxime, $(CH_3 \cdot C{:}NOH)_2$]. The reagent is a remarkably specific precipitant for nickel and is most frequently used for that purpose, although it can be used for the precipitation of palladium.[7]

Cupferron (ammonium nitrosophenylhydroxylamine, $C_6H_5N \cdot NO \cdot ONH_4$). This compound is especially valuable in that it can be used for the precipitation in an acid solution of iron, vanadium, zirconium, titanium, and tin, and their separation from aluminum, chromium, beryllium, and manganese.

[7] The monograph by Diehl, *Applications of the Dioximes to Analytical Chemistry* (Columbus, Ohio: G. Frederick Smith Chemical Company, 1940), contains a comprehensive treatment of the subject.

8-*Hydroxyquinoline* ($C_9H_6N{\cdot}OH$). A large number of metals are precipitated by this compound, but by proper control of the *p*H a considerable specificity can be attained. The precipitate can be weighed or else dissolved and titrated with a bromate in the presence of bromide. This latter procedure is useful in that it permits a volumetric determination of aluminum.

REFERENCE

I. M. Kolthoff, E. B. Sandell, E. J. Meehan, and S. Bruckenstein (*Quantitative Chemical Analysis*, 4th ed., New York: Macmillan, 1969, pp. 259–280) give an excellent summary of organic reagents and their application to qualitative methods, as well as references to sources giving more comprehensive treatments.

GAS–LIQUID AND GAS–SOLID SEPARATIONS

Gas–liquid and gas–solid separations occur by means of two closely related and even overlapping processes: volatilization and distillation. *Volatilization* is the more general term, applied to the conversion of a substance to the gaseous state. The term *distillation* means a volatilization process but usually implies the vaporization of a substance having a relatively low partial pressure above a solution by boiling that solution and removing the substance in the resultant gas phase.

Volatilization processes are usually considered to be processes in which the constituent is separated as a pure gas. The constituent may be volatilized without change (as when heat is used to remove mercury from an alloy or water from a hydrate), or the constituent may be changed by a chemical reaction prior to its volatilization (thus the sulfur in a sulfide can be separated by treatment with an acid and volatilized as hydrogen sulfide, and the carbonate in a limestone determined by similar treatment with acid and collection of the carbon dioxide). When a substance changes directly from a solid to a gas phase, as when solid iodine is volatilized without the formation of a liquid phase, the process is termed a *sublimation.*

Distillation implies a somewhat more complicated process than *volatilization.* The constituent of interest is present or is formed in a liquid phase and is extracted and removed from the liquid phase by a second gas phase. In distillations from aqueous solutions the second gas phase consists of bubbles of water vapor. Usually the constituent being removed has a relatively low partial pressure in the gas phase but is continuously extracted into the water bubbles and eventually is removed from the liquid phase. When desired, the constituent of interest can be collected by condensing the water vapor or by passing it through a suitable collecting agent. Distillation methods are extensively used in the preparation and separation of organic compounds.

Volatilization

As mentioned above, mercury can be separated from alloys or other nonvolatile constituents by volatilizing the metal; the mercury can then be collected on a gold foil, and a determination can be made if desired.

The anions of volatile acids can be eliminated from a solution by the addition of an excess of a much less volatile acid and heating the solution. Thus the halides can be volatilized from a solution by adding perchloric or sulfuric acid and heating to fuming. This process serves to dissolve such insoluble salts as calcium fluoride and the silver halides.

The iron in an ammonia precipitate can be separated from such elements as aluminum, zirconium, beryllium, and chromium by heating the precipitate at 200°C–300°C in a stream of dry chlorine and hydrogen chloride.[8] The inert gases can be volatilized from silicates by vacuum fusion; this process is used in radioactive rock-dating techniques.

Silica, SiO_2, is frequently a source of difficulty in the analysis of native materials, but it can be removed by being converted to the tetrafluoride, SiF_4, which is then removed by volatilization. Carbonates and sulfides are converted to CO_2 and H_2S by acid, and these can be removed by boiling.

Sublimation, mentioned above, is the term applied to a volatilization process in which there is direct vaporization of a solid and subsequent condensation of the gas to the solid without an intermediate liquid state. The primary uses of sublimation for separations have been in the purification of substances. For example, both I_2 and As_2O_3 can be separated from nonvolatile impurities by this process.

Distillations

Various elements and groups of elements can be separated by distilling their compounds from solutions. For example, arsenic can be separated from the other common metallic elements by distilling the trichloride or the bromide from hydrochloric acid solutions. In their *System of Analysis for the Rare Elements*, Noyes and Bray (Footnote 1) separate a group composed of arsenic, germanium, and selenium by distilling these elements as bromides from a hydrobromic acid solution; they also separate osmium and ruthenium from other elements and from each other by successive distillations of their tetroxides from nitric and from fuming perchloric acid solutions.

Procedures have also been developed for the separation of selenium and tellurium by distillation of their chlorides from sulfuric acid solution.[9] The

[8] Gooch and Haven. *Am. J. Sci.*, **4**, 7, 370 (1899).

[9] V. Lehner and D. P. Smith. *Ind. Eng. Chem.*, **16**, 837 (1924); H. C. Dudley and H. G. Byers. *Ind. Eng. Chem., anal. ed.*, **7**, 3 (1935).

chlorides of arsenic, antimony, and tin can be separated from each other and other elements by distillation.[10] Chromium has also been distilled as CrO_2Cl_2 from perchloric-acid–sodium-chloride solutions.

LIQUID–LIQUID SEPARATIONS

The phase-distribution law (which states that when the same substance is present in two phases at equilibrium, the ratio of the activities of the substance in the two phases at a given temperature will be a constant) was discussed in Chapter 4. A special case of this law, applicable to the same substance present in two *liquid* phases, states that at a given temperature, the ratio of the equilibrium concentrations of a substance in two immiscible solvents is a constant, called the *distribution ratio.* The use of carbon tetrachloride, or other organic solvents, for the extraction of iodine or bromine from aqueous solutions is one of the most familiar examples of the use of a liquid-phase distribution for an analytical separation. Since the distribution ratios for iodine and bromine between carbon tetrachloride and water are known, the distribution law can be used for predicting the completeness of such separations under a given set of conditions.

Separations of this kind, which are sometimes called *solvent separations* or *solvent extractions*, have certain desirable features. First, even though the distribution ratio may be such that significant amounts of the substance being extracted are left after one extraction with equal volumes of the extracting solvent, the process can be repeated until the desired completeness has been achieved. Thus, if even one-tenth of the substance is left after the first extraction, the fraction left after n similar extractions will be $(0.1)^n$; after three extractions only one-thousandth or 0.1% of the initial quantity will remain unextracted. The volume of solvent required to achieve a certain completeness of extraction can be reduced by using smaller portions of solvent and making a larger number of extractions. Secondly, solvent-extraction methods eliminate the coprecipitation phenomena that are always present when precipitation methods are used. Thirdly, one can extract a quantity of a substance so small that it would not produce a visible precipitate; this feature of solvent-extraction methods has been of particular advantage in recent work with small quantities of radioactive isotopes. Finally, such methods often permit separations that could not be effected by precipitation methods. This advantage is being extended by the fact that many cations combine with weak organic acids to form compounds that are soluble in organic solvents; therefore in many cases it has been found preferable to separate such a

[10] J. A. Sherrer. *Natl. Bur. Std. J. Res.*, **16**, 253 (1936); **21**, 95 (1938); W. D. Mogerman. *Natl. Bur. Std. J. Res.*, **33**, 307 (1944).

compound by extraction rather than to separate the precipitate by filtration or centrifugation. In the past twenty years there has been a great increase in the use of solvent-extraction methods for effecting chemical separations. Only a few applications of these methods can be cited below; they are chosen to illustrate the general features of such procedures.

The extraction of iodine and bromine from aqueous solutions has been mentioned above. This method is used for the separation of these elements from each other, after selective oxidation, and from other elements. The distribution involved is a relatively simple one in which the substance is present in both phases as the elementary compound, and the distribution is relatively independent of the hydrogen-ion concentration, at least at *p*H values where hydrolysis of the halogens is not significant.

The extraction of ferric iron from hydrochloric acid solutions by ether is an example of a distribution that is much more complicated. This method, first proposed by Rothe,[11] has been used extensively for analytical purposes, and the equilibria involved are still being investigated.[12]

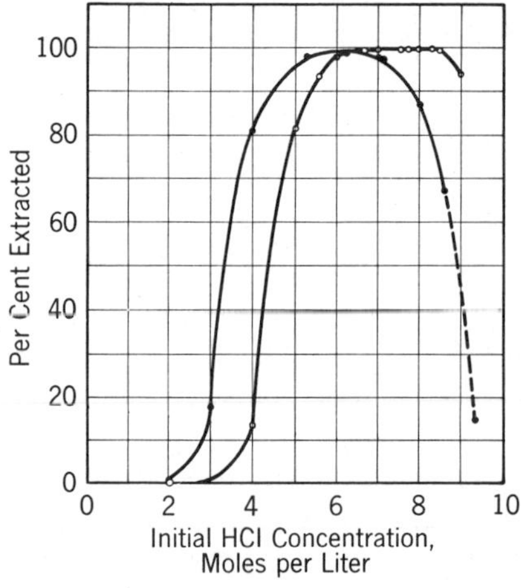

FIGURE 9-2
Extraction of ferric chloride from aqueous hydrochloric acid solutions by an equal volume of ethyl and of isopropyl ether. ●, ethyl ether; ○, isopropyl ether.

The iron is extracted by various ethers, the compound extracted has the formula $HFeCl_4$ and carries varying amounts of water with it into the ether, and the percentage extracted by a given volume of ether is dependent upon the acid concentration and to some extent upon the iron concentration.[13] The effect of the acid concentration is shown in Fig. 9-2.

[11] Rothe. *Stahl Eisen*, **12**, 1052 (1892).

[12] Nachtrieb and Conway. *J. Am. Chem. Soc.*, **70**, 3547 (1948); Nachtrieb and Fryxell. *J. Am. Chem. Soc.*, **70**, 3552 (1948); Myers, Metzler, and Swift. *J. Am. Chem. Soc.*, **72**, 3764 (1950); Myers and Metzler. *J. Am. Chem. Soc.*, **72**, 3776 (1950).

[13] Dodson, Forney, and Swift. *J. Am. Chem. Soc.*, **58**, 2573 (1936); Axelrod and Swift. *J. Am. Chem. Soc.*, **62**, 33 (1940).

TABLE 9-3

Behavior of Various Elements on Shaking a 6 *F* Hydrochloric Acid Solution with an Equal Volume of Ether

Element and Oxidation State	Percentage Extracted	Reference No.* and Remarks	Element and Oxidation State	Percentage Extracted	Reference No.* and Remarks
Al(III)	0	1	Hg	<1	4
Sb(V)	99+	2. Isopropyl ether	Mo(VI)	80–90	7a
Sb(III)	2.5	2. Isopropyl ether	Ni(II)	0	8
As(V)	<1	3. Isopropyl ether	Os(VIII)	0	9
As(III)	80	3. 8*F* HCl. Isopropyl eher	Pd(II)	0	4
Bi(III)	0	3	Pt(IV)	<1	4
Cd(II)	0	3	Rh(III)	0	9
Cr(III)	0	1	Se(IV)	<1	3
Co(II)	0	1	Ag(I)	0	4
Cu(II)	<1	4	Te(IV)	34	4
Ga(III)	97	5	Th(IV)	0	7b
Ge(IV)	40–60	3	Tl(III)	95	10. Isopropyl ether
Be(II)	0	3	Sn(IV)	17	4
Au(III)	95	4	Sn(II)	15–30	3
In(III)	<1	6	Ti(IV)	0	11
Ir(IV)	5	4	W(VI)	0	12
Fe(III)	99	7. Also isopropyl ether	V(V)	<1	3. 15–25% with 7.7 F HCl and isopropyl ether (Refs. 13 and 14)
Fe(II)	0	3	V(IV)	<1	13 and 14. None with isopropyl ether
Pb(II)	0	4	Zn(II)	<1	4
Mn(II)	0	1	Zr(IV)	0	11

NOTE: Unless otherwise stated, the results are for ethyl ether.

NOTE: In many cases the values given above were obtained under different experimental conditions and are often only approximate.

* REFERENCES: 1. Speller, *Chem. News*, **83**, 124 (1901).

2. Edwards and Voight. *Anal. Chem.*, **21**, 1204 (1949); Bonner. *J. Am. Chem. Soc.*, **71**, 3909 (1949).
3. Swift. unpublished experiments. California Institute of Technology.
4. Mylius and Huttner. *Ber.*, **44**, 1315 (1911).
5. Swift. *J. Am. Chem. Soc.*, **46**, 2375 (1924); Nachtrieb and Fryxell. *J. Am. Chem. Soc.*, **71**, 4035 (1949).
6. Wada and Ato. *Sci. Papers Inst. Phys. Chem. Res.* (*Tokyo*), **1**, 70 (1922).
7. (a) Sammet and (b) Wada. Unpublished experiments with A. A. Noyes. Massachusetts Institute of Technology.
8. Langmuir. *J. Am. Chem. Soc.*, **22**, 102 (1900).
9. Dalton. Unpublished experiments. California Institute of Technology.
10. Thomas Lee. Unpublished experiments. California Institute of Technology.
11. Noyes, Bray, and Spear. *J. Am. Chem. Soc.*, **30**, 515, 558 (1908).
12. Carter. Unpublished experiments. California Institute of Technology.
13. Dodson, Forney, and Swift. *J. Am. Chem. Soc.*, **58**, 2576 (1936).
14. Lingane and Meites. *J. Am. Chem. Soc.*, **68**, 2443 (1946).

The extent to which various other elements are extracted is shown in Table 9-3; in many of these cases the distribution phenomena have not been critically studied.

The compounds formed by the metallic elements with organic reagents frequently can be extracted into organic solvents. Thus, 8-hydroxyquinoline, one of the organic precipitants mentioned previously, forms compounds that can be extracted into chloroform. By adjustment of the *p*H of the aqueous solution, various separations are possible. For a review of this subject, as well as of the use of other organic compounds in such extraction methods, with special reference to trace quantities, see Sandell. *Colorimetric Determination of Traces of Metals* (3rd ed.) New York: Wiley-Interscience, 1959. A monograph that treats the analytical aspects of solvent extraction is Morrison and Freiser, *Solvent Extraction in Analytical Chemistry* (New York: Wiley, 1957).

MULTIPLE OPERATIONS AND VARIANTS

Multiple Operations

In gas–liquid separations, small differences in boiling points may make a single distillation relatively ineffective in providing a separation. Similarly, in liquid–liquid extraction an unfavorable distribution ratio may necessitate a large number of extractions. Various multiple-operation techniques have been developed in order to handle such separations.

A simple distillation apparatus such as is shown in Figure 9-3 may be adequate for the separation of two substances of widely different boiling points. However, if their boiling points are close to each other, even the first portion of the distillate may have almost the same composition as the original solution.

An improved separation can be achieved if the vapor phase is condensed and then redistilled, for in a two-component system, such as alcohol and water, the vapor phase is richer in the more volatile component than is the liquid phase. Therefore, each distillation and condensation results in an improved separation. A fractionating column achieves multiple condensation and redistillation without the necessity of a sequence of separate operations. Figure 9-4 shows one type of fractionating column, made by packing the vertical arm of the distillation flask with short lengths of glass tubing upon whose large surface the vapors condense and from which the redistillation occurs. Such a column may have the effect of from 20 to 40 theoretical plates per meter of length; that is, it behaves as if the vapor had gone through from 20 to 40 separate condensation-distillation cycles. The technique of fractional distillation is of great importance in organic chemistry, in which compounds of very similar properties often occur in mixtures and must be separated. A

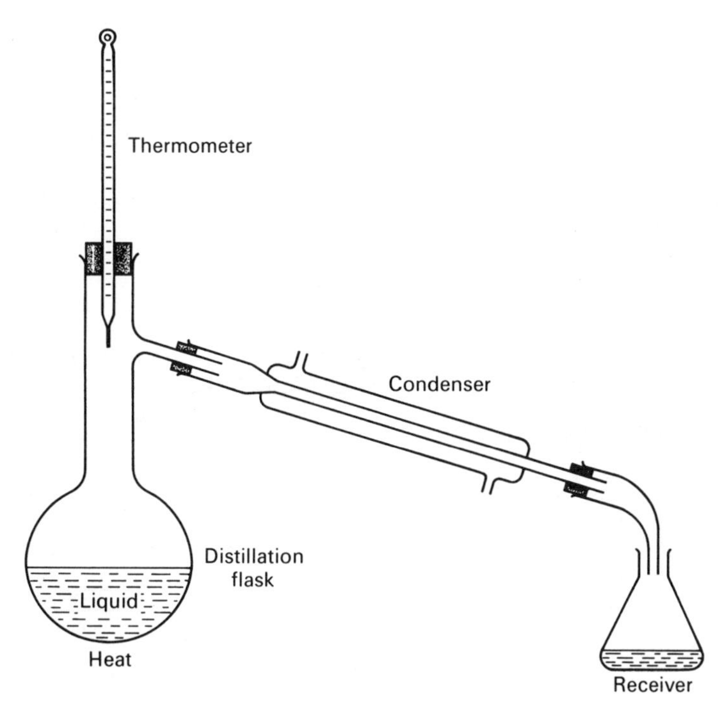

FIGURE 9-3
Simple distillation apparatus.

brief theoretical discussion of distillation and a bibliography of monographs on the subject are given by E. W. Berg (*Physical and Chemical Methods of Separation.* New York: McGraw-Hill, 1963).

Continuous extraction techniques can be used to eliminate the need for multiple extractions. In these methods the organic solvent, if heavier than water, is allowed to fall dropwise through the water phase in a long column. If the solvent is lighter than water it is released in drops at the bottom of the aqueous column. In either case an overflow arrangement collects the solvent with its solute after it has passed through the aqueous phase. Other devices have been designed that automatically equilibrate the aqueous phase with successive portions of an immiscible solvent, thus permitting multiple extractions with minimal effort. Detailed discussions of these and other multistage operations are given in *Solvent Extraction in Analytical Chemistry* by G. H. Morrison and H. Freiser (New York: Wiley, 1957) and *Physical and Chemical Methods of Separation* by E. W. Berg (New York: McGraw-Hill, 1963).

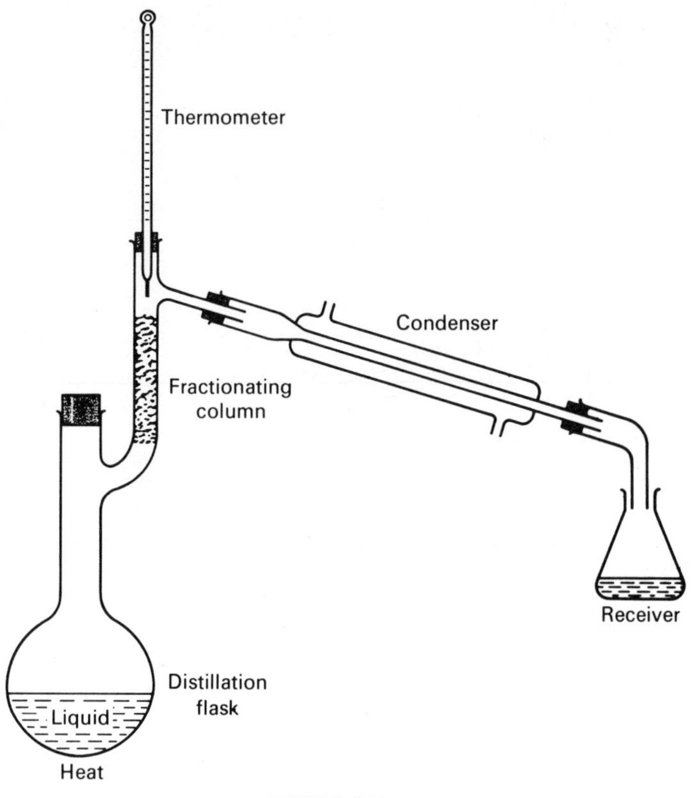

FIGURE 9-4
Fractionating still.

Electrolytic Oxidation and Reduction

Earlier we discussed electrolytic separations in which oxides were deposited. Many electrolytic methods are used in which a metal or salt is deposited. These methods serve both for separations and for quantitative determinations. Copper and silver are readily separated from zinc by being deposited as solids on a platinum cathode from dilute nitric acid solutions. Perhaps the most useful type of electrolytic separation is that involving the mercury cathode.

The mercury cathode has two distinct advantages. First, the evolution of hydrogen is decreased because of the unusually high overvoltage at a mercury electrode (see Chapter 28). Second, many metals dissolve in mercury, forming amalgams or intermetallic compounds. Consequently the activity of the metal is lowered in amalgams to a value determined by the mole fraction of the metal in the amalgam or, where compounds are formed, by the stability of the compound. Because of these two effects it is possible to deposit the very reactive alkaline earth and alkali metals at mercury cathodes.

The mercury cathode is of especial value in metallurgical analysis, since by its use large amounts of iron, chromium, zinc, nickel, cobalt, tin, molybdenum, copper, bismuth, and silver can be deposited and separated from aluminum, phosphorus, titanium, arsenic, vanadium, and uranium. The purpose of the separation is the removal of the deposited elements in order to facilitate the analysis for those remaining in the solution.[14]

In *controlled potential electrolysis*, the potential of the cathode is controlled at a predetermined value. If the deposition potentials of two metals are sufficiently different, it is possible to deposit one quantitatively without commencing deposition of the other. The usefulness of this technique was recognized over 60 years ago[15] but received little application until electronic potentiostats were developed, since manual control of the potential required constant and tedious adjustment. These subjects are considered in much more detail in Chapter 28.

Separation by electrolytic deposition is not restricted to the formation of metals or of their oxides. The silver halides, for example, can be deposited at a silver anode and can be separated from each other by control of the anodic potential.

Chromatographic Methods

These methods involve systems in which the phase in which the sample is dissolved passes through a stationary phase that has a very large surface area. The name *chromatographic* comes from Greek words meaning "color" and "to write" and applied well to the earliest applications of the technique, since these usually involved colored compounds. The name scarcely fits a large number of present applications but is so well established that change is unlikely. If you remember the requirement of a moving phase that contains the sample and a large-surfaced stationary phase you will be able to identify chromatographic methods. For ease in discussion we will consider two subdivisions of these methods: (1) adsorption chromatography and (2) gas chromatography.

Adsorption Chromatography. The most commonly used technique involves passing a solution of the substance or substances to be separated through a

[14] Hillebrand and Lundell. *Applied Inorganic Analysis.* New York: Wiley, 1929, p. 105; Cain. *Ind. Eng. Chem.*, **3**, 476 (1911); Lundell, Hoffman, and Bright. *Ind. Eng. Chem.*, **15**, 1064 (1923); Brophy. *Ind. Eng. Chem.*, **16**, 963 (1924).

[15] H. J. S. Sand. *J. Chem. Soc.*, **91**, 401 (1907); A. Z. Fischer. *Angew. Chem.*, **20**, 134 (1907).

vertical column of an adsorbent; alumina, calcium carbonate, magnesium oxide, and silica gel have been frequently used. As the solvent moves down the column, the various solutes, including the substances to be separated, will do likewise; but the rate at which any one of these solutes migrates will be determined by the adsorption equilibrium that is set up between the solute and the adsorbing material. That is, the more firmly a solute is adsorbed, the greater the fraction of the time it will be held on the adsorbent and the more slowly it will move. Consequently, the solutes will become separated on the column or will pass from the column at different times. The separation is effected either by discontinuing the flow of solvent and recovering the various solutes by removing sections of the adsorbent column, or by continuing the flow of solvent until the solutes migrate completely through the column.

It is observed experimentally that adsorption is an equilibrium process and can be represented by the empirical equation

$$\frac{x}{m} = Kc^{1/n}$$

in which x is the weight of material adsorbed by the mass, m, of adsorbent; K and n are characteristic constants for the system; and c is the equilibrium concentration of solute. (We will encounter this same relationship in Chapter 12, in which coprecipitation effects are considered.)

As is seen from the adsorption equation, the amount of a given substance adsorbed under given conditions is determined by the concentration of the species being adsorbed; therefore specific effects can be obtained, especially in the separation of inorganic substances, by the use of pH control and of complex-ion formation.

The technique for adsorption chromatography described above is called column chromatography. A modification that permits the separation of species in a very small sample is *thin-layer chromatography*, in which the adsorbent stationary phase is prepared as a uniform layer. This technique permits increased separation of species since separation can occur in each of two dimensions. The sample is placed at one corner of the layer of adsorbent, and the layer, still supported on the flat plate (glass, for example) upon which it was prepared, is suspended vertically with its bottom edge in solvent. This ascends the layer by capillary action. The components being separated form a series of spots or bands (the *chromatogram*). If the separation is incomplete, the layer can be rotated 90° (so that the chromatogram first obtained is at the bottom edge) and suspended in a second solvent. As this solvent ascends the layer, further separation may result. With proper selection of adsorbent material and solvents, very complex separations can be achieved.

A widely used variation of adsorption chromatography is *paper chromatography*, in which adsorbent paper strips are used as the stationary phase. This technique and that for thin-layer chromatography are quite similar; both have been used extensively for analysis of minute quantities of complex organic and biological mixtures. Inorganic separations can also be made by these methods. For example, microgram quantities of platinum, palladium, ruthenium, and iridium have been separated by paper chromatography.[16]

Gas Chromatography. The stationary phase used in gas chromatography may be either a solid or a nonvolatile liquid adsorbed on a solid. The moving phase is a gas stream—the gaseous sample in a carrier gas. If a solid is used as the stationary phase, the process is called *gas–solid chromatography* (GSC); if a liquid on a solid is used the name given is *gas–liquid chromatography* (GLC). The latter is more versatile, is better understood in terms of theory, and is used much more than GSC. In usual practice, the sample is introduced into a continuous flow of carrier gas that is passing through a column packed with a solid, such as crushed fire-brick, upon which is adsorbed the immobile liquid phase. Columns ordinarily have inner diameters (ID) of from 5 to 8 mm and are up to 30 feet long. The gas emerging from the column is either analyzed continuously or is separated into portions containing the various components. One simple method of analysis is to measure the thermal conductivity of the effluent gases, since the thermal conductivity is different for the components of the sample than for the carrier gas.

A very effective and important modification of the packed column is the capillary column, which is about 0.25-mm ID, up to 300 feet long, and is coated with a thin film of the liquid stationary phase. Exceptional separations on very small samples of complex mixtures can be achieved with capillary columns.

For reviews of chromatographic methods, see the biennial Fundamental Reviews section of *Analytical Chemistry.* For example, *Anal. Chem.*, **40**, 33R, 490R (1968); **38**, 31R, 61R (1966), and each 2 years previously. Many monographs are available: R. L. Pecsok (Ed.) *Principles and Practice of Gas Chromatography*, New York: Wiley, 1959; R. W. Moshier and R. E. Sievers. *Gas Chromatography of Metal Chelates.* New York: Pergamon, 1965; E. Lederer and M. Lederer. *Chromatography.* Amsterdam: Elsevier. 1957.

Ion-Exchange Methods

These methods are similar in principle and technique to column chromatographic methods and are sometimes treated as a special field of that general

---

[16] F. H. Pollard. *Chromatographic Methods of Inorganic Analysis With Special Reference to Paper Chromatography.* New York: Academic Press, 1953.

topic. In ion-exchange methods, use is made of inorganic or organic substances of large molecular weight that can be considered as being acids, bases, or salts that are capable of exchanging their cation or anion constituents with other ions in solutions.

A familiar example of substances of this type is the zeolites, which are the minerals used in water-softening apparatus. The zeolites are a class of minerals that comprises various sodium aluminosilicates such as NaH_6AlSiO_7 and $Na_2Al_2Si_4O_{12}$. The sodium ions in these minerals are loosely fixed and are capable of being replaced in the framework of the crystal by other cations. Thus, in the water-softening process the reaction taking place can be written as follows, where the zeolite is represented by Z^-:

$$Na_2{}^+Z^- + Ca^{2+} \rightleftarrows Ca^{2+}(Z^-)_2 + 2Na^+$$

Since this reaction is reversible, the original sodium compound can be regenerated by treating the zeolite with a concentrated solution of a sodium salt. More recently, organic compounds termed *ion-exchange resins* have been developed; these are ionic solids in which either the cation or the anion is polymeric, multiple-charged, and of high molecular weight, and in which the ion of opposite sign is relatively small and is freely diffusible and replaceable. The cation-exchange resins are frequently essentially organic acids and the anion resins essentially organic amines. The reaction with a cation-exchange resin can be represented as

$$2RCOOH + Ca^{2+} \rightleftarrows (RCOO)_2Ca^{2+} + 2H^+$$

and that with an anion resin as

$$2RNH_3{}^+OH^- + SO_4{}^{2-} \rightleftarrows (RNH_3{}^+)_2SO_4{}^{2-} + 2OH^-$$

By passing tap water through columns containing first one type of a resin and then the other, the ionic constituents can be reduced to such a small concentration that the product can be used to replace distilled water for most laboratory uses.

Cation-exchange resins have been extensively used for separating the rare-earth elements and other products of atomic disintegration. A very comprehensive treatment of these applications, as well as of the various principles involved, is given in the series of papers in the *Journal of the American Chemical Society*, **69**, 2769–2881 (1947). For a monograph on the subject see O. Samuelson. *Ion Exchange Separation in Analytical Chemistry*. New York: Wiley, 1963. In Chapter 26, ion-exchange reactions are discussed in more detail, and a procedure is given for the determination of total cations in a solution by the use of a cation-exchange resin.

QUESTIONS AND PROBLEMS

1. *Solubility Effects.* (The answers to this problem will require a consideration of the principles involved in Chapters 4 through 9. Logical application of principles rather than specific facts is required.)

   Explain fully but concisely the following solubility effects:

   (*a*) $Ni(OH)_2$ is more soluble in 1 *F* NH_4ClO_4 than in water.
   (*b*) $Mn(OH)_2$ is less soluble in 0.1 *F* NH_3 than in water.
   (*c*) $Mg(NH_4)AsO_4$ is more soluble in 0.1 *F* NH_4NO_3 than in 0.1 *F* NH_3.
   (*d*) AgCl is more soluble in 1 *F* K_2SO_4 than in water.
   (*e*) $PbSO_4$ is less soluble, but $BaCrO_4$ more soluble, in acetic acid than in magnesium acetate solution.
   (*f*) Calcium oxalate is more soluble in a magnesium nitrate solution than in water.
   (*g*) $Pb(OH)_2$ is less soluble in 0.1 *F* NH_3 and more soluble in 0.1 *F* NaOH than in water.
   (*h*) Silver sulfide is more soluble in 9 *F* HCl than in 9 *F* $HClO_4$.
   (*i*) Silver sulfide dissolves more readily in hot 3 *F* HNO_3 than in hot 3 *F* HCl.

2. *Hydrous Oxide Separations. The Use of Ammonia and Ammonium Salts.* Kolthoff and Kameda (*J. Am. Chem. Soc.*, **53**, 832, 1931) have found that soluble zinc salts can be maintained in solution (up to 0.1 *F*) if the *pH* is not greater than approximately 6. In attempting the separation of ferric iron from zinc by an ammonia precipitation, an excess of 1 drop (0.04 ml) of 1 *F* NH_3 over that required to precipitate the iron is added; the volume of the solution is 100 ml. Calculate what must be the ammonium-ion concentration in order to avoid the precipitation of any zinc. Is ammonia and ammonium salt an efficient buffering system for this separation? What additional effect contributes towards maintaining the zinc in solution?

3. *The "Basic Acetate" Separation.* This classical process depends upon buffering a solution with an acetic acid and acetate buffer system. The Fe(III) precipitate is a mixture of hydrous oxide and basic acetates. In a solution 0.1 *F* in both ferric and zinc salts, it is desired to precipitate the iron so completely that not more than 0.01% remains in the solution without at the same time causing the precipitation of any zinc.

   (*a*) Calculate the maximum and minimum hydrogen-ion concentrations that will give the desired result. (Assume the Fe(III) precipitate to be $Fe(OH)_3$.) *Ans.* 2.2×10^{-4}, 1×10^{-6}.
   (*b*) Calculate the geometrical mean of the maximum and minimum values and the concentrations of acetic acid and acetate ion necessary to buffer the solution initially at this value and not have the $[H^+]$ increase to over 10^{-4} after precipitation of the iron.

4. *Hydrous Oxide Separations.* Kolthoff, Stenger, and Moskovitz (*J. Am. Chem. Soc.*, **56**, 812, 1934) have recommended that benzoic acid and ammonium benzoate be used as a buffer system for the separation of iron(III), aluminum(III), and chromium(III) from bipositive Zn, Ni, Co, and Mn. If it is desired to retain in

100 ml of solution as much as 500 mg of zinc, calculate the maximum ratio of the molar concentrations of benzoate to benzoic acid which can be used. (Assume the tripositive elements precipitate as hydroxides.) *Ans.* 66.

Sulfide Separations

5. A solution at room temperature and buffered at a *p*H of 4.0 is saturated with H_2S at atmospheric pressure. Calculate the sulfide-ion concentration. (The solubility of H_2S under these conditions is 0.10 *F*.)

6. Hydrogen sulfide gas was passed into 1 liter of a solution buffered at a *p*H of 9.0 until 0.10 mole of the gas was absorbed. Calculate the molar concentrations of the H_2S, HS^-, and S^{2-}. *Ans.* $[H_2S] = 9 \times 10^{-4}$.

7. Show that the solubility of a sulfide of the type $M^{2+}S^{2-}$ in a solution saturated with H_2S is directly proportional to the solubility product of the sulfide and to the square of the hydrogen ion concentration.

8. Calculate how many milligrams of lead would remain in 10 ml of a solution at room temperature and in which the hydrogen-ion and hydrogen sulfide concentrations were 0.10 *M* and 1.0×10^{-3} *M*, respectively. Discuss briefly the uncertainties involved in using such calculations to predict experimental behavior.

9. (*a*) Calculate the molar concentration of the manganous ion required to saturate the solution of Problem 5 with MnS.
   (*b*) Repeat the calculation for the solution of Problem 6.

10. At what *p*H would a solution be saturated with FeS if the H_2S were 0.1 *F* and the Fe^{2+} were 10^{-3} *M*? *Ans.* *p*H = 3.5.

11. At what *p*H would 10 ml of a solution which contained 0.1 mg of Pb(II) be saturated with PbS if the H_2S were 0.1 *F*? *Ans.* $[H^+] = 2.3$, *p*H = −0.36.

12. Two elements, each having atomic weights of approximately 60, are to be separated by means of a sulfide precipitation. It is required that not more than 0.1% of the less soluble element be left in 100 ml of a solution that initially contained 600 mg of each element in the dipositive state and as the perchlorate. The solubility products of the sulfides of these elements are 10^{-19} and 10^{-15}, respectively.
    (*a*) Calculate whether the desired separation is possible, assuming that equilibrium is attainable.
    (*b*) State any liabilities that you think should be subjected to experimental verification.

13. *Preparation of an Ammoniacal Magnesium Nitrate Solution.* An ammoniacal solution of a magnesium salt is frequently used for the precipitation of phosphate or arsenate. In preparing a magnesium nitrate solution, 130 g of $Mg(NO_3)_2 \cdot 6H_2O$ were dissolved in water and 35 ml of 6 *F* NH_3 were added, with the result that a precipitate formed. It is desired to dilute the final solution to 1 liter. Calculate how much solid NH_4NO_3 should be added so that a clear solution will result. *Ans.* 67 g.

14. *Precipitation of Arsenate and Phosphate by an Alkaline Magnesium Nitrate Reagent.* The precipitation of arsenate and phosphate is made from a solution that contains 120 meq of ammonium ion and 20 meq of magnesium nitrate in a volume of approximately 70 ml. Calculate the maximum number of milliequivalents of ammonia that could be present without causing a precipitate of magnesium hydroxide. To how many milliliters of 6 F NH_3 does this number correspond? *Ans.* 9.5 ml.

SECTION II

EQUIPMENT AND TECHNIQUES

CHAPTER **10**

Instructions Regarding Laboratory Equipment and Operations

PRELIMINARY LABORATORY INSTRUCTIONS

Equipment

Obtain an apparatus list and check the items in your desk against this list. Report any missing, defective, or dirty equipment to your instructor. Sign the list when your equipment is complete and give it to the instructor.

Obtain a permanently bound notebook before your first laboratory period. Consult your instructor as to type and size.

Safety Precautions

Obtain a laboratory apron and wear it whenever you are working. You and your neighbors will be repeatedly pouring, shaking, and heating solutions. Accidents are inevitable. Many of these solutions are so corrosive or stain-producing that a few drops are adequate to cause personal injury or to ruin your clothing. Bare feet, or even sandals, are a liability. If corrosive solutions come into contact with your person, *immediately* wash the area repeatedly with water. Acid or alkaline solutions should be neutralized with sodium bicarbonate, concentrated solution or solid, followed by water. Chemicals spilled on desks or floors should be similarly removed, and the area of the spill should be dried.

Eyeglasses, prescription or safety type, are a very desirable protection when in a chemical laboratory. In many laboratories their use is required at all times. *Consult your instructor regarding the laboratory policy on eye protection.*

Instructions Regarding the Notebook

Your notebook should provide a complete and permanent record of exactly what you do and the results you obtain. No set of instructions is perfect, nor is it likely that many experiments or measurements will be carried out perfectly. As a result, you may frequently obtain anomalous results. If you have kept a complete record of your operations in your notebook there is a good chance that either you or your instructor can interpret such anomalies so that you can salvage the work and time already expended. Your notebook record will at least help you to avoid a subsequent occurrence of the same kind. These considerations apply to any research, experimental, or developmental work that you may do in future laboratory courses or in your subsequent career.

Observe the following rules in regard to the notebook.

(1) *Use a permanently bound notebook.* Never tear a sheet from the book. If an experiment fails or an accident happens, note that fact.

(2) *Record in ink all data and observations as soon as possible.* Memory is uncertain and tricky. Do not use loose scraps of paper; they may be lost or errors made in transcribing to the notebook. Pencil marks are not permanent, can be erased or altered, and may inspire doubt as to the integrity of the work. Original records are of more value than those that have been transcribed for the sake of neatness.

(3) *Begin each new experiment on a separate page and give it a descriptive title.*

(4) *Show the date of each entry clearly.* The date is often of importance in indicating the reliability of standard solutions, in interpreting subsequent effects, or in determining the sequence of experiments; it is of vital importance in experimental work that may lead to patent proceedings.

(5) *Make a complete record of the experiment.* State the purpose of the experiment. Record not only all measurements but also all experimental conditions, such as temperature, however insignificant they may seem, that may in any way affect the results or conclusions. Many experiments have had to be repeated because some conditions or observations not considered worth recording were later found to be necessary in interpreting the final results. *The method of procedure, the observation, and the results obtained should be so clearly and explicitly stated that any other reasonably skilled person could repeat the experiment and obtain the same results or know at once if his results deviated in any way from those originally obtained.* A convenient form for the laboratory record in quantitative analysis is provided by ruling the page of the notebook into three columns and heading them

"Procedure," "Observations," and "Conclusions." Also enter all reactions under the column headed "Conclusions." Although a complete record is necessary, it should be as brief and concise as is consistent with clarity.

Figures and diagrams of apparatus are valuable and space-saving. When a mass of data is to be recorded, it is more easily comprehended and space is conserved if a tabular form is provided and the measurements are entered directly therein. Clearly indicate the significance of all data; do not rely on the memory to interpret them at some later date.

(6) *At the completion of an experiment, record at once a summary of the results and the conclusions that can be drawn.* This will save much time if it is necessary to refer to or interpret the data at a later time.

(7) *Record all experiments; never tear pages from a notebook.* Faulty experiments should be labeled as such, together with an explanation of causes or mistakes. This type of record causes one to have confidence in the integrity of the notebook.

(8) *Prepare an index.* Leave sufficient space at the front of the book so that the titles of all experiments can be indexed for ready reference.

CONSTRUCTION OF LABORATORY EQUIPMENT

Construction of Useful Glass Equipment

A supply of simple glass equipment (such as stirring rods of various sizes and types, droppers, and wash bottles) should be constructed before beginning your experimental work. Construction of glass equipment requires that the glass be raised to a temperature at which it becomes soft enough to be bent or otherwise formed into the desired shape. The equipment described below can be made from "soft glass" (a glass that softens at a low temperature); a simple laboratory gas burner is used for obtaining the desired temperature. Since gas burners are used continuously in analytical work, and since they are frequently used improperly, instructions regarding their operation are given below.

Instructions for Using the Gas Burner

A gas burner is an instrument for producing heat by means of an oxidation-reduction reaction. The gas contains various reducing agents such as hydrocarbons, carbon monoxide, and hydrogen. The oxygen of the air is the oxidizing agent. The products are carbon dioxide and water; the reactions are exothermic, that is, heat is evolved. When higher temperatures are needed, as when working with Pyrex glass, the air is replaced by pure oxygen. This increases the concentration of one of the reactants and eliminates the cooling effects of the inert nitrogen in the air.

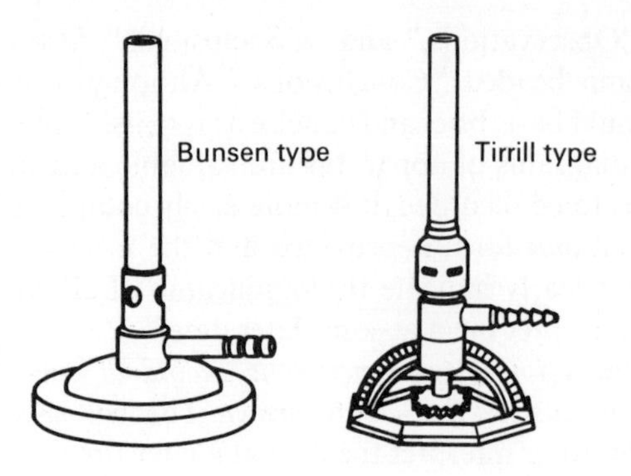

FIGURE 10-1
Gas Burners.

The simple Bunsen burner is shown in Figure 10-1, together with the modification known as the Tirrill burner. With the simple Bunsen type the gas supply and therefore the height of the flame are controlled by the stopcock on the gas line. The Tirrill type should be operated with the line stopcock *fully opened*; the gas supply should be controlled only by the needle valve in the base.

The air supply is controlled on the Bunsen by the rotating sleeve just above the base, on the Tirrill type by rotating the upper portion of the burner.

Always use as small a flame as will accomplish your purpose. For most purposes use an oxidizing flame; that is one in which there is a slight excess of air and therefore of oxygen. Such a flame has a very pale bluish inner cone with a violet color at the tip as shown in Figure 10-2. A reducing flame has an

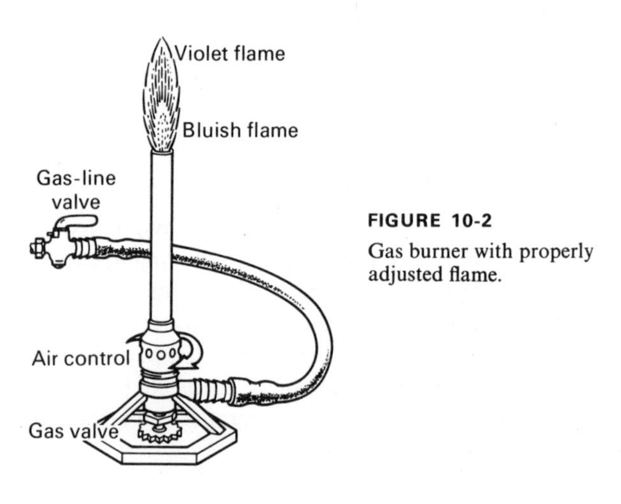

FIGURE 10-2
Gas burner with properly adjusted flame.

excess of gas and has a yellow color; such a flame will coat a cold surface with a dark layer of carbon particles.

A Tirrill burner can be used to attain temperatures up to 1000°C. For temperatures up to 1200°C, a Meker burner (Figure 10-3) is used. This burner is designed to give complete combustion of the gas without "striking back," that is, without the gas igniting at the base.

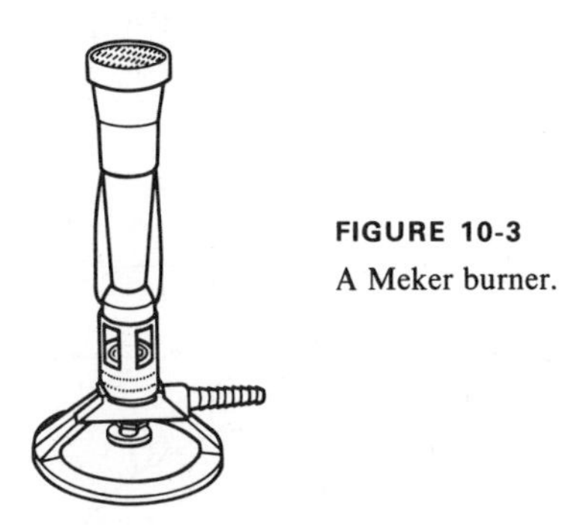

FIGURE 10-3
A Meker burner.

When you have finished with the burner, *turn it off at the gas line stopcock.* Rubber hoses deteriorate and gas leaks can be dangerous!

Instructions for Making Stirring Rods

Stirring rods are needed for stirring solutions; they are used when transferring solutions or precipitates, and when washing precipitates or treating them with solvents.

Obtain 50 cm–60 cm of 3- or 4-mm soft-glass rod. Cut the rod into lengths from 15 cm to 25 cm as directed below.

Cutting Glass Rod (or Tubing). Hold the rod as shown in Figure 10-4*A* and by means of a short triangular file *draw a perpendicular scratch* at the desired breaking place.

Grasp the rod with both hands as shown in Figure 10-4*B*; turn the thumbs inward and place them on the opposite side from the scratch; hold the elbows outward and the forearms horizontally and in toward the chest.

Pull outward with the hands and simultaneously apply pressure with the thumbs.

The rod or tubing should break smoothly. If protruding glass splinters remain, stroke them against a square of wire gauze as shown in Figure 10-4*C*. Fire-polish the rods as instructed below.

Fire-Polishing Glass Rod (or Tubing). Grasp the rod between the thumb and forefinger as shown in Figure 10-5. Incline the rod and rotate it between

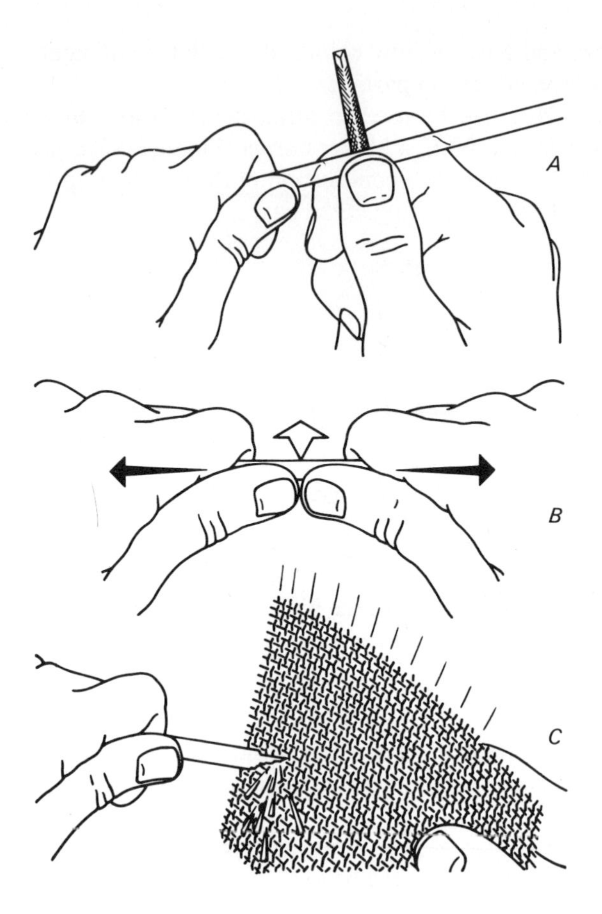

FIGURE 10-4
Cutting glass rod or tubing.

the thumb and finger with the upper end being supported and held by the little finger. Practice this operation until you can continuously rotate the rod in one direction. Slowly introduce the end of the rod into the hottest part of the flame—just above the inner cone. Continue to rotate the rod in this position until the end has rounded and enlarged slightly. Withdraw the rod from the flame and hold it vertically with the polished end downward until the glass becomes firm—otherwise the end of the rod may sag to one side.

Rotation of the piece being heated is essential in glass-working. It assures even heating and prevents distortion and sagging of the glass. With practice you can learn to rotate the glass continuously and evenly in one direction while maintaining it in a fixed position in the flame.

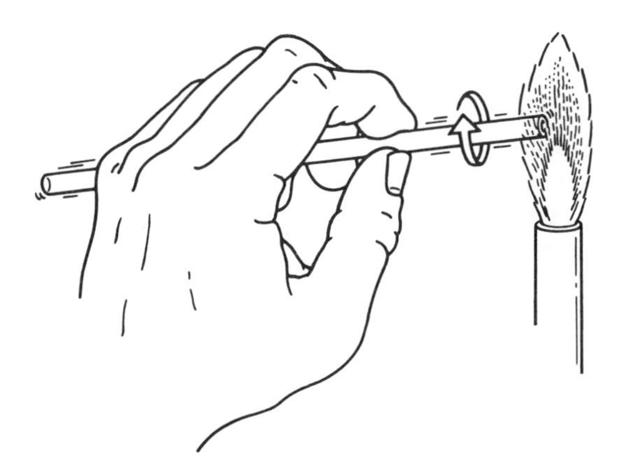

FIGURE 10-5
Rotating and fire-polishing glass rod or tubing.

Making Stirring Rods with Shaped Ends. A rod with a broad flat end (Figure 10-6), made by heating the end and pressing it against a clean surface of metal or "transite," is useful for breaking up solids that are being treated with solvents. The transite or metal surface should be heated momentarily just before the hot end of the rod is touched to it.

A rod useful for transferring precipitates from a filter to other vessels is made by bending 15 mm–20 mm of the lower end of the rod to roughly a 40° angle and then flattening this end (Figure 10-6).

FIGURE 10-6
Stirring rods.

Instructions for Making Droppers

Droppers, frequently termed "medicine droppers," are actually small pipets with attached rubber bulbs. They are extensively used for adding reagents dropwise, for adding measured small volumes of reagents, for washing precipitates, and for removing solutions from centrifuged precipitates.

The usual dropper available commercially is so short that it cannot be used to withdraw solution from a flask and has such a wide tip that it delivers very large drops. Since good droppers are easily constructed, each student should prepare at least two.

The glass tubing of a dropper should be of such bore that the rubber bulb fits firmly. The tubing should be of such length that solution cannot be drawn up into the bulb; when this happens contamination of the reagent is almost inevitable. In order to avoid such contamination a dropper containing a solution should *never be inverted or left in such a position that the solution enters the bulb.*

Cut a 30 cm–35 cm length of 6- or 7-mm OD (outside diameter) glass tubing. Rotate the tubing and, holding it horizontally, heat the center portion until the wall shrinks and thickens to almost twice its normal thickness (see Figure 10-7). Rotate it with both hands simultaneously (not alternately) and take care not to pull the ends outward during this period. Support the tubing with the little fingers and use the thumbs and forefingers for obtaining rotation.

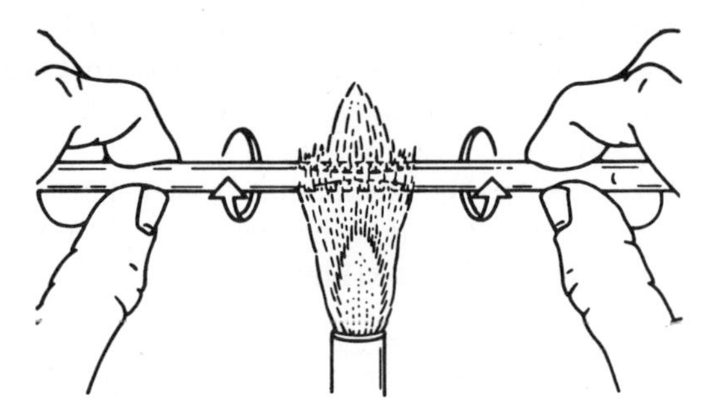

FIGURE 10-7
Rotating and heating a glass tube prior to drawing a dropper tip.

Remove the tubing from the flame, hold it *vertically* for a few seconds, then pull firmly and steadily downward on the lower end until the center portion has been reduced to about 1 mm in internal diameter; hold in a vertical position until the glass becomes firm.

Better control of the capillary portion is attained by allowing the glass to cool for a few seconds so that a firm pull is required. A straighter capillary portion is obtained by holding the tube vertically. If the pull is made too rapidly a thin-wall capillary will result.

Allow the tube to cool, then cut the capillary to the desired length and fire-polish it. For most purposes a short stout tip is satisfactory and is less likely to be broken. A long capillary is useful for removing small volumes of solution from a centrifuged precipitate (Figure 10-8).

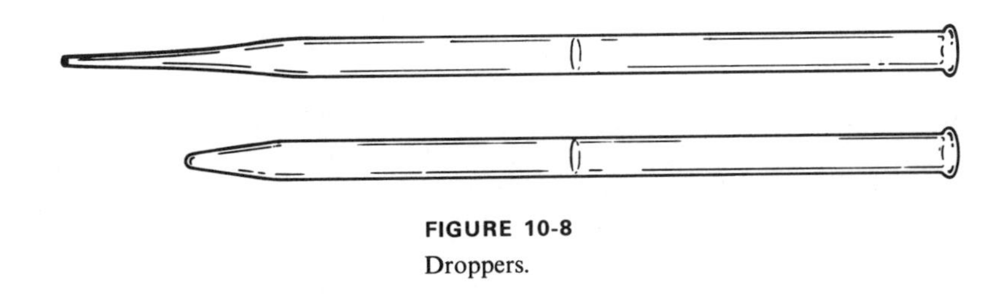

FIGURE 10-8
Droppers.

Fire-polish and flare the large end of the dropper. (If 7-mm tubing is used, flaring may be unnecessary, but the end must still be fire polished.) To make the flare, rotate the tube and rest the inner side lightly against the handle end of a file (see Figure 10-9). First determine the extent to which the end needs to be flared in order to provide a firm fit for the rubber bulbs that are available.

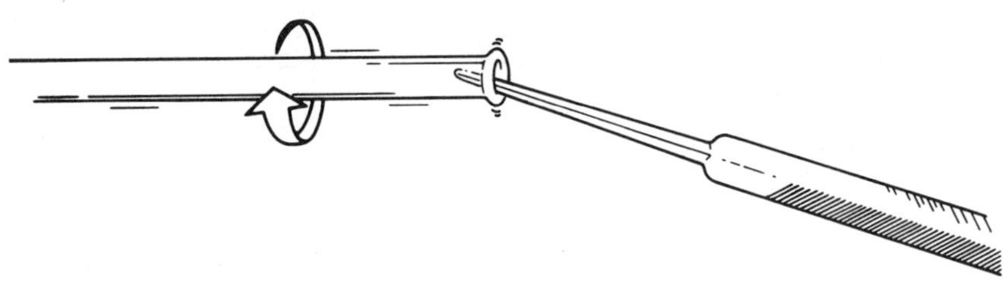

FIGURE 10-9
Flaring the end of a glass tube.

Instructions for Calibrating Droppers

Add water to a 10-ml graduated measuring cylinder or "graduate." Attach the rubber bulb to the dropper tube, insert the tip just below the surface of the water, and withdraw water until the *lower* meniscus of the water in the graduate is slightly below any one of the ml marks; raise the dropper until the tip is just above the water surface. Now slowly add water until the lower meniscus *just touches* the ml mark; a ml mark is used because it extends around the graduate and can be used to obtain a perpendicular line of vision. Hold the graduate so that the line of vision is perpendicular to the graduate (see Figure 10-10). Eject the water from the dropper.

Now withdraw slightly more than 0.5 ml of water. Raise the dropper and slowly return water to the graduate until the lower meniscus just touches the 0.5-ml mark. Withdraw the dropper and by means of your thumbnail fix the position of the lower meniscus of the water in the dropper when the tip is completely filled. By means of a file make a light but plainly visible scratch perpendicular to the dropper tube at this position of the meniscus.

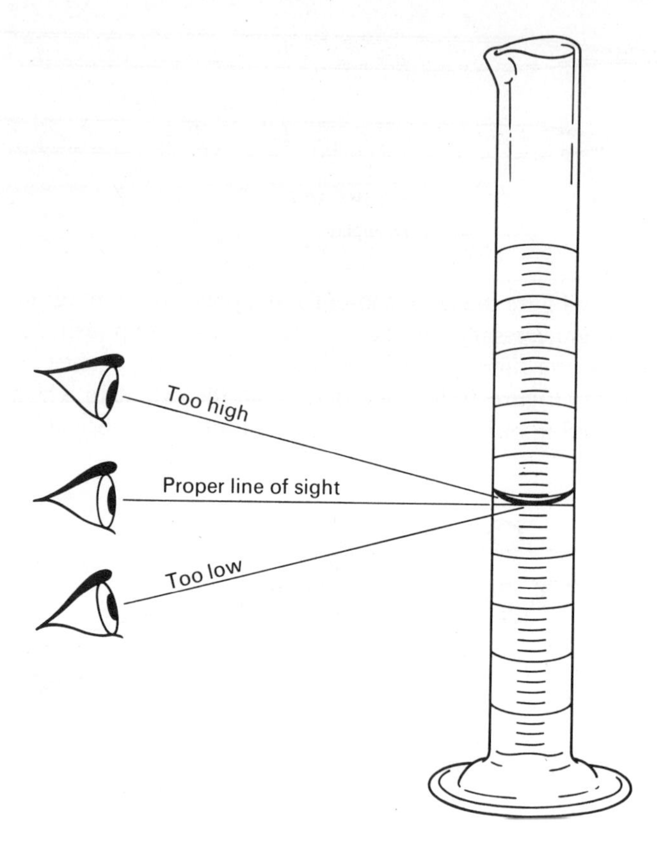

FIGURE 10-10
Proper method of reading a meniscus.

Similarly withdraw a 1-ml portion of water and scratch the dropper tube at the position of the meniscus.

Fill the dropper until the lower meniscus just touches the 1-ml mark. Hold it in a *perfectly vertical* position and cause the water to flow out in drops. Count and record in your notebook the number of drops per ml delivered by your dropper.

Repeat this measurement until you obtain successive drop counts that agree within 1 or 2 drops. The dropper should deliver from 25 drops to 30 drops per ml if the tip is of proper size. Repeat the count with the dropper held at a 45° angle. (This demonstrates the necessity for holding the dropper vertically when you want to add known volumes of a reagent.)

You now have a means of accurately delivering volumes of reagents of 1 ml or less.

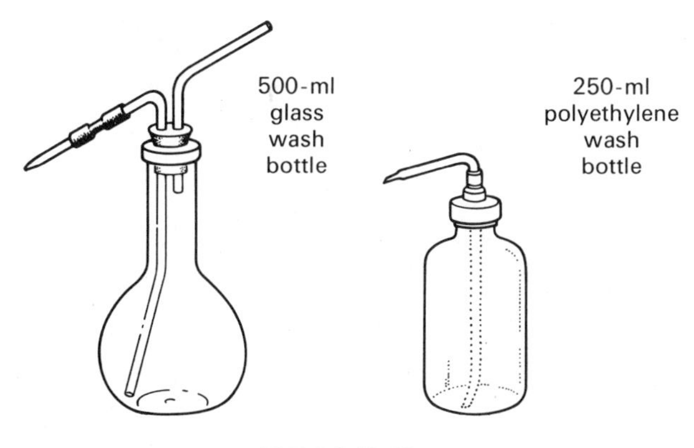

FIGURE 10-11
Wash bottles.

Instructions for Making Wash Bottles

Wash bottles are a necessary piece of equipment for the chemist. They serve as containers for distilled water, and as a means of washing precipitates and equipment. Figure 10-11 shows a 500-ml glass wash bottle of conventional design; this size is convenient for the work of this text.

If a larger or smaller size flask is provided, it can be used, or a conical flask can be substituted for the flat-bottom flask shown. Consult your instructor.

If the bottle is to be used with hot solutions it should be of Pyrex glass, and the neck should be wrapped with heavy twine, asbestos sheet, or cord. Also, a rubber mouthpiece is advisable, since it is less likely to burn the mouth and is more flexible.

Polyethylene wash bottles of 250-ml capacity (Figure 10-11) are extremely convenient, and their use is recommended *for other than hot solutions.*

Assembly of Materials. The following instructions call for a 500-ml flat-bottom glass flask as shown in Figure 10-12; you can easily modify these directions for other sizes or for the use of a conical flask. Assemble the following items:

1 flat-bottom Pyrex flask, 500 ml
1 rubber stopper, no. 5, 2 hole
60-cm glass tubing, 6 mm OD
8-cm rubber tubing, $\frac{1}{4}''$ ID, $\frac{1}{16}''$ wall

New tubing and stoppers are desirable; wash the stoppers free of talc or sulfur.

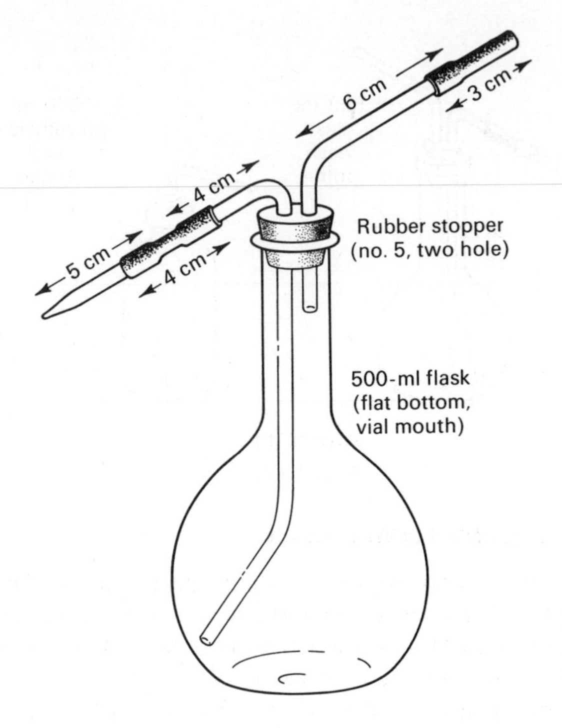

FIGURE 10-12
The dimensions and construction of a wash bottle.

If only solid stoppers are available, bore the holes with cork borers of appropriate size. Lubricate the borer with water or preferably glycerine; dilute sodium hydroxide can be used if care is taken not to spill it and if both the stopper and the borer are thoroughly washed afterwards. When boring the hole, *support the stopper on a block of wood* in order to avoid cutting the desk tops.

Making and Fitting the Outlet Tube. Measure and cut a piece of the tubing of proper length to serve as the outlet tube (Figure 10-12); use the flask in making this measurement.

Fire-polish the ends of this tubing.

Use the instructions given below to bend one end of this tubing at the proper position to form the upper end of the outlet tube.

Instructions for Bending Glass Tubing. Obtain and fit a "wing top," shown in Figure 10-13, to your burner. Light the burner and adjust the flame to a

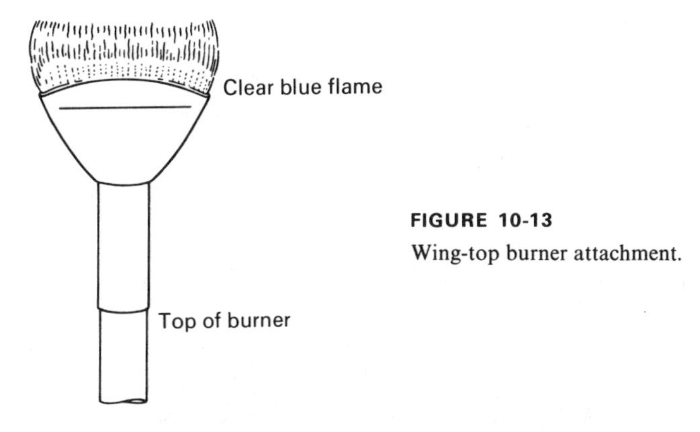

FIGURE 10-13
Wing-top burner attachment.

height of 1 cm to $1\frac{1}{2}$ cm over the center of the burner. A good wing top provides a horizontal flame across the entire top of the burner.

Rotate the tubing as directed under "Instructions for Making Droppers" and heat the portion to be bent in the flame. The length of the bend is controlled by the length of the heated portion. When the tube is sufficiently flexible to bend slowly of its own weight, remove it from the flame for a few seconds, then bend it, while holding it in a vertical plane, to the desired angle. Keep the bent lengths in the same plane. Exert a slight outward pull as the bend is made in order to avoid collapsing the inner wall of the bend.

Allow this piece to cool. (Begin making the tip meanwhile.)

Insert the straight end of the tube in one of the holes of the stopper as described in the next paragraph.

Forcing Glass Tubing Through Stoppers. CAUTION: *Forcing glass tubing through stoppers is a dangerous operation. Should the tube break, your hands can be seriously cut.* First, always wet the stopper hole in order to reduce the force required to insert the stopper. Second, wrap the tube with a folded towel and grasp it as close as possible to the end to be inserted, as shown in Figure 10-14. Third, insert the tube with a rotary motion; push only when the tube is rotating in the stopper. If excessive force is required, enlarge the hole with a cork borer. Finally, as soon as possible wrap and grasp the emerging end, then pull the tube into the desired position with a rotary motion; continue to grasp the tube as closely as possible to the stopper.

After inserting the straight end in the stopper, bend the lower end to the angle shown in Figure 10-12. Fit the stopper in the flask and note if the lower end of the tube has the position shown in Figure 10-12. If necessary, reheat the tube until the bend can be properly adjusted.

Making the Tip. Fire-polish one end of the remaining piece of tubing, then allow it to cool. (Complete the outlet tube meanwhile.)

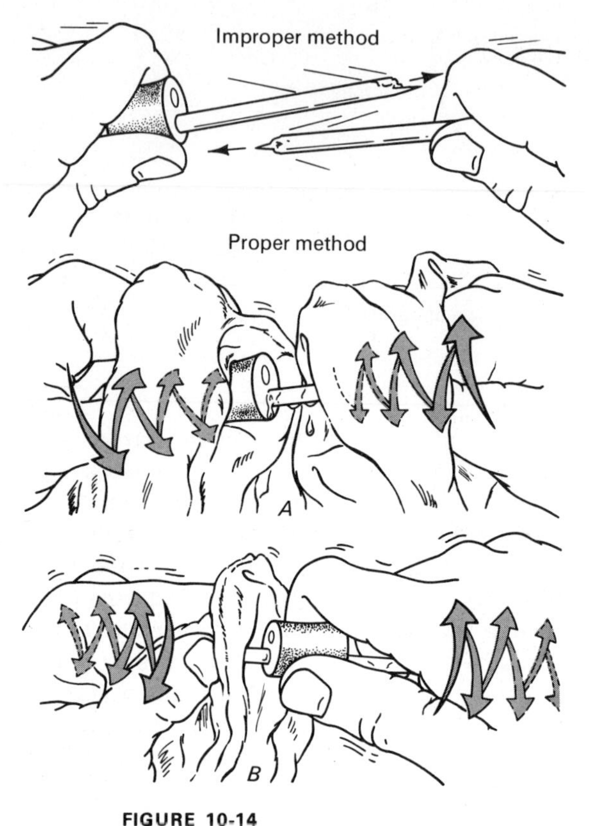

FIGURE 10-14
Forcing glass tubing through stoppers.

Rotate the tubing (as directed in "Instructions for Making Droppers") at the proper length from the fire-polished end to give a tip of the length shown in Figure 10-12. Then pull out the capillary tip according to the instructions given for making droppers. Cut the capillary, then fire-polish the tip so that it will deliver a fine jet of water.

Cut a piece of the rubber tubing to the proper length and use it to connect the tip to the outlet tube.

The tip should be so positioned that the forefinger of the hand holding the wash bottle can be used as shown in Figure 10-15 to direct the jet of water as desired.

Making the Inlet Tube. Cut away the capillary portion of the tube from which the tip has been drawn. Then cut it to the proper length and bend it to form the inlet tube. Fire-polish the ends.

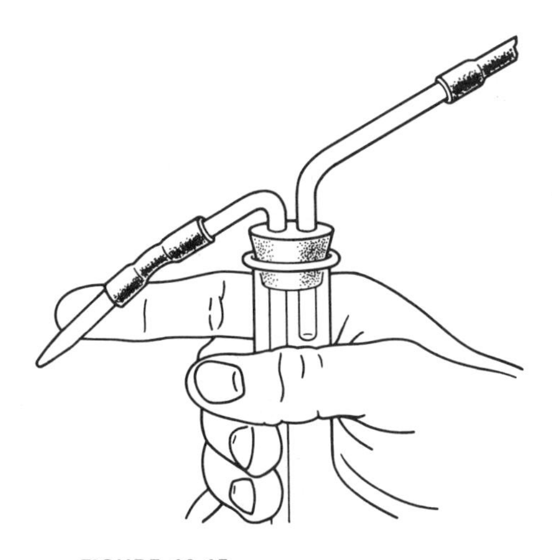

FIGURE 10-15
Directing the stream from a wash bottle.

Allow the tube to cool and insert it in the stopper. (CAUTION: Follow the method described above.)

When the wash bottle is used for hot water affix a 4-cm length of rubber tubing to the outer end of the inlet tube.

The proper method of controlling the direction of the stream from a wash bottle when washing precipitates is shown in Figure 10-15. Pressure bulbs do not permit flexible control of the rate of flow and should not be used. Longer lengths of tubing used on the inlet tube or for connecting the tip are not only clumsy to use but are hazardous to apparatus on the bench.

Preparation and Use of Certain Equipment

Desiccators

A desiccator is a container made of glass or aluminum in which objects, especially crucibles, may be placed and allowed to cool in a dry atmosphere protected from dust, carbon dioxide, and the fumes common to a laboratory. Dried materials, such as samples and primary standards, should also be kept in the desiccator. A common type of desiccator is shown in Figure 10-16. The dessicating agent most frequently used is anhydrous calcium chloride; for special purposes sulfuric acid, phosphorus pentoxide or anhydrous magnesium perchlorate may be used. The desiccator should be opened by sliding the cover to one side only so far as is necessary. It should be *opened only when necessary*, *closed as soon as possible*, and kept tightly closed at all

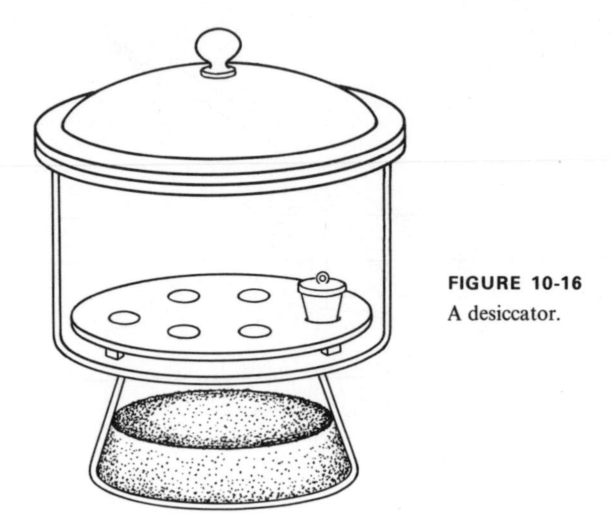

FIGURE 10-16
A desiccator.

other times;[1] the desiccant should be replaced as soon as it begins to lose its efficacy. After being heated to a high temperature (commonly spoken of as being "ignited"), an object should be allowed to cool to approximately 100°–150° before it is placed in the desiccator, and should be allowed to remain there for 20 to 30 minutes before it is weighed.

The cover fits the desiccator with a ground-glass joint that should be greased with a thin film of vaseline. When hot objects are placed in a desiccator the air expands; when the cover is replaced and the object is allowed to cool, the air contracts, creating a partial vacuum that may cause the removal of the cover to be difficult. In such cases the desiccator should be firmly supported and the cover *slowly* moved to one side; otherwise the sudden inrush of air may blow precipitates from crucibles, or a sudden jerk may overturn crucibles or weighing bottles (it is a worthwhile precaution to place the latter in small beakers).

Cleaning and Charging the Desiccator. Remove the cover as directed above, then remove the supporting plate. (Consult your instructor as to the disposition of any spent desiccant.) Wipe off most of the old grease (a paper towel can be used). If the desiccator is visibly dirty, wash it. Wipe the surfaces and the plate clean with a dry lintless towel.

---

[1] Booth and McIntyre (*Ind. Eng. Chem.*, anal. ed., **8**, 148, 1936) give data indicating that as much as 90 minutes may be required to dry the air in a desiccator; Smith, Bernhart, and Wiederkehr (*Anal. Chim. Acta*, **6**, 42, 1952) subsequently concluded that 15 minutes was adequate. Theirs and Beamish (*Anal. Chem.*, **19**, 434, 1947) have shown that there is some advantage in allowing platinum crucibles and nonhygroscopic precipitates to cool prior to being weighed under conditions of humidity approaching that of the balance room.

Apply a thin film of vaseline (stopcock grease is too viscous) to the *dry* ground surface, replace the cover, and rotate it until the vaseline becomes transparent.

Fill the lower compartment of the desiccator about two-thirds full of the desiccant specified by the instructor. Calcium chloride is recommended for student use; a small container of indicating Drierite can be used to show when the desiccant has lost its effectiveness. In filling the desiccator pour the dessicant through a sheet of glazed paper rolled into a large funnel to avoid getting particles or dust of the disiccant on the upper walls of the desiccator or the plate. Such particles will become moist on subsequent exposure to the air and may adhere to objects placed in the desiccator.

If sulfuric acid is used as a desiccant, the bottom of the desiccator should first be filled with glass beads to the desired level of the acid to minimize splashing.

Weighing Bottles. These glass bottles usually vary in volume from 10-to 100-ml capacity and are designed to contain substances, either solid or liquid, that are to be weighed. Samples of materials that are to be analyzed are placed in them, either before or after being dried. When accurate weighings are to be made, the weighing bottle should not be handled with the bare fingers. A piece of clean lintless cloth or chamois can be used; a folded strip of paper, long enough to encircle the bottle so that the two ends can be grasped, is quite useful.

Three types are shown in Figure 10-17. Type *C*, with the ground surface on the outside of the bottle, is used to advantage with finely powdered materials and also reduces the danger of contamination of the sample by dirt picked up by the stopper.

Wash the weighing bottles provided with your equipment and wipe them with a clean lintless towel. Remove the stoppers, place them crosswise of the mouth of the bottles, put the bottles in a beaker having an identifying mark—not a paper label. Place the beaker and bottles in an oven at 110°C–150°C for 1 hour. Then remove and transfer them to the desiccator.

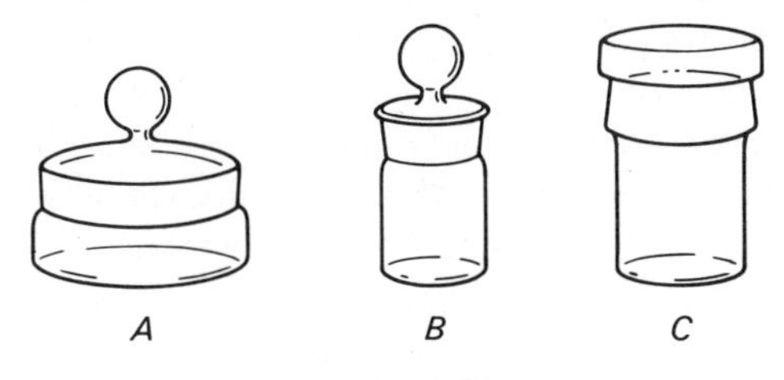

FIGURE 10-17
Weighing bottles.

If an oven is not available, the bottles can be dried as follows:

Secure an iron or nickel crucible of approximately 75-ml capacity and place in the bottom a square of wire gauze with the corners turned down so that there is an air space between the gauze and the bottom of the crucible. Support the crucible on a triangle and heat it with the full flame of a burner until any grease or other volatile matter has been removed. Allow the crucible to cool and suspend a thermometer with the bulb well down in and slightly away from the side of the crucible. Place the weighing bottle on the gauze in the crucible and again heat the crucible until the thermometer registers 125°C–150°C. After 15 minutes allow the crucible to cool somewhat and remove the bottle to the dessicator.

ANALYTICAL OPERATIONS

Most of the analytical operations involving solutions are carried out in glass vessels. This is true, first, because glass, compared with metals or alloys, is relatively inert to attack by acids (except HF), oxidizing agents, and other reagents. Second, glass is more resistant to thermal shock than most other chemically inert ceramic materials. Finally, the transparency of glass is a tremendous advantage in observing the course and results of reactions.

Instructions for Cleaning Glassware

Glassware that is to be used for analytical purposes should be perfectly clean. Not only should it be free from visible dirt but when rinsed with distilled water the surface should drain cleanly without leaving streaks or droplets.

Usually glassware can be effectively cleaned by treatment with a soap or detergent solution; when necessary a suitable brush can be used advantageously. After such a treatment the apparatus should be rinsed *copiously* with tap water, then *sparingly* with distilled water. Finally, it should be inverted and left to dry on a cloth or paper towel. *Do not wipe with a towel;* this is likely to leave a greasy film. Such films cause water to leave streaks and droplets on the glass.

Do not use the conventional sulfuric acid-chromic acid "cleaning solution" unless so-advised by your instructor. Most inorganic stains, such as manganese dioxide and ferric oxide, are more effectively dissolved by hot concentrated HCl than by cleaning solution. In addition, the cleaning solution is dangerous, especially when hot, if spilled or spattered on your skin or clothing. It is corrosive to equipment, is difficult to remove completely by rinsing, and can cause serious trouble if inadvertently introduced into an analysis.

When using brushes in test tubes or flasks be careful not to scratch the surface with exposed wires. Such scratches induce subsequent cracks when the vessels are heated or cooled rapidly. In addition, precipitates tend to form in and on such cracks or scratches and are then difficult to remove.

Handling and Measuring Solid Chemicals

One of the most frequent problems you will encounter is that of obtaining a small predetermined quantity of a solid reagent from a relatively large stock bottle without contamination of the stock supply and without waste.

First, carefully check the label on the stock bottle to be sure that the formula or name is that of the compound required. A hasty glance can easily mistake Na_2SO_3 for Na_2SO_4, or Mg for Mn.

Solid reagents are usually dispensed from wide-mouth bottles with stoppers made of glass or plastic. If these stoppers are hollow, as is usually the case if the stopper is glass, the transfer from the bottle may be made by partly inverting, then shaking or rolling the bottle until approximately the desired quantity is caught in the hollow of the stopper. Remove the stopper and transfer the desired quantity from it to a watch glass or paper. Tap or roll the stopper to control the quantity removed. Excess material in the stopper can be returned to the bottle provided it has not been touched or contaminated. Never return material from a paper or other container to the bottle.

If the bottle has a solid glass stopper, tilt or roll the bottle with the stopper held loosely in place until the approximate quantity needed is caught in the neck of the bottle. *Partly* withdraw the stopper and shake or roll this material directly on to the watch glass or paper. Partial withdrawal of the stopper permits control of the quantity taken and minimizes danger of the mass of material in the bottle suddenly spilling forth.

Material can be obtained from wide-mouth bottles with plastic screw caps by removing the cap, inclining the bottle, then tapping or rolling the required quantity of material into the cap.

Consult your instructor in case the material in a bottle has become "caked" so that the above methods cannot be used.

Spatulas of stainless steel or porcelain are useful for transferring small quantities of solids. With practice one can measure quantities by the area of the spatula tip that is covered or by the quantity taken in the spoon of a porcelain spatula.

Instructions for Transferring Solids from Containers

(*a*) Ascertain or calculate the exact quantity required and then make an estimate as to the bulk represented. Many chemicals are expensive—do not waste them by taking an unnecessary excess!

(*b*) Transfer this quantity from the container to a clean dry watch glass—if a weighed quantity is to be taken the watch glass should first be weighed or counterbalanced. A piece of glazed paper can be used if the material is a dry, nonabsorbent, nonreactive solid. Take your container to the reagent shelf, *never take the stock bottle to your desk. Read the label carefully before you open the bottle and again as you replace it on the shelf.*

Certain substances, such as $CaCl_2$, absorb water so rapidly that they will stick to paper. Na_2O_2 is such a vigorous oxidizing agent that paper in contact with it may start to burn.

(*c*) Close the container and return it to its proper place.

(*d*) Weigh or measure (as described below) the precise quantity required.

(*e*) Discard the excess. Never return it to the stock bottle—there is great danger of contamination or of returning it to the wrong bottle. Immediately clean up and put in the waste jars any solid that is spilled in taking it from the bottle or in weighing it.

Instructions for Weighing Solids

Balances are instruments for making weighings. The analytical balance, which is discussed in detail in Chapter 11, is the most accurate and highly developed of such instruments. It is restricted to loads of less than 200 g and should be used only for highly accurate measurements. The type to be used for other purposes will depend upon the total weight and the accuracy desired. Masses up to 2000 g can be weighed to within 0.1 g on a platform or "trip" balance such as is shown in Figure 10-18.

So-called "triple-beam" balances can be used to weigh to within 0.01 gram masses up to 100 g or even 200 g provided they are in perfect condition and adjustment and if they are carefully used. A low-sensitivity analytical balance, a "general laboratory" balance, or a "pulp" balance can be used to weigh to within 1 mg masses up to between 100 and 200 g.

A rugged balance having high capacity is used for the calibration of volumetric flasks and for the preparation of solutions of known weight-concentration (gravimetric solutions). The balance is constructed with wide bows and pans, and usually has a capacity of between 2 kg and 3 kg with a precision of 0.01 g.

Top-loading balances are constructed with various capacities and precisions and are now generally replacing pulp balances and such balances as

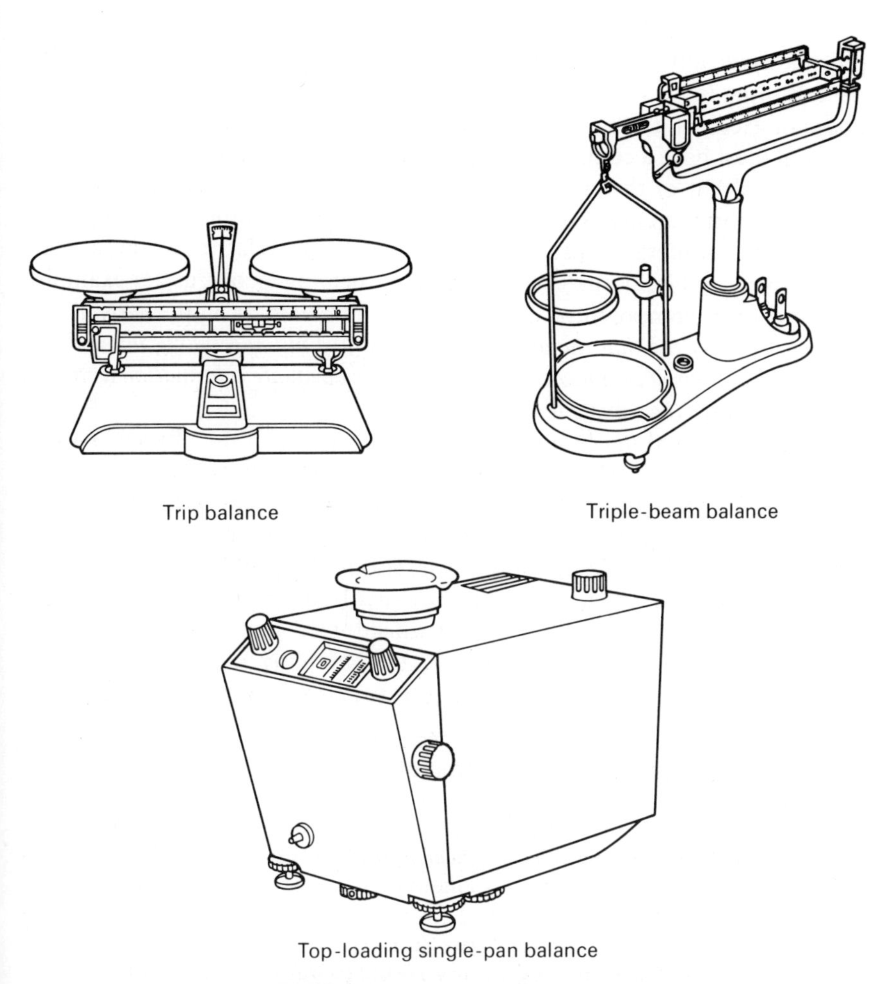

FIGURE 10-18
Balances and weights for analytical work.

those discussed in the preceding paragraph. Although more expensive, they are also more rapid and convenient to use (see Figure 10-18).

Instructions for (*a*) Obtaining, (*b*) Drying, (*c*) Weighing, and (*d*) Transferring Samples

(*a*) *Obtaining Representative Samples.* The importance that attaches to the process called "sampling," that is, the method of securing a representative sample of the original material, is often not sufficiently realized by an inexperienced analyst. In order to appreciate the difficulties involved in this process, especially when the material is a nonhomogeneous solid, consider the following facts: (1) the weight of the sample used for an analysis usually does not exceed 1 g; (2) if the analysis is to be of value, this small amount of material must have a composition that is the same, within the accuracy of the analysis, as that of the entire mass of the material being analyzed; (3) the mass of the original material—it may be a carload or a shipload or a mineral deposit of considerable size—is usually very large relative to the size of the sample; and (4) this material may be quite heterogeneous, as, for example, an ore deposit composed of widely differing minerals. *It must, therefore, be strongly emphasized that the analyst is not justified in undertaking an analysis unless he has himself sampled the material or has adequate information as to the method by which the sample has been obtained.*

The method used in obtaining samples will vary with the physical nature of the material. A detailed discussion of the procedures that have been developed for sampling various types of materials is beyond the scope of this book but the principles involved in securing a sample of a nonhomogeneous solid are simple and are illustrated by the problem below. Where the materials of a particular industry are involved, one should always refer to the standard methods that have been adopted and published by that industry. More detailed treatments of the sampling of various types of materials will be found in the following texts and reference works:

American Society for Testing and Materials, ASTM Standards. *Methods for Chemical Analysis of Metals.* (2nd ed.) Philadelphia: American Society for Testing and Materials, 1950.

American Society for Testing and Materials, ASTM Standards. *Official Method of Analysis of the Association of Official Agricultural Chemists.* Washington, D.C.: American Society for Testing and Materials, 1955.

Hillebrand, W. F., Lundell, G. E. F., Bright, H. A., and Hoffman, J. I. *Applied Inorganic Analysis.* (2nd ed.) New York: Wiley, 1953.

Laitinen, H. A. *Chemical Analysis.* New York: McGraw-Hill, 1960, pp. 579–594.

Problem: Preparation of the Sample. A 1000-g sample composed of particles of pyrite (80%) and quartz (20%) is to be crushed and quartered. (*a*) Calculate the maximum permissible weight of the largest particle in order that the addition or removal of this particle from the reserved portion will not cause an error of more than 0.1% in the determination of the sulfur in the sample. (*b*) The material is to be passed through a sieve to ensure that no larger particles are present. Calculate approximately what should be the size of the sieve openings. *Ans.* (*a*) 0.5 g, (*b*) 0.58 cm for cubes.

(*b*) *Drying Samples.* No general rule can be given for drying samples. The temperatures and times to be used depend upon the type of material and the conventions used to obtain a uniform basis upon which to report the analyses. The analyses of some samples are reported on an "as received" basis; more commonly samples are dried at 100°C–110°C for a specified time, usually

1 hour, or until constant weight is obtained. Make sure that you know the proper drying conditions for the sample to be analyzed.

In the procedures of this book specific directions will be given for drying each sample.

(c) *Weighing Samples.* After being properly dried, samples for analysis, primary standards, and similar substances are usually kept in glass-stoppered weighing bottles. Samples may be weighed from these by either of two general methods, here termed the *direct* and *difference* methods.

In using the direct method, an empty weighing bottle, a watch glass, a piece of glazed paper, or any other suitable container is weighed, the desired amount of material from the weighing bottle is transferred to it, and the weight of the container plus the material is determined. (Tared watch glasses or a scoop with a counterweight are convenient for reducing the time required for making weighings by this method.) The material is usually transferred to the weighed container by means of a spatula or by holding the sample bottle directly over the weighed receiver, removing the stopper, and then tipping the bottle and slowly rotating it, holding the top close to or against the receiving vessel, until approximately the desired amount of material has been transferred. If an exact weight of sample is desired, the direct method is preferable, since finely powdered material can be added or taken by means of a pointed steel spatula until this is obtained. Regardless of method, extreme care must be taken not to spill any sample on the balance pans and not to lose any in the subsequent transfers.

In the difference method the weighing bottle containing the dry material is first weighed, the desired amount of material is transferred from it directly to the beaker or flask in which it is to be treated, and the weighing bottle is closed and again weighed; the loss, or difference, in weight represents the sample taken. The transfer of the material is accomplished by the process described above, care being taken that all of the material leaving the bottle is caught in the beaker or flask. Since it is not good practice to return material from the receiving vessel (which may not be entirely dry or clean) to the sample vessel, it is not convenient to take exact sample weights by this method. Approximate sample weights can be taken by removing slightly less than the weight of the desired sample from the right pan and then intermittently removing material from the bottle until an approximate balance is again obtained; however, frequent removal and insertion of the stopper may cause loss of material by "dusting," especially with finely powdered dry material.

The direct method is usually more convenient when a large amount of material is to be weighed, presents less chance of loss of finely powdered material, and permits taking an exact sample weight; the difference method is more rapid when several successive weighings are to be made or when the sample is likely to absorb moisture during weighing, and permits weighing out the sample directly into the beaker in which it is to be treated.

(*d*) *Transferring Samples.* Two methods are used for transferring weighed material to another container for subsequent treatment. If the material is finely ground, crystalline, and readily soluble, it may be transferred directly by means of a funnel as directed above. Any solid remaining in the funnel is readily dissolved by the water that is to be added. If the material is of an amorphous nature, tending to lump together, or is not readily soluble, it is advisable to transfer it first to a beaker where it can be more readily brought into solution (heated if necessary); this solution is then poured through the funnel into the flask. When transferring weighed samples from a watch glass to a beaker, the latter should be of adequate size to permit lowering the glass into the inclined beaker before beginning the transfer.

Handling and Measuring Solutions

Most of the operations of analytical chemistry require the handling of solutions, for volumes of solutions can be measured rapidly and accurately. However, practice is needed to acquire dexterity in pouring, filtering, or in the other manipulations necessary with liquids. When such dexterity has been acquired laboratory work becomes neater and more rapid, and more enjoyable. For these reasons detailed instructions are given below to help you learn the proper techniques.

Instructions for Transferring Solutions

Pouring from Bottles into Graduates or Test Tubes. First, *carefully check the label!* Be sure the name or formula is the one desired. Next, *check the concentration.* Many procedures are critically dependent upon concentration adjustment. In some cases you will have to dilute the reagent to the specified concentration.

Two problems arise when a solution is to be poured from a reagent bottle: first, preventing contamination of the reagent; second, controlling the flow of the solution so as to avoid spilling the solution or getting it on your hands.

Bottles standing in a laboratory often accumulate a visible white deposit composed primarily of ammonium salts. Such a deposit should be washed from the stopper and neck of the bottle before the stopper is removed; finally rinse with distilled water and dry the bottle with a clean towel. Chloride is likely to be introduced into analyses if this precaution is not taken.

In order to prevent possible contamination, the ground-glass surface of the stopper *should never come in contact with the desk or other surface.* Also, the bottle should be left open for as short a time as possible.

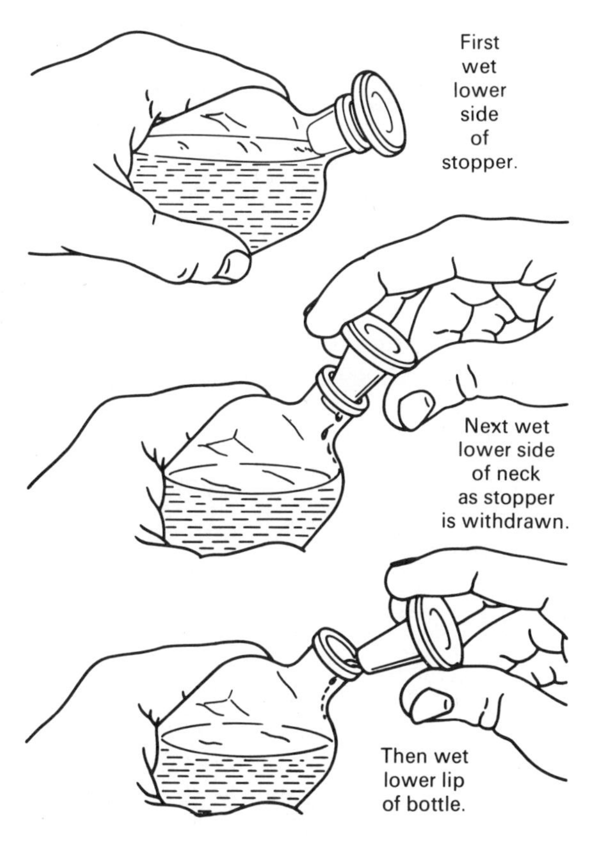

FIGURE 10-19
Pouring from a bottle.

When first poured over the dry glass of the neck of a bottle, a solution tends to flow in an erratic manner both as to path and volume. Much better control is possible if the ground glass neck and the lip are first wetted as shown in Figure 10-19.

The type of stopper in the bottle will determine the preferable method of handling the stopper while pouring from the bottle. Various types of stoppers are shown in Figure 10-20. Type *A* is designated as a *penny* or *coin-headed* stopper. The flat vertical disc is convenient in that the stopper, after being loosened, can be withdrawn with the back of the fingers and the stopper and bottle held in the same hand while pouring from the bottle (see Figure 10-21A and B). Be sure that the wet stopper is drained against the inside neck of the bottle; otherwise the reagent may drain down on your fingers and in some cases this may cause discomfort. However, this stopper has the disadvantage

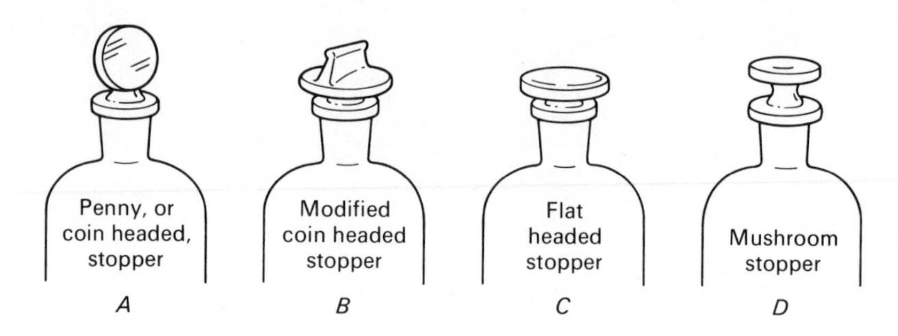

FIGURE 10-20
Types of reagent bottles.

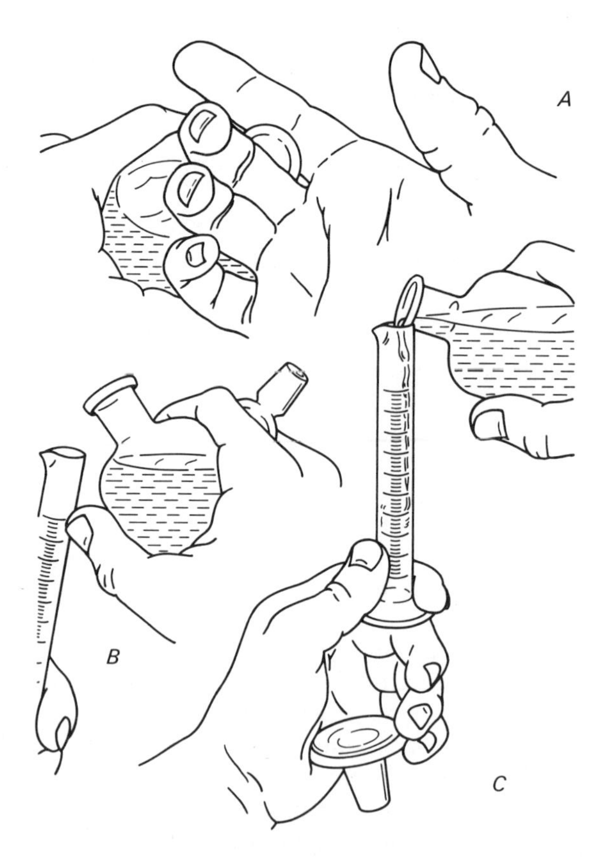

FIGURE 10-21
Handling bottle stoppers.

that it cannot be placed on a desk without the ground glass surface touching the desk. In addition, dust can settle upon the lip of the bottle and thus contaminate solution flowing over it. These disadvantages have been overcome in Type *B* stoppers.

Types *C* and *D* are *flat-headed* stoppers. Type *D* is known as a *mushroom head.* This stopper protects the neck from dust and can be inverted and placed on the desk. This invites the possibility that you may forget to replace it, or if more than one bottle is opened, that you may replace it in the wrong bottle. For this reason, whenever possible, hold this stopper (with the last two fingers) against the palm of the hand which is holding the graduate or test tube, as shown in Figure 10-21*C*. Always hold the bottle lip against the edge of the graduate or test tube into which the solution is poured.

Pouring into Beakers or Flasks. Use of Stirring Rods. The stirring rod is used in this process to guide the first portions of solution over the lip of the pouring vessel, to minimize sudden spurts, to prevent solution from running down the outside of the container from which it is poured, to guide the solution to the desired place, to prevent spattering in the receiving vessel, and to remove the last drops of solution from the lip of the pouring vessel.

FIGURE 10-22
Proper use of a stirring rod.

When a stirring rod is not used always hold the lip of the bottle or container *against* the inner surface of the receiving vessel. The proper use of the stirring rod for transferring a solution from a test tube to a conical flask is shown in Figure 10-22; use a stirring rod in the same way when pouring from flasks or beakers.[2]

Instructions for the Use of Droppers

Measuring and Transferring Reagents. The construction and calibration of droppers have been described earlier, pages 167–170.

The calibration marks are used for adding 1- and $\frac{1}{2}$-ml volumes; smaller volumes are measured by counting the drops delivered. *Never put your dropper directly into the reagent bottle;* this presents too great a chance of contamination. First take a slightly larger volume than is required in a small graduate or test tube.

Never put the tip of the dropper into the receiving solution; to prevent spattering, deliver the drops near the surface of the solution to which the reagent is being added or near the wall of the receiving vessel.

Washing or Dissolving Precipitates. Wash solution can be effectively applied to a centrifuged precipitate by means of a dropper. The solution can be added so as to expose and mix all portions of the precipitate with the wash solution.

When a precipitate on a filter is to be washed, use a dropper to apply the wash solution dropwise as shown in Figure 10-23. If the precipitate is lumped or compacted, use the tip of the dropper to gently disintegrate the material so that all parts are washed. Finally, be sure that the entire surface of the filter is washed, especially, that over the triply folded part of the filter.

When a precipitate on a filter is to be treated with a *minimum of a solvent*, the latter should be applied dropwise and the tip of the dropper used to disintegrate the precipitate. The solution should then be drawn through the filter, the vessel containing the filtrate removed, another vessel put in place, and the same solvent passed through the filter as before. When practical, heat the solution before again passing it through the filter. This process can be repeated as many times as is required.

Instructions for Diluting Concentrated Acids

When water is added to anhydrous acids the hydration reaction evolves heat. With sulfuric acid this effect is of such magnitude that *the addition of water to the acid* may cause spattering because the water is converted to

[2] So-called dripless beakers, which have a strip of Teflon applied to the rim, are now available. The Teflon minimizes the tendency of solutions to run down the outside surface of the beaker. Such beakers are also more resistant to breakage.

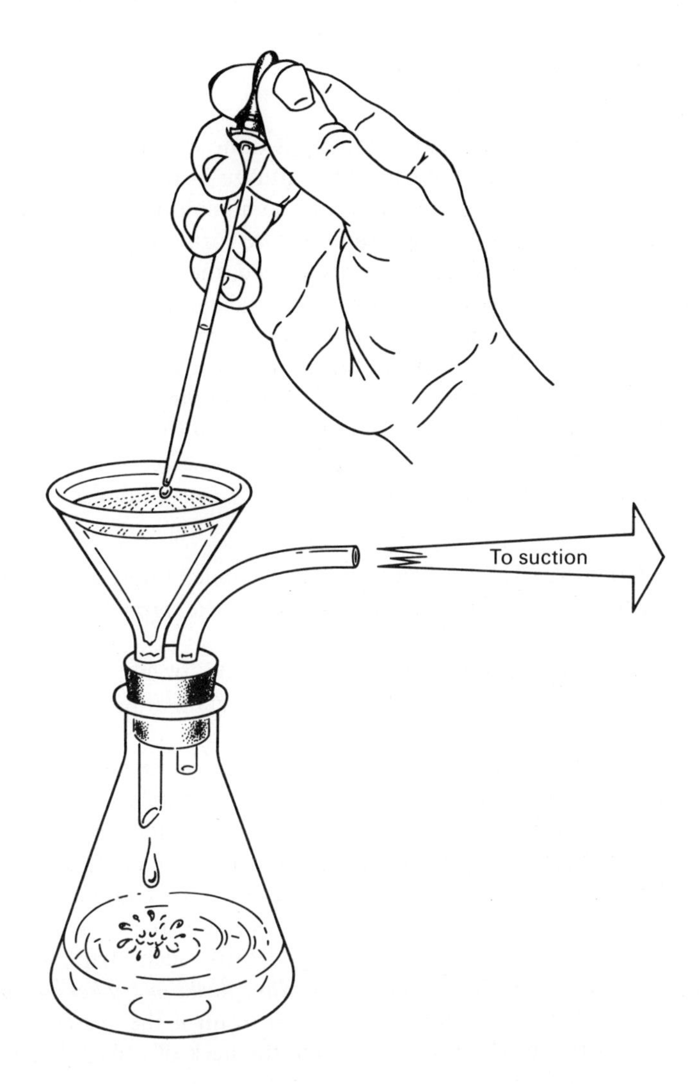

FIGURE 10-23
Washing a filtered precipitate with a dropper.

steam. When the acid is added slowly to a large excess of water, not enough heat is evolved to raise the temperature of the water to its boiling point. Therefore when mixing or diluting anhydrous acids, especially H_2SO_4, *always add the acid slowly to the water, be sure that the solution is constantly stirred, and cool it if necessary.* The latter is especially necessary when the excess of water is not large.

Instructions Regarding Heating and Evaporating Solutions

Much of the work in analytical chemistry is done with hot or boiling solutions. Just as all solutions tend to supersaturate in the absence of crystallization nuclei before precipitation occurs, so all solutions have a tendency to superheat above their boiling points before formation of the gas phase takes place. Then, once the gas phase is formed, the boiling frequently proceeds so rapidly that solution is thrown from the container by this rapid gas formation (steam in aqueous solution). An occurrence of this kind, which is called *bumping*, may be dangerous, since hot corrosive chemicals may be spattered on your or your neighbor's skin or clothing. Moreover such an occurrence is likely to be discouraging; it may mean the loss of the time and effort already expended on that analysis. For these reasons the following suggestions are presented in the hope that they can minimize such occurrences.

Heating Solutions in Flasks. Solutions can be boiled safely and evaporated rapidly in conical flasks held *directly over the flame, provided the following rules are observed:*

First, the flask initially must not contain over 30% or 40% of its maximum capacity; that is, a 50-ml flask should not contain more than 20 ml of solution.

Second, the solution in the flask *must be kept in continuous, rapid motion.* This is most effectively done by rapidly swirling the solution, as shown in Figure 10-24. Test-tube holders of the large wooden clothespin type are most convenient for this purpose. A satisfactory holder can be made by repeatedly folding lengthwise a sheet of firm glazed paper so that a compact strip about 1 cm in width is formed. This strip is looped around the neck of the flask, as shown in Figure 10-24. If moistened and dried such strips will hold a circular shape. *Do not use your tongs for handling flasks or beakers.* Corrosion of the tongs and contamination of solutions result if the vessel is grasped with one tip on the inside surface. Holding the neck of a flask between the bend in the tongs is hazardous.

Third, the flame used must be carefully controlled and the flame never applied above the solution level. A 50-ml flask should not be heated with a flame more than about 3 cm in height.

A solution in a 50-ml flask can be evaporated to 1 ml or 2 ml without danger of cracking the flask, *provided the swirling is so continuous that the portion of the flask exposed to the flame is kept continuously covered with solution.*

A solution in a flask is almost certain to bump, with danger of spattering, if heated to boiling on a wire gauze without being swirled or being stirred

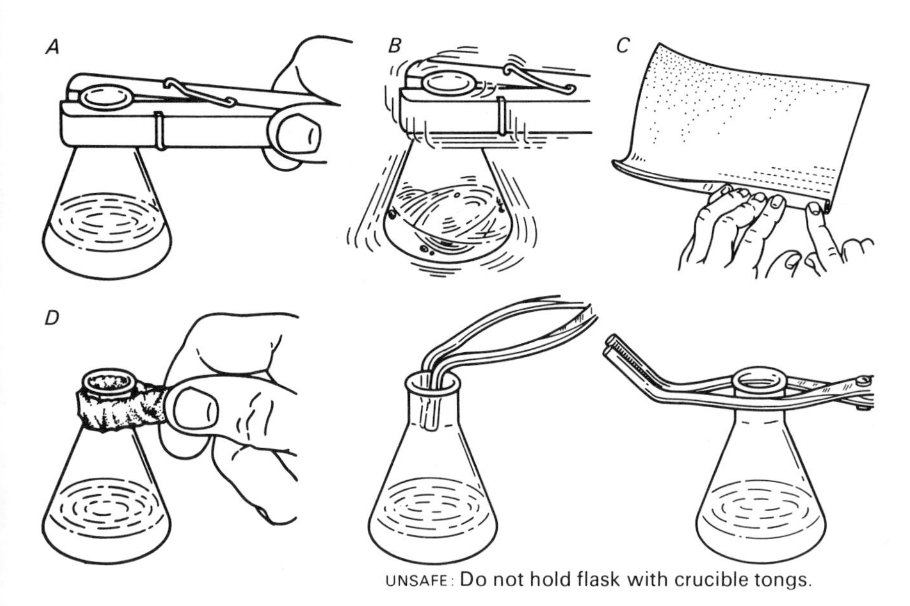

FIGURE 10-24
Boiling solutions in flasks.

continuously. This tendency is minimized, but not eliminated, by placing a stirring rod in the flask and applying the flame directly under the rod.

Use of an Ebullition Tube. An effective way to minimize bumping is to provide a gas phase, a bubble, in the solution at the point where heat is applied. A simple device that accomplishes this is the *ebullition tube*, Figure 10-25. This can be constructed from a 20-cm length of 3 mm–5 mm outer diameter (O.D.) soft-glass tubing. The tube is rotated slowly as it is heated at a point about 5 cm from one end until it is closed by the constriction of the glass. The tube should be caused to thicken slightly during the closing and then pulled to a uniform outer size afterward. After the tube has cooled, the shorter end is cut so it extends about 1 cm past the sealed section, the cut end is fire polished, and the other end is heated until it is closed and rounded.

The tube is placed in a beaker or flask with the open end down, and heat is applied directly under the end of the ebullition tube. If boiling is stopped and is to be resumed, the tube must be removed and shaken to remove the liquid from the open end. The tube must never be inserted into a solution that may be superheated.

Do not place a hot flask or beaker directly on the top of your desk. The paint or other finish is likely to soften and stick to the glass. Allow the container to cool on a wire gauze.

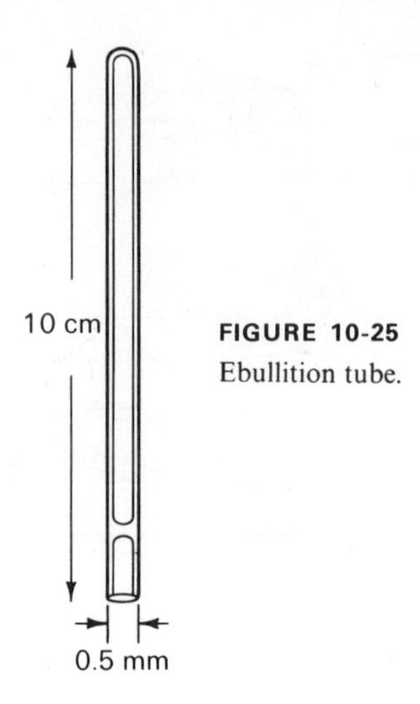

FIGURE 10-25
Ebullition tube.

Heating Solutions in Beakers. Heating solutions to boiling in open beakers involves a greater danger of loss from spattering than boiling in flasks. The solution cannot be kept in as vigorous motion, and solution is more easily lost even when the beaker is covered with a watch glass. You invite danger of loss if a solution is boiled in a beaker that is filled to over 20% the rated capacity. Danger of bumping can be decreased by placing a stirring rod in the beaker and applying the tip of the flame directly under the end of the rod. The area of contact serves as a center for gas phase (steam) formation. An ebullition tube gives much better protection against bumping.

Heating Solutions in Test Tubes. Boiling solutions in small test tubes or centrifuge tubes is especially dangerous and should be avoided whenever possible—one large steam bubble can cause violent expulsion of most of the solution from a tube! *Never heat a solution in a test tube while the tube is pointed toward another person; never look in the mouth of a test tube that is being heated!*

Do not fill the test tube to more than 20% of its capacity if you wish to boil the solution. Even in this case, the solution *must be kept in vigorous motion*, the tube *must be inclined*, and the flame *must be applied near the top of the liquid*, as shown in Figure 10-26.

Instructions Regarding the Use of Volumetric Flasks and Storage Bottles for Standard Solutions

A volumetric flask (Figure 16-1) is a device designed to *contain* a known and accurately reproducible volume of solution. Detailed instructions concerning its calibration are given in Chapter 16.

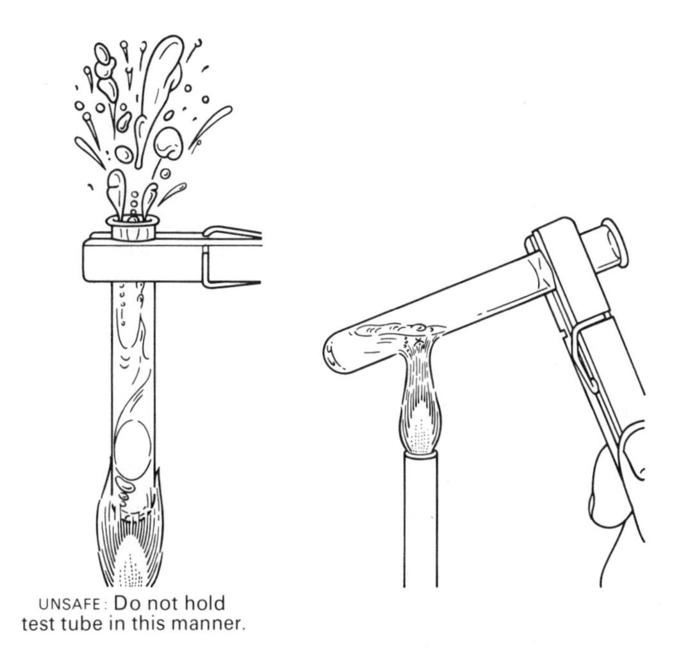

FIGURE 10-26
Boiling solutions in test tubes.

When a standard solution—a solution of accurately known concentration—is prepared in a volumetric flask an accurately weighed portion of the dry reagent-grade substance is transferred to the flask and dissolved in water. Water is then added with gentle swirling so that the solution is well mixed before the final adjustment of the volume is made. The solution of a solid or the mixing of solutions may result in an appreciable volume and temperature change. A dropper or pipet is used to add the last portions of water to bring the lower meniscus just coincident with the calibration mark.

If, as sometimes happens, the calibration mark is overrun, the solution may be saved by the following expedient:

Paste against the neck of the flask a thin strip of paper and mark on it with a sharp pencil the position of the meniscus. After removing the thoroughly mixed solution from the flask, fill the flask with water until the meniscus is tangent to the calibration mark. By means of a buret (or, if the volume to be added is small, a 1- or 2-ml measuring pipet) add water to the flask until the meniscus is raised to the mark on the strip of paper. Note and record the volume so added.

One of the most common sources of error in volumetric analysis is the incomplete mixing of solutions following their preparation in volumetric flasks.

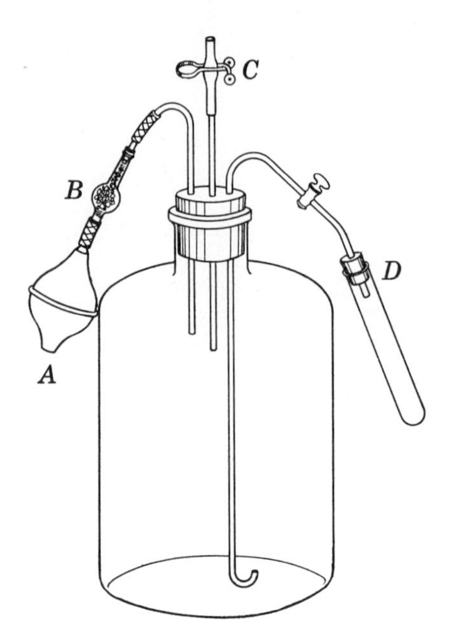

FIGURE 10-27

A storage bottle for standard solutions. *A* is an atomizer bulb; *B* is a soda-lime bulb packed with sterile cotton; *C* is a relief tube fitted with a pinch clamp; *D* is a guard tube fitted over the outlet and containing a small volume of the standard solution in order to prevent evaporation. By removing *D* and pumping with *A*, solution is forced out as needed. The flow may be quickly stopped by releasing the pinch clamp at *C*. The relief tube may be eliminated by providing a stopcock on the outlet tube.

Such solutions can be effectively mixed by swirling the solution, inverting, and again swirling, repeating this sequence at least 10 times.

Standard solutions should not be kept in volumetric flasks, since the latter are relatively thin-walled and easily broken, and are also likely to be etched by solutions on long standing. When transferring a standard solution to a storage bottle one first rinses the bottle with several small portions of the solution thus avoiding the necessity of drying the bottle.

Standard solutions that contain nonvolatile compounds should not be poured from storage bottles, since the solution adhering to the neck of the bottle evaporates and the residue may fall back into the bottle when the stopper is withdrawn. If large volumes of such solutions are to be used frequently, some type of delivery tube should be arranged (see Figure 10-27). For occasional use the solution may be removed by means of a *clean, dry* pipet. Because of the excessive amount of time required to dry a pipet, an acceptable technique is to rinse the pipet thoroughly with distilled water, carefully drain it, wipe the outside of the tip dry with a clean towel, and while applying gentle suction lower the tip into the solution. When 2 ml or 3 ml of solution have been drawn into the pipet, withdraw it, rinse the inside, and drain it. Repeat this with a second portion and then the pipet can be used for transferring the solution.

When not in use the stoppers of the bottles should be protected from dust and other foreign matter by inverting over them small beakers.

Instructions Regarding the Preparation of Standard Gravimetric Solutions

These solutions have known weight concentration—moles or equivalents per kilogram of solution. If such a solution is prepared directly from a primary standard reagent the technique is as follows:

A suitable clean container (flask, beaker, or bottle) is thoroughly dried; then it is weighed on a top-loading or other rugged balance having a capacity greater than that of the combined weight of the solution and the container, and also having an appropriate sensitivity.

The dry reagent is weighed on an analytical balance to the required accuracy and transferred to the previously weighed container. The desired quantity of water is added, the mixture is swirled until the reagent is dissolved, and the container and solution are weighed. From the weight of the reagent and the weight of the solution one calculates the weight formality of the solution.

If the reagent dissolves readily, the solution is conveniently prepared in the bottle in which it is to be stored. If heating is required in order to dissolve the reagent, the solution should be prepared in a beaker or flask and then transferred to the storage bottle. The same precautions in storage are required here as for standard volumetric solutions (see page 194).

THE SEPARATION OF PHASES

As has been explained in Chapter 9, many of the separations made in analytical and industrial work are based upon the separation of two phases. In these separations the constituents of interest are converted either into compounds that form a new phase, such as a gas or an insoluble precipitate, or into compounds that can be made to pass into a separate phase, such as an immiscible liquid.

The three types of phase separations most commonly used are:

(1) *Solid–Liquid Phase Separations.* The most frequently occurring example of this type involves the formation of an insoluble precipitate and the separation of the solid phase from the liquid phase. Separations of this type are discussed below.

(2) *Gas–Liquid Phase Separations.* Such separations are based upon differences in the volatilities, or more fundamentally the vapor pressures, of various compounds. If a constituent in a solution can be converted into a compound that is relatively volatile at ordinary temperatures, then this compound can be removed from the solution in a gas phase and collected in some suitable manner. Thus carbonate carbon can be converted to carbon

dioxide, which can be swept from the solution by steam or another inert gas and collected in an alkaline solution.

(3) *Liquid–Liquid Phase Separations.* The constituent of interest is converted into a compound that is more soluble in a second immiscible liquid. Thus iodide iodine in an aqueous solution can be oxidized to elementary iodine and extracted into a liquid such as carbon tetrachloride.

Solid–Liquid Phase Separations

The type of separation most frequently used in analytical work involves the formation of a solid phase, called the precipitate, and its separation from a liquid phase, usually an aqueous solution. Two methods are used to accomplish these separations: *filtration* and *centrifugation.* Of these two, filtration is most commonly used in quantitative methods and is discussed in detail below.

Filtering of Precipitates

As stated above filtering has for its object the separation of the precipitate from the solution in which it has been formed. This separation is accomplished by passing the solution through a porous medium that will be capable of allowing the solvent and the constituents that are in true solution to pass but that will retain very finely divided precipitates—even those with particles having diameters of approximately 1 μ to 2 μ (1 μ, called *one micron,* is 0.001 millimeter). With particles of diameters of less than 0.2 μ to 0.5 μ, colloidal properties become evident; and although it is possible to devise filtering media that will retain even colloidal suspensions, special apparatus is required, and the rate of filtration is so slow that it is not practical for analytical purposes to try to retain particles of less than approximately 1 μ. Certain restrictions are imposed upon the material used for a filtering medium. (1) It must be relatively inert to the various solvents and solutions to be filtered. This implies that it must not be weakened or disintegrated during the filtering process, and that it must not introduce any contamination into the solutions passed through it. (2) It must not retain, either by adsorption or by absorption, the soluble constituents of the solution. (3) For quantitative purposes (where the precipitate is to be subsequently weighed), it must either remain constant in its weight or be capable of complete removal—for example, by being burnt or volatilized.

Filtering Media

The filtering media most extensively used in analytical work are paper and asbestos; alternative media are glass fiber mats, and sintered-glass and porous porcelain crucibles. The techniques involved in the use of these media and their relative advantages are discussed below.

Paper Filters. The medium used almost exclusively for qualitative work, and still used most frequently for the separations involved in quantitative analysis, is pure cellulose paper. Because of the use of asbestos and the development of filtering crucibles of glass and porcelain, paper filters are now used for gravimetric determinations only with precipitates that must be heated before being weighed to temperatures causing deterioration of other media.

Paper filters have several disadvantages: (1) They are not inert, being attacked by concentrated solutions of both alkalies and acids, and by many oxidizing agents. (2) They are lacking in mechanical strength, are not readily adapted to vacuum filtration, and often disintegrate, introducing fibers into otherwise clear solutions. (3) They have the property of adsorbing constituents from the solutions passed through them. (4) They cannot be dried to a constant weight for precise quantitative work. Their chief advantages lie in their cheapness, availability, and superior filtering efficiency, especially in the filtration of gelatinous precipitates. This advantage over other filtering media is partly due to the larger surface exposed and a larger ratio of pore space to total surface.

Paper filters are obtainable in various degrees of porosity. Those of more open texture are suited for the rapid filtration of easily retained precipitates, the more dense ones being adapted for finer precipitates; the finer filters will retain particles of approximately 2 μ in diameter, and the coarser ones will retain particles of about 6 μ in diameter.

When used for gravimetric work, paper filters have a serious disadvantage in that they cannot be dried to a constant weight and therefore have to be ignited or burned before the precipitate can be weighed. This process is often unsatisfactory for the following reasons: (1) It may require a higher temperature than is necessary to dry the precipitate and one at which the precipitate may be unstable. (2) Many precipitates (for example, the silver halides and barium sulfate) are appreciably reduced during the burning of the paper and have to be treated subsequently to correct this reduction. Where this reduction is serious, it is necessary to remove most of the precipitate from the paper and burn it separately, a process that invites mechanical loss. (3) A nonvolatile residue, or ash, always remains after the paper is burned.

Quantitative (Ashless) Paper Filters. In order to reduce this error, paper filters that are intended for quantitative gravimetric work are washed in the process of manufacture with hydrochloric and hydrofluoric acids to remove

TABLE 10-1
Weight of the Ash of Various Types of Paper Filters

Diameter of Filter (cm)	Qualitative (mg)	Quantitative (mg)
7	0.2–0.9	0.02–0.05
9	0.4–1.3	0.03–0.08
11	0.6–3.3	0.05–0.11

as completely as possible the nonvolatile inorganic salts and the silicon compounds. This process increases the cost of these filters but reduces the weight of the ash to a very low value. The average weight of this ash in various types of filters is shown in Table 10-1.

Table 10-1 shows that the ash of the quantitative filters has been reduced to such a small value that it can be precisely corrected for or even neglected in most routine work. Because of the development of several types of satisfactory filtering crucibles, the use of filters papers is no longer recommended for precipitates that can be readily collected on filtering crucibles and that can be dried at low temperatures (100°C–500°C), or that have to be separated from the filter before the paper is burned.

Quantitative filters cost approximately five times as much as qualitative filters of the same filtering characteristics; therefore they should be used only when a low ash is necessary. A good grade of paper is desirable; the cheaper papers are likely either to be too stiff to fit properly in the funnel or else to disintegrate too easily.

In order to obtain rapid and effective filtration with paper filters, care must be taken to select the proper type of filter and to fit the paper to the funnel properly. In special cases gentle suction may be used. The type of filter to be used, quantitative or qualitative, should be determined by the subsequent treatment of the precipitate.

The size of the filter to be used should be determined largely by the size of the precipitate (or residue) that is to be collected and not by the volume of the solution to be filtered. The sizes most commonly used are 7 cm, 9 cm, and 11 cm in diameter. In general, the precipitate should not fill more than one-third of the capacity of the filter. The funnel used should be of such size that the paper never comes closer than 0.5 cm or 1 cm to the top; papers that are too large should be cut to size after being folded.

General Instructions for the Use of Paper Filters

Selection of the Paper. (a) Select a paper of the proper type—*quantitative* ("ashless," if the precipitate is to be weighed, otherwise qualitative). (*b*)

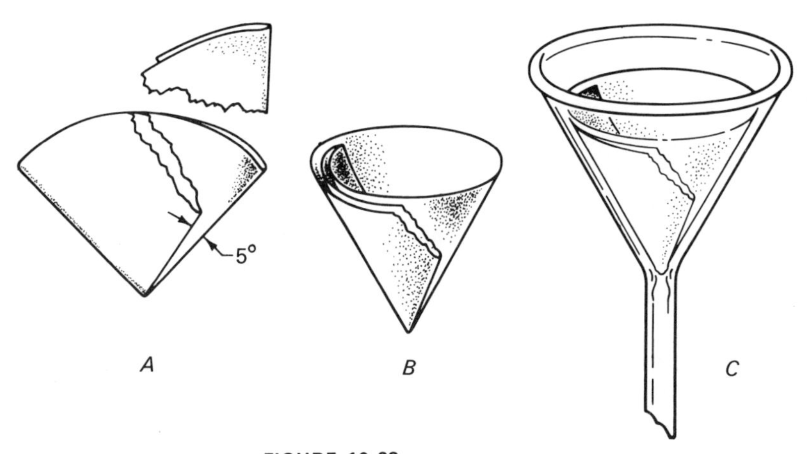

FIGURE 10-28
Folding and fitting a filter paper.

Select a paper of the proper porosity—the finer and more dispersed the precipitate, the less porous and more dense the paper required. (*c*) Select a paper of the proper size—never fill a paper over one-third full with precipitate.

Fitting the Paper to the Funnel. Use a funnel of such size that the rim will extend $\frac{1}{4}''$ to $\frac{1}{2}''$ above the top of the fitted paper. Fold a filter exactly in half (see Figure 10-28). Crease the fold by pressing it firmly between the fingers; do not rub the crease. Fold again, but with the two straight edges making an angle of approximately 5°. Again crease firmly. Tear an irregular segment from the smaller fold, as shown in Figure 10-28 (Note 1).

Open the filter between the untorn folds and place it in the funnel. Fill the paper with water and fit it into place by gently pressing downward with the fingers on the upper edges. *The paper should fit so that the stem retains a column of water after the filter is empty*, otherwise air will leak around the folds and will prevent efficient filtration. This column of water, by the suction it produces, greatly increases the rate of filtration. When filtering, the beveled tip of the funnel should always be made to touch the inside of the receiving vessel (Note 2).

NOTES

1. The instructions given here are to prepare a filter for a 60° funnel. If a 58° funnel is used, the second fold should be made with the two straight edges together rather than at an angle of 5°. The slightly smaller funnel angle causes this folded filter to fit tightly at the top edge.

2. When the above precautions are observed, the use of suction with paper filters is usually unnecessary, unless the precipitate is large and bulky. When suction

Asbestos Fiber Filters

Asbestos is a calcium-magnesium silicate, and that used for quantitative purposes should be of a white, silky, long-fibered variety, free from magnetite and iron. It has been found that even the asbestos purchased and designated specifically "for quantitative analysis" should be acid-treated and washed before being used. Asbestos can be safely heated only to moderately high temperatures. When used for gravimetric work, Gooch crucibles (and other filtering crucibles) have the great advantage over paper that they can be dried to constant weight, and the disadvantages connected with the burning of a paper filter are eliminated. Asbestos filters are intended for use with suction. With coarse crystalline precipitates and with a thin, properly made mat, they afford a more rapid filtration than paper. However, asbestos and other filtering media such as glass and porcelain are not so satisfactory as paper for gelatinous precipitates, since these precipitates tend to clog the relatively small filtering surface. Also, asbestos is not so satisfactory for precipitates that have to be heated to very high temperatures, since some specimens tend to lose weight, and the asbestos insulates the precipitate from the heat of the burner; studies by Duval (*Anal. Chim. Acta*, **3**, 163, 1949) made with a thermobalance show that asbestos slowly loses weight above 283°C. Asbestos filters (and filtering crucibles in general) should be heated to a constant weight at the temperature subsequently to be used for drying the precipitate. They possess the disadvantage that they require more time in their preparation than a paper filter; also, some specimens of asbestos tend to absorb water so rapidly as to cause difficulty in weighing. For qualitative work or in the removal of a precipitate (as in Procedure 18-1, in which MnO_2(s) must be removed from the $KMnO_4$ solution), an asbestos filter can be made upon a circular bevel-edged perforated porcelain plate or upon a wad of glass wool supported in an ordinary funnel (see Figure 10-29).

NOTES

is used, the lower portion of the filter must be supported by a smaller filter of hardened paper, a perforated platinum cone, or a small circular piece of cloth; the cloth may be purchased or cut from surgical gauze or similar material. Only a moderate suction can be used, and care must be taken in applying it in order to avoid breaking the filter. Precipitates, such as some hydrous oxides, which are so colloidal as to pass through the filter, can often be retained by the addition of paper pulp to the mixture before the filtration is begun. This pulp may be purchased in the form of readily disintegrating tablets or may be made by tearing a piece of quantitative paper into small bits and violently shaking it with water in a test tube. Precipitates, such as silver chloride or barium sulfate, which have been properly coagulated, should be retained by a medium or rapid filtering paper and should not require the use of either paper pulp or suction filtration.

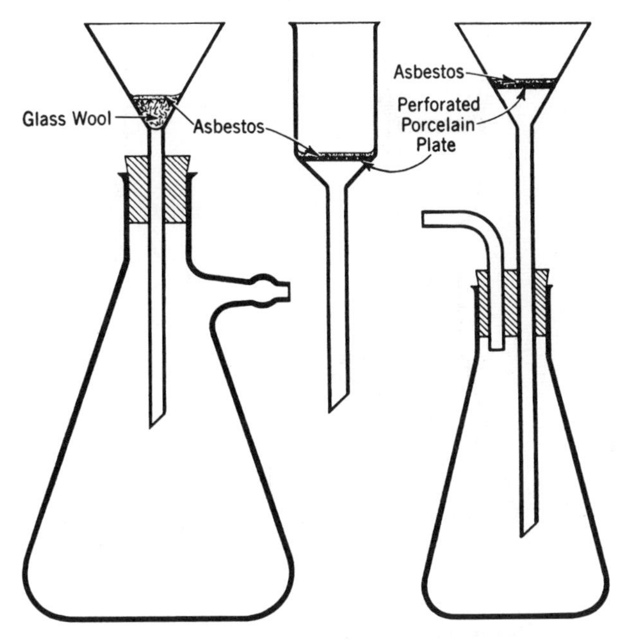

FIGURE 10-29
Funnel-type asbestos filters for qualitative work or for quantitative separations.

Funnel-Type Asbestos Filters

Instructions for Preparing Funnel-Type Asbestos Filters. (*a*) *Asbestos Filter with Porcelain-Plate Support.* Obtain a 20-mm perforated porcelain plate, preferably one with a 60° beveled edge, and place it with the glazed side uppermost in the funnel of the suction filtration tube. Check that the plate is seated firmly and does not rock. With no suction applied, pour the asbestos suspension on the plate until the perforations are thinly covered. Apply light suction until the fibers are drawn against the plate; only light suction is required to draw the asbestos into place. Examine the mat to see if all the perforations are covered. The tendency at first is to use too thick a mat. Usually a 1-mm or at most 2-mm mat is adequate. Add water, taking care not to disrupt the mat, until the funnel is two-thirds full; draw this water through the mat. Repeat this washing process until no more asbestos fibers are drawn through the mat. The filter is ready for use. (In case the filter is to be used with solutions containing silver salts, wash until the wash water is free of chloride.)

(*b*) *Asbestos Filter with Glass-Wool Support.* Obtain a supply of glass wool. (This must be fine, soft fiber, not springy strands. Pyrex No. 800 fiber or

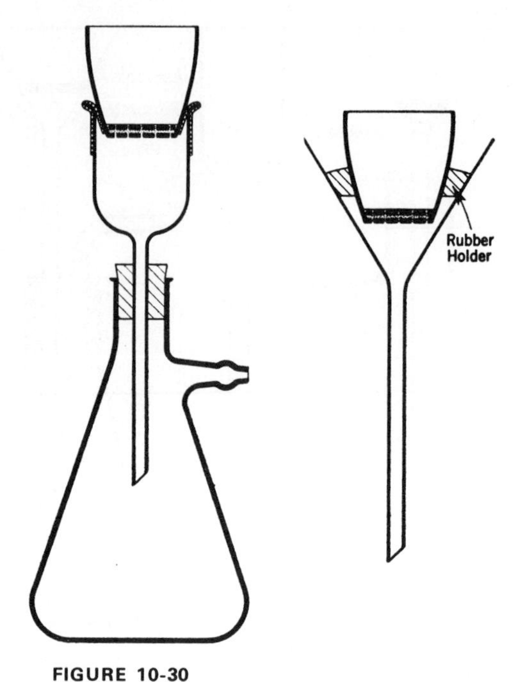

FIGURE 10-30
Asbestos filters in Gooch crucible with holders.
For gravimetric determinations.

equivalent should be used.) By means of a stirring rod with a flattened end, loosely tamp a quantity of the glass wool into the base of the funnel (see Figure 10-29). The upper surface of the mat should have a diameter of from 10 mm to 15 mm. Adjust the quantity of wool to the size and nature of the precipitate; the larger quantity will be required only for large voluminous precipitates. With no suction applied, pour the well-shaken asbestos suspension on the mat until a 1-mm to 2-mm layer of asbestos is formed. Apply light suction to draw the asbestos into place. Wash as directed above.

Crucible-Type Asbestos Filters

For quantitative work, asbestos is almost universally used in porcelain crucibles with perforated bottoms, called Gooch crucibles, although the form used by Gooch[3] was of platinum (see Figure 10-30).

Instructions for Preparing a Gooch-Type Asbestos Filter. Obtain a clean crucible of the proper size, and fit it into a filter crucible holder (Figure

[3] F. A. Gooch, *Proc. Am. Acad.*, **13**, 342 (1878).

10-30). Adjust the crucible and holder so that the rubber does not extend beyond the bottom of the crucible and the filtrate does not come into contact with the rubber. Fit the stem of the crucible holder or of the funnel into a one-hole rubber stopper and support it in a suction filter flask as shown in Figure 10-30 (Note 1).

Shake up a water suspension of acid-washed asbestos (see Appendix 8) and pour it through the crucible until a uniform mat of asbestos of about 1-mm or 2-mm thickness is obtained (Note 2). Place a perforated porcelain plate (a "Witte" plate) on the mat and, while applying very gentle suction, pour into the crucible just enough of the asbestos suspension to fill around the edges and the perforations of this plate (Note 3). Wash the mat until the water passing through it is free from asbestos fibers. Guide the first portions of water carefully so that the mat is not disturbed; thereafter do not allow one portion of liquid to run out entirely before adding the next.

NOTES

1. Suction flasks are conical flasks made with a side neck and with thick walls to withstand the external pressure when a vacuum is produced. It is unsafe to heat solutions in them.

When a water aspirator is used for suction, eliminate the possibility of contamination of the filtrate by tap water being sucked back into the filter flask by fitting a conical flask with a two-hole stopper carrying an inlet and outlet tube and then connecting the inlet tube to the filter flask and the outlet tube to the aspirator; such an arrangement also safeguards against solution in the filter flask overflowing into a vacuum system.

2. The thickness of mat to be used will vary with the type of precipitate to be retained. For example, with silver chloride or barium sulfate precipitates that have been properly coagulated, a thin, moderately fast filtering mat may be used. When such a mat is held up to the light, the perforations of the crucible should be dimly visible; by such an inspection, imperfections in the mat may often be detected. If the holes are not discernible, or if, after the top plate with its film of asbestos is applied, water runs through the filter slowly with full suction applied, the mat had better be remade. The thinner the mat (which will retain the precipitate), the more rapid the filtration and washing of the precipiate, and the more quickly the crucible can be dried and brought to constant weight.

3. This plate and the additional asbestos serve to prevent the disruption of the mat, not as an aid to filtering. If care is taken when pouring solutions upon the mat, the plate may be dispensed with and more rapid filtering and washing of the precipitate obtained. When adding a solution to the empty crucible, always apply a slight suction and guide the solution with a stirring rod held against the plate or, if the plate is not used, with the stirring rod held just above the mat. Avoid too strong a suction, since it may break the mat and, in many cases, will so pack the precipitate against the mat as to clog it and slow the rate of flow.

Other Types of Filters

Glass-Fiber Mats. There are discs of glass fibers commercially available that can be used in the place of the asbestos layer in a Gooch crucible. A Witte plate or two mats should be used; in this case, extreme care must be taken to prevent the mats from rising at the edges because precipitate will pass around them. These mats are very satisfactory for filtering solutions that would attack paper, such as permanganate or concentrated acids. For gravimetric methods they can be used for precipitates that can be dried at 500°C or lower temperatures. They are sold in various sizes and need only to be placed in the crucible and covered with a Witte plate. They provide the convenience of sintered-glass and porous porcelain crucibles (though not the temperature range of the latter) and avoid the problems of cleaning inherent with those crucibles.

Filtering Crucibles. Filtering crucibles of glass, quartz, porcelain, platinum, and other materials are now offered that have a filtering element of the same material fabricated in place as an integral part of the crucible. Such crucibles possess the advantage over paper that they can be dried to constant weight, as can the Gooch-type asbestos filter. They possess the advantages over the asbestos filter that they are ready for immediate use, that the filtering elements can be obtained with various pore sizes, that they are more quickly brought to a constant weight, and that they are less hygroscopic thereafter. They possess the disadvantage that special methods are necessary in order to clean the precipitate from them. The characteristics of these crucibles are discussed briefly below.

Sintered-Glass Crucibles. The filtering element in the bottom of these crucibles is composed of glass particles that have been powdered and graded as to size, and then sintered together (in place) to form a filtering crucible

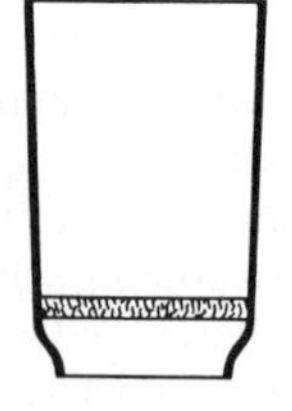

FIGURE 10-31
Sintered-glass filtering crucible.

of the desired porosity (see Figure 10-31). They possess the advantage of being transparent, of quickly coming to constant weight, and of being easy to clean. They are made of borosilicate glass and, with care in heating and cooling, they can be used up to 500°C. They can be used with strong acids (except hydrofluoric) and with moderately concentrated bases. A tendency of some of these crucibles to lose weight, apparently from disintegration of the filtering mat, after being heated much above 150°C has been noted; otherwise they are excellent.

Sintered-Quartz Crucibles. These crucibles are similar to the glass crucibles but are constructed of clear quartz. They possess the advantage that they can be used for temperatures to 1200°C. They are expensive.

Porous Porcelain Crucibles. These crucibles are similar to sintered-glass crucibles but are constructed of porcelain and have a porous porcelain filtering element in the bottom. They can be obtained in any desired porosity and can be heated to between 1000°C and 1200°C.

Platinum Filtering Crucibles (Munroe type). These crucibles are similar in design to a Gooch crucible except that they are constructed of platinum, and the filtering element is a porous layer of platinum; this is prepared by building up a layer of ammonium chloroplatinate, $(NH_4)_2PtCl_6$, of the desired thickness and then igniting it. These crucibles have many advantages. They can be made to retain fine precipitates, yet filter rapidly; they quickly attain constant weight and can be heated to high temperatures; because of their superior heat conductivity, precipitates in them are not insulated as they are by asbestos mats. They are usually provided with a cap that fits over the bottom and protects the precipitate from possible action of gases from burners. They are expensive, however, and the mats have to be frequently renewed, especially when used at high temperatures. They are recommended for use in highly precise work. Details of the preparation, use, and cleaning of these crucibles are available in the chemical literature.[4]

Cleaning Filtering Crucibles. Before being used, sintered-glass and porcelain crucibles may be cleaned by any acid-cleaning solution except HF; if the sulfuric acid-dichromate mixture is used, it should be free of solid material and the crucible should be then rinsed with water until free of acid. Prolonged treatment with concentrated alkali is not advisable, although ammonia and cyanide solutions can be used. The proper solution to use will be determined by the substances it is desired to remove from the crucible.

---

[4] Snelling. *Chem. News*, **99**, 229 (1909); Snelling. *J. Am. Chem. Soc.*, **31** 456 (1909); Swett, *J. Am. Chem. Soc.*, **31**, 928 (1909).

Miscellaneous Operations

Instructions Regarding the Use of *p*H Papers

So-called *p*H papers are strips of paper impregnated with organic compounds called "indicators." These indicators can be assumed to be organic acids or bases that change color upon being converted to their corresponding salts.

Litmus paper is the classical *p*H paper. The essential constituent of litmus paper is azolitmin; this substance changes from red to blue as the *p*H changes from about 5 to 8. This paper is very convenient to use but is restricted in its *p*H range.

Wide-range *p*H papers have been developed in recent years and have found general use. These papers have been impregnated with several indicators having *p*H ranges that differ so that a continuous color change is observed on changing from a strongly acid solution, *p*H less than 2, to a strongly alkaline solution, *p*H greater than 11. The change of colors with changing *p*H is usually as follows:

*p*H 2 or less—*orange-red*
*p*H 3 to 5—*orange-yellow*
*p*H 5 to 7—*yellow*
*p*H 7 to 9—*green to purplish blue*
*p*H 10 or greater—*dark blue*

Before making a pH *test be sure that the solution is thoroughly mixed.* Many an analysis has failed because the test was made without mixing the solution, or the test paper came in contact with acid or base left on the walls of the container.

After mixing the solution insert a clean stirring rod, remove it, and touch the wall of the container just above the solution in order not to remove too large a drop. Touch the rod to the extreme end of the paper. Wait 1 or 2 seconds and compare the color produced with the color chart that is usually on the bottle from which the papers are dispensed. Tear off the moistened portion of the strip.

A single strip of paper can be used for repeated tests since only a small area need be moistened.

Instructions for Drying Substances

All solid materials that have been exposed to the air will contain a certain amount of water. The substance may combine with the moisture in the air to form a stable hydrate, or (if the vapor pressure of water above its saturated solution is less than that in the air) it will become moist (deliquesce). Even though no hydrate formation occurs, water may be adsorbed on the surface to a significant extent, especially if the material is a finely divided powder.

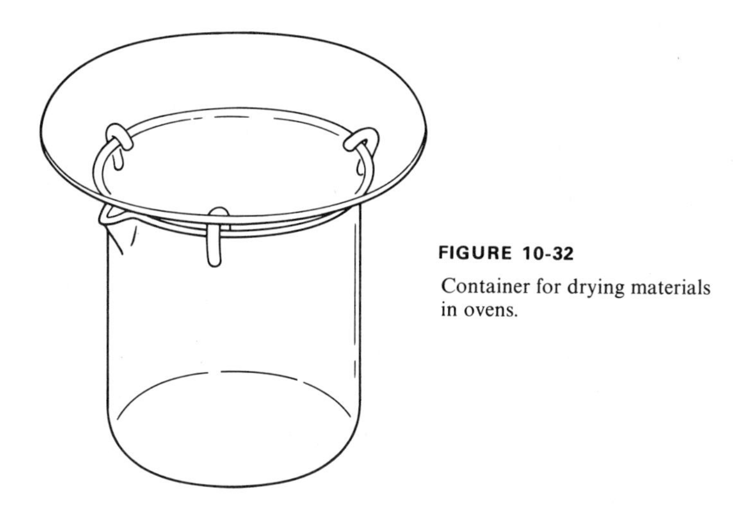

FIGURE 10-32

Container for drying materials in ovens.

Therefore, for quantitative work, samples and solid materials have to be dried before being weighed. Samples are usually dried at 105°C–110°C for a definite period of time or until they reach a constant weight. The time and temperature required for complete drying of materials such as primary standards will vary. For example, silver nitrate crystals that are not too large or that have been crushed will contain less than 0.1% of water after being dried at 110°C for 2 hours; the last traces of water can be removed only by fusing the salt and keeping it at a temperature between 220°C and 240°C for 10 minutes. Many salts can be made anhydrous only by being heated above their melting points.

A sample that is to be dried must be protected from dust and other substances that might react with it. Therefore, to dry a sample, transfer it to a weighing bottle and place this in a small beaker that you have identified by marking with pencil on the ground glass or opaque portion provided for that purpose. If more than one sample is to be dried at once, the contents of each weighing bottle should be indicated in pencil on the ground-glass seal. Paper labels should never be used on articles placed in the drying oven, since the adhesive material chars and gives off volatile reducing substances (tarry materials) that may react with the sample. Cover the beaker with a watch glass and place it in the oven. The rate of drying is increased if the watch glass is supported on glass hooks as shown in Figure 10-32, since air circulation around the sample is improved.

Instructions Regarding the Use of Distilled Water

Because of dissolved impurities, ordinary tap water is never used in the preparation or dilution of solutions for analytical work. When a procedure in this text calls for "water" to be added, distilled water is meant. The mere fact that water has been taken from a tap or container labeled "Distilled Water" is no guarantee of its purity. In fact, distilled water will always contain dissolved volatile constituents, such as carbon dioxide and ammonia, and small amounts of suspended material, usually organic dust particles. Because of faulty still operation (foaming, and so on), appreciable amounts of the original constituents of the water or of water conditioning compounds may be present; in addition, corrosion of the still, connecting lines, or containers frequently causes the presence in the water of significant amounts of certain constituents, such as tin, copper, or silica.

It is good practice to check a sample of the distilled water that is to be used for the preparation of certain standard solutions. For example, water used to prepare $AgNO_3$ solutions should not cause a precipitate when the silver salt is added.

The Purity of Chemicals

Various grades of chemicals can be purchased. The *technical* grade is usually intended for large-scale industrial use and may be quite impure. The *commercial* grade is intended for use where relatively small amounts of various impurities can be tolerated. The designation "pure" is less commonly used than formerly and should be viewed with suspicion. The label "U.S.P." implies that the chemical conforms to the standards of the United States Pharmacopoeia and in general does not indicate as high a state of purity as does "C.P." (*chemically pure*). "C.P." does not mean that the substance is absolutely pure, since such a goal is unobtainable; nor does it mean that the chemical is so pure that no foreign material can be detected by chemical means. In fact, the designation has had no specific meaning and because of this fact the best grades of chemicals are now usually designated as *reagent*, *analyzed*, or *guaranteed*. Such products usually carry on the label an analysis showing the percentages of the most common impurities. The Committee on Analytical Reagents of the American Chemical Society has begun to set up definite standards for such reagents, and frequently the label will also carry the statement "Conforms to ACS Specifications" or simply the letters "A.C.S."[5] In general, only the analyzed or reagent grade should be used for the preparation of analytical reagents. It is to be emphasized, however, that implicit trust *cannot be placed on the analyses accompanying such chemicals.* Impurities not mentioned in the analysis have been found, foreign

[5] The standards are published by the American Chemical Society in the book *Reagent Chemicals—A. C. S. Specifications* (4th ed., 1968). The publication is available from the Special Issue Sales of that Society, 1155 16th St., N.W. Washington D.C. The first supplement to the 4th edition appeared in the December, 1969, issue of *Analytical Chemistry*.

substances may be introduced in packaging, and, if the bottle has been once opened, subsequent contamination is an ever-present possibility. Therefore, where the accuracy or importance of the work justifies it, tests assuring the purity of the chemical, or at least the absence of injurious impurities, should be made.[6]

Reagents of primary standard quality can be obtained from the National Bureau of Standards. *N.B.S. Miscellaneous Publication 260* lists the reagents that are available.

---

[6] Rosin (*Reagent Chemicals and Standards.* New York: Van Nostrand, 1937) and Murray (*Standards and Tests for Reagent Chemicals.* 2nd ed., New York: Van Nostrand, 1927) give useful information and methods for testing chemicals. Recent operations, such as the analysis of lunar rocks, of materials for solid state devices, and of foods for trace metals have required reagents exceeding "reagent grade" purity. M. A. Zeif, *Industrial Research,* **13** (April), discusses the preparation, handling, and containment of such reagents.

SECTION III

GRAVIMETRIC MEASUREMENTS AND METHODS

CHAPTER 11

Measurement of Mass

In this and the following three chapters we will discuss gravimetric measurements and methods; that is, methods in which the constituent being determined is converted into an element or compound of definite composition, which is then weighed. The measurement of mass, or *weighing* as the operation is commonly called, is the fundamental operation in gravimetric methods.

The *mass* of an object is a measure of the quantity of matter in that object and is defined in terms of the inertia. The *weight* of an object is a force. Consequently in a gravitational field, g, the weight, W, is related to the mass, M, by the expression,

$$W = Mg$$

From this equation it can be seen that an object of mass M has different weights when it is subjected to different gravitational fields. A steel bar weighs more at sea level than high on a mountain; it weighs considerably more on the earth than it would on the moon, since the latter has a much smaller gravitational field. Yet the mass of the steel bar is not different in one place than it is in the other place. The observed or apparent weight of an object is changed as it is immersed in fluids of different density. Thus the observed weight of the steel bar is less if the bar is in water than if it is in air. Moreover, its observed weight is less in air than in a vacuum. This smaller observed weight results from the buoyant force that the fluid medium exerts upon the object.

To determine the *weight* of an object one could use a spring balance, a device that measures the gravitational force on the object by showing the extension of a spring when the object is hung from it.

To determine the *mass* of an object one could use the equal-arm balance. The origin of this type of balance is unknown. Szabadváry in his *History of Analytical Chemistry*[1] states, "The balance has been known since very early times and is so old that the peoples of the ancient civilizations attribute its origin to the Gods." Stone weights dating from the twenty-sixth century B.C., when Gudea was king of Babylon, have been found. To use this balance one places the object on one pan and balances it with weights (of known masses) on the other pan.

Let us assume first that the balance is in a vacuum, so that there are no buoyant forces acting upon either the unknown or the known masses. Now, since the two masses just balance each other it is evident that the moments or torques about the central point of support are equal. But the balance arms are equal, so the forces or weights must be equal:

$$W_1 = W_2$$

Each weight is the product of the mass and the acceleration due to gravity, g:

$$W_1 = M_1 g$$

$$W_2 = M_2 g$$

Therefore,

$$M_1 = M_2$$

because both unknown and known masses were at the same location and in the same gravitational field. It is seen from this reasoning that the operation of weighing on a simple balance is essentially the operation of making a comparison of the mass of an unknown object with the mass of known objects.

In the example just considered, the balance was assumed to be in a vacuum at the time the comparison of masses was made. This assumption permitted us to ignore the problem of buoyant forces acting upon the unknown and known masses. The buoyant force of air is small but must be calculated under some circumstances; we will discuss the subject in detail in a later section of this chapter.

[1] Ferenc. Szabadváry. *History of Analytical Chemistry*. New York: Pergamon, 1966.

THE EQUAL-ARM BALANCE

While the past two decades have seen extensive replacement of the simple equal-arm balance in the laboratory by much more rapid balances, it is well worthwhile to consider the principles of its operation. The more complex modern balances are readily understood by one who is well acquainted with the principles of the equal-arm balance. There is no other relatively inexpensive laboratory instrument that is capable of such high precision over as wide a range as is a laboratory balance. A student can use an analytical balance that retails for less than $300 to weigh a 100-g object to a precision of 0.1 mg. (This is an uncertainty of only 1 part in a million.)

The equal-arm balance consists in principle of a horizontal lever (the beam) supported at its center on an agate knife-edge (the central knife-edge) and carrying at each end edges (the terminal knife-edges) that support the pans on which the objects and weights are placed. Since the pans that carry the object and weights are flexibly supported on the terminal knife-edges, it follows that applying a load to the pan is the same as concentrating this load at the terminal knife-edge. Therefore, when an object, O, is placed on a pan, a rotational moment, OL_1, is produced (where L_1 is the distance from the point of support, or central knife-edge, to the terminal knife-edge supporting the applied load). When the object is balanced by applying weights to the opposite pan, an equal and opposing moment, WL_2, will have been produced, and

$$OL_1 = WL_2 \tag{1}$$

If L_1 and L_2 are made equal, it follows that $O = W$. In order to protect the knife-edges from damage when the instrument is not in use, means are provided for raising the beam and pans from the knife-edges; this arresting mechanism is controlled by means of the milled head labeled "beam-arrest control" shown in Figure 11-1. Means are also provided for arresting the motion of the pans, and this mechanism is controlled by the button labeled "pan-arrest control." In some balances these controls are combined.

It is obvious that the essential parts of the balance are the beam and knife-edges, and that the design and construction of these largely determine the qualities of the balance. The principal requirements of a good beam are rigidity and strength with minimum weight, and these conflicting qualifications have led to much research both as to the design and the material used in its construction. Various designs can be noted by reference to the catalogues of balance manufacturers. The materials most commonly used at present are aluminum alloys of the "dural" type.[2]

---

[2] For a discussion of some of the problems of beam manufacture, see Ainsworth. *Ind. Eng. Chem.*, anal. ed., **11**, 572 (1939).

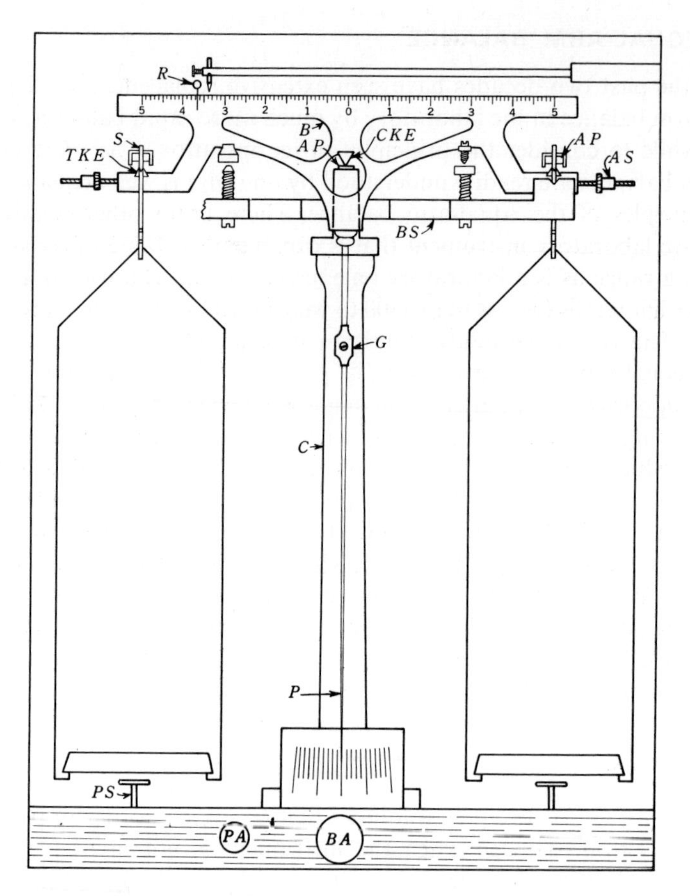

FIGURE 11-1

The essentials of the analytical balance: *BA*, beam-arrest control; *C*, column or post; *B*, beam; *P*, pointer; *G*, movable weight for adjusting sensitivity; *R*, rider; *TKE*, terminal knife edge; *CKE*, central knife edge; *S*, stirrup; *AP*, agate plate; *BS*, beam support; *PS*, pan support; *AS*, adjusting screw; *PA*, pan arrest control.

Knife-edges are of a special grade of agate, a silicon dioxide mineral, or of synthetic sapphire; these materials are used because of their hardness, durability, and resistance to corrosion. The central and terminal knife-edges bear against agate plates set in the "post" and in the pan "stirrups," respectively.

The relative positions of the knife-edges are of fundamental importance in determining the performance of the balance. First, and obviously, the central knife-edge or point of support should be above the center of gravity of the

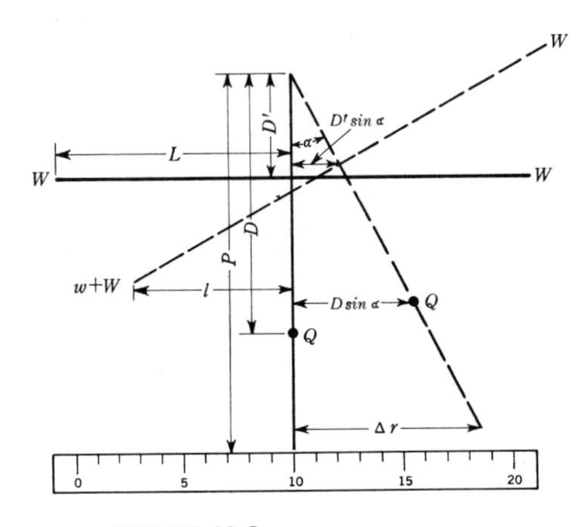

FIGURE 11-2
Principles of the equal-arm balance.

beam system in order for it to be in stable equilibrium; also, as will be shown later, the distance between these two points is an important factor in determining the *sensitivity*, that is, the response of the balance to a given difference in the load on the pans. The sensitivity is conventionally measured in terms of the deflection of the pointer produced by an excess in weight of 1 mg on either pan. By reference to Figure 11-2, it is seen that the sensitivity, S, can be expressed in more general terms as

$$S = \frac{\Delta r}{w} \tag{2}$$

where Δr is the change in the point of rest upon the addition of an excess of weight, w, to either pan. Second, the knife-edges should be parallel to each other and perpendicular to the plane of rotation of the beam in order to avoid excessive frictional effects and possible variation in arm length as the beam oscillates. Third, the knife-edges should be in the same horizontal plane in order that the sensitivity of the balance remain constant with varying loads. That this is true can be seen from the following analysis of the factors determining the sensitivity of the balance.

Figure 11-2 shows the fundamentals of a balance system. Let Q represent the weight of the beam and unloaded pan system with the center of gravity in the position shown; W is the weight applied to each pan and w is the excess of weight applied to one pan; α is the angle of deflection made by the rotating beam; L is the arm length; D is the distance from the point of support to the center of gravity of the beam system, and D' is the distance from the point of support to the center of gravity of the applied weights. This latter center of gravity is on a line joining the two terminal knife-edges, since, owing to the fact that the pans are flexibly suspended from the terminal knife-edges, the pan load can be considered as centered at the suspending knife-edge. It is apparent that as w is applied, a *moment* equal to w times its horizontal distance from the point of support will be set up, which will cause displacement of the beam system until the moment due to the weight Q times its horizontal distance from the point of support plus that due to $2W$ times its horizontal distance from the point of support becomes equal to the displacing moment. From Figure 11-2 the displacing moment due to w is seen to be

$$wl \tag{3}$$

and the restoring moments will be

$$QD \sin \alpha \tag{4}$$

and

$$2WD' \sin \alpha \tag{5}$$

At equilibrium the displacing moments must equal the restoring moments, and therefore

$$wl = QD \sin \alpha + 2WD' \sin \alpha \tag{6}$$

For very small angles, which can be justifiably assumed in this case because the angular deflections of a balance are relatively small, l approaches L and the value of $\sin \alpha$ approaches that of $\tan \alpha$. Therefore Equation 6 can be rewritten as follows:

$$wL = \tan \alpha(QD + 2WD') \tag{7}$$

As previously shown, the sensitivity, S, of the balance can be expressed in terms of the applied excess weight and the deflection, r, of the pointer as follows:

$$S = \frac{\Delta r}{w}$$

and from Figure 11-2 it is seen that

$$\tan \alpha = \frac{\Delta r}{P} \tag{8}$$

where P is the length of the pointer. Therefore it follows that

$$\tan \alpha = \frac{Sw}{P} \tag{9}$$

Substituting this expression for $\tan \alpha$ in Equation 7, we obtain the following general formula for the sensitivity of the balance:

$$S = \frac{LP}{(QD + 2WD')} \tag{10}$$

An inspection of this equation shows that if the terminal knife-edges are in the same plane with the central knife-edge, then D' becomes zero, and the equation is simplified to the form

$$S = \frac{LP}{QD} \tag{11}$$

This shows that the sensitivity is directly proportional to the arm length and to the pointer length, inversely proportional to the distance from the point of support to the center of gravity and to the weight of the beam system, *but independent of the applied load.* If the terminal knife-edges are below the central knife-edge, then as shown in Equation 10, the sensitivity becomes less as the load ($2W$) is increased. If the terminal knife-edges are above the central knife-edge or point of support, Equation 10 has the form

$$S = \frac{LP}{(QD - 2WD')} \tag{12}$$

and it is seen that as $2W$ is increased, the sensitivity increases, until $2WD'$ becomes equal to QD, when it becomes infinite and the balance becomes unstable.

From these considerations it would appear that if the balance is constructed with the terminal knife-edges in a plane with the central knife-edge, the sensitivity would remain constant; however, owing to bending of the beam with the applied load and to frictional forces, the sensitivity would be likely to decrease. Because of this fact, the terminal edges are sometimes placed slightly above the central one, and occasionally balances will be found in which the sensitivity may actually increase with applied load. A means of changing the sensitivity is provided by the adjustable weight on the pointer, which by being shifted changes the value of D. Increasing the sensitivity of balances by increasing the arm length, L, is limited practically by the mechanical difficulty in building a rigid beam of great length and light weight, and by the fact that the period of oscillation of the balance increases as the arm length is increased; this last factor increases the time required for making a weighing.

Devices for Expediting Weighings

Chain-Weights and Notched Beams. Figure 11-3 shows an equal-arm balance that has been modified to speed the weighing process by the addition of a fine gold chain, one end of which is supported by the beam and the other by a movable support. The latter is operated from outside the balance case, varying the fraction of the weight of the chain carried by the beam. In this way fractional weights up to 100 mg are replaced by the chain. The effective weight applied to the beam is read to 0.1 mg by means of a vernier scale. The balance in Figure 11-3 also has a notched beam, which permits a cylindrical weight or rider to be placed at fixed positions on the beam thus applying effective weights up to 1 g in 100 mg increments.

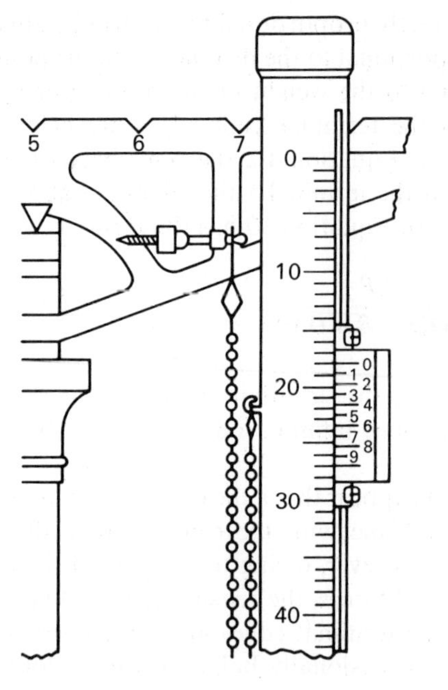

FIGURE 11-3
A chain-weight balance attachment.

Damping Devices. Balances provided with damping devices are available. These serve to bring the beam system to rest within 10 sec–20 sec and thus effect a considerable saving of time in determining the point of rest. Both magnetic dampers and air dampers are used. Present-day two-pan balances

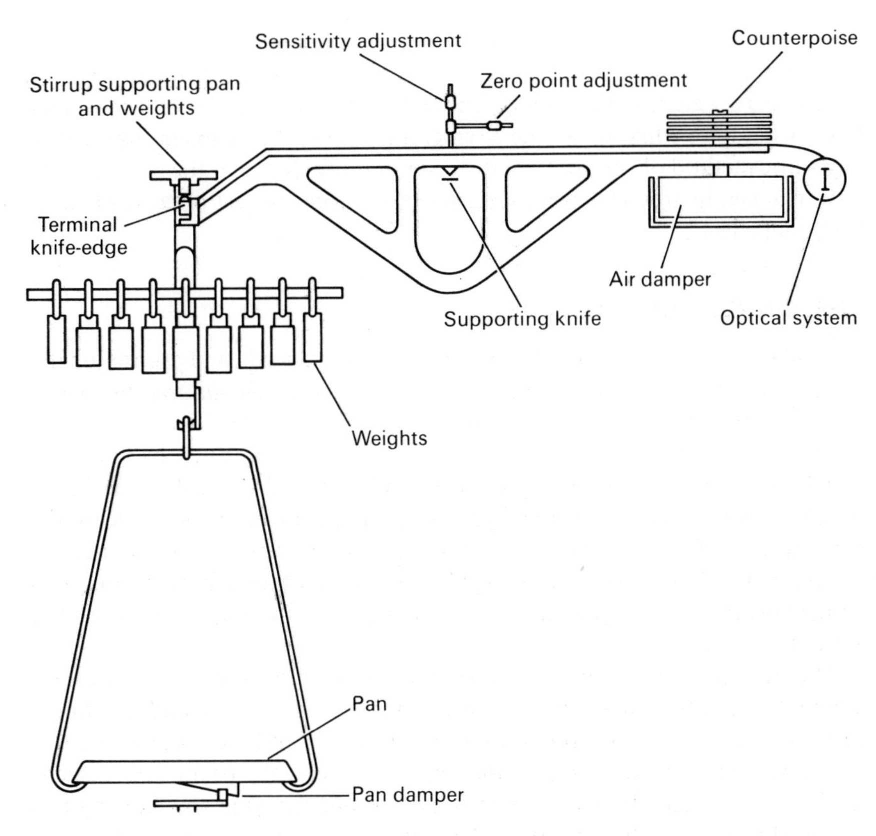

FIGURE 11-4
Essential features of a single-pan balance equipped with air damper. (The beam-arrest mechanism is not shown.)

often have magnetic dampers. The damping effect is obtained by suspending an aluminum plate from the stirrup between the poles of a magnet (which is supported by the balance post). The eddy currents induced by the vertical movement of this plate cause the damping. Air dampers consist of cylinders that are closed at the top and are suspended from the beam by the stirrups and move vertically inside cylinders closed at the bottom and supported in a fixed position by the central post. The two cylinders do not touch but fit sufficiently closely for the friction of the displaced air to give the desired damping. An adjustable vent allows the desired amount of damping to be obtained. A single-pan balance with an air damper is shown in Figure 11-4.

One should recognize that the above devices do not improve the precision or accuracy of the simple balance but are intended to decrease the time required for weighing. The chain, while made carefully, consists of discreet links; and minute particles of dust or grease may cause sticking of the links at the bottom of the loop. Any error in the placement of the notches for the rider (a 0.5 g weight in the balance shown) would result in an error in weighings. A damper, while very time-saving, introduces the problem inherent in any static method of determining the rest point of a balance: the balance may have reached its true rest point *or* it may have come to rest because of small frictional effects.

THE SINGLE-PAN BALANCE

In 1946 Erhard Mettler of Switzerland introduced and subsequently developed a balance that can be called a *single-pan, unequal-arm, two-knife-edge, substitution-weighing, constant-load balance*; the significance of these terms will become apparent from the discussion below. Such balances are usually provided with damping devices, with external control of the larger weights, with an accompanying digital scale, and with an optical scale for the small weights. Balances of this type have now largely replaced the equal-arm, double-pan balance and have changed weighing an object from a tedious and time-consuming art to a much more rapid procedure, requiring much less skill and experience.

The essential features of the single-pan balance are shown in Figure 11-4 where it is seen that there are only two knife-edges—a central and a terminal one. The terminal one supports both the single pan and the set of weights. A counterweight is rigidly fixed to the opposite side of the beam and this side of the beam is usually made longer than the one carrying the pan; by this means the total weight carried by the central knife edge is decreased.

The balance is initially adjusted to its zero point of rest with all weights suspended from the terminal knife-edge and the pan empty. Then the object to be weighed is placed upon the pan, and weights are removed from their suspension until the beam is again balanced; in effect, the object is substituted for the weights—hence the term *substitution weighing*. The beam always carries the same total load and therefore has a constant sensitivity, since any distortion is constant. This maximum constant load demands special structural characteristics in the beam and knife-edges. An equal-arm analytical balance may be designed for 160 g but in ordinary use will seldom carry a load of over 100 g and will most commonly carry much smaller loads. The beam of a constant-load balance designed for 160 g is always loaded at 160 g. For this reason the knife-edges and bearing plates of such balances are made from synthetic sapphire rather than the agate generally used with equal-arm balances.

The need for very small weights is avoided by the use of an optical pointer. Light passes through an optical scale that is attached to the end of the beam and the image of the scale is projected upon a ground glass screen at the front of the balance. This permits direct reading of the last 0.1 g or 1 g, depending upon the balance. This use of the optical system is possible only because of the constant sensitivity of the balance. The balance shown in Figure 11-4 is air-damped by means of a cylinder and piston arrangement.

Once again, it should be clearly understood that the modifications make the weighing of an object much less time-consuming but do not make that operation more precise or accurate. The readout on a modern constant-load balance is presented in such a clear and readable manner that we are tempted to believe that special trust can be placed in the numbers. We must remember that all the problems of weight calibration, friction, finger prints, unlevel balance, and air currents from thermal nonequilibrium are still present. Unfortunately, the calibration of the weights on a single-pan balance can be accomplished only with the use of a complete auxiliary set of calibrated weights.[3]

Constant-load balances lend themselves especially well to automatic processes. Some are now available that print the weight onto a gummed strip of paper, which can be placed directly in a notebook thus eliminating reading and recording errors. Others record the weights in punched cards for computer use. Taring devices are built into some balances, which permit the container to be balanced before the weighing of container-plus-sample so that the read-out shows only the sample's weight.

OTHER BALANCE SYSTEMS

Two other systems are used in the construction of modern balances. One of these makes use of torsional forces. Figure 11-5 shows a schematic diagram of a quartz-fiber torsion microbalance, which can weigh up to 25 mg with a precision of 0.02 μg. The same principle has been applied to macrobalances, in which special alloy fulcra replace the quartz fiber.

Another system uses an accurately measured electrical force to oppose the torque caused by the object being weighed. This system is especially easily adaptable to use with automatic data-processing devices.

ELIMINATION OF ERRORS IN WEIGHING

The assumption has been made above that when one places an object on the pan of a balance and then restores the zero point by adding weights to the

[3] When the constant-load balance was first introduced, there was doubt, apparently unjustified, as to its durability and resultant accuracy. A discussion of the relative merits of the two types of balances is given in the *Treatise on Analytical Chemistry* by Kolthoff and Elving (Part 1. Vol. 7. New York: Wiley–Interscience, 1967, p. 4247).

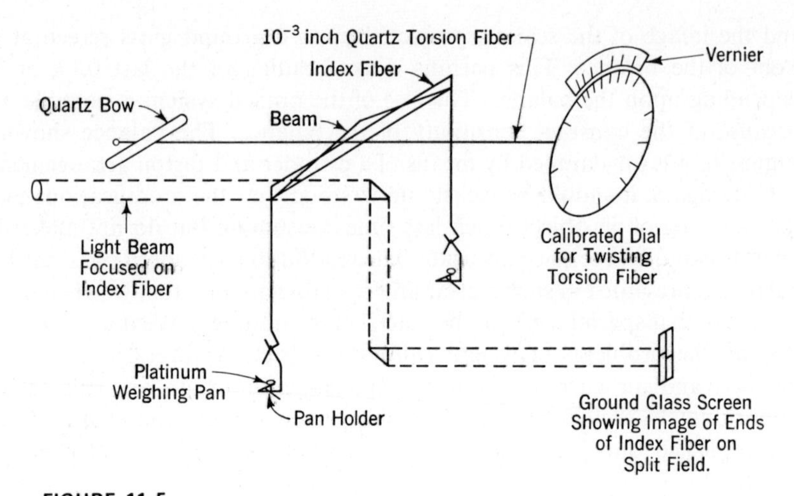

FIGURE 11-5

Schematic diagram of a torsion quartz fiber microbalance. Sensitivity, 0.02 μg; weighing range, 300 μg; capacity, 25 mg. (Courtesy *Chemical and Engineering News*.)

opposite pan of the equal-arm balance or removing weights from the single-pan balance, the mass of the object is given by the face values of the weights. There are three possible sources of error this assumption overlooks: (1) the arm lengths of the equal-arm balance may be unequal, (2) the weighing is usually made in air, a buoyant medium, and (3) the values assigned to the weights may be incorrect. These errors and methods for eliminating them are discussed below.

Correction for Unequal Arm Lengths

When an object is balanced by weights on a supposedly equal-arm balance the moments arising from the object and the applied weights are equal; that is,

$$Q \times L = R \times W,$$

where Q is the weight of the object, L is the length of the left arm, W is the applied weights, and R is the length of the right arm. Obviously, only when $L = R$ is the above assumption justified, and this may not be true within the desired experimental limits. Accordingly, for precise work, it is necessary either to determine the effect of this inequality and then correct for it, or to carry out the weighing in such manner that the effect is eliminated. The latter procedure is more commonly used and may be accomplished by either of two methods.

1. *Method of Substitution.* The method of substitution, commonly called *Borda's method*, consists in placing the object on one pan and balancing it with any suitable material T (commonly called a *tare*), preferably weights from an old set, and then removing the object and balancing the tare with precise weights W. From the principle of moments, the following relation applies to the first weighing:

$$QL = RT$$

and for the second,

$$WL = RT$$

from which it is seen that L and R can be eliminated from the equations and that $Q = W$.

2. *Method of Transposition or Double Weighing.* The method of transposition, commonly called *Gauss's method*, consists in placing the object, Q, on one pan and balancing it with weights, W_1, and then transferring the object to the other pan and again balancing with weights, W_2. For the first weighing,

$$QL = RW_1$$

and for the second,

$$W_2L = RQ$$

from which it is seen that

$$Q = \sqrt{W_1 W_2}$$

and the true weight is the geometric mean of the observed weights. Since the difference between W_1 and W_2 is relatively small, the arithmetical mean,

$$Q = \frac{W_1 + W_2}{2}$$

will give the true weight well within the desired experimental limits and is generally used. This latter method is usually more convenient, since it does not require an additional set of tare weights and is more generally used for precise weighings.

Correction for Buoyancy

A body immersed in a fluid is buoyed up by a force equal in magnitude to the weight of the fluid displaced. Since weighings in the laboratory are done in air, the question of buoyancy by the air must be considered.

It should be evident that on an equal-arm balance there will be no net buoyant force if the object and the weights are of the same density. The larger the difference in the densities of the object and the weights the greater will be the net buoyant effect. Consider that a sample having a density, d_s, is weighed on an equal-arm balance against weights of density, d_w. If d_s is smaller than d_w, which is usually the case, the vacuum weight, W_v, is greater than the observed weight, W_o, by the net buoyant force:

$$W_v = W_o + \text{net buoyant force}$$

This net buoyant force is the net weight of the air displaced and is the product of the density of the air, d_a, and the net volume of air displaced.[4] Hence,

$$W_v = W_o + d_a(V_s - V_w) \tag{13}$$

where V_s and V_w are the volumes of object and of weights. If the density of the sample is d_s and that of the weights is d_w, then V_s and V_w can be replaced by W_s/d_s and W_o/d_w. However, the second term on the right side of Equation 13 is very small and need not be known to the same relative accuracy in order to be added with the much larger first term. Therefore, $V_s \approx W_o/d_s$ is a good approximation for substitution into Equation 13. This equation now becomes

$$W_v = W_o + d_a\left(\frac{W_o}{d_s} - \frac{W_o}{d_w}\right) = W_o\left[1 + d_a\left(\frac{1}{d_s} - \frac{1}{d_w}\right)\right]$$

This easily derived equation permits us readily to calculate the vacuum weight or mass of a sample if the densities of weights and sample are known.

Although the density of air varies with barometric pressure, temperature, humidity, and carbon dioxide content, it is usually sufficiently accurate to assume that 1 ml of air weighs 0.0012 g. Weights for use on balances are commonly made of brass ($d = 8.3$ g/cm^3) or stainless steel ($d = 7.8$ g/cm^3). Because of the small relative magnitude of the term in which this density is used an average value of 8 g/cm^3 can ordinarily be used for the density of the weights.

On first consideration, the buoyancy problem may appear to be very different in the case of a constant-load balance, since weights are removed from the arm on which the sample is placed rather than added to the opposite arm. However, the large counterbalancing weight on the opposite arm of the balance may be imagined to be divided into two portions. One of these just balances the remaining weights, which hang on the stirrup along with the pan and sample. Since all the weights are of the same material, there is no net

[4] This is from the definition of density as the mass per unit volume; $d = m/v$.

buoyant force. The other portion of the counterbalancing weight balances the sample; unless the sample is of the same material as the counterbalancing weight, there is a buoyant force just as with the simple balance.

W. R. Burg and D. A. Veith (*J. Chem. Ed.*, **47**, 192, 1970) have derived equations showing the essential similarity of corrections for buoyancy for a single-pan balance and for an equal-arm balance. J. E. Lewis and L. A. Woolf (*J. Chem. Ed.*, **48**, 639, 1971) have pointed out a minor discrepancy that results from the method of weight adjustment in some single-pan balances.

Correction of Weights

Equal-Arm Balances. It is obvious that regardless of the type of balance, a weighing can be no more accurate than the values assigned to the weights involved. The weights in the sets used with equal-arm balances are not protected and may have been used, or misused, by others. Therefore, unless such weights have been specifically certified by the makers or by the Bureau of Standards, and their previous history known, they cannot be relied upon for precise work.[5] The process of determining the true value of each weight in a set is known as calibration. The principle involved in the method discussed on page 239 for this calibration is due to Richards[6] and consists in assuming a value for one of the smaller weights, or the rider, and obtaining the value of all the other pieces in the set in terms of the arbitrary standard. These relative values may then be used where the absolute weight of the object is not necessary. However, in order to obtain absolute weighings (as, for instance, in calibrating volumetric vessels in true metric units)[7] it is necessary to compare one of the larger pieces with a piece of known value, usually one certified by the Bureau of Standards. Once an absolute value has been established for this piece, the values of the remaining pieces can be calculated by means of the ratio of the absolute and relative values. In order to eliminate an accumulation of errors due to inequality of arm lengths, the weighings should be made by substitution or by double weighing.

Single-Pan Balances. The original limits of error of the weights of a single-pan balance are usually stated by the maker. In addition, the weights are enclosed and are not handled manually; therefore, they are much less subject to errors from misuse. Nevertheless, they should be checked periodically,

---

[5] A brief discussion of some causes of variability in weights is given by Craig (*Anal. Chem.*, **19**, 72, 1947).

[6] T. W. Richards. *J. Am. Chem. Soc.*, **22**, 144 (1900).

[7] Volumetric vessels can be calibrated with the relative values of a set of weights; the volume so obtained will not be in metric units but can be used for measurements that are based on the relative values of that particular set of weights.

especially if weighings of high accuracy or of a critical nature are to be made. A set of weights of known values is necessary for this purpose, and each weight is compared with one of the same value. A quick check can be made by checking the sum of all of the weights against the standard set. If there is satisfactory agreement of the sum of the weights it is unlikely that any single weight is in serious error.

USE OF THE ANALYTICAL BALANCE

Procedures are given below for both the equal-arm and the single-pan balance. Although the latter is rapidly displacing the former as an analytical tool, there are still many excellent equal-arm balances in service.

THE EQUAL-ARM BALANCE

There are four fundamental operations that are involved in the use of an equal-arm balance. These are (1) determination of the rest point, (2) determination of the sensitivity, (3) determination of the weight of an object, and (4) calibration of the weights. Procedures for performing these operations are presented below.

PROCEDURE 11-1 (A and B)

Determination of the Point of Rest of an Equal-arm Balance

Outline. Two methods are presented here for determining the point of rest. Procedure 11-1A uses the method of long swings; Procedure 11-1B uses the method of short swings.

DISCUSSION

The first operation in the use of the balance is the determination of the *point of rest* of the empty balance,[8] that is, the reading of the pointer when the beam assembly has been released and has been allowed to assume its equilibrium position. For convenience and brevity, this equilibrium position of the pointer with zero applied load will be designated the zero point of rest.[9]

The obvious method of determining the point of rest would be to set the beam in motion, allow it to come to rest, and then carefully observe the position of the pointer on its scale. Such a static method is obviously too time-consuming for practical use, and, in addition, small frictional effects might cause the beam to stop at other than its exact equilibrium position. Consequently, dynamic methods of determining the point of rest are used. The method most commonly described, which may be termed the "long-swing method," consists in setting the beam in motion so that the pointer swing covers from 5 to 10 divisions on the scale, and then making a series of readings of the extreme points of the swings. An odd number of readings, from three to five, are taken on one side, and an even number, from two to four, are taken on the other side. The average of each of these two sets is

[8] The procedures used for testing new balances are beyond the scope of this text. Reference to the following is suggested: Craig. *Ind. Eng. Chem.*, anal. ed., **11**, 581 (1939); Kreider. *Ind. Eng. Chem.*, anal. ed., **13**, 117 (1941); Corwin. *Ind. Eng. Chem.*, anal. ed., **16**, 258 (1944); MacNevin. *The Analytical Balance.* Sandusky, Ohio: Handbook Publishers, 1951. For suggestions regarding the maintenance of balances see Niederl and Niederl. *Micro-methods of Quantitative Organic Elementary Analysis.* (2d ed.) New York: Wiley, 1942.

[9] This position is sometimes called the "zero point," because the pointer scales of balances were often calibrated from zero in the center to 10 divisions on each side; the term "zero point" is also sometimes used to designate the equilibrium position of the pointer with zero load, and the "rest point" the equilibrium position with a load.

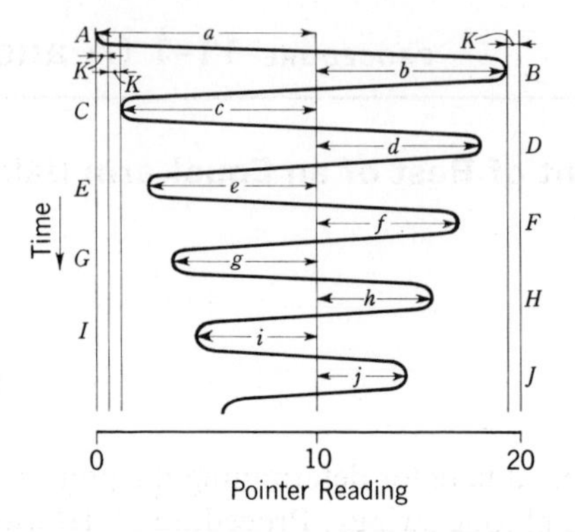

FIGURE 11-6
The effect of the decrement on the oscillations of a balance.

taken, and the point of rest is the mean of these two values. For example, if the successive readings are *a, b, c, d, e, f*, and *g* (see Figure 11-6), the point of rest would be

$$\text{P.R.} = \frac{\dfrac{a + c + e + g}{4} + \dfrac{b + d + f}{3}}{2}$$

By taking an odd number of readings on one side, the error due to the decrease in the amplitude of successive swings (called the *decrement*) caused by friction and air resistance is eliminated. This result will be understood by a consideration of Figure 11-6, where $A, B, C \ldots$ represent the swings of an ideal balance with no decrement, and $a, b, c \ldots$ represent the swings of a balance with a decrement K. The value of K can be assumed to be constant for the limited number of swings considered here, since its value is small in comparison to the amplitude of the swing. With no decrement it is obvious that the true point of rest would be the mean of the first two observations, thus:

$$\text{P.R.} = \frac{(A + B)}{2}$$

The real observation, b, is equal to $B - K$; therefore $(A + B - K)/2$ differs from the true point of rest by $-K/2$. However, if two swings to the left and one to the right are taken, we have, for the ideal balance,

$$\text{P.R.} = \frac{\dfrac{A + C}{2} + B}{2}$$

and for the real observations,

$$\text{P.R.} = \frac{\dfrac{A + C + 2K}{2} + B - K}{2}$$

in which K is eliminated and the true point of rest is obtained. The same treatment can be extended to a larger number of readings.

The method of long swings requires considerable time and computation, and in order to avoid this the so-called *short-swing method* of determining the point of rest may be used with a precision usually well within the experimental limits desired. In this method the swing is limited to one or, at the most, two divisions, and the point of rest is determined by visually estimating the center of the pointer swing; for very precise work the extremes of one of the short swings may be recorded and the mean taken. Since the pointer is moving much more slowly, the period of oscillation being constant, and since the decrement with such a short swing is hardly appreciable, the center can be very precisely placed.[10]

INSTRUCTIONS (Procedure 11-1: Determination of the Point of Rest of an Equal-Arm Balance. A. Method of Long Swings)

Examine the balance to ascertain that it is in proper operating condition (Notes 1, 2). Close the case. Release the pan rests by pushing in the button controlling them; this usually may be locked in by a slight turn. If the pans begin to swing, bring them to a stationary position by repeated momentary application of the pan rests. Now lower the beam by very slowly turning the beam-rest control in a counterclockwise direction as far as it will go (Note 3). By means of the rider control momentarily lower the rider to the beam near the center so that the pointer is caused to swing from five to six divisions to one side of the center. After one or two swings (Note 4) begin recording in the notebook (Note 5) the successive extreme positions of the pointer, taking five such readings, three on one side and two on the other (Notes 6, 7). Arrest the beam by a gradual application of the beam rest, taking special care that *the pointer is near the center of the scale* (that is, that the beam is in a horizontal position) when it is lifted. Finally, release the pan-rest control. From the readings so obtained calculate the zero point of rest of the balance (Note 8). Repeat the determination (Note 9).

---

[10] For a discussion of the relative merits of the long and short swing methods see Wells. *J. Am. Chem. Soc.*, **42**, 411 (1920).

NOTES

1. The balance should be situated in a position that is as free as possible from the corrosive fumes common to a chemical laboratory, from air movement, and from sudden changes in temperature, such as would be caused by an oven in the vicinity or by direct sunlight. Adequate illumination should be provided. The balance should be solidly mounted so as to avoid vibrations.

2. Before using the balance, examine it to see that it is level (a leveling instrument is usually provided in the case), that the beam and pan arrests are operating smoothly, and that the "rider" is properly controlled by its mechanism throughout the entire range of the beam calibrations.

Students should never attempt adjustments or repairs but should report the condition to the instructor. The balance should be kept scrupulously clean at all times; any evidence of any material spilled on the pans or in the case, or of corrosion on the pans, should be reported immediately. Even if apparently clean, the pans should be gently brushed off with a small soft camel's-hair brush (provided for that purpose only) before each series of weighings.

Keep the balance case closed at all times except when placing or removing objects or weights.

3. It cannot be too strongly emphasized that this release of the beam must be done with the utmost care, since by it the knife-edges with their loads are brought into contact with the bearing plates. The operator of a balance can tell by the feel as the beam control is turned at what point the knife-edges contact the bearing plates and must use special care to bring them into contact as gently as possible.

4. The first swings of the balance may be somewhat erratic because of air currents in the balance case or oscillation of the pans; therefore one should not immediately begin recording the swings but should wait until the extremes of the swings show an orderly decrement.

5. The instructions relating to notebooks on pages 162–163 should be read and carefully observed.

6. In order to avoid errors due to parallax, the observer should seat himself squarely in front of the case and attempt to have his line of sight perpendicular to the scale as the reading is made.

7. See the discussion above for the reason for taking an odd number of readings, or "an even number on one side and an odd on the other"; seven or even nine readings may be taken if desired.

8. In the recording of observations and measurements and in calculations involving observations and measurements, attention should be paid to the proper use of *significant figures*. Review the discussion on pages 24–26. Since the observation of the extreme of a swing is usually uncertain to approximately 0.2 of a division, it would be recorded to one decimal place and not beyond; an observation thought to be exactly 10 would be recorded as 10.0, not merely 10, and not 10.00.

9. The rest points obtained by a series of three or more such determinations should not show a maximum deviation of more than 0.2 of a division. A constant shift in the point of rest usually indicates uneven temperature conditions.

INSTRUCTIONS (Procedure 11-1: Determination of the Point of Rest of an Equal-Arm Balance. B. Method of Short Swings)

Release the pan and beam arrests and, if necessary, set the beam in motion by momentarily touching it near the center with the rider so that a swing of not over 1 to 1.5 divisions is obtained (Note 1). Determine the point of rest by visually estimating the central point of these swings. Record this value of the "zero point of rest" of the balance and, without arresting the beam, repeat the determination by recording the extremes of a swing and taking the mean. Raise the beam and release the pan rests. Repeat the determination (Note 2).

NOTES

1. Usually, unless the point of rest is exactly at the center of the scale, a swing of from 0.5 to 1 division can be obtained without the use of the rider by carefully releasing the beam.
2. The values obtained by this method should be compared with those obtained by the method of long swings. It is recommended that the discussion of the relative merits of the two methods by Wells (*J. Am. Chem. Soc.*, **42**, 411, 1920) be read. From the data obtained, the analyst can draw his own conclusions as to the relative precision of the two methods and decide which method he is justified in using.

PROCEDURE **11-2**

Determination of the Sensitivity of an Equal-arm Balance

Outline. The response of the balance to an applied unbalance of 1 mg is determined as a function of the total load on the balance. A plot of sensitivity versus total load is prepared from these data.

DISCUSSION

The structural factors influencing the sensitivity of the balance have been discussed above. It is advantageous to know the sensitivity of the balance at various loads, because this knowledge enables one to estimate the weight to add or subtract in order to restore the original point of rest when making weighings, because it is useful in making weighings by the sensitivity method (see Procedure 11-3B), and because it indicates the maximum load that the balance should carry. It is generally stated that a balancc should not be made to carry an excess of weight over that which causes the sensitivity to decrease to 40% of its maximum value. The conventional analytical balance is usually designed to carry a maximum of 100 g or, for better instruments, 200 g on each pan. Obviously, the above criterion cannot be applied to the somewhat rare case in which a balance shows a continued increase in sensitivity with loading. A convenient sensitivity for making weighings to 0.1 mg is from 2 to 5 divisions per milligram.

INSTRUCTIONS (Procedure 11-2: Determination of the Sensitivity of a Balance)

Determine and record the zero point of rest of the balance. Place upon the center of one of the pans (Note 1) a 1-mg weight, or if a 1-mg weight is not provided in the set of weights being used, place the rider upon the division on the beam indicating 1 mg (Note 2). Determine the point of rest with this applied load. The number of scale divisions through which the pointer has been deflected is the sensitivity of the balance with zero load.

Repeat the determination with 10-, 20-, and 50-g loads on each pan (Note 3). Construct a curve in the notebook by plotting the sensitivity as ordinate against the load as abscissa.

NOTES

1. When weighings are made, the object is usually placed on the left pan, since the manipulation of weights is more conveniently made upon the right pan.

2. Since balance beams are calibrated differently, care should be taken that the weight of the rider used should correspond to the value of the calibration directly above the terminal knife-edge; the exact weight of the rider should also be checked.

3. The point of rest with weights of equal face value on each pan may not agree with that of the empty balance, owing to inequality of length of the arms of the balance or small errors in the weights. If the pointer is deflected from the central portion of the scale with the heavier loads because of these effects, it should be brought back by adjusting the rider before the initial point of rest is determined.

PROCEDURE **11-3 (A and B)**

Weighing an Object on an Equal-arm Balance

Outline. Two methods of weighing are provided. In Procedure 11-3A weights are added until the zero point of rest is restored. In Procedure 11-3B weights are added until the total is within 1 or 2 mg of the required mass; then the weight that would need to be added or removed is calculated from the sensitivity of the balance.

DISCUSSION

The simplest method of weighing an object would be to determine the zero point of rest of the balance, then to place the object on one pan, and by means of adding weights to the other pan and adjusting the rider, to restore the point of rest to the zero value. This is the method commonly used by chemists for routine work and is employed in Procedure 11-3A, below. An alternative method, which avoids having to make the weight adjustment to the exact mass of the unknown object, known as the method of *weighing by sensitivity*, is often resorted to, especially when making precise weighings. In this method the weights are applied until the total is within 1 or 2 mg of the required amount, and the point of rest is determined. Then, if the zero point of rest and the sensitivity of the balance are known, the weight to be added or subtracted in order to restore the zero point of rest can be calculated. In making weighings by this method, it is convenient to have the sensitivity expressed in milligrams per division of the pointer scale. Since the sensitivity of the balance may vary with the load, it is better practice for precise work to determine the sensitivity under the actual load conditions by again taking a point of rest after adding or subtracting a milligram from the load on the pan. This method is employed in Procedure 11-3B. When weighing by sensitivity you should avoid using the extreme outer portions of the pointer scale. Theoretically the sensitivity decreases with increase of the angle of deflection; in addition, the effects of irregularities of knife-edges and plates become more pronounced.

The above methods have assumed that the arms are of equal length and have neglected the buoyant effect of the air on the weights and objects.

Means for eliminating these sources of error have been discussed on page 224 and page 225.

INSTRUCTIONS (Procedure 11-3: Weighing an Object on an Equal-Arm Balance. A. Method of Restoring the Original Point of Rest)

Obtain an object of known weight from the instructor.[11] Determine the point of rest of the empty balance, or zero point of rest (Procedure 11-1). Place the object in the center of the left pan (Note 1) and place on the center of the right pan a weight thought to be approximately equal to that of the object. Release the pan rests. If there is any swinging of the pans, arrest this motion by successive application of the pan rests (Note 2). Slowly lower the beam until the first motion of the pointer indicates whether the applied weight is too large or too small. Then systematically add or remove weights, arresting the beam and pans between each change and taking care to prevent oscillation of the pans, until the object is balanced within the range of the rider (Note 3). Close the balance case and visually estimate the point of rest by noting the extremes of the first full swing. From this approximate value of the point of rest and the value of the zero point of rest, make a mental calculation of that position of the rider that will restore the point of rest to the zero value. Shift the rider to this calculated position and repeat this process until a final point of rest is obtained that is coincident with the zero point of rest within the limits of accuracy desired for the weighing (Note 4). Raise the

NOTES

1. Extreme care must be taken to make sure that any object placed directly on the pan of the balance is perfectly clean and dry. If there is cause for doubt, the object should be wiped with a clean lintless cloth or with a chamois skin kept for the purpose.
2. The pans must remain motionless before the beam is released; otherwise, erratic swings of the pointer will be obtained. Sidewise oscillation of the pans is minimized by placing objects and grouping weights as near the center as possible.
3. Always add the weights in the order in which they occur in the weight box, beginning with the largest apparently required and then in decreasing order of mass.
4. It is usually a waste of time to attempt to restore the point of rest to the *exact* zero value. Assuming that one wishes to weigh to 0.1 mg and that the balance has a sensitivity of 4 divisions per milligram, an agreement of the points of rest to within

---

[11] Weights with removable tops (so that their weight can be adjusted at will) make excellent objects for this determination.

beam and release the pan arrests. Record in the notebook the individual weights on the pan and the value of the rider. Check this list against the weights missing from the box. Return the weights to the box, again checking the original list. Add the total value of the weights (Note 5).

INSTRUCTIONS (Procedure 11-3B: Weighing an Object on an Equal-Arm Balance. B. Method of Using the Balance Sensitivity)

Follow Procedure 11-3A until the point of rest has been restored to within the limits of the pointer scale (preferably within five divisions of the zero point of rest). Determine and record the point of rest. Shift the rider by 1 mg in that direction, which will cause the point of rest to come closer to the center of the scale. Determine and record this point of rest. Lift the beam and release the pan arrests.

The difference between these two points of rest represents the sensitivity of the balance under the existing load conditions. By the use of this sensitivity value, calculate the weight to be added or subtracted in order to restore the original point of rest. Add this value to, or subtract it from, the value of the applied weights and rider in order to obtain the weight of the object.

NOTES

0.4 division is all that is required. The instructor should state the accuracy desired in the above weighing.

One can save time in making weighings by considering relative errors. The errors in most analytical procedures are such that there is no justification for reducing the weighing error to less than 1 part per 1000; this value is attained if an object weighing 1 g is weighed to 1 mg. However, if a crucible weighing 10 g is to be used to contain a 0.2-g precipitate, the weight has to be within 0.2 mg in order to determine the weight of the precipitate to within 1 part per 1000.

5. If the weights have been calibrated, the correction for each weight should be listed; these corrections should then be totaled algebraically and added to the total face value.

Calibration of a Set of Weights

A method for the calibration of a set of weights is presented in abbreviated form here; the references cited in Footnotes 6 and 12 should be consulted for detailed instructions.

The method of Richards (page 227) consists in assuming or assigning a value for one of the smaller weights in the set, and obtaining the values for all other pieces in the set relative to this arbitrary standard. As was mentioned, this method can lead to a serious accumulation of errors if the balance arms are of unequal length; therefore the weighings should all be done by either substitution or double weighing. The latter method avoids the use of a second set of weights, and the modification suggested by Weatherill[12] is particularly well adapted for the calibration process. This method can be illustrated as follows. Suppose that the two pieces W_1 and W_2 are to be compared; W_1 is placed on the left pan and W_2 on the right, and the point of rest is determined. The pieces are exchanged, and the difference in weight, Δw, necessary to restore the first point of rest is determined. For the first point of rest,

$$W_1 L = R W_2$$

and for the restored point of rest,

$$W_2 L = R(W_1 + \Delta w).$$

By properly combining these equations it is seen that

$$W_2 - W_1 = \frac{\Delta w R}{L + R},$$

and since Δw is small, it can be safely assumed that L and R are equal; therefore

$$W_2 - W_1 = \frac{\Delta w}{2}$$

If an arbitrary value has been assigned to W_1, then $W_2 = W_1 + (\Delta w/2)$. If weight has to be added to W_1 to restore the first point of rest, Δw is positive; if weight has to be removed, Δw is negative.[13]

---

[12] Weatherill. *J. Am. Chem. Soc.*, **52**, 1938 (1930).

[13] It is to be noted that this method is fundamentally the same as double weighing. In the equation (page 225).

$$Q = \frac{W_1 + W_2}{2}$$

W_2 is equal to $W_1 + \Delta w$, and therefore

$$Q = W_1 + \frac{\Delta w}{2}$$

TABLE 11-1

Calibration of a Set of Weights

1 Face Value	2 Weights on Left Pan	3 Weights on Right Pan	4 Point of Rest	5 $\Delta w/2$ (mg)	6 Rel. Value (Based on 5-mg wt.)	7 Absolute Value	8 Corr. (mg) (Rounded)
0.005	0.005	rider	10.2		0.00500	0.00501	
rider	rider	0.005	9.1	−0.11	0.00489	0.00490	−0.1
0.01	0.005, rider	0.01	8.7				
	0.01	0.005, rider	10.5	+0.18	0.01007	0.01008	+0.1
0.01*	0.005, rider	0.01*	9.1				
	0.01*	0.005, rider	10.2	+0.11	0.01000	0.01002	—
0.02	0.01, 0.01*	0.02	9.5				
	0.02	0.01, 0.01*	10.0	+0.05	0.02012	0.02015	+0.2
0.05	0.02 + Σ[1]	0.05	10.4				
	0.05	0.02 + Σ	8.8	−0.16	0.04992	0.04998	—
0.1	0.05 + Σ	0.1	10.5				
	0.1	0.05 + Σ	8.9	−0.16	0.09984	0.09997	—
0.1*	0.05 + Σ	0.1*	10.4				
	0.1*	0.05 + Σ	9.0	−0.14	0.09986	0.09999	—
0.2	0.1, 0.1*	0.2	9.7				
	0.2	0.1, 0.1*	9.7	0	0.19970	0.19995	—
0.5	0.2 + Σ	0.5	9.7				
	0.5	0.2 + Σ	9.4	−0.03	0.49937	0.50000	—
1	0.5 + Σ	1	9.9				
	1	0.5 + Σ	9.1	−0.08	0.99869	0.99996	—
1*	1	1*	9.7				
	1*	1	9.7	0	0.99869	0.99996	—
1**	1	1**	9.7				
	1**	1	9.7	0	0.99869	0.99996	—
2	1, 1*	2	6.7				
	2	1, 1*	12.4	+0.57	1.99795	2.00049	+0.5
5	2, 1, 1*, 1**	5	10.8				
	5	2, 1, 1*, 1**	7.4	−0.34	4.99368	5.00003	—
10	5, 2, 1, 1*, 1**	10	10.1				
	10	5, 2, 1, 1*, 1**	6.3	−0.38	9.98732	10.00002	—
10*	10	10*	7.7				
	10*	10	8.7	+0.10	9.98742	10.00012	+0.1
20	10, 10*	20	7.0				
	20	10, 10*	6.3	−0.07	19.97467	20.00007	+0.1
50	20 + Σ	50	4.7				
	50	20 + Σ	0.5	−0.42	49.93669	50.00019	+0.2
1	1 standard	1	9.8				
	1	1 standard	9.4	−0.04	—	0.99996	—

[1] Σ indicates the sum of the smaller weights of the series, in this case the 0.01-, 0.01*-, and 0.005-g weights and the rider.

This method is illustrated by the actual data taken in calibrating a set of weights and shown in Table 11-1. The 0.005-g weight was taken as the arbitrary basis on which to obtain the relative values. The sensitivity of the balance did not vary appreciably from 0.20 mg per division throughout the load applied. It should be noted that the arms of the balance used for obtaining these data were not exactly equal, the effect caused by this becoming quite appreciable at the heavier loads. Column 1 shows the face value of the weight being compared, Columns 2 and 3 the weights on each pan, and Column 4 the point of rest with the weights in the original and transposed positions. Column 5 shows $\Delta w/2$ (in milligrams), and in Column 6 the values of the pieces on the basis of the 0.005-g weight are summed up. After the 1-g piece has been compared with a 1-g weight from the Bureau of Standards, the absolute values of the other pieces can be calculated by correcting each of the pieces by the ratio of the absolute weight to the relative weight, that is, 0.99996:0.99869, and the values shown in Column 7 are obtained.[14] For convenience in making weighings, the value of the correction in milligrams to be applied to the face value of each weight to give its absolute weight has been rounded off and is shown in Column 8.

GENERAL RULES FOR THE USE OF BALANCES

(The rules regarding weights apply to equal-arm balances only.)

1. The balance should be isolated from other laboratory operations and placed on a vibration-free support.
2. The balance should be level; check the indicator periodically.
3. Before using your balance, see that it is clean, free of dirt or corrosion, and is operating normally. Report any abnormal condition of balance or weights to your instructor. *Once you begin use of the balance you are responsible for its condition.*
4. Keep the balance case and pans clean. Brush off pans with camel's-hair brush before making a weighing.
5. Sit directly in front of the balance to avoid error from parallax.
6. Release the beam with a gradual steady motion.

---

[14] The labor involved in the calculations required for the calibration can be greatly reduced as follows. In the data above we find that the relative value for the weight of face value 1 g is 0.99869, whereas the absolute value is 0.99996; therefore for this 1-g weight a correction of +0.00127 g has to be applied to the relative value in order to obtain the absolute value. Since the relative values are consistent among themselves, the correction to be added to any other weight can be assumed to be +0.00127 g multiplied by the ratio of the face value of that weight to that of the 1-g weight: in the case above, the correction to be added to the relative value of the 0.5-g piece will be $+0.00127 \times 0.5/1$, which is 0.00063. The advantage of applying a ratio to the correction rather than to the relative value is that only the same number of significant figures have to be carried in the ratio as are necessary in the correction. For a discussion of the assumptions involved in the calibration process, see the articles by Hurley (*Ind. Eng. Chem.*, anal. ed., **9**, 238, 1937) and by Blade (*Ind. Eng. Chem.*, anal. ed., **11**, 499, 1939).

7. Arrest the swing of the beam only when the pointer is at the center of the scale. Always raise the beam before leaving the balance.

8. Do not leave weights or objects on pans after weighing. *Keep the case closed, except when making weighings.*

9. Place weights and objects as near as possible to the center of the pan. Damp oscillation of the pans before determining the point of rest.

10. As a general rule, substances should not be placed directly upon the pans. No liquids, unless in stoppered bottles, are to be brought into the balance case.

11. Allow heated objects to cool to balance-room temperature before weighing.

12. Do not vigorously rub glass or plastic objects (possibly producing an electrostatic charge) just before weighing. Use counterpoises for large glass objects.

13. Ascertain rated capacity of the balance, or determine it from the change in sensitivity, and do not exceed it.

14. *Record weighings at once in your notebook.* Do not carry scraps or loose sheets of paper into the balance room.

15. Report any damage to the balance to the instructor at once. Do not attempt to make adjustments or repairs yourself.

GENERAL RULES FOR THE USE OF WEIGHTS

1. Keep the weight box closed except when making weighings.

2. Periodically inspect and check weights.

3. Use only your assigned weights. Handle the weights with the ivory-tipped forceps provided for that purpose; never use your hands.

4. When making successive weighings, use the same larger weights whenever possible.

5. *Triple-count the weights to avoid error*: (1) count the weights on the pan; (2) count the spaces in the box; and (3) check the weights as they are removed from the pan.

THE SINGLE-PAN BALANCE

The general features of the single-pan balance are discussed on pages 222–223, and its mechanical construction is shown in Figure 11-4. Because several manufacturers make single-pan balances with various methods of control, the details of operation for the specific balance being used will be given by your instructor or can be obtained from the manufacturer's manual. There are, however, certain general instructions and operations given below that are applicable to any of the single-pan balances.

Weighing an Object on a Single-pan Balance

Outline. An unknown object is weighed in order to obtain practice in making preliminary checks and adjustments, and in making an accurate weighing.

INSTRUCTIONS

(Read the general rules for the use of balances, pages 241–242. Read Notes 1, 2, 3, 5, and 6 of Procedure 11-1A)

Pre-Weighing Checks

Check to see that the balance is level; if not, use the foot-screws to adjust it. Ascertain that the beam is arrested. Gently brush the pan with the camel's-hair brush (to be used for this purpose only). Ascertain that all weight-control knobs are set at zero. (Report any evidence of pan corrosion or malfunctioning of the balance to your instructor.)

Zero-Point Adjustment

With all weight-control knobs set at zero, no weight on the pan, and the doors closed, carefully rotate the arrest knob to the fully released position (Note 1). Then rotate the zero adjustment knob until the optical scale reads zero (Note 2).

Weighing an Object

Obtain an object to be weighed. With the balance arrested place the object carefully in the center of the pan by means of tongs, forceps, or a band of paper. Close the doors of the balance. Partially release the beam. Rotate the weight-control knobs gently in the following order: First rotate the 10-g control adding weight in 10-g increments until the optical scale indicates overweight (Note 3).

Now rotate the dial back one position (to remove 10 g). Next rotate the 1-g control (adding weight in 1-g increments), again until the optical scale indicates overweight. Rotate this control back one position (to remove 1 g). Arrest the beam, then set it in the *fully released* position.

When the optical scale has reached its equilibrium position rotate the fine-control knob until, depending upon the balance being used, the digital or micrometer read-out is properly adjusted. Record the weight directly in the notebook in ink. Carefully recheck your observation.

Arrest the balance. Remove the object from the pan. Close the doors of the weighing chamber and adjust all controls to zero.

NOTES

1. This release of the beam should be done with care, since by it the knife-edges with their loads are brought into contact with the bearing plates. Remember that this type of balance is always fully loaded. The operator can tell by the optical scale at what point contact is made and should rotate the control with special care until this point is reached.

2. The zero-adjustment knob rotates a prism set in the path of the beam of light that passes through the reticle at the rear of the beam and that is projected on the optical scale. It permits small adjustments of the zero point. If the optical scale cannot be adjusted to zero, the balance should be arrested and the level of the balance adjusted carefully. If there is still difficulty in adjusting the balance to zero, the instructor should be consulted. *Before each use of the balance the zero point should be rechecked. At the conclusion of each use the balance should be arrested, and all of the knobs should be returned to zero.*

3. With a single-pan balance, weights are actually removed from the front knife-edge; the dial reads the weight removed. With this type balance, weights equal to the weight of the object are removed in order to restore the original point of rest.

QUESTIONS AND PROBLEMS

1. Derive the relation $\tan \alpha = wL/DQ$, where α is the angle of rotation of the beam of a balance from its horizontal position, w is a small weight applied to one side, L is the arm length, D is the distance from the point of support to the center of gravity and Q is the weight of the beam system.

2. Derive the equation $S = LP/(QD - 2WD')$ for the case where the terminal knife-edges are above the central knife-edge.

3. What conclusions regarding the construction and condition of balances A and B can be drawn from the following observations?
   (*a*) Balance A has a sensitivity of 1.8 div./mg with no load, and 2.5 div./mg with a 20-g load.

(*b*) Balance *B* has a sensitivity of 3.6 div./mg with a 1-g load, and 1.4 div./mg with a 50-g load.

4. Three balances and the accompanying sets of weights were tested by determining (*a*) the point of rest of the empty balance, (*b*) the point of rest with the weight marked "10" (g) on the left pan and the one marked "10*" on the right, and (*c*) the point of rest with these weights transposed. The data obtained were tabulated as follows:

Balance	Point of Rest		
	Empty Balance	10 on Left, 10* on Right	10* on Left, 10 on Right
C	9.8	10.7	8.6
D	10.2	11.2	9.2
E	9.9	10.6	10.3

What conclusions can be drawn regarding these balances and weights?

5. The distance from the central knife-edge to the left terminal knife-edge of a balance is 10.0003 cm, and the distance from the central knife-edge to the right terminal knife-edge is 9.9998 cm. The apparent weight of a weighing bottle on the left pan is 24.7562 g. What is its correct weight? *Ans.* 24.7549 g.

6. In constructing a balance, the two terminal knife-edges are mounted in the beam 8 cm apart. It is desired that the error of the balance not exceed 0.5 mg with a 50-g load (on each pan). What is the maximum distance from the center that the central knife-edge can be placed? *Ans.* 0.00002 cm.

7. The point of rest of an unloaded balance was 9.8; with a weight of exactly 100 g on each pan, it was 6.4; when a 1-mg weight was added to the left pan, it became 8.9. Calculate the ratio of the arm lengths, L/R, of the balance. *Ans.* 0.999986.

8. A 10-mg rider is being used on a balance having its right arm calibrated in 12 equal divisions, the twelfth calibration being exactly above the terminal knife-edge. With a crucible on the left pan, the original point of rest is restored when weights totaling 24.2400 g are placed on the right pan and the rider position is read as 6.25. What is the crucible weight?

9. The specifications and characteristics of a balance are as follows: weight of beam assembly, 420 g; beam length, 18 cm; pointer length, 20 cm; divisions on pointer scale, 0.1 cm; sensitivity, 4 div./mg; knife-edges in same horizontal plane. Calculate the distance from the central knife-edge to the center of gravity of the beam assembly.

10. A 100-ml volumetric flask is balanced by a tare and then filled to the mark with water at 20°C. It requires 99.902 g of brass weights with a specific gravity of 8.3 to balance the water. What is the weight of the water (*in vacuo*)? (Assume that 1 liter of air weighs 1.2 g.) What is the volume of the flask in milliliters? *Ans.* 100.188 ml.

11. What would be the percentage error caused by failure to correct for buoyancy when weighing out the following substances? (*a*) Sodium oxalate ($Na_2C_2O_4$), (*b*) benzoic acid (C_6H_5COOH), and (*c*) copper foil.

12. The data shown below were obtained in the calibration of a set of weights. Prepare a table similar to Table 11-1 page 240, and complete Columns 5, 6, 7, and 8.

Data Obtained in the Calibration of a Set of Weights

1 Face Value	2 Weights on Left Pan	3 Weights on Right Pan	4 Point of Rest	1 Face Value	2 Weights on Left Pan	3 Weights on Right Pan	4 Point of Rest
0.005 rider	0.005	rider	10.0	1	0.5 + Σ	1	10.3
	rider	0.005	10.1		1	0.5 + Σ	9.5
0.01	0.005, rider	0.01	10.1	1*	1	1*	10.0
	0.01	0.005, rider	10.0		1*	1	9.8
0.01*	0.005, rider	0.01*	10.1	1**	1	1**	9.8
	0.01*	0.005, rider	10.2		1**	1	10.0
0.02	0.01, 0.01*	0.02	9.8	2	1, 1*	2	9.8
	0.02	0.01, 0.01*	10.2		2	1, 1*	9.8
0.05	0.02 + Σ[a]	0.05	10.2	5	2, 1, 1*, 1**	5	7.5
	0.05	0.02 + Σ	9.6		5	2, 1, 1*, 1**	11.7
0.1	0.05 + Σ	0.1	10.2	10	5 + Σ	10	9.4
	0.1	0.05 + Σ	9.7		10	5 + Σ	8.5
0.1*	0.05 + Σ	0.1*	10.2	10*	10	10*	7.4
	0.1*	0.05 + Σ	9.7		10*	10	10.2
0.2	0.1, 0.1*	0.2	10.2	20	10, 10*	20	10.2
	0.2	0.1, 0.1*	9.8		20	10, 10*	9.8
0.5	0.2 + Σ	0.5	10.0	50	20 + Σ	50	10.2
	0.5	0.2 + Σ	9.8		50	20 + Σ	9.6
				1	1 standard	1	10.0
					1	1 standard	9.6

NOTE: Sensitivity of the balance: 2.2 div./mg for the loads applied.

[a] Σ indicates the sum of the smaller weights of the series.

CHAPTER **12**

General Principles of Gravimetric Measurements

In Chapter 11 we discussed the measurement of mass, which is the fundamental measurement of a gravimetric method. A gravimetric analysis usually involves weighing the sample, dissolving it in an aqueous solvent, separating the constituent of interest as a relatively insoluble precipitate, drying or heating until an element or compound of known and definite composition is obtained, and then weighing that substance.[1] In the development of a gravimetric method certain factors have to be considered. These may be classified into groups as they affect the following:

I. The solubility of the precipitate.
II. The physical characteristics of the precipitate.
III. The composition and purity of the filtered precipitate.
IV. The composition and stability of the weighed precipitate.

[1] There are some gravimetric methods that do not involve the formation of a precipitate. For example, the loss in weight after heating can be used to determine volatile constituents such as water of hydration and carbonate content. In other methods a constituent can be volatilized, collected by a suitable absorbent, and the gain in weight determined. The carbon content of an organic compound is often determined by burning it in a stream of oxygen and collecting the carbon dioxide.

I. FACTORS AFFECTING THE SOLUBILITY OF THE PRECIPITATE

The factors affecting the amount of a given constituent that will remain in the filtrate when the precipitate is filtered may be classified as follows:

1. Common ion and activity effects (the solubility-product principle).
2. Formation of complex ions.
3. Hydrogen (and hydroxide) ion effects.
4. Effects of the solvent.
5. Temperature effects.
6. Effect of time on the completeness of precipitation (supersaturation and rate effects). These factors are discussed briefly below.

1. Common Ion and Activity Effects (the Solubility-Product Principle)

It is a fundamental necessity for the development of a gravimetric method that the precipitate be so insoluble (under the conditions of the precipitation) that the amount left in the saturated filtrate and in the wash solution be negligible in comparison with the other errors of the method (or, as is possible in certain cases, that a reliable correction can be made for the amount so lost). In some cases calculations can be made from solubility data as to whether the amount of the compound lost in the filtrate and wash solution will be within the accuracy that is desired, or whether it can be reduced to the desired limits by the addition of a common ion; for making such calculations, the solubility-product principle is employed.

This principle has been discussed in Chapter 4, where it was applied in predicting the course of certain precipitation reactions, and reference should be made to that discussion. There it was pointed out that the solubility-product principle (and the mass-action law in general) is obeyed experimentally only when the activities of the substances are used instead of their concentrations, or, if the latter are used, only under the following limitations: (1) with quite insoluble precipitates, (2) when the total ion concentration of the solution is low, and (3) with salts yielding singly charged ions. In Table 12-1 the experimental data of Jahn[2] have been used to show (1) the solubility of silver chloride in dilute solutions of potassium chloride and (2) the constancy of the calculated solubility-product value when the assumption is made that the potassium chloride is completely ionized and that the activities of the ions are equal to their formal concentrations. Note that the values for the solubility product agree reasonably closely in solutions which are below

[2] Jahn. *Z. physik. Chem.*, **33**, 454 (1900).

TABLE 12-1
Solubility of Silver Chloride in Potassium Chloride Solutions

KCl (moles/liter)	Ag (moles/liter)	K_{sp}
0.00670	1.7×10^{-8}	1.14×10^{-10}
0.00833	1.39×10^{-8}	1.16×10^{-10}
0.01114	1.07×10^{-8}	1.19×10^{-10}
0.01669	0.738×10^{-8}	1.23×10^{-10}
0.03349	0.388×10^{-8}	1.30×10^{-10}

0.01 F in KCl. This agreement is to be compared with the values (shown in Figure 4-2, page 50) for the solubility of silver sulfate, a more soluble salt and one containing a binegative ion, where—especially in the presence of an excess of sulfate ion—the solubility varies considerably more from the predicted value; in fact, *the common ion causes very little decrease in the solubility of the salt.* As previously stated, these deviations are due to the attractive forces existing between the ions in the solution, and these forces are responsible for the marked increase in the solubility of silver sulfate in solutions of salts not having a common ion. An indication of the magnitude of the solubility increase caused by these forces can be obtained from the activity coefficients in Table 4-1, page 52. For example, at ionic strength 0.01 the activity coefficients for Ag^+ and Cl^- are both 0.90. The solubility product for AgCl is

$$a_{Ag^+}a_{Cl^-} = 1.8 \times 10^{-10} \tag{1}$$

where a_i represents the activity of the ith species. As noted in Chapter 4, $a_{Ag^+} = \gamma_{Ag^+}[Ag^+]$ so, substituting into Equation 1 we obtain

$$\gamma_{Ag^+}[Ag^+]\gamma_{Cl^-}[Cl^-] = 1.8 \times 10^{-10} \tag{2}$$

The formal solubility of the silver chloride in a solution of ionic strength 0.01 is calculated from Equation 2 to be $1.4_9 \times 10^{-5}$, 11% higher than the $1.3_4 \times 10^{-5}$ calculated for activity coefficients equal to unity.

2. Formation of Complex Ions

A still different effect from those mentioned above is shown in Figure 12-1 where data of Forbes and of Jonte and Martin[3] for the solubilities of silver

[3] Forbes. *J. Am. Chem. Soc.*, **33**, 1939 (1911); Jonte and Martin. *J. Am. Chem. Soc.*, **74**, 2052 (1952).

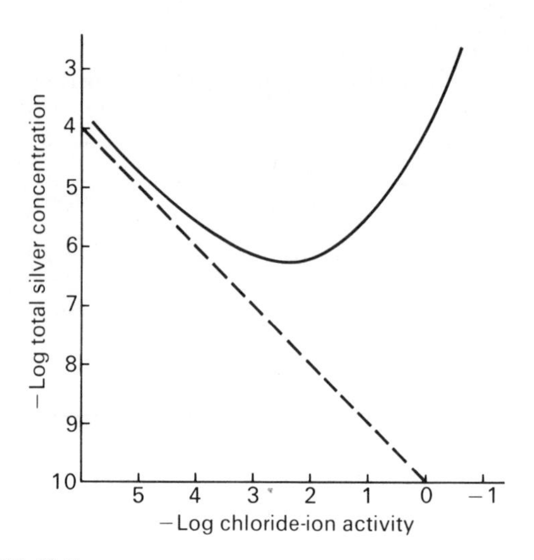

FIGURE 12-1

The solubility of silver chloride as a function of the chloride-ion activity. (Data from Jonte and Martin and Forbes.)

chloride in solutions of various chloride concentrations are plotted. The dotted line in the figure is the plot which would have been obtained if the only silver-containing species in the solution were Ag^+. If this were the case, then the solubility product,

$$[Ag^+][Cl^-] = K_{sp}$$

would permit one to calculate not only the silver-ion concentration in equilibrium with a given chloride-ion concentration but also the total silver in solution. In fact, however, the total silver in solution is made of Ag_2Cl^+, Ag^+, AgCl(aq), $AgCl_2^-$, $AgCl_3^{2-}$, and $AgCl_4^{3-}$. It should be noted that the plot is logarithmic and that strikingly large deviations from the solubility predicted by the solubility product are encountered. For example, if the activity of chloride ion is unity, the total silver in solution is approximately a million times higher than that calculated from the solubility product. Thus it is seen that if these effects are ignored, calculations on the basis of the solubility-product principle would lead to erroneous and misleading predictions. Furthermore, the *haphazard addition of a large excess of a precipitant may defeat the purpose sought*; it is entirely possible that a large excess of even a common ion may cause an increase in the solubility of the precipitate.

One must not conclude that the solubility-product principle does not apply to cases where complex formation exists, but merely that the additional equilibria involved in their formation must also be considered. One should note that, since in the large majority of cases complex ions are formed as the result of the addition of an excess of a negative ion to a positive one, less pronounced effects will follow the addition of an excess of a positive ion as precipitant to a negative ion than the reverse. Thus, the solubility of silver chloride in 2 *F* silver nitrate is only 5×10^{-5} *F*, which is greater than it is in water but is less than one-tenth that in 2 *F* sodium chloride solution.

Other obvious applications of the tendency of certain elements to form complex ions are demonstrated in the solubility of silver chloride and of copper oxide in the presence of ammonia (because of the formation of the ammine ions of silver(I), such as $Ag(NH_3)_2^+$, and of copper(II), such as $Cu(NH_3)_4^{2+}$). Where the values of the dissociation constants of these compounds have been determined and are available, calculations can be made as to the effect of such compound formation on the solubility of a precipitate. Thus, since the value for the dissociation constant, K_D, for the expression

$$\frac{[Ag^+][NH_3]^2}{[Ag(NH_3)_2^+]} = K_D$$

has the value 6.8×10^{-8}, it can be predicted that silver chloride will be very soluble in a solution 0.5 *F* in ammonia, but that silver iodide will be relatively insoluble and that a separation of iodide from chloride can be thus effected; this separation is an experimental fact.

3. Hydrogen (and Hydroxide) Ion Effects

The effects of the hydrogen-ion concentration on the solubility of compounds are due to two types of reactions, namely, neutralization and displacement. The general principles underlying these reactions have been discussed in Chapter 7 in connection with acid–base titrations. Because of reactions of the first type, those substances, such as hydroxides, basic oxides, and basic salts, that ionize in aqueous solutions to produce hydroxide ion become more soluble as the hydrogen-ion concentration is increased.

Because of reactions of the second type, formalized as

$$H^+ + A^- = HA$$

all salts of weak acids become more soluble as the hydrogen-ion concentration is increased. The adjustment of the hydrogen-ion concentration is also of importance in gravimetric precipitations because it often affects the physical character of the precipitate (this effect is discussed later) and because it can be used to prevent the precipitation of undesired compounds. Thus, if an attempt were made to precipitate silver chloride from a neutral solution, there would also precipitate any carbonate, phosphate, arsenate, or other anions of weak acids that form insoluble silver salts; the same would be true of the attempt to precipitate barium sulfate from a neutral solution.

When the hydroxides are precipitated by addition of hydroxide ion, each specific case has to be considered separately; thus, a high concentration of hydroxide ion is required for the complete precipitation of magnesium hydroxide, whereas with aluminum hydroxide, because of its amphoteric nature, an excess of hydroxide ion must be carefully avoided.

4. Effect of the Solvent

In many cases the solubility of a salt can be reduced by altering the properties of a particular solvent or by changing to a different medium. In general, inorganic salts, especially if they are highly ionized, are less soluble in organic solvents, such as alcohol and ether, than they are in water. Because of this effect, strontium chromate, which is appreciably soluble in water solutions, can be quantitatively precipitated from an aqueous solution containing a considerable proportion of alcohol. Likewise, the quantitative precipitation of potassium perchlorate has to be made from an alcoholic or other non-aqueous medium; and nickel chloride, which is very soluble in water, can be precipitated by passing hydrogen chloride gas into an ether solution.

5. Temperature Effects

Where the precipitate is sufficiently insoluble and stable, and where other undesirable effects (such as the hydrolysis of certain salts or the oxidation of some of the constituents present) are not encountered, there are definite advantages in carrying out the precipitation, filtering, and washing of precipitates at elevated temperatures, since, in general, the precipitates are more readily coagulated and brought into a filterable form. As an example, barium sulfate that has separated in such a finely dispersed condition as to pass through even a dense filter can be converted into a much coarser and more readily filtered form by keeping the solution near the boiling temperature for a short period. The higher temperature increases the rate at which the larger particles increase in size at the expense of the more soluble smaller particles.

Solutions are more rapidly filtered when hot, mainly because their viscosity is much less; the specific viscosity of water is 0.637 at 15°C and only 0.158 at 100°C. Experimentally, it has been found that hot solutions will pass through a paper filter from 5 to 10 times as fast as those at room temperature. Another very important advantage arises from the fact that the substances contaminating the precipitate are usually more soluble in hot solutions and are thus more readily extracted.

6. Effect of Time on the Completeness of Precipitation

It is not generally realized that in many precipitation reactions there is a considerable lapse of time before equilibrium is reached. There may be two general causes for this delay. The first is the familiar phenomenon of supersaturation, which is most obvious with moderately soluble compounds and occurs even when the constituent ions of a salt are present in high concentrations. In the second case the substance may be extremely insoluble, and yet the precipitation may be very slow because the constituent ions are present in such very minute concentration or because they react very slowly. The slow precipitation of arsenic sulfide, As_2S_5, from acid solution is probably an example of both causes. This slowness can be due to the very low concentration of simple pentapositive arsenic ions in the solution and also to a low rate of formation of these ions from the other molecular species in which the arsenic exists. Another example of this type of rate effect is involved in the precipitation of cobalt as potassium cobaltinitrite, $K_3Co(NO_2)_6$, where the precipitation has to be preceded by the oxidation of the cobaltous ion to the cobaltic state and by the formation of the complex cobaltinitrite ion.

Although no general rule can be given, it is extremely important that sufficient time elapse between the adding of the precipitant and the beginning of the filtration, not only for the attainment of equilibrium between solution and precipitate but also to enable the precipitate to develop into a filterable form. Precipitation can be induced in supersaturated solutions by vigorous agitation, such as stirring or shaking; the expedient (often used in preparative work) of adding a crystal of the precipitate is usually not practical in analytical procedures. In cases of the second effect, precipitation can be hastened by providing a high concentration of the reactants, by increasing the temperature, and occasionally by mechanical agitation.

II. FACTORS AFFECTING THE PHYSICAL CHARACTERISTICS OF THE PRECIPITATE

It would seem that the physical characteristics of a precipitate—that is, whether it is crystalline or amorphous, granular or gelatinous, whether it separates in a highly hydrous condition, and even its color—would be determined by its chemical composition. Thus it would seem that sulfides are of an amorphous nature, and that hydroxides are gelatinous and sulfates crystalline because of their chemical composition. However, owing in a large measure to the work of von Weimarn,[4] it has been shown that *these characteristics are influenced more by the conditions under which the precipitate forms than by its chemical composition.*

---

[4] Von Weimarn. *Zur Lehre von den Zuständen der Materie.* Dresden: Steinkopff, 1913; Von Weimarn. *Chem. Rev.*, **2**, 217 (1926).

From the mechanism of precipitation, it seems reasonable to expect that if a precipitate forms in a solution in which its normal solubility is only slightly exceeded, only a relatively few crystal nuclei will be formed initially, and that after these are present, the subsequent precipitation will consist mainly of an enlargement, or "growth," of these crystals. This subsequent growth of crystals is in accord with the experimental fact that the solubility of extremely small particles is appreciably greater than that of larger ones. This behavior is predicted from theoretical considerations and has been studied experimentally by various workers,[5] who have found solubility increases ranging from 15% to 80% when studying small particles (ranging from 0.0001 mm to 0.0002 mm in diameter) of barium sulfate and of calcium sulfate monohydrate, $CaSO_4 \cdot H_2O$. Therefore a solution that is saturated with respect to the very small particles is obviously supersaturated with respect to the larger ones. As a result, precipitation takes place on the larger ones, and the smaller ones tend to pass into solution.

If the initial precipitation takes place in a solution that is greatly supersaturated, then there will be formed a large number of very small primary crystal nuclei, and the precipitate will separate in a highly dispersed condition. With very insoluble precipitates, ordinary methods of mixing the solutions will always result in a relatively high degree of supersaturation, and therefore the precipitate will appear as a large number of very small particles; in fact, these particles may be so small that they will remain colloidally dispersed. Furthermore, if a precipitate is extremely insoluble, the concentration of the saturated solution will be so small that the growth of large crystals at the expense of the more soluble smaller ones will be slow.

According to von Weimarn, this degree of supersaturation, which can be expressed as the ratio of the initial supersaturation of the substance (before precipitation begins) to its equilibrium solubility (or $[S - s]/[s]$, where S represents the initial concentration of the substance before precipitation begins and s the equilibrium solubility, each expressed in equivalents per liter), is a major factor in determining the physical characteristics of precipitates. From this principle the generalization is made that the physical characteristics of two precipitates will be in general the same, irrespective of their chemical nature, if they are produced under corresponding conditions, and the most important factor determining these conditions is the value of $[S - s]/[s]$. It follows that if the same substance is precipitated under conditions where $[S - s]/[s]$ has widely varying values, its physical characteristics will be quite different.

[5] Ostwald. *Z. Physik. Chem.* **34**, 495 (1900); Hulett. *Z. Physik. Chem.*, **37**, 385 (1901); **42**, 581 (1903); **47**, 357 (1904); Hulett and Allen. *J. Am. Chem. Soc.*, **24**, 667 (1902); Dundon and Mack. *J. Am. Chem. Soc.*, **46**, 2479 (1923); Dundon. *J. Am. Chem. Soc.*, **45**, 2658 (1923); R. Becker and W. Doering, *Ann. Physik.*, 5. Folge, **24**, 719 (1935).

As an example, barium sulfate, which normally appears as a finely divided but distinctly crystalline precipitate, can be made to separate in a colloidal or gelatinous form when it is caused to precipitate from a highly supersaturated solution, or one where the value of $[S - s]/[s]$ is large. An increase in this ratio can be attained in two ways: (1) by decreasing the solubility of barium sulfate (for example, by the addition of alcohol to the aqueous solution), thus decreasing s; or (2) by mixing highly concentrated solutions, thus increasing S. As specific examples, it is found that the precipitate resulting from mixing equal volumes of 0.025 N cobalt sulfate and 0.025 N barium thiocyanate in 50% alcohol has at first a distinctly colloidal appearance, and on coagulating might easily be mistaken for aluminum hydroxide, whereas the precipitate resulting from rapidly mixing equal volumes of 7 N manganous sulfate and 7 N barium thiocyanate is a viscous gel.[6] Since colloidal solutions consist of finely dispersed particles (usually with diameters ranging from 10^{-4} mm to 10^{-6} mm), it would be expected that a high value of the ratio of $[S - s]/[s]$ would be favorable to their formation; this is true, as is shown by the fact that barium sulfate (a typically crystalline precipitate) can be produced as either a suspensoid or even as a gel. On the other hand, by precipitating barium sulfate from a 4 F hydrochloric acid solution, where s is considerably greater than it is in a neutral solution, the precipitation takes place quite slowly, and relatively large crystals can be obtained.[7]

Apparent exception to this rule can be noted in the fact that substances of approximately the same solubility will precipitate in distinctly different physical form; barium sulfate and silver chloride furnish an example of this anomaly. These two substances have about the same solubility, yet they normally appear as precipitates of a very different type—silver chloride as a curdy flocculated colloid and barium sulfate in a definitely crystalline form.

One partial explanation is given by the "supersolubility" curves plotted by Miers,[8] (Figures 12-2 and 12-3) in which three distinct regions are apparent. The *unsaturated* region represents concentrations lower than and up to the equilibrium saturation concentration. The solid line is the equilibrium solubility curve, showing for the substance being considered an increase in solubility with increased temperature. Between the solid and the dotted lines is the *metastable* region representing conditions in which a concentration above the equilibrium solubility of the substance is present, but in which the

[6] These particular compounds are used because of their solubility, and not because of any specific effects of the cobalt, manganese, or thiocyanate.

[7] A more detailed discussion of these effects is given by H. A. Laitinen (*Chemical Analysis.* New York: McGraw-Hill, 1960) and by H. F. Walton (*Principles and Methods of Chemical Analysis.* Englewood Cliffs, N.J.: Prentice-Hall, 1964). See also A. G. Walton. *The Formation and Properties of Precipitates.* New York: Wiley-Interscience, 1967.

[8] H. A. Miers and F. Isaac. *J. Chem. Soc.*, **89**, 413 (1906); Miers and Isaac. *Proc. Roy. Soc.* (*London*), **A79**, 322 (1907).

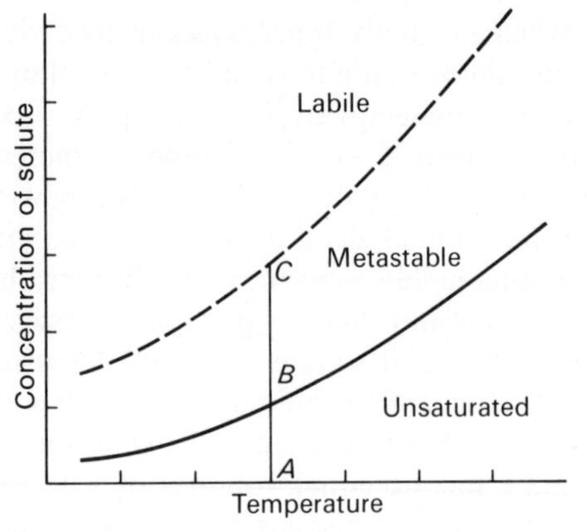

FIGURE 12-2
Supersolubility curve for a solute with a large metastable region.

solute and solvent may remain homogeneous unless in some way nuclei for crystal growth are introduced. That is, crystal growth will occur in this region but crystal nuclei are not formed under the conditions represented here. The dotted line represents the limit of supersaturation; at concentrations above this nucleation occurs spontaneously. The region above is referred to as the labile region.

Consider the slow addition of solute starting at point A in the unsaturated region with the temperature maintained at a constant value. No precipitation will occur until the labile region is reached at point C.

Now, since there are crystal nuclei present, precipitation will continue spontaneously until the concentration at point B is reached. The addition of more precipitant, if done slowly, may not take the mixture into the labile region; if not, new nuclei do not form, but the existing crystals grow larger. Therefore, one sees that a large separation of the solubility curve and the limit of supersaturation curve favors crystal growth rather than formation of new nuclei. A small separation, as in Figure 12-3, makes improbable the addition of precipitant without the labile region being reached; hence new crystal nuclei are produced and a large number of small crystals is the result. Barium sulfate has a relatively broad region of supersaturation and can be

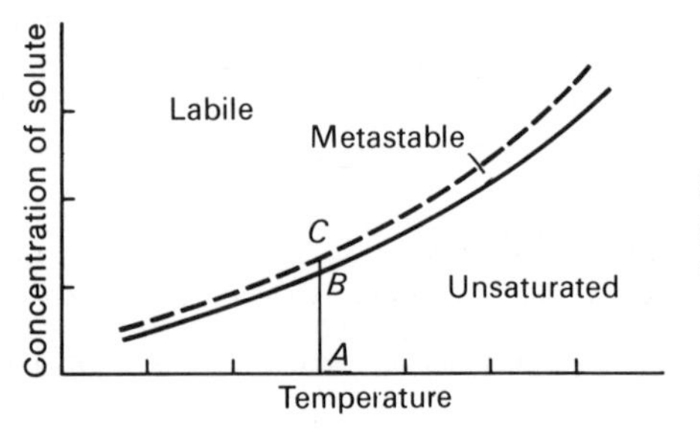

FIGURE 12-3
Supersolubility curve for a solute with a small metastable region.

represented by a plot such as Figure 12-2; from very dilute solutions it tends to precipitate as rather coarse uniform particles. Silver chloride has a narrow region of supersaturation, characteristic of Figure 12-3; the primary particles of silver chloride are very small when precipitated from very dilute solution. The curdy precipitates obtained with the silver halides are aggregates of these very small, almost colloidal, particles.

A related explanation[9] is that the solubility of barium sulfate increases much more rapidly as the particle size is decreased than does silver chloride, and therefore the initial supersaturation is much less with barium sulfate than with silver chloride.

Thus it is seen that larger and more readily filterable particles can usually be obtained (1) if the initial supersaturation of the solution during the precipitation process is kept as small as possible, and (2) if time is given for the extremely small particles first formed to increase in size. This "growth" can be attributed to the increased solubility of the small particles. In addition, if a crystalline precipitate is caused to form very rapidly from relatively concentrated solutions, it is probable that many imperfections and strains will result in the process, and that if such a precipitate is allowed to age, a recrystallization and adjustment will take place. This process, which is sometimes denoted by the term "aging" or "perfecting," will usually result in a decrease in the total surface and gives a more filterable precipitate. Frequently it is responsible for the changes taking place upon "digesting" (that is, maintaining the solution for a period of time at or near its boiling point) a very fine or colloidal precipitate in order to convert it into a form that can be filtered.[10]

Precipitation from Homogeneous Solution

It is obviously difficult to avoid local supersaturation when the precipitant is added in the form of a relatively concentrated solution. To avoid this difficulty and to cause an increase in the concentration of the precipitant at a low and controlled rate the process called precipitation from homogeneous solutions has been developed. In this process the precipitant is not added manually or mechanically but instead is generated in the solution by means of a reaction whose rate can be controlled.

As an example, consider the precipitation of hydrous aluminum oxide by addition of ammonia to an acid solution of an aluminum salt. The precipitate

[9] Kolthoff. *J. Phys. Chem.* **36**, 860 (1932). An extensive discussion of precipitation and coprecipitation phenomena is given in this article.

[10] For a discussion of this subject of the "aging" of precipitates see I. M. Kolthoff, E. B. Sandell, E. J. Meehan, and S. Bruckenstein. *Quantitative Chemical Analysis.* (4th ed.) New York: Macmillan, 1969, pp. 220–225. See also Laitinen, op. cit., Chap. 9; H. F. Walton, op. cit., pp. 37–40.

produced by the conventional addition of an ammonia solution is usually very dispersed and difficult to filter and wash, even though the NH_3 is added very slowly and with constant stirring. In 1937 Willard and Tang[11] showed that the physical characteristics of the precipitate were greatly improved if the ammonia were generated in the solution by the hydrolysis of urea, as follows:

$$CO(NH_2)_2 + H_2O = 2NH_3 + CO_2$$

Subsequent investigations by Willard and his co-workers also showed that the process could be applied to the precipitation of a large number of other hydrous oxides and basic salts. In addition, the use of urea was extended to the precipitation of slightly soluble salts of weak acids (for example, calcium oxalate) held in solution by a strong acid.

The hydrolysis of esters of various acids has also been used as a means of generating the anion of the acid as a precipitant. As an example, instead of precipitating barium as sulfate by addition of a sulfuric acid solution, one can add dimethyl sulfate, which slowly hydrolyzes and produces sulfate homogeneously throughout the solution as follows:

$$(CH_3O)_2SO_2 + 2H_2O = 2CH_3OH + SO_4{}^{2-} + 2H^+$$

Precipitation from homogeneous solutions has found extensive applications because not only can precipitates with more favorable physical characteristics be produced but in many cases there is less contamination of the precipitate by other constituents of the solution.

To summarize the above discussion:

In most cases the physical characteristics of a precipitate can be affected by the method of precipitation. During analytical precipitations the filtering and washing can be facilitated by producing precipitates that are crystalline or granular rather than colloidal and voluminous. The following precautions will aid in obtaining this objective.

1. The mixing of the reagents should be done very slowly and with effective stirring.
2. Dilute solutions should be used.
3. In many cases the solubility of the precipitate should be increased, usually by the addition of an acid, or by working in hot solutions.
4. The precipitate should be allowed to stand, preferably at a higher temperature, until the particle size is such that it will be retained by the filter.

---

[11] H. H. Willard and N. K. Tang. *J. Am. Chem. Soc.*, **59**, 1190 (1937). For a general review of this method of precipitation see L. Gordon, M. L. Salutsky, and H. H. Willard. *Precipitation from Homogeneous Solutions.* New York: Wiley, 1959.

III. FACTORS AFFECTING THE COMPOSITION AND PURITY OF THE PRECIPITATE

Coprecipitation

It is a universally observed phenomenon that when a precipitate separates from a solution, it will carry with it in varying amounts the soluble constituents of the solution; hereafter the term *coprecipitation* will be used to designate this phenomenon without implying any specific cause, mechanism, or time of occurrence. Coprecipitation effects are one of the most important factors involved in precipitation methods and have to be considered regardless of whether a separation or a gravimetric determination is desired. In general, coprecipitation is not the result of any one effect but may result from the operation of any one or more of the four following causes:

1. adsorption
2. compound formation
3. solid-solution formation
4. mechanical inclusion.

A brief discussion of these effects and of means for minimizing them is given below.

1. *Adsorption as a Cause of Coprecipitation.* Adsorption may be broadly defined as the process that causes an increase in the concentration of a gas, liquid, or dissolved substance at an interface—that is, at the surface between two phases. In analytical chemistry there is special interest in the process whereby constituents of a solution are concentrated on the surfaces of precipitates. Charcoal is probably the best-known example of an effective adsorbing agent, and consequently is extensively used scientifically and industrially for collecting gases and for removing many substances from dilute solutions; an example of the latter is the commercial use of special charcoals for the recovery of iodine from oil-well brines.

The adsorption of acetic acid from aqueous solutions by charcoal was investigated quite early in the study of adsorption phenomena, and data showing the relation between the amount of adsorption taking place on a given amount of charcoal and the concentration of the acetic acid are shown in Table 12-2. If the adsorption process were of a chemical nature—due conceivably to the formation of insoluble compounds—it would be expected that definite saturation values would be found that would limit the amount of material taken up by a given amount of adsorbing agent. If the process were of a physical nature, it would be expected to be more analogous to the distribution of a substance between two phases, and such a process should obey the distribution law discussed in Chapter 4.

TABLE 12-2
The Adsorption of Acetic Acid from Aqueous Solutions by Charcoal*

Equilibrium Concentration (c) of HC_2H_3O (moles/liter)	Mmoles HC_2H_3O per g of charcoal (x/m)
0.0181	0.467
0.0309	0.624
0.0616	0.801
0.1259	1.11
0.2677	1.55
0.4711	2.04
0.8817	2.48
2.785	3.76

* Freundlich. *Kapillarchemie*. 1909.

This law is formulated as follows:

$$\frac{C_2}{C_1} = K$$

where C_1 and C_2 represent the concentrations of a common substance in two phases that are in equilibrium and where the substance exists in the same molecular form in each phase. Should the substance exist in different molecular aggregations in the two phases—for example, be polymerized in one phase—the equation would have the more general form

$$\frac{(C_2)^n}{(C_1)} = K$$

where n is the number of molecules associated in the polymer, in agreement with the general mass-action law.

It has been found experimentally that adsorption is an equilibrium process and that it follows an empirical equation, called the *adsorption isotherm*, which has the following form:

$$\frac{x}{m} = Kc^{1/n}$$

where x represents the weight of material adsorbed by the weight m of adsorbing material, c represents the equilibrium concentration of the adsorbed material, and K and n are constants. The constant K is highly dependent upon the specific conditions to which the equation is being applied, and

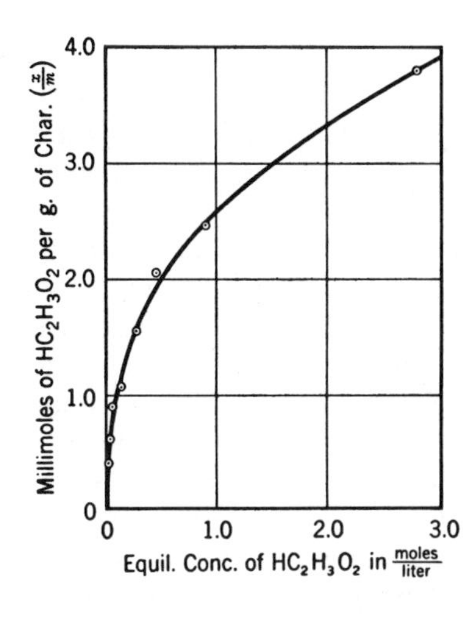

FIGURE 12-4

The adsorption of acetic acid from aqueous solutions by charcoal.

n is generally greater than 1, usually less than 5, and frequently approximately 2. It is seen that this equation bears a formal resemblance to the distribution law. Figure 12-4 shows the curve obtained by plotting the data of Table 12-2. It is also evident that the equation may be written

$$\log \frac{x}{m} = \frac{\log c}{n} + \log K$$

and that a straight line should result from plotting the data logarithmically; this is also shown in Figure 12-5. This graphical method is frequently used to determine whether a given process is due to adsorption or to the presence of other effects that would cause a departure from the straight line.

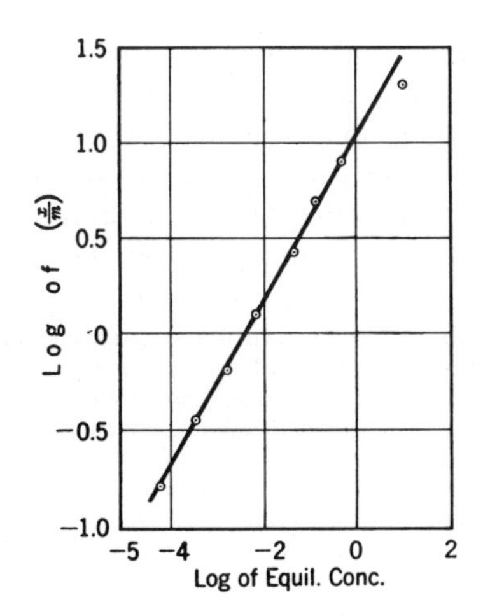

FIGURE 12-5

The adsorption of acetic acid from aqueous solutions by charcoal (logarithmic plot).

Not only are molecular compounds adsorbed from solutions by solids, but it is also found that ions are similarly concentrated at the surface of precipitates; the charge associated with colloidal particles is in general due to this effect, and the presence of this charge contributes to the stability of the suspension.[12]

It is found that, as a general rule, precipitates tend to adsorb the ions of which they are composed; and the less soluble the precipitate, the greater this tendency. A colloidal suspension of silver chloride, bromide, or iodide that has been formed in the presence of an excess of the halide ion will be found to be negatively charged because of the adsorbed negative ions; in the presence of an excess of silver ions such a precipitate will be positively charged. The adsorption of silver nitrate by silver iodide has been extensively studied, and the results of one such study are shown in Figure 12-6.

The silver halide first formed by the addition of silver nitrate to a solution containing halide ions is negatively charged and usually remains in suspension until quite near the equivalence point; then as the concentration of the halide ion in the solution is reduced to a very small value, the adsorbed halide ions are withdrawn from the precipitate and coagulation takes place.[13] By addition of a large excess of silver nitrate, the precipitate can be peptized, that is, again colloidally dispersed, and the particles will now have a positive charge. Such a suspension may be coagulated by the addition of negative ions to the solution; however, the coagulated precipitate will be found to have carried down with it some of the negative ions, and these may be an undesirable contamination for quantitative purposes. Since it is probable that all precipitates pass through the colloidal state to some extent in the mechanism of their formation, it is not surprising to find that marked adsorption may occur even with typically crystalline precipitates. Thus, it has been found that the coprecipitation of potassium nitrate and of sodium chloride by a barium sulfate precipitate appears to follow the adsorption equation.

The magnitude of the adsorption taking place in precipitation processes will vary from an amount not exceeding the usual quantitative errors to cases where even qualitative separations cannot be carried out. Adsorption phenomena are probably at least partly responsible for the striking coprecipitation effects that have been found when the "ammonia precipitation"

[12] It should be pointed out that the solution does not acquire an electrostatic charge because of the adsorption of ions of a particular sign. This effect is prevented by the formation of a second loosely attached and more mobile layer of ions of opposite sign (thus producing the so-called "electrical double layer"). Upon coagulation or filtration, the double layer is carried with the precipitate, and the solution remains electrically neutral.

[13] In many precipitation reactions this effect can be used to indicate when an approximately equivalent amount of the precipitant has been added and an unnecessary excess thus avoided. In certain cases the effect is so pronounced that a titrimetric end point can be taken with moderate precision.

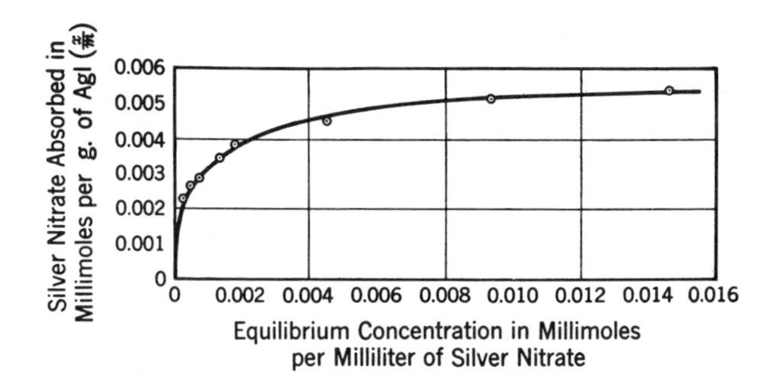

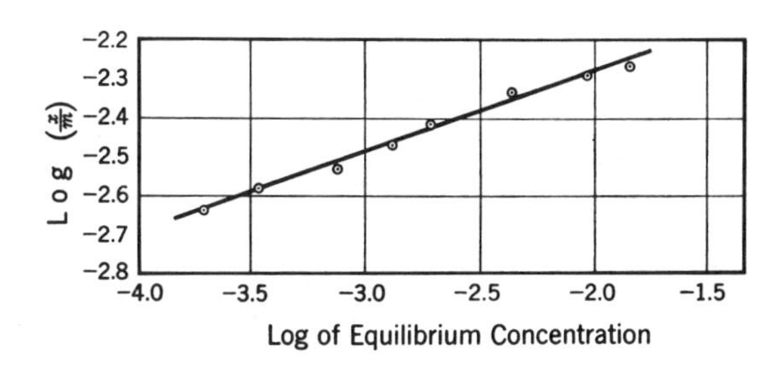

FIGURE 12-6
The adsorption of silver nitrate by silver iodide.

is used for the separation of tripositive iron, aluminum, and chromium from dipositive manganese, nickel, cobalt, and zinc.[14] It is also found that the coprecipitation of the dipositive elements by a given tripositive was usually greatest with zinc, the least soluble of the dipositive hydroxides, and least with manganese, the most soluble. This result is in conformity with a general rule that, other conditions being comparable, the less soluble the compound adsorbed, the greater the adsorptive tendency. In many cases it is difficult to distinguish between adsorption and compound formation. Thus, in the hydroxide separations just mentioned, it is probable that the precipitates adsorb hydroxide ion on their surfaces and that this adsorption results in the formation of the hydroxides of the soluble elements. A somewhat similar effect has been noted in a study of the coprecipitation of zinc sulfide by copper sulfide and has been attributed to the induced precipitation of zinc sulfide caused by the adsorption of hydrogen sulfide on the copper sulfide precipitate.[15]

[14] Swift and Barton. *J. Am. Chem. Soc.*, **54**, 2219 (1932).
[15] Kolthoff and Pearson. *J. Phys. Chem.*, **36**, 549 (1932).

Space does not permit an adequate treatment of even the most important of those adsorption and colloidal phenomena that are encountered in analytical precipitations and separations. For a more detailed treatment reference should be made to the texts of Footnote 10; or to reference books and textbooks of colloidal and surface chemistry.

2. *Compound Formation as a Cause of Coprecipitation.* Compound formation was once thought to be a much more important cause of coprecipitation than at present. As examples, the coprecipitation of zinc by chromium hydroxide when the latter is precipitated by ammonia solution was attributed to the formation of a zinc chromite. This is unlikely because of the fact that in the hydroxide concentration provided in an ammonia precipitation there would hardly be a significant concentration of chromite formed; in fact, present evidence indicates that the solution of chromic hydroxide in more concentrated hydroxide solutions is due not to chromite formation but to the formation of a stable colloidal suspension in which there is a strong adsorption of hydroxide ion. Such a suspension would tend to be coagulated by, and to carry down with it, a positive ion, such as zinc; in general, the greater the charge, the more effective the coagulating action. Many cases formerly attributed to "double-salt" formation are now believed to be due to adsorption; the adsorption of cadmium salts by cadmium sulfide precipitates is an example.[16]

Compound formation is a more probable cause of coprecipitation in cases where partially hydrolyzed products are formed, where complex compounds may exist, or where a preliminary adsorption process may be followed by compound formation. The attempted precipitation of bismuth hydroxide by ammonia solution is likely to result in a mixture of hydroxide and basic salts in varying proportions depending upon the anions present, the hydroxide-ion concentration of the solution, the temperature, and other factors such as the method and rate of mixing the solutions. The coprecipitation of sulfate with a ferric hydroxide precipitate has been attributed to the formation of a basic ferric sulfate.

3. *Solid Solutions (Mixed Crystals) as a Cause of Coprecipitation.* It is possible to have a constituent uniformly distributed throughout a solid, and thus to form what is known as a *solid solution*. For example, it was observed quite early[17] that if a solution of iodine in benzene was cooled until solid benzene separated, iodine would be uniformly distributed throughout the

[16] Weiser and Durham. *J. Phys. Chem.*, **32**, 1061 (1928); for a study of the coprecipitation of cadmium with mercuric sulfide, see Ritter and Schulman. *J. Phys. Chem.*, **47**, 537 (1943).

[17] Van't Hoff. *Z. Physik. Chem.*, **5**, 322 (1890).

solid phase and that *the concentration of the iodine in the solid phase was proportional to the concentration of the iodine in the liquid phase.*

The occurrence of isomorphism was also observed quite early. Isomorphous substances are those having so nearly the same crystal configuration that the constituents of one can be substituted in the crystal lattice of the other without causing an appreciable change in the crystal. Thus, evaporation of a solution containing two isomorphous substances produces homogeneous crystals of each substance; however, each will be found to contain a certain proportion of the other substance. Since it seemed reasonable to assume that isomorphous substances should have analogous molecular formulas, this rule was used by Mitscherlich as early as 1820 in deciding which multiple of the combining weight of an element should be its atomic weight. An experiment that strikingly shows mixed crystal formation can be performed by crystallizing potassium perchlorate from a solution containing even a very low concentration of permanganate ions. The crystals will remain pink even after washing (preferably with a saturated solution of potassium perchlorate) until the wash solution is colorless.[18]

It has been found, however, that isomorphism is not necessary for a limited amount of solid-solution formation to occur. The striking coprecipitation of nitrate by a barium sulfate precipitate, which previously had been attributed to compound formation [compounds of the type $Ba(NO_3)_2 \cdot BaSO_4$ having been postulated], and the coprecipitation of the alkali metals by barium sulfate have been studied by means of x-ray diffraction experiments with results that indicate that limited solid solutions are formed in these cases.[19]

4. *Mechanical Inclusion (Occlusion) as a Cause of Coprecipitation.* In many cases it has been made evident, especially from microscopic studies, that portions of the solution have been either entrapped by crystals or entrained in a mass of crystal capillaries and in this manner carried out with the precipitate. Such a process is properly termed *occlusion*, even though this term is sometimes used to designate coprecipitation in general. This effect can often be minimized by proper adjustment of conditions and method of precipitation and is probably the least serious of the causes of coprecipitation discussed. An example of it is found in the precipitation of potassium perchlorate, where it has been observed[20] that sodium perchlorate and perchloric acid are entrained by the precipitate.

---

[18] For a study of the distribution of arsenate between the solution and a $MgNH_4PO_4$ precipitate, see Kolthoff and Carr. *J. Phys. Chem.*, **47**, 148 (1943).

[19] Walden and Cohen. *J. Am. Chem. Soc.*, **57**, 2591 (1935); **68**, 1742, 1750 (1946).

[20] Smith and Ross. *J. Am. Chem. Soc.*, **47**, 774 (1925).

General Observations on Coprecipitation Effects

1. *Coprecipitation Effects Often Appear to be Specific.* At the present time the general principles governing coprecipitation effects are better understood than formerly. However, in many cases the systems involved are so complex, and the conditions affecting them are so varied, that only very general predictions can be made in regard to the occurrence or magnitude of the phenomena; in many cases the effects appear to be controlled entirely by specific conditions. However, certain general tendencies have been noted and are discussed below. It should be mentioned that a program of intensive work that has been under way for the past twenty years is being extended by various workers in regard to the mechanism of the formation of precipitates, their changes on aging, the role of adsorption in coprecipitation, and other related topics; a much better general understanding of the subject has already resulted from these investigations, and additional progress can be expected.

2. *Effect of the Concentration of the Coprecipitated Substance.* Where coprecipitation results from adsorption alone, the effect of the concentration of the adsorbed substance can be predicted from the adsorption equation. Where only isomorphous solid-solution formation is involved, and where equilibrium is attained, it has been shown that a constant can be obtained for the distribution between solid and liquid phases. Thus, in a study of the coprecipitation of arsenate by magnesium ammonium phosphate[21] it was found that the equilibrium could be expressed by the relationship

$$\frac{C_{AsO_4}}{C_{PO_4}} = D\frac{N_{AsO_4}}{N_{PO_4}}$$

where D is a distribution constant, C represents the concentration in the liquid phase of the respective constituents, and N is the mole fraction in the solid phase. Regardless of the cause, it is qualitatively apparent that the greater the concentration of the coprecipitated substance, the greater the amount coprecipitated, other factors being constant.

3. *Effect of Temperature on Coprecipitation.* Since adsorption processes are exothermic, it is to be predicted that the higher the temperature, the less the amount of adsorption taking place. On the other hand, where hydrolysis is involved in the coprecipitation process, raising the temperature is likely to increase coprecipitation. In certain cases where separations are being made that depend upon maintaining some substance in a supersaturated condition, raising the temperature appears to break this state of supersaturation and

[21] See Footnote 18.

results in extensive coprecipitation. This effect is noted in the precipitation of zinc sulfide at a *p*H of 2 in the presence of cobalt; in a cold solution a precipitate can be obtained that will remain white for hours but that rapidly darkens if the solution is heated.

4. *Effect of the Solubility of the Coprecipitated Substance.* It has been observed that where other conditions are constant, the amount of coprecipitation occurring increases as the conditions of the precipitation more nearly approach the saturation value of the coprecipitated substance. Thus, in a determination of the sulfate in a solution that was approximately 0.02 *F* in calcium chloride, a barium sulfate precipitate weighing approximately 1 g showed a loss in weight of 16 mg (replacement of barium by calcium), whereas under the same conditions with magnesium chloride present the error was 3 mg positive (coprecipitation of barium chloride).[22] It was mentioned above that in the separation of the bipositive elements manganese, cobalt, nickel, and zinc from the tripositive elements aluminum, iron, and chromium by a carefully controlled ammonia precipitation, it was found that coprecipitation increased in the order named; zinc, the least soluble of the bipositive hydroxides, was the most extensively carried out with the tripositive hydroxides.

5. *Coprecipitation and Complex-Compound Formation.* It is frequently observed that where a constituent of the precipitate tends to form a complex compound with some constituent of the solution, there will be coprecipitation of that compound. When sulfate is precipitated as barium sulfate, ferric iron, which forms the complex compound $FeSO_4^+$ with sulfate, is much more extensively coprecipitated than aluminum (where any such complex is much less stable) or than ferrous iron.

6. *Coprecipitation Varies with the Methods and Conditions of Precipitation.* Since the physical characteristics of a precipitate, especially its surface, can be varied so extensively by the conditions of precipitation, it would be expected that such effects as adsorption would be similarly influenced. Also, since the concentration of the adsorbed ion is a factor in coprecipitation effects, it is possible to vary the effective concentration of such an ion during the formation of the precipitate (when coprecipitation is usually most pronounced) by the order of mixing the solutions. Thus, in the precipitation of sulfate by addition of an excess of barium chloride, the coprecipitation of barium chloride will be less if it is added to the sulfate solution than if the mixing is made in the reverse order. However, if alkali metal ions are present

[22] Blasdale. *Quantitative Analysis.* Princeton, N.J.: Van Nostrand, 1928, pp. 140, 188.

in the solution, more alkali sulfate will be precipitated when the barium chloride is added to the sulfate solution; consequently, Popoff and Neumann[23] have recommended the reverse order of precipitation. The possible effects of the temperature at which the precipitation is made have been already mentioned above.

General Methods for Minimizing Coprecipitation Effects

Certain general methods for avoiding or minimizing coprecipitation effects are mentioned below.

1. *Avoid Introducing or Remove Substances Known to Cause Coprecipitation.* It is preferable to avoid a fusion process in the solution of the sample because such processes introduce large amounts of sodium or potassium salts. Sometimes species used in the dissolving of the sample may need to be removed in order to decrease the extent of coprecipitation. For example, in the determination of sulfur in a pyrite (FeS_2) material, nitric acid is often used in the dissolving solution along with a powerful oxidizing agent such as Br_2. The sulfur is oxidized to sulfate and then precipitated as $BaSO_4$ by the addition of a barium chloride solution. The coprecipitation of barium nitrate causes serious errors, so nitric acid is removed by repeated evaporations with hydrochloric acid. The iron is oxidized to the Fe(III) state by the dissolving process and it must be removed or reduced to the Fe(II) state or it will be coprecipitated as ferric sulfate.

2. *Keep the Concentration of Coprecipitated Substances Low.* The concentration of coprecipitated substances can be kept low by avoiding large excesses of precipitants or of acids and bases that have to be subsequently neutralized, and by making the precipitation from dilute solutions.

3. *Properly Control the Method and Conditions of Precipitation.* Factors involved would be (*a*) the adjustment of the volume of the solution, (*b*) the order of mixing the reagents, (*c*) the temperature at which the precipitation is carried out, (*d*) the rate of addition of the precipitant, (*e*) the stirring of the solution, (*f*) the time allowed before filtering of the precipitate, and (*g*) the selection of a suitable medium for the precipitation.

Addition of the precipitant as a relatively concentrated solution always produces local high concentration. For this reason the use of precipitations from homogeneous solution, discussed above, offers distinct advantages with

[23] Popoff and Neumann. *Ind. Eng. Chem.*, anal. ed., **2**, 45 (1930).

many substances that normally appear as colloidal or voluminous precipitates.

4. *Reprecipitation.* As a last resort, it is often necessary to dissolve the precipitate and to repeat the precipitation. Since in the second solution the concentration of the coprecipitated substance is relatively small, the coprecipitation is correspondingly decreased. It is obvious that a single reprecipitation is effective only when a small proportion of the adsorbed substance is carried out in the first precipitation.

IV. FACTORS AFFECTING THE COMPOSITION AND STABILITY OF THE WEIGHED PRECIPITATE

In order to effect a gravimetric determination, it is necessary, after filtering and washing the precipitate, that it be (1) completely dried, (2) of definite composition, and (3) sufficiently stable to be accurately weighed.

Assuming that volatile contaminants are not present (usually as coprecipitated material), precipitates can be divided into two types: (1) those that are weighed in the same form as they were filtered and (2) those that have to be converted into a more stable and uniform compound before they can be weighed.

With precipitates of the first type it is necessary only to remove the superficial water (adsorbed or entrained) or other solvent medium. Examples of such precipitates are silver chloride, bromide, iodide, or thiocyanate, lead sulfate, barium sulfate, bismuth oxychloride, and nickel dimethylglyoxime. With such precipitates the drying can usually be accomplished by heating at a relatively low temperature, in some cases as low as 110°C. In certain cases where the water may be more firmly held, or where a high order of precision is desired—as in atomic-weight determinations—higher temperatures must be used. For example, in order to remove the last traces of moisture (usually less than 0.01%) from a silver chloride precipitate, it must be heated until fused.

There are certain advantages in the drying of precipitates of this first type by treating them with some volatile solvent that has a great tendency to take up water, such as ethyl alcohol or a combination of alcohol and ether, and then removing this solvent by placing the crucible and precipitate in a desiccator and evacuating for a short time. A considerable saving of time is effected, since the whole process is carried out at room temperature, and there is no need to wait 20 to 30 minutes for the crucible to cool before it can be weighed. Also, certain precipitates that are unstable if dried by heat can be dried by this process; magnesium ammonium phosphate hexahydrate

($MgNH_4PO_4{\cdot}6H_2O$) is such a substance.[24] The use of alcohol and ether followed by the aspiration of air through the precipitate or a brief treatment in a vacuum desiccator has been proposed by Dick[25] for the rapid drying of quite a number of precipitates. The precision of the method has been questioned[26] and confirmed by others.[27]

Precipitates of the second type can be divided into two different subgroups. In drying those of the first subgroup a relatively high temperature is necessary because what may be called *water of constitution* must be eliminated completely. This group includes the hydrous oxides (ferric and aluminum hydroxides, and silicic and "metastannic" acids are examples), which are of very uncertain composition and which have to be converted to the oxides before a stable compound of definite composition is obtained. The second subgroup of precipitates is composed of those that may be partly decomposed upon drying by heat, and therefore it is necessary to convert them into a compound of more uniform and stable composition before they can be weighed. Examples are magnesium ammonium phosphate ($MgNH_4PO_4{\cdot}6H_2O$), which is converted by heat to magnesium pyrophosphate ($Mg_2P_2O_7$), and calcium oxalate ($CaC_2O_4{\cdot}H_2O$), which is weighed as either the carbonate or the oxide.

The treatment of these precipitates will vary greatly. For example, in order to convert calcium oxalate to calcium carbonate, the temperature must be controlled between 375°C and 525°C;[28] or if the ignition is carried out in an atmosphere of carbon dioxide, the temperature can be raised to 700°C.[29] The complete dehydration of aluminum hydroxide or of silicic acid requires heating above 1000°C for some time.[30]

It should be emphasized that certain precipitates that are weighed in the same form as they separate are heated at higher temperatures because of the presence of coprecipitated compounds; thus, barium sulfate, which can be dried at a relatively low temperature, is heated to temperatures varying from

---

[24] Fales. *Inorganic Quantitative Analysis.* London: Century, 1925, p. 222; Worsham. The Stability of Magnesium Ammonium Phosphate. Thesis, Columbia University, 1923.

[25] Dick. *Z. Anal. Chem.*, **77**, 352–363 (1929); **78**, 414 (1929); **83**, 105 (1931).

[26] Moser and von Żombory. *Z. Anal. Chem.*, **81**, 95 (1930).

[27] Wassiljew and Sinkowskaja. *Z. Anal. Chem.*, **89**, 262 (1932).

[28] Willard and Boldyreff. *J. Am. Chem. Soc.*, **52**, 1888 (1930).

[29] Foote and Bradley. *J. Am. Chem. Soc.*, **48**, 676 (1926).

[30] An extensive series of investigations of the limiting temperatures required in order to attain constant weight has been made by C. Duval and co-workers in which use is made of a sensitive thermobalance. Such a balance enables one to record the change in weight of precipitate as a function of time as the temperature is increased. This type of balance was originally developed by Chevenard, Wache, and Tullaye, (*Bull. Soc. Chim.*, **11**, 41, 1944) and has since found extensive use. For information regarding this technique, see C. Duval. *Thermogravimetric Analysis.* Trans. Oesper. (2nd ed.) Amsterdam: Elsevier, 1963; P. D. Garn. *Thermoanalytical Methods of Investigation.* New York: Academic Press, 1965; W. W. Wendlandt. *Thermal Methods of Analysis.* New York: Interscience, 1964.

TABLE 12-3
Precipitates Frequently Used in Gravimetric Analysis

Compound Precipitated	Compound Weighed	Temperature Used (°C)	Remarks
$AgCl$	$AgCl$	130–150	Estimation of Ag or Cl
$PbSO_4$	$PbSO_4$	500–600	Precipitate washed with dilute H_2SO_4
$BiPO_4$	$BiPO_4$	200–300	Decomposes above 650°
$BiOCl$	$BiOCl$	105–110	
CdS	$CdSO_4$	450–500	CdS treated with H_2SO_4
HgS	HgS	105–110	
Hg_2Cl_2	Hg_2Cl_2	105–110	
As_2S_3	As_2S_3	105–110	
$MgNH_4AsO_4 \cdot 6H_2O$	$Mg_2As_2O_7$	850–950	
Sb_2S_3	Sb_2S_3	280–300	
$SnO_2(H_2O)_x$	SnO_2	1100–1200	
$Fe_2O_3(H_2O)_x$	Fe_2O_3	1100–1200	In oxidizing atmosphere
$ZnNH_4PO_4 \cdot 6H_2O$	$ZnNH_4PO_4$	110–135	
	$Zn_2P_2O_7$	850–950	
ZnS	$ZnSO_4$	450–500	ZnS treated with H_2SO_4, decomposes above 700°
	ZnO	850–950	
$NiC_8H_{14}N_4O_4$	$NiC_8H_{14}N_4O_4$	110–120	
CoS	$CoSO_4$	450–500	CoS treated with H_2SO_4
MnS	$MnSO_4$	450–500	May decompose above 550°
$MnNH_4PO_4 \cdot H_2O$	$Mn_2P_2O_7$	850–950	
$Al_2O_3(H_2O)_x$	Al_2O_3	1050–1200	
$BaSO_4$	$BaSO_4$	120–900	For Ba or S
$SrSO_4$	$SrSO_4$	450–500	
$CaC_2O_4 \cdot H_2O$	$CaCO_3$	475–525	
	CaO	1200	
$MgNH_4PO_4 \cdot 6H_2O$	$Mg_2P_2O_7$	1000–1100	For Mg or P
$KClO_4$	$KClO_4$	350	
$NaCl$	$NaCl$	600	
$SiO_2(H_2O)_x$	SiO_2	1200	

350°C to 900°C in order to reduce the error caused by coprecipitated material. Table 12-3 shows some of the precipitates most frequently used in gravimetric analysis, the form in which they are weighed, and the temperatures to which they are usually heated.

RELATIVE ADVANTAGES OF GRAVIMETRIC AND TITRIMETRIC METHODS

Gravimetric methods, depending as they do upon measurements of weight, lend themselves to work of exceedingly high accuracy. For example, the extremely precise atomic-weight measurements made by T. W. Richards[31]

[31] T. W. Richards, *Chem. Rev.*, **1**, 1 (1924).

and his co-workers were based on the conversion of the carefully purified chlorides of various elements to AgCl, which was determined gravimetrically. Some of the atomic weights obtained by this technique agree within 6 parts per hundred thousand with the best results obtainable today by physical mass methods. Such results are obtainable, however, only if extraordinary care is used at every stage of the operation, and even then only for relatively few substances. If a titrimetric method is available, and especially if repetitive determinations are to be made, a chemist will usually choose to avoid the much more time- and technique-demanding gravimetric determination.

QUESTIONS AND PROBLEMS

(Problems involving the application of mass-action principles to solubility effects are given at the end of Chapters 4, 5, 9, and 14.)

1. Why is it desirable analytically to cause precipitates to form in relatively slightly supersaturated solutions? Discuss practical methods whereby supersaturation can be minimized during precipitations.

2. The determination of calcium in limestones, cements, and other products is of considerable technological importance. The calcium is usually precipitated as calcium oxalate monohydrate, $CaC_2O_4 \cdot H_2O$, and determined gravimetrically by igniting the precipitate to $CaCO_3$ or CaO, or volumetrically by dissolving the precipitate in sulfuric acid and titrating the oxalate with permanganate. If the precipitation is made from an ammoniacal solution, as is necessary to reduce the solubility to acceptable limits, the precipitate tends to be finely divided and difficult to filter.
   Suggest as many means as you can to obtain a more filterable precipitate.

3. An analyst weighed out (at 20°C) 10.000 g of sulfuric acid (density 1.0661), introduced it into water, and diluted the solution to approximately 100 ml. He then added 100 ml of 0.25 *N* $BaCl_2$, diluted the resulting mixture to exactly 500 ml, let it stand until equilibrium was attained, and filtered out, washed, dried, and heated the precipitate, which was found to weigh 2.380 g.
   (*a*) From the weight of precipitate found, calculate: (*1*) the percentage of SO_3 in the original sulfuric acid solution, (*2*) the formal concentration, (*3*) the neutralization normality, and (*4*) the weight formality of the sulfuric acid in the original solution. *Ans.* (*1*) 8.17%.
   (*b*) Calculate the percentage error caused by the solubility of the precipitate in the 500 ml of filtrate.
   (*c*) State the effect upon the physical characteristics of the precipitate, other conditions being kept constant, of:
      (*i*) diluting the H_2SO_4 solution to 400 ml before adding the $BaCl_2$ solution;
      (*ii*) neutralizing the sulfuric acid solution with NH_3 to *p*H 7 before adding the $BaCl_2$;
      (*iii*) adding 10 ml of 12 *N* HCl to the sulfuric acid solution;
      (*iv*) adding the $BaCl_2$ rapidly rather than adding it slowly.

(*d*) State the effect upon the accuracy of the analysis (sign and approximate magnitude, parts per thousand, of any error) of the following:

(*i*) substituting $Ba(NO_3)_2$ for the $BaCl_2$;

(*ii*) adding the H_2SO_4 to the $Ba(NO_3)_2$;

(*iii*) the presence in the sulfuric acid solution of 10 mmoles of $CaCl_2$;

(*iv*) the presence of 10 mmoles of $FeCl_3$.

A discussion of various substances coprecipitated by a barium sulfate precipitate and the effect of precipitation methods and conditions is given in Procedure 14-2, pages 291–301.

4. To a solution saturated with lead chloride, concentrated hydrochloric acid was added in increments such that each increment increased the molar chloride concentration by a factor of ten. What would be the effect on the $[Pb^{2+}]$ of each increment (neglect activity effects)? State qualitatively what effect these increments might have on the formal solubility of lead chloride.

CHAPTER **13**

Operations of Gravimetric Measurements

A large majority of the operations of gravimetric methods are concerned with precipitates—their formation, separation from the solution, conversion to a known composition, and the final weighing.

In Chapter 10 certain of the fundamental operations of the chemical laboratory were discussed and instructions given for constructing useful apparatus and for performing the more frequently required of these operations.

In Chapter 12 the factors affecting the solubility, physical characteristics, composition, purity, and stability of a precipitate were considered. In this chapter we will discuss in more detail the principles and techniques involved in the operations of gravimetric analysis: filtering, washing, drying, and weighing precipitates. Following this discussion is a summary of operating suggestions and precautions covering these topics.

(1) PREPARATION OF FILTERS

A discussion of the relative advantages of various types of filters for general purposes and instructions for their preparation has been given in Chapter 10, pages 196–203; this material should be reviewed. Additional instructions are given below for the use of these filters for gravimetric determinations.

Paper Filters

Select an ashless quantitative filter of appropriate porosity for the nature of the precipitate, that is, a retentive one for finely dispersed precipitates, a more porous one for coarsely crystalline ones. Select a size that will cause the filter to be not over one-third filled with precipitate. Select a funnel of such size that the fitted paper does not come closer than about 1 cm to the top, Fit and test the filter as directed on pages 197–199.

Filtering Crucibles

Select the proper size of crucible for the precipitate to be filtered—the precipitate should form a layer of not over 1 mm or filtration and washing will be slow. Clean, prepare, and wash the filter as directed in Chapter 10, pages 200–203. Dry and heat the filter as directed in section 3 below.

(2) FILTERING AND WASHING PRECIPITATES

Do not begin the filtration until the precipitate has been brought into a filterable condition. Most precipitates will not be retained by filtering media when first formed; such finely divided precipitates will clog the filter and in many cases will make completion of the filtration and washing difficult if not impossible. The following means of increasing the filterability of a precipitate should be considered: (*a*) method and conditions of the precipitation, such as rate of addition of reagents, volume of solution, *p*H of the solution, and temperature; (*b*) subsequent heating of the mixture; (*c*) stirring; (*d*) addition of electrolytes; (*e*) long standing.

Filter and wash by decantation if the nature of the precipitate permits. To the extent possible, coagulate precipitates and then allow them to settle before beginning the filtration. Then decant as much as possible of the clear solution through the filter, taking care not to stir up the precipitate. In this manner it is frequently possible not only to filter a mixture completely but also to wash the precipitate without removing it from the precipitating vessel. This operation, termed *filtering and washing by decantation*, should be practised whenever the nature of the precipitate makes it feasible, since it permits these processes to be done more rapidly and efficiently than when the precipitate is allowed to pass to the filter. When this happens the precipitate usually partly clogs the filter, making the filtration much slower; also, the precipitate can be much more thoroughly treated with the wash solution in the vessel than on the filter.

After a portion of wash solution has been added and thoroughly mixed with the precipitate, it is an advantage to allow the vessel to stand in an inclined position. The precipitate then tends to settle compactly into the space at the juncture of the bottom and side of the vessel and is less likely to be stirred up when the vessel is further tilted to pour out the clear solution.

Transferring Solutions

When pouring a solution from one vessel to another or to a filter, guide the liquid by means of a stirring rod placed against the lip of the vessel (Figure 13-1). This operation not only prevents splashing of the solution on the filter or in the receiving vessel but also prevents the solution from running down the underside of the lip of the vessel from which it is poured. With some vessels the lip is so constructed that this spilling may take place even with the use of a stirring rod; in such cases rub the underside of the lip with the thinnest possible film of vaseline or other grease. (See Footnote 2, page 188, regarding "dripless' beakers.)

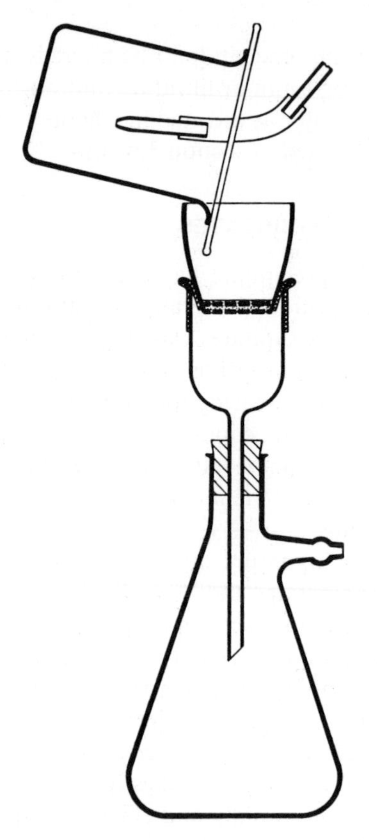

FIGURE 13-1
Method of transferring solution and precipitate from beaker.

Transferring Precipitates

When transferring a precipitate from a vessel, take extreme care that none is lost by spattering. First suspend the precipitate in a small amount of the wash solution and slowly drain it from the beaker into the crucible, using the stirring rod to guide it to the bottom of the crucible. Then, while holding the inclined beaker with the stirring rod held against the lip in one hand (Figure 13-1), direct a gentle stream of the wash solution (contained in a small wash bottle) against the bottom and sides of the beaker until all the precipitate has been washed out. If precipitate adheres to the sides of the beaker (and as a general precaution), gently rub the entire inner surface of the beaker with a "policeman," a stirring rod capped at one end with a short piece of soft gum tubing specially made for the purpose by being sealed at one end. (Policemen should not be used as stirring rods, since they are attacked by some solutions, may disintegrate, or may be hard to clean if left in the solution when the precipitation takes place.)

Washing Precipitates

In washing a precipitate, the first consideration is the selection of the most effective washing medium. It is not often that this will be pure water. In making this selection, the analyst must consider the following factors: (1) the solubility of the precipitate, (2) the stability of the precipitate, (3) the nature of the material to be washed out, (4) the tendency of the precipitate to become colloidal, and (5) the effect of the wash solution or its constituents on the drying and heating of the precipitate.

If the precipitate is appreciably soluble, it will be necessary to reduce its solubility in the wash medium. This reduction is often accomplished by (1) the addition of a common ion; (2) the use of some special solution, an example being the washing of a magnesium ammonium phosphate or arsenate precipitate with dilute ammonia; (3) the addition of organic liquids (frequently alcohol); and (4) the use of cold wash solutions (especially where the solubility of the salt has a high temperature coefficient).

Certain precipitates may undergo changes if washed with water; thus, bismuth phosphate may be partly converted into a basic salt. Other precipitates, certain sulfides for example, may be partly oxidized unless the wash solution is freed of oxygen, or an inhibiting substance (such as hydrogen sulfide) is present.

The nature of the ions and compounds in the filtrate must also be considered in the selection of the washing medium. This factor is especially important in separations, since, if the precipitation has taken place in a strongly acid solution and the precipitate were washed with water, some of the constituents of the filtrate might be partly precipitated. Thus, arsenic can be separated from antimony by precipitation as sulfide from a 9 F hydrochloric acid solution, but if the solution adhering to the precipitate were diluted by addition of wash water, precipitation of antimony sulfide would be likely to result.

The method of washing will be determined by the nature of the precipitate and the method of filtering. The advantages of decantation apply to the washing as well as the filtering process; settling the precipitate by centrifuging will often facilitate a decantation process. The type of filter used and the nature of the precipitate will indicate whether suction can be used to advantage.

In washing a precipitate, it is usually desirable, because of solubility effects and for prevention of undue dilution of the filtrate, to use the smallest possible volume of wash solution. For these reasons it is usually more effective to wash with repeated small portions of wash solution than with a smaller number of larger portions.

The effect of multiple washings upon any material in the solution is shown as follows: The concentration, C_1, of some species left in the solution with the precipitate after a single washing can be seen to be related to the original concentration of that species, C_o, by the expression

$$C_1 = \left(\frac{V_r}{V + V_r}\right) C_o$$

where V is the volume of wash solution added and V_r is the volume of solution that had remained with the precipitate before the washing. If n washings are made, each with the same volume, V, of wash solution, the concentration after the nth washing is

$$C_n = \left(\frac{V_r}{V + V_r}\right)^n C_o$$

Thus, if a precipitate weighing 0.2 g has left with it and the filter 0.5 ml of a solution and 0.1 g of contaminating material, and is washed with 20 ml of water, added in one case in 10-ml portions and the other in 5-ml portions, it is seen that in the first case there will be left 2.3×10^{-4} g of material and in the second only 6.8×10^{-6} g, these corresponding to an error of 0.1% and 0.004%, respectively.

These calculations indicate that precipitates can be washed with relatively small volumes of water; however, the assumption has been made that the foreign material has been merely held in the solution remaining on the precipitate and filter and that it has been uniformly distributed through each portion of wash water. In many cases, especially when bulky precipitates are dealt with, it may be difficult to get the wash water into effective contact with all portions of the mass. Still more important, the foreign material is likely to be adsorbed on the precipitate. Because of this latter effect, calculations such as the above are likely to be extremely misleading, and *the only safe way to determine when to discontinue washing a precipitate is to make tests on the wash solutions until some substance known to be present in the original filtrate in appreciable amounts, and for which a sensitive test can be made, is shown to be absent.* A negative test in the wash water does not guarantee that coprecipitated material is completely removed from the precipitate.

A precipitate should never be allowed to dry, or cake and crack, before or during a washing.

When one has decanted most of a solution from a precipitate or has successfully filtered a precipitate from the solution in which it was formed, it is usually an advisable precaution to remove this solution from the receiving vessel, or to substitute another receiving vessel, before beginning the washing process. Many precipitates that filter perfectly from their original solution tend to become colloidal (or, as it is termed, to *peptize*) and pass through the filter when treated with a wash solution; should this occur, or should a break subsequently occur in the filter, much less solution will have to be refiltered.

(3) DRYING AND HEATING OF PRECIPITATES

Certain precipitates will require only that they be heated sufficiently to remove superficial water; others will require a temperature adequate to remove water of constitution or to convert them to a different compound of known composition before they can be weighed. The factors that determine the temperature and time for which a precipitate should be dried have been discussed in section IV of Chapter 12. Data regarding the compounds precipitated, the compounds weighed, and the temperatures required are shown in Table 12-3, page 271.

Specific instructions for the use of various filtering crucibles with precipitates of silver chloride and barium sulfate are given in Procedures 14-1 and 14-2.

Sources of Heat

Electric ovens, provided with thermostatic controls, can be obtained with temperature ranges extending up to 150°C or in some cases 325°C. They are used for drying samples (usually at 110°C), for the drying of precipitates that can be rendered anhydrous at these temperature ranges, or for the preliminary drying of certain precipitates prior to ignition at a higher temperature.

Substances that would evolve corrosive fumes should not be heated in such ovens, nor should they be used for the evaporation or digestion of solutions.

Electric furnaces, preferably equipped with a temperature-indicating device and having a maximum temperature of at least 1000°C (preferably 1200°C), are used for ignitions requiring temperatures above the range of the ovens. They are indispensable for cases in which a close control of high temperatures is necessary.

Gas burners are used in the absence of electric equipment or when paper filters are to be ignited separately from the precipitate. They are not so satisfactory as the electric ovens, since the flame creates air currents that may cause loss of finely divided, fluffy precipitates, and in certain cases the products of combustion may contaminate the precipitate. Furthermore, control of the temperature of the heated object is much more difficult, since it has to be based upon experience or judged by the color of the heated object. The following general observations are useful:

A properly adjusted Bunsen burner can produce a temperature of 900°C–1050°C in a covered platinum crucible; a Tirrill burner, 1050°C–1150°C; a Meker burner, 1150°C–1250°C; and a blast burner, 1100°C–1300°C. The temperature attained in a porcelain crucible will be from 200°C to 300°C less (Hillebrand, Lundell, Bright, and Hoffman. *Applied Inorganic Analysis*, 2nd ed. New York: Wiley, 1953).

In judging temperatures, one should remember that at approximately 500°C a perceptible red glow is produced; this becomes a bright red glow at around 1000°C and an intense white glow at 1500°C.

For general purposes, care should be taken to adjust the burner so that an oxidizing (nonluminous) flame is produced. A luminous yellow flame causes carbon deposition on vessels and crucibles, does not produce as high temperatures, and *is very injurious to platinum-ware.*

The triangles used for supporting crucibles over gas flames for high-temperature ignitions are preferably made of fused silica, of clay tubes on heavy metal wire, or of platinum. Triangles of nichrome alloy are satisfactory for medium temperatures but are not recommended for high-temperature ignitions, especially with platinum crucibles. Triangles should be kept scrupulously clean, since many materials tend to fuse to the crucible.

(4) WEIGHING PRECIPITATES

After being heated precipitates should be allowed to cool to balance-room temperature. After the first weighing, they should be again heated and weighed, and this process continued until constancy of weight is obtained. Many precipitates after being heated to a high temperature will tend to absorb some moisture even from the air in a desiccator. Such precipitates should be weighed as soon and as rapidly as possible. Others, such as calcium oxide, rapidly absorb both water and carbon dioxide. In such cases they should be cooled and weighed in stoppered weighing bottles; allow them to cool somewhat before placing them in the container.

FILTERING AND WASHING OF PRECIPITATES. SUMMARY OF SUGGESTIONS AND PRECAUTIONS

1. Do not begin the filtration until the precipitate has been brought into a filterable condition. Most precipitates will not be retained by filtering media when first formed; such finely divided precipitates will clog the filter and in many cases will make completion of the filtration and washing difficult if not impossible.
2. Filter and wash by decantation if the nature of the precipitate permits. Where the precipitate is not to be weighed but is to be dissolved or to be treated with a solvent to extract some constituent, retain as much as possible of it in the original vessel and carry out the subsequent treatment there.
3. Keep both the solution and the wash solution hot during the filtration if the solubility and stability of the precipitate permit.
4. *Always* use a stirring rod when transferring a solution from one vessel to the filter or to another vessel.
5. Remove the film of precipitate from the surface of the precipitating vessel by means of a policeman; do not use a stirring rod for this purpose.
6. Remove the original filtrate before beginning to wash the precipitate.
7. Wash the precipitate immediately after it is filtered; otherwise it is likely to dry into a caked mass, with cracks, and to become resistant to solvents.
8. Washing is more efficient with repeated small portions of wash water (see page 278).
9. Where necessary, add acid or other electrolytes to prevent the precipitate from becoming colloidal.
10. *All filtrates or the clear supernatant solution should be properly tested* to be sure an excess of precipitant is present.
11. *All wash solutions should be properly tested* to be sure washing is completed.
12. Cover and plainly label all vessels containing solutions.
13. Clean apparatus and remove it from the working desk. Vessels left containing solutions to be discarded are likely to cause confusion.
14. *Keep filter stands, apparatus, and desks dry and clean.* A sloppy, cluttered desk usually indicates an inefficient, ineffective, untrustworthy workman.

USE OF VARIOUS FILTERING MEDIA

I. Paper Filters

(Most of the suggestions that follow are discussed above under Paper Filters)

1. If the precipitate is to be weighed (or ignited), use a quantitative, "ashless" filter.
2. Select a paper of the *proper type* for the precipitate (or the solution) to be filtered.
3. Select a paper of the proper *size* for the precipitate. Do not fill a paper filter over one-third full of precipitate or two-thirds full of solution.
4. Fit the paper to the funnel, *and test it*, before beginning the filtration.
5. Cause the beveled tip of the funnel to touch the side of the receiving vessel.

II. Filtering Crucibles

1. In general, gelatinous or flocculent and bulky precipitates are more effectively filtered through paper.
2. Use the minimum amount of suction (strong suction tends to pack the precipitate against the filtering element and thus to clog it).
3. To remove a precipitate, choose the proper solvent for that precipitate and type of crucible.

A. *Asbestos Filters* (with particular reference to Gooch-type filtering crucibles).

1. Use only acid-washed asbestos (see Asbestos Filters, page 200, and Asbestos, under Reagents in the Appendix).
2. Use the thinnest mat that will retain the precipitate.
3. Thoroughly wash the mat before using it; watch the filtrate for asbestos fibers.
4. Add solution so as not to stir up the mat (protective plates are advisable).

B. *Other Filtering Crucibles.* Other filtering crucibles are made of (*a*) sintered glass, (*b*) porcelain, (*c*) quartz, (*d*) platinum (Munroe type).

1. Do not use these crucibles with solutions that will attack them, especially HF solutions with (*a*), (*b*), or (*c*), or oxidizing solutions containing halide ions with (*d*).
2. Do not collect precipitates in these crucibles that cannot be readily removed by treatment with the proper solvent.
3. Do not heat sintered-glass crucibles above 450°C–500°C; they should be heated and cooled with care.

4. Precipitates that are easily reduced to elements alloying or forming compounds with platinum (such as silver, lead and tin salts, and sulfides) should not be heated in platinum crucibles.

CHAPTER 14

Gravimetric Methods

The two preceding chapters have presented principles and operations of gravimetric analysis. In this chapter procedures for the gravimetric determination of chloride and of sulfate are presented. The samples for analysis are assumed to be soluble and not to contain interfering constituents. A general discussion of methods for separating and eliminating such constituents has been given in Chapters 8 and 9. The discussions of Chapters 12 and 13 concerning the general principles and operations of gravimetric methods should be reviewed in connection with these procedures.

PROCEDURE **14-1**

Gravimetric Determination of Chloride as Silver Chloride

Outline. In this procedure (*a*) a sample of a soluble chloride is dissolved and the solution is acidified with nitric acid, (*b*) silver chloride is precipitated by addition of silver nitrate in small excess and the precipitate is coagulated, (*c*) the precipitate is filtered through a weighed filter and washed, and (*d*) the precipitate is dried and weighed.

DISCUSSION

The chloride present in a solution can be very accurately determined gravimetrically by precipitating it as silver chloride and weighing the precipitate. This method is made use of here because, first, it can be carried out with exceptional accuracy. This is true because silver chloride is remarkably free from the coprecipitation phenomena discussed in Chapter 12, so that with reasonable care even an inexperienced analyst should obtain satisfactory results. Second, it can be used to illustrate exceptionally well many of the principles and much of the technique involved in gravimetric analysis.

Certain features of the successive steps in this procedure are discussed below.

Preparation of the Solution

The solution of a chloride sample is usually made acidic with nitric acid prior to the addition of silver nitrate. This is done to make the precipitation more specific by preventing the possible precipitation of the silver salts of weak acids, such as carbonate and phosphate, which are insoluble in neutral solutions (see Table 17-2). A moderate hydrogen-ion concentration is also an effective aid in the coagulation of the colloidal precipitate that is first obtained, and acid is added to the wash water to prevent the precipitate from reverting to a colloidal condition.

This procedure for determining chloride cannot be applied to solutions or materials containing (1) anions that form silver salts insoluble in acid solutions (see Table 17-2); (2) cations, such as mercuric ion, that form stable un-ionized or complex ions with chloride; or (3) constituents capable of reducing silver ion to metallic silver.

Precipitation of the Silver Chloride

The solubility of silver chloride in pure water, 1.8 mg per liter at 25°C and 21 mg at 100°C, is such that by the addition of an excess of silver nitrate, which should not exceed 0.01 *F*, the solubility loss can be reduced to a value that is inappreciable for most work.[1] Because of interionic attraction effects and the formation of complex ions such as Ag_2Cl^+, a large excess of silver nitrate may lead to an *increase* in solubility, especially if other salts are present in the solution. In addition, there will be more adsorption of silver nitrate and more effort required in adequately washing the precipitate.

As was noted above, the initial silver chloride precipitate tends to be colloidal. This colloidal precipitate undoubtedly contains adsorbed silver ions, but according to Kolthoff and Sandell[2] a silver chloride precipitate "ages" so rapidly and with such a decrease in total surface that only a few minutes after the precipitation, even at room temperature, the precipitate does not exert noticeable adsorptive properties toward silver nitrate. This "aging" effect takes place more rapidly when the mixture is heated.

Filtering and Washing the Precipitate

A silver chloride precipitate can be filtered through a paper filter or any of the filtering crucibles discussed in Chapter 10. Paper filters have the disadvantage that there is danger of loss of fine particles and of volatilization of precipitate during the burning ("ignition") of the paper. In addition there is partial reduction to metallic silver during this process and a treatment with nitric and hydrochloric acid is required to convert the reduced silver to the chloride. For these reasons the use of paper filters is to be avoided if possible.

If a coagulated precipitate, and especially silver chloride, is washed with pure water, it will tend to "peptize," that is, revert to a colloidal state, and to pass through the filtering medium. This peptizing process results from removal of a coagulating electrolyte, in this case predominantly silver nitrate. By providing a volatile electrolyte in the wash water, the nonvolatile silver

---

[1] For exceedingly accurate work, such as atomic-weight determinations, care is taken to obtain the optimum excess concentration and to make corrections for the dissolved silver chloride. For reports on such work see Richards and Wells. *J. Am. Chem. Soc.*, **27**, 459 (1905); Richards and Willard. *J. Am. Chem. Soc.*, **32**, 4 (1910); and Baxter and Hilton. *J. Am. Chem. Soc.*, **45**, 695 (1923).

[2] Kolthoff, Sandell, Meehan, and Bruckenstein. *Quantitative Chemical Analysis.* (4th ed.) New York: Macmillan, 1969, p. 582.

nitrate can be replaced by nitric acid, which is volatilized when the precipitate is dried.

Photochemical Effect. Silver chloride is photochemically decomposed by light into metallic silver and chlorine. If this process takes place after the precipitate has been filtered and washed, there is a loss in weight. If this process occurs in the presence of a solution that contains an excess of silver ions (as is the case when a chloride sample is precipitated as silver chloride), additional precipitate is formed because of the hydrolysis and disproportionation to chloride and chlorate:

$$2AgCl(s) \underset{h\nu}{=} 2Ag(s) + Cl_2$$

$$3Cl_2 + 3H_2O + 5Ag^+ = 5AgCl + ClO_3^- + 6H^+$$

If prolonged exposure to bright daylight is avoided, these errors are usually not serious, even though the precipitate may become purplish violet to gray in color.[3]

Drying the Precipitate

A silver chloride precipitate can be heated to the fusion point (455°C) without decomposition or significant volatilization, and fusion is recommended when extreme accuracy is required, as in atomic-weight determinations. It has been found that precipitates dried at 280°C retain only 0.01% of water,[4] and at 100°C–120°C only 0.03%–0.04%.[5]

INSTRUCTIONS (Procedure 14-1: Gravimetric Determination of Chloride as Silver Chloride)

Precipitation of the Chloride

In case the sample is a solution, pipet an appropriate volume (Note 1) of the solution into a 400-ml beaker (Note 2). Dilute the solution to 100 ml (Note 3).

NOTES

1. Consult your instructor regarding the number of determinations to be made. The volume or weight of sample to be taken will be determined by the concentration of the chloride in the solution or the percentage of chloride in the solid. If less than 25 ml is pipeted, the precision of that volumetric measurement becomes less

[3] Lundell and Hoffman (*J. Res. Natl. Bur. Std.*, **4**, 109, 1930) found a positive error of 0.2% after 3 hours' exposure of a precipitate and solution to bright daylight.

[4] W. F. Hillebrand, G. E. F. Lundell, H. A. Bright, and J. J. Hoffman. *Applied Inorganic Analysis.* (2nd ed.) New York: Wiley, 1953, p. 207.

[5] Kolthoff, Sandell, Meehan, and Bruckenstein, op. cit., p. 583.

In case the sample is a solid, accurately weigh out a sample that is thought to contain 0.15 g–0.25 g of chloride (Note 1) and transfer it to a 400-ml beaker (Note 2). Add 100 ml of water and dissolve the solid (Note 3).

Add to the solution 1 ml of 6 *F* HNO_3. Calculate the volume of a 0.2 *F* $AgNO_3$ solution thought to be equivalent to the chloride present, and add it dropwise to the solution (Notes 4 and 5); stir the mixture vigorously (do not spatter or splash it) during the addition, and then add 1 ml in excess (Notes 6 and 7). Do not carry out this precipitation or the subsequent operations in sunlight or in very intense artificial illumination. Heat the mixture

NOTES

than is desirable; it is also desirable that the weight of the silver chloride precipitate be not less than 0.3 g (to reduce the weighing error) and not larger than 1 g (to avoid difficulty in the filtering and washing of the precipitate). Duplicate (for special work, triplicate) determinations should be made.

2. A beaker, rather than the conical flasks generally used for titrations, is employed to advantage when it is desired to transfer the resulting precipitate to a filter or to another vessel.

3. If the solution of the sample is alkaline, indicating a soluble base or salts of weak acids, make it just acid by careful addition of 6 *F* HNO_3.

If the sample is a solid that does not dissolve in water, dissolve it in an excess of 6 *F* HNO_3, then neutralize the excess with 6 *F* NaOH, and finally make the solution just acid with the HNO_3. Consult more advanced reference works in regard to appropriate methods for decomposing insoluble or organic materials and for removing interfering constituents.

4. The amount of silver nitrate to be added cannot be accurately calculated except in the analysis of a sample whose chloride content is already closely known. In other cases the equivalent amount can be determined by noting when no more precipitation takes place upon addition of a portion of the reagent. In this operation add the $AgNO_3$ in successively smaller portions as the rate of precipitation is seen to be decreasing, until finally the effect of a few drops is noted. As the equivalence point is approached, shake or stir the mixture vigorously after each addition of $AgNO_3$ and allow the precipitate to settle so that the effect of the next portion can be observed in the supernatant liquid. Usually the precipitate will remain colloidally dispersed until near the equivalence point, when a distinct coagulation will be observed. This is termed the *clear point* and is used in certain titrations as an end-point method. After the equivalent amount of silver nitrate required has been noted, the 1-ml excess is added.

5. The calculated volume of the $AgNO_3$ may be measured in a graduate and added from a dropper, or it may be added from a graduated pipet or buret. In order to avoid splashing, deliver the solution very close to the surface of the solution or flow it down the side of the beaker.

6. Stirring rods used for this purpose should have their ends fire-polished and should not be rubbed against the walls of the vessel; otherwise, small particles of glass may be chipped off or the inside of the vessel may be scratched. Precipitates tend to form in such cracks and are difficult to remove.

to approximately 80°C–90°C and stir it frequently until the precipitate settles rapidly, leaving the upper portion of the solution clear. Keep the beaker covered when not stirring the solution.

Add to the clear supernatant solution 3 drops of 0.2 *F* $AgNO_3$. If a turbidity is produced, add 1 ml of the $AgNO_3$, and again heat, stir, and test as before. Repeat this process until the solution remains clear upon addition of the 3 drops of $AgNO_3$ (Note 8).

Continue heating and stirring the solution or allow it to stand in the dark until the entire solution is clear (Note 9).

Filtering the Precipitate

Read pages 200–205, regarding the preparation of filtering crucibles and pages 274–277 regarding the filtering of precipitates.

Prepare and dry to constant weight at 100°C–120°C filtering crucibles of the type suggested by your instructor (Note 10).

Insert the crucible in the holder and, applying a very slight suction (Note 11), decant the clear solution through the filter (Note 12). Carefully guide the first portion of the solution onto the filter with a stirring rod, and thereafter

NOTES

7. Vigorous stirring is very effective in coagulating a silver chloride precipitate, although extreme care must be taken to avoid splashing the solution.

8. It is a general rule that a test should be made to ascertain if an excess of the precipitating reagent has been used by adding a small additional amount to the clear supernatant liquid, or to the first portion of the filtrate coming through the filter, and then, after allowing sufficient time, to note if any additional precipitate forms.

9. Do not begin filtration until the solution is clear. If possible, make the precipitation so that the solution can be allowed to stand overnight, since usually at least 1 or 2 hours of heating and stirring are required before a clear solution is obtained.

10. Instructions for the preparation of the various types of crucibles are given in Chapter 10. The pages for specific instructions are as follows: asbestos filters, page 202; glass fiber mats, page 204; sintered glass and porous porcelain, page 205.

11. The slight suction is used to hold asbestos fibers in place when the solution first hits the mat. With fabricated mats the suction should be dispensed with for as long as the solution flows through the filter at a reasonable rate. *Avoid too strong a suction*; it may break the mat and will tend to pack the precipitate so tightly as to clog the filter and slow the flow.

12. See pages 275–279 regarding filtering and washing by decantation. Decantation should be used whenever the nature of the precipitate makes it feasible, since it permits these processes to be done much more rapidly and efficiently than when the precipitate is allowed to pass to the filter.

do not allow the crucible to become empty (Note 13). Note carefully if any turbidity or any asbestos fibers (if an asbestos filter is used) appear in the first portions of the filtrate; if so, it must be refiltered (Notes 14 and 15).

Washing the Precipitate

Read pages 277–279 regarding the washing of precipitates.

Prepare a wash solution by adding 5 ml of 6 F HNO_3 to 500 ml of water and put this in a small wash bottle. (Label this special wash bottle.) Wash the precipitate by decantation with three 25-ml portions of this wash solution, transfer it to the crucible (Note 16), and wash with the dilute acid until 2–3 ml of the washings give no precipitates upon the addition of a single drop of HCl. Finally, wash with 5 ml of water, added dropwise; if the last drops through the filter turn blue litmus red, continue washing (Note 17).

Drying and Weighing the Precipitate

Place the crucible in a covered beaker and heat it at 120°C for 1 hour (Note 18). Cool the crucible for 30 minutes in a desiccator and weigh. Repeat the heating and weighing until the weight is constant to 0.2 mg.

Calculate the concentration of chloride in a solution sample or the percentage of chloride in a solid sample.

NOTES

13. See Figure 13-1, page 276, and accompanying material regarding transferring solutions and precipitates.

14. Remove the clear solution from the receiving vessel, or substitute another receiving vessel, before beginning the washing process.

15. Treat the original filtrate and the first portion of wash water with a slight excess of hydrochloric acid, allow the precipitate to settle, decant the solution away, and transfer the residue to the "silver residue" container. After finishing this procedure, also add the weighed silver chloride precipitates to the same container for later recovery of the silver.

16. See pages 277–279 regarding the transferring and washing of precipitates.

17. Add this water in drops so that the *entire inner surface* of the *crucible* and *filter* is washed. Let each drop flow or be sucked away before applying the next one. Use a dropper to control the drops properly.

Nitric acid not removed from the asbestos or bottom of the crucible may not be completely volatilized by one period of heating, especially at around 110°C–120°C. When the crucible is weighed, this acid causes the corrosion rings often seen on balance pans.

18. Dry the precipitate at the same temperature as the crucible was dried. If final temperatures of above approximately 120°C are to be used, the precipitate

Gravimetric Determination of Sulfate as Barium Sulfate

Outline. The sulfate sample is dissolved in water and the solution acidified with hydrochloric acid. Barium sulfate is precipitated by the addition of barium chloride. The precipitate is washed, dried at 500°C–700°C, and weighed.

DISCUSSION

Precipitation of the $BaSO_4$

Barium sulfate was one of the first compounds used as the basis of a gravimetric determination. Also, some of the earliest observations regarding coprecipitation phenomena were made in connection with this precipitate,[6] and it is probable that more work has been done on the conditions and methods for precipitating this compound than on any other used for gravimetric methods. This is true, first, because of the commercial and technical importance attached to this method for determining sulfur, and second, because barium sulfate, even though usually precipitated in a distinctly crystalline form, coprecipitates both cations and anions to such an extent as not only to cause serious errors but also to present favorable opportunities for the study of the causes of the coprecipitation. Some of these studies are discussed later.

NOTES

should be first dried for 20 minutes at 100°C–110°C and then for the same length of time at the higher temperature. The precipitate should be gradually heated at first, or it tends to "cake" and enclose the water, so that a fusion is necessary thereafter to complete its expulsion. If the precipitate is heated to this higher temperature, the crucible should have been similarly treated before the initial weighing.

[6] Berzelius (*Ann. Chim. Phys.*, **14**, 376, 1820) discussed the state of combination of the impurities, and Turner (*Phil. Trans. Roy. Soc. London*, **119**, 295, 1829) spoke of the "adhesion of potassium sulfate" to the barium sulfate precipitate.

Solubility of $BaSO_4$

The solubility of barium sulfate, $BaSO_4$, has been found to be 0.00023 g at 18°C and 0.00041 g at 100°C per 100 ml of water. The formal solubility of barium sulfate in water at room temperature is approximately the same as that of silver chloride. It seems anomalous that under comparable conditions barium sulfate will appear in a granular crystalline form whereas silver chloride appears as a colloidal suspension, coagulating into an amorphous solid. As discussed in Chapter 12, this difference in form has been attributed by Kolthoff[7] to the fact that the solubility of silver chloride (which forms soft crystals of low surface tension) remains relatively constant, whereas that of barium sulfate (which forms hard crystals of high surface tension) increases greatly as the particle size is decreased. Thus Kolthoff[8] has calculated that the solubility of barium sulfate particles of radius of 0.02 micron (one micron is 10^{-4} cm) is 1000-fold greater than that of large crystals.

The solubility of barium sulfate in water is increased by the addition of strong acids. This result would be expected both because of interionic attraction effects and because the second proton of sulfuric acid is not completely ionized. The data shown below have been taken from various sources and show the effect of hydrochloric and nitric acid on the solubility of barium sulfate in aqueous solutions:

Acid	—	HCl	HCl	HCl	HNO_3	HCl	HNO_3	HNO_3
Formality	0.00	0.10	0.30	0.50	0.50	1.0	1.0	5.0
$BaSO_4$ dissolved (mg/100 ml)	0.23	1.0	2.9	6.7	7.0	8.9	10.7	24

A certain amount of acid is desirable in the solution from which the precipitation is made, since it prevents the precipitation of barium salts of weak acids, such as barium carbonate or phosphate, and since the increase in solubility causes the precipitate to be less dispersed and more readily filtered. The precipitation is usually made in solutions from 0.01 to 0.1 F in hydrochloric acid; nitric acid is to be avoided because of coprecipitation effects. The addition of a small excess of barium chloride serves to reduce the solubility at the above acid concentrations to quantities that can be neglected except in work of exceptional accuracy.

As would be expected, the presence of cations that form complex ions with sulfate increases the solubility of barium sulfate. This effect is so pronounced with tripositive chromium that the precipitation should not be attempted

[7] Kolthoff. *J. Phys. Chem.*, **36**, 860 (1932).
[8] Ibid.

in the presence of chromic ion; 100 ml of a solution containing 1 g of ferric chloride dissolves 12 mg of barium sulfate; a similar solution of aluminum chloride dissolves 9 mg.

Coprecipitation by Barium Sulfate

The use of barium sulfate in the analysis of iron pyrites, FeS_2, has been one cause for the commercial and technical importance attached to gravimetric barium sulfate determinations. In this analysis the crude pyrites is treated with either a "dry" or "wet" oxidizing medium. The "dry way" consists of a fusion with an alkaline flux, such as sodium carbonate, to which an oxidant, such as sodium peroxide, is added. The fusion product is then dissolved in hydrochloric acid. The "wet way" consists of treatment with a strong acid, such as hydrochloric, to which an excess of an oxidant, such as liquid bromine, is added.

As a result of these treatments the sulfur is quantitatively converted to sulfate, and the iron to Fe(III). It was realized quite early in the history of analytical chemistry that ferric iron is extensively coprecipitated by barium sulfate.[9] An interesting series of studies, extending from 1889 to 1900, led to various conclusions as to the cause of this coprecipitation. Jannasch and Richards[10] attributed the effect to the formation of a double salt; Schneider[11] proposed that limited solid solution formation was responsible; Kuster and Thiel[12] postulated that complex formation was responsible and stated that compounds such as $Ba[Fe(SO_4)_2]_2$ were formed; Richards[13] in 1900 argued that the formation of a basic salt such as $FeOHSO_4$ was the predominant factor. Other explanations have involved the adsorption of negative complex ions, such as $Fe(SO_4)_2^-$, on the positively charged colloidal barium sulfate resulting from the precipitation in the presence of an excess of barium chloride. The exact mechanism of this coprecipitation is still uncertain. Several effects may be involved in varying degrees, depending upon the conditions, such as *p*H, temperature, and concentrations, prevailing at the time of the precipitation.

Aside from attempts to minimize the coprecipitation of iron by modifying the conditions of the precipitation, various methods have been suggested for eliminating the effect and three methods have found extensive use. They are (*a*) reduction to ferrous iron, (*b*) precipitation of ferric hydroxide, and (*c*) removal by a cation-exchange resin.

---

[9] Turner, op. cit., p. 291.
[10] Jannasch and Richards. *J. Prakt. Chem.* (2), **39**, 321 (1889); **40**, 233 (1889).
[11] Schneider. *Z. Physik. Chem.*, **10**, 425 (1892).
[12] Kuster and Thiel. *Z. Anorg. Chem.*, **19**, 97 (1899); **21**, 73 (1899); **22**, 424 (1900).
[13] Richards. *Z. Anorg. Chem.*, **23**, 383 (1900).

(*a*) *Reduction of Ferric Iron.* The reduction of the ferric iron to the ferrous state was proposed quite early, since it was found that ferrous iron is much less coprecipitated than is ferric, and various reducing agents have been suggested for this purpose.[14] A procedure involving the use of aluminum metal is extensively used at the present time and is capable of giving very reproducible results.

(*b*) *Precipitation of Ferric Iron with Ammonia.* Lunge, in 1880, suggested that the ferric iron be removed before the barium sulfate precipitation by precipitation of "ferric hydroxide," actually a hydrous ferric oxide, with ammonia;[15] this method has been extensively used. Because the precipitate tends to coprecipitate sulfate, for very accurate work it is dissolved in hydrochloric acid and the precipitation repeated.

(*c*) *Removal of Ferric Iron by a Cation-Exchange Resin.* The iron and other cations can be quickly and conveniently removed by a cation-exchange resin,[16] and this process offers the advantage that the precipitation can be made in a solution free from all interfering metal ions. This method is recommended if repetitive determinations are being made and when a cation-exchange column is available.

Coprecipitation of Other Ions

As stated above, barium sulfate coprecipitates both foreign cations and anions to a very appreciable extent; and in spite of the tremendous amount of experimental work that has been done on the subject, in many cases the exact causes and mechanism of these effects are not clearly established.

Anions are coprecipitated largely as the barium salts and will cause positive errors. In accord with the general rules applying to coprecipitation effects discussed in Chapter 12, the coprecipitation of a given anion increases with the concentration of that anion; therefore it is advantageous to use dilute solutions and to avoid undue salt concentrations. With anions of a given type, such as the halides, the coprecipitation varies inversely as does the solubility of the barium salt of the anion. Thus, the order for the barium halides is $BaCl_2 > BaBr_2 > BaI_2$; the coprecipitation of nitrate is more pronounced

---

[14] Johnson (*Chem. News*, **70**, 212, 1894) suggested the use of a hypophosphite; Allen and Bishop (*Eighth Intern. Congr. Appl. Chem.*, **1-2**, I-II, 48, 1912) and Moore (*Ind. Eng. Chem.*, **8**, 26, 1916) have used aluminum foil and powder, respectively; Virgin (*Tefinsk. Tids., Kemi*, **53**, 1, 1923) used hydroxylamine; and Liebenberg and Leith (*J. S. African Chem. Inst.*, **14**, 47, 1931) used potassium iodide.

[15] Lunge. *Z. Anal. Chem.*, **19**, 419 (1880); Weiser (*Inorganic Colloid Chemistry*. Vol. 2. The Hydrous Oxides and Hydroxides. New York: Wiley, 1935, pp. 27–85) gives a discussion of the formation and properties of the hydrous oxides and hydrates of ferric iron.

[16] The use of these resins is discussed in Chapter 26 where such a resin is used in Procedure 26-1 for the determination of the total cations in a solution.

than that of these halides. Coprecipitation will be greater if the sulfate solution is added to a barium solution containing the anion than with the reverse order. The coprecipitation of cations takes place primarily as the sulfates and in general follows these same rules. However, cation coprecipitation will tend to cause negative errors; the cation is coprecipitated as sulfate and most cations have a lower equivalent weight than does barium. In addition, the coprecipitation appears to be increased when sulfate complex formation is involved, being greater with ferric iron than with either ferrous or aluminum ion.

The causes and mechanisms involved in this coprecipitation of anions and cations by barium sulfate have been the subject of many investigations. Thus, Kolthoff and Sandell[17] state that under analytical conditions the barium sulfate precipitate is microcrystalline and that the coprecipitated materials are largely present in the interior of the crystals as the result of the incorporation in the crystals of ions or molecules adsorbed during the growth of the precipitate; studies by others have also produced evidence that adsorption is a considerable factor.[18] However, x-ray diffraction studies of barium sulfate crystals containing adsorbed alkali metal cations, and also those containing adsorbed barium nitrate, have produced evidence that limited solid solutions are formed.[19]

The general magnitude of the coprecipitation effects and the variations obtained when the precipitation of the barium sulfate is made under uniform conditions but in the presence of equivalent concentrations of various chlorides and nitrates are shown in Table 14-1.

Means of Minimizing Coprecipitation by Barium Sulfate. In the procedure below, nitrate and ferric iron are assumed to have been absent from the sample or are eliminated before the precipitation is made, and the precipitation is then made from a large volume of hot solution. The conventional method of precipitation consists of the slow addition of the barium chloride to the sulfate solution containing hydrochloric acid. This method tends to cause low results because of coprecipitation of ammonium or alkali metal sulfates, since the formation of most of the precipitate takes place in the presence of an excess of sulfate. For this reason the suggestion has been made that, especially when high concentrations of alkali metal salts are present, the sulfate solution be added to the barium chloride solution; this procedure tends to give positive errors because of coprecipitation of barium salts, but

[17] Kolthoff, Sandell, Meehan, and Bruckenstein, op. cit., p. 603.

[18] McHenry, Ampt, and Heyman. *J. Am. Chem. Soc.*, **62**, 448 (1940); de-Brouchère. *Bull. Acad. Roy. Belg.*, **13**, 827 (1947); Gaicolone and Russo. *Gazz. Chim. Ital.*, **66**, 631 (1936).

[19] Walton and Walden. *J. Am. Chem. Soc.*, **68**, 1742, 1750 (1946); Walden and Cohen. *J. Am. Chem. Soc.*, **57**, 2591 (1935); Averell and Walden. *J. Am. Chem. Soc.*, **59**, 907 (1937).

TABLE 14-1
Coprecipitation of Various Chlorides and Nitrates by Barium Sulfate

Chloride added	None	HCl	NaCl	KCl	NH_4Cl	$CaCl_2$	$MgCl_2$
Error (mg)	+1.7	+2.1	−2.2	−5.9	−3.2	−16.0	+3.0
Nitrate added	None	HNO_3	$NaNO_3$	KNO_3	NH_4NO_3	$Ca(NO_3)_2$	$Mg(NO_3)_2$
Error (mg)	+1.7	+27.0	+27.3	+18.2	+12.3	+4.1	+21.7

SOURCE: The data for this table are taken from Blasdale. *Quantitative Analysis*. Princeton, N.J.: Van Nostrand, 1918, p. 138.

NOTE: In all cases 25 ml of a standard solution containing sulfuric acid equivalent to 1.0118 g of barium sulfate were taken. Ten milliequivalents of the compounds shown were added, the solution was diluted to 200 ml and heated to boiling, 50 ml of a solution containing 1.3 g of $BaCl_2$ were added, the solution was allowed to stand for 16 hours, and then the precipitate was filtered, washed, and ignited. The errors shown are the average of at least two experiments.

in the presence of alkali metals gives more nearly accurate values.[20] A procedure that represents a compromise between these two methods, which was proposed by Hintz and Webber[21] and subsequently investigated by Rieman and Hagen,[22] involves adding the barium chloride *rapidly* to the sulfate solution; the time during the formation and growth of the precipitate that the sulfate is in excess is minimized and, according to Rieman and Hagen, the positive and negative errors are more nearly balanced. For this reason, and because of its quickness and convenience, this procedure is used below. The initial precipitate is finely divided but ages so rapidly that filtration can be started after the mixture is kept hot for 1 hour.

Filtering and Washing the Precipitate

If the precipitate has been allowed to stand or is "digested" hot for a sufficient length of time, there is usually no difficulty in filtration.

Filtering crucibles or paper filters can be used, each having certain advantages and liabilities. Asbestos (Gooch) filters may be slow in filtering and in attaining constant weight; sintered-glass filtering crucibles do not permit drying at temperatures much above 350°C–400°C; porous porcelain and platinum sponge crucibles can be heated to higher temperatures. Removal of the precipitate from the fabricated crucibles may be difficult. The filtering

---

[20] Popoff and Neuman. *Ind. Eng. Chem.*, anal. ed., **2**, 45 (1930).

[21] Hintz and Webber. *Z. Anal. Chem.*, **45**, 31 (1906).

[22] Rieman and Hagen. *Ind. Eng. Chem.*, anal. ed., **14**, 150 (1942). This article gives an interesting comparative study of various precipitation procedures; that by Fales and Thompson (*Ind. Eng. Chem.*, anal. ed., **11**, 206, 1939) has an extensive bibliography and also gives the results of studies of various factors such as the order of precipitation, concentrations, temperature, *p*H, and methods of drying.

and washing can be done rapidly and effectively with paper filters, but the subsequent ignition and treatment of the residue to eliminate errors caused by reduction of the sulfate is time-consuming and is accompanied by the danger of accidents. Optional procedures for the use of these various filtering methods are provided below.

Hot or warm water can be used for the washing of the precipitate, provided a minimum amount is used.

Drying and Heating the Precipitate

In heating a barium sulfate precipitate, two problems have to be considered: (1) the dehydration or removal of water and (2) the volatilization or decomposition of coprecipitated impurities. There is evidence that some of the water carried down by the precipitate is present as a limited solid solution, and its removal requires temperatures above 500°C.[23]

Although pure $BaSO_4$ has been shown to be stable and nonvolatile at the maximum temperatures ordinarily used for analytical work (1000°C–1200°C), this is not true of the impurities that may be present. Thus, sulfuric acid, acid sulfate, ammonium sulfate, and certain metal sulfates would be decomposed or volatilized at such temperatures, with a resultant increase in the negative error caused by their coprecipitation. This effect has caused Fales and Thompson to recommend that the precipitate be dried at 110°C–120°C instead of being ignited at 600°C–800°C as more commonly recommended. It is probable that by this method of drying there is a compensation of the negative errors caused by sulfate coprecipitation with the positive errors due to incomplete removal of water.

Weighing the Precipitate

Although it has been shown that a barium sulfate precipitate will take up water after having been dried at a high temperature, this effect is not so pronounced as to require special precautions in weighing the ignited precipitate.

Other Gravimetric Methods Involving the Precipitation of Barium Sulfate

Essentially the procedure below can be used for determining the sulfur in organic and inorganic compounds containing sulfur that can be converted to sulfate. Provided that loss by volatilization is guarded against, these

---

[23] Walton and Walden, op. cit., p. 1750; see also Kolthoff and McNevin. *J. Phys. Chem.*, **44**, 921 (1940).

would include sulfides, sulfites, and, in fact, most inorganic sulfur compounds except the insoluble sulfates such as barium and lead.

Barium can be precipitated and weighed as the sulfate. This determination is less susceptible to coprecipitation errors, since most anions that are co-precipitated as barium salts can be corrected for by adding sulfuric acid to the precipitate and again igniting.

INSTRUCTIONS (Procedure 14-2: Gravimetric Determination of Sulfate as Barium Sulfate)

(Consult your instructor regarding the number of determinations to be made.)

Precipitation of the Sulfate

In case the sample is a solution, pipet appropriate volumes of the solution into 400-ml beakers. Dilute each to 300 ml with water. To each add 3 ml 6 *F* HCl.

In case the sample is a solid, accurately weigh out 0.3- to 0.5-g portions into 400-ml beakers. Dissolve each in water and dilute to 300-ml. To each add 3 ml of 6 *F* HCl.

Heat the sulfate solutions almost to boiling. Dilute 13 ml of 0.5 *F* $BaCl_2$ to 50 ml in a graduate cylinder. Stir the sulfate solution and add the $BaCl_2$ as rapidly as possible without causing spattering. Stir the hot mixture continuously for five minutes.

Cover the beakers and either (*a*) let the mixture stand at least 12 hours or (*b*) keep it boiling for 1 hour, stirring at frequent intervals (Note 1).

Optional Instructions

Optional instructions are given below for the use of filtering crucibles (Option *A*) or paper filters (Option *B*).

Filtering the Precipitate. Read pages 275–277 regarding the filtering of precipitates.

NOTES

1. As soon as the precipitate settles, leaving a clear supernatant solution, add by means of a dropper held at the surface of the solution 1 ml more of the 0.5 *F* $BaCl_2$; then note if more precipitate forms. If a precipitate is obtained, continue the addition of $BaCl_2$ in 1-ml portions until no more precipitate forms; then add 3 ml in excess.

Option A. The Use of Filtering Crucibles

Read pages 202–205 regarding the preparation of filtering crucibles.

Preparing the Crucible. Obtain, prepare, and dry to constant weight filtering crucibles of the type directed by your instructor (Note 2; also read Note 10, Procedure 14-1).

Filtering and Washing the Precipitate. Insert the crucible in the holder and, applying a very slight suction, decant the clear solution through the filter (Notes 11 and 12, Procedure 14-1). Carefully guide the first portion of the solution onto the filter with a stirring rod, and thereafter do not allow the crucible to become empty. (Notes 13 and 14, Procedure 14-1.) Wash the precipitate by decantation with three 25-ml portions of water at 60°C–80°C (see pages 277–279 regarding the washing of precipitates). Transfer the precipitate to the crucible, and wash with the hot water until a 2- to 3-ml portion of the washings gives no precipitate upon the addition of a drop of 1 F $AgNO_3$.

Drying and Weighing of the Precipitate. Dry the crucible and precipitate by the same method and at the same temperature used for drying the empty crucible. Cool the crucible for 30 minutes in a desiccator and weigh. Repeat the heating and weighing until the weight is constant to 0.2 mg.

From the weight of the precipitate calculate the concentration of sulfate in the solution or the percentage of sulfur in the solid sample.

Option B. The Use of Paper Filters

These instructions are provided for the use of a paper filter in case suitable filtering crucibles are not available or that it is desired to dry the $BaSO_4$ precipitate at a higher temperature than is permitted by the kind of filtering crucibles available.

Read pages 196–199 regarding the characteristics of paper filters and instructions for their use.

NOTES

2. If an asbestos filter in a Gooch-type crucible or a porous porcelain crucible is used both the empty crucible and the crucible with precipitate can be dried at 500°C–700°C in an electric furnace or with a Bunsen or Tirrill burner. Sintered-glass and glass-fiber filters must not be used above 500°C; these can be dried at 120°C.

Filtering and Washing the Precipitate. Select a quantitative filter of the proper size and porosity and fit it to a funnel. Decant the solution through the filter, then wash the precipitate by decantation as directed in Procedure 14-1 (Note 3).

Burning ("Igniting") the Paper. Cover the funnel with a watch glass or by moistening the outer edge of a filter paper and crimping it over the edge; place it in a drying oven (at 100°C–110°C), and leave until dry.

Obtain and clean a porcelain crucible and cover. Mark the crucible (with a file or china-marking pencil), and place it on a clean triangle (Note 4). Warm the crucible gently at first, and then heat it and its cover at 500°C–700°C for 10 minutes (Note 5). Do not cause the crucible to glow brightly. Allow the crucible to cool somewhat, place it in a desiccator for 20 minutes, and weigh. Repeat the heating (for 5 minutes), cooling, and weighing until the weight is constant to 0.2 mg.

Remove the filter paper from the funnel (Note 6), then fold the filter paper over the precipitate and place it in the crucible. Cover the crucible. Dry the paper in an oven or over a *very low flame* (too rapid heating will cause spattering). When the paper is dry, partly uncover the crucible and heat it over a flame until all "smoking" ceases. Do not cause a flame; immediately cover the crucible if flaming occurs. Incline the crucible, partly remove the cover, and then gradually raise the temperature to the full heat of the burner (not

NOTES

3. When washing a precipitate on a paper filter, take care finally to wash down the upper portions of the paper, and especially the folded portions. Dropwise addition of the wash water is most efficient.

4. Triangles for high-temperature ignitions are preferably made of fused silica, clay tubes on heavy metal wire, or of platinum. Triangles of nichrome alloy are satisfactory for medium temperatures but are not recommended for high-temperature ignitions, especially with platinum crucibles. Scrupulously keep triangles clean, since any material is likely to fuse to the crucible.

5. Take care to adjust the burner so that an oxidizing (nonluminous) flame is produced. A luminous yellow flame causes carbon dsposition on vessels and crucibles, does not produce as high temperatures, and *is very injurious to platinum ware.* See page 280 in regard to the temperature produced by various burners and the judging of temperature by the glow of the heated object.

6. To remove a filter from a funnel, the paper may be raised, by inserting a clean spatula between it and the funnel, and then folded forward so that only the outside of the filter is handled. An alternative method is to tear a small piece of paper from a quantitative filter and to use this against the inside of the filter when lifting it; this paper is then burned with the filter. In case any precipitate has been spattered or has "crept" onto the funnel above the filter, it should be first wiped onto this paper.

a Meker), and maintain that temperature until the precipitate is white (Note 7). Allow the crucible and precipitate to cool partly, place them in a desiccator for at least 30 minutes, and weigh. Replace the crucible on the triangle, add *1 drop* of 9 *F* H_2SO_4, heat *very cautiously* until the acid is volatilized (Note 8), and then again heat to the maximum temperature of the burner for 10 minutes. Cool and weigh as before. Repeat the heating (not the treatment with H_2SO_4) until the weight is constant.

NOTES

7. In case a resistant carbon deposit remains on the cover, invert it separately on a triangle and heat it until the deposit is removed.

8. The sulfuric acid is added to convert to sulfate any barium sulfide resulting from reduction of sulfate by carbon. A preliminary weighing is made before this treatment, since the subsequent evaporation of the sulfuric acid is likely to cause loss unless carried out with skill and patience.

QUESTIONS AND PROBLEMS

Stoichiometric Calculations

1. A 0.5-g precipitate of silver chloride was heated in a 30-ml crucible, and the crucible was covered and allowed to cool in a desiccator until it and the air in the desiccator were at 25°C. Crucible and contents were then weighed in a balance case in which the air was at 20°C, dry, and at a barometric pressure of 740 mm. Calculate in milligrams and as a percentage the error in the weight of the precipitate caused by the warm air in the crucible (neglect the effects of convection currents and of possible unequal expansion of the balance arms caused by the warm crucible). Is the error negative or positive? The density of dry air at 20°C and 740 mm is 0.00117 g/cm^3. *Ans.* 0.6 mg; 0.12%.

2. A hydrochloric acid solution was standardized by Procedure 14-1. A pipet delivering 25.01 ml at 20°C was used at that temperature, and the silver chloride precipitate was found to weigh 0.3584 g. Calculate the normality of the acid. If a quantitative paper filter with a stated ash content of 0.04 mg had been used, should a correction for the ash be made?

3. In a gravimetric standardization of an HCl solution, a 57.26-g sample of the solution was weighed out and diluted to 150 ml, a small excess of silver nitrate was added, the mixture was allowed to stand in the dark for 24 hours, and the precipitate was collected on a Munroe crucible, washed free of silver with a 1% HNO_3 solution, washed free of acid with water, and dried at 130°C. The weight of the precipitate was 0.8188 g. The filtrate and washings were evaporated and the dissolved silver chloride was determined nephelometrically, the equivalent of 0.2 mg of HCl being found. Calculate the grams of HCl per gram of the solution. Calculate the weight normality (eq./100 g of solution).

4. The bismuth in an alloy was determined by precipitating and weighing it as BiOCl. Calculate the factor (called the *gravimetric factor*) by which the weight of the precipitate must be multiplied in order to obtain the weight of bismuth. *Ans.* 0.8024.

5. Calculate the gravimetric factor (see Problem 4 above) for each of the cases given in the accompanying table.

Substance to be Reported	Form in Which Weighed
Al_2O_3	$AlPO_4$
$(NH_4)_2SO_4$	$BaSO_4$
FeS_2	$BaSO_4$
S	$BaSO_4$
As_2O_3	$MgNH_4AsO_4 \cdot 6H_2O$
Mg	$MgNH_4AsO_4 \cdot 6H_2O$
P	$Mg_2P_2O_7$
P_2O_5	$(NH_4)_3PO_4 \cdot 12MoO_3$

6. A 0.2500-g sample of pure NaCl is dissolved in water, 50.0 ml of a 0.100 *N* $AgNO_3$ solution are added, and the mixture is diluted to 500 ml. Calculate the formal concentration of the excess $AgNO_3$ (neglect solubility effects). *Ans.* 0.00144.

7. An 0.800-g sample of pyrite (containing 87.50% FeS_2) was decomposed with aqua regia and bromine, the silica was filtered out, the iron was precipitated with ammonia, and the filtrate was neutralized. Calculate the volume of 0.50 *N* $BaCl_2$ required to precipitate the sulfate present in the solution (neglect solubility effects).

8. The sulfur in a crude pyrite (FeS_2) is being determined by converting it to sulfate and then precipitating and weighing the $BaSO_4$. If it is desired that each milligram of $BaSO_4$ obtained represent 0.1% sulfur in the sample, what weight of the sample must be taken? *Ans.* 0.1374 g.

9. A 0.1936-g sample consisting of a mixture of pure KCl and KBr is dissolved and the halides are precipitated with $AgNO_3$. The weight of the precipitate was 0.3311 g. Calculate the percentage by weight of chlorine as chloride and of bromine as bromide in the mixture. How many significant figures are justified in the percentage values? Explain. What would be the relative percentage error caused in each of these values by an error of 0.3 mg in weighing the silver precipitate?

10. In the analysis of an 0.5000-g sample of a mixture of soluble salts, the chloride and iodide present gave a silver precipitate weighing 0.3781 g. After the precipitate was heated in a stream of chlorine, the weight was 0.2867 g. Calculate the percentage of chloride and iodide in the original sample. *Ans.* 7.09% Cl^-.

11. During the course of a chloride determination, a silver chloride precipitate was allowed to stand in bright light until, when it was weighed, 10% of it could be represented by the formula Ag_2Cl. What would be the relative percentage error introduced into the analysis?

12. Calculate the normality of a hydrochloric acid solution that was standardized as follows (Andrew, *J. Am. Chem. Soc.*, **36**, 2089, 1914): A small beaker was carefully cleaned, dried at 120°C, and weighed. Very pure silver nitrate crystals were then added, and the beaker and crystals were again weighed. The crystals were dissolved in a minimum amount of water, and then 50.00 ml of the hydrochloric acid solution were slowly added by means of a pipet. The beaker was put into a larger beaker and placed on a hot plate, and the mixture was evaporated to dryness without being allowed to boil. The residue and beaker were then dried at 120°C and weighed. *Data*: Weight of beaker, 32.4265 g; weight of beaker and $AgNO_3$, 34.9427 g; weight of beaker and residue, 34.6755 g. *Ans.* 0.2015 *N*.

    Discuss the advantages and liabilities of the above method.

Calculations Involving Chemical Equilibria

13. One liter of water at 100°C dissolves 0.0211 g of silver chloride. A 0.2923-g sample of sodium chloride was dissolved in 100 ml of water, and to this was added 100 ml of 0.1000 *N* $AgNO_3$. The mixture was heated to boiling and filtered. Calculate the weight of silver chloride in the filtrate (assume filtrate at 100°C and at equilibrium with solid).

14. In a chloride determination, the solution had a volume of 250 ml, the filtration was made at 100°C, and the weighing of the silver chloride precipitate was made accurate to 0.05 mg. Calculate the required molar concentration of the excess silver ion in the solution in order that the loss of silver chloride in the filtrate should not be greater than the weighing error. How many milliliters of 0.2 *N* $AgNO_3$ would this quantity represent? *Ans.* 19 ml.

15. In a gravimetric determination, a 0.2500-g barium sulfate precipitate was obtained. Calculate the percentage error caused by the solubility of the precipitate if only an exactly equivalent amount of the precipitant had been added, the volume was 500 ml, and the temperature during the filtration was 100°C. (The solubility of barium sulfate at 100°C is 0.0039 g/liter.) Calculate the molar concentration of the excess precipitant necessary to reduce the error to 0.1%. What volume of 0.1 *N* precipitant does this quantity represent? What would be the effect of the acidity of the solution on the solubility of the precipitate?

16. A precipitate of strontium sulfate weighing 0.5 g is washed with 250 ml of water at room temperature. If it is assumed that the wash water becomes saturated in the process, calculate the loss in weight of the precipitate and the percentage error caused. (A saturated solution of $SrSO_4$ at 20°C is 8.2×10^{-4} *F*.) Calculate what the molar concentration of sulfate ion in the wash water would have to be in order to prevent a loss of over 0.1%.

17. In the preceding problems no *p*H data have been given, and the solutions have been assumed to be neutral. This is usually not the case in gravimetric precipitations, and the effect of the acidity of the solution is illustrated in this problem.

    In making a sulfate determination, a 0.7103-g sample of pure Na_2SO_4 was dissolved in water, 25 ml of 0.50 *N* $BaCl_2$ and 10.0 ml of 5.0 *N* HCl were added, and the mixture was diluted to 500 ml, allowed to stand until equilibrium was attained, and filtered. Calculate the weight of the precipitate dissolving in the filtrate. *Ans.* 0.044 mg of $BaSO_4$.

18. In order to check the accuracy of Procedure 14-2, a 50.00-ml portion of a solution which was 0.02000 *F* in ferric ferrous alum, $Fe_2(SO_4)_3 \cdot FeSO_4 \cdot 24H_2O$ and 0.1000 *F* in ferric chloride was put through a cation-exchange column (containing resin that had been regenerated with HCl), and the column was effectively washed with 100 ml of water. (Assume that there is complete exchange of all metal cations.)

    The effluent solution was collected, 18.0 ml of 0.500 *N* $BaCl_2$ were added, and the mixture was diluted to 500 ml, allowed to stand until equilibrium was attained, and filtered at room temperature; then the precipitate was washed, ignited (500°C), and weighed.

    (*a*) Calculate the stoichiometric weight of $BaSO_4$ that should be obtained.
    (*b*) Calculate the weight of barium sulfate remaining in the solution in equilibrium with the precipitate.
    (*c*) Make a prediction as to the sign and magnitude (parts per 1000) of the experimentally found error in such a procedure (neglecting any errors caused by technique). Explain the basis of your prediction.
    (*d*) State the effect upon (1) the physical characteristics of the precipitate, and (2) any errors in the determination, of (*a*) very rapid, and (*b*) very slow (dropwise) addition of the barium chloride.

19. Calculate the normality (metathetical) of a sulfuric acid solution that was standardized as follows. A platinum dish was dried and weighed. Pure barium chloride crystals, $BaCl_2 \cdot 2H_2O$, were added and the dish and barium chloride heated at 200°C until a constant weight was attained. The crystals were dissolved in water, and 50.00 ml of the sulfuric acid solution were slowly added. The mixture was then evaporated to dryness and the dish and residue were again heated at 200°C until constant weight was obtained. Data:

Weight of platinum dish	50.2645 g
Weight of dish and barium chloride	51.5141 g
Weight of dish and residue	51.6398 g

    Ans. 0.1998 *N*.

20. A determination of the silver in a silver solder (Ag, Cu, Zn) was made by dissolving a 0.5000-g sample of the solder in hot concentrated nitric acid, evaporating to 2 ml, diluting to 250 ml, adding 250 ml of 0.01000 *N* NaCl, filtering (at room temperature), washing, drying, and weighing the precipitate. The weight of the precipitate was 0.2867 g.

    (*a*) Calculate the percentage of silver in the alloy (neglect any solubility loss). *Ans.* 43.16%.
    (*b*) Calculate the weight of silver remaining in the solution under equilibrium conditions. (State assumptions made.)
    (*c*) State the effect of the HNO_3 upon the solubility of the silver chloride.

(*d*) Predict the result of applying this method to a 0.5-g sample of a silver-mercury amalgam of approximately 20% silver.

(*e*) Would photochemical decomposition of the silver chloride precipitate cause positive or negative errors (1) in the above procedure? (2) in the analogous determination of Cl by precipitation of silver chloride?

SECTION IV

TITRIMETRIC MEASUREMENTS AND METHODS

CHAPTER 15

Titrimetric Methods

GENERAL PRINCIPLES

The application of precipitation reactions to both gravimetric and titrimetric methods and the distinction between the end point and equivalence point have been discussed briefly in Chapter 5, pages 58–60; that material should be reviewed. As was explained in more detail in Chapter 12, a gravimetric analysis involves weighing the sample, separating the constituent of interest as an element or compound of known and definite composition, and then weighing that substance. It was also explained there that in order for a reaction to serve as the basis for a gravimetric determination, certain requirements must be met.

In titrimetric methods the determination of a substance is made by measuring the quantity of a standard solution (one having an accurately known concentration) that reacts directly or indirectly with the substance. For example, the amount of chloride in a sample of drinking water can be determined by measuring the quantity of a standard silver nitrate solution required for the precipitation reaction

$$Cl^- + Ag^+ + AgCl(s)$$

This operation is known as a *titration.*

Similarly, the quantity of iron in an ore can be determined by titration with a standard permanganate solution, the iron being first brought into solution and reduced to the ferrous state by means of a suitable reductant. In this case the titration reaction is represented by the equation

$$5Fe^{2+} + MnO_4^- + 8H^+ = 5Fe^{3+} + Mn^{2+} + 4H_2O$$

The solution added in a titration, permanganate in this case, is called the *titrant*.

There are two general techniques whereby the quantity of a titrant solution used in a titration is determined. In the first of these the *volume* of the solution is measured and the process is termed a *volumetric titration*. In the second the *weight* of the titrant solution is measured and the process is termed a *gravimetric titration*. The same general principles apply to both titrations and are discussed first below. Following that, certain special features and advantages of each of these types of titrations are presented.

Titrimetric methods are more extensively used than are gravimetric because if standard solutions are available, titrimetric determinations usually can be carried out much more rapidly and require less experience and special technique.

Preparation of Standard Solutions

These can be prepared in various ways. The preferred approach is to dissolve a weighed portion of a pure dry substance and then dilute the solution to a fixed weight or to a fixed volume at a known temperature. Substances of high purity that can be used for the direct preparation or standardization of solutions are called *primary standards*. In the majority of cases a substance of adequate purity cannot be obtained; therefore, an indirect approach must be used in which a solution of approximately the desired concentration is prepared and then standardized by titration against a weighed quantity of a primary standard. A more detailed discussion of these methods of obtaining standard solutions is given in Chapter 17.

Storage of Standard Solutions

The precautions required to prevent concentration changes during storage vary with the nature of the solution. Silver nitrate and permanganate solutions must be protected from excessive exposure to light; sodium hydroxide solutions attack glass containers and also absorb carbon dioxide from the air; thiosulfate solutions must be sterilized to prevent bacterial action; some of the most effective reducing solutions are oxidized by oxygen of the air; and permanganate solutions are reduced by organic matter such as particles of paper or cloth. Any standard solution must be protected from evaporation and from contamination by foreign substances. Bottles with ground-glass stoppers are the preferred storage containers for most standard solutions. Glass bottles with polyethylene stoppers find wide use now because of the lower cost and the lower tendency for the stopper to stick when alkaline solutions are used.

Requirements of a Titrimetric Method

As was true for a gravimetric method, in order for a given reaction to be developed into a titrimetric method, certain requirements must be met.

1. The fundamental reaction involved, that is, the reaction between the standard solution and the titrated substance, must, when it reaches an equilibrium, be complete to within the desired accuracy. The equilibrium conditions can usually be predicted from available data such as solubility measurements, ionization constants, or potential values.
2. The reaction must proceed with practical rapidity. The reaction rate cannot be predicted from the equilibrium calculations and usually must be measured experimentally. Thus, one would predict from equilibrium conditions that chlorate ion would be quantitatively reduced to chloride by iodide ion in even dilute acid solutions according to the equation

$$ClO_3^- + 6I^- + 6H^+ = Cl^- + 3I_2 + 3H_2O$$

   Experimentally, the rate at which the reaction proceeds is so low that the reaction cannot be made the basis of a quantitative volumetric method unless a catalyst is added.
3. The reaction must be stoichiometric; that is, the reaction must proceed according to some definite equation or equations, so that from the volume or weight and the concentration of the standard solution used, the exact amount of the titrated substance present can be calculated. These criteria are not met by titrations involving side reactions or *induced reactions.* An example of each of these follows.
   (*a*) When iodine is titrated with thiosulfate, the stoichiometric equation

$$I_2 + 2S_2O_3^{2-} = 2I^- + S_4O_6^{2-}$$

   is assumed to represent the change taking place; however, if the titration is made in a solution that is slightly alkaline, part of the thiosulfate is oxidized to sulfate and more than the calculated amount of iodine is used.

(*b*) When stannous tin in a hydrochloric acid solution is titrated with permanganate, the resulting change is assumed to be represented by the stoichiometric equation

$$2MnO_4^- + 5SnCl_4^{2-} + 10Cl^- + 16H^+ = 2Mn^{2+} + 5SnCl_6^{2-} + 8H_2O$$

However, if the titration is made in the presence of air, the oxidation of stannous tin by oxygen is induced, and such a large error is introduced that this particular determination is not practical. Often the existence and magnitude of these effects can be determined only by experiment.

4. Some means must be available for determining to the desired accuracy the end point of the titration, that is, the point at which it is assumed that an amount of the standard solution equivalent to the substance being determined has been added. Various means are available:
   (*a*) One of the reactants may have such a distinctive and intense color that the appearance or disappearance of the color serves as an indication of the end point. In titrations with permanganate the first pink permanganate color that persists is commonly used as the end point. The iodine in solutions can be reduced in titrations with thiosulfate to the colorless iodide ion. The disappearance of the iodine color, or of the color iodine gives in the presence of starch, is taken as the end point.
   (*b*) Some product of the titration may have properties that permit its use as an indication of the end point. Certain titrations are continued until additional titrant causes no more visible precipitation; the substance being titrated has been effectively removed by the titrant. In other cases the titration is stopped when a new product with different properties is first formed. For example, when cyanide ion is titrated with silver ion the reaction first produces soluble $Ag(CN)_2^-$; additional Ag^+ causes precipitation of $Ag_2(CN)_2(s)$ and the appearance of the first turbidity serves as the end point.
   (*c*) *Indicators*, substances that cause an observable change in the region of the equivalence point, may be used. As we shall discuss in subsequent chapters, these changes are caused by the formation or disappearance of a precipitate, or by a color change. The majority of indicators are substances that change color as the *p*H of the solution changes, or as they are oxidized or reduced by one of the constituents of the titration reaction.
   (*d*) A large and growing number of physical methods is available. Some of these depend upon optical properties and are closely

related to the indicator methods of (*a*), (*b*), and (*c*) above. Instruments having greater sensitivity and selectivity than the unaided eye are used to detect the changes in light absorption or in turbidity that serve as an indication of the end point. Others of the physical methods make use of properties of light that are ordinarily undetectable by the eye, such as refractive index or ultraviolet absorption.

Many physical methods depend upon electrical properties of the solution. For example, changes in the conductivity or the potential of the solution can serve for detection of the end point. The heat of a reaction is used in a thermometric titration as an indication of its completeness. Even such unlikely sounding properties as the freezing point and the viscosity have been used as end-point methods for titrations. By these means the end point of the titration is taken. As stated before, this must be clearly distinguished from the *equivalence point*, which is the point at which an exactly equivalent amount of the titrating substance has been added. The end point is that point at which the titration is stopped, and the error in the titration is measured by the difference between the equivalence point and the end point.

The imposition of the requirements listed above greatly limits the number of reactions upon which accurate titrimetric methods can be based.

Types of Titrimetric Methods

The methods used in titrimetric analyses may be divided into three general types, as follows:

1. precipitation methods;
2. oxidation and reduction methods;
3. ionization methods (including neutralization, complex-ion formation, and un-ionized compound formation).

This classification is based upon the type of reaction taking place between the substance being titrated and the titrant. With methods of the first type the completeness of the reaction is dependent upon the solubility of the precipitate formed, and solubility-product data are required for calculating the conditions obtaining at the equivalence point.

With methods of the second type the completeness of the reaction is dependent upon the potentials of the half-cell reactions involved. However, these potentials *do not* determine reaction rate, and oxidation–reduction reactions are more likely than the other two types to proceed so slowly that practical titrations are not possible.

Methods of the third type are dependent upon the formation of un-ionized compounds, and the completeness of the titration reaction is dependent upon the degree of ionization of the compound formed. Where the ionization constants of these substances are available, the equilibrium conditions can be calculated. Specific examples of these general types will be discussed in detail in the subsequent chapters.

VOLUMETRIC-TITRATION METHODS

In a volumetric method the standard solution is prepared to have a known concentration in moles (or more frequently, equivalents) of solute per liter of solution (volume concentration). Titrations are then performed with the use of a volumetric buret, which enables one to make an accurate measurement of the *volume* of standard solution used.

If a standard solution is prepared from a primary standard, a quantity of the latter is weighed into a volumetric flask (see page 321). Water is added to dissolve the substance, and the addition of water continued until the solution reaches the mark on the flask. The temperature of the solution at the time the adjustment to the mark is made must be known; the volume that the flask will contain and the volume of the solution both depend upon the temperature. That is, both flask and solution expand as the temperature is raised, but the coefficients of expansion of the two are different; only at the temperature of preparation will the meniscus of the solution and the mark on the flask coincide. To calculate the concentration of the solution, you must know the weight of the primary standard and in addition the volume of the flask at the temperature at which the solution is prepared. From this you can then calculate the concentration of the solution at other temperatures, by using thermal expansion data for dilute solutions. Since the volume concentration depends upon expansion and contraction of the solution, temperature changes result in concentration changes.

The volume of titrant used in a volumetric titration is measured by means of a buret that is calibrated to deliver a measured volume of solution at a given temperature. In ordinary macrotitrations, a 50-ml buret is most often used, and has calibration marks for each 0.10 ml. Volumes delivered are estimated to within 0.01 ml–0.02 ml. The walls of the buret must be scrupulously cleaned; otherwise, uneven drainage of the solution from the walls will cause errors much larger than the uncertainty in reading the volume. These problems are discussed in detail in Chapter 16.

Volumetric-titration methods have been very extensively used because they are much more rapid than gravimetric determinations and require less manipulative skill. Until recently weight measurements were so time-consuming that the increased accuracy inherent in gravimetric titrations was more than offset by the time required by their use. With the development of more rapid balances this limitation has been removed. Gravimetric titrations are discussed below.

GRAVIMETRIC (WEIGHT) TITRATION METHODS

In a gravimetric titration the standard solution is prepared to have a known concentration in moles (or equivalents) of solute per kilogram of solution (weight concentration). The weight of titrant used in a reaction is then measured; from this weight and the known concentration, the number of moles or equivalents used is calculated.

Gravimetric titrations have been used for highly accurate or specialized determinations for many years but have gained little general acceptance primarily because of the much greater time they have required. Today, however, gravimetric titrations can be performed so rapidly that their special benefits are made easily available.

A gravimetric titration consists of the following steps: (1) The titrant solution is prepared, or standardized, so that its weight concentration is known (see Chapter 2). (2) An appropriate weight of standard solution is put into a gravimetric buret (commonly called a *weight buret*) and the buret and solution are weighed. (3) The solution of the unknown is titrated. (4) The gravimetric buret with remaining titrant is weighed.

From the weight of solution used, the number of moles or equivalents of titrant used is calculated.

A gravimetric titration is free from problems of temperature effects upon the volume of the glassware and upon the concentrations of the solutions. Drainage problems, which are critical in volumetric titrations, are of no consequence in gravimetric titrations. Glassware calibrations are not required. In addition, because the weight of a solution can be measured much more accurately than its volume, gravimetric titrations can be made on smaller samples without loss of accuracy. For the same reason, more concentrated solutions can be used when this is advantageous.

The concentration units used in gravimetric titrations are weight formal, weight molar, and weight normal. Each of these units depends upon a weight of solute (translated to a number of moles or equivalents) and a weight of solution, therefore the concentrations are temperature-invariant. Although the volume and density of solutions change with temperature the measurements are all weight measurements; consequently, there is no change in weight concentrations as temperatures change.

With a gravimetric buret there is no requirement that the walls drain uniformly, for only the weight change is required. Indeed, in the gravimetric titrations given in this text, a small inexpensive polyethylene wash bottle is suggested as the gravimetric buret, and though droplets often adhere to the inner wall of the polyethylene container they cause no error. Such droplets would cause serious errors if they occurred on the walls of a volumetric buret or pipet.

The perceptive student will see that there may be a buoyancy problem involved in the preparation of a standard solution for gravimetric titrations and again in its use in the titration. Consider the preparation and use of a 0.1 *WF* (weight formal) $AgNO_3$ solution. The solution can be prepared by accurately weighing about 17 g (0.1 mole) of $AgNO_3$ into a weighed beaker or bottle and adding water until the solution weighs about 1000 g. Because of the buoyancy of the air, the apparent weight of the solution is less than its vacuum weight by about 0.1 %. If buoyant forces are ignored, the calculated concentration is, therefore, about 0.1 % greater than the actual concentration. Then, in a titration, a weighed portion of this solution is used, but an error in the weight results from the buoyancy of the air. The actual weight taken is about 0.1 % higher than that indicated by the balance. Thus one has taken 0.1 % more of the solution which is 0.1 % less concentrated than he calculated it to be—and the errors cancel. The buoyancy error in weighing silver nitrate crystals (density 4.3) is approximately 0.01 % and usually can be neglected.

CHAPTER 16

Titrimetric Methods. Operations

As was stated in Chapter 12, the constituent of interest in a gravimetric determination is converted to an element or compound of definite composition and then weighed. In a titrimetric determination the quantity of a standard solution that reacts directly or indirectly with the constituent of interest is measured by a titration. The quantity of solution used can be measured volumetrically or gravimetrically; in Chapter 15 the relative advantages of these two types of measurement were discussed. The general operations and techniques involved in titrimetric methods are described below, and instructions are given for the use of certain essential equipment.

VOLUMETRIC TITRATIONS

The accuracy with which volumetric measurements can be made depends upon your ability to use flasks, pipets, and burets properly. In this section we will discuss the use and calibration of such volumetric apparatus.

Types of Volumetric Apparatus

Volumetric apparatus may be classified into two general types, depending upon whether it is used *to contain* or *to deliver* a certain volume of solution. Flasks are usually designed to contain a definite volume of solution, whereas pipets and burets are usually designed to deliver definite volumes. Uniform drainage is a critical factor in the use of delivery apparatus.

Cleaning of Volumetric Apparatus

The vessels used in volumetric measurements must be scrupulously cleaned to avoid contaminating standard solutions and erratic drainage. A vessel may be considered suitable for use when it is visually clean and when a solution drains from it without leaving perceptible streaks or droplets. The most common contaminant is a grease film. Various agents and methods have been suggested for cleaning glassware. Among these are grease solvents such as alcohol, ether, and acetone; concentrated acid solutions such as sulfuric, nitric, and hydrochloric, either alone or in various combinations; acids with oxidizing agents added, such as concentrated sulfuric and chromic acids, or dilute sulfuric acid and sodium permanganate; soaps, either powder or liquid; and various laboratory cleaning preparations (such as sodium hexametaphosphate), which are commercially available.

It is recommended that, wherever possible, glass apparatus be cleaned by scouring it with soap powder or solution or with a cleaning preparation,[1] and then by thoroughly rinsing it with distilled water. To economize with distilled water, the beakers, flasks, and similar apparatus should be first liberally rinsed with tap water and then with a smaller amount of distilled water. When a resistant film of grease is encountered, one that cannot be removed by the above treatment, or when the construction of the vessel makes the use of soap powder ineffective, recourse may be made to the so-called "cleaning solution." This is a solution of an alkali metal dichromate, or of chromic acid, in concentrated sulfuric acid.[2] Such a solution, though effective, should not be used indiscriminately because of its corrosive nature and because of the difficulty with which it is completely removed from glass surfaces.[3] It is frequently necessary to leave the solution in contact with the surface being cleaned for an hour or longer. Often more effective and rapid action can be obtained by treating the vessel with 6 N NaOH for 5 to 10 minutes and then by following this with the cleaning solution. Calibrated glassware should not be steamed or treated with hot solutions if this can be avoided, nor should such glassware be dried with hot air; the thermal readjustments take place slowly.

All volumetric glassware should be calibrated on the basis of the liter. Because there has been some confusion in regard to units of volume and

[1] Cleaning or "scouring" powders containing abrasives should not be used even for flasks. If the glass surface of a vessel is scratched, it is often difficult to remove precipitates. Moreover, a badly scratched surface is much more rapidly attacked by alkaline solutions and may cause appreciable amounts of the constituents of the glass to be introduced into a solution that is being analyzed.

[2] An effective cleaning solution can be made by dissolving 30 g of sodium dichromate in 500 ml of concentrated sulfuric acid. Sodium dichromate is less expensive and is more soluble than potassium dichromate.

[3] E. P. Lang. *Ind. Eng. Chem.*, anal. ed., **6**, 111 (1934).

because glassware as purchased is often not precisely calibrated or may have been altered in its capacity by previous use or abuse, all volumetric glassware that is to be used for precise measurements should be checked before use. In principle this checking usually involves finding the weight of the water that such apparatus will contain or deliver, as the case may be, and then calculating the equivalent capacity from this weight, with the liter as the unit of capacity.

The National Bureau of Standards has adopted 20°C as the standard laboratory temperature to which all apparatus should be calibrated. In order to check the calibration of a piece of volumetric glassware—say a pipet—one weighs the water delivered by the pipet at a known temperature and from the weight so obtained calculates the effective volume of the pipet at 20°C. The weight of the water is obtained in air, and this air exerts a buoyant force upon both water and weights. The magnitude of the buoyant force depends upon the density of the air, and the density in turn depends upon the temperature and pressure of the air. Therefore, the data that must be obtained if one is to *calibrate* (check the calibration of) a pipet are the weight of water delivered, the temperature, and the barometric pressure.

The calculation of the volume of the apparatus at 20°C from the apparent weight of the water at the prevailing temperature and barometric pressure is outlined below. The path from the weight of water at T°C delivered by the pipet to the effective volume of the pipet at 20°C can be seen as a sequence of steps:

$$W_o \xrightarrow[A]{} W_v \xrightarrow[B]{} V_{H_2O} = V_{P(T)} \xrightarrow[C]{} V_{P(20)}$$

where

W_o is the observed weight of water delivered by the pipet at T°C,
W_v is the vacuum weight of the water delivered by the pipet at T°C,
V_{H_2O} is the volume of the water at T°C,
$V_{P(T)}$ is the effective volume of the pipet at T°C, and
$V_{P(20)}$ is the effective volume of the pipet at 20°C.

First the vacuum weight of the water must be calculated. This step (step A) is essential since a volume is wanted, and a simple relation is known between *mass* and volume of water. When corrections are made for buoyancy, *weights* obtained with a simple balance become *masses*.[4] Hence the first step is to correct for the buoyant effect of the air upon the water and the weights.

[4] In fact, the vacuum weight, which is obtained by applying buoyancy corrections to the observed weight, differs from the mass by a constant factor g, the acceleration due to gravity at any specified location. Since this constant factor would appear in all steps it is omitted for simplicity.

ILLUSTRATIVE PROBLEM

The calibration of a pipet was being checked. The apparent weight in air of the water delivered at 25°C and 1 atmosphere pressure was 24.89 g. The density of air under these conditions was 1.18 g/liter; the density of the weights used was 8.3 g/cm^3. Calculate the vacuum weight of the water. (Buoyancy corrections were discussed in Chapter 11.)
Ans. 24.92 g.

From this value for the vacuum weight and the density of water at 25°C, the volume of the water (and hence of the pipet) at 25°C is calculated to be 24.99 ml (step *B*).

The volume of the pipet at 20°C is calculated from its volume at 25°C (step *C*) by means of the general expression

$$V_{P(20)} = V_{P(T)}(1 + 1 \times 10^{-5}\,\Delta T)$$

where ΔT is the difference in temperature between the standard (or any other desired value) and the observed temperature, and the constant, 1×10^{-5}, is the average cubical coefficient of expansion of borosilicate laboratory glass per degree for the range from 0°C to 100°C. It is the increase in volume per unit of volume per degree centigrade.

In this case

$$V_{P(20)} = V_{P(25)}[1 + 10^{-5}(20 - 25)]$$

and the volume remains 24.99 ml since the change of -0.00005 is too small to be significant.

PROCEDURE **16-1**

Calibration of a Flask

Outline. The water *contained* by a volumetric flask is weighed and from this weight the volume of the flask at 20°C is calculated.

DISCUSSION

A volumetric flask is usually designed to contain a specified volume and is constructed with a relatively long, narrow neck on which is placed the calibration mark (see Figure 16-1). This construction ensures that any variation in adjusting the meniscus of the solution to the calibration mark will represent a relatively small volume compared with the total volume of the flask, and thus will cause only a small relative error.

Flasks are calibrated by adding a definite mass of water and suitably marking the neck of the flask at the meniscus. A previously calibrated flask is tested by weighing the water that it contains when filled to the calibration mark at a definite temperature; this latter operation is also frequently spoken

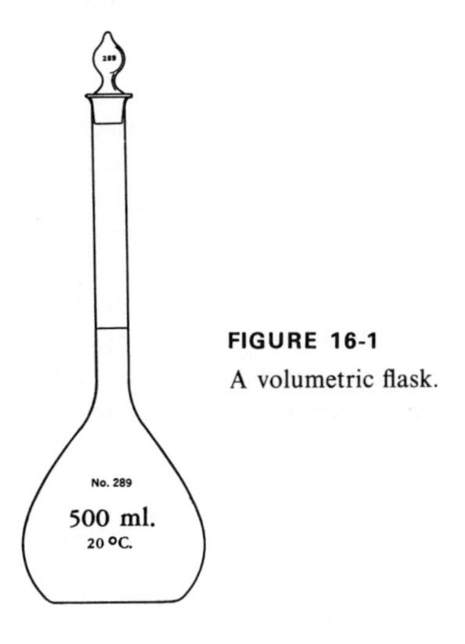

FIGURE 16-1
A volumetric flask.

of as a calibration and is the procedure most commonly required for routine analytical work. It is described in the procedure below.

The gravimetric calibration methods described above are direct or primary calibrations in that the volume is determined by directly weighing the water. In general, a measurement of mass can be made more accurately than can a measurement of volume. It is also possible to calibrate a flask volumetrically by measuring the volume of water that is delivered into it from a specially designed calibrating bulb. This method, though more rapid, especially if a large number of flasks are to be checked, is a secondary calibration, since the calibrating bulb itself has first to be calibrated directly by weighing the water that it delivers.

The tolerances, or permitted deviations from the indicated volume, that have been established by the Bureau of Standards for flasks are shown in Table 16-1.[5] Much of the volumetric apparatus that may be purchased will be found to exceed the Bureau of Standards tolerances; however, all apparatus should be tested before being used.

INSTRUCTIONS (Procedure 16-1: Calibration of a Flask)[6]

Clean the flask with soap solution or powder, with cleaning solution, or, if necessary, with 6 *F* NaOH followed by cleaning solution (see the discussion above, also Note 1); finally rinse it thoroughly with distilled water (Note 2). Dry the flask (Note 3). Weigh the stoppered flask on a top loading balance or a rugged balance (see page 180) of suitable capacity. (With flasks of 100-ml capacity or greater, ± 50 mg accuracy is adequate.) Record this apparent weight. Remove the flask and by means of a funnel, which must extend below

NOTES

1. It is especially necessary that the inner surface of the flask in the region of the calibration mark be free of grease; otherwise, the meniscus of the solution may be so distorted as to cause a serious error.
2. *Vessels should not be rinsed at the distilled-water tap*; carry water to the desk.
3. The flask may be dried by allowing it to stand in an inverted position until the inside surface is dry. The evaporation of the water from the surface is hastened by inserting a tube nearly to the bottom and *drawing* air through. *Do not heat the flask.* In case of urgent haste, the drying process can be expedited by rinsing the inside walls with several successive small portions of alcohol, or, better, acetone (grease-free).

[5] Consult *National Bureau of Standards Circular* No. 9 (1916) for more complete details and specifications concerning volumetric apparatus.

[6] In general, it has been found preferable for students to defer the calibration procedures until the apparatus is actually needed.

TABLE 16-1
Capacities and Limits of Error for Flasks

Capacity (ml) Less Than and Including	Limit of Error (ml)	
	If to Contain	If to Deliver
25	0.03	0.05
50	0.05	0.10
100	0.08	0.15
200	0.10	0.20
300	0.12	0.25
500	0.15	0.30
1000	0.30	0.50
2000	0.50	1.00

the calibration mark, fill it almost to the mark with water at room temperature (Note 4). Then remove the funnel; with a dropper or pipet, add water until the lower meniscus is just tangent with the calibration mark. Before making the final adjustment remove any droplets of water remaining against the side of the flask above the calibration mark with filter paper wrapped around a glass rod. (See the discussion of Procedure 16-3 for precautions to be taken in adjusting or reading the position of a meniscus.) Stopper the flask and again weigh it. Record this apparent weight (Note 5). Note the room and water temperature (Note 6). From the data thus obtained calculate the volume of the flask at the standard room temperature of 20°C. Duplicate calibrations should be made (Notes 7 and 8).

NOTES

4. The apparatus to be calibrated and the water to be used should have been allowed to come to room temperature so that there will be no thermal volume changes during the calibration. For most volumetric work the temperature of the water need be known only to within one degree; the volume change of water at these temperatures is about 0.02% per degree. Ordinary distilled water may be used; the effect of the dissolved air and carbon dioxide is negligible for calibration purposes.

5. The value of the weights originally applied minus the weights added to balance the filled flask gives the apparent weight in air of the water in the flask.

6. For highly precise work the barometric pressure and humidity of the air in the laboratory also would have to be noted in order to calculate the buoyancy correction.

7. The number of the flask and the correction should be recorded in the notebook (see page 162). If the piece of calibrated apparatus is to be used exclusively by one person, it is convenient to affix a small paraffined label noting the correct volume, date of calibration, and initials of the person making the calibration.

8. Duplicate calibration values that do not agree well within the limits of error shown in Table 16-1 should be repeated.

Calibration of a Pipet

Outline. The water *delivered* by a pipet is weighed, and from this weight the volume of the pipet at 20°C is calculated.

DISCUSSION

Pipets are in general used "to deliver" certain volumes of solution. *Transfer pipets* are used to deliver a fixed volume; *measuring pipets* are marked at regular intervals and can be used to deliver various volumes (see Figure 16-2).

It is obvious that transfer pipets are the more accurate of these instruments, since the bore of the stem in the vicinity of the calibration is made small in comparison with the total volume of the pipet in order to minimize the error caused by a given variation in fixing the meniscus.

Pipets are calibrated or tested by determining the weight of pure water that they will deliver at a given temperature. The volume of a liquid that is delivered from a pipet (or buret) will be affected by the "drainage," which controls the amount of the liquid adhering to the glass surface. Therefore, first, it is imperative that the inner surface of any delivery vessel be thoroughly clean; otherwise, irregular droplets of the solution will form on the walls. In addition, the volume of the film of solution that remains even on clean walls must be constant. Because the volume of this film will vary with the time and the method of delivering the solution from the pipet, these two factors must be carefully fixed. This rate of flow of a liquid from a pipet is controlled within the proper limits by the size of the tip orifice. The specifications of the Bureau of Standards are as follows:

"The outlet of any transfer pipet must be of such size that the free outflow shall last not more than one minute and not less than the following for the respective sizes:

Capacity (ml) up to and including	5	10	50	100	200
Outflow time (seconds)	15	20	30	40	50"

Various methods of emptying pipets have been used, differing mainly in the treatment of the solution that remains in the tip. Experiments have shown that more reproducible results are obtained by leaving this solution

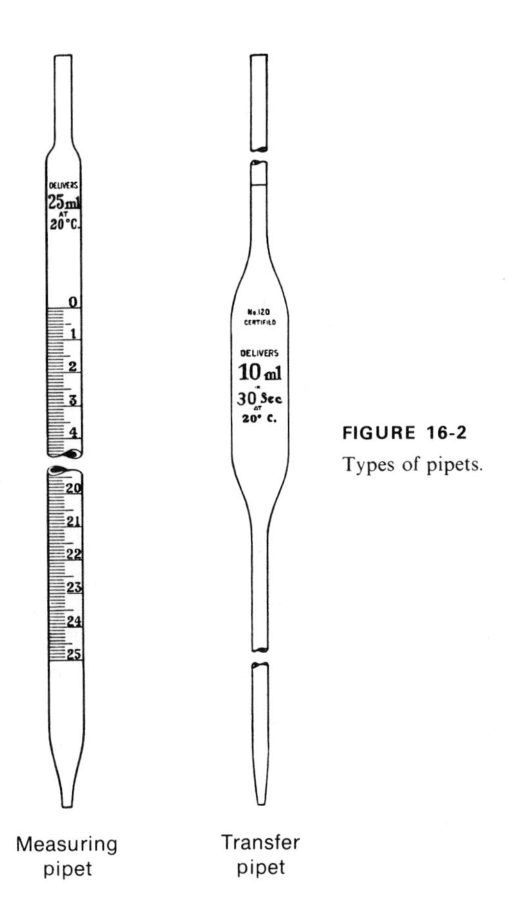

FIGURE 16-2
Types of pipets.

in the tip, after touching the tip against the wet surface of the receiving vessel, than by blowing or otherwise expelling the solution. The method given in the procedure below conforms closely to that recommended by the Bureau of Standards.

The tolerances for transfer pipets are shown in Table 16-2.

Experiments have shown that the difference between the volume of pure water and that of dilute aqueous solutions, even up to one formal concentration, delivered from a pipet (or buret) is negligible for most volumetric measurements.[7] It is therefore permissible to calibrate delivery apparatus with distilled water and to use such apparatus with the dilute solutions usually employed in volumetric work. Organic solvents and concentrated acids or bases may show large drainage differences from water.

[7] Schloesser. *Chemiker Z.*, **30**, 1074 (1906); Stott. *Volumetric Glassware*. London: Witherby, 1928, pp. 95–99.

TABLE 16-2
Capacities and Limits of Error for Transfer Pipets

Capacity (ml) Less Than and Including	Limit of Error (ml)
2	0.006
5	0.01
10	0.02
30	0.03
50	0.05
100	0.08

INSTRUCTIONS (Procedure 16-2: Calibration of a Pipet)

Clean the pipet, then test it to be sure that water drains uniformly from it without leaving streaks or droplets on the inside walls.

Note the time required for the pipet to empty (see the discussion above).

Weigh a weighing bottle or a thin-walled conical flask covered with a small watch glass (Note 1). The vessel need not be dry on the inside.

Immerse the tip of the pipet into water at room temperature and draw the liquid into the pipet until it is slightly above the calibration mark; use a short length of rubber tubing or a pipet bulb to apply suction (Notes 2 and 3). Remove the bulb or tubing and quickly place the forefinger (*not the*

NOTES

1. A consideration of the measurements involved will show that unless an accuracy of greater than 1 part per 1000 is desired, the vessel need be weighed to only 5 mg or 10 mg when pipets of greater than 10 ml capacity are being calibrated.

2. Vigorous suction should not be used in drawing solutions into the pipet; this causes bubbles to form that may adhere to the walls of the pipet. Never use direct oral suction when pipetting corrosive or poisonous solutions. Volatile solutions should be forced into the pipet by pressure

3. When the pipet is being used to measure *standard solutions* or solutions to be analyzed, certain additional precautions are necessary. First, the pipet should have been carefully drained, the tip should have been wiped dry with a clean towel, and gentle suction should be applied as the tip is immersed in the solution; otherwise, a standard solution may be diluted or contaminated. The tip should be immersed in the solution only so far as to ensure that it will not be uncovered as the solution is withdrawn. Then, in order to avoid having to dry the inside of the pipet completely, a small portion of the solution should be drawn into the pipet, used to wet the inner walls of the pipet, drained out, and discarded; at least two such portions should be used.

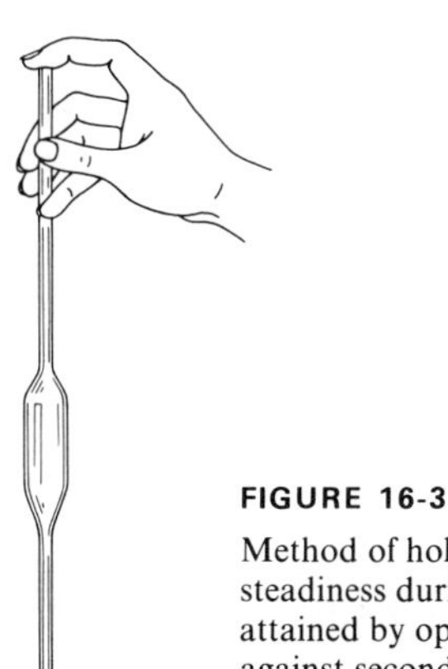

FIGURE 16-3

Method of holding pipets. Additional control and steadiness during adjustment of meniscus is attained by opposing thumb and little finger against second and third fingers.

thumb) over the upper end of the pipet so that the solution is held above the calibration mark.

Wipe off any liquid adhering to the outside of the lower stem with a clean towel or piece of absorbent paper. While holding the pipet vertically (see Figure 16-3), allow the solution to escape very slowly (by regulating the pressure of the finger on the top of the pipet or by slightly rotating the pipet against the finger) until the lower meniscus of the solution just reaches the calibration mark. Maintain the column of liquid in this position and carefully touch the tip of the pipet against a glass surface (Note 4).

While holding the pipet vertically, lower the tip sufficiently close to the receiving surface to prevent spattering, then allow the solution to run into the previously weighed vessel. Fifteen seconds after continuous flow has ceased, lightly touch the tip of the pipet against the wet side of the receiving vessel. Do not blow out the portion remaining in the tip.

Again weigh the receiving vessel. From the weight of the water thus delivered, calculate the volume of solution delivered by the pipet at 20°C. Duplicate calibrations should be made (Notes 5 and 6).

NOTES

4. Once the meniscus has been adjusted, any sudden vertical movement of the pipet must be carefully avoided; otherwise, loss may result, since the solution is suspended in the pipet by the elastic column of air above it.
5. In repeating the calibration, the vessel must be reweighed after being emptied.
6. Duplicate calibrations of 25-ml pipets should not differ by more than 0.02 ml.

Calibration of a Volumetric Buret

Outline. A buret is cleaned and the calibrations checked at a series of volume readings. The water delivered at these intervals is weighed and from these weighings the volume of the buret at 20°C is calculated for each interval.

DISCUSSION

Burets are constructed with various types of stopcocks (see Figure 16-4); Teflon stopcocks have the advantage of not needing lubrication and of not

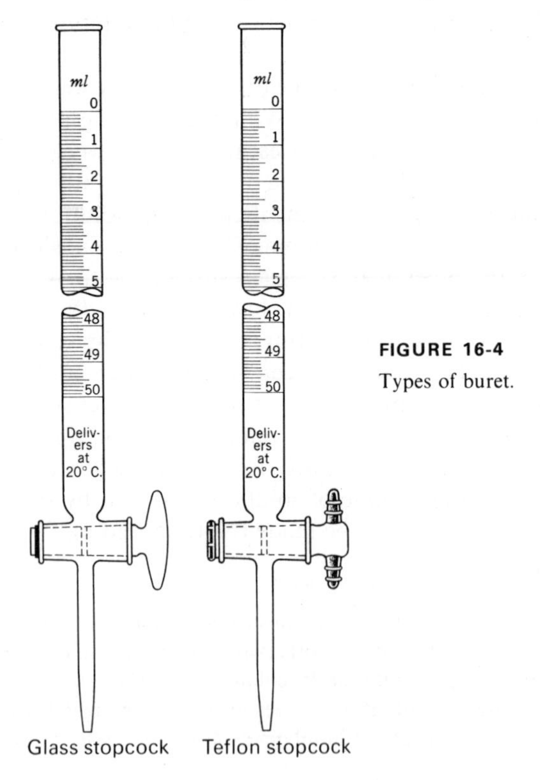

FIGURE 16-4
Types of buret.

TABLE 16-3
Minimum Outflow Time for Burets

Length Graduated (cm)	Time of Outflow (sec)
65	140
60	120
55	105
50	90
45	80

sticking when left in contact with alkaline solutions. Glass stopcocks are somewhat less expensive and will give excellent service if properly lubricated and cared for.

Burets are almost invariably used *to deliver* certain volumes of solutions. Even greater precautions must be taken to obtain uniform drainage when using burets than when using pipets because of the greater surface to volume ratio of the buret. For this reason it is essential that the tip of the buret be so constricted that it does not permit too rapid a flow. The minimum outflow time established by the Bureau of Standards for burets of different lengths is shown in Table 16-3.[8]

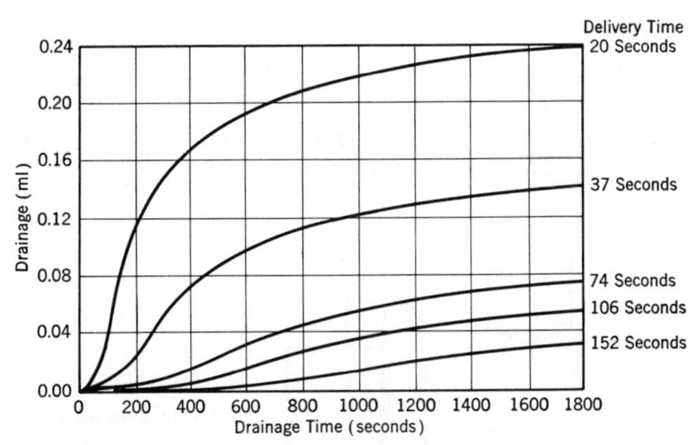

FIGURE 16-5
Effect of delivery time on drainage of a buret. The volume of water draining from the walls of a 50-ml buret after delivery, in the times shown, is plotted against the time of drainage.

The effect of the rate of flow, the so-called delivery time, on the volume of water left on the walls of a 50-ml buret and on the rate of drainage is shown in Figure 16-5. It is seen that with a delivery time of 20 seconds, not only is

[8] *National Bureau of Standards Circular* No. 9, 1916.

there over 0.24 ml of water to drain from the walls but also that this drainage persists at an appreciable rate for about 10 minutes and that the rate of drainage *increases* after approximately 40 or 50 seconds. With a delivery time of 106 seconds, the drainage is hardly appreciable even after 4 or 5 minutes.

Since the bore of a buret is much larger than that of a transfer pipet at the calibration mark, serious errors will result unless great care is taken in adjusting and reading the meniscus of the solution. To a large extent these errors are caused by parallax and by variable shadows on the meniscus.

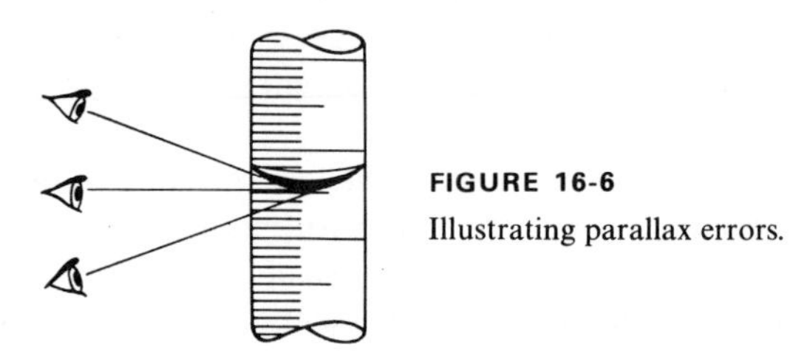

FIGURE 16-6
Illustrating parallax errors.

Parallax errors are illustrated in Figure 16-6. They can be avoided if you make sure that the calibration mark on the front and rear of the buret is coincident. All calibration marks on a buret should extend *at least* half way around the buret. When using a buret not thus calibrated, you should use an aid, such as that shown in Figure 16-7, to ensure that the line of vision is perpendicular to the buret. Mirrors and buret-reading lenses are necessary only for special purposes; buret floats are not reliable.

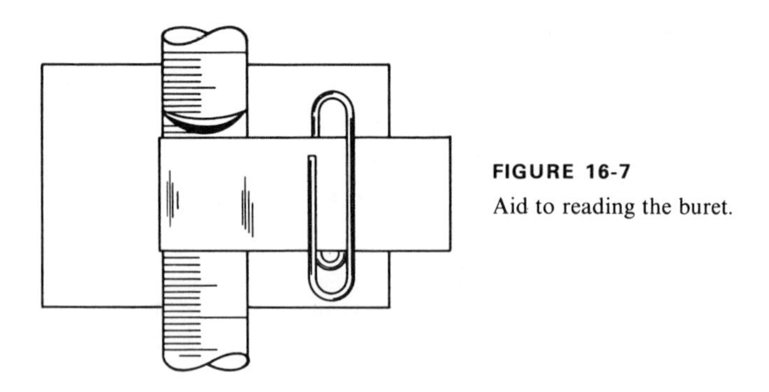

FIGURE 16-7
Aid to reading the buret.

With other than opaque solutions, the lowest meniscus should be read when the position of solutions in volumetric apparatus is being determined. The shadows on a meniscus, and therefore the apparent position of the meniscus, will vary with the height of the liquid and the lighting. Therefore, it is helpful to darken the meniscus uniformly by encircling the buret just below the meniscus with a strip of darkened paper or a piece of black rubber tubing, while a white card or piece of filter paper held back of this makes a contrasting background on which to view the darkened meniscus (see Figure 16-7). Even a piece of white cardboard with the lower half blackened, preferably with India ink, is quite useful, though not so satisfactory as the above aids.

Burets are gravimetrically calibrated (or, more properly, the calibrations are tested) by weighing the amount of water delivered at definite intervals, say 5 or 10 ml, and calculating from the weights of the water the corresponding volumes. Burets may also be calibrated volumetrically by measuring with a special calibrating pipet the volume of water delivered from certain positions on the buret. Directions for the gravimetric calibration are given in Procedure 16-3.

INSTRUCTIONS (Procedure 16-3: Gravimetric Calibration of a Buret)

Clean the buret and ascertain that it drains uniformly (Note 1). Test the stopcock for leakage and ease of operation (Note 2). Test the time of delivery of the buret with stopcock full open (see the discussion above).

NOTES

1. Before the buret is used, it should be cleaned with a buret brush and soap powder or cleaning solution and then thoroughly rinsed with water; it should then drain without leaving any streaks or droplets.

Care should be taken in using the buret brush that the inside of the buret is not scratched.

If the buret does not drain uniformly, cleaning solution (or sodium hydroxide followed by cleaning solution) should be used. The cleaning solution attacks the stopcock grease of burets with glass stopcocks; therefore, the buret should be inverted, a piece of rubber tubing fitted to the tip, and the cleaning solution drawn from a beaker until the calibrated portion of the buret is covered. Closing the stopcock enables the solution to be kept in the buret as long as desired. After allowing the cleaning solution to drain, wash most of the acid from the buret by first drawing successive portions of water into it.

2. A stopcock can be tested for leakage as follows. Remove all grease from the plug and bore. Wet the plug with water, replace it firmly in the closed position, fill the buret to near the zero mark with water at room temperature, and carefully note the reading. After 10 minutes again note the reading. The leakage should not have exceeded 0.1 ml. Rotate the plug through 180 degrees and repeat the test.

Lubricated stopcocks cannot be tested effectively because grease can prevent leakage temporarily even in badly defective stopcocks. If a perceptible drop of

Weigh to within 5 mg a 50-ml weighing bottle or a lightweight conical flask covered with a small watch glass (Note 3, also Note 1 of Procedure 16-2).

Fill the buret with water at room temperature; make sure that no air bubbles remain around the stopcock or in the tip (Note 4). Draw the solution down to within 0.02 to 0.03 ml of the zero mark, allow 30 seconds for drainage, then carefully adjust the meniscus to the zero mark (Notes 5, 6, and 7). Touch the tip of the buret to a moist glass surface to remove the hanging drop.

NOTES

water forms on the tip of a filled and lubricated buret within 3 minutes, the stopcock should be dried and relubricated or the buret tested as indicated above.

Unless glass stopcocks are lubricated properly, they cannot be operated smoothly. Vaseline or, if there is a tendency to leakage, stopcock grease should be used; the latter can be prepared or purchased. Before the grease is applied, both the plug and the bore should be dried. Grease may be removed from the tips of burets by the use of a fine wire, by a solvent such as carbon tetrachloride or benzol, or by immersing the tip in boiling water. Apply a thin film of lubricant to the plug on either side of the opening; otherwise, it is likely to collect in the tip and cause stoppage during a titration. Insert the plug firmly and rotate it until a uniform transparent film is obtained between the moving surfaces.

3. This container should be dry on the outside; it is not necessary to dry the inside.

4. Buret funnels are useful when filling burets from beakers, bottles, or other such vessels. Burets should not be filled with water by reversing the flow from the wash bottle. The mouthpiece of the wash bottle may not be clean and breakage frequently results from catching it in the top of the buret.

Air bubbles can be forced out by opening the stopcock full; if necessary tap the buret while the stream of water is flowing.

5. The solution should be withdrawn slowly until the lowest meniscus is *just tangent* to the zero mark. It is preferable to begin the calibration (and titrations) from a fixed reference mark; the position of the solution is more accurately determined when the meniscus is tangent to a calibration than when its position between two calibration marks has to be estimated visually. Record all readings to 0.01 ml. As a general rule, if the meniscus is flattened by a calibration line, the reading is recorded as 0.01 ml more than the value of the line; 0.02 ml if the meniscus is just visible below the line.

6. As stated in the discussion above, all calibrations on the buret should extend halfway around the buret. The calibrations should be thin lines. Thick lines, expecially when they are filled with opaque coloring material, are a liability, since they interfere with noting the position of the meniscus.

7. Make it a habit to push in slightly on the plug of a glass stopcock as it is turned; otherwise, it may slip out of the bore sufficiently to allow leakage. Rubber bands or spring clamps are sometimes used to prevent this slippage. If it does occur, the stopcock should be removed and dried and again greased. Water emulsified with the stopcock grease decreases its lubricating and sealing properties and also tends to cause the grease to flake off and cause stoppages.

Draw water slowly from the buret into the weighed container until the meniscus is just above the 5-ml mark (or other interval—consult your instructor), and allow 15–20 seconds for the delivery of 5 ml (Note 8). After 30 seconds adjust the meniscus to the unit mark; it is convenient, but not absolutely necessary that the meniscus be stopped exactly on a calibration (see Note 5). Touch the tip of the buret to the inside of the receiving vessel. Close and weigh the receiving vessel.

Refill the buret and again adjust the meniscus to the zero mark as directed above (Note 9). Withdraw and weigh a volume twice that of the first portion of water, observing the same precautions as with the first portion. Repeat this process until the entire length of the buret has been calibrated (Note 10). Empty the receiving vessel when the total weight of vessel and water exceeds approximately 100 g. Refill the buret with the distilled water and close it with a clean cork stopper, or by inverting a test tube over the top (Note 11).

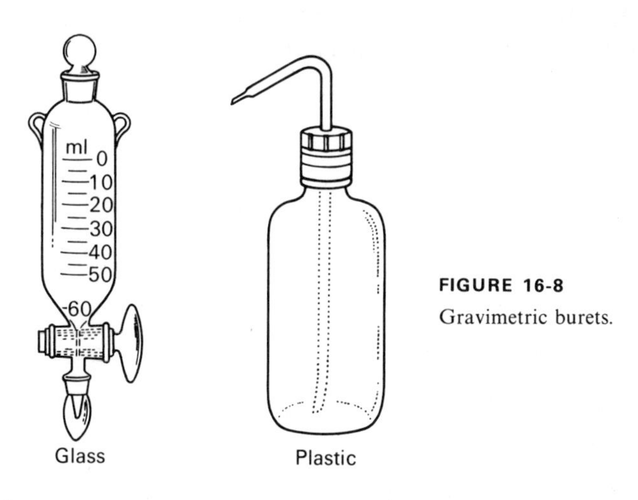

FIGURE 16-8
Gravimetric burets.

NOTES

8. Since the buret empties much more rapidly when filled to the top, the stopcock should not be opened wide when solutions are being withdrawn from the upper half of the buret.

9. If the highest accuracy is not required, the calibration may be continued by withdrawing successive portions of water from the buret without refilling. This method is somewhat more rapid but does not so closely reproduce the drainage conditions that generally obtain in titrations and is more likely to be affected by stopcock leakage.

10. Duplicate tests (which check to within 0.02 ml) should be made for each calibration point.

11. If the buret is left empty, or not closed, the walls rapidly become filmed with grease and cleaning is necessary in order to obtain proper drainage

From the weight of water obtained from each buret interval calculate the volume of water delivered and the corresponding correction for that interval. Make a plot of the corrections as abscissae against the volumes as ordinates.

GRAVIMETRIC TITRATIONS

In gravimetric-titration methods the measuring instrument is the balance. The device for containing and delivering the standard solution is the gravimetric buret, or as it is commonly called the "weight buret"; examples of such burets are shown in Figure 16-8. The glass burets are designed to be hung from the stirrup of a balance; in a titration these burets are mounted in a buret clamp and used in the same way a volumetric buret is used. The volume markings are approximate and are for convenience in replicate determinations. The plastic buret offers the advantages of very low cost, excellent durability, one-hand operation, and convenience in weighing. It, too, can have approximate volume markings.

PROCEDURE **16-4**

Use of a Gravimetric Buret

Outline. Practice is given in the use of a gravimetric buret. Sample titrations are made in order to provide experience in the titrating operation.

DISCUSSION

There is no calibration necessary for a gravimetric buret, for the buret serves only to contain and deliver the standard solution during the titration and weighing. The same cleanliness is required here as for flasks, beakers, and bottles that are used for titrations and for storing standard solutions. There is no requirement of uniform drainage of solution from the walls of the gravimetric buret as there is for volumetric burets.

The instructions that follow are designed to give practice in the use of the buret in order that it can be used effectively in subsequent quantitative procedures. These instructions are applicable to the use of a plastic wash-bottle type of gravimetric buret; glass burets of the designs shown in Figure 16-8 can be used also. If the glass buret is used, a light-weight plastic beaker can serve to hold the buret upright while it is being weighed; this avoids the necessity of hanging the buret from the balance stirrup.

There appear to be no plastic gravimetric burets on the market at the present time.[9] The 2-oz (60-ml) wash bottle recommended here works very

[9] The Mettler Instrument Corporation, Princeton, N.J., has recently introduced a convenient gravimetric-titration apparatus designed for research or industrial use by experimenters who have unshared use of a top-loading balance. The titrant reservoir remains on the balance throughout the titration; titrant is delivered by means of tubing and a flow valve.

well if the tip is modified and if approximate volume markings are added as is explained in the Instructions below.

INSTRUCTIONS (Procedure 16-4: The Use of a Gravimetric Buret)

Wash the buret thoroughly (Note 1); check it to be certain that the top and delivery tube fit tightly so that air does not leak from the buret when water is forced through the delivery tube.

If the buret has no approximate calibration marks, they are easily applied by the use of a felt-tipped marking pen in the following way: remove the top and empty the buret of water. Add 5 ml of water from a volumetric buret, measuring pipet, or 10-ml graduated cylinder. At the water level, carefully make a narrow mark one-half inch long on the buret with a marking pen. Add 5 ml more water and again mark the water level; repeat the process over the height of the buret.

Practice delivering one drop at a time from the buret. Place the buret firmly upon the desk top so that its tip is just inside the receiving vessel (Figure 16-9). If the buret needs to be raised, a small inverted beaker or a block of wood makes a satisfactory rest for the buret (Note 2).

NOTES

1. The buret recommended here is a 2-oz (60-ml) plastic wash bottle having thin walls. The bottles manufactured by the Nalge Company of Rochester, New York are suitable. Wash bottles having a one-piece molded cap and delivery tube assembly are free from leakage problems, but the samples tried have been so stiff-walled that precise control of solution delivery is difficult. If a two-piece cap and delivery tube assembly is used and is found to leak, the most satisfactory remedy is to obtain a new plastic cap and, with a cork borer, make a hole in which the delivery tube fits very tightly.

Because of the nonadhering properties of the plastic, a test-tube brush used with laboratory detergent is usually effective in cleaning the buret. If stubborn residues from previous use coat the buret or delivery tube, use appropriate solvent solutions. For example, a deposit of silver is readily dissolved by warm dilute nitric acid; addition of a small amount of a soluble nitrite increases its effectiveness. Manganese dioxide films are removed easily by being reduced; acidified nitrite or sulfite solutions are effective. The plastic buret is very resistant to attack by most inorganic reagents, however, one should use care not to expose it to strong oxidizing acids for extended periods of time. Hot solutions should in general be avoided.

2. It is essential that the buret have a flexible wall, otherwise difficulty will be experienced in obtaining single drops or in splitting drops.

It is much easier to control the delivery of single drops if the buret is resting upon its base and if your hand and forearm are resting upon the laboratory bench than if the buret is held free from the desk. Moreover, if the buret it held above

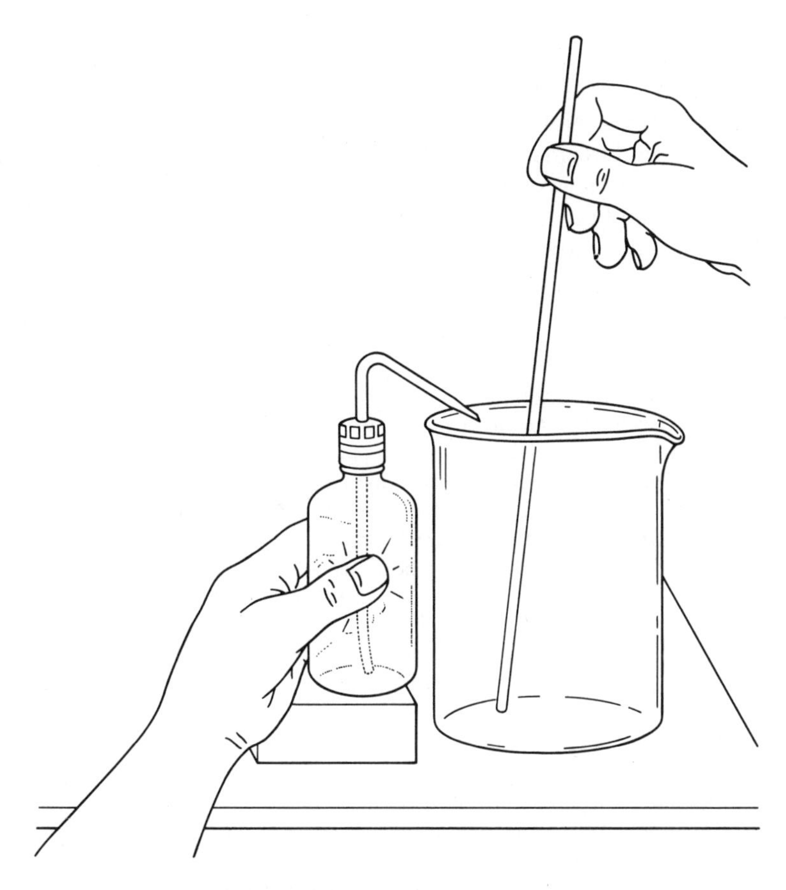

FIGURE 16-9
Method of using a plastic gravimetric buret.

NOTES

the desk there is the possibility that it will be tipped so that the solution enters the cap while pressure is being applied. If there are minute leaks where the cap and delivery tube or the cap and bottle join, solution may be lost. If the bottle is resting on its base, only air can escape from such leaks. For this same reason *never fill the buret completely to the shoulder.*

Better control of a single drop is often obtained if the hand is relaxed between drops.

Titrations with a volumetric buret are usually made by swirling or stirring the solution with the right hand and controlling the stopcock with the left. The student who is accustomed to using a volumetric buret will probably find it easier to continue to use his left hand for operating the gravimetric buret while stirring with his right. The student should find and use the combination most convenient for himself.

When you have become proficient in controlling the delivery of drops from the buret, carefully blot (Note 3) the outside of the buret with a dry towel; then weigh the buret (Note 4). Remove it from the balance, deliver one drop from the buret, and weigh it again (Note 5). Repeat this process several times until you know the range of drop size delivered by your buret. Now practice taking a split-drop (fractional drop) from the tip by means of a stirring rod until you can consistently remove a portion that weighs less than 5 mg.

NOTES

3. The polyethylene buret easily acquires an electrostatic charge if it is rubbed; this makes accurate weighing difficult. The proper practice for the use of such a device is to clean it thoroughly once, and from then on keep it clean and dry. The buret should always be placed on a clean part of the laboratory bench or a clean towel, and the student should always wipe his finger tips on a dry towel before touching the buret. Weighings in the titrations made here need to be made to ± 5 mg; if care is used that fingers are dry before the buret is touched there should be no appreciable error. If smaller portions were to be titrated so that more accurate weighings had to be made, special handling techniques would be required.[10]

4. The weighing operations with a gravimetric buret can be done on a single-pan analytical balance or on a top-loading balance that has 1-mg sensitivity. The operations could, of course, be done on an equal-arm balance, but the time required for the weighings makes the titrations very slow compared to volumetric titrations. With most single-pan analytical balances an optical reading to 0.1 mg will be obtained, but the 0.1 mg value is not significant for the purpose of this experiment.

5. If your gravimetric buret delivers drops that weigh more than about 15 mg, the tip should be redrawn (consult your instructor): Obtain a microburner and adjust this to the smallest flame that will burn continuously. The air valve should be closed completely. Remove the delivery tube from the buret and heat a section about $\frac{3}{4}''$ from the tip by rotating it slowly while holding it approximately 1″ above the top of the flame. The plastic becomes transparent as it is heated; when this transparent section has thickened about 50 %, remove the tube from the heat and, holding it in a vertical plane, slowly draw the tip to the desired diameter. Hold it in the drawn position until the portion that was heated has again become opaque. (Blowing gently on the tubing speeds this cooling process.) Cut the tip with a pair of scissors or a sharp knife on an angle, as shown in Figure 16-8.

If a buret delivers less than about 4 ml in 10 seconds when a continuous stream is flowing, titrations made with it will be unnecessarily time-consuming. The tip on such a buret should be trimmed until the flow-rate is about 4 ml in 10 seconds with moderate hand pressure; this requirement is easily compatible with the requirement above that a drop not weigh more than about 15 mg.

---

[10] C. A. Seils, R. J. Meyer, and R. P. Larsen (*Anal. Chem.*, **35**, 1673, 1963) give one such technique to prevent touching the buret.

Sample Titrations

(Note 6) Pipet a 5.00- or 10.00-ml-portion of 0.20 F NaOH (Note 7) into a 250-ml conical flask, add water from a graduated cylinder to bring the volume to about 50 ml, and add 2 drops of methyl-orange indicator. Fill the gravimetric buret (no higher than $\frac{1}{4}''$ below the shoulder) with 0.20 F HCl and weigh it. While continuously swirling the NaOH solution, titrate with the HCl until the first distinct change in color of the indicator from yellow toward reddish yellow or pink is obtained.

Weigh the buret and record this weight in your notebook.

Add and titrate two more portions of the HCl in the same manner. The weights of titrants used should agree to within the limit of error of the pipet (Note 8).

Continue making these titrations until you obtain weights within this range.

NOTES

6. These titrations are intended to provide experience and to build confidence in the use of the gravimetric buret. A very simple titration is performed in which there is a distinct color change. No calculation of concentration is to be made. You are to make several titrations of samples having identical size and check the reproducibility that you can obtain.

7. If only a 25-ml transfer pipet is available it is convenient to use a hydroxide solution that is less concentrated than the HCl so that the practice titrations can be done quickly with small portions of titrant. If the HCl is about three times as concentrated as the NaOH, the latter can be pipetted with a 25-ml pipet and only about 8 g of titrant will be required

8. The water delivered from the buret can be measured more accurately than water can be delivered from the pipet. Table 16-2 indicates that the limit of error for a 5-ml pipet is ± 0.01 ml.

QUESTIONS AND PROBLEMS

1. What factors determine whether the apparent weight of an object is greater or less than the vacuum weight?

2. What is the weight (*in vacuo*) of 1 liter of water at 20°C? *Ans.* 998.20 g.

3. Solution A is 0.1000 F at 20°C. What is its formality at 25°C? Solution B is 0.1000 WF at 20°C. What is its weight formality at 25°C?

4. Calculate to 0.01 g the difference in the weight of brass weights required to balance 1 liter of water at 20°C (a) in a vacuum and (b) in air with the barometric pressure 760 mm. What would be the percentage error caused by neglecting the buoyancy correction when weighing water in air under the above conditions? *Ans.* 1.05 g.

5. It requires 24.925 g of weights to balance in air (760 mm) the water delivered by a 25-ml pipet at 20°C. What is the volume of the pipet in milliliters at that temperature? *Ans.* 24.995 ml.

6. What volume of water will this pipet deliver at 30°C? *Ans.* 24.998 ml.

7. It is desired to calibrate a liter flask for use at 20°C, but the laboratory temperature at the time is 25°C and the barometric pressure is 760 mm. Calculate to 0.01 g the apparent weight of water to be used for the desired calibration, the empty flask being balanced with a tare.

8. You are required to calibrate a buret marked "20°C" for use at that temperature but find the laboratory temperature at the time of the calibration to be 25°C and the barometric pressure to be 755 mm. Calculate the factor by which any apparent weight of water delivered by the buret can be converted to the required volume. *Ans.* 1.00395.

9. A silver nitrate solution was prepared to be 0.10175 *N* at 17°C. What is its normality at 24°C?

10. A permanganate solution was standardized at 25°C and found to be 0.09982 *N* at that temperature. The solution was used without correction in a titration at 18°C to determine the iron in an ore. What percentage error was thus introduced into the determination? *Ans.* 0.16% negative error.

11. The calibration of a glass pycnometer was made by determining the weight of specially purified mercury (density at 25°C: 13.5340), which it contained. The following experimental data were obtained:

Apparent weight of pycnometer in air at 25°C and 760 mm	25.8276 g
Apparent weight of pycnometer filled with mercury under same conditions	93.5998 g

Calculate to 0.0001 ml the volume of the pycnometer in ml at 20°C. *Ans.* 5.0070 ml.

12. The data given below were obtained in checking the calibration of a liter flask. Calculate the volume of the flask ($\pm$0.01 ml) when used at 20°C.

Apparent weight in air, empty flask	215.72 g
Apparent weight in air, flask filled	1213.66 g

The room temperature was 20°C, the barometric pressure, 760 mm.

13. A liter flask was filled to the calibration mark with water at 15°C. The weight (*in vacuo*) of this water was 998.87 g.

A standard silver nitrate solution was prepared by dissolving 16.989 g of the pure salt in water and diluting the solution to the calibration mark in the above flask when the room temperature was 25°C. Calculate the formal concentration of the silver nitrate solution (*a*) at this temperature, and (*b*) at 20°C.

14. A volumetric flask was found to contain exactly 1000.00 ml of pure water at 20°C. What volume of 0.1 *F* $AgNO_3$ will it contain at 25°C? *Ans.* 1000.05 ml.

15. A liter flask was filled to the mark with water at 15°C. The apparent weight of this water was 998.90 g. The inner diameter of the exactly-cylindrical neck of the flask measured 1.74 cm at 20°C. It is desired to scratch a new mark on the neck such that the volume of the flask, when filled to the new mark, is exactly 1 liter at 20°C. How far below the old mark would the new one have to be placed? *Ans.* 4 mm.

16. The water in a volumetric flask filled to the calibration mark at a room temperature of 15°C had an apparent weight of 99.925 g weighed with brass weights in air at 15° and 760 mm. Calculate the volume of water (to 0.01 ml) which the flask will contain when similarly filled (*a*) at 20°C and (*b*) at 25°C.

17. Where approximate methods of calculation do not introduce errors greater than the precision of the measurements involved, such approximate methods should always be used; they not only eliminate much tedious arithmetical work but frequently greatly reduce the chance of errors in the calculation. A large number of the cases where approximations are useful can be treated by the general formula

$$(1 \pm \delta)^m \approx 1 \pm m\delta$$

where m is any number and δ is small in comparison with unity. The particular cases most often encountered are $1/(1 \pm \delta) \approx 1 \mp \delta$; $\sqrt{1 \pm \delta} \approx 1 \pm \frac{1}{2}\delta$; and $(1 \pm \delta)^2 \approx 1 \pm 2\delta$.

A solution that has been standardized at 20°C is being used at 23°C in a buret that was calibrated for use at 20°C. Calculate what factor should be applied to the volume readings thus obtained in order to convert them to a 20°C basis. *Ans.*

$$\begin{aligned} V_{20} &= V_{23} \text{ (observed)} \times [1 + 0.000010(23 - 20)] \times \frac{(0.99754)}{(0.99820)} \\ &= V_{23} \text{ (observed)} \times (1 + 0.000030) \times \frac{(1 - 0.00246)}{(1 - 0.00180)} \\ &= V_{23} \text{ (observed)} \times (1 + 0.000030 - 0.00246 + 0.00180) \\ &= V_{23} \text{ (observed)} \times 0.99937 \end{aligned}$$

CHAPTER 17

Precipitation Methods

As was explained in Chapter 5, a titrimetric precipitation method is one in which the insolubility of a precipitate causes the titration reaction to be complete within the desired quantitative limits. For example, silver thiocyanate is very slightly soluble and the reaction $Ag^+ + SCN^- = AgSCN(s)$ is the basis for a very well known volumetric method.

However, not all precipitation reactions can be used for volumetric determinations. As was stated in Chapter 5, if a reaction is to be so used, certain requirements must be fulfilled; these are discussed in more detail below.

REQUIREMENTS FOR PRECIPITATION METHODS

The first requirement is that the solubility of the precipitate must be sufficiently low to cause the reaction to be complete within the desired experimental accuracy. This solubility should not be greater than approximately 10^{-5} formal. The critical factor is not so much the solubility of the precipitate *per se*, as that the solubility controls the rate of change in the vicinity of the end point of the concentrations of the two ions involved in the titration reaction. This was discussed in Chapter 5, pages 60–63, in connection with the titration of chloride with silver nitrate. Methods have been suggested that make use of more soluble precipitates by taking advantage of the common-ion effect; an excess of the standard precipitant is added, the precipitate coagulated and filtered, then the excess precipitant determined. Thus, iodate can be determined by adding an excess of standard silver nitrate, filtering the silver iodate ($K = 2 \times 10^{-8}$), then titrating the excess silver with a thiocyanate solution. Such methods are used infrequently.

A second requirement is that the precipitation takes place rapidly and without the formation of supersaturated solutions. The formal solubility of silver chloride is slightly greater than that of barium sulfate, but when the latter is precipitated supersaturation occurs to such an extent that titrimetric methods based on its formation are impractical.

Third, the precipitate must be of exactly known composition. Ferric hydroxide, or more properly, hydrous ferric oxide, is a very insoluble compound, but the titration of a ferric salt with a standard solution of a soluble hydroxide is not possible because adsorption effects and the formation of basic salts cause the precipitate to be of uncertain and varying composition. Precipitates tend to adsorb the ions of which they are composed and the less soluble the precipitate and the greater the charge on an ion the greater this tendency. For these reasons most carbonates, phosphates, arsenates, and sulfides of the so-called heavy metals, although quite insoluble in neutral solutions, are infrequently used as the basis for volumetric precipitation methods.

ELEMENTS INVOLVED IN PRECIPITATION METHODS

Precipitation reactions are restricted to a surprising extent to reactions between singly charged cations and anions; the cation used almost exclusively is Ag^+. This situation can be attributed to the fact that, with one exception, singly charged cations are limited to the elements of columns Ia and Ib of the periodic table; the exception is thallium. The cations of the so-called *alkali metals*, column Ia, form relatively few insoluble salts and none of these meets the criteria for a titrimetric precipitation process.

The periodic positions of the elements of column Ib and of three other elements of row 6 having related analytical properties are shown in Figure 17-1. Also shown are their electronic configurations and common oxidation states. The symbols of certain nearby elements are shown for purposes of orientation. The three elements of column Ib all have the $(n-1)d^{10}ns^1$ configuration of their outer electrons; as would be expected from the single *s* electron, all can exist in the unipositive oxidation state, but only the unipositive silver ion is stable in aqueous solutions.

Both unipositive copper and gold tend to disproportionate to give the metal and a higher oxidation state. For example, cuprous copper reacts as follows:

$$2Cu^+ = Cu + Cu^{2+}$$

Ib	IIb	IIIa	IVa
29 +1 +2 Cu [Ar] $3d^{10}\,4s^1$	Zn	Ga	Ge
47 +1 +2 +3 Ag [Kr] $4d^{10}\,5s^1$	Cd	In	Sn
79 +1 +3 Au [Xe] $4f^{14}\,5d^{10}\,6s^1$	80 +1 +2 Hg [Xe] $4f^{14}\,5d^{10}\,6s^2$	81 +1 +3 Tl [Xe] $4f^{14}\,5d^{10}\,6s^2\,6p^1$	82 +2 +4 Pb [Xe] $4f^{14}\,5d^{10}\,6s^2\,6p^2$

FIGURE 17-1
Elements forming insoluble halides.

The unipositive states of both metals can be stabilized by the formation of an insoluble compound, such as CuI, or a stable complex ion, such as $Au(CN)_2^-$. Selected half-cell potentials of these elements are shown in Table 17-1 and they can be used to confirm the statement made above. Thus, by making use of the following half-cells and their potentials

$$Cu(s) = Cu^+ + e^- \qquad -0.52\ v$$

$$Cu(s) = Cu^{2+} + 2e^- \qquad -0.34\ v$$

one finds that the constant for the reaction

$$2Cu^+ = Cu^{2+} + Cu(s)$$

is approximately 10^6. The two potential values given above also indicate that the reduction of cupric ion will proceed to the metal unless the Cu(I) is stabilized by the formation of an insoluble compound or stable complex ion. Unipositive gold is also unstable and tends to form the metal and tripositive gold.

The three elements to the right of gold are of analytical interest because unipositive mercury and thallium and dipositive lead form relatively insoluble compounds with the halide anions (except fluoride), and with the pseudo halide, thiocyanate. In systems of qualitative analysis for the common elements, the first group of elements to be separated is precipitated by the addition of chloride to a nitric acid solution of the sample. The group

TABLE 17-1
Representative Potentials Involving Cu, Ag, and Au

	$E°$ (v)		
	Cu	Ag	Au
$M(s) = M^+ + e^-$	−0.52	−0.799	−1.69
$M(s) = M^{2+} + 2e^-$	−0.34	−1.39	−1.5
$M(s) = M^{3+} + 3e^-$			
$M^+ = M^{2+} + e^-$	−0.15	−1.98	−1.4
$M^{2+} = M^{3+} + e^-$		−2.1[1]	−1.3
$M(s) + Cl^- = MCl(s) + e^-$	−0.14	−0.22	−1.17
$M(s) + I^- = MI(s) + e^-$	+0.18	+0.15	−0.50
$M(s) + 2CN^- = M(CN)_2^+ + e^-$	+0.43	+0.31	+0.60

[1] $Ag^{2+} + H_2O = AgO^+ + 2H^+ + e^-$.

precipitate consists of silver chloride, mercurous chloride, and lead chloride and is usually called the *silver group.* Lead chloride is moderately soluble, is only partially precipitated, and is usually removed from the precipitate by extraction with hot water. Only mercurous mercury is precipitated with this group and the precipitate has the formula Hg_2Cl_2. The unipositive mercury ion is unique in being diatomic; this results from the single remaining 6*s* electrons on two ions forming an electron-pair bond. Gold and thallium are not included in conventional systems of qualitative analysis for the common elements; cuprous compounds are normally oxidized to the cupric state before the precipitation of the silver group.

The elements forming uninegative anions are found in column VIIa and those of analytical interest are shown in Figure 17-2. Chloride, bromide, and iodide all form insoluble compounds with the unipositive cations discussed above—the solubilities of these compounds decrease in the order given. Conversely the solubilities of the calcium halides increase with increasing atomic number. The difference can be attributed to the type of bonds formed. Those formed by the halide ions with calcium and certain other cations having inert gas structures are predominantly ionic; the electrostatic forces will be greatest and the bonds strongest with the smaller ions, and the compounds formed will tend to be less soluble. Thus, calcium fluoride is relatively insoluble, and the other calcium halides are very soluble. This size effect is seen also with the cation; the solubilities of the fluorides of the alkali metals in column Ia increase with increasing atomic number; the formal solubility of sodium fluoride is about ten times that of lithium fluoride, and that of potassium fluoride about twenty times that of sodium fluoride. Fluoride also forms complex ions with small multicharged cations, examples are FeF_6^{3-} and AlF_6^{3-}, but not with large cations such as Ag^+, Au^{3+}, or Hg^{2+}.

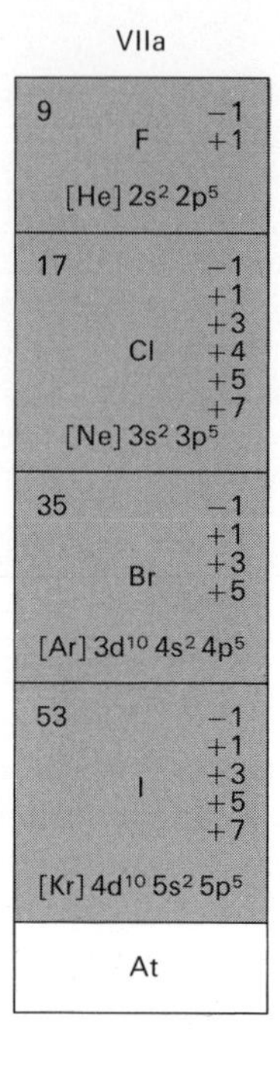

FIGURE 17-2

Elements forming uninegative anions of analytical interest.

Fluorine, like the other first-row elements, differs more in its properties and compounds from the elements below it than they do from each other. This difference can be attributed to the relatively much smaller size of the fluoride ion and to a much greater charge density.

However, the unipositive cations of the column Ib elements—copper, silver, and gold—have 11 outer electrons and form bonds with chloride, bromide, and iodide, which increase in covalent character with increased ionic radius of the halide. As the size of the halide ion increases there is a greater ease of ion polarization, that is, deformation of the electronic structure, and of covalent bond formation. Increase in such bonding results in a decrease in solubility with increase in atomic radius of the halide, and a corresponding increase in stability of the silver halide complex ions. The solubility products and instability constants given below show this trend,

X^-	Cl^-	Br^-	I^-
$[Ag^+][X^-]$	1.8×10^{-10}	5×10^{-13}	1×10^{-16}
$\dfrac{[Ag^+][X^-]^2}{[AgX_2^-]}$	6×10^{-6}	8×10^{-8}	10^{-13}

Among other effects that influence solubilities, but that are beyond the scope of this text, are the formation energies of the crystals and the energy of solvation of the ion or molecules dissolved.

With the exception of the fluoride, the precipitates of the silver halides, as well as the thiocyanate and cyanide, are not only insoluble but are uniform in composition, and thus a large majority of titrimetric precipitation methods involve the use of a standard silver nitrate solution.

METHODS FOR DETERMINING THE END POINTS OF PRECIPITATION TITRATIONS

One should understand that there are usually various means that can be used to determine the end point of any given titration (regardless of whether it is a precipitation titration). Frequently the choice of the particular end point to be used will be affected by the presence of other ions in the solution; for example, these ions may be intensely colored or may require that the titration be carried out in an acid rather than a neutral solution.

The general methods that are most commonly used for determining the end points of precipitation titrations are as follows:

I. Cessation of observable precipitate formation.
II. Use of internal indicators. These may form either (*a*) a colored precipitate, or (*b*) a colored soluble compound, or (*c*) a colored adsorbed compound.
III. Use of external indicators.
IV. Instrumental methods based upon physicochemical measurements.

Method I. Cessation of Observable Precipitate Formation

In using this method the analyst titrates until he observes that no more precipitate forms upon addition of the standard solution of the precipitant.

The classical example of this method is the Gay-Lussac titration of silver with a standard solution of sodium chloride, the reaction being

$$Ag^+ + Cl^- = AgCl(s)$$

This method, used in certain mints for assaying silver, depends upon the fact that as the equivalence point is approached the amount of precipitate forming upon addition of a given increment of the standard solution decreases very rapidly to an amount that can no longer be observed. It is not correct to say that no more precipitate is formed; theoretically, increasing the chloride-ion concentration of the saturated solution (by addition of more standard solution) should continue to cause further precipitation, since the solubility product is thereby exceeded. Applications of this method are obviously restricted to very insoluble salts that precipitate rapidly and do not form stable supersaturated solutions; also, though the method is capable of giving exceedingly exact results under carefully controlled conditions, it is laborious and time-consuming. For these reasons it does not find extensive application.

A variation of this method is based upon an adsorption phenomenon. Precipitates tend to adsorb, hold closely on their surfaces, their constituent ions. Therefore when a precipitate forms in a solution containing an excess of one of its ions, the particles tend to be similarly charged and mutually repellent, consequently, they remain dispersed. Thus if one titrates a silver-ion solution with chloride, the initially finely dispersed precipitate remains colloidally dispersed because of the adsorbed silver ions. The adsorption process is a reversible effect (see pages 259–264) and is dependent upon the concentration of the adsorbed ion. As one continues the addition of chloride the silver-ion concentration decreases very rapidly near the equivalence point, the adsorbed silver ions pass into the solution, and the particles lose their charges. As a result the precipitate coagulates and settles from the solution. This effect is very useful as an indication that an equivalent quantity of a precipitant has been added. In some cases the effect is so pronounced that, if highly accurate results are not required, it can be used as the end point; in such cases the method is known as a *clear-point* titration.

Method II. Use of Internal Indicators

An internal indicator is a substance that is added to the solution being titrated for the purpose of obtaining some visual evidence in the solution of the point at which an equivalent amount of the standard solution has been added.

These visual effects are usually caused by the formation of a characteristic precipitate or colored compound as the result of a reaction between the indicator and the reactive constituent of the standard solution. Internal indicators can be classified into three sub-classes as shown above. An example of subclass (*a*) results when a chloride solution is being titrated with a solution of silver nitrate, and a soluble chromate is used as an indicator. Since silver chloride is less soluble than silver chromate, none of the characteristic red silver chromate precipitate will be formed until an amount of the silver substantially equivalent to the chloride present has been added. The sequence of reactions taking place will be

$$Cl^- + Ag^+ = AgCl(s)$$
(Titration reaction)

$$CrO_4^{2-} + 2Ag^+ = Ag_2CrO_4(s) \quad \text{(reddish brown)}$$
(Indicator reaction)

An example of subclass (*b*) is obtained when silver ion is titrated with thiocyanate, with a ferric salt used as indicator. The solubility of silver thiocyanate is such that the characteristic soluble red ferric thiocyanate compound is not formed in visible amounts until substantially all the silver has been precipitated. In this case the reactions can be represented as follows:

$$Ag^+ + SCN^- = AgSCN(s)$$
(Titration reaction)

$$Fe^{3+} + SCN^- = FeSCN^{2+} \quad \text{(red color)}^1$$
(Indicator reaction)

The third subclass of end-point methods makes use of internal indicators but is restricted to precipitation titrations, since it depends upon the adsorption of the indicator and a resulting color change on the surface of the precipitate that is being formed during the titration.

The first two of these subclasses of end points will be discussed in detail in connection with the procedures that follow.

Method III. Use of External Indicators

In some cases satisfactory internal indicators are not available and recourse is made to the expedient of removing a small fraction of the solution being titrated and adding it to an external indicator. For example, zinc ion forms an insoluble zinc potassium ferrocyanide in the presence of potassium ions, and zinc can be estimated volumetrically by making use of the following reaction:

$$3Zn^{2+} + 2K^+ + 2Fe(CN)_6^{4-} = K_2Zn_3[Fe(CN)_6]_2(s)$$

---

[1] There has been some uncertainty as to the formula of the colored compound. Compounds such as $K_3Fe(SCN)_6$ can be prepared; and Schlesinger and Van Falkenburgh (*J. Am. Chem. Soc.*, **53**, 1212, 1931), and Schlesinger (*J. Am. Chem. Soc.*, **63**, 1765, 1941) attributed the color to anions of the type $Fe(SCN)_6^{3-}$ or $Fe(SCN)_4^-$. However, Møller (*Chem. Zentr.*, **1**, 3176, 1938), Bent and French (*J. Am. Chem. Soc.*, **63**, 568, 1941), Edmonds and Birnbaum (*J. Am. Chem. Soc.*, **63**, 1471, 1941), and Frank and Oswalt (*J. Am. Chem. Soc.*, **69**, 1321, 1947) have obtained substantiating evidence that the color is caused by the cation $Fe(SCN)^{2+}$ when an *excess of iron* is present.

Cupric ferrocyanide is red, but cupric ion cannot be used as an internal indicator for the above titration because cupric ferrocyanide is less soluble than the zinc potassium ferrocyanide, and therefore a red precipitate would be obtained at the beginning of the titration. If, however, drops of the solution are removed during the course of the titration and added to a drop of a cupric solution, no perceptible precipitate of the red cupric ferrocyanide will be obtained until substantially all of the zinc has been precipitated. Such methods are time-consuming and are very seldom used in modern laboratories.

Method IV. Instrumental Methods Based Upon Physicochemical Measurements

In many cases certain physicochemical properties of the solution being titrated, such as its electrical potential or conductivity, will show rapid and characteristic changes in the neighborhood of the equivalence point. By measuring these changes with suitable instruments, very accurate determinations of the end points of titrations can be made.

In the procedures that follow there will be found examples of Method II accompanied by detailed discussions of the factors, both experimental and theoretical, that affect the agreement between the observed end point and the stoichiometric equivalence point.

Preparation of a Standard Solution of Silver Nitrate

Outline. In this procedure a standard solution of silver nitrate is prepared directly from a weighed portion of dry, reagent grade $AgNO_3$.

If the solution is to be used for volumetric titrations, this $AgNO_3$ is dissolved in distilled water of a known temperature, and the solution diluted to a known volume in a volumetric flask.

If the solution is to be used for gravimetric titrations, the weighed $AgNO_3$ is transferred to a dry weighed beaker, dissolved in water, and the solution diluted to a known weight.

DISCUSSION

A standard solution for titrimetric work can be prepared by two methods: (1) Directly. The required amount of a pure substance is weighed precisely and dissolved. For volumetric titrations the solution is diluted at a definite temperature to an exactly known volume in a calibrated flask. For gravimetric titrations the solution is diluted to a known weight. (2) Indirectly. The solution of approximately known concentration is prepared and then standardized by measuring the volume or weight of it which reacts with a precisely weighed amount of some pure substance (called a *primary standard*).

The direct method is used when the substance from which the standard solution is made can be obtained in a high degree of purity, can be precisely weighed, and does not change upon being dissolved in water or when the solution is diluted (for example, does not react with the gases or traces of organic matter usually present even in distilled water). The indirect method is more frequently employed because many of the substances used for standard solutions are not readily obtained in a sufficiently pure form or are not stable when first dissolved in distilled water.

Standard Silver Nitrate Solutions

Very pure silver nitrate can be prepared by recrystallization or can be purchased, and can be dried and weighed (even fused at 220°C to 250°C) without change, provided organic matter, reducing gases, or hydrogen sulfide are avoided. Standard solutions of silver nitrate are usually prepared by the direct method described above. Such solutions are then stable, provided they are protected from organic matter, reducing gases, and also from light.

When the silver nitrate solution with its concentration precisely known is available, it can be used for standardizing other solutions that can not be directly prepared.

Concentrations of Standard Solutions

The silver nitrate solution prepared below and a majority of the standard solutions used in macromethods of volumetric analysis are approximately 0.1 *N*. The choice of the concentration to be used is governed by several factors. The solution should be so dilute that the volume or weight required for a titration will be such that the errors in the measurements will be relatively small; it should not be so dilute as to require refilling a buret during a titration. The solution should not be diluted beyond the point where the measurement errors are relatively less than the uncertainty in the determination of the end point.

Optional procedures are given below for preparing solutions for volumetric titrations and for gravimetric titrations. Consult your instructor regarding use of these procedures.

INSTRUCTIONS (Procedure 17-1: Preparation of a Silver Nitrate Solution)

In this procedure for the first time you will encounter certain nomenclature and will use several general operations that have been described previously. These are listed below and should be studied before coming to the laboratory and, when necessary, reviewed while carrying out this procedure:

Operation	Page
Designation of purity of chemicals	208
Drying substances	206
Use of weighing bottles	177
Evaluation of measurements	27
Weighing and transferring samples	183
Use of distilled water	207
Use of volumetric flasks	192
Handling solutions	184
Use of various equipment	175

Use a trip, top-loading, or similar balance to weigh into a clean, dry weighing bottle approximately 17.0 g of *reagent grade* silver nitrate (10.0 g if preparing a gravimetric solution). (Notes 1 and 2.)

Place the open bottle in a small beaker loosely covered with a watch glass, and heat for at least 2 hours in an electric oven at 105°C to 110°C (Note 3).

Transfer the weighing bottle and contents to a desiccator and allow them to cool. Weigh the bottle and contents.

If the solution is to be used for volumetric titrations, continue by following optional procedure *A* below.

If the solution is to be used for gravimetric titrations, continue by following optional procedure *B* below.

Optional Procedure A. Preparation of a Solution for Volumetric Titrations

Insert a large-stem funnel in the neck of a clean, calibrated (see Procedure 16-1) 1-liter flask and carefully transfer the silver nitrate to the funnel. Allow any particles of the silver nitrate adhering to the bottle to remain. Close the bottle and return it to the desiccator. (*Do not place the weighing bottle on the desk.*) Dissolve the silver nitrate remaining on the funnel by slowly flowing *distilled* water from a wash bottle over the crystals. Fill the flask

NOTES

1. These instructions are for preparing a liter of the volumetric solution and 500 g of the gravimetric solution. Since $AgNO_3$ is an expensive chemical, a smaller quantity should be prepared if less solution is required (consult your instructor). By approximately weighing out the amount of silver nitrate taken, no excess is dried, and the solution will be close to the desired formality.

When laboratory time is limited, the student can be given the required weight of dry $AgNO_3$ in a weighing bottle; this procedure also eliminates some loss of $AgNO_3$.

2. Because measurements of weight can be made more accurately than those of volume, smaller quantities of more concentrated solutions can be used for gravimetric than for volumetric titrations.

In general, gravimetric solutions are prepared to be approximately 20% more concentrated than volumetric solutions because the more concentrated solutions enable one to make titrations more expeditiously. Accuracy is not lost; a weight measurement is substituted for one of a volume and smaller drops can be added.

3. Since silver compounds are easily decomposed by organic matter and reducing agents, the oven should be protected from dust and from reducing gases. Paper labels should not be used on articles placed in the drying oven, because the adhesive material used chars, giving off volatile reducing substances (tarry materials), which cause the reduction of silver compounds. Weighing bottles should be identified by marking them with a pencil on the ground-glass seal around the stopper. Most commercial silver nitrate will darken slightly on being heated, but with the better grades this decomposition is not significant.

almost to the mark with water at room temperature. (The solution in the flask should be gently swirled as the water is added so that it is well mixed before the final adjustment of the meniscus is made. The dissolving of a solid or the mixing of solutions may result in an appreciable volume and temperature change.)

Use the funnel so as to avoid wetting the flask above the calibration mark. Finally, remove the funnel, being sure to rinse off any solution adhering to it; then with a dropper or pipet add water until the lower meniscus just coincides with the calibration mark. Any droplets of water spilled against the side of the flask above the calibration mark should be removed with filter paper before this final adjustment is made. Stopper the flask and mix the contents by swirling the solution, inverting, and again swirling, repeating this sequence at least 10 times. (*One of the most common sources of error in volumetric analysis is the incomplete mixing of solutions.*)

Remove the weighing bottle from the desiccator and weigh it.

Transfer the solution to a clean bottle with a ground-glass stopper, first rinsing the bottle with several small portions of the silver nitrate solution. Calculate the formality of the solution at the standard laboratory temperature (20°C) and label the bottle. (Standard solutions should not be kept in volumetric flasks, since the latter are relatively thin-walled and easily broken, and are also likely to be etched by solutions left in them for long periods.)

Keep this light-sensitive solution in a brown glass bottle or, if these are not available, in a bottle that is covered with paper or that has been painted with black lacquer.

(All standard solutions, reagents, and solutions for analysis that are to be reserved for future use should be carefully labeled, dated, and, if for general use, initialed by the person preparing them.)

Optional Procedure B. Preparation of a Solution for Gravimetric Titrations

Weigh a clean, dry 1-liter (or smaller—see Note 1 above) glass-stoppered bottle, without the stopper, on a top-loading or "solution balance" to the nearest 0.5 g (or 0.05% of solution weight). The outside of the bottle must be dry and free from any loose particles. It should be placed on a clean towel or paper after it is weighed. Insert a large-stem funnel in this weighed bottle and carefully transfer the silver nitrate to the funnel. Allow any particles of the silver nitrate adhering to the weighing bottle to remain. Close the weighing bottle and return it to the desiccator. (*Do not place the weighing bottle on the desk.*) Dissolve the silver nitrate remaining on the funnel by slowly flowing distilled water from a wash bottle over the crystals. Remove the funnel, being sure to rinse off any solution adhering to it. Return the bottle to the balance on which it was weighed, and add distilled water until the bottle and solution weigh 500 g (or desired weight) more than did the empty bottle. Record the weight of the bottle and solution, stopper the bottle, and mix the contents by swirling the solution, inverting, and again swirling, repeating the sequence at least 10 times. (One of the common sources of errors in titrimetric analysis is the incomplete mixing of solutions.)

Remove the weighing bottle from the desiccator and weigh it. Calculate the weight-formality of the solution and label the bottle. Keep this light-sensitive solution in a brown-glass bottle or, if these are not available, in a bottle that is covered with paper or that has been painted with black lacquer.

(All standard solutions, reagents, and solutions for analysis which are to be reserved for future use should be carefully labeled, dated, and, if for general use, initialed by the person preparing them.)

Titration of Chloride with Silver Ion. Chromate as an Internal Indicator

Outline. In this procedure (*a*) the weighed chloride sample is dissolved in water, (*b*) the resulting solution is adjusted to a *p*H range of 6.3 to about 10, (*c*) potassium chromate is added as indicator, and (*d*) the solution is titrated with standard silver nitrate solution to the appearance of silver chromate precipitate. The equations involved are

$$Ag^+ + Cl^- = AgCl(s) \quad \text{(white precipitate)}$$
(Titration reaction)

and

$$2Ag^+ + CrO_4{}^{2-} = Ag_2CrO_4(s) \quad \text{(reddish-brown precipitate)}$$
(Indicator reaction)

DISCUSSION

A discussion and calculation of the end-point error of this titration, which is known as the Mohr method,[2] has been given in Chapter 5. That material should be reviewed.

Weighing the Sample

In this titration procedure a soluble chloride sample is used in order that the sample can be prepared quickly and easily for the titration. Replicate portions of the sample are weighed into a conical flask; this measurement must be made with an accuracy within that desired in the final result, usually between 1 and 2 parts per thousand or between 0.1% and 0.2%.[3] Thus, since the quantitative measurements made in the entire procedure are *weight* of $AgNO_3$ used to prepare the standard solution, *volume* or *weight* of $AgNO_3$ solution prepared, *weight* of unknown chloride sample, and *volume* or *weight*

[2] See Footnote 1, page 59.
[3] See Chapter 3, pages 23–27, for a more detailed discussion of precision of measurements.

of standard $AgNO_3$ solution used in the titration, each of these measurements should be made to at least the accuracy of the poorest measurement of the group. Let us consider first the volumetric titration. The volumetric flask is calibrated and can be used to an accuracy of $\pm 0.1\%$ (one part per thousand). In the case of a 1-liter flask the manufacturer needs only to place the calibration mark so that, in volume units, it is within 1 ml of exactly 1 liter. Since this is easily done, a flask of this size is seldom found that varies more than 0.1% from its stated value. The pipet and buret have both been calibrated to 0.1% or better, so it is evident that the weighings must be made accurate to $\pm 0.1\%$. If a 0.2-g sample is taken, it must be weighed to 0.2 mg. (While the weighing bottle and sample may weigh 18 g together, it is the *sample* that must be weighed to 0.1%. If the weighing bottle plus sample were weighed accurately to 18 mg, which is 0.1% of the total weight, there would be a 0.9% error in the weight of the sample.)

In the case of the gravimetric titration of a solid sample, all measurements are weight measurements. Since each of these can be made with an error of considerably less than 0.1%, it becomes evident that the accuracy of the final results may well be determined by the inherent error in the titration (that is, the "titration error," see page 64) and upon the experimenter's ability.

Factors Affecting the End Point

The discussion of titration error in Chapter 5 showed that the experimental accuracy of the titration will be largely determined by the concentrations of the chromate and the hydrogen ions; the latter control the chromate–dichromate equilibrium. These conditions are discussed separately below:

The Chromate-Ion Concentration. This is the most important factor involved, since it can be controlled over a considerable range and it determines where in the titration the end point (the reddish Ag_2CrO_4) appears. For example, an increase in the concentration of chromate ion causes a decrease in the silver-ion concentration and an increase in the chloride-ion concentration at the end point. Practically, the chromate concentration cannot be increased beyond the point where its color becomes objectionable by masking the first appearance of the precipitate of silver chromate. Concentrations ranging from 4×10^{-4} to 1.5×10^{-2} F have been recommended; the concentration used in the procedure below is approximately 1.3×10^{-3} F.

The Amount of the Ag_2CrO_4 Precipitate. This quantity is variable, being dependent upon the individual observer, the concentration of the chromate (and therefore the color of the solution), the dispersion of the precipitate, and the conditions obtaining at the time of the observation. Under favorable conditions it appears to range from 0.08 mg to 0.2 mg of silver chromate, but if care is not taken, it will be much larger. Thus, in a series of observations made with 100 ml of a solution that was 6×10^{-3} *F* in potassium chromate, it was found that 0.28, 0.24, and 0.22 ml—or an average of 0.25 ml—of 0.01 *M* silver nitrate was required to give the first perceptible color change.

In the laboratory, this error is effectively reduced by applying an end-point correction. This consists in subtracting from the silver nitrate required for the titration a volume equal to that required to produce a similar silver chromate precipitate in a solution similar to the one being titrated but containing no chloride. Such a correction is made in this procedure.

The Hydrogen-Ion Concentration. This controls the chromate concentration, and should be between 5×10^{-7} and 10^{-10} *M*.[4] The upper limit is fixed by the solubility of silver chromate in acid solutions, and the lower limit by the low solubility of silver oxide in alkaline solution.

Two effects cause an increase in the solubility of silver and other metal chromates in acid solutions. First, chromic acid is not a highly ionized acid. The second hydrogen ion is ionized to about the same extent as is the first hydrogen ion of carbonic acid or the second one of phosphoric acid; the equilibrium constant for the reaction

$$HCrO_4^- = CrO_4^{2-} + H^+$$

has been estimated[5] to be 3.2×10^{-7}. In addition to this, there is an equilibrium between hydrogen chromate ($HCrO_4^-$) and dichromate ion, as follows:

$$2HCrO_4^- = H_2O + Cr_2O_7^{2-}$$

The equilibrium constant for this reaction is 98.[6] From these two expressions one can derive the expression for the equilibrium between the chromate and the dichromate ions,

$$2CrO_4^{2-} + 2H^+ = Cr_2O_7^{2-} + H_2O$$

The equilibrium constant for this reaction can be calculated from those given above to be 9.5×10^{14}. It is to be noted that the hydrogen ion enters

[4] Doughty (*J. Am. Chem. Soc.*, **46**, 2707, 1924) has claimed that the $[H^+]$ for this titration can be as high as 10^{-5}, and recommends that the solution be buffered by acetate and acetic acid, provided that the acetate–acetic acid ratio is greater than 2. Better end points are obtained in the more alkaline solutions specified above.

[5] Neuss and Rieman. *J. Am. Chem. Soc.*, **56**, 2238 (1934).

[6] Tong and King. *J. Am. Chem. Soc.*, **75**, 6180 (1953).

into this equilibrium as the square of its concentration. Calculations will show that at *p*H 7, with the total chromate 6×10^{-3} *F*, chromate ion is the predominant species, but that at *p*H 6 this is no longer true. The dichromates are, in general, much more soluble than the chromates; therefore, the chromate-ion concentration must be above the minimum value cited in the discussion above if the end-point precipitate is to appear at an acceptable point in the titration; moreover, the presence of dichromate ion, which has an intense orange color, tends to obscure the end point even if enough chromate ion is present to cause precipitation of silver chromate.

The Titration with Standard Silver Nitrate Solution

The mixture must be vigorously swirled during titration with the $AgNO_3$ solution; otherwise, a premature end point will result from adsorption of chloride ions upon the solid silver chloride. Another benefit of the swirling is the rapid coagulation of the silver chloride precipitate, which occurs just a few tenths of a milliliter prior to the end point. This coagulation is a convenient indication that the end point is near; however, one may miss the coagulation or clear point entirely unless he swirls the mixture vigorously.

The appearance of the mixture during the titration might be described in this way: the first drop of standard silver nitrate solution added to the yellow solution of chloride and chromate ions causes a milky white precipitate of AgCl(s) to form. This mixture becomes more dense with addition of the silver solution, and the precipitate coagulates into curdlike flocs just before the end point. During the titration a reddish color will be seen in the region where the stream of $AgNO_3$ solution enters the mixture, but this color will not spread so long as there is a large excess of chloride ion present. When the end point is very near, the reddish color will be seen to spread through most of the mixture with the addition of a single drop of $AgNO_3$ solution, until finally at the end point the color persists. This reddish color is Ag_2CrO_4(s), which disappears if the chloride-ion concentration remains high enough to force the equilibrium

$$Ag_2CrO_4(s) + 2Cl^- = 2AgCl(s) + CrO_4{}^{2-}$$

significantly to the right.[7]

[7] Fortunately for the purposes of this titration, the freshly precipitated Ag_2CrO_4 is rapidly metathesized by the above reaction. This ability of newly precipitated solids, which are highly dispersed and hence have large surface areas, to adjust rapidly to equilibrium conditions is in sharp contrast to the behavior of solids that have had time for development and perfection of their lattice structures. If crystals of Ag_2CrO_4 were added to the mixture early in the titration the disappearance of the red Ag_2CrO_4 would be slow compared with the observed rate of disappearance.

Two other factors should be considered in performing this titration. First, the solubility of silver chromate increases rapidly with temperature; therefore, the mixture should not be much above 20°C during the titration. Second, silver chloride is sensitive to light (see page 287), and exposure of the mixture to direct sunlight should be avoided.

Properly performed titrations by this method will give a reproducibility of within 2 parts to 3 parts per thousand.

Applications and Limitations of the Method

This *Mohr titration*, that is, the titration with silver ion with chromate as the indicator, can be used for the determination of bromide and (if the solution is kept slightly alkaline) of cyanide; it is not suitable for the determination of iodide or thiocyanate because of adsorption effects. The application of this method to any of the above determinations is limited by the fact that its use is precluded by the presence of any anion that forms a precipitate with silver ion in either an acid or neutral solution that is less soluble than silver chloride or silver chromate. If the anion forms a precipitate that is more soluble than silver chromate, it may be present in moderate concentrations. The solubilities of the more common silver salts are shown in Table 17-2.

Cations that are hydrolyzed in neutral solutions (such as ferric and aluminum ions) or cations that form insoluble chromates (such as lead and barium) cannot be present.

TABLE 17-2
Solubility of Silver Salts

Insoluble in Neutral and Acid Solutions	Insoluble in Neutral Solutions	Soluble in Neutral Solutions
$AgBr$[a,c]	Ag_3AsO_4[b]	$AgC_2H_3O_2$[b]
$AgCN$[a,c]	$Ag_2B_4O_7$[b]	$AgClO_3$[b]
$Ag_3Fe(CN)_6$[a,c]	Ag_2CO_3	AgF[b]
$Ag_4Fe(CN)_6$[a,c]	Ag_2CrO_4[b]	$AgNO_3$[b]
$AgIO_3$	$Ag_2C_2O_4$[b]	$AgNO_2$[a,d,e]
AgI[a,c]	Ag_3PO_4[b,f]	$AgClO_4$[b]
Ag_2S[a,c]	Ag_2SO_3[b,d]	$AgMnO_4$[b]
$AgSCN$[a,c]	—	Ag_2SO_4[b,e]

[a] Less soluble than silver chloride.
[b] More soluble than silver chloride in neutral solution.
[c] Less soluble than silver chromate in neutral solution.
[d] Causes reduction of chromate ion.
[e] Only moderately soluble.
[f] This compound has been used (instead of silver chromate) as the indicator precipitate; the results are not as satisfactory.

Colored compounds cannot be present in sufficient concentration to obscure the end point; nor can substances that would cause reduction of chromate. The method cannot be used for the direct titration of silver with chloride, since the silver chromate precipitate becomes so coagulated and compacted as to redissolve slowly near the equivalence point; it is possible to add an excess of standard chloride, then to add the indicator and back-titrate with silver.

The procedure below provides for the analysis of solutions and readily soluble solids that do not contain any of the interfering constituents mentioned above. More advanced texts or reference works should be consulted in regard to the preliminary treatment required for the analysis of insoluble materials and the elimination of interfering constituents.

Optional procedures are given below for volumetric titrations (Procedure 17-2A) and for gravimetric titrations (Procedure 17-2B). These procedures are identical in principle, but use different methods for measuring the titrant. They have been separated in this first procedure for clarity of presentation. Consult your instructor regarding the use of these procedures.

INSTRUCTIONS (Procedure 17-2A: Volumetric Titration of Chloride with Silver Ion. Chromate as an Internal Indicator)

(Early procedures are given in detail in order to aid in development of technique. In later procedures detail is minimized.)

I. Preparation of Samples for Titration

If the sample is a solid, transfer 1 or 2 g to a weighing bottle and dry at 110°C until the weight is constant to 0.2 mg–0.3 mg. Weigh out into 250-ml beakers or 200-ml flasks (Note 1) two or three samples (consult your instructor) that will require from 20 ml to 40 ml of standard 0.1 *F* silver nitrate solution (Note 2). Dissolve each sample in 50 ml of water.

NOTES

1. Either beakers or conical flasks can be used for titrations. Samples can be transferred to beakers somewhat more easily than to flasks; with flasks, solutions can be effectively mixed by swirling, and there is less danger of loss by splashing.

2. Information is usually available as to the approximate chloride content of the sample. On this basis the weight or volume of the sample should be so adjusted that the volume of silver nitrate used will be within the above limits in order that the errors in reading the buret will be small relative to the volume of solution used, and that refilling the buret during the titration will be unnecessary. In case information

If the sample is a solution, pipet (read Procedure 16-2 in regard to the proper technique) into 250-ml beakers or 200-ml flasks (Note 1) that volume of solution that will require from 20 ml to 40 ml of standard 0.1 *F* silver nitrate solution (Note 2). Dilute each solution to 50 ml.

Add to the neutral solution (Note 3) approximately 8 drops (0.25 ml) 0.5 *F* K_2CrO_4.

Fill a Geissler-type buret with the previously prepared (Procedure 17-1) standard silver nitrate solution (Note 4). Withdraw the solution until the meniscus is just above the zero mark. Allow 30 seconds for drainage and adjust the meniscus to the mark, then remove any hanging drop from the tip.

II. Preparation of Comparison Solutions

Grind 0.1 g of reagent-grade, precipitated, *chloride-free* calcium carbonate (see the last paragraph of Note 3) with 2 ml–3 ml of water in a clean mortar. To each of two vessels similar to that to be used in the titration add 0.25 ml (8 drops) of 0.5 *F* K_2CrO_4 indicator and 100 ml of water (approximately the same volume of water as will be present in the titrated mixture at the end point). To each solution add 3 drops–5 drops of the carbonate suspension (Note 5). To the second solution add the silver nitrate solution in fractions

NOTES

as to the chloride content of the sample is not available, it is recommended that a preliminary titration be made on a small portion in order to decide the proper volume or weight to use for subsequent titrations.

3. If it is not known that the solution of the sample will be neutral, dissolve a preliminary 0.1-g portion in 10 ml–20 ml of water and test the solution with litmus or a wide-range *p*H test paper. In case the solution is acid, add powdered *chloride-free* calcium carbonate in 0.1-g portions to the solution to be titrated and stir for 15 sec–30 sec, until a *small excess* of the carbonate is present. In case the solution is basic, add 1 drop of phenolphthalein indicator and then add 6 *F* CH_3COOH until the solution becomes colorless.

The calcium carbonate used for this purpose should be tested for chloride by dissolving 0.1 g–0.2 g in 5 ml of 3 *F* HNO_3, diluting the solution to 10 ml–15 ml, and adding 1 ml of 1 *F* $AgNO_3$; no significant precipitate should result.

4. Review Procedure 16-3 in regard to the use and calibration of burets. If you have not performed that procedure, it should be referred to constantly during this titration. Draw the distilled water from the buret and rinse the entire bore with at least two small portions (3 ml–5 ml) of the silver nitrate before filling it. Also review pages 192–194 regarding storage and handling of standard solutions. *Keep storage bottles closed except when withdrawing solutions!*

5. Only enough of the carbonate mixture should be added to produce a *suspended turbidity* in the titrated solution; from 0.03 ml to 0.15 ml is usually required. The carbonate *must be free of chloride and oxide, and only a minimum amount should*

of a drop until the first perceptible change from a greenish yellow toward a brownish yellow is obtained (Note 6). Note and record the volume of silver nitrate required for this end-point correction. Keep these solutions for comparison during the subsequent titration of the sample.

III. Preliminary Titrations

(Optional. Consult your instructor.) Pipet 5.00 ml of a standard 0.1 F NaCl solution (provided by your instructor) into a 250-ml beaker or 200-ml flask. Dilute the solution to 50 ml. Add 0.25 ml (8 drops) of 0.5 F K_2CrO_4.

Note and record the buret reading. Gently swirl or stir the chloride solution and slowly add the standard silver nitrate to it until the transient reddish-brown color produced locally by the silver solution begins to disappear somewhat slowly. (When a beaker is used, keep the buret tip close enough to the surface of the solution to prevent loss from splashing, and use a stirring rod constantly during the titration in order to mix the solution; when a flask is used, mix the solution by swirling.) Thereafter add the silver nitrate solution dropwise until the first perceptible color change is produced that remains after the mixture is shaken vigorously for 30 seconds and that corresponds to that of the comparison solution used for the end-point correction (Notes 7 and 8). Allow time for drainage, then read the buret. Record the volume of the $AgNO_3$ used.

NOTES

be added; otherwise, an abnormally large amount of the silver nitrate will be needed. The carbonate should not contain significant amounts of calcium oxide or hydroxide, or a precipitate of silver oxide may result.

6. If over 0.08 ml of the 0.1 F $AgNO_3$ is required for this end-point correction, too much color has been produced, too much of the carbonate mixture was used, or the carbonate is contaminated by chloride, oxide, or hydroxide. In any case, repeat the procedure and omit the carbonate.

This end-point change is best obtained by reflected, not transmitted, light, and by incandescent lighting rather than by bright daylight or fluorescent lighting. The change is from the greenish yellow of the chromate solution to the first detectable *change* toward a brownish yellow. The appearance of the brownish-yellow color is more easily perceived if the solution being titrated is observed against a uniform white background, such as white paper or tile. For this reason the titration is sometimes made in a large casserole.

7. The approach to the end point is shown by the slower disappearance of the local reddish-brown color caused by each drop of silver nitrate. This lag is partly due to adsorption of chloride on the precipitate and partly to the slowness with which the local precipitate of Ag_2CrO_4 is metathesized by the low concentration of chloride remaining in the solution. Because of these effects the mixture should be vigorously shaken so that equilibrium is established. A person making this

Add a 1.0-ml portion of the 0.10 *F* NaCl and repeat the titration (Note 9).

IV. Titration of the Samples

Refill the buret with the standard silver nitrate solution. Adjust the solution to the zero mark as directed in the last paragraph of I above.

Titrate one of the chloride samples as directed in the second paragraph of III above.

Allow time for drainage, read the buret, and *immediately record this reading in ink* in a permanent notebook (see page 162).

Similarly titrate the second sample of the chloride. From the formality of the silver nitrate solution, the net volumes of silver nitrate used (that used for the titration minus that for the end-point correction), and the weights of the samples, calculate and report the percentage or formality of chloride found in each sample and the average of these values (Notes 10 and 11).

NOTES

titration for the first time should add the silver nitrate solution in full drops in order to obtain a definite recognition of the end point. On subsequent titrations, as the end point is approached the silver nitrate should be added by allowing a fraction of a drop to form on the tip of the buret, removing this, and adding it to the solution by means of a stirring rod.

Near the end of a titration, the inside walls of the flask or other titration vessel should always be washed down by means of a jet of water (only a few milliliters are needed) from the wash bottle; any of the standard solution or of the solution being titrated adhering thereto is thus recovered.

8. A common procedure that allows rapid titration of a solution without overrunning the end point is to reserve a small fraction of the solution and to add it after a rapid titration to a preliminary end point. This operation can be performed by drawing up a small portion, in this case approximately 0.5 ml, of the solution into a long "dropper" and leaving this in the solution until a preliminary end point is obtained. The reserved portion is then expelled, and the dropper is washed out by alternately drawing up and expelling the solution being titrated. An end point is again obtained, and the dropper is again flushed out to ascertain whether the end point is permanent.

9. These preliminary titrations are done in order to familiarize you with the approach to the end point and thus minimize the danger of overrunning end points when titrating the samples.

10. See page 18 for a note on calculations.

11. All silver solutions and residues should be put into a "silver residues" bottle. In the above case, first allow the precipitates to settle and then decant away most of the clear solution.

INSTRUCTIONS (Procedure 17-2B: Gravimetric Titration of Chloride with Silver Ion. Chromate as an Internal Indicator)

I. Preparation of the Samples for Titration

If the sample is a solid, transfer 0.5 g–1 g to a weighing bottle and dry at 110°C until the weight is constant to 0.2 mg–0.3 mg. Weigh out into 250-ml beakers or 200-ml flasks [Note 1 (all references are to the notes in Procedure 17-2A unless otherwise specified)] two or three samples (consult your instructor), which will require from 10 g to 20 g of standard 0.12 *WF* silver nitrate solution (Note 2). Dissolve each sample in 80 ml of water.

If the sample is a solution, pipet (read page 326 in regard to the proper pipetting technique) into 250-ml beakers or 200-ml flasks (Note 1) that volume of solution that will require from 10 g to 20 g of standard 0.12 *WF* silver nitrate solution (Note 2). Dilute each solution to 80 ml (Note 1, below).

Add to the neutral solution (Note 3) 0.25 ml (approximately 8 drops) of 0.5 *F* K_2CrO_4.

Fill a weight buret (pages 333–334) with the gravimetric standard silver nitrate solution (Procedure 17-1B) and weigh it.

II. Preparation of Comparison Solutions

Grind 0.1 g of reagent-grade, precipitated, *chloride-free* calcium carbonate (see the last paragraph of Note 3) with 2 ml–3 ml of water in a clean mortar. To each of two vessels similar to that to be used in the titration add 0.25 ml (8 drops) of 0.5 *F* K_2CrO_4 indicator and 100 ml of water (approximately the same volume of water as will be present in the titrated mixture at the end point). To each solution add 3 drops–5 drops of the carbonate suspension (Note 5). To the second solution add the silver nitrate solution in half drops until the first perceptible change from a greenish yellow toward a brownish yellow is obtained (Note 6). Note and record the weight of silver nitrate required for this end-point correction. Keep these solutions for comparison during the subsequent titration of the sample.

NOTES

1. If the sample is a solution its volumetric concentration is determined by a titration with a silver solution of known gravimetric concentration. Therefore a measured *volume* of the chloride sample is taken and the *weight* of the silver solution that reacts with it is measured.

If the chloride unknown had been a gravimetric solution, then a known weight of it would have been taken for the titration.

III. Preliminary Titrations

(If the preliminary titrations of Procedure 17-2A have been performed, the preliminary titration below can be omitted. Consult your instructor.)

Pipet 5.0 ml of a 0.10 *F* NaCl solution to a 250-ml beaker or 200-ml flask. Dilute the solution to 80 ml. Add 0.25 ml (8 drops) of 0.5 *F* K_2CrO_4.

Note and record the weight of the buret and solution. Gently swirl or stir the chloride solution and slowly add the silver nitrate to it until the transient reddish-brown color produced locally by the silver solution begins to disappear somewhat slowly. (When a beaker is used, keep the buret tip close enough to the surface of the solution to prevent loss from splashing, and use a stirring rod constantly during the titration in order to mix the solution; when a flask is used, mix the solution by swirling.) Thereafter add the silver nitrate solution dropwise until the first perceptible color change is produced that remains after the mixture is shaken vigorously for 30 seconds and that corresponds to that of the comparison solution used for the end-point correction (Notes 7 and 8). Weigh the buret and record the weight of the $AgNO_3$ solution used.

Add about 1 ml of the 0.10 *F* NaCl and repeat the titration (Note 9).

IV. Titration of the Samples

Refill the buret with the standard silver nitrate solution.

Titrate one of the chloride samples as directed in the second paragraph of III above. Weigh the buret and *immediately record this data in ink* in a permanent notebook (see page 162).

Similarly titrate the second sample of the chloride. From the weight formality of the silver nitrate, the net weights of silver nitrate used (that used for the titration minus that for the end-point correction), and the weights of the samples, calculate and report the percentage or weight formality of chloride found in each sample and the average of these values (Notes 10 and 11).

CALCULATION OF THE END-POINT ERROR FROM EXPERIMENTAL DATA

Experimental Determination of the End-Point Error

After completing your titrations do the following. (Consult your instructor.)

(*a*) Calculate the initial formal and normal (for the indicator reaction) concentrations of the K_2CrO_4 in the "comparison solutions" prepared above. (Assume the volume to be 100 ml).

(*b*) Calculate the initial molar concentrations of the K^+ and CrO_4^{2-} in the "comparison solutions." (Assume complete ionization of the K_2CrO_4; neglect hydrolysis of the chromate and the dichromate equilibrium.)

(*c*) Calculate the molar concentration of silver ion required to saturate the "comparison solution" with Ag_2CrO_4.

(*d*) Calculate the volume of 0.10 *N* $AgNO_3$ that the above concentration of Ag^+ represents.

(*e*) From the volume of the standard $AgNO_3$ required to give a perceptible precipitate in the second comparison solution, calculate the total milliequivalents of the $AgNO_3$ added.

(*f*) From the data obtained calculate the milliequivalents of Ag_2CrO_4 precipitated. (Assume equilibrium between solid and liquid phases.)

(*g*) Calculate the chloride-ion concentration that would be in equilibrium with the silver-ion concentration of (*c*) above. (Assume the solution is just saturated with AgCl.)

(*h*) Calculate the end-point error expressed in milliequivalents and in milligrams of chloride in a titration made under the conditions of the second "comparison solution."

(*i*) Calculate the percentage error if 100 mg of chloride were titrated without application of the end-point correction.

EXPERIMENT 17-2: Reactions of Silver Compounds

The reagents needed for this experiment are listed below and methods for their preparation are given in Appendix 7. They are:

0.10 *F* $AgNO_3$	1 *F* KI
0.1 *F* $NaHCO_3$	1 *F* Na_2S
1 *F* NaCl	6 *F* NaOH
0.2 *F* KCN (which is also 0.1 *F* in KOH)	

These experiments—on a larger scale—can be used for class demonstrations.

1. (*a*) Add 2 ml of 0.1 *F* $AgNO_3$ and 3 ml of H_2O to a 50-ml conical flask. Add 0.1 *F* $NaHCO_3$, dropwise, shaking after each drop, to the $AgNO_3$ solution until the first permanent precipitate is obtained. (Note initial appearance of precipitate and any change.) Finally, warm the mixture to 40°C–60°C. (Note any change.)

NOTE: A solution of $NaHCO_3$ is slightly alkaline because hydrolysis predominates over ionization, that is, $HCO_3^- + HOH = H_2CO_3 + OH^-$. Heating such a solution causes an increase in alkalinity because of loss of CO_2, $H_2CO_3 = CO_2 + H_2O$. Silver hydroxide tends to lose water to give the oxide; the latter will be in equilibrium with silver hydroxide in the solution.

Illustrative Problem

Calculate the equilibrium constant for the reaction

$$Ag_2CO_3(s) + 2OH^- = 2AgOH(s) + CO_3^{2-}$$

Predict the result of adding an excess of Ag_2CO_3 to a solution 0.1 F in NaOH. Justify your prediction by calculation.

(*b*) To the mixture from (*a*) add 1 ml 1 F NaCl. Note the result. Write an equation for the reaction.

Illustrative Problem

A solution having a volume of 10 ml and a pH of 10 was in equilibrium with 0.1 mmole of Ag_2O. One mmole of NaCl was added. Predict the composition of the precipitate after equilibrium was established. What would be the effect of adding NH_4Cl instead of NaCl?

(*c*) To the mixture from (*b*) add 1 ml 6 F NH_3. Note the result. Shake the mixture vigorously until no further change is observed and the precipitate settles. Decant the supernatant solution into a second flask. Add 2 ml 6 F NH_3 to the precipitate. Shake the mixture. Comment on the statement found in the literature that "silver chloride is soluble in ammonium hydroxide." Combine the two solutions.

Illustrative Problem

Calculate how many millimoles of silver chloride will dissolve in 10 ml of a solution 0.6 F in NH_3 and 0.1 F in NaCl. Is your calculated result consistent with your experiment?

(*d*) Add 1 ml of 1 F KI to the solution. Note the color of the precipitate.

Illustrative Problem

Calculate how many milligrams of AgI would dissolve in 10 ml 2 F NH_3.

(*e*) Swirl the mixture from (*d*) and add 0.2 F KCN dropwise until the precipitate just disappears.

Illustrative Problems

(*i*) Calculate the ratio of $[Ag(CN)_2^-]$ to $[Ag(NH_3)_2^+]$ in a solution 1 M in NH_3 and 10^{-2} M in CN^-. (*ii*) To 10 ml of a solution 1.0 F in NH_3 and 0.1 F in $AgNO_3$ was added 2 mmoles of KI. The mixture was then titrated with 2.0 F KCN until a barely perceptible precipitate remained. Calculate the molar concentrations of the Ag^+, the $Ag(NH_3)_2^+$, the $Ag(CN)_2^-$, and the CN^-. (Neglect volume change and the AgI precipitate at the end of the titration.)

Comment on the possible accuracy of a titration of silver under similar conditions.

(*f*) To the solution from (*e*) add 1 ml of 0.20 F KCN; now add 1 ml 1 F Na_2S.

Illustrative Problem

Calculate the $[Ag^+]$ in equilibrium with a solution 0.1 *M* in S^{2-}. Calculate the $[S^{2-}]$ in equilibrium with a solution 1 *M* in Ag^+. Comment on this latter result (remember Avogadro's number).

2. Add 0.25 ml 0.1 *F* $AgNO_3$ and 1 ml H_2O to a test tube. Add 1 drop 6 *F* HCl. Note the size of the precipitate. Shake the mixture and pour it, in small portions, into 5 ml 12 *F* HCl in a second test tube. Swirl the contents of the second tube after each addition. Note and explain the behavior of the precipitate. Warm the mixture to 40°C–60°C if particles of precipitate remain.

   Finally pour the solution into 50 ml of water in a 100-ml conical flask. Note the result and explain it.

3. *Titration of cyanide with silver* (*Liebig method*).
   (*a*) Pipet (using a bulb) 5 ml of a standard 0.20 *F* KCN solution (which is also 0.1 *F* in KOH) into 25 ml of water in a 100-ml flask. Add 0.5 ml 6 *F* NaOH. Add dropwise from a measuring pipet or calibrated dropper standard 0.10 *F* $AgNO_3$ until the first permanent turbidity is obtained.

   The volumetric ratio of the two solutions, but not necessarily their concentrations, should be known so that undue time will not be consumed in the first approach to the end point. Use a background behind the flask to facilitate perception of the precipitate.

   Back-titrate with KCN to the disappearance of the precipitate; use fractions of a drop; then titrate with the $AgNO_3$ to the reappearance of the precipitate. Note the reversibility and sensitivity of the end point. Finally cause the precipitate to *just* disappear.
   (*b*) Add to the *clear* solution from a dropper 0.1 ml (3 drops) of a 1 *F* KI solution. Swirl the solution until the precipitate forms.
   (*c*) Again titrate with the KCN, using split drops, until the solution is clear. Note the volume used. Again back-titrate as in (*a*). Finally leave a *barely perceptible* precipitate.
   (*d*) Add to the turbid solution from (c) 1 ml of 6 *F* NH_3. Swirl until the precipitate dissolves. Again titrate with the $AgNO_3$, using split drops, until a dispersed precipitate is again perceptible.

Illustrative Problem

In a determination of the KCN in a solution by the modified Liebig method, 2.0 ml of a KI solution (2.0 g of KI in 100 ml of solution) were added as indicator and the end point was recognized when 0.10 mg of precipitate formed. The total amount of KCN present in the sample taken was 0.20 g. The volume of the solution at the end point was 100 ml. (The KCN solution contained sufficient KOH to make the hydrolysis of cyanide negligible.)
(*i*) Calculate the silver-ion concentration at the end point.
(*ii*) Calculate the cyanide-ion concentration at the end point.
(*iii*) Calculate the error (deviation in g of KCN) in the determination. *Ans.* 6.7×10^{-5} g of KCN, minus error.

(*iv*) Taking the results obtained in (*iii*), calculate what would have to be the formal concentration of ammonia in the solution in order to obtain correct results. *Ans.* 2.8 *F*.

NOTE: See page 104 regarding NH_3 and NH_4OH. Pages 101–103 contain a discussion of other anhydrides and their hydration.

PROCEDURES 17-3, 17-4 (A AND B), AND 17-5

Determination of the Silver in an Alloy

GENERAL DISCUSSION

When silver ion is titrated with thiocyanate the very slightly soluble AgSCN(s) is precipitated,

$$Ag^+ + SCN^- = AgSCN(s) \quad \text{(white)}$$

The indicator reaction involves the formation of a red-colored complex of thiocyanate with ferric ion

$$Fe^{3+} + SCN^- = FeSCN^{2+} \quad \text{(red)}$$

These reactions are the basis of what is known as the Volhard method[8] for the determination of silver. Procedure 17-3 contains instructions for the preparation of a thiocyanate solution and its standardization against silver nitrate; Procedure 17-4 gives the application of this method to the determination of silver in an alloy.

[8] Volhard. *J. Prakt. Chem.*, **117**, 217 (1874). Jacob Volhard (1834–1910) was one of the group, which included Mohr and Liebig, that contributed to the development of analytical chemistry during the latter half of the 19th century. The "Volhard method" was actually discovered by Charpentier in 1870, but was rediscovered and better publicized by Volhard.

PROCEDURE **17-3**

Preparation of a Thiocyanate Solution

Outline. In this procedure a thiocyanate solution of approximately known concentration is prepared by weighing and dissolving potassium thiocyanate crystals, then diluting the solution to the approximate volume desired.

DISCUSSION

Since both ammonium and potassium thiocyanate are somewhat hygroscopic, are too soluble to be readily purified by recrystallization, and cannot be easily dried, standard solutions of these substances are not usually prepared by direct weighing as was done with silver nitrate.[9] Solutions of approximately the desired concentration are therefore prepared and then standardized, usually against pure metallic silver or silver nitrate. If the thiocyanate solution is to be standardized against pure silver, Procedure 17-5 should be used.[10]

Thiocyanate solutions are quite stable; a change of only 0.1% in concentration was observed in an experimental study extending over a period of 164 days.[11]

INSTRUCTIONS (Procedure 17-3: Preparation of a Thiocyanate Solution)

Weigh out 10.0 g of potassium thiocyanate (12.0 g if for gravimetric titrations —consult your instructor) on a suitable balance. Transfer the crystals to a

[9] Kolthoff and Lingane (*J. Am. Chem. Soc.*, **57**, 2126, 1935) have prepared KSCN suitable for use as a primary standard by recrystallization from water, drying over P_2O_5, and, finally, melting for a short time at 200°C. The dried material is not hygroscopic at a relative humidity of less than 45% but deliquesces rapidly at relative humidities greater than 50%.

[10] Very pure silver, known as proof silver, can be purchased from the United States Mint, Philadelphia, Pa.

[11] Campbell and Hook. *Proc. Soc. Chem. Ind. Victoria*, **36**, 1106 (1936).

beaker and dissolve them in water. Dilute the solution to a liter (Note 1), transfer it to a clean bottle with a ground-glass stopper, and thoroughly mix the solution by swirling it repeatedly in the bottle (if possible, avoid wetting the ground-glass surfaces of the stopper). Label the bottle.

NOTE

1. Since the solution is to be standardized subsequently, this dilution does not have to be made precisely. It may be accomplished by adding to the crystals 1 liter of distilled water measured by a graduate.

Standardization of a Thiocyanate Solution

Outline. The thiocyanate solution (from Procedure 17-3) is standardized by titrating volumetrically (Procedure 17-4A) or gravimetrically (Procedure 17-4B) an accurately measured portion of the standard silver nitrate solution (prepared in Procedure 17-1). Weighed samples of silver nitrate or metallic silver can be dissolved and used also.

The titration reaction is

$$SCN^- + Ag^+ = AgSCN(s)$$

Ferric iron is used as indicator and the reaction is

$$SCN^- + Fe^{3+} = FeSCN^{2+} \quad \text{(red)}$$

DISCUSSION

In the Volhard titration, standard thiocyanate is added to silver ion in a nitric acid solution that also contains a soluble ferric salt to serve as the indicator. As long as there is an appreciable amount of silver in the solution being titrated, the concentration of the thiocyanate remains so small, owing to the precipitation of silver thiocyanate,

$$Ag^+ + SCN^- = AgSCN(s)$$

that no perceptible amount of the red compounds that thiocyanate forms with ferric iron[12] is formed. When the amount of thiocyanate added to the solution becomes approximately equivalent to the silver present, further addition of thiocyanate causes its concentration to increase very rapidly (see the discussion and figures on pages 60–67); this concentration increase

[12] There has been some uncertainty as to the formula of the colored compound. Compounds such as $K_3Fe(SCN)_6$ can be prepared; and Schlesinger and Van Falkenburgh (op. cit., and Schlesinger, op. cit.) attributed the color to anions of the type $Fe(SCN)_6^{3-}$ or $Fe(SCN)_4^-$. However Møller (op. cit.), Bent and French (op. cit.), Edmonds and Birnbaum (op. cit.), and Frank and Oswalt (op. cit.) have obtained substantiating evidence that the color is caused by the cation $FeSCN^{2+}$ when an *excess* of iron is present.

results in the reaction

$$Fe^{3+} + SCN^- = FeSCN^{2+} \quad \text{(red)}$$

producing perceptible amounts of the red-colored compound. The appearance of this color is taken as the end point.

Studies of this method have shown that there is a tendency for the silver thiocyanate precipitate that is first formed to adsorb silver ions on its surface; therefore, unless the mixture is vigorously stirred, the first end point obtained may be premature. In addition, the solution must be approximately 0.3 F in nitric acid in order to repress the partial hydrolysis of the ferric ion into products such as $Fe(OH)^{2+}$; these products impart a brownish-yellow color to the solution, making the end point less easily detected. Chloride, like thiocyanate and hydroxide, forms complex ions with ferric ion; they are yellowish.[13] The solution should be cold, since ferric ion is more hydrolyzed in hot solutions and the ferric thiocyanate ion is more dissociated; furthermore, thiocyanate ion is oxidized by nitric acid and by ferric ion at elevated temperatures.[14] Thiocyanate is also oxidized by nitrous acid (which may be present after the dissolving of an alloy in nitric acid), a transitory red color being formed during the oxidation that may be mistaken for the end point.

An experimental study of the ferric thiocyanate end point in titrations of Hg^{2+} with thiocyanate has been discussed on pages 107–110. There it is stated that in order to produce a perceptible red color in 100 ml of a solution 0.013 F in $Fe(NO_3)_3$ the total concentration of thiocyanate—that is $SCN^- + Fe(SCN)^{2+}$—required was 1.0×10^{-5} F; under these conditions the SCN^- concentration was 3.6×10^{-6} M. Since the solubility product for silver thiocyanate is approximately 10^{-12}, the silver-ion concentration in such a saturated solution would be 2.8×10^{-7} M.

The titration error, expressed in equivalents (eq.), is given by the equation

$$\text{T.E.} = \Sigma \text{ eq SCN} - \Sigma \text{ eq Ag}$$

where

$$\Sigma \text{ eq SCN}$$

represents the total equivalents of thiocyanate added at the end point and

$$\Sigma \text{ eq Ag}$$

[13] The following references describe measurements of the dissociation constants of ions such as $FeOH^{2+}$ and $FeCl^{2+}$: Bray and Hershey. *J. Am. Chem. Soc.* **56**, 1889 (1934); Lamb and Jacques. *J. Am. Chem. Soc.*, **60**, 977, 1215 (1938); Rabinowitch and Stockmayer. *J. Am. Chem. Soc.*, **64**, 335 (1942); and Frank and Oswalt, op. cit.

[14] For an experimental study of this method consult the article by Kolthoff and Lingane (*J. Am. Chem. Soc.*, **57**, 2126, 1935).

the total equivalents of silver present. At the end point,

$$\Sigma \text{ eq SCN} = \text{eq AgSCN} + \text{eq SCN}^- + \text{eq FeSCN}^{2+}$$

and

$$\Sigma \text{ eq Ag} = \text{eq AgSCN} + \text{eq Ag}^+$$

therefore,

$$\text{T.E.} = \text{eq SCN}^- + \text{eq FeSCN}^{2+} - \text{eq Ag}^+$$

Assuming that the titration is made in 100 ml of solution, we can obtain the equivalents of these substances by dividing the molar concentrations obtained above by 10; then, when these values are substituted, the equation becomes

$$\text{T.E.} = (3.6 \times 10^{-7}) + (6.4 \times 10^{-7}) - (2.8 \times 10^{-8}) = 9.7 \times 10^{-7} \text{ eq}$$

If we were titrating 25 ml of 0.1 N $AgNO_3$, there would be present 2.5 milliequivalents of silver, or 2.5×10^{-3} equivalents. Therefore the percentage error would be only

$$\frac{9.7 \times 10^{-7} \times 100}{2.5 \times 10^{-3}} = 0.038\,\% \text{ (positive error)}$$

As is indicated by these experiments and calculations, this titration is capable of giving highly accurate results.

In the discussion of the titration of a chloride with silver nitrate with chromate as indicator, it was shown that the point at which the precipitate appeared, and at which the end point was taken, depends upon the relative solubilities of silver chloride and silver chromate, and upon the concentration of the chromate indicator. Analogous considerations with respect to this thiocyanate titration of silver show that the point at which the color becomes perceptible and the end point is taken depends upon the relative solubility of silver thiocyanate and the *degree of dissociation* of the colored compound. If silver thiocyanate were much more soluble or the ferric thiocyanate compound were much less ionized, the end point would be taken before the equivalence point. If the ferric thiocyanate compound were highly dissociated, it would require a larger concentration of thiocyanate to force the equilibrium

$$Fe^{3+} + SCN^- = FeSCN^{2+}$$

to the right and thus produce a perceptible color; the end point would then occur after the equivalence point. It is also apparent that the end point can be shifted by varying the concentration of the ferric salt added as indicator. We have just seen that the titration of silver with thiocyanate under the conditions specified above is well within the accuracy of the experimental apparatus commonly used; therefore, we will consider below a determination that makes use of thiocyanate but that, under the above conditions, would have a very large error.

The Determination of Chloride by the Volhard Titration. Elimination of the Titration Error

The Volhard titration can be used for the determination of chloride. This is done by adding a known quantity of silver ion (in excess of the chloride) to the solution of the unknown chloride. The mixture then contains AgCl and the excess Ag^+. Ferric nitrate is added to the mixture and the excess silver ion is titrated with standard thiocyanate solution to the appearance of the $FeSCN^{2+}$ color.

If the determination is made under the conditions discussed above, relatively large errors can result. Therefore it is of interest to see if the titration-error expression can be used to calculate the conditions that would result in coincidence of the end point and the equivalence point. The reactions are

$$Cl^- + Ag^+ = AgCl(s)$$

$$Ag^+ \text{ (excess)} + SCN^- = AgSCN(s)$$

and

$$Fe^{3+} + SCN^- = FeSCN^{2+}$$

We calculate the equivalents of chloride from the difference of the equivalents of silver and of thiocyanate. However, the method can lead to large errors because the solubility of silver chloride is greater than that of silver thiocyanate. When thiocyanate is added to a solution that is in equilibrium with solid silver chloride, the following reaction tends to occur:

$$AgCl(s) + SCN^- = AgSCN(s) + Cl^- \tag{1}$$

The equilibrium constant for this reaction is

$$\frac{[Cl^-]}{[SCN^-]} = K$$

and is seen to be just the ratio of the solubility products of AgCl and AgSCN; the numerical value is 180.

Because of the above reaction of SCN^- with AgCl the end point of the titration tends to fade rapidly, an excess of titrant is used, and *low* results are obtained for the chloride. Even though an excess of titrant is used, a low value for the chloride is obtained, since this is a back-titration. Various approaches have been used to decrease the error caused by the metathesis of AgCl to AgSCN. The AgCl can be removed by filtration before the back-titration, the mixture can be boiled for several minutes in order to make the AgCl more resistant to dissolving, or an organic substance such as nitrobenzene can be added to coat the precipitate and prevent its coming to equilibrium with the solution. A procedure that avoids the delays and difficulties of these techniques and that is based upon equilibrium considerations[15] makes use of the fact that if the ferric-ion concentration is increased, the thiocyanate-ion concentration at the end point is lowered, thus decreasing the tendency of Equation 1 to proceed to the right. The titration-error expression for this chloride determination is

$$\text{T.E.} = \Sigma \text{ eq Ag} - \Sigma \text{ eq Cl} - \Sigma \text{ eq SCN}$$

If there is no error, T.E. = 0, and

$$0 = \Sigma \text{ eq Ag} - \Sigma \text{ eq Cl} - \Sigma \text{ eq SCN} \quad (2)$$

At the end point of the titration

$$\Sigma \text{ eq Ag} = \text{eq Ag}^+ + \text{eq AgCl} + \text{eq AgSCN} \quad (3)$$

$$\Sigma \text{ eq Cl} = \text{eq Cl}^- + \text{eq AgCl} \quad (4)$$

and

$$\Sigma \text{ eq SCN} = \text{eq SCN}^- + \text{eq AgSCN} + \text{eq FeSCN}^{2+} \quad (5)$$

Substitution of Equations 3, 4, and 5 into 2 gives

$$0 = \text{eq Ag}^+ - \text{eq Cl}^- - \text{eq SCN}^- - \text{eq FeSCN}^{2+} \quad (6)$$

We observed in a preceding section (page 108) that in a 100 ml solution, 6.4×10^{-7} eq of $FeSCN^{2+}$ gave the first detectable color. Equation 6 becomes

$$0 = \text{eq Ag}^+ - \text{eq Cl}^- - \text{eq SCN}^- - 6.4 \times 10^{-7} \quad (7)$$

and this can be simplified appreciably if one observes that multiplication of the entire equation by 10 gives

$$0 = [\text{Ag}^+] - [\text{Cl}^-] - [\text{SCN}^-] - 6.4 \times 10^{-6} \quad (8)$$

[15] Swift, Arcand, Lutwack, and Meier. *Anal. Chem.*, **22**, 306 (1950).

The solution volume was 100 ml, so the factor of 10 changes *eq* SCN^- (in 100 ml) to the *number of equivalents of* SCN^- *in 1 liter*. Since SCN^- has 1 equivalent per mole, this is also the number of moles of SCN^- per liter, which is, of course, the molar concentration. Identical considerations apply to the Ag^+ and Cl^- terms.

Equation 8 relates the concentrations of Ag^+, Cl^-, and SCN^- to each other; these concentrations are also related independently through the solubility products:

$$[Ag^+][SCN^-] = 1 \times 10^{-12} \tag{9}$$

and

$$[Ag^+][Cl^-] = 1.8 \times 10^{-10} \tag{10}$$

Substitution of Equations 9 and 10 into 8 gives

$$0 = [Ag^+] - \frac{1.8 \times 10^{-10}}{[Ag^+]} - \frac{1 \times 10^{-12}}{[Ag^+]} - 6.4 \times 10^{-6}$$

Hence

$$[Ag^+]^2 - 6.4 \times 10^{-6}[Ag^+] - 1.8 \times 10^{-10} = 0$$

From this

$$[Ag^+] = 1.7 \times 10^{-5}$$

$$[SCN^-] = 6 \times 10^{-8}$$

and

$$[Cl^-] = 1.1 \times 10^{-5}$$

From the dissociation constant for $FeSCN^{2+}$ and the known concentrations of $FeSCN^{2+}$ and SCN^-,

$$[Fe^{3+}] = 0.8\ M$$

This, then, is the ferric-ion concentration that would meet the condition demanded by Equation 7, that just 6.4×10^{-7} eq of $FeSCN^{2+}$ (the first perceptible quantity) forms when the equivalence point is reached.

It is found experimentally that in 1 *F* nitric acid, dilute ferric nitrate solutions have no perceptible color, but if the ferric nitrate concentration is 0.15 *F* a grayish-purple color can be detected. This color becomes distinct with 0.3 *F* ferric nitrate. Therefore, the 0.8 *M* Fe^{3+} calculated above for zero titration error would obscure the ferric thiocyanate color at the end point. Calculation shows that if -0.1% error can be tolerated in a titration of about 2.5 mmoles of chloride, the ferric-ion concentration can be lowered to 0.2 *M*. At this concentration the ferric solution is not sufficiently colored to cause serious difficulty in detection of the end point.

It is seen that by a combination of theoretical and experimental considerations a procedure can be developed that permits accurate determination of chloride under equilibrium conditions.

Instructions are given below for standardizing the thiocyanate solution from Procedure 17-3 against the standard silver nitrate solution from Procedure 17-1. The titrations can be made volumetrically (Procedure 17-4A) or gravimetrically (Procedure 17-4B). The standardization can also be made against weighed samples of silver nitrate or pure silver.

INSTRUCTIONS (Procedure 17-4A: Volumetric Standardization of a Thiocyanate Solution)

I. Preliminary Experiments and Questions

(Optional. Consult your instructor.)

Experiments. Effect of Acid and of Chloride on the Color of Fe(III) *Solutions and on the End Point.* Add 5 ml of water to each of three 15-ml test tubes. Designate the test tubes as *A*, *B*, and *C*.

(*a*) To *A* add 1 ml of water and 1 ml of 0.3 *F* $Fe(NO_3)_3$.
(*b*) To *B* add 1 ml of 6 *F* HNO_3 and 1 ml of 0.3 *F* $Fe(NO_3)_3$.
(*c*) To *C* add 1 ml of 6 *F* HCl and 1 ml of 0.3 *F* $Fe(NO_3)_3$.

Questions

1. What chemical species causes the color in tube *A*?
2. What causes the lack of color in tube *B*?
3. What causes the color in tube *C*?
4. Would you expect the *p*H of a $Fe(NO_3)_3$ solution to be greater, approximately equal to, or less than 7? Explain.
5. How does the fraction of the iron existing as Fe^{3+} vary as a $Fe(NO_3)_3$ solution is diluted? Explain.

II. Volumetric Standardization of a Thiocyanate Solution

Use a pipet to transfer three 25-ml portions of the silver nitrate solution into separate 200-ml flasks (Notes 1 and 2). Add to each portion 10 ml of 6 *F* HNO_3 (Note 3), 5 ml of chloride-free 0.3 *F* $Fe(NO_3)_3$ solution (Note 4) and 25 ml of water.

NOTES

1. Read the suggestions made in pages 328–333 regarding the use of burets and on pages 194 and 326 for the removal of standard solutions from storage bottles.

Fill a buret with the thiocyanate solution. Withdraw the solution until the meniscus approaches the zero mark; allow 30 seconds for drainage, again adjust the meniscus to the mark, and remove any hanging drop from the tip. Gently swirl the silver nitrate solution and add the thiocyanate to it until the local and transient pink color first produced by the thiocyanate shows a tendency to spread throughout the solution. Thereafter add the thiocyanate in drops, and then in fractions of a drop, until a color is obtained that remains permanent when the mixture is shaken vigorously for 30 seconds (Note 5). After allowing time for drainage, read the buret, and *immediately record this reading in ink in a permanent notebook.* Similarly titrate the other portions of silver nitrate. From the normality of the silver nitrate solution and the volumes of silver nitrate and thiocyanate used, calculate the normality of the thiocyanate solution (Note 6).

NOTES

2. If solid silver nitrate is to be used, weigh accurately 0.40-g–0.45-g samples, transfer them to 200-ml flasks, add 25 ml of water, dissolve the crystals and proceed as directed for the pipetted portions. If pure silver is to be used, weigh 0.22-g–0.28-g samples and follow Procedure 17-5.

3. The HNO_3 should be free from nitrous acid and from chloride (see the Discussion). If it is at all yellowish (indicating nitrous oxides), boil it vigorously for several minutes, replacing the water lost by evaporation, and then cool it to room temperature.

4. If chloride-free $Fe(NO_3)_3$ is not available, 1 ml–2 ml of a saturated solution of ferric ammonium sulfate, $Fe(NH_4)(SO_4)_2 \cdot 12H_2O$, can be used.

5. When unknown amounts of silver are titrated, the approach to the end point is shown by the slower disappearance of the local red color caused by each drop of thiocyanate. As mentioned in the Discussion, the first apparently permanent color may fade; this fading is partly due to adsorption of silver on the precipitate, and the mixture should be vigorously shaken so that this may be removed. At this stage of the titration the thiocyanate should be added by allowing only a fraction of a drop to form on the tip of the buret, removing this, and adding it to the solution by means of a stirring rod. Also, near the end of a titration, the inside walls of the flask or other titration vessel should always be washed down by means of a jet of water from the wash bottle; any of the standard solution, or of the solution being titrated, adhering thereto is thus recovered. When very precise measurements are desired, an end-point correction should be made by adding the thiocyanate solution to a solution of the same volume as that at the end point of the titration, containing the same amount of acid and ferric nitrate, until a color matching that used for the end point is obtained. This volume is subtracted from the volume used in the titration. The end point of this titration is so sharp, and the agreement with the equivalence point so close, that the end-point correction is usually omitted.

6. Properly carried out, these titrations should give results which agree to within 1 part–2 parts per 1000.

INSTRUCTIONS (Procedure 17-4B: Gravimetric Standardization of a Thiocyanate Solution)

Follow procedure 17-4A, the volumetric standardization above, with the following changes:

1. If the standard volumetric $AgNO_3$ solution (from Procedure 17-1, Option A) is to be used, pipet 25-ml portions.

   If the standard gravimetric $AgNO_3$ solution (from Procedure 17-1, Option B) is to be used, weigh 10 g–15 g portions (to 10 mg–15 mg).
2. Use a weight buret for the titration.
3. Calculate the weight normality of the thiocyanate solution.

PROCEDURE **17-5**

Determination of the Silver in an Alloy

Outline. The alloy is cleaned, dissolved in nitric acid, the oxides of nitrogen boiled from the solution, the solution cooled and titrated with standard thiocyanate solution as in Procedure 17-4.

DISCUSSION

When nitric acid is used to dissolve metals or alloys, the reduction products are capable of oxidizing thiocyanate and must be removed from the solution. The course of the reaction and the products resulting from the reduction of nitric acid are determined largely by the concentration of the solution and the potential of the reducing agent. The two principal reduction reactions with their potentials are as follows:

$$H_2O + NO_2 = NO_3^- + 2H^+ + e^- \qquad (-0.79 \text{ v})$$

and

$$2H_2O + NO = NO_3^- + 4H^+ + 3e^- \qquad (-0.94 \text{ v})$$

In dilute solutions, except with very powerful reducing agents, nitric oxide is the product; in concentrated solutions, nitrogen dioxide is the invariable product. The latter is true because, even though the direct product of the reduction reaction is nitric oxide, it can be oxidized by concentrated nitric acid, the equilibrium constant for the reaction

$$NO + 2NO_3^- + 2H^+ = 3NO_2 + H_2O$$

being 5×10^{-9}. Other equilibria are involved as follows:

$$2NO_2 + H_2O = HNO_2 + NO_3^- + H^+$$

$$3HNO_2 = NO_3^- + 2NO + H^+ + H_2O$$

Fortunately the oxides are volatile and thus all of the reduction products can be removed by boiling such a solution.

Interfering Constituents

Cobaltous and nickelous ions impart interfering colors to the solution; cupric ions are permissible except in high concentrations, when not only is the color objectionable but cupric and cuprous thiocyanates may also be formed. With these exceptions, the method may be applied in the presence of most of the common cations except mercury; the behavior of this element is discussed below.

The Titration of Mercuric Ion

The diatomic mercurous ion, $Hg_2{}^{2+}$, forms an insoluble precipitate with thiocyanate and with all of the halides except fluoride. Mercuric ion, Hg^{2+}, forms moderately soluble but extremely *un-ionized* salts with thiocyanate, cyanide, and all the halides except fluoride. It is, therefore, possible to titrate a highly ionized mercuric salt (such as the nitrate, sulfate, or perchlorate) with a standard thiocyanate solution and to use ferric ion as the indicator under conditions identical with those of the silver titration.[16] It should be understood that the titration of mercuric ion is an example of a titrimetric ionization process. Although a precipitate of $Hg(SCN)_2$ usually forms near the end of the titration, the main reaction is quantitative not because of the formation of this precipitate but because of the formation of the un-ionized $Hg(SCN)_2$. The equilibria and conditions obtaining in the region of the end point of this titration have been discussed in Chapter 7, pages 107–110.

The procedure given in Note 5 below can be used for the analysis of certain mercury amalgams, oxides, and most other mercury compounds or mixtures not containing the halogens.

INSTRUCTIONS (Procedure 17-5: Determination of Silver in an Alloy)

Clean the alloy (Note 1). Weigh out into 200-ml flasks duplicate samples that should each require from 25 ml to 35 ml of the 0.1 *F* thiocyanate solution (Note 2). Add 15 ml of chloride-free 6 *F* HNO_3, cover the flask with a small

NOTES

1. Flat pieces of metal may be scrubbed with cleaning powder until free of tarnish and then dried; cuttings or drillings should be washed with ether to remove grease or oil.

2. If the approximate silver content of the alloy is not known, conduct a preliminary trial, using a sample of about 0.5 g.

[16] J. Volhard. *Liebigs Ann. Chem.*, **190**, 57 (1877). Volhard appears to have been the first to apply the Volhard titration to mercuric ion—see Footnote 8, page 371.

watch glass, and heat the mixture almost to boiling (Note 3) until the alloy is completely dissolved (Note 4). Finally, boil the mixture until any nitrous compounds are completely expelled (see the discussion, Procedure 17-4), and cool the solution to room temperature (Note 5). Add 25 ml of water and 3 ml of chloride-free 0.3 *F* $Fe(NO_3)_3$ solution (Note 4, Procedure 17-4), and proceed as directed in the last paragraph of Procedure 17-4, Optional Method A (volumetric titration), or Optional Method B (gravimetric titration).

NOTES

3. When directed to heat a solution "almost to boiling," a small flame may be used directly, or the container may be put on a water bath. In the above procedure, should the reaction become so vigorous as to cause possible spattering, discontinue the heating and, if necessary, cool the mixture by holding the flask under running tap water.

4. Most silver alloys will dissolve completely in the HNO_3. Gold will remain as a brownish powder; tin will be partially precipitated as the hydrous stannic oxide (the so-called metastannic acid). The latter can be quantitatively precipitated by evaporating the mixture *almost* to dryness on a water bath.

5. If this procedure is used for the determination of the mercury in amalgams or in mercurous compounds, complete oxidation to the mercuric state may not be obtained by the above treatment with HNO_3. Therefore proceed as follows:

Add 0.1 *F* $KMnO_4$ dropwise until the first pink color is observed that persists after the solution is swirled for a few seconds (an excess of permanganate should be avoided; otherwise, a precipitate of MnO_2 may form). Then add 0.1 *F* $FeSO_4$ dropwise until the solution is colorless; avoid an excess.

The permanganate oxidizes the mercurous ion and is reduced to manganous ion; any nitrous acid is also oxidized. The excess of permanganate must be reduced; if it were not reduced, it would oxidize the thiocyanate.

When large amounts of mercury are titrated, a coarse crystalline precipitate of $Hg(SCN)_2$ may separate during the latter part of the titration.

EXPERIMENT 17-5: Titration of Mercuric Ion with Thiocyanate

(It is recommended that this experiment be done if the titration of Hg(II) described in Note 5 of Procedure 17-5 has not been done. Read the instructions given in that Note. The equilibria involved in this titration have been discussed in Chapter 7, pages 107–110.)

Transfer with a pipet 5 ml of 0.05 *F* $Hg(NO_3)_2$ into a 100-ml conical flask. Add 1.0 ml 0.3 *F* $Fe(NO_3)_3$ and then add 1.0 ml of 6 *F* HNO_3.

What causes the color of the solution to disappear?

Add 0.1 *F* KSCN dropwise from a measuring pipet until the first permanent pink color is observed. Swirl the solution during this titration. Keep the solution cool.

What is the precipitate that forms near the end of the titration? What causes the pink color? Is this titration an example of a volumetric precipitation process?

Add 1 drop of 6 *F* HCl to a test tube, add 9 drops of water, and mix the solution. Add 1 drop of this solution to the titrated solution.

Explain the result.

Problem. Calculation of the End-Point Error when Titrating Mercuric Ion with Thiocyanate. A titration of 2.5 meq of $Hg(NO_3)_2$ was made under the conditions of the experimental study of the end-point error of the titration of Ag^+ with thiocyanate (page 375). Calculate the end-point error to be expected.

QUESTIONS AND PROBLEMS

Stoichiometric Calculations

1. An analyst weighed out and dissolved 15.462 $\pm$ 0.002 g of dry silver nitrate in water and then diluted the solution to the mark in a flask, which had been found by calibration to contain 1001.40 ml. Calculate the normality of the solution when used for procedures involving the precipitation of insoluble silver salts. *Ans.* 0.09090 *N*.

2. The silver nitrate solution of Problem 1 was used for the standardization of a sodium chloride solution by the Mohr titration (Procedure 17-2A); 50.00 ml of the sodium chloride solution required 45.00 ml of the silver nitrate.

   (*a*) Write the equation for the indicator reaction.
   (*b*) What factors set the *p*H limits within which the titration is practicable?
   (*c*) Calculate the normality of the sodium chloride solution.

3. The silver nitrate solution of Problem 1 was used for the standardization of a thiocyanate solution by the method of Procedure 17-4A; 25.00 ml of the silver solution required 27.50 ml of the thiocyanate.

   (*a*) Calculate the normality of the thiocyanate.
   (*b*) Explain the result of attempting to carry out this titration in a neutral solution. *Ans.* (*a*) 0.08263 *N*.

4. *The Determination of the Silver in an Alloy.* A section of a clean silver coin weighing 0.3426 g was dissolved in 10 ml–15 ml of 6 *F* HNO_3, and the oxides of nitrogen were expelled by boiling.

   Write equations showing the predominant reactions taking place when metallic silver is dissolved in (*a*) concentrated and (*b*) dilute nitric acid.

   Explain the effect of failure to remove the oxides of nitrogen before making the titration.

   The solution was cooled, 1 ml of chloride-free 1 *F* $Fe(NO_3)_3$ was added, the solution was diluted to 50 ml and then titrated with the thiocyanate solution of Problem 3 as directed in Procedure 17-5, 34.70 ml being required. Calculate the percentage of silver in the coin.

NOTE: This titration of silver can be carried out in the presence of copper unless the ratio of copper to silver is greater than approximately 7.

5. A sample of pure sodium chloride was dissolved and titrated with standard $AgNO_3$ as directed in Procedure 17-2A, chromate being used as indicator (the Mohr method). The titration required 12.62 ml of 0.1016 *N* $AgNO_3$. Calculate the milligrams of sodium in the sample. *Ans.* 29.49 mg.

6. Bismuth can be quantitatively precipitated as BiOCl, bismuthyl chloride, and separated from such elements as copper, small amounts of lead, cadmium, zinc, nickel, manganese, and the alkaline-earth metals by proper dilution or neutralization of a chloride solution. After such a procedure, the BiOCl precipitate was dissolved in HNO_3; 12.00 ml of 0.1000 *N* $AgNO_3$ were added; and 4.00 ml of 0.0500 *N* KSCN were used for the final titration. Calculate the milligrams of bismuth present.

   NOTE: This problem illustrates one of the uncertainties attached to the use of normal concentrations. Why would one expect BiOCl to dissolve in HNO_3?

7. A mixed precipitate of sodium and potassium chlorides weighing 0.2076 g was dissolved and titrated by the Mohr method; 28.50 ml of 0.1055 *N* $AgNO_3$ were used. Calculate the weight percentages of the two salts in the mixture. *Ans.* 29.0% NaCl.

8. A thiocyanate solution was standardized by treating 0.4562 g of metallic mercury by Procedure 17-5 (see Note 5 of that Procedure); 42.56 ml were required for the titration.

   (*a*) Calculate the normality of the thiocyanate solution.
   (*b*) Write the equation for the titration reaction.
   (*c*) Can this titration be classified as a volumetric precipitation process? Explain.
   (*d*) Predict the probable effect of the presence of the following ions in the above titration in amounts approximately equivalent to that of the mercury: (*i*) bromide, (*ii*) iodate, (*iii*) perchlorate, (*iv*) phosphate, (*v*) unipositive silver.

9. The silver in various alloys is to be determined by dissolving the alloys in HNO_3 and titrating with a standard KSCN solution (the Volhard titration). Calculate the sample weight of the alloy in order to have the buret reading in milliliters give directly the percentage of silver in the alloy when an exactly 0.1 *N* thiocyanate solution is used. *Ans.* 1079 mg.

10. The mercury in cinnabar (HgS) ores is to be determined by titration with standard thiocyanate solutions. If a 1-g (± 1 mg) sample of the cinnabar is to be taken, calculate what the normality of the thiocyanate should be for the buret reading in milliliters multiplied by 2 to give directly the percentage of mercury in an ore.

11. A potassium ferrocyanide solution was standardized by dissolving pure ignited zinc oxide in hydrochloric acid and titrating the resulting solution with the ferrocyanide solution until a drop of the titrated solution produced a brownish coloration

when added to a drop of a cupric salt solution on a spot plate (see page 349). A 0.5420-g portion of ZnO required 24.20 ml of the ferrocyanide. Calculate the formal concentration of the potassium ferrocyanide solution. *Ans.* 0.1835 *F*.

NOTE: With reactions involving mixed precipitates (such as the potassium zinc ferrocyanide in this method) or complex ions (such as $Ag(CN)_2^-$ or $Ni(CN)_4^{2-}$) less uncertainty is involved if formal, rather than normal, concentrations are used.

12. The Liebig method for determining cyanide (see page 110) can be utilized for the determination of the silver in a silver chloride precipitate. This procedure involves dissolving the precipitate in ammonia with the aid of an excess of standard KCN, then adding iodide and back-titrating with standard silver nitrate to the first appearance of a silver iodide precipitate.

    In such a determination, a silver chloride precipitate was treated with 24.75 ml of 0.2024 *F* KCN, and 0.24 ml of 0.1020 *F* $AgNO_3$ was used for the back-titration.

    (*a*) Calculate the milligrams of silver present.
    (*b*) Can this determination be classified as a volumetric precipitation process?

13. A 0.2350-g sample of a mixture of crude KCN and KSCN was dissolved in water and titrated with standard silver nitrate. The end point was taken when no more precipitate formed on addition of the silver nitrate (see page 347); 30.00 ml of 0.1000 *F* solution were required. Then 1.00 ml of 1 *F* KI and 5 ml of 6 *N* NH_3 were added, and the mixture was titrated with standard potassium cyanide solution until all the precipitate dissolved, 40.00 ml of 0.1000 *F* solution being used (see page 110).

    (*a*) Write equations for all reactions.
    (*b*) Calculate the percentages of KCN and of KSCN in the sample. *Ans.* 41.35% KSCN.

14. A standard silver nitrate solution was prepared by dissolving 16.989 g of the pure salt in water and diluting in a volumetric flask to a volume of 1000 ml at 20°C.

    This solution was used in the standardization of a thiocyanate solution, and 25.00 ml of the silver nitrate required 20.00 ml of the thiocyanate. The room temperature was 15°C.

    A bismuthyl chloride, BiOCl, precipitate was dissolved in 10 ml of cold 6 *N* HNO_3.

    (*a*) Write an equation for the solution reaction (no oxidation–reduction reactions were involved).

    A 50.00-ml portion of the silver nitrate was pipetted at 20°C into the HNO_3 solution of the BiOCl precipitate, and the resulting precipitate was filtered, washed, and discarded. To the filtrate and washings were added 2.0 ml of 1 *F* $Fe(NO_3)_3$, and the solution was titrated with the thiocyanate solution until the appearance of the first perceptible pink color; 20.00 ml of thiocyanate were required at 20°C.

(*b*) Write the equation for the indicator reaction.
(*c*) What factor sets the minimum hydrogen-ion concentration at which this titration is practicable?
(*d*) Calculate the weight of the bismuthyl chloride precipitate. *Ans.* 651.3 mg.
(*e*) What data would you need to predict the effect of the bismuth on this titration?

15. A solution of $AgNO_3$ is 0.10015 *N* at 20°C. A 33.00-ml sample of this solution at 15°C is titrated with a thiocyanate solution at the same temperature, 30.00 ml being required. At 25°C, a 44.00-ml portion of a mercuric nitrate solution is titrated with this thiocyanate solution (also at 25°C); 40.00 ml were required.

(*a*) What is the normality of the mercury solution at 20°C?
(*b*) What is the normality at 15°C?

16. State the effects taking place (compounds, etc., formed and results on the analysis) upon making or attempting to make the titrations listed below in (*a*) an acid solution ($HNO_3 = 10^{-2}$ *F*), (*b*) a neutral solution (*p*H 7), and (*c*) an alkaline solution ($NH_3 = 10^{-2}$ *F*).

(*1*) The titration of bromide with standard silver nitrate, chromate being used as indicator.
(*2*) The titration of mercury with standard thiocyanate, ferric sulfate being used as indicator.
(*3*) The titration of cyanide with standard silver nitrate, the silver cyanide precipitate being used as indicator.

17. To a 25.00-ml portion of a nickel nitrate solution were added 5 ml of 6 *F* NH_3, 25 ml of water, 1.25 ml of 1.010 *F* KI, and 1.00 ml of 0.1000 *F* $AgNO_3$. The resulting mixture was then titrated until no more precipitate was visible with a solution that was 0.01000 *F* in KOH and 0.1000 *F* in KCN; the volume required for the titration was 12.00 ml. Calculate the grams of nickel in the solution. *Ans.* 0.0147 g.

Write equations for the reactions taking place (*a*) upon addition of (1) the NH_3, (2) the $AgNO_3$, and (*b*) during the titration.

18. A sample of pure $Ag_2Cr_2O_7$ was added to a dilute solution of HNO_3, then 50.0 ml of 0.1000 *N* HCl were added, and the precipitate was removed by filtration. A slight excess of calcium carbonate was added to the filtrate and this solution was then titrated with 0.1000 *N* $AgNO_3$ to the appearance of the first reddish turbidity; 30.0 ml of the $AgNO_3$ were used and the final volume was 200 ml.

(*a*) Write equations for the reactions taking place.
(*b*) Calculate the weight of the $Ag_2Cr_2O_7$ sample.
(*c*) Calculate the concentration of the chloride ion at the end point of the titration.
(*d*) What factors would determine if the end point coincided with the equivalence point?

Calculations Involving Chemical Equilibria

19. A soluble chloride is titrated with a standard silver nitrate solution. Chromate is used as indicator, and just 1 ml of a solution containing 50 g of K_2CrO_4 per liter has been added. Calculate the concentration of chloride ion in the titrated solution if the volume was 100 ml when the solution became just saturated with Ag_2CrO_4. *Ans.* $[Cl^-] = 3.6 \times 10^{-6}$.

    What factors set the maximum and minimum chromate-ion concentrations that can be used under experimental conditions for this titration?

20. In making an experiment to determine the end-point error of a Mohr titration, 2.0 ml of 1.0 *N* K_2CrO_4 were diluted to 100 ml and then 0.010 *N* $AgNO_3$ was added until the first perceptible precipitate was observed, 0.50 ml being required. Calculate the error (grams of chloride, plus or minus) in a determination made under the above conditions. (Assume that equilibrium is attained between the solution and precipitate.) *Ans.* 1.3×10^{-4} g, plus error.

    To how many ml of 0.10 *N* $AgNO_3$ does this error correspond?

21. In order to determine the end-point error of a Volhard titration, an analyst diluted 1.0 ml of 1.0 *F* $Fe(NO_3)_3$ and 6 ml of 6 *F* HNO_3 to 100 ml and then added 0.010 *F* KSCN from a 10-ml buret until he could detect a perceptible pink-red color; 0.087 ml of the KSCN was required.

    (*a*) Calculate the concentration of the red compound required for perception of the end point by this analyst.
    (*b*) Calculate the end-point error (grams of silver, plus or minus). *Ans.* 9.2×10^{-5} g (+ error).
    (*c*) To how many ml of 0.1 *N* KSCN does this correspond?
    (*d*) To what percent error would this correspond if 0.25 g of silver were titrated?
    (*e*) What would be the effect of omitting the nitric acid?

22. A chloride determination was made by adding an excess of silver nitrate to the unknown chloride solution, then adding nitric acid and ferric nitrate indicator solution and titrating with a standard solution of potassium thiocyanate until the first perceptible red-pink color appeared. Experiments had shown that the red $Fe(SCN)^{2+}$ ion was detected when its concentration was 5.0×10^{-6} *M*. In the above titration the ferric ion was 3.0×10^{-3} *M* and the volume was 100 ml when the end point was taken.

    (*a*) Calculate the error in the determination (mg of chloride, plus or minus).
    (*b*) What means would you suggest to minimize the calculated error?
    (*c*) Explain what factors would set the *p*H limits between which the above titration would be practical.

CHAPTER 18

Oxidation-Reduction Reactions. Permanganate Methods

In the preceding chapter, we discussed titrimetric procedures that depend upon the formation of slightly soluble substances. In the next four chapters we will consider titrations in which redox reactions are used. The general principles of redox reactions are covered in Chapter 6; this material should be reviewed. This chapter will present procedures in which standard permanganate solutions will be used to titrate reducing substances.

Permanganate was first used as a volumetric reagent by Margueritte[1] in 1846 for the determination of iron in ores. He described the process in the conventions of that period as follows:

"The reaction of iron protoxide and chamaeleon [the name for permanganate] can be expressed by the following equation:

$$Mn^2O^7, KO = Mn^2O^2 + O^5 + KO$$

$$Mn^2O^2 + O^5 + KO + 5Fe^2O^2 = Mn^2O^2 + 5Fe^2O^3 + KO$$

It can be seen that one equivalent of permanganate oxidizes 10 equivalents of iron protoxide."

According to Szabadváry,[2] this was the first paper in which a redox process was expressed by a chemical equation.

---

[1] Frédéric Margueritte. *Ann. Chim. Phys.*, **18**, 244 (1846).

[2] Ferenc Szabadváry. *History of Analytical Chemistry*. New York: Pergamon, 1966, p. 230.

Standard solutions of permanganate are now used extensively in titrimetric methods involving oxidation and reduction reactions. This widespread use is due, first, to the large negative value, -1.45 v, of the standard potential of the half-cell reaction,

$$Mn^{2+} + 4H_2O = MnO_4^- + 8H^+ + 5e^- \quad (E^\circ = -1.45 \text{ v}) \qquad (1)$$

and the resultant pronounced oxidizing tendency of the permanganate ion; second, to the intense purple color of the permanganate ion, which enables it to serve as its own indicator; and third, to the fact that, properly prepared and kept, permanganate solutions are stable over long periods of time.

In order to demonstrate, first, the effectiveness of permanganate as an oxidizing agent and, second, the use of the Nernst equation for calculating the conditions existing at the end point of a titration involving an oxidation-reduction reaction, let us consider the calculation of the potential of a solution 10^{-3} M in Mn^{2+}, 1×10^{-5} M in MnO_4^-, and 1 M in H^+. These concentrations approximate those that frequently exist at the end of a permanganate titration.[3] By substituting these concentrations into the Nernst equation, we can calculate the potential existing in such a solution as follows:

$$E = -1.45 - \frac{0.059}{5} \log \frac{(1 \times 10^{-5})(1)^8}{(10^{-3})} = -1.43 \text{ v}$$

This potential is only slightly less negative than the standard potential for the manganous-permanganate half-cell; consequently, one predicts from a survey of the half-cell potentials shown in Appendix 4 that under the conditions permanganate would be an excellent oxidizing agent. The rate at which permanganate would react with a given reductant cannot be predicted from potential values.

Permanganate solutions are extensively used for the titration of solutions of ferrous salts; and, since at the end point such solutions have approximately the above potential value, we can calculate what the ratio of ferric iron to ferrous iron would be at the end point of a titration made under the above conditions and thereby predict how completely ferrous iron would be oxidized under equilibrium conditions when titrated by a permanganate solution. Therefore, by substituting the above value into the Nernst equation

---

[3] Thus, if 1 meq of a reducing agent had been titrated and the final volume is 200 ml, 0.2 mmole of manganous ions would have been formed, since in the half-cell reaction shown above there are five oxidation equivalents for each mole of permanganate. The concentration of permanganate ion required for perception of its color, and therefore for the end point, is approximately 10^{-5} M.

for the following half-cell

$$Fe^{2+} = Fe^{3+} + e^- \qquad E° = -0.782 \text{ v} \tag{2}$$

$$-1.43 = -0.782 - \frac{0.059}{1} \log \frac{[Fe^{3+}]}{[Fe^{2+}]}$$

we obtain

$$\frac{[Fe^{3+}]}{[Fe^{2+}]} = 1.0 \times 10^{11}$$

This result indicates, as is found experimentally, that the reaction would be complete well within the usual analytical limits. The possibility that the rate at which the reaction proceeds might be too slow for it to be of practical use is not precluded. Also, should there be any ions in the solution capable of forming complex ions with ferric ion, this additional equilibrium would have to be considered.

Another way to predict the completeness of the permanganate-ferrous iron reaction is to calculate the equilibrium constant for the reaction from the half-cell potentials. If we multiply Equation 2 by five and subtract Equation 1 from it we obtain

$$MnO_4^- + 8H^+ + 5Fe^{2+} = Mn^{2+} + 5Fe^{3+} + 4H_2O \quad E°_{cell} = +0.67 \text{ v}$$

The relatively large positive value of the standard cell potential, $E°_{cell} = 0.67$ v, indicates that the reaction should proceed as written. Substituting into Equation 11, Chapter 6,

$$E°_{cell} = \frac{0.059}{n} \log K$$

we obtain

$$+0.67 = \frac{0.059}{5} \log K$$

The value for n is 5 because each half-cell reaction had 5 electrons in it when the two were combined. A whole reaction can only be obtained from two half-cell reactions if both have the same number of electrons; that number is the value to substitute for n in the equation for the equilibrium constant. Solution of the above equation gives $K = 10^{57}$, which suggests that the reaction should proceed very far to the right.

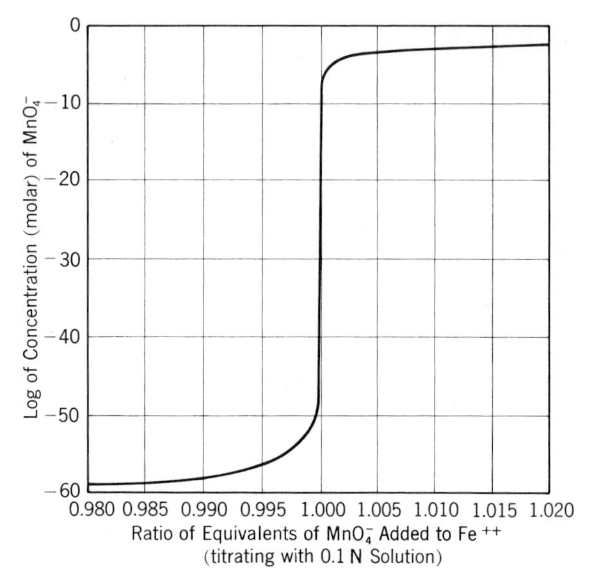

FIGURE 18-1

Changes in permanganate-ion concentration during the titration of a ferrous salt solution.

CHANGE OF CONCENTRATIONS DURING OXIDATION-REDUCTION TITRATIONS

In Chapter 5 a series of curves was given showing the calculated changes in the concentrations of the ions involved during the titrations of a soluble silver salt with various anions that form insoluble silver salts. In these curves the predominant feature was the very rapid change in the concentra-

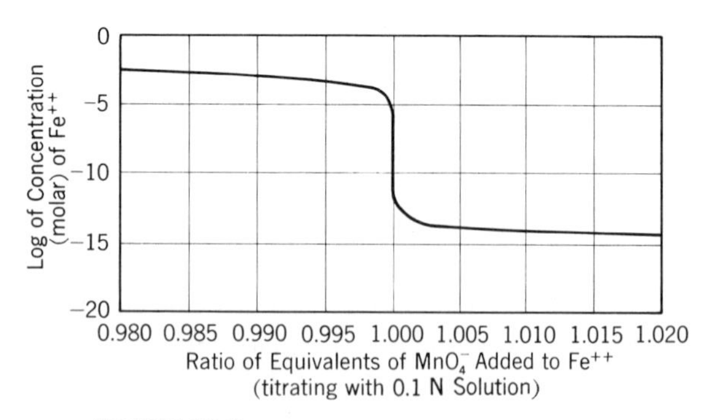

FIGURE 18-2

Changes in ferrous-ion concentration during titration with permanganate.

tions of the silver ion and of the anion near the equivalence point, and it was pointed out that these large concentration changes could be used to cause some perceptible effect that would serve as the end point of the titration. This same principle is illustrated in Figure 18-1, where the changes in the permanganate concentration during the titration of ferrous iron with permanganate have been calculated and are shown as a function of the two substances present. Note that as long as an appreciable amount of ferrous iron is present, the concentration of the permanganate is practically negligible, but that near the equivalence point, there is a very rapid rise in the permanganate concentration, which causes its color to become visible; therefore, the appearance of this color can be taken as the end point of the titration. It would also be possible to determine the end point by testing for the ferrous ion, because, as is seen from Figure 18-2, its concentration very rapidly decreases to a negligible quantity near the equivalence point.[4]

CHANGE OF POTENTIAL DURING AN OXIDATION-REDUCTION TITRATION. POTENTIOMETRIC TITRATIONS

In Figure 18-3 there has been plotted for the same titration of ferrous iron with permanganate, not the concentration of the ions, but the potential existing in the solution as a function of the ratio of the equivalents of the oxidizing and reducing substances present. As would be predicted, since the potential in the solution is a function of these concentrations, the rate of change of the potential is also much greater near the equivalence point. This fact is the basis of the potentiometric method of determining the end point, in which the potential set up by the ions or compounds present in the solution is experimentally measured by suitable means at frequent intervals during the course of the titration, and a curve similar to those of Figure 18-3 obtained, from which the end point can be determined. The principle of the method used in making these potential measurements is shown in Figure 18-4 and the accompanying explanatory material.[5]

---

[4] Such a method was used in the so-called Penny method for titrating ferrous iron with standard dichromate solution (which is not so intensely colored as to be used as its own indicator). It involves the use of an external or "outside" indicator. As the titration proceeds, drops of the titrated solution are transferred to a potassium ferricyanide indicator solution. As long as there is an appreciable concentration of ferrous iron present, a blue color is obtained. For the details of the titration see Treadwell and Hall. (*Analytical Chemistry.* Vol. 2, *Quantitative.* 9th ed., New York: Wiley, 1942, p. 575) or Scott (*Standard Method of Analysis.* Vol. 1. 5th ed., Princeton, N.J.: Van Nostrand, 1939, p. 470).

[5] A more complete treatment of this subject will be found in textbooks of quantitative analysis or in special reference books dealing with this subject, such as *Electroanalytical Chemistry* by J. J. Lingane (2nd ed., New York: Interscience, 1958).

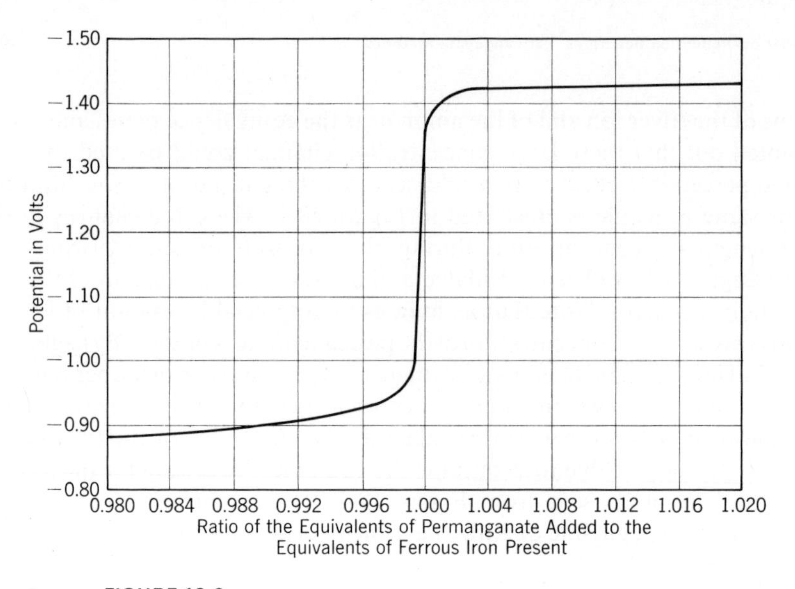

FIGURE 18-3

Changes in potential during the titration of ferrous ion with permanganate.

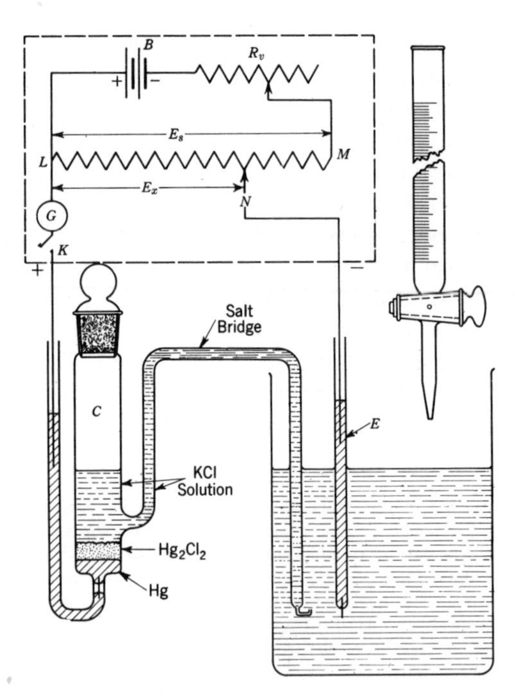

FIGURE 18-4

See facing page for description.

EFFECT OF REACTION RATE ON OXIDATION-REDUCTION TITRATIONS

It has been mentioned above that one cannot predict from calculations of the equilibrium conditions how rapidly a reaction will proceed. In general it is necessary to determine by experiment the effect that rate phenomena may have upon a given reaction or method. This factor of rate is especially important in permanganate titrations. Referring to the table of standard potentials, we see that there are three half-cell reactions that might be involved in such titrations in acid solutions, namely,

$$MnO_2(s) + 2H_2O = MnO_4^- + 4H^+ + 3e^- \qquad (-1.59\text{ v}) \qquad (3)$$

$$Mn^{2+} + 4H_2O = MnO_4^- + 8H^+ + 5e^- \qquad (-1.45\text{ v}) \qquad (4)$$

$$Mn^{2+} + 2H_2O = MnO_2(s) + 4H^+ + 2e^- \qquad (-1.24\text{ v}) \qquad (5)$$

FIGURE 18-4

Principles of potentiometric titrations: B, battery of constant e.m.f.; LM, uniform slide curve resistance; N, sliding contact; R_v, variable resistance; G, galvanometer; K, tapping key switch; C, calomel electrode; E, platinum-wire indicator electrode.

Explanation of method: Within the broken lines are shown the essential features of the potentiometer method of measuring an unknown electromotive force. The battery B is connected through a variable resistance to the terminals of the uniform slide wire resistance LM, and should maintain a constant potential drop through this resistance; this potential drop can be properly adjusted for the electromotive force to be measured by the variable resistance R_v. The cell whose unknown electromotive force is to be measured is connected as shown. In this case this cell is composed of the reference calomel electrode and the platinum electrode, which assumes the potential set up by the constituents in the titrated solution. As thus connected, the unknown electromotive force tends to oppose that of the battery; in making a measurement, the sliding contact N is adjusted until a position is found at which no current exists in the galvanometer circuit as evidenced by no deflection of the galvanometer, G, being obtained upon closing the tapping key, K. At this point the opposing electromotive forces through the resistance are equal. As the ratio of the lengths, LN/LM, is equal to the ratio of the resistances of these two lengths and to the potential drop across them, it is seen that if the potential drop across LM has been fixed at a definite value, E_s, the unknown potential drop, E_x, across LN can be found; that is,

$$E_x = E_s \frac{LN}{LM}.$$

By this means no current need be drawn from the unknown cell as the measurement is made; this avoids polarization effects, which would result if the cell were directly connected to a voltmeter.

As the electromotive force of the reference calomel half-cell is constant, the change during the course of a titration of the measured electromotive force, E_x, represents the change in the potential of the solution in which the platinum indicator electrode is immersed. By the substitution of a reference standard cell of known electromotive force in place of the measured cell the electromotive force E_s can be accurately determined, and thus a measurement of the actual value of E_x can be obtained; this is the basis of the electrical method of determining the standard and formal potential values.

When a reducing agent is titrated with a standard permanganate solution, it is usually desirable for the reduction reaction to proceed from right to left according to Equation 4 and not according to Equation 3 with the resultant formation of manganese dioxide. Since the hydrogen-ion concentration enters to the eighth power in the Nernst equation for Equation 4, one would expect that as long as an excess of the reducing agent is present and an acid solution is maintained, no appreciable amount of manganese dioxide would be formed. However, as the end point is approached the permanganate-ion concentration increases rapidly and the manganous-ion concentration has become appreciable; therefore one might expect that Equation 3 would proceed to the left, forcing Equation 5 to the right according to the reaction

$$3Mn^{2+} + 2MnO_4^- + 2H_2O = 5MnO_2(s) + 4H^+ \qquad (6)$$

In addition, by the methods illustrated above, one can calculate that the equilibrium expression for this reaction has an exceedingly large constant

$$\frac{[H^+]^4}{[Mn^{2+}]^3[MnO_4^-]^2} = 10^{36}$$

By substituting into this expression reasonable values for the manganous-ion concentration at the end of a titration, and for the permanganate-ion concentration necessary to give a perceptible end point, one can calculate that the hydrogen-ion concentration would have to be raised to an impracticable value in order to prevent the precipitation of manganese dioxide. Fortunately, it is observed experimentally that the rate of the reaction shown by Equation 6 is very low. It is found that if the hydrogen-ion concentration is kept above approximately 0.5 *M* when permanganate titrations are being made, manganese dioxide is not likely to form sufficiently rapidly to interfere with observing the end point color. If titrated solutions are allowed to stand, the permanganate color slowly fades and eventually a brownish turbidity will be observed. These effects are illustrated by the experiment below.

EXPERIMENT: Stability of the Permanganate End Point

To each of two 125-ml conical flasks, marked *A* and *B*, add 20 ml of water, 2 ml of 3 *F* H_2SO_4, and 0.5 ml of 1 *F* $FeSO_4$ or $FeSO_4{\cdot}(NH_4)_2SO_4$. (The Fe(II) is added to simulate a titration and provide Mn(II).)

By means of a dropper add 0.02 *F* (0.1 *N*) $KMnO_4$ dropwise to flask *A* until the first persistent pink color is obtained. Repeat the titration with flask *B*.

Write an equation for the titration reaction. Heat solution *A* just to boiling and allow it to stand a few minutes. Compare it with the solution in flask *B*.

Write equations for any reactions. Note and explain your observations.

Rate effects are also an important factor in permanganate titrations that are carried out in hydrochloric acid solutions; again it would be predicted that chloride would be oxidized by permanganate, but the rate at which this process takes place is so low that, under properly controlled conditions, such titrations are experimentally possible. An extended discussion of these effects is given in Procedure 18-2.

The reaction between oxalic acid, $H_2C_2O_4$, and permanganate in acid solutions provides another interesting example of the effect of reaction rates on the reactions of permanganate. The titration was first proposed by Hempel[6] almost 120 years ago. Subsequently oxalic acid and sodium oxalate have been used extensively as primary standards for permanganate solutions, and, as a result, the titration has been the subject of many investigations.

The reaction between permanganate and oxalate when taking place in an acid solution, is represented by the equation

$$2MnO_4^- + 5H_2C_2O_4 + 6H^+ = 2Mn^{2+} + 10CO_2 + 8H_2O$$

and is, moreover, an example of what is called an *irreversible* reaction, since it has not been found possible experimentally to cause the reduction of carbon dioxide to oxalate.

However, the rate of the reaction between oxalate ion, or oxalic acid, and permanganate is somewhat low at room temperature, and it is necessary to heat the solution in order to cause the reaction to proceed to completion at a practical rate. It will be found experimentally that the first portion of the permanganate added will be decolorized very slowly even at 90°C, but that thereafter it disappears more rapidly. This behavior is due to the fact that manganous ion catalyzes the reaction. Therefore, since manganous ion is one of the products, the reaction proceeds more rapidly as soon as an appreciable amount of the permanganate has been reduced. This is an example of an *autocatalytic reaction*, that is, one that is catalyzed by one of the products of the reaction.

The mechanism of this reaction and its catalysis is very complicated and, although extensively investigated, is not yet entirely clear.[7] In very general terms it seems that, as stated, the reaction between oxalate, or oxalic acid, and permanganate ion is quite slow, but that if manganous ion is present, it can be oxidized by permanganate to tripositive or quadripositive manganese compounds and that these compounds in turn can rapidly oxidize the oxalate. This simple picture is complicated by the formation of complex com-

[6] W. Hempel. *Memoire sur l'emploi de l'acide oxalique dans les dosages á liqueurs titrées.* Lausanne, 1853.

[7] One of the first to note these effects was Skrabal. (*Z. Anorg. Chem.*, **42**, 1, 1904); H. A. Laitinen (*Chemical Analysis.* New York: McGraw-Hill, 1960) gives a detailed review of the more recent literature on this subject.

pounds between the various manganese ions and oxalate ion, and the possible formation from the oxalate of unstable oxidation products, such as the ion CO_2^-.

The question of the stoichiometric nature of the reaction has been exhaustively studied by McBride[8] and later by Fowler and Bright,[9] who showed that by proper control of the conditions of titration the stoichiometric error should not exceed 0.1 %, and if weight burets are used, the instrumental errors should also be less than that value.

The catalytic effect of manganous ion on the titration of oxalate with permanganate is shown by the following experiment.

EXPERIMENT: Effect of Mn(II) on the Permanganate-Oxalate Reaction

Dissolve two 0.25-g samples of $Na_2C_2O_4$ in 300 ml of water and 30 ml of 3 *F* H_2SO_4, contained in each of two 400-ml beakers. Designate the beakers *A* and *B*. To beaker *B* add 1 ml of 0.5 *F* $MnSO_4$. Heat both solutions to 60°C.

Add to beaker *A* 1 ml of 0.01 *N* $KMnO_4$. Note the time required for the solution to decolorize.

To beaker *B* add 1 ml of the $KMnO_4$. Note the time required for the color to fade.

Add successive 1-ml portions of the $KMnO_4$ to beaker *A*. Note the time required for decolorization after each addition.

PERMANGANATE METHODS INVOLVING OTHER OXIDATION STATES

Titrimetric methods are available in which permanganate is reduced to other than Mn(II). In one of these the product is Mn(III). The tripositive ion is unstable because of the reaction

$$2Mn^{3+} + 2H_2O = MnO_2(s) + Mn^{2+} + 4H^+$$

However, in the presence of dihydrogen pyrophosphate, $H_2P_2O_7^{2-}$, the stable complex $Mn(H_2P_2O_7)_3^{3-}$ is formed and titrations based on this fact have been developed.

There are also certain titrations with permanganate that are carried out in essentially neutral solutions and in which the permanganate is reduced to Mn(IV) and precipitated as the hydrous manganese dioxide. In fact, the reaction of Equation 6 is the basis for what is known as the *Volhard method* for the titrimetric determination of manganese; the dipositive manganese is oxidized in a neutral solution by permanganate to manganese dioxide.[10]

---

[8] McBride. *J. Am. Chem. Soc.*, **34**, 393 (1912).

[9] Fowler and Bright. *J. Res. Natl. Bur. Std.*, **15**, 493 (1935).

Methods have also been proposed that are carried out in strongly alkaline solutions in which the permanganate is reduced only to manganate, the half-cell reaction being

$$MnO_4^{2-} = MnO_4^- + e^- \qquad (E° = -0.54 \text{ v})$$

Further reduction of the manganate is prevented by providing barium ion and thus precipitating the slightly soluble $BaMnO_4$.[11] These titrations are discussed in more detail in Chapter 22, pages 491–493.

---

[10] The reaction was first used by Guyard (*Bull. Soc. Chim.*, **6**, 89, 1863) and later modified by Volhard (*Ann. Chem.*, **198**, 314, 1879). This is the same Volhard who developed the method for the titration of silver with thiocyanate.

[11] This method was proposed by Stamm (*Angew. Chem.*, **47**, 191, 1943; **48**, 150, 710, 1935).

Preparation of a Permanganate Solution

Outline. Potassium permanganate crystals are dissolved in water. The solution is boiled to rapidly oxidize any reducing agents present. The resultant manganese dioxide is filtered out because it would catalyze subsequent reduction of the permanganate.

DISCUSSION

It has already been stated that permanganate solution, if properly prepared and stored, is stable over long periods of time. It is not practical to prepare standard solutions of permanganate by direct weighing, for the best grades of potassium permanganate commercially obtainable almost invariably contain some manganese dioxide that has formed on the surface; although it is possible by elaborate means to prepare a very pure product, the distilled water used for the solution usually contains sufficient reducing gases or organic material to cause the production of some manganese dioxide upon standing. This manganese dioxide then serves as a catalytic agent for the further decomposition of the permanganate, and the solutions rapidly decrease in strength, especially if they are exposed to light, which induces the decomposition reaction. Therefore, it is customary to prepare the solution, to heat it or allow it to stand until the reducing substances in the water are oxidized, and then to filter the solution through asbestos or sintered-glass filters into the storage bottle. Thereafter the solution must be protected from light and from contact with dust, organic matter, or reducing gases.

From the standard potential values of the two half-cell reactions,

$$MnO_2 + 2H_2O = MnO_4^- + 4H^+ + 3e^- \qquad (E° = -1.59 \text{ v})$$

and

$$2H_2O = O_2 + 4H^+ + 4e^- \qquad (E° = -1.23 \text{ v})$$

one would predict that aqueous permanganate solutions would not be stable because of the tendency of permanganate to oxidize water and from these half-cells one can obtain the following cell reaction and potential

$$4MnO_4^- + 4H^+ = 4MnO_2(s) + 3O_2 + 2H_2O \qquad E^\circ_{cell} = 0.36 \text{ v}$$

From the value of the cell potential one concludes that the reaction should proceed from left to right, and calculations will show that even in dilute neutral aqueous solutions permanganate should be reduced to MnO_2. Fortunately, the rate of this reduction is so low that standard permanganate solutions are practical volumetric reagents and are extensively used.

If acid permanganate solutions are heated, however, reduction takes place at a significant rate. Therefore, you cannot make an accurate determination if you add an excess of permanganate to a reducing agent in an acid solution, heat the solution in order to accelerate a slow oxidation reaction, and then titrate the excess of permanganate.

The cell equation given above appears unusual because water is not shown as a reactant, although the equation was derived from a half-cell reaction involving the oxidation of the oxide oxygen of water to oxygen. This apparent anomaly results from more water being produced in the reduction of permanganate than is lost in the oxidation of the water. The equation represents the stoichiometry of the reaction and not the mechanism by which it occurs. The equation also indicates that permanganic acid is not a stable compound, and such is the case—the addition of crystals of a permanganate to concentrated sulfuric acid is likely to result in an explosion.

When neutral permanganate solutions, even those prepared from reagent-grade potassium permanganate, are heated as described in the procedure below, they become alkaline, and experiments have shown that this effect is largely caused by impurities in the permanganate, since it could be minimized by use of carefully recrystallized reagent. It is possible that this decrease in the hydrogen-ion concentration contributes to the stability of the resultant solutions; some workers have recommended that small quantities of sodium hydroxide be added to standard permanganate solutions.

INSTRUCTIONS (Procedure 18-1: Preparation of a Permanganate Solution)

Weigh out 3.2 g (take 4.0 g if the solution is to be used for gravimetric titrations) of the best grade of $KMnO_4$ obtainable (Note 1), transfer it to a large

NOTES

1. Even though the solution is later to be standardized, it is an advantage to use a good grade of $KMnO_4$, because it is less likely to contain substances that will later cause a slow reduction of the permanganate and the formation of MnO_2. Sodium permanganate is deliquescent and not readily purified.

beaker, add 1 liter of water, and heat the solution to boiling, stirring the mixture until the crystals have dissolved. Cover the solution with a clock glass and keep it *just* boiling for *at least* 20 minutes; do not allow the solution to concentrate by evaporation (Note 2). Allow the solution to cool, and filter it through a sintered-glass filter (Note 3). Before use, the filter should be thoroughly rinsed with distilled water and then with small portions of the permanganate solution. Collect the solution in a bottle with a ground-glass stopper (see page 194) that has just been cleaned with cleaning solution, then rinsed with distilled water and next with small portions (3 ml–6 ml) of the permanganate solution. Swirl the solution, without wetting the neck of the bottle, until it is thoroughly mixed (Note 4).

NOTES

2. If time is available, the solution should be kept hot, preferably on a water bath for an hour, or heated and left to cool overnight. In this way, slowly oxidized organic substances are more effectively eliminated.

3. See Figure 10-30, page 202, regarding support for this filter. If sintered-glass filters are not available, any one of the following filters can be used; (*a*) porous porcelain filters (page 205), (*b*) glass-fiber discs (page 204), (*c*) asbestos mats (see page 200).

If the filter is supported by a rubber holder, take care that the $KMnO_4$ does not come in contact with the rubber.

4. It is desirable that the neck and stopper of the bottle not be wet with the permanganate, since this then evaporates, and a deposit of $KMnO_4$ and MnO_2 results, which may fall into the solution when the bottle is opened again. For this reason, when filling a buret do not pour the solution but remove it by means of a pipet.

PRIMARY STANDARDS

A substance that is directly weighed and used for the standardization of a solution is called a *primary standard*. For very precise work it is preferable that a solution be directly standardized against a primary standard rather than compared with another standard solution.

A primary standard should meet the following qualifications:

1. It should be a pure substance. It may be an elementary substance such as iodine, copper, or iron, or it may be a compound such as sodium oxalate, arsenious oxide, or potassium iodate; but it must be capable of purification, usually by recrystallization or sublimation, to a definitely known composition.
2. It should be stable. Hydrated substances, though frequently used, are subject to question, since they are difficult to free of extraneous moisture and may change their moisture content upon storage.

Compounds subject to surface oxidation, such as $FeSO_4 \cdot (NH_4)_2SO_4 \cdot 6H_2O$, and Fe wire, or to reduction, such as $KMnO_4$, are to be avoided if possible.

3. It should not be hygroscopic. Difficulty in storage and in weighing is likely with hygroscopic materials.
4. It should react stoichiometrically. Critical conditions difficult of attainment, complex apparatus, or special techniques should not be necessary in order to make the reaction stoichiometric. A particular standard may be useful for only one solution of a general class. Arsenious oxide is an excellent primary standard for iodine solution; but, as will be discussed later, it can be used for permanganate only under very special conditions.
5. It should have a large equivalent weight. It is desirable that the weight required for a titration should be so large that the weighing errors will be small relative to the other errors involved in the standardization.

Certain primary standards (sodium oxalate, benzoic acid, arsenious oxide, potassium dichromate, and potassium hydrogen phthalate) can be obtained from the National Bureau of Standards and are accompanied by certificates showing their purity. Others may be obtained from commercial sources or may have to be prepared as needed.[12]

The primary standards that have been most commonly used for permanganate solutions are ferrous ammonium sulfate, $FeSO_4 \cdot (NH_4)_2SO_4 \cdot 6H_2O$; oxalic acid, $H_2C_2O_4 \cdot 2H_2O$; iron wire; sodium oxalate, $Na_2C_2O_4$; and arsenious oxide, As_2O_3. In this procedure the permanganate solution will be standardized against arsenious oxide, since it fulfills the qualifications listed above.

---

[12] The Mallinckrodt Chemical Works, St. Louis, Mo. 63107, offers the following primary standards with assays from 99.95% to 100.05%: benzoic acid, arsenious oxide, potassium hydrogen phthalate, potassium dichromate, sodium oxalate, and sodium carbonate.

PROCEDURE **18-2**

Standardization of a Permanganate Solution. Arsenious Oxide as Primary Standard

Outline. (*a*) The weighed sample of arsenious oxide is dissolved, (*b*) iodine monochloride is added as catalyst, and (*c*) the resultant solution is titrated with permanganate; ortho-phenanthroline ferrous sulfate is used as indicator. The titration reaction is

$$2MnO_4^- + 5H_3AsO_3 + 6H^+ = 2Mn^{2+} + 5H_3AsO_4 + 3H_2O$$

and the indicator reaction is

$$MnO_4^- + \underset{\text{red}}{5Fe(C_{12}H_8N_2)_3^{2+}} + 8H^+ = Mn^{2+} + \underset{\text{pale blue}}{5Fe(C_{12}H_8N_2)_3^{3+}} + 4H_2O$$

DISCUSSION

(*a*) Dissolving the Arsenious Oxide

Arsenic is classified as a nonmetal (note its position in the periodic table) and therefore arsenious oxide is an acidic oxide; that is, it reacts with water to form an acid, H_3AsO_3, as follows

$$As_2O_3 + 3H_2O = 2H_3AsO_3$$

This hydration reaction takes place slowly, therefore the As_2O_3 is first dissolved in sodium hydroxide and the solution is then acidified. Hydrochloric acid is used because the catalytic reaction involved in the titration and discussed below requires a high chloride concentration. In the procedure 60 mmoles of NaOH are used (for the dissolving reaction) initially and 124 mmoles of HCl are added subsequently (including that in the solution of the catalyst). The volume at the start of the titration is 120 ml so the concentrations of chloride and hydrogen ions are about 1 *M* and 0.5 *M*, respectively. Under these conditions the arsenic is present primarily as arsenious acid, H_3AsO_3.

(*b*) Catalysis of the Titration Reaction

The standard potential for the half-cell

$$H_3AsO_3 + H_2O = H_3AsO_4 + 2H^+ + 2e^-$$

is -0.556 v, which suggests that arsenious acid would be quantitatively oxidized by permanganate ion, and, therefore, that the solid As_2O_3 would be a suitable primary standard for permanganate solutions. However, the reaction is neither rapid nor stoichiometric if the titration is carried out in sulfuric acid solutions at room temperature. The first portions of the permanganate appear to be rapidly reduced, but the solution becomes yellow, then green-orange, and finally dark brown upon further addition of permanganate, and these colors disappear slowly even when the solution is heated. The colors are attributed to the formation of complex substances containing tripositive manganese and arsenic acid. If the titration is made in a hot hydrochloric acid solution, these effects are minimized and a fairly satisfactory titration can be made by titrating very slowly and observing the first appearance of a uniform pink color. However, the end point is not permanent because of the reduction of permanganate by chloride in hot acid solutions.

Apparently Lang[13] was the first to discover that the oxidation of arsenious acid by permanganate is catalyzed by the presence of even very minute quantities of an iodate. Subsequently it has been shown that the catalysis is independent of the oxidation state of the added iodine. A mechanism for this catalysis is suggested from the experimental observations that (1) arsenious acid in hydrochloric acid solutions is rapidly oxidized by iodine monochloride and (2) under the same conditions elementary iodine is rapidly oxidized to iodine monochloride by permanganate.

The equations for these two reactions can be written as follows:

$$2ICl + H_3AsO_3 + H_2O = I_2 + H_3AsO_4 + 2H^+ + 2Cl^- \qquad (7)$$

and

$$5I_2 + 2MnO_4^- + 10Cl^- + 16H^+ = 10ICl + 2Mn^{2+} + 8H_2O$$

If these two equations are added, with Equation 7 first multiplied by 5, the equation

$$2MnO_4^- + 5H_3AsO_3 + 6H^+ = 2Mn^{2+} + 5H_3AsO_4 + 3H_2O$$

---

[13] Lang. *Z. Anal. Chem.*, **45**, 649 (1906).

is obtained, and the net result of these reactions is seen to be the oxidation of the arsenious acid by permanganate, with the iodine acting only as a catalyst.

The above series of reactions is an example of the catalysis of oxidation-reduction reactions in single-phase or homogeneous systems. Catalytic action is required in such systems when, even though there is adequate potential difference, the oxidant of one half-cell reaction,

$$\text{reductant } A = \text{oxidant } A + ne^-$$

reacts slowly with the reductant of a second half-cell reaction,

$$\text{reductant } B = \text{oxidant } B + ne^-$$

The slow reaction can be generalized as

$$\text{oxidant } A + \text{reductant } B \xrightarrow{\text{slow}} \text{reductant } A + \text{oxidant } B$$

In order to catalyze this reaction, a third half-cell reaction is used, generalized as

$$\text{reductant}_{\text{cat.}} = \text{oxidant}_{\text{cat.}} + ne^-$$

The qualifications of this third half-cell reaction are as follows: (1) The reductant ($\text{reductant}_{\text{cat.}}$) must react rapidly with primary oxidant A. (2) The oxidant ($\text{oxidant}_{\text{cat.}}$) must react rapidly with primary reductant B. (3) The standard potential should preferably be intermediate in value between the standard potentials of the other two half-cells; otherwise one of the reactions specified in (1) and (2) would not proceed to a significant extent.

The mechanism of the catalysis can be represented as follows:

$$\begin{array}{l} \text{oxidant } A + \text{reductant}_{\text{cat.}} \xrightarrow{\text{fast}} \text{reductant } A + \text{oxidant}_{\text{cat.}} \\ \underline{\text{reductant } B + \text{oxidant}_{\text{cat.}} \xrightarrow{\text{fast}} \text{oxidant } B + \text{reductant}_{\text{cat.}}} \\ \text{oxidant } A + \text{reductant } B \xrightarrow{\text{fast}} \text{reductant } A + \text{oxidant } B \end{array}$$

Since the catalyst is regenerated, it is obvious that a small quantity can catalyze the reaction of an indefinite quantity of the primary reactants. Catalytic action will be just as effective regardless of whether reductant catalyst or oxidant catalyst is initially added to the solution. The *final* state of the catalyst will be determined by whether there is an excess of primary oxidant or primary reductant. The statement that such a catalyst does not undergo change is not rigorously correct, since its catalytic activity is dependent upon the fact that it does react rapidly with the primary reactants, and hence its initial and final states may or may not be the same.

If the iodine needed for the catalysis of the permanganate–arsenious acid reaction is added as iodine monochloride, no correction for the amount of catalyst added will be necessary, since at the end point substantially all of any of the iodine monochloride reduced during the titration will be re-oxidized. A study of the use of ICl as the catalyst in the standardization of MnO_4^- against As_2O_3 has shown that the method gives both high precision and excellent agreement with other standard methods.[14] A demonstration of the catalytic effects of both chloride and iodine monochloride is provided in experiment 18-2 below.

(*c*) Detection of the End Point. Use of Redox (Potential) Indicators

Two methods are available for obtaining the end point of the arsenious acid–permanganate titration. First, the color of the permanganate may be used, or, second, an oxidation–reduction (potential) indicator may be added as a means of eliminating the transient end-point color that permanganate gives in the presence of large concentrations of hydrochloric acid. A *potential indicator* may be defined as a substance that changes from a colorless to a colored form, or from one color to another, when it is oxidized or reduced. Such a potential indicator is used in the procedure below. Since these indicators are becoming increasingly important in oxidation–reduction titrations, they are discussed in some detail.

Unlike permanganate, many of the standard solutions used in titrimetric analyses are not sufficiently colored to serve as their own indicators. This fact formerly required the use of outside indicators or of potentiometric titrations when such solutions (for example, ceric salts, iodate, or dichromate) were employed. To eliminate this disadvantage, the use of oxidation–reduction, or potential, indicators has been developed. As will be explained below, this color change will be observed within a definite potential range, which is commonly termed the *transition range*, of that indicator; the analogy with acid-base, or *p*H, indicators will be apparent later.

A potential indicator is employed in the titration of arsenious acid with permanganate in the procedure given below; the theory of these indicators can be illustrated by a consideration of its use. This indicator is a complex ion formed by ferrous iron with three molecules of an organic compound having the empirical formula $C_{12}H_8N_2$ and known as orthophenanthroline (or, 1,10-phenanthroline). The structure of this organic compound can be

[14] Metzler, Myers, and Swift. *Ind. Eng. Chem.*, anal. ed., **16**, 625 (1944); for investigations of this titration with iodide or iodate being used as the catalyst see Bright. *Ind. Eng. Chem.*, anal. ed., **9**, 577 (1937) or *J. Res. Natl. Bur. Std.*, **19**, 691 (1937); and Kolthoff, Laitinen, and Lingane. *J. Am. Chem. Soc.*, **59**, 429 (1937).

represented as follows:

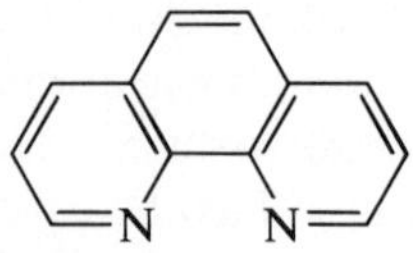

Because each of the two nitrogen atoms has two unshared electrons the compound tends to form stable complexes with the positive ions of such metals as iron, cadmium, copper, cobalt, and zinc. The ferrous complex, which we will indicate as $(\text{phen.})_3\text{Fe}^{2+}$, has an intense red color, but the ferric complex, $(\text{phen.})_3\text{Fe}^{3+}$ is pale blue; this color has an intensity about one-tenth that of the ferrous compound. When the ferrous compound in an aqueous solution is oxidized the half-cell reaction taking place can be written as

$$(\text{phen.})_3\text{Fe}^{2+} = (\text{phen.})_3\text{Fe}^{3+} + e^- \qquad (E^\circ = -1.06 \text{ v})$$

The formal potential of the arsenious acid–arsenic acid half-cell is -0.58 v in 1 F hydrochloric acid. This value indicates that if a small amount of this indicator were added to an acid solution of arsenious acid, substantially all of the arsenious acid could be oxidized by a suitable oxidant such as permanganate before a significant fraction of the indicator was oxidized.

The agreement of the end point with the equivalence point that could be obtained when using permanganate for the titration of arsenious acid can be predicted by the following calculations. When within 0.1% of the equivalence point the ratio of the equivalents of permanganate added to the arsenious acid initially present has the value 0.999 and, assuming quantitative oxidation of the arsenious acid by the permanganate, the ratio of arsenic acid to arsenious will be 0.999 to 0.001.

The potential existing in the solution can be calculated by use of the Nernst equation as follows:

$$E = E^\circ - \frac{0.059}{2}\log\frac{[\text{H}_3\text{AsO}_4][\text{H}^+]^2}{[\text{H}_3\text{AsO}_3]}$$

If the $[\text{H}^+]$ is assumed 1 M

$$E = -0.58 - \frac{0.059}{2}\log\frac{(0.999)(1)^2}{(0.001)} = -0.67 \text{ v}$$

At equilibrium all half-cell reactions in a solution must have the same value; therefore the ratio of the indicator in the oxidized form to that of the reduced form can be calculated from the equation

$$-0.67 = -1.06 - \frac{0.059}{1}\log\frac{[(\text{phen.})_3\text{Fe}^{3+}]}{[(\text{phen.})_3\text{Fe}^{2+}]}$$

to be approximately 10^{-7}. This value indicates that no significant amount of the indicator will have been oxidized and no color change observed.

When the equivalence point has been exceeded by 0.1% the ratio of the equivalents of permanganate added to the tripositive arsenic initially present will be 1.001 and the molar ratio of excess permanganate to manganous ion formed during the titration will be 0.001/1.000. (Quantitative reduction of the permanganate to Mn^{2+} by the arsenious acid is assumed.) Again, if the hydrogen-ion concentration is assumed to be one molar, the potential in the solution is

$$E = E^\circ - \frac{0.059}{5} \log \frac{[MnO_4^-][H^+]^8}{[Mn^{2+}]}$$

$$E = -1.45 - \frac{0.059}{5} \log \frac{0.001}{1.000} = -1.40 \text{ v}$$

In a solution having this potential value, calculations similar to those above will show that the ratio of $(phen.)_3Fe^{3+}$ to $(phen.)_3Fe^{2+}$ will now be almost 10^6. This shows that the indicator will have been substantially all oxidized and a color change from red to pale blue observed. Thus 0.1% short of the equivalence point, no significant amount of the indicator will have been oxidized, and no color change observed, but before 0.1% excess of permanganate has been added, substantially all of the indicator has been oxidized. Therefore a very sudden color change will be obtained, which should give a very accurate end point. It has been assumed above that the amount of the indicator added was so small that the amount of the standard solution required to convert the indicator to the oxidized form was insignificant. If this is not true, an end-point correction should be made.

From the above discussion it can be seen that a potential indicator changes in color when the potential of the solution changes through such a range that a sufficient fraction of the indicator is converted from one form to the other to result in a visible color change. The percentage conversion required to cause a visible color change depends upon the specific nature of the indicator and upon whether the change is from colorless to colored or from one color to another. In general, the human eye cannot detect the presence of less than about 10% of one colored form of an indicator in the presence of another colored form. Hence the color change can usually be recognized during the range of ratios of the two forms from approximately 1/10 to 10/1; a change in ratios of 100. Therefore, if the indicator half-cell reaction can be written as $In_{red.} = In_{ox.} + ne^-$, then from

$$E_{In} = E^\circ_{In} - \frac{0.059}{n} \log \frac{In_{ox.}}{In_{red.}}$$

it can be seen that this range of ratios through which a perceptible color change can be observed covers a change in potential of the solution of $(0.059 \times 2)/n$ or approximately $(0.12/n)$ v. With the orthophenanthroline ferrous complex the change from the red of the reduced form to the less intense blue of the oxidized form is usually perceived when the ratio

$$In_{ox.}/In_{red.} \approx 10$$

or at a potential of about -1.12 v.

There are many organic compounds that are capable of being oxidized and reduced, and with which process there is associated an appearance, disappearance, or change of color. Unfortunately, many of these compounds cannot be used as potential indicators because (1) the color change is not sufficiently pronounced; (2) they are often so easily oxidized as to be of no value for most titrations; and (3) the oxidation-reduction reaction may not be readily reversible, one of the oxidation states reacting only slowly or being further changed by excess of the titrating agent. The study and development of compounds having the above qualifications has been an important contribution to analytical research.

Potential indicators that have been extensively used are the following.

(1) Diphenylamine, which has a transition range around -0.76 v. This indicator has been widely used for the titration of ferrous salts with dichromate.[15] It is not a strictly reversible reaction. When diphenylamine is oxidized, the first product is *colorless* diphenylbenzidine, which is further oxidized to the colored diphenylbenzidine violet.[16] Diphenylbenzidine has been recommended instead of diphenylamine for certain titrations because a smaller end-point correction is required.[17]

(2) Diphenylamine sulphonic acid[18] turns a reddish-violet color at about -0.83 v. It therefore gives a sharper end point in the titration of ferrous salts in the absence of phosphoric acid and also is more soluble in aqueous solutions than is diphenylamine.

(3) Orthophenanthroline ferrous complex,[19] which has been discussed above, has been extensively used with ceric sulfate solutions. The transition occurs at a potential of about -1.12 v; this value indicates, and experiments

---

[15] Knop. *J. Am. Chem. Soc.*, **46**, 263 (1924).

[16] Kolthoff and Sarver. *J. Am. Chem. Soc.*, **52**, 4179 (1930).

[17] Cone and Cady. *J. Am. Chem. Soc.*, **49**, 356 (1927); Willard and Young. *Ind. Eng. Chem.*, **20**, 764 (1928); Kolthoff and Sandell. *Ind. Eng. Chem.*, anal. ed., **2**, 140 (1930); Willard and Young. *Ind. Eng. Chem.*, anal. ed., **5**, 154 (1933).

[18] Sarver and Kolthoff. *J. Am. Chem. Soc.*, **53**, 2902, 2906 (1931); Willard and Young, op. cit.

[19] Walden, Hammett, and Chapman. *J. Am. Chem. Soc.*, **55**, 2649 (1933); Hume and Kolthoff. *J. Am. Chem. Soc.*, **65**, 1895 (1943).

have shown, that this indicator is not very satisfactory for use with dichromate solutions.[20]

In Procedure 18-2 below, orthophenanthroline ferrous complex can be used to advantage as an indicator, instead of the permanganate color, because of the previously mentioned tendency of the excess permanganate to be reduced by chloride ion, with consequent fading of the end point.

The experiment below demonstrates the rate effects in the permanganate–arsenious acid reaction and the catalytic effects of chloride and iodine monochloride.

EXPERIMENT 18-2: Reactions of Permanganate with Arsenious Acid

To each of three 125-ml flasks, designated *A*, *B*, and *C*, add 10 ml of a 0.05 *F* (0.1 *N*) solution of H_3AsO_3 (prepared as directed in Note 2, Procedure 18-2), and 20 ml of water.

(*a*) To flask *A* add 2 ml of 6 *F* H_2SO_4. Using a calibrated dropper titrate this solution with 0.02 *F* (0.1 *N*) $KMnO_4$ until the resultant brownish coloration would obviously obscure a permanganate end point. Note the appearance of a yellowish color (usually after 0.2 to 0.6 ml of the $KMnO_4$), which changes to a yellowish brown (1 ml–2 ml), followed by a greenish brown (3 ml–4 ml), and finally, after passing the end point, an orange brown.

(*b*) To flask B add 2 ml of 6 *F* HCl. Repeat the titration, and note the rate at which the resultant brownish color is bleached. This becomes apparent after approximately 3 ml of the $KMnO_4$ have been added, and is emphasized by rapidly adding a 0.5 ml–1 ml portion, then allowing the color to fade.

(*c*) To flask *C* add 2 ml of 6 *F* HCl and 0.3 ml of 0.0025 *F* ICl in 4 *F* HCl. Titrate as in (*b*) and compare the rate of bleaching at the same points in the titration with that observed in (*b*).

Explain your observations.

INSTRUCTIONS (Procedure 18-2: Standardization of a Permanganate Solution. Arsenious Oxide as Primary Standard)

Dry approximately 0.7 g of As_2O_3 (Note 1) at 100°C–105°C for 1 hour and weigh out 0.2-g portions into each of three 500-ml flasks (Note 2). Dissolve

NOTES

1. As_2O_3 for standardization purposes can be purchased from the Bureau of Standards; it is recommended that this product be used for highly accurate work;

[20] For more extensive discussions of potential indicators and the applications, see Whitehead and Wills. *Chem. Rev.*, **29**, 69 (1941); Oesper. *Newer Methods of Volumetric Chemical Analysis.* Princeton, N.J.: Van Nostrand, 1938; Kolthoff and Stenger. *Volumetric Analysis.* Vol. 1. New York: Wiley-Interscience, 1942.

the As_2O_3 in 10 ml of cold 6 *F* NaOH (Note 3); stir or swirl the mixture intermittently until complete solution of the solid is obtained. As soon as the As_2O_3 has dissolved, add 90 ml of water, 20 ml of 6 *F* HCl, and 1 ml of 0.0025 *F* ICl in 4 *F* HCl (Note 4).

NOTES

for other purposes, material of "primary standard" grade can be purchased commercially. Pure As_2O_3 is not hygroscopic and need not be dried unless extreme accuracy is desired or it has been exposed to considerable moisture. Since As_2O_3 tends to sublime, it is often dried by being left over sulfuric acid in a desiccator for 12 hours.

Commercial As_2O_3 often contains chloride, sulfide, and water. The product may be purified by recrystallization followed by sublimation.[21]

2. Weighing out three or more separate samples of arsenious oxide can be avoided by preparing a stock solution in a volumetric flask and taking aliquot portions of this with a pipet (consult your instructor).

A 250-ml flask and a 50-ml pipet are convenient for this purpose. If they have not been calibrated, a relative calibration is entirely satisfactory for this purpose, since it is desired only that an exactly known fraction of the total be taken. Such a calibration is made as follows:

Clean and dry a 250-ml volumetric flask. (To dry the flask or similar apparatus, rinse it with several small portions of alcohol, invert it, insert a glass tube almost to the bottom, attach the tube to a vacuum line, and draw air through it. Compressed air is not desirable, since it is likely to contain oily material picked up from the pumps.) Carefully deliver 5 pipets of water into the flask and mark the position of the meniscus (a small gummed label may be used).

To prepare the arsenious acid solution, proceed as follows. Weigh out accurately about 1 g of As_2O_3 into a small beaker, dissolve it in 5 ml to 10 ml of 6 *F* NaOH, add 50 ml of water, and transfer the solution to the 250-ml flask with the aid of a funnel. Wash out the beaker repeatedly with portions of water and dilute the solution to the calibration mark. Mix the solution, pipet 50 ml into a 500-ml flask, and treat as directed in the above Procedure.

The solution should be used at once. Alkaline arsenite solutions are slowly oxidized on exposure to air; neutral or acid solutions are quite stable.

The analyst should realize that this method is neither as reliable nor as accurate as weighing out separate portions of As_2O_3, since there is no check on the weight of the one portion and since several volumetric measurements are involved.

3. The NaOH used for this purpose should be free of either oxidizing or reducing substances. NaOH solutions that have been stored in bottles with rubber stoppers will contain reducing substances and should *not* be used. See that all solid particles are dissolved. Arsenious oxide is only moderately soluble in water and dissolves slowly; in fact, the dry powder may dissolve slowly even in the sodium hydroxide solution.

---

[21] See Chapin (*Ind. Eng. Chem.*, **10**, 522, 1918) for a discussion of the preparation and testing of pure As_2O_3.

Prepare an end-point correction solution by adding to 100 ml of water the same quantities of NaOH, HCl, and iodine compound as used above. Reserve this solution until the first titration is completed (Note 5).

Volumetric Titrations

Titrate the As_2O_3 solution with the permanganate solution, *stirring continuously*, until a lag in the rate of decolorization of the permanganate is noticed. Then add the permanganate dropwise (still with continuous stirring) until the pink color first spreads *almost* throughout the solution. Add to the solution 1 drop of 0.025 *F* orthophenanthroline ferrous sulfate indicator (Note 6). Now add the permanganate dropwise until the first transient *fading* of the *pink color of the indicator* is observed, usually within 1–2 drops from the end point. Thereafter add the permanganate in fractions of a drop until the *pink color of the indicator* is no longer perceptible (Note 7). Observe and record the volume of permanganate used.

Obtain an end-point correction by adding 1 drop of the 0.025 *F* orthophenanthroline indicator to the prepared end-point correction solution. Then add the permanganate in fractions of a drop until the pink color of the indicator is no longer perceptible. Subtract this volume of permanganate from that used for the titration (Note 8). From the corrected volume of permanganate used and the weight of As_2O_3 taken, calculate the formality

NOTES

4. If the ICl solution is not available, 0.04 ml of 0.0025 *F* KI or KIO_3 can be used. With this small quantity of catalyst, transient and premature end points will be obtained near the end point, and the titration should be made more slowly. Larger quantities of these compounds should not be added unless standard solutions are used and corrections made for the volume added.

5. Since the end point obtained in the titration may fade, the solution to be used for the end-point correction should be prepared before the end point is reached.

6. The indicator is advantageously withheld until this preliminary indication is obtained, since it permits a more rapid titration to this point; otherwise the pink color of the indicator prevents observation of this transient permanganate color.

7. A more sensitive end point is obtained by noting the disappearance of the pink color, rather than attempting to titrate to the appearance of a distinct blue color. This blue color is so pale that one not familar with the end point may miss it because of the subsequent appearance of a pink color caused by excess permanganate. The end point should be stable for at least 30 seconds. If the titration is carried to the appearance of a blue color, this same color should be reproduced in obtaining the end-point correction.

8. This correction should not exceed 0.05 ml and is usually 0.03 ml or less.

9. Results obtained from standardizations by this method should agree to within 1 part to 2 parts per 1000.

[also the normality for Mn(VII)–Mn(II) titrations] of the permanganate solution (Note 9, page 415).

Gravimetric Titrations

Proceed as directed above for volumetric titrations, but use a weight buret. From the corrected weight of permanganate used and the weight of As_2O_3 taken, calculate the weight formality [also the weight normality for Mn(VII)–Mn(II) titrations] of the permanganate solution.

Determination of the Iron in an Ore. Reduction by Stannous Chloride and Titration with Permanganate

Outline. (*a*) The ore is dissolved and the ferric iron is reduced to the ferrous state by treatment with hydrochloric acid and stannous chloride, (*b*) the excess stannous chloride is oxidized with mercuric chloride, (*c*) a manganous salt and phosphoric acid are added to minimize oxidation of chloride and to decolorize the solution, and (*d*) the ferrous iron is titrated with standard permanganate. The titration reaction is

$$MnO_4^- + 5Fe^{2+} + 8H^+ = Mn^{2+} + 5Fe^{3+} + 4H_2O$$

The permanganate color is used as the indicator.

DISCUSSION

(*a*) Dissolving Iron Ores

The industrially important iron ores usually contain ferric iron as Fe_2O_3 (hematite), $Fe_2O_3 \cdot nH_2O$ (brown ore, including limonite, $2Fe_2O_3 \cdot 3H_2O$), and Fe_3O_4 (magnetite) in which both Fe(II) and Fe(III) are present. These ores are much more readily dissolved by treatment with hydrochloric acid than by treatment with sulfuric, nitric, or perchloric acids. This difference is usually attributed to the formation of stable complex compounds by ferric iron in hydrochloric acid solutions. The time required for solution of the ore is greatly diminished if a reducing agent such as stannous tin is added to the hydrochloric acid, and many iron ores can be dissolved in a few minutes by such a treatment. This decrease in time is probably caused by a surface reduction of the ferric iron.

Some more resistant oxide and silicate ores may leave a residue insoluble in hydrochloric acid–stannous chloride solutions. Such a silicious residue can be treated with hydrofluoric acid in order to ensure that all of the iron is dissolved. The hydrofluoric acid dissolves silica and silicates with the formation of SiF_4 (or of H_2SiF_6 with an excess of HF) and is also an effective solvent for ferric compounds because of the formation of various complex ions.[22] Occasionally a very resistant ore will require a fusion process in order to dissolve all of the iron. The procedure below is designed for use with ores from which the iron is readily dissolved by a hydrochloric acid–stannous chloride solution; however, by the procedure in Note 3 a residue that may contain iron can be treated by a fusion process in case such a treatment seems necessary.

Reduction of Ferric Solutions. In the procedure below, the iron in the ore is assumed to be in the ferric state (as is usually the case) and must be reduced before the titration. This reduction can be accomplished by (1) metals (such as zinc, cadmium, or aluminum), (2) reducing gases (such as hydrogen sulfide or sulfur dioxide), or (3) solutions of a reducing agent (usually stannous chloride). After such a reduction, the excess of the reducing agent must be removed. If metals are used, this operation may involve dissolving the excess of metals, using the metal in the form of a spiral that can be lifted from the solution, or employing special reduction apparatus that permits the solution to flow through a bed of the finely divided metal. (The use of metals and amalgams as reductants is discussed in Chapter 22.) If reducing gases are used, they have to be expelled by long boiling or else swept out with an inert gas such as carbon dioxide or nitrogen. In case the iron is in a hydrochloric acid solution (as is frequently the case because of the effectiveness of this acid in dissolving iron ores), the reduction can be very easily carried out with a stannous chloride solution.

It would be a gross oversimplification to write the equation for the reduction of ferric iron in a hydrochloric acid solution by stannous tin as

$$2Fe^{3+} + Sn^{2+} = 2Fe^{2+} + Sn^{4+}$$

In such solutions the Fe(III) forms relatively stable complex ions ranging from $FeCl^{2+}$ to $FeCl_6^{3-}$ depending upon the chloride concentration;[23] the Sn(IV) is present as the very stable $SnCl_6^{2-}$. The chloride complexes of Fe(III) are yellowish in color and these colors can be used to indicate when an excess of stannous has been added to the ferric solution.

The rate of the reaction between Fe(III) and Sn(II) is of interest because, first, it has been shown that in relatively concentrated perchloric acid solutions, in which there is little complex formation or hydrolysis, the rate is

---

[22] This solvent effect is commonly attributed to the formation of the anion FeF_6^{3-}. Work by Dodgen and Rollefson (*J. Am. Chem. Soc.*, **71**, 2600, 1949) indicates that species such as FeF^{2+}, FeF_2^{+}, and FeF_3 predominate in acid solutions.

[23] Axelrod and Swift (*J. Am. Chem. Soc.*, **62**, 33, 1940) have shown that a compound having the formula $HFeCl_4$ is extracted by dichloroethyl ether from hydrochloric acid solutions; the extraction is most efficient from about 6 *F* acid.

extremely low. Second, the addition of chloride increases the rate and this rate increase is proportional to the fourth power of the chloride concentration, indicating that the reacting species could range from $FeCl_4^- \cdot Sn^{2+}$ to $Fe^{3+} \cdot SnCl_4^{2-}$.[24] Even in the presence of chloride, the reduction reaction is perceptibly slow; therefore, it is essential that a high chloride-ion concentration be provided, and that the solution be kept hot.

(*b*) Oxidation of Excess Stannous Chloride with Mercuric Chloride

Mercuric chloride, which is one of the few relatively un-ionized salts, is uniquely fitted for the oxidation of the excess chloride present after reduction of the iron (1) because mercuric chloride does not oxidize ferrous ion, and (2) because mercurous chloride (the reduction product) has such a low solubility —especially if it is caused to precipitate in a crystalline form—that it is not appreciably oxidized by either ferric ion or permanganate during the course of the titration. Mercurous ion is unusual in that it is diatomic, Hg_2^{2+}; the insoluble chloride is Hg_2Cl_2.

(*c*) Titration of Ferrous Salts in Hydrochloric Acid Solutions with Permanganate

As was mentioned in the general discussion of permanganate methods, one would calculate from the potentials involved that permanganate, even in the small concentration necessary to give an end point, would be reduced by chloride in a strongly acid solution. Fortunately, under most conditions the rate at which this reduction takes place is so low that such titrations can be carried out; thus, oxalate and ferrocyanide can be titrated with permanganate in the presence of relatively high concentrations of hydrochloric acid. However, investigators discovered quite early that there is more rapid oxidation of chloride when ferrous salts in hydrochloric acid solutions are titrated with permanganate.[25]

Later investigations have shown that this is the result of an induced reaction, since ferric ion does not catalyze the reaction between permanganate and chloride.

One is likely to find unusual rate effects when an oxidant such as permanganate reacts with a reductant, such as ferrous iron, where there is a large

---

[24] This reaction has been studied by F. R. Duke and R. C. Pinkerton (*J. Am. Chem. Soc.*, **73**, 3045, 1951). A general discussion of rate phenomena is given by Duke in Part 1, Vol. 1 of *Treatise on Analytical Chemistry* (Kolthoff and Elving, Eds., New York: Wiley-Interscience, 1959).

[25] The first mention of the effect of chloride was made by Lowenthal and Lensson (*Z. Anal. Chem.*, **1**, 329, 1862).

difference between the electrons involved; it is kinetically unlikely that five ferrous ions will simultaneously react with a permanganate ion. In such cases the over-all reaction is likely to proceed in steps, with reactive intermediate oxidation states being formed. These reactive intermediates can be a source of trouble in that they may react rapidly with some constituent of the solution other than the reductant of interest and thus cause what is known as an induced reaction.

The reduction of permanganate could result in several intermediate oxidation states, namely Mn(VI), (V), (IV), and (III), which, while unstable, could react more rapidly with chloride than with Fe(II) to produce chlorine or hypochlorous acid. The lower the concentration of the Fe(II) and the higher the chloride concentration the more pronounced the induction is.

Alternatively, an induced reaction can occur when a reductant is oxidized to a higher and unstable oxidation state that can react rapidly with a constituent of the solution. There is some evidence that the induction of the oxidation of chloride by permanganate in the titration of Fe(II) is caused by the permanganate first oxidizing the ferrous iron to an unstable higher oxidation state, probably ferrate (FeO_4^{2-}),[26] as follows

$$MnO_4^- + Fe^{2+} = FeO_4^{2-} + Mn^{3+}$$

The ferrate then oxidizes chloride ion to hypochlorous acid; experiments have shown that the oxidized chlorine is largely present as this compound.[27] As yet the specific mechanism of this induced chloride oxidation is an unsolved problem.

It was also observed quite early[28] that this induction is decreased by the presence of a relatively high concentration of manganous salts.[29] Zimmermann[30] appears to have been the first to make practical use of this pheno-

[26] Manchot. *Ann. Chem. Pharm.*, **325**, 105 (1902); Bohnson and Robertson. *J. Am. Chem. Soc.*, **45**, 2493 (1923); Hale. *J. Phys. Chem.*, **33**, 1633 (1929). For suggestions regarding the analytical use of ferrate solutions see Schreger, Thompson, and Ockerman. *Anal. Chem.*, **22**, 691, 1426 (1950); **23**, 1312 (1951).

[27] Baxter and Frevert. *Am. Chem. J.*, **34**, 109 (1905).

[28] Kessler. *Ann. J. Chem.*, **118**, 17 (1863).

[29] The mechanism of the inhibiting effect of the manganous ion has not been clearly established. In partial explanation the suggestion has been advanced that raising the concentration of the manganous ion decreases the oxidizing potential of the permanganate–manganous ion half-cell. However, the change in potential that can be thus produced is so limited that the validity of this explanation is doubtful. It is more probable that the effect is one in which the mechanism of the reaction is changed by the presence of manganous ion. Thus, manganous ion may react rapidly with the permanganate to give an intermediate oxidation state that does not rapidly oxidize chloride or the Fe(III) to FeO_4^{2-}. Alternatively, the reactive manganese intermediate, or the ferrate (or the hypochlorous acid that it forms) may react very rapidly with manganous ion to give tripositive or quadripositive manganese, which then rapidly oxidizes ferrous iron to the ferric state.

[30] Zimmermann, *Ann. Chem.*, **213**, 305 (1882).

menon by the addition of large amounts of manganous salts to the solution to be titrated.

As stated above ferric iron imparts to hydrochloric acid solutions an intense yellow color because of the formation of complex compounds. This color would obscure the permanganate color and would necessitate the addition to such solutions of a higher concentration of permanganate in order to obtain an end point. This yellow color can be bleached by the addition of phosphoric acid, which forms more stable (but colorless) complex compounds with ferric iron;[31] sulfuric acid also forms complex compounds but these are less stable.[32]

Thus, by the addition of phosphoric acid, a colorless solution and a much sharper and more stable end point are obtained. The increased stability of the end-point color is caused by the fact that in a colorless solution the concentration of permanganate required for recognition of the end point is decreased; thus the tendency toward oxidation of chloride and of the mercurous chloride precipitate is minimized. Reinhardt[33] suggested that manganous sulfate, phosphoric acid, and sulfuric acid be combined in one reagent; this solution, still universally used, is known as the Zimmermann–Reinhardt "preventive" solution.

By the use of this "preventive" solution, by avoidance of an unnecessary excess of stannous chloride and therefore of a large amount of mercurous chloride precipitate, and by slow titration in a cold dilute solution, the method can be made to give precise results. For this reason, and because of the rapidity with which it can be carried out, this method is very extensively employed, especially for the analysis of iron ores.

Elements such as vanadium, molybdenum, tungsten, and platinum would cause errors in the procedure given below by being reduced by the stannous chloride and reoxidized by the permanganate. These elements can be assumed to be absent from the ore samples submitted for analysis.

EXPERIMENT: Titration of Fe(II) with Permanganate

(Consult your instructor regarding performing these experiments.)

---

[31] These were thought by Weinland and Ensgraber (*Z. Anorg. Allgem. Chem.*, **84**, 340, 1931) to be of the type $H_3Fe(PO_4)_2$, and by Bonner and Romeyn (*Ind. Eng. Chem.*, anal. ed., **3**, 85, 1931) to be $Fe(H_2PO_4)_3$. Later work by Langford and Kiehl (*J. Am. Chem. Soc.*, **64**, 291, 1942) and by Yamane and Davidson (unpublished experiments, California Institute of Technology, 1958) indicates that the complex cation $Fe(HPO_4)^+$ is the predominant species.

[32] These have been shown by Whiteker and Davidson (*J. Am. Chem. Soc.*, **75**, 3081, 1953) to be $FeSO_4^+$ and $Fe(SO_4)_2^-$.

[33] Reinhardt. *Chem. Ztg.*, **13**, 323 (1889). For other information on this titration see Hough. *J. Am. Chem. Soc.*, **32**, 539 (1910); Barneby. *J. Am. Chem. Soc.*, **36**, 1429 (1914).

1. Effect of *p*H on the Titration

(*a*) Pipet 10 ml of a solution 0.10 *N* (0.10 *F*) in ferrous sulfate and 0.03 *F* in H_2SO_4 (prepared as directed in Appendix 7) into a 300-ml conical flask containing 75 ml of water. Titrate the solution with 0.1 *N* $KMnO_4$ until the coloration and precipitate will make the first pink of the end point difficult to detect.

(*b*) Add to the solution 10 ml of 3 *F* H_2SO_4. Swirl until any precipitate is dissolved and the solution is essentially colorless. Continue the titration to a permanganate end point.

What two precipitates may have formed in (*a*)?

Which of these sets the lower hydrogen-ion concentration permissible in permanganate titrations?

Write equations for the reactions occurring when these precipitates dissolved upon addition of the acid.

2. Effect of Chloride on the Titration

(*a*) To 75 ml of water and 10 ml of 6 *F* HCl in a 300-ml flask add 0.1 *N* (0.02 *F*) $KMnO_4$ until a perceptible pink color is obtained (usually 0.02 ml–0.04 ml). Note the stability of the color.

(*b*) To 75 ml of water and 10 ml of 6 *F* HCl, add 2.5 ml of 1 *F* $FeCl_3$ (free of Fe(II)). Note and explain the color.

Add 0.1 *N* (0.02 *F*) $KMnO_4$ until a perceptible pink color matching as closely as possible to that in (*a*) is obtained.

Note the stability of the color.

(*c*) Pipet a 25-ml portion of the same ferrous sulfate solution used in 1(*a*) into a 300-ml conical flask containing 75 ml of water; then add 10 ml of 6 *F* HCl. Titrate with the permanganate until a color matching that originally present in (*b*) is obtained. Note any odor over the solution.

Compare the volume of $KMnO_4$ used for the titration in Experiment 1 with that required for this titration. Explain any difference in the volume required.

(*d*) Repeat (*c*) but add 10 ml of Zimmermann-Reinhardt reagent before beginning the titration.

Explain the effect of this reagent.

INSTRUCTIONS (Procedure 18-3: Determination of the Iron in an Ore. Reduction by Stannous Chloride and Titration with Permanganate)

Volumetric Titrations

Weigh out an amount of the dried ore (Note 1) that should require from 25 to 35 ml of the standard permanganate, add to it in a 150-ml beaker 20 ml

NOTES

1. Most iron ores should be dried at 100°C–105°C for 1 hour. If the ore is especially hygroscopic, take an air-dried sample for the analysis; take a separate

of 6 F HCl and 1 ml of 0.5 F $SnCl_2$, cover the beaker (Note 2), and heat the mixture almost to boiling until the ore is dissolved or only a white silicious residue remains; if the solution becomes yellow, add additional $SnCl_2$ in 0.1-ml portions until it is decolorized (Note 3). When the solution of the ore is complete, add 0.2 F $KMnO_4$ dropwise until the first yellow color is obtained (Notes 4 and 5).

Evaporate the solution to about 15 ml. While keeping it hot and swirling it, add 0.5 F $SnCl_2$ dropwise until the solution becomes colorless (Note 6); then add 1 drop excess $SnCl_2$.

Cool the solution to room temperature and add rapidly 10 ml of a saturated mercuric chloride solution (Note 7). Allow the mixture to stand 2 minutes (Note 8); then transfer it to a 600-ml beaker containing 400 ml of water and 25 ml of Zimmermann-Reinhardt solution (Note 9). Immediately titrate the mixture with standard $KMnO_4$. Do not add the $KMnO_4$ rapidly at any time;

NOTES

sample at the same time and determine the moisture by heating the ore at 105°C for 3 hours–4 hours. Correct the iron analysis for the moisture present and report it on the dry basis.

2. Vessels from which acids or other corrosive, poisonous, or unpleasant gases are being expelled should *always be placed under a hood.*

Always cover a beaker in which a solution is being evaporated or boiled (Figure 10-32, page 207). When removing the cover, rinse it; collect the rinse in a beaker.

3. The presence of undissolved iron is indicated by a dark or reddish residue that remains after the solution has been heated with an excess of stannous chloride for 15 or 20 minutes.[34] In such a case proceed as follows:

Filter the residue on a small filter and wash it with 5 ml–10 ml of 6 F HCl; collect the wash solution with the main solution. Transfer the filter to a small porcelain crucible and burn off the paper (see page 300). Mix 0.5 g–0.7 g of anhydrous Na_2CO_3 with the residue; then heat the mixture until a liquid melt is obtained. Allow the melt to cool, add 5 ml of water, and then *slowly* add 5 ml of 6 F HCl. Warm the crucible until the melt is completely disintegrated; then transfer the mixture to the main solution.

4. A large excess of stannous tin is undesirable because later it causes a large precipitate of mercurous chloride and may cause the formation of metallic mercury (see Note 7 below). Unless organic matter may be present, only sufficient $KMnO_4$ is added here to oxidize the excess stannous tin; the yellow color of the ferric iron is used as an indicator, since it will not become apparent until the tin has been oxidized to the stannic state.

[34] Some "student" or "analyzed" samples that are sold as "acid-soluble" iron ores may leave a grayish residue that usually can be neglected. It should be emphasized, however, that in the exact analysis of iron ores, any residue that is other than white and silicious should not be neglected.

stir the solution continuously, and approach the end point drop by drop. Take the end point when the first perceptible pink tinge spreads uniformly throughout the solution and persists for at least 15 seconds (Note 10). Use a white background for viewing the solution and avoid overrunning the end point. Make an end-point correction (Note 11). From the corrected volume of standard permanganate used, calculate the amount of iron present.

NOTES

5. If there is reason to suspect that the ore may contain organic matter, the addition of the $KMnO_4$ should be continued at this point until an excess is present as shown by the appearance of a distinct pink or brownish color. Troublesome organic matter is thus oxidized.

6. The solution may have a pale greenish-yellow tinge after reduction of the ferric iron. This color will not decrease with addition of $SnCl_2$ and should not be confused with the intense yellow color of ferric iron. If more than the one drop excess of $SnCl_2$ is added inadvertently, the treatment with the $KMnO_4$ and $SnCl_2$ should be repeated.

7. A white precipitate of Hg_2Cl_2 should be apparent within 30 seconds. The solution is cooled and the $HgCl_2$ solution is added all at once in order to prevent the formation of metallic mercury. If a large excess of $SnCl_2$ is present or if the solution is warm, a black precipitate of finely divided mercury is likely to result.[35] In this case the analysis must be discarded, since mercury in this form rapidly reduces both permanganate and ferric iron.

8. Allow the mixture to stand for 2 minutes in order that complete precipitation of the mercurous chloride may occur and that it may change from the finely divided, often colloidal, form in which it first separates to a more compact crystalline form. In this state it is oxidized much more slowly by ferric iron or permanganate. Lack of a precipitate indicates that an excess of stannous chloride has not been added and that a quantitative reduction of the iron has not been obtained. Do not let the solution stand much longer than the 2 minutes; otherwise, oxidation of the ferrous iron by the oxygen of the air will become appreciable. For the same reason, titrate the solution *immediately* after adding the Zimmermann-Reinhardt solution.

9. The Zimmermann-Reinhardt solution is approximately 0.3 F in $MnSO_4$, 3 F in H_3PO_4, and 2 F in H_2SO_4. See Appendix 7 for directions for its preparation.

10. The end point will slowly fade because of the oxidation of the mercurous chloride, but if the above conditions have been adhered to, it will persist for at least 15 seconds and usually considerably longer.

11. If the purity of the reagents is in question a combined correction for the end point and for reducing impurities in the reagents can be made as follows:

Take the volume of hydrochloric acid and of stannous chloride originally used, oxidize the stannous chloride with 0.2 F $KMnO_4$. Then carry out the remainder of

[35] Crowell, Luke, and Mastin (*Ind. Eng. Chem.*, anal. ed., **13**, 94, 1941) found no serious error as long as the solution was not over 30°C at this point and not over 40°C during the titration with permanganate. They recommend that an amount of ferric chloride equivalent to that present at the end point be added to the solution used for making an end-point correction.

Gravimetric Titrations

Proceed as instructed above for volumetric titrations, but use a weight buret and the standard gravimetric permanganate solution.

SUPPLEMENTARY REFERENCES

Laitinen, H. A. *Chemical Analysis*, New York: McGraw-Hill, 1960. Chapter 18 presents a discussion of the reversibility of the MnO_4^--Mn^{2+} half-cell and of some of the rate phenomena involved in permanganate methods.

Kolthoff, I. M. and Belcher, R. *Volumetric Analysis*. Vol. 3. *Oxidation-Reduction Reactions*. New York: Wiley-Interscience, 1957. Chapter 2 contains a comprehensive survey of permanganate methods.

NOTES

the operations as directed in the second and third paragraphs of the procedure above, with the following exceptions: add 2 ml–3 ml of 1 *F* $FeCl_3$ (free of $FeCl_2$) after the addition of the Zimmermann-Reinhardt solution, and then titrate with the $KMnO_4$ 1 minute later.

The correction will usually vary between 0.03 ml and 0.05 ml of 0.1 *N* $KMnO_4$.

QUESTIONS AND PROBLEMS

Stoichiometric Calculations

1. What weight of $KMnO_4$ should be taken to prepare a liter of 0.100 *N* solution? (The titrations are to be carried out in acid solutions.) *Ans.* 3.16 g.

2. Manganous ion can be oxidized to manganese dioxide by permanganate in a neutral solution, and this reaction is the basis of the so-called Volhard titration of manganese. Calculate the weight of $KMnO_4$ to be taken to prepare a liter of 0.100 *N* solution for use in the Volhard titration.

3. Certain organic compounds can be oxidized to carbon dioxide in strongly alkaline solutions by permanganate. In the presence of barium ion the manganate formed precipitates as barium manganate. Calculate the weight of $KMnO_4$ to be taken to prepare a liter of 0.100 *N* solution for such titrations (see page 401 and page 492).

4. *Standardization of permanganate solutions against ferrous ammonium sulfate.* A permanganate solution was standardized by weighing pure ferrous ammonium sulfate, $Fe(NH_4)_2(SO_4)_2 \cdot 6H_2O$, dissolving the crystals in 0.5 *F* sulfuric acid, and titrating with the permanganate solution. A 1.5684-g sample of the ferrous ammonium sulfate required 38.46 ml of the $KMnO_4$ solution.

   (*a*) Calculate the number of oxidation-reduction equivalents of ferrous sulfate taken. (Note that the equivalents of such a compound are determined by the initial and final oxidation states of the element undergoing change.)

(*b*) Calculate the normality of the above permanganate solution. (Note that the normal concentration of the $KMnO_4$ can be calculated without the necessity of writing a balanced equation for the titration reaction.) *Ans.* 0.1040 *N*.

(*c*) Predict probable effects on the titration of dissolving the ferrous salt in (*i*) 0.05 *F* H_2SO_4, (*ii*) 0.5 *F* HCl.

5. An approximately 0.1 *N* $KMnO_4$ solution is to be standarized against sodium oxalate (pages 399–400). Calculate the maximum and minimum limits of the weight of sample that should be taken. Explain.

6. A 0.2526-g sample of Bureau of Standards sodium oxalate was used in the standardization of a permanganate solution (see page 399). The titration required a total volume of the permanganate of 36.25 ml; the end-point and blank correction required 0.04 ml.
   (*a*) Calculate (*i*) the formal concentration of the $KMnO_4$; (*ii*) the normal concentration for the titration reaction. *Ans.* 0.02082 *F*.
   (*b*) Calculate the normal concentration of the solution when it is used in a Volhard titration (see Problem 2) of manganese.

7. A 0.2120-g sample of Bureau of Standards As_2O_3 was used in the standardization of a permanganate solution by Procedure 18-2. The titration required a net volume, after the end-point and blank correction, of 40.26 ml of the permanganate.
   (*a*) Write equations for the reactions taking place (*i*) when the As_2O_3 is dissolved in NaOH, (*ii*) when the NaOH solution is made acid with HCl, (*iii*) when the ICl solution is added, and (*iv*) when I_2 is oxidized by $KMnO_4$.
   (*b*) Calculate the normality of the permanganate solution (for the titration reaction). *Ans.* 0.1065 *N*.

8. A sample of iron ore weighing 0.4621 g was treated by Procedure 18-3. The final titration required a net volume of 26.87 ml of 0.1021 *N* $KMnO_4$. Calculate the percentage of iron in the ore.

9. What weights of $KMnO_4$ should be taken to prepare 1 liter of exactly 0.1 *N* solutions for the following titrations:
   (*a*) Strongly alkaline solutions containing a soluble barium salt?
   (*b*) Neutral solutions?
   (*c*) Acid solutions?

10. The following statements apply to Procedure 18-3 for the determination of iron. Write "T" if the statement is true, "F" if it is false. Wrong answers will be given a negative value.

    (*1*) T F HCl is used in dissolving iron ores because it is a reducing acid.
    (*2*) T F H_2SO_4 could not be used for dissolving the ore because it forms complex ions with ferric ion.
    (*3*) T F Perchloric acid could not be used because of its oxidizing characteristics.
    (*4*) T F The addition of too large an excess of $SnCl_2$ during the final reduction of the Fe(III) would tend to cause positive errors.

(*5*) T F Slow addition of the $HgCl_2$ to a hot solution would tend to give negative errors.

(*6*) T F The formation of a "black precipitate" on addition of the $HgCl_2$ indicates that the ratio of $SnCl_2$ to $HgCl_2$ is too low.

(*7*) T F The "black precipitate" is darkened by the same chemical species as is the "black precipitate" obtained on treating a precipitate of mercurous chloride with ammonia.

(*8*) T F Not diluting the solution either before adding the Zimmermann-Reinhardt solution or before titrating would tend to cause negative errors.

(*9*) T F Addition of the Zimmermann-Reinhardt solution results in a smaller end-point correction.

(*10*) T F Addition of phosphoric acid causes the Mn^{2+}-MnO_4^- half-cell potential at the equivalence point to be less negative (oxidizing).

(*11*) T F Allowing the solution to stand after adding the preventive solution tends to give low results.

(*12*) T F Ferric chloride is more rapidly reduced by metallic mercury than by mercurous chloride.

11. A volumetric determination of calcium can be made by precipitating the calcium as $CaC_2O_4 \cdot H_2O$, washing excess oxalate from the precipitate, dissolving it in sulfuric acid, and titrating the resulting solution with standard permanganate.

    The calcium oxalate precipitate resulting from the analysis of a 0.2250-g sample of a limestone was dissolved in sulfuric acid and the solution titrated with permanganate, 44.22 ml of 0.0982 *N* $KMnO_4$ being used for the titration. Calculate the percentage of CaO in the sample. *Ans.* 54.12%.

    NOTE: In the statement of analyses it is often conventional to report the elements as their oxides. This is advantageous, especially in silicate and carbonate rock analyses, since the sum of the basic and acid oxides, corrected by the oxygen equivalent of any nonoxygen anions, such as fluoride or sulfide, should give the total weight of the sample.

12. In the determination of the manganese in a steel by the so-called Williams method, a 2.00-g sample was dissolved in nitric acid, an excess of solid $KClO_3$ was slowly added, and the mixture was boiled until yellow fumes of ClO_2 were no longer evolved. The resulting MnO_2 precipitate was dissolved in a sulfuric acid solution into which 50.00 ml of 0.1014 *F* ferrous sulfate solution had been pipetted. The resulting solution was titrated with 0.01042 *F* $KMnO_4$, 20.27 ml being required.

    (*a*) Write an equation for (*i*) the oxidation of Mn(II) in a nitric acid solution to MnO_2 (assume the principal product to be ClO_2), (*ii*) the action of the Fe(II) solution on the MnO_2 precipitate.

    (*b*) Calculate the percentage of manganese in the steel.

13. Certain metals can be determined by being precipitated as their sulfides. The sulfide precipitate is treated with an acid solution containing an excess of a ferric salt, whereupon the following type of reaction takes place:

$$MS + 2Fe^{3+} = M^{2+} + S(s) + 2Fe^{2+}$$

When the resulting mixture is titrated with permanganate, the ferrous iron is oxidized to ferric; the sulfur does not react at a significant rate with either the permanganate or the ferric iron.

(*a*) In the analysis of a 0.4762-g sample of an aluminum solder that contained zinc, aluminum, and cadmium, the alloy was dissolved in HCl, the solution made 0.3 F in HCl and saturated with H_2S. The sulfide precipitate was treated as described above. The titration required 13.55 ml of a 0.06242 N $KMnO_4$ solution.

(*b*) The *p*H of the HCl filtrate was adjusted to 5, the solution again saturated with H_2S, and this precipitate treated similarly; 48.70 ml of a 0.2242 N $KMnO_4$ solution were required for the titration. Calculate the percentages of cadmium and of zinc in the solder.

14. *Titration of ferrocyanide.* Potassium ferrocyanide, $K_4Fe(CN)_6 \cdot 3H_2O$, because of its high equivalent weight and the ease with which it can be purified by recrystallization, has been suggested as a primary standard for permanganate solutions. The reaction taking place can be represented as follows:

$$5\,Fe(CN)_6{}^{4-} + MnO_4{}^{-} + 8H^+ = 5Fe(CN)_6{}^{3-} + Mn^{2+} + 4H_2O$$

The titration must be carried out in a relatively large volume of a solution that should be approximately 1 F in sulfuric or hydrochloric acid. Otherwise the reaction will be complicated by the formation of a precipitate of potassium manganese ferrocyanide, $K_2MnFe(CN)_6$; this compound is so insoluble that the oxidation of the precipitated ferrocyanide proceeds very slowly, leading to an uncertain end point. Since the oxidation of chloride by permanganate is not induced by the permanganate-ferrocyanide reaction, the titration can be made in hydrochloric acid solutions. Because the yellow color of the ferricyanide obscures the end point, titrations are made in large volume, and a comparison solution containing a comparable amount of ferricyanide is used.

Procedure. Four 25.00-ml portions of a ferrocyanide solution containing 40.256 g of $K_4Fe(CN)_6 \cdot 3H_2O$ per liter were pipetted into 400-ml flasks, and 200 ml of water and 40 ml of 6 N H_2SO_4 were added. The solutions were titrated with a permanganate solution until the color changed from a light greenish yellow to yellowish pink. The solution being titrated was compared with one containing an equivalent amount of potassium ferricyanide (ferrocyanide-free) in the same volume of water and acid; after the titrations were finished, permanganate was added to this solution until it matched the titrated solutions; 0.06 ml was required.

The titrations required 30.10, 30.14, 30.23, and 30.12 ml of the permanganate. Calculate and report the normality (oxidation-reduction for this titration) of the permanganate solution. *Ans.* 0.0793 N.

15. Ferrocyanide can be quantitatively oxidized to ferricyanide by permanganate (see Problem 14).

A permanganate solution was standardized against sodium oxalate, 0.2680 g of oxalate requiring 40.00 ml of the $KMnO_4$. This solution was then used in titrating a ferrocyanide solution, 25.00 ml being required to titrate a 100-ml portion of the ferrocyanide. Calculate the weight of $K_4Fe(CN)_6 \cdot 3H_2O$ per liter required to prepare the ferrocyanide solution.

16. A 0.2102-g sample of dry As_2O_3 was dissolved in 10 ml of cold 6 *N* NaOH. To this solution were added 90 ml of H_2O, 20 ml of 6 *N* HCl, and 1.00 ml of 0.0200 *F* KIO_3. The solution was titrated with $KMnO_4$ solution until the color almost spread throughout the solution. One drop of orthophenanthroline ferrous sulfate indicator was added; then the $KMnO_4$ was added until the pink color was not perceptible. A blank of the same volume, having the same amounts of NaOH, HCl, and indicator, was prepared. The titration required 41.28 ml of the $KMnO_4$ solution; the blank, 0.03 ml. What was the normality of the $KMnO_4$?

17. A 3.00-g sample of steel was dissolved in nitric acid. When $KClO_3$ was added and the solution was heated, manganese dioxide precipitated and was filtered out. The precipitate was dissolved in a sulfuric acid solution containing 25.00 ml of 0.1320 *F* ferrous sulfate, and the resulting solution was titrated with 0.01000 *F* $KMnO_4$, 22.80 ml being required.
    (*a*) Calculate the percentage of manganese in the steel. *Ans.* 1.98%.
    (*b*) Why is the Zimmermann-Reinhardt solution not used for the titration of iron in this procedure?

18. A 0.7842-g sample of $(NH_4)_2Fe(SO_4)_2 \cdot 6H_2O$ was dissolved in 100 ml of 0.5 *F* H_2SO_4 and titrated with a permanganate solution, 24.00 ml being required.

    (*a*) Write the equation representing the titration reaction.
    (*b*) Calculate the formal concentration of the $KMnO_4$ solution.
    (*c*) Solutions of (*i*) $NaHC_2O_4$, (*ii*) HBO_3 (peroxyboric acid), and (*iii*) HCOOH were titrated with the permanganate according to the following equations:

    $$(1)\ 5HC_2O_4^- + 2MnO_4^- + 11H^+ = 10CO_2 + 2Mn^{2+} + 8H_2O,$$

    $$(2)\ 3HBO_3 + 2MnO_4^- + H_2O = 3HBO_2 + 3O_2 + 2MnO_2 + 2OH^-,$$

    $$(3)\ HCOOH + 2MnO_4^- + 2Ba^{2+} + 2OH^- = CO_2 + 2BaMnO_4(s) + 2H_2O.$$

    If 25.00 ml of each of these solutions required 50.00 ml of the permanganate solution, calculate the formal concentration and the normal concentration of each solution in the reaction involved.

19. The calcium oxalate precipitate obtained from the analysis of a 0.3000-g sample of a limestone was added to 200 ml of a hot 0.5 *F* H_2SO_4 solution; then 50.00 ml of a 0.02000 *F* $KMnO_4$ solution were added. After this solution was heated to 80°C–90°C for 5 minutes, 25.00 ml of a 0.1000 *F* solution of $(NH_4)_2SO_4 \cdot FeSO_4$ were added. The resulting solution was then titrated with the 0.02000 *F* $KMnO_4$ to the first pink color; 15.00 ml were required.

    (*a*) Write balanced equations for the reactions described above.
    (*b*) Calculate the percentage of lime (expressed as CaO) in the limestone. *Ans.* 37.40%.
    (*c*) State any sources of error in the procedure.

20. The following procedure was designed for the analysis of an "acid-soluble" iron ore (iron content 10%–15%). State where this procedure might lead to error and explain your statement. Where an error is involved, state its probable magnitude and sign.
Weigh out 0.25 g–0.30 g of the air-dried ore, add to it in a 150-ml beaker 50 ml 12 *N* HCl, 5 ml 36 *N* H_2SO_4, and 1 ml 0.1 *N* $SnCl_2$, and heat until only a white residue remains. Evaporate the solution to a volume of about 25 ml, cool, add 0.1 *N* $SnCl_2$ until the yellow color disappears, and then add 10 drops 0.1 *N* $SnCl_2$ in excess. Warm the solution to 80°C–90°C and add slowly 10 ml of saturated $HgCl_2$ solution. Allow the mixture to stand 15 minutes, stirring it frequently. Add to the mixture 25 ml of 3 *F* NaH_2PO_4 and transfer it to a large beaker containing 400 ml of water to which has been added 25 ml of 0.3 *F* $MnSO_4$. Immediately and rapidly titrate the mixture with standard (0.1 *N*) $KMnO_4$ until an end point is obtained that is permanent for 3 minutes. Calculate the percentage of iron in the sample from the volume of the $KMnO_4$ used in the titration.

21. *Volhard Titration of Bipositive Manganese with Permanganate.* The following procedure was used to check the accuracy of the Volhard titration: A 25.00-ml portion of a 0.02000 *F* permanganate solution was pipetted into a flask containing an excess of concentrated HCl, and the solution was boiled, evaporated to fuming with an excess of sulfuric acid, cooled, and diluted. State the reactions taking place.
An excess of solid zinc oxide was added to the solution. Explain why this produces a buffered solution having a *p*H of approximately 6.
The mixture was heated to boiling and titrated with another portion of the original permanganate solution until a pink color was perceptible. Calculate how many milliliters would theoretically be required for the titration.

NOTE: In the above titration the precipitate that is formed tends to coprecipitate appreciable amounts of the manganous ion, either by adsorption or by formation of a manganous manganite. By providing a high concentration of a zinc (or calcium) salt, selective coprecipitation of the latter is caused, and the manganese in the precipitate more nearly approaches the quadripositive state.

Calculations Involving Chemical Equilibria

22. In a titration of a certain reducing substance, 25 ml of 0.020 *F* $KMnO_4$ were added, the permanganate being reduced quantitatively to the bipositive state. Then 0.030 ml of the permanganate was added before the pink color was observed and the end point was taken. Assuming that the final volume was 200 ml and the final hydrogen-ion concentration was 1.0 *M*, calculate the potential set up by the manganous-permanganate half-cell in the solution. *Ans.* -1.42 v.

23. If the titration in Problem 22 were carried out in a solution that was maintained 1 *M* in chloride ion, calculate what the partial pressure of chlorine above the solution would have to be in order for the chloride-chlorine half-cell to be in equilibrium with the solution. How are the results of this calculation to be reconciled with the conditions of the titration in Procedure 18-3? *Ans.* 100 atm.

24. (*a*) A 0.010 *F* solution of ferric perchlorate is shaken with an excess of solid silver until equilibrium is attained. Calculate the percentage of the iron not reduced. (*Note:* Because of a rate effect, perchlorate ion is not significantly reduced under these conditions.)
    (*b*) The process is repeated in a solution that is maintained 1.0 *F* in HCl. Repeat the calculation.

25. It was desired to determine the value of the standard potential of the following half-cell: $Fe^{2+} = Fe^{3+} + e^-$. For this determination, a study was made of the equilibrium of the following reaction: $Fe^{3+} + Ag(s) = Ag^+ + Fe^{2+}$. The $Ag = Ag^+ + e^-$ half-cell potential was known to be -0.799 v. An experiment was made in which an excess of metallic silver was added to 250 ml of a solution 0.1000 *F* in Fe^{3+} and the mixture was rotated in a thermostat at 25°C for 10 days. One hundred ml of the mixture were then pipetted out through a filter and titrated with 0.1000 *N* $KMnO_4$ solution, 62.00 ml being required.

    (*a*) Calculate the desired standard half-cell potential. *Ans.* -0.740 v.
    (*b*) State any assumptions made in the calculations involved.

26. Hydrogen sulfide is frequently used to reduce ferric iron solution preliminary to titrations with permanganate or other oxidizing agents such as dichromate or ceric sulfate. The excess of H_2S is removed by boiling or passing an inert gas such as CO_2 through the solution.

    (*a*) Calculate the equilibrium value of the ratio $[Fe^{2+}]/[Fe^{3+}]$ in a solution above which the partial pressure of the H_2S had been reduced to 10^{-6} atmospheres and in which the hydrogen-ion concentration was 2 *M*. (Assume no complex formation or hydrolysis of the ferric iron.) *Ans.* Ratio $= 1.3 \times 10^{-7}$.
    (*b*) State qualitatively the effects upon the $[Fe^{2+}]/[Fe^{3+}]$ ratio of carrying out the above reduction in a solution 1 *F* in H_2SO_4 and 1.5 *F* in Na_2SO_4.

CHAPTER **19**

Oxidation-Reduction Reactions. Cerimetric Methods

In the preceding chapter titrimetric methods involving the reduction of Mn(VII) to Mn(II) have been considered. In this chapter methods involving the reduction of Ce(IV) to Ce(III) will be presented.

Cerium is the first of a period of fourteen elements extending from atomic number 58 to 71. They are known as the rare earth metals or the lanthanide series. These elements are quite similar in their properties and differ in their electronic structure by the successive addition of fourteen electrons into the seven 4*f* orbitals; their outer electronic structures remain constant, which accounts for the similarity of their properties. These elements, their electronic structures and stable oxidation states are shown in Figure 19-1. Only the tripositive oxidation state is common to all these elements and is the most stable. Cerium, praseodymium, and terbium form quadripositive compounds, which are strong oxidants. Only Ce(IV) is stable in acid solutions; Pr(IV) and Tb(IV) compounds are so unstable that they tend to evolve oxygen in such solutions. Samarium, europium, and ytterbium form dipositive compounds, which are such strong reductants that in acid solutions they tend to evolve hydrogen.

The analytical use of solutions of ceric compounds as oxidants was first suggested over one hundred years ago,[1] but comprehensive investigations as to their application were not begun until after Martin in 1927 proposed a

[1] L. T. Lange. *J. Prakt. Chem.*, **82**, 129 (1861).

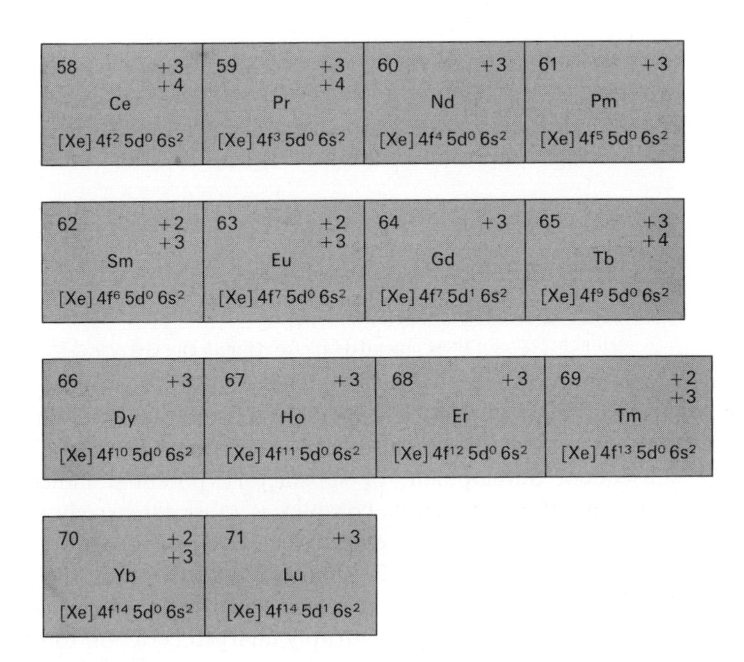

FIGURE 19-1

Electronic structures and stable oxidation states of the lanthanide series.

method for determining azide in which an excess of a standard ceric sulfate solution was used to oxidize the azide and the excess ceric sulfate determined iodometrically.[2] This work was followed in 1928 by two independent series of investigations, one by Furman and the other by Willard and their collaborators.[3]

These studies not only demonstrated the wide applicability of ceric sulfate as an oxidant but also showed that its solutions afford certain advantages over permanganate solutions. Among these advantages are the following: (1) such solutions appear to be stable indefinitely; (2) there is only one reduction product, tripositive cerium; and (3) titrations can be made in hydrochloric acid solutions without significant oxidation of chloride. While the oxidizing power of such solutions is not quite as great as that of permanganate, they can be used to replace permanganate in most cases.

Ceric sulfate solutions are yellow, but the intensity of the color is so low that visual end points based on that color are not practical for quantitative

---

[2] J. Martin. *J. Am. Chem. Soc.*, **49**, 2133 (1927).

[3] One interested in this subject should consult the series of papers by Furman and co-workers beginning *J. Am. Chem. Soc.*, **50**, 755 (1928), and extending through 1932; and by Willard and co-workers beginning in the same journal, **50**, 1322 (1928), and extending through 1934. For subsequent summaries see the Supplementary References at the end of this chapter.

TABLE 19-1

Formal Potentials of the Ce(III)–Ce(IV) Half-Cell in Certain Acids

Acid (*F*)	$HClO_4$	HNO_3	H_2SO_4	HCl
1	−1.70 v	−1.61 v	−1.44 v	−1.28 v
2	−1.71 v	−1.62 v	−1.44 v	—
4	−1.75 v	−1.61 v	−1.43 v	—
6	−1.82 v	—	—	—
8	−1.87 v	−1.56 v	−1.42 v	—

SOURCE: Data for these potentials are from Smith and Getz (*Ind. Eng. Chem.*, Anal. Ed., **10**, 191, 1938).

work; therefore, originally, resort was made to potentiometric end points. However, the necessity for potentiometric methods has been minimized by the subsequent development of satisfactory potential indicators; of these, orthophenanthroline ferrous complex is most extensively used. The characteristics of this indicator have been discussed in Procedure 18-2.

The potential of the Ce(III)-Ce(IV) half-cell, and therefore the effectiveness of ceric solutions as standard oxidants, is strongly influenced by two factors: first, the tendency of the quadruply charged ceric ion to hydrolyze in aqueous solutions, and second, the tendency of ceric ion to form complex ions. Both of these reactions are caused by the tendency of the ceric ion, with its large positive charge, to form electron-pair bonds with other ions or compounds. These two factors are discussed below and their effect on the cerous-ceric potential in certain acids of various concentrations is shown in Table 19-1.

HYDROLYSIS OF CERIC ION

Compounds of quadripositive cerium hydrolyze in water to such an extent that the standard solutions are usually made 0.2 *F* to 0.5 *F* in sulfuric acid in order to prevent precipitation of the hydrous oxide or basic salts. The standard potential of the half-cell

$$Ce^{3+} = Ce^{4+} + e^-$$

is not accurately known because of uncertainty regarding both this hydrolysis and the extent to which complex-ion formation occurs.

Measurements by Sherrill and co-workers[4] produced no evidence of complex-ion formation in perchloric acid solutions; in general, perchlorate shows the least tendency toward complex formation of the common acid

---

[4] Sherrill, King, and Spencer. *J. Am. Chem. Soc.*, **65**, 170 (1943).

anions. They did find a dependence of potential on hydrogen-ion concentration and attributed this to the hydrolysis of Ce^{4+} to $Ce(OH)^{3+}$ and $Ce(OH)_2{}^{2+}$. (No evidence of hydrolysis of Ce(III) was found when the perchloric acid was above 0.2 formal.) Subsequently, evidence for dimeric species such as $(CeOCe)^{6+}$, $(CeOCeOH)^{5+}$, and $(OHCeOCeOH)^{4+}$ has been found.[5] A tendency to polymerize upon hydrolysis has been observed with other highly charged cations such as Fe^{3+} and Cr^{3+}.

COMPLEX FORMATION OF CERIC ION

An inspection of the formal potential values shown in Table 19-1 shows a decrease in negative values—thus in oxidizing tendency—as one proceeds from perchloric to nitric, to sulfuric, and to hydrochloric acid. These values indicate that the ceric ion is forming complex ions of increasing stability in these acids; studies by various methods have shown this to be the case. Studies of the nitrate complex ions have been made by Garner.[6]

Both Ce(III) and Ce(IV) form sulfate complexes. Hardwick and Robertson[7] calculated the following values for the formation constants of certain of these:

$$Ce^{4+} + HSO_4{}^- = Ce(SO_4)^{2+} + H^+ \qquad 3{,}500$$

$$Ce(SO_4)^{2+} + HSO_4{}^- = Ce(SO_4)_2 + H^+ \qquad 200$$

$$Ce(SO_4)_2 + HSO_4{}^- = Ce(SO_4)_3{}^{2-} + H^+ \qquad 20$$

Above about 0.5 F sulfate, migration to the anode only was observed. The value of the potential in hydrochloric acid would indicate very stable chloride complexes, but doubt that the potential represents an equilibrium situation has been expressed.[8] There is slow oxidation of chloride in acid solution, but titrations of reductants in hydrochloric acid solutions with ceric sulfate can be carried out if the acid is less than about 3 F. Standard solutions of quadripositive cerium in nitric or perchloric acid are less stable than those in sulfuric acid. In spite of this relative instability, the perchloric acid solutions have been used for certain titrations in which a more oxidizing medium than the sulfate could be used to advantage.[9]

---

[5] Heidt and Smith. *J. Am. Chem. Soc.*, **70**, 2476 (1948); Hardwick and Robertson. *Can. J. Chem.*, **29**, 818, 828 (1951); Duke and Parchen. *J. Am. Chem. Soc.*, **78**, 1540 (1956).

[6] C. S. Garner. unpublished experiments; Yost, Russell, and Garner. *The Rare Earth Elements and Their Compounds.* New York: Wiley, 1947, p. 61.

[7] See Footnote 5.

[8] Wadsworth, Duke, and Goetz. *Anal. Chem.*, **29**, 1824 (1957).

[9] Smith and Duke. *Ind. Eng. Chem.*, anal. ed., **13**, 558 (1941); Smith. *Cerate Oxidimetry.* Columbus, Ohio: G. Frederick Smith Chemical Co., 1942, pp. 88–123.

Preparation of Ceric Sulfate Solutions

Outline. In this procedure instructions are given for two alternative preparations of ceric sulfate solutions. *Option A:* Primary standard $(NH_4)_2Ce(NO_3)_6$ is dried, weighed accurately, dissolved in 18 F H_2SO_4, and diluted to known volume or weight. *Option B:* $(NH_4)_4Ce(SO_4)_4 \cdot 2H_2O$ is dissolved in 1 F H_2SO_4 and is then diluted to the concentration desired. This solution is subsequently standarized for volumetric or gravimetric titrations (Procedure 19-2).

DISCUSSION

The lack of commercially available compounds of cerium suitable for the preparation of standard solutions was one factor that inhibited the rapid adoption of ceric methods of volumetric analysis. Today such compounds are available, and even solutions of 0.5 F ceric sulfate in 0.5 F–1 F sulfuric acid can be purchased.[10] Ammonium ceric nitrate, $(NH_4)_2Ce(NO_3)_6$ [more exactly: ammonium hexanitratocerate(IV)], of primary standard grade can be prepared or purchased and can be used to prepare standard solutions of known concentrations; when dissolved in sulfuric acid, stable solutions equivalent to those prepared from the sulfate are obtained and neither the ammonium nor nitrate ions interfere with subsequent titrations. For the preparation of primary standard material, see the directions of Smith and Fly.[11] Optional Procedure 19-1A below can be used to prepare a standard solution directly by exact weighing of primary standard grade $(NH_4)_2$-$Ce(NO_3)_6$. Ammonium ceric sulfate, $(NH_4)_4Ce(SO_4)_4 \cdot 2H_2O$ [more exactly: ammonium tetrasulfatocerate(IV)], is a satisfactory solid material for the preparation of solutions to be subsequently standardized. Procedure 19-1B below is provided for its use, is shorter, and is recommended unless the $(NH_4)_2Ce(NO_3)_6$ solution is desired for some specific purpose. Ceric sulfate solutions have been shown to be stable for periods of several years. Thermodynamically, there should be oxidation of water, but as with permanganate

[10] The G. Frederick Smith Chemical Company of Columbus, Ohio, has pioneered in the commercial preparation and distribution of a wide variety of these compounds.

[11] Smith and Fly. *Anal. Chem.*, **21**, 1233 (1949).

the rate appears to be extremely low at room temperatures. Unlike permanganate, the solutions are not light-sensitive.

INSTRUCTIONS (Procedure 19-1A: Preparation of a Standard Ceric Sulfate Solution from $(NH_4)_2Ce(NO_3)_6$)

I. *Solution for Volumetric Titrations.* Dry an adequate supply of the $(NH_4)_2Ce(NO_3)_6$ at 85°C for at least an hour. Weigh out exactly 54.826 g of the dried material into a 1500-ml beaker (Note 1). Add 60 ml of 18 F H_2SO_4 and stir the mixture. Do not stop stirring throughout the remainder of the procedure. After 2 or 3 minutes, slowly and *cautiously* add 50 ml of water (Note 2). After another 2 or 3 minutes, slowly add 100 ml of water; 2 or 3 minutes later, add another 100 ml of water. Repeat this process until the volume is 500 ml–600 ml and a clear orange-colored solution is obtained. Cool the solution and transfer it to a 1-liter volumetric flask. (Read the instructions for transferring solution, page 354.) When the solution has reached room temperature, dilute it to the calibration mark. Mix the solution thoroughly; then transfer it to a clean bottle with a ground-glass stopper. Label the bottle.

II. *Solution for Gravimetric Titrations.* Proceed as above for a volumetric solution, except (*a*) weigh the dry $(NH_4)_2Ce(NO_3)_6$ into a *dry, weighed* beaker and (*b*) after the final dilution, weigh the beaker and solution; weigh to an accuracy of 1 part per 1000 (Note 3).

NOTES

1. If an exactly 0.1 F solution is not required, time can be saved by taking approximately 55 g, but weighing to ± 1 mg.

2. Normally one dilutes concentrated sulfuric acid by adding it *to* cold water—this avoids danger of spattering because of the heat generated. The order of addition is reversed here in order to convert the ceric nitrate into sulfate by action of the hot sulfuric acid. Unless this is done a precipitate may form when the solution is diluted to the final volume.

3. Although weight concentrations are independent of temperature, hot solutions should not be placed on the balance pan.

INSTRUCTIONS (Procedure 19-1B: Preparation of a Standard Ceric Sulfate Solution from $(NH_4)_4Ce(SO_4)_4 \cdot 2H_2O$)

Take 500 ml of water in a 1500-ml beaker or flask. While constantly swirling or stirring the contents, slowly add 30 ml of 18 F H_2SO_4. Weigh out 63 g of

$(NH_4)_4Ce(SO_4)_4 \cdot 2H_2O$ and add it to the sulfuric acid solution; stir during this addition, and continue to stir thereafter until the solid is dissolved (Note 1).

Dilute the solution to approximately 1 liter, and transfer it to a clean, ground-glass-stoppered bottle. Label the bottle.

NOTE

1. If a turbidity is obtained, the solution should be allowed to stand as long as possible and then should be filtered through a sintered-glass or asbestos filter before use. Impure salts may contain small amounts of phosphate or fluoride, which are exceedingly troublesome because they may cause precipitation over a period of several days. If such a precipitate forms subsequent to the standardization, the solution should be filtered and restandardized.

Some supposedly reagent-grade $(NH_4)_4Ce(SO_4)_4 \cdot 2H_2O$ has been found to contain much less than the theoretical amount of ceric cerium. In order to avoid preparing solutions more dilute than desired, it is suggested that an approximate standardization be made before the dilution to the final volume.

Standardization of Ceric Sulfate Solutions Against Arsenious Oxide

Outline. In this procedure (*a*) the As_2O_3 is dissolved in NaOH as was done in Procedure 18-2, (*b*) the solution is acidified with sulfuric acid, (*c*) osmium tetroxide, OsO_4, is added as a catalyst, (*d*) orthophenanthroline ferrous sulfate is added as the indicator, and (*e*) the solution is titrated with the ceric sulfate solution.

DISCUSSION

Procedures have been developed for the standardization of ceric sulfate solutions against sodium oxalate,[12] against pure iron wire, and against arsenious oxide.[13] Since the titration of oxalate with quadripositive cerium has to be made under rather special conditions because of the slowness of the reaction, and since iron wire of the necessary purity may be difficult to obtain and to dissolve, arsenious oxide is the most practical primary standard.

The reaction between arsenious acid and quadripositive cerium is slow but can be catalyzed by either osmium or iodine compounds. The procedure below illustrates both the use of an osmium compound as a catalyst and the use of a potential indicator for obtaining the end point.

Gleu appears to have been the first to apply the catalytic action of osmium tetroxide, OsO_4, to the titration of arsenious acid by ceric salts in sulfuric acid solutions, while Willard and Young showed that the potential indicator orthophenanthroline ferrous complex could be used for obtaining the end point. The mechanism of the catalysis by the osmium compound has been investigated by Habig, Pardue, and Worthington[14] who found evidence that two catalytic cycles were involved. In one, the reduction of the Os(VIII) to Os(VI) by As(III) takes place by a two-electron step, and reoxidation of the Os(VI) occurs by two successive one-electron steps. In the second cycle the Os(VI) is further reduced by As(III) to Os(IV) by a two-electron step, followed by reoxidation of the Os(IV) to Os(VI) by successive one-electron steps.

[12] Willard and Young, *J. Am. Chem. Soc.*, **50**, 1322 (1928); **55**, 3260 (1933).
[13] Gleu. *Z. Anal. Chem.*, **95**, 305 (1933).
[14] R. L. Habig, H. L. Pardue, and J. B. Worthington. *Anal. Chem.*, **39**, 600 (1967).

There is a lack of information as to the various osmium species present in sulfuric acid and other acid solutions. OsO_4 hydrates to some extent to form the weak acid H_2OsO_5 ($K_{A_1} = 10^{-13}$); salts of the hypothetical acid H_2OsO_4 are known, and oxy ions such as OsO^+ and OsO^{2+} have been postulated; in HCl solutions complex ions such as $OsCl_6{}^{3-}$ and $OsCl_6{}^{2-}$ are formed. There is also a lack of information as to the half-cell potentials of the various oxidation states involved in the catalysis.[15]

One would expect that osmium, and also ruthenium, which is just above it in the periodic table and similar in its properties, would be able to act as homogeneous catalysts because they have such a large number of oxidation states.

The studies by Gleu and by Willard and Young (Footnotes 12 and 13) indicated that titrations by essentially the procedure given below are capable of giving an accuracy of 1 part per 1000. However, a subsequent study by Zielen[16] indicated that this accuracy was due to compensating errors that were dependent upon concentrations of catalyst and indicator, and upon the order of the titration. For this reason the procedure given below should be followed closely.

The use of orthophenanthroline ferrous complex as a potential indicator has been discussed in Procedure 18-2.

INSTRUCTIONS (Procedure 19-2: Standardization of Ceric Sulfate Solutions against Arsenious Oxide)

Weigh out samples of As_2O_3 and dissolve them in NaOH as directed in the first paragraph of Procedure 18-2 (also see Note 1, Procedure 18-2). To the NaOH solution of the As_2O_3, add 75 ml of water and 25 ml of 3 F H_2SO_4. Then add 2 drops of 0.01 F OsO_4 reagent (see Appendix 7) and 1 drop of 0.025 F orthophenanthroline ferrous sulfate.

A. *Volumetric Titrations.* Titrate the prepared solutions with the ceric sulfate solution from Procedure 19-1 until there is a pronounced local decrease in the pink color upon dropwise addition of the ceric sulfate (Note 1); thereafter add the ceric sulfate in fractions of a drop until the pink color of the indicator is no longer perceptible (Note 7, Procedure 18-2).

Obtain an end-point correction by similarly titrating a solution having the same volume as did the titrated solution and prepared with the same amounts

---

[15] Cartledge (*J. Phys. Chem.*, **60**, 1468, 1956) gives the following:

$$Os(OH)_4(s) = OsO_4(aq) + 4H^+ + 4e^- \qquad E° = -0.96 \text{ v}$$

[16] A. S. Zielen. *Anal. Chem.*, **40**, 139 (1968). He recommended that the Ce(IV) be titrated *with* the As(III) and a potentiometric end point be used; data showing an accuracy of 0.1% were presented.

of NaOH, H_2SO_4, OsO_4, and indicator (Note 2). Subtract this volume from that used in the titration. Calculate the formality of the ceric sulfate solution from the corrected volume of ceric sulfate used and the weight of As_2O_3 taken (Note 3).

B. *Gravimetric Titrations.* Proceed as above for volumetric titrations, except use a weight buret. Calculate the weight formality of the ceric sulfate solution from the corrected weight of ceric sulfate used and the weight of As_2O_3 taken (Note 3).

NOTES

1. If the ceric solution is added rapidly, a transitory green color may appear in the solution; this is caused by combination of the yellow color of the ceric salt and the blue of the oxidized form of the indicator.

2. The end-point correction should not exceed 0.05 ml and is usually 0.03 ml or less.

3. Values obtained by this method should agree to within 1 or 2 parts per 1000.

Determination of the Iron in an Ore. Reduction with Stannous Chloride and Titration with Ceric Sulfate

Outline. Iron ore is dissolved in hydrochloric acid and the Fe(III) reduced to Fe(II) with stannous chloride. The excess Sn(II) is oxidized to Sn(IV) by $HgCl_2$. The Fe(II) is titrated to Fe(III) with a standard ceric sulfate solution. Orthophenanthroline ferrous complex is used as indicator.

DISCUSSION

Iron ore is dissolved, and the iron reduced as in Procedure 18-3 (see Discussion, Procedure 18-3). Then ceric sulfate is substituted for permanganate in the titration of the ferrous iron. Since ceric sulfate does not oxidize chloride under the conditions of the titration, the Zimmermann-Reinhardt reagent (H_3PO_4, $MnSO_4$, and H_2SO_4) is omitted. However, since orthophenanthroline ferrous complex is used as a visual potential indicator, an amount of phosphoric acid equivalent to that in the Zimmermann-Reinhardt reagent is added to bleach the yellow color of the ferric chloride complexes so that it will not obscure the indicator transition at the end point.

INSTRUCTIONS (Procedure 19-3: Determination of the Iron in an Ore. Reduction with Stannous Chloride and Titration with Ceric Sulfate)

Proceed as directed in the first and second paragraphs of Procedure 18-3. Take special care to avoid adding more than 1 drop excess of the stannous chloride (Note 1).

Cool the solution to room temperature and add rapidly 10 ml of a saturated mercuric chloride solution (Note 7, Procedure 18-3). Allow the mixture

NOTES

1. A large precipitate of mercurous chloride, caused by an undue excess of stannous chloride, makes the end point less distinct, apparently because of adsorption of the indicator on the precipitate. This effect is partially eliminated by subsequent addition of more HCl than was used in the permanganate titration.

to stand 2 minutes (Note 8, Procedure 18-3); then transfer it to a 600-ml beaker containing 300 ml of water, 20 ml of 6 *F* HCl, and 10 ml of 15 *F* (85%) H_3PO_4. Add 2 drops of 0.025 *F* orthophenanthroline ferrous sulfate.

Then titrate with either the volumetric or gravimetric standard ceric sulfate until the first transient fading of the pink color of the indicator is observed; thereafter titrate progressively more slowly until the pink color of the indicator is no longer perceptible (Note 2).

Make a combined end-point and reagent-blank correction as follows: Take the volume of HCl and of $SnCl_2$ originally used for dissolving the ore; then carry out the remainder of the operations as was done in this procedure (Note 3).

From the corrected volume or weight of ceric sulfate used, calculate the percentage of iron in the ore.

NOTES

2. If a large precipitate of mercurous chloride is present, the end point may be a transition from a pink to a greenish color.

The pink color may return after the mixture has stood for several minutes because of a slow reduction of the oxidized indicator by the mercurous chloride.

3. If over 0.05 ml of ceric sulfate is required, either the end point was overrun or some of the reagents used contain reducing constituents.

APPLICATIONS OF THE USE OF STANDARD SOLUTIONS OF CERIC COMPOUNDS

Methods for the determination of arsenic, iron, oxalate, and antimony have been mentioned previously. In addition investigations have shown that ceric sulfate can be used for the determination of the following metals by means of the changes in oxidation state indicated: copper, Cu(I) to Cu(II); mercury, Hg(I) to Hg(II); molybdenum, Mo(III) or Mo(V) to Mo(VI); thallium, Tl(I) to Tl(III); titanium, Ti(III) to Ti(IV); uranium, U(IV) to U(VI); vanadium, V(IV) to V(V); tellurium, Te(IV) to Te(VI); and tin, Sn(II) to Sn(IV). Calcium and sodium are indirectly determined after precipitation as $CaC_2O_4{\cdot}2H_2O$, and $NaHg(UO_2)_3(CH_3COO)_9{\cdot}6.5H_2O$ (or the analogous zinc compound), respectively.

The use of solutions of ceric sulfate and ceric perchlorate for the oxidation of organic compounds has been extensively studied by Smith and his co-workers. Consult more comprehensive texts for specific procedures and the sources cited below for more detailed descriptions of the above applications and for references to the original investigations.

SUPPLEMENTARY REFERENCES

Kolthoff, I. M. and Belcher, R. *Volumetric Analysis.* Vol. 3. New York: Wiley-Interscience, 1957.

Laitinen, H. A. *Chemical Analysis.* New York: McGraw-Hill, 1960.

Oesper. *Newer Methods in Volumetric Analysis.* New York: Van Nostrand, 1938.

Smith, G. F. and Duke, F. R. Cerate Oxidimetry. *Ind. Eng. Chem.*, anal. ed., **13**, 558 (1941).

Smith, G. F. and Duke, F. R. Cerate and Periodate Oxidimetry. *Ind. Eng. Chem.*, anal. ed., **15**, 120 (1943).

Smith, G. F. *Cerate Oxidimetry.* Columbus, Ohio: G. Frederick Smith Chemical Co., 1942.

[See also the series of articles cited in Footnote 3.]

QUESTIONS AND PROBLEMS

1. State the effect to be expected of adding a sample of $(NH_4)_2Ce(NO_3)_6$ to a large volume of water.

2. The formal potential of the Ce(III)-Ce(IV) half-cell was measured in a solution 0.1 F in Ce $(ClO_4)_3$, 0.1 F in $Ce(ClO_4)_4$, and 6 F in $HClO_4$. State qualitatively any potential shifts to be expected and the reasons therefor when portions of the solution were treated as follows: (*a*) diluted 10-fold; (*b*) added to an equal volume of 3 F $NaHCO_3$; (*c*) added to an equal volume of 3 F Na_2SO_4; (*d*) added to an equal volume of 6 F HNO_3.

3. State the relative advantages involved in the use of standard solutions of potassium permanganate as compared with solutions of ceric sulfate as volumetric oxidizing agents.

4. The formal cerous-ceric half-cell potential is -1.61 v in 1 F HNO_3, -1.56 v in 8 F HNO_3, -1.70 v in 1 F $HClO_4$, and -1.87 v in 8 F $HClO_4$. Explain.

Stoichiometric Calculations

5. What weights of (*a*) $(NH_4)_4Ce(SO_4)_4{\cdot}2H_2O$ and (*b*) $(NH_4)_2Ce(NO_3)_6$ should be taken in order to prepare one liter of solutions that will be exactly 0.1 N with respect to (*a*) the half-cell reaction $Ce^{III} = Ce^{IV} + e^-$, and (*b*) the displacement reaction $NH_4^+ + OH^- = NH_3 + H_2O$? *Ans.* (*a*) 63.259 g.

6. A ceric sulfate solution was standardized against arsenious oxide. A 0.2024-g portion of As_2O_3 required a net volume of 34.26 ml of the ceric sulfate. Calculate the normality of the ceric sulfate solution.

7. *Standardization of a ceric sulfate solution against ethylenediamine ferrous sulfate.* The compound ethylenediamine ferrous sulfate tetrahydrate, $(CH_2NH_3)_2SO_4{\cdot}FeSO_4{\cdot}4H_2O$, has been proposed as an oxidimetric primary standard [Oesper.

J. Chem. Ed., **24**, 235 (1947)]. The ferrous iron is quantitatively oxidized to the ferric state by such standard reagents as ceric sulfate and permanganate.

In the standardization of a ceric sulfate solution a 1.0246-g portion of the ferrous compound was taken and transferred to a 400-ml beaker, then dissolved in 150 ml of water to which had been added 30 ml of 3 *F* H_2SO_4. Two drops of 0.025 *F* orthophenanthroline ferrous sulfate were added and the solution was titrated with the ceric sulfate solution until there was a pronounced local decrease in the pink color upon dropwise addition of the ceric sulfate solution. Thereafter the ceric sulfate was added in fractions of a drop until the pink color of the indicator was no longer perceptible; 32.46 ml of the ceric solution were required.

An end-point correction was made by similarly titrating a solution having the same volume and the same amount of sulfuric acid; 0.04 ml of the ceric solution was required.

Calculate the normality of the ceric solution for this titration. *Ans.* 0.0946 *N*.

8. A 0.3824-g sample of impure arsenic trichloride was weighed into a 500-ml flask containing 10 ml of 6 *F* NaOH. The solution was then treated by Procedure 19-2, beginning with the second sentence of the first paragraph (page 440). The titration with ceric sulfate required 40.22 ml of 0.1003 *N* ceric sulfate.

   (*a*) Calculate the percentage by weight of $AsCl_3$ in the sample. *Ans.* 95.6%.
   (*b*) What products are formed when $AsCl_3$ is added to an excess of NaOH solution?
   (*c*) Give an explanation of the catalytic action of the OsO_4 in the procedure.
   (*d*) Under what conditions could ICl be used as a catalyst (see Procedure 18-2)? Explain.

9. *Standardization of Ceric Sulfate Solutions Against Pure Iron.* In order to standardize a ceric sulfate solution, a 0.2042-g sample of electrolytic iron was transferred to a 400-ml beaker containing 20 ml of 9 *F* HCl. The solution was heated until the iron was dissolved; then 0.5 *F* $SnCl_2$ reagent was added dropwise until 1 drop caused the color of the solution to change from yellowish to a faint green. One additional drop of $SnCl_2$ was then added, followed by an excess of $HgCl_2$ (see Procedure 18-4), and the solution was titrated. The titration with the ceric sulfate solution required 30.64 ml; the end-point correction, 0.03 ml.

   Calculate the normality of the ceric sulfate solution.

10. A 0.3264-g sample of a stibnite ore (crude Sb_2S_3) was dissolved in HCl, the H_2S removed by boiling, and the solution titrated with standard Ce(IV) solution. The titration reactions are similar to those with arsenic (Procedure 19-2). (Note the relative positions of antimony and arsenic in the periodic table.) The subsequent titration required a net volume of 35.42 ml of 0.09872 *N* ceric sulfate.

    Calculate the percentage by weight of antimony in the sample. *Ans.* 65.23%.

    Write the equation for the reaction involved when Sb_2S_3 is dissolved in concentrated hydrochloric acid.

Calculations Involving Chemical Equilibria

11. Although there is some doubt as to the exact species present, the half-cell reaction involving osmium in the +4 and +8 oxidation states can be formulated as $OsO^{2+} + 4H_2O = H_2OsO_5 + 6H^+ + 4e^-$, and the formal potential has been calculated to have a value of approximately −1.0 v.

    On this basis calculate the ratio of Os(VIII)/Os(IV) that would be present under equilibrium conditions in a solution in which the hydrogen-ion concentration was 1.0 *M* and in which the ratio of H_3AsO_4 to H_3AsO_3 was 1000.

    Repeat the calculation of the Os(VIII)/Os(IV) ratio for a solution that is 1.0 *M* in sulfuric acid and in which the ratio of Ce(III)/Ce(IV) is 1000.

    State what predictions can be made from these calculations, and state the assumptions involved.

12. A ferrous solution is titrated with ceric sulfate in a solution that is kept 1.0 *F* in sulfuric acid. Calculate the potential of (*a*) the ferrous-ferric half-cell when 99.8% of the iron is oxidized and (*b*) the cerous-ceric half-cell when a 0.2% excess of the ceric sulfate has been added.

    If the potential indicator orthophenanthroline ferrous complex is present in very small quantity in the solution, calculate the equilibrium ratio $In_{ox.}/In_{red.}$ in each of the above cases. *Ans.* (*a*) −0.84 v, (*b*) −1.27 v.

    NOTE: Usually a 10% transformation of an indicator will cause a perceptible color change.

CHAPTER 20

Oxidation-Reduction Reactions. Iodometric Methods (I)

In Chapters 18 and 19 we discussed the use of manganese(VII) and cerium(IV) as oxidants. In this chapter we will consider the general properties and reactions of iodine and its compounds of analytical interest with emphasis on the use of the half-cell reaction $3I^- = I_3^- + 2e^-$. Chapter 21 will present typical methods illustrating the use of this half-cell. Iodine solutions can be used for the titration of reductants, whereas iodide can be used for the reduction and determination of oxidants. In addition there are methods involving iodine compounds in which it has oxidation numbers $+1$, $+5$, and $+7$. Because of the use of this large number of oxidation states and the relative ease with which iodine in several of these oxidation states can be made to react with other oxidants or reductants, iodine and its compounds find widespread use. In addition to the usefulness of iodometric methods, iodine exhibits in its compounds and reactions certain unusual features that make it of special interest to chemists.

In this and the following chapter we will consider properties of the iodide-iodine half-cell and its use in analytical methods. In Chapter 22 the analytical applications of iodine in certain other oxidation states, as well as of chlorine and bromine are discussed.

THE TRI-IODIDE ION

The first of these unusual features involves the solubility of iodine in water and the formation of the tri-iodide ion. An aqueous solution saturated with iodine at a temperature of 25°C is only 1.34×10^{-3} M in $I_2(aq)$, which indicates that aqueous iodine solutions are too dilute for most analytical purposes; for convenience standard solutions are usually prepared to be approximately 0.1 N. In addition, the partial pressure of iodine above a saturated solution would be the same as that of solid iodine at the same temperature, which is approximately 0.4 mm at 25°C; this value means that there would be danger of significant loss of iodine in handling such solutions. (Note the odor of iodine above a bottle containing the solid.)

Fortunately, iodine solutions of convenient formal concentrations can be prepared if advantage is taken of the increased solubility of iodine in iodide solutions that results from the tendency of iodine to react with iodide to form the tri-iodide ion:

$$I_2(aq) + I^- = I_3^- \tag{1}$$

The equilibrium for this reaction lies well to the right as indicated by the formation constant for I_3^-,

$$K_f = \frac{[I_3^-]}{[I_2(aq)][I^-]} = 768 \text{ at } 25°\text{C}^1 \tag{2}$$

Hence, while the concentration of $I_2(aq)$ is limited to the low equilibrium molar solubility (1.34×10^{-3} M), the *formal* solubility ($[I_2(aq)] + [I_3^-]$) in iodide solutions can be made relatively high. Equation 2 indicates that in a saturated solution with the equilibrium iodide concentration at 0.10 M, the molar I_3^- concentration would be about 77 times greater than that of the I_2. When used in titrations such solutions react towards reducing agents as though they were iodine solutions; either the equilibrium is rapidly established or both iodine and tri-iodide react rapidly with reducing agents.

Standard "iodine" solutions are, therefore, prepared by dissolving iodine crystals in iodide solutions and are predominately tri-iodide solutions.[2] These solutions are found experimentally to be quite stable when prepared to be about 0.05 F in iodine and 0.15 F in potassium iodide, and if several precautions are observed in storage and use. Such solutions must be protected from light because of the light-induced oxygen error (discussed below), must be kept stoppered as much as possible because of the still significant partial pressure of iodine above the solutions, and must be protected from dust and reducing gases.

---

[1] M. Davis and E. Gwynne. *J. Am. Chem. Soc.*, **74**, 2748 (1952).

[2] There is evidence of the formation of ions such as I_5^- and I_7^- in very concentrated iodide-iodine solutions (Grace. *J. Chem. Soc.*, **1931**, 594; Briggs, Clack, Ballard, and Sassaman. *J. Phys. Chem.*, **44**, 350, 1940).

It is of interest to note that iodine is also more soluble in aqueous bromide and chloride solutions because of the formation of analogous ions such as BrI_2^- and ClI_2^-; these ions are less stable than is tri-iodide, the formation constants being 12 and 3.5 respectively.

THE STANDARD POTENTIALS OF THE IODIDE-IODINE HALF-CELLS AND THE REVERSIBILITY OF CERTAIN TITRATION REACTIONS

The iodide-iodine half-cell can be formulated in three ways,

$$2I^- = I_2(s) + 2e^- \qquad E° = -0.536 \text{ v}$$

$$3I^- = I_3^- + 2e^- \qquad E° = -0.54 \text{ v}$$

$$2I^- = I_2(aq) + 2e^- \qquad E° = -0.62 \text{ v}$$

The first value is applicable to a solution in which the iodide concentration is 1 molar (more exactly, the iodide has an activity of one) and that is saturated with solid iodine; the second value is applicable to a solution 1 *M* in iodide and 1 *M* in tri-iodide ion. The two potentials are essentially the same because a solution 1 *M* in both iodide and tri-iodide is essentially a saturated solution and the molar concentration of the $I_2(aq)$ in both solutions is essentially that of a saturated solution, 1.34×10^{-3} *M*; that this is true is just a remarkable coincidence. The third value is applicable to a hypothetical half-cell that is 1 *M* in iodide and also 1 *M* in iodine ($I_2(aq)$); this latter concentration is over 700 times greater than that of a saturated solution and not attainable experimentally.

An inspection of the table of half-cell potentials given in Appendix 4 will show that a value of −0.54 v is intermediate between those of half-cells in which the reductant can be classified as a strong reducing agent (for example, stannous tin) or where the oxidant is a strong oxidizing agent (for example, permanganate). This indicates that many of the strong oxidants can oxidize iodide quantitatively, and many of the strong reductants can reduce iodine quantitatively. The second half-cell reaction shown above can, therefore, serve as the basis for a very large number of titrations. The determination of reducing agents involves a titration with a standard iodine solution. The determination of oxidizing agents usually *does not*, as one would expect, involve a titration with a standard iodide solution. Such a titration is possible by using a potentiometric end point; otherwise the color of the iodine produced interferes with the detection of the end point by visual indicators. Largely for this reason, the determination is usually made by an indirect method that involves the addition of an excess of a soluble iodide to a solution of the oxidizing agent with the result that an equivalent amount of iodine is produced; this iodine is then titrated with a standard solution of a reducing agent, usually thiosulfate, which quantitatively reduces the iodine to iodide.[3] Because of the intermediate value of the iodide-iodine half-cell, iodine is not a powerful oxidant nor is iodide a powerful reductant; consequently there are many cases where attempts to titrate a reductant with iodine or to reduce an oxidant with iodide will result in the reaction reaching an equilibrium before the titration proceeds quantitatively to within the desired limits. The possibility of controlling titration conditions so as to cause such reactions to proceed quantitatively in the desired direction presents an interesting challenge to analytical chemists. Examples of such reactions that can be caused to proceed quantitatively in either direction are discussed below.

THE REVERSIBILITY OF IODOMETRIC REACTIONS

The potentials of the iodide-iodine half-cells are independent of the hydrogen-ion concentration, while the potentials of half-cell reactions involving compounds that contain oxygen (and a majority of strong oxidants do contain oxygen) are in general highly dependent upon the hydrogen-ion concentration. Examples of such half-cell reactions are

$$H_3AsO_3 + H_2O = H_3AsO_4 + 2H^+ + 2e^- \qquad E^\circ = -0.56 \text{ v}$$

$$HSbOC_4H_4O_6 + H_2O = HSbO_2C_4H_4O_6 + 2H^+ + 2e^- \qquad E^\circ = -0.7 \text{ v}$$

$$VO^{2+} + 3H_2O = V(OH)_4{}^+ + 2H^+ + e^- \qquad E^\circ = -1.00 \text{ v}$$

$$2Cr^{3+} + 7H_2O = Cr_2O_7{}^{2-} + 14H^+ + 6e^- \qquad E^\circ = -1.3 \text{ v}$$

In solutions having pH values of about 7 or higher, any of the reductants on the left of the above half-cell reactions can be oxidized quantitatively by iodine; however, in very acidic solutions any of the oxidants on the right can be quantitatively reduced by an excess of iodide ion.

THE REACTION OF ARSENIOUS ACID WITH IODINE

This reaction is of special interest for several reasons. It is of historical interest because it was one of the first analytical procedures for which

[3] Methods of the first type, involving the use of a standard solution of iodine, are sometimes designated as "iodimetric" methods, and those of the second type—where iodine is liberated and titrated with thiosulfate—as "iodometric" methods; this convention will not be used in this book.

theoretical calculations were used to establish the conditions—in this case, especially the upper and lower hydrogen-ion concentrations—within which highly accurate titrations could be made.[4] Secondly, because of its accuracy, the titration is still used for the standardization of iodine solutions against primary standard arsenious oxide, As_2O_3.[5] Finally, by raising the hydrogen-ion concentration sufficiently, the reaction can be quantitatively reversed and used for the determination of pentapositive arsenic.

A summary of the assumptions and calculations involved in predicting the *p*H limits for the titration of tripositive arsenic with iodine is given below.

TITRATION OF ARSENIOUS ACID WITH IODINE

The equilibrium constant for the reaction

$$H_3AsO_3 + I_3^- + H_2O = H_3AsO_4 + 3I^- + 2H^+ \tag{3}$$

has been found experimentally[6] to have the value 5.5×10^{-2}. Consequently, as mentioned previously, this reaction can be forced in either direction by controlling the hydrogen-ion concentration. When an aqueous arsenious acid solution is titrated with iodine, the reaction is incomplete if the acid formed by the reaction is allowed to accumulate in the solution; therefore, some means must be provided to prevent this increase in the hydrogen-ion concentration. However, if the solution is made alkaline, iodine reacts with hydroxide ion

$$I_3^- + OH^- = 2I^- + HIO \tag{4}$$

causing an excess of iodine to be consumed before the starch-iodine end point

---

[4] E. W. Washburn. *J. Am. Chem. Soc.*, **30**, 31 (1908). This report was entitled "The Theory and Practice of the Iodometric Determination of Arsenious Acid" and is of interest not only for the calculations but because of the detailed account of the experimental details taken in demonstrating the highly precise results attainable. Washburn and Bates (*J. Am. Chem. Soc.*, **34**, 1341, 1912) subsequently used the titration in a study of the use of the iodine coulometer in establishing the value of the faraday. Friedrich Mohr (1806–1879), who developed the chloride titration, appears to have been the first to titrate arsenious acid with iodine. He noted that the reaction was reversible in acid solutions and he first used sodium carbonate, then later the "bicarbonate," to control the acid concentration.

[5] The formula of arsenious oxide is usually given as As_2O_3 with a formula weight of 197.84. Structural studies have shown that the compound exists as As_4O_6 molecules (molecular weight 395.68); the more descriptive name is *diarsenic trioxide*. The "trioxides" of phosphorus and antimony are dimerized also.

[6] Washburn and Strachan. *J. Am. Chem. Soc.*, **35**, 681 (1913). This constant can be calculated from the values given in Appendix 4 for the standard potential of the I^--I_3^- half-cell and the formal potential for the H_3AsO_3-H_3AsO_4 half-cell. The calculated value is in reasonable agreement with the earlier experimental constant.

is obtained. Therefore, the maximum and minimum hydrogen-ion concentrations under which the reaction can be made the basis of a precise iodometric method are limited; an approximate calculation of these limits can be made as follows:

In order to calculate the upper limit for the hydrogen-ion concentration, let it be assumed (1) that the final volume will be 200 ml, (2) that 5 milliequivalents of arsenious acid are to be titrated, (3) that the standard iodine solution is 0.05 *F* in iodine (I_2) and 0.15 *F* in potassium iodide, and (4) that an accuracy of 0.001% is desired; this small value is arbitrarily taken to ensure a reasonable factor of safety.

Since 50 ml of the iodine solution will be used, the final iodide concentration will be 0.0625 *M*; to give the accuracy desired, the ratio of the concentration of H_3AsO_3 to H_3AsO_4[7] will have to be 10^{-5}, and, if the minimum detectable starch-iodine color is used as the end point, the concentration of the tri-iodide ion may be as low as 2×10^{-7} *M*. (This value was found by Washburn experimentally.) By substitution of these values in the equilibrium expression for Equation 3 the maximum hydrogen-ion concentration is found to be approximately 2×10^{-5} *M*.[8]

The factors limiting the lowest permissible hydrogen-ion concentration are the reaction of tri-iodide with hydroxide

$$I_3^- + OH^- = 2I^- + HIO \qquad K = 1.3 \times 10^2 \tag{5}$$

and the fact that HIO is unstable and tends to decompose in alkaline solutions into iodate and iodide as follows:

$$3HIO + 3OH^- = 2I^- + IO_3^- + 3H_2O \qquad K = 1.2 \times 10^{13} \tag{6}$$

The two above equations can be combined to give an equation which represents the hydrolysis or reaction with hydroxide of iodine or tri-iodide, to

[7] More exactly, the ratio of H_3AsO_3 to $H_3AsO_4 + H_3AsO_3$. For this approximate calculation the assumption is made that H_3AsO_4, rather than $H_2AsO_4^-$, will be a predominant species under the final conditions; as noted later, this is not the case but the final *p*H limit is not changed greatly.

[8] The first ionization constant of H_3AsO_4 is 5.6×10^{-3}, therefore in a solution having this maximum hydrogen-ion concentration, the arsenic acid will exist primarily as $H_2AsO_4^-$. A recalculation on this basis gives the maximum hydrogen-ion concentration to be approximately 3×10^{-4} *M*. Park (*Ind. Eng. Chem.*, anal. ed., **3**, 77, 1931) obtained a similar value for the maximum $[H^+]$ that would not cause significant reduction of As(V) by iodide. He was interested in the iodometric determination of copper in ores by the reaction $2Cu^{2+} + 5I^- = 2CuI(s) + I_3^-$. Such ores often contain arsenic, which would be present in the solution being analyzed as H_3AsO_4, and its reduction to give I_3^- would cause errors in the copper determination.

form iodate and iodide,[9] as follows

$$3I_3^- + 6OH^- = 8I^- + IO_3^- + 3H_2O \qquad K = 2.6 \times 10^{19} \tag{7}$$

This equation can be rewritten in terms of the hydrogen-ion concentration as follows:

$$3I_3^- + 3H_2O = 8I^- + IO_3^- + 6H^+ \tag{8}$$

The equilibrium expression for this reaction is

$$\frac{[I^-]^8[IO_3^-][H^+]^6}{[I_3^-]^3} = K$$

and the value of K is calculated to be 2.6×10^{-65}.

For calculation of the minimum hydrogen-ion concentration, it can be assumed that the final iodide and tri-iodide ion concentrations are essentially the same as in the above calculation, and since, in order to obtain the accuracy assumed above, not over 0.001% of the iodine used can be converted into IO_3^-, the final concentration of this substance should not exceed 4.2×10^{-8} M.[10] When these values are inserted in the equilibrium expression given above, the minimum permissible *hydrogen*-ion concentration is calculated to be approximately 4×10^{-12} M.[11] Therefore, from an *equilibrium* point of view it is predicted that if the hydrogen-ion concentration is controlled between the limits 2×10^{-5} and 5×10^{-12} M, the reaction could be made the basis of an accurate volumetric method.

EFFECT OF REACTION RATES ON THE TITRATION

The rates of the various reactions involved in the titration have not been considered in these calculations. Experimentally it has been found that it is

---

[9] Reaction 5 results from the tendency of all halogens to react with water and disproportionate; the type reaction is $X_2 + H_2O = HXO + X^- + H^+$. The equilibrium constant for the reaction with Cl_2 is 5×10^{-4}, for Br_2, 5×10^{-9}, and for I_2, 3×10^{-13}. Note that the reaction is more complete the lower the atomic number; F_2 oxidizes water so rapidly that the hydrolysis equilibrium cannot be measured. Reaction 6 results from the tendency of the unipositive halogen acids to disproportionate, the type reaction being $3HXO = HXO_3 + 2I^- + 2H^+$.

[10] Five meq of arsenious acid would require 5 meq of iodine or 2.5 mmoles of iodine. Therefore only 2.5×10^{-5} mmole of iodine should be permitted to hydrolyze (or react with hydroxide). Since in this hydrolysis three I_2 give one IO_3^- only 8.3×10^{-6} mmole of IO_3^- should be allowed to form. Since the final volume is assumed to be 200 ml, the maximum permissible molar concentration of iodate is 4.2×10^{-8}. The HIO and IO^- that would be present under equilibrium conditions have been neglected for these approximate calculations.

[11] McAlpine (*J. Chem. Ed.*, **26**, 362, 1949) reviews these calculations and presents an experimental discussion of the lower limits of the permissible hydrogen-ion concentration.

better to limit the maximum hydrogen-ion concentration to approximately 10^{-7} M, since in more acid solutions the rate at which the iodine is reduced by the arsenious acid becomes so low as to make the titration quite tedious. On the other hand, if titrations are made in solutions as alkaline as that calculated for the lower limit, there is a tendency for any local accumulation of iodine to hydrolyze, or to react with hydroxide ion, to form iodate and iodide. Since the iodate thus formed is extremely slow in reacting with either arsenious acid or iodide, the titration is usually made in solutions that are maintained between the *p*H limits of 7 and 9.

Some means must be provided by which the hydrogen-ion concentration of the solution can be controlled within this favorable range. It is obvious that these concentrations are so small that this control cannot be achieved by the manual addition of a solution of a base or acid as the titration reaction progresses. Recourse is therefore made to what is known as a chemical *buffer system* for accomplishing this control; such systems are discussed below.

BUFFER SYSTEMS

As was shown in the discussion of buffer systems in Chapter 7, in any solution containing an incompletely ionized acid and its salt the hydrogen-ion concentration is determined by the ionization constant of the acid and the ratio of the molar concentration of the un-ionized acid to that of the salt, thus:

$$[H^+] = K\frac{[HA]}{[A^-]}$$

Therefore, first, if an acid with a constant having a value close to that of the desired hydrogen-ion concentration is selected and, second, if relatively large amounts of both the acid and the salt are provided, the hydrogen-ion concentration of the solution will be fixed and, furthermore, will be little changed by addition or formation of either acid or base. Such a solution is said to be *buffered*. It is desirable that the constant of the acid be approximately equal to the desired hydrogen-ion concentration; this value permits a ratio of acid to salt of near unity and provides the most effective buffering action against either acid or base for a given amount of the buffering material.[12]

For the present case, acids will have to be selected for the buffer system that will not be oxidized by iodine or reduced by iodide and that have constants with values near that of the desired hydrogen-ion concentration

---

[12] See Park (*J. Chem. Ed.*, **19**, 171, 1942) for a simple mathematical demonstration of this statement.

(approximately 10^{-7}). Such acids are dihydrogen phosphoric acid ($H_2PO_4^-$, $K_A = 2 \times 10^{-7}$) and carbonic acid (H_2CO_3, $K_A = 3 \times 10^{-7}$), and these acids and their salts were experimentally found to be satisfactory by Washburn. If sodium hydrogen carbonate ($NaHCO_3$) is added to an acid solution, carbonic acid is set free; and since the solubility of carbon dioxide in water has a fixed value at a given temperature and at a given pressure of the carbon dioxide above the solution, the concentration of the carbonic acid (H_2CO_3) in a solution saturated with carbon dioxide at atmospheric pressure is fixed. Thus by neutralizing an acid solution with a soluble hydrogen carbonate (usually $NaHCO_3$) and then adding a definite amount in excess one obtains a solution buffered at a value determined by the excess of the hydrogen carbonate added.

The molar concentration of carbonic acid (H_2CO_3) in a solution saturated with carbon dioxide at atmospheric pressure and at 20°C can be taken to be 3.4×10^{-2}; see page 103 regarding the use of H_2CO_3 to represent $[CO_2 + H_2CO_3]$. Therefore, if a sufficient excess of hydrogen carbonate is added to make its molar concentration about 0.34, the hydrogen-ion concentration of the solution would be 3×10^{-8} *M*. During the titration of 5 meq of arsenious acid with iodine, 5 meq of hydrogen ion are produced;[13] therefore, if the solution volume is 200 ml, this acid would cause the hydrogen carbonate concentration to decrease to 0.315 *M* (because of the reaction $H^+ + HCO_3^- = H_2CO_3$), and the hydrogen-ion concentration would be increased proportionally to 3.2×10^{-8} *M*, approximately a 7.5% change. If the same amount of acid is added to 200 ml of pure water, the hydrogen-ion concentration would change from 10^{-7} to 2.5×10^{-2}, a 200,000-fold change.

In addition, we should note that if carbon dioxide were continuously passed through the hydrogen carbonate solution, it would also be buffered against a decrease in the hydrogen-ion concentration. This is because the addition of 5 meq of hydroxide ion would result merely in the hydrogen-carbonate-ion concentration increasing to 0.365 *M*, and proportionally decreasing the hydrogen-ion concentration to 2.8×10^{-8}—again only a 7.5% change as compared with one of 200,000-fold if the hydroxide ion were added to the same volume of pure water.

It is obvious that in any solution containing an incompletely ionized base and its salt, similar considerations apply, thus:

$$[OH^-] = K_B \frac{[BOH]}{[B^+]}$$

---

[13] For this approximate calculation, the partial ionization of H_3AsO_4 to $HAsO_4^{2-}$ is neglected. If this ionization is assumed to be complete, the hydrogen carbonate concentration would be lowered to only 0.29 *M* and the resulting $[H^+]$ would be 3.5×10^{-8}.

Therefore, when it is desired to buffer a solution to a basic range, a base with the proper ionization constant and its salt can be used effectively.

Buffer solutions have again been discussed at some length here because they are of great importance in analytical, commercial, and biological processes. Their application should be thoroughly understood. The principles relating to them may be summarized from the practical side by recalling that in order to buffer a solution effectively, two important considerations should be observed. First, an acid (or base) should be selected whose constant is of the same order of magnitude as that of the hydrogen-ion (or hydroxide-ion) concentration at which it is desired to maintain the solution, since for a given total amount of acid and salt the buffering is most effective when the ratio of acid to salt is unity. Secondly, the amount of the buffer material provided must be large in comparison to the amount of hydrogen ion or hydroxide ion which may be formed.

EXPERIMENT: Reversibility of the Iodine-Arsenious Acid Reaction

Have available a standard 0.050 *F* (0.10 *N*) arsenious acid solution and a standard 0.050 *F* (0.10 *N*) iodine solution. See Appendix 7 for the preparation of these solutions. The arsenious acid solution was prepared by dissolving the As(III) oxide in NaOH, adding phenolphthalein, then neutralizing with HCl until the indicator became colorless. (If these experiments are to be done as a class demonstration, use 5 times the quantities specified.)

(*a*) Pipet 5 ml of the 0.050 *F* H_3AsO_3 into a 200-ml conical flask and add 45 ml of water. Calculate the volume of the standard iodine solution required for the titration. Add the iodine solution in 0.5-ml portions (do not add starch) to the flask until an iodine color is obtained which persists for 30 seconds–60 seconds. Reserve the solution for use in (*c*) below.

Note the rate effect as shown by the increasing time required for bleaching of the iodine color.

The initial *p*H of the As(III) solution has been adjusted to approximately 8. Which of the following species will be predominant: H_3AsO_3, $H_2AsO_3^-$, $HAsO_3^{2-}$, AsO_3^{3-}?

Calculate the approximate *p*H of the titrated solution when 50% of the As(III) has been oxidized. What will be the major As(V) species? Predict the observable effects of adding to the titrated solution a quantity of iodine equivalent to the As(III) present.

(*b*) Take 30 ml of 1.5 *F* Na_2CO_3 and, using a dropper, add dropwise about 0.25 ml of the standard iodine solution. Note and explain the effect. Write equations for possible reactions taking place. Predict the effect of adding 10 ml of 6 *F* HCl to the solution.

(*c*) To the partially titrated solution from (*a*) add 1 g of $NaHCO_3$. Make an approximate calculation of the *p*H. Predict the result of adding additional iodine solution.

Again add 0.5 ml–1 ml of the iodine solution. Note and explain the effect. Add 1 ml of a "starch solution." Complete the titration. Compare the calculated and found end-point volumes.

(*d*) Add to the titrated solution from (*c*) 5 g of KI, swirl until dissolved, and then add (slowly until violent effervescence ceases) 10 ml of 12 *F* HCl. Explain the effect. Discuss the implications of this effect as regards the possibility of making an iodometric determination of As(V).[14]

Other Reversible Iodometric Reactions

The reaction of iodide with ferric iron according to the equation

$$2Fe^{3+} + 3I^- = 2Fe^{2+} + I_3^-$$

represents a reversible reaction where the oxidant is not an oxygen anion and where the hydrogen-ion concentration is apparently not involved. However, if one attempts the titration of the ferric salt without addition of acid, an equilibrium is reached and the reaction is incomplete because of hydrolysis of the ferric ion. Addition of acid minimizes the hydrolysis and an accurate determination can be made.

One can take advantage of the reversibility of the reaction and titrate Fe(II) with iodine if the $[Fe^{3+}]$ is kept low by the addition of a complexing agent, such as fluoride or pyrophosphate.

INDICATORS FOR IODOMETRIC TITRATIONS

The use of potential indicators for redox titrations has been discussed in Chapters 18 and 19. These indicators are rarely used in iodometric titrations because several methods that depend upon certain specific properties of iodine are available for determining the end point. First, although the color of the iodine molecule, I_2, in aqueous solutions is not of sufficient intensity for it to serve as a sensitive indicator, the color of the tri-iodide ion can be detected in colorless solutions, with good lighting, when its concentration is from 2×10^{-5} to 5×10^{-6} *M*. A sensitivity of 2×10^{-5} *M* is equivalent to about 0.04 ml of a 0.1 *N* iodine solution, which is adequate for many titrations.

---

[14] A discussion of this method is given by Kolthoff and Belcher (*Volumetric Analysis*. Vol. 3. New York: Wiley-Interscience, 1957, p. 316).

Second, the nonpolar iodine molecule is much more soluble in certain organic solvents (carbon tetrachloride, CCl_4; chloroform, CH_3Cl; benzene, C_6H_6) than in water and in such solvents has an intense purple-to-violet color. Therefore, a very sensitive, if not so convenient, end point can be obtained by shaking a small amount of one of these solvents with the solution being titrated and observing when the color appears or disappears in the organic solvent. As would be expected, the sensitivity of this method is decreased by the presence of iodide in the aqueous phase; only iodine is extracted, and the formation of the tri-iodide ion decreases its concentration. Thus, in experiments in which 200 ml of an aqueous phase were shaken with 10 ml of carbon tetrachloride, a perceptible color was detected when the aqueous phase was initially from 2 to 2.5×10^{-6} F in I_2 and no KI was present; when the aqueous phase was 3×10^{-2} F in KI, a color was detected in the carbon tetrachloride when the I_2 in the aqueous phase was initially from 3 to 5×10^{-6} F; when the KI was 0.15 F, the sensitivity of detection of the I_2 dropped to the range 1.0 to 2.5×10^{-5} F.

The third and most frequently used indicator method is made possible by the almost unique reaction whereby iodine imparts to aqueous starch suspensions an intense blue color. The nature of this blue color has been the subject of much research and controversy. Investigations by Rundle and co-workers[15] have indicated that a complex is formed between iodine and β-amylose, the latter being a constituent of starch. According to these investigators, the amylose molecules are in the form of long helical chains, with the helix of proper size to permit iodine molecules to enter with their long axes coincident with the axis of the helix. It was also found that iodide or tri-iodide could replace iodine in the helix; Lambert[16] showed that little color is formed in the absence of iodide. Others[17] have suggested that the iodine is present in a linear complex such as $3I_2 \cdot 2I^-$, and that the color is related to the length of this chromophore or light-absorbing complex.

The color of the complex is so intense that solutions 4 to 8×10^{-6} F in iodine and at least 4×10^{-5} F in iodide give an easily visible blue color; some experimeters have detected concentrations as low as 2×10^{-7} F in iodine. The intensity of the color decreases with less than this concentration of iodide, and in the absence of iodide an iodine concentration of from 2 to 4×10^{-5} F is required to give a detectable color. The end point is more sensitive in slightly acid than in neutral solution; above 25°C, the sensitivity

[15] R. E. Rundle, J. F. Foster, and R. R. Baldwin. *J. Am. Chem. Soc.*, **66**, 2116 (1944); **73**, 4321 (1951); R. S. Stein and R. E. Rundle. *J. Chem. Phys.*, **16**, 195 (1948). Numerous references to previous articles and discussions are given in these articles.

[16] Lambert. *Anal. Chem.*, **23**, 1251 (1951).

[17] G. A. Gilbert and J. V. Marriot. *Trans. Faraday Soc.*, **44**, 84 (1948); S. Ono, S. Tsuchihashi, and T. Kuge. *J. Am. Chem. Soc.*, **75**, 3601 (1953).

is markedly decreased. The color is also somewhat dependent upon the presence of other salts, changing to purple or brown in concentrated salt solutions.

IODOMETRIC DETERMINATION OF OXIDANTS

It was stated at the beginning of this chapter that iodometric methods could be divided into two classes: First, the determination of reductants by titration with standard iodine solutions and, second, the determination of oxidants by addition of excess iodide and titration of the iodine produced by a standard reductant. This indirect process is made necessary because, as stated previously, iodide is not a strong reducing agent. Therefore, the reduction reactions, which can be generalized as

$$\text{oxidant} + I^- = \text{reductant} + I_2$$

tend not to be quantitative with most oxidants unless a large excess of iodide is added, and unless the concentration of the iodine is lowered to a very small value by titration with a reductant. One might raise the question, why not titrate the oxidants directly with the reductant? There are two reasons: First, most standard solutions of strong reductants, such as tin(II), chromium(II), and titanium(III), are quite inconvenient to use because they are easily oxidized by atmospheric oxygen and therefore elaborate precautions are required in their storage and use. Second, there is a reductant whose solutions are surprisingly stable in the air, but that is unique in that iodine (or triiodide) is the only oxidant that reacts with it *rapidly*, *stoichiometrically*, and *quantitatively*. This reductant is thiosulfate and its properties and its reaction with iodine are discussed below.

THIOSULFATE AND ITS REACTION WITH IODINE

The thiosulfate ion, $S_2O_3^{2-}$, is of interest because the two sulfur atoms are in different oxidation states. It can be made by treating an alkaline sulfite solution with elementary sulfur:

$$SO_3^{2-} + S(s) = S_2O_3^{2-}$$

In this reaction it appears that the elementary sulfur has been reduced to S(−II) and the sulfite sulfur oxidized to S(VI). This view is supported by experimental evidence from various sources, which indicates that the two sulfur atoms are not equivalent and that the electronic structure of the ion can be represented as

$$\left[\begin{array}{ccc} & :\ddot{\underset{\cdot\cdot}{O}}: & \\ :\ddot{\underset{\cdot\cdot}{O}}: & \ddot{\underset{\cdot\cdot}{S}} & :\ddot{\underset{\cdot\cdot}{S}}: \\ & :\ddot{\underset{\cdot\cdot}{O}}: & \end{array}\right]^{2-}$$

The prefix *thio* (from *theion*, Greek for sulfur) implies that thiosulfate can be considered as sulfate in which an oxide oxygen is replaced by a sulfide sulfur. These considerations are consistent with the reaction of thiosulfate in acid solutions whereby sulfur is produced

$$S_2O_3^{2-} + H^+ = HSO_3^- + S(s)$$

It is seen that this reaction is essentially the reverse of the one given above for the formation of thiosulfate from sulfite and sulfur.

When iodine reacts with thiosulfate the following reaction occurs:

$$I_2 + 2S_2O_3^{2-} = 2I^- + S_4O_6^{2-} \tag{9}$$

Since the iodine solutions being titrated contain predominantly tri-iodide ion, the titration reaction may be better represented as

$$I_3^- + 2S_2O_3^{2-} = 3I^- + S_4O_6^{2-} \tag{10}$$

The electronic structure of tetrathionate can be written as

$$\left[\begin{array}{ccccc} & :\ddot{\underset{\cdot\cdot}{O}}: & & :\ddot{\underset{\cdot\cdot}{O}}: & \\ :\ddot{\underset{\cdot\cdot}{O}}: & \ddot{\underset{\cdot\cdot}{S}} & :\ddot{\underset{\cdot\cdot}{S}}:\ddot{\underset{\cdot\cdot}{S}}: & \ddot{\underset{\cdot\cdot}{S}} & :\ddot{\underset{\cdot\cdot}{O}}: \\ & :\ddot{\underset{\cdot\cdot}{O}}: & & :\ddot{\underset{\cdot\cdot}{O}}: & \end{array}\right]^{2-}$$ [18]

The answer to the interesting question as to why iodine should be uniquely effective in removing electrons from two thiosulfate ions so as to rapidly

[18] This ion with 2 thiosulfate groups connected by two sulfur atoms is analogous to the peroxydisulfate ion

$$\left[\begin{array}{ccccc} & :\ddot{\underset{\cdot\cdot}{O}}: & & :\ddot{\underset{\cdot\cdot}{O}}: & \\ :\ddot{\underset{\cdot\cdot}{O}}: & \ddot{\underset{\cdot\cdot}{S}} & :\ddot{\underset{\cdot\cdot}{O}}:\ddot{\underset{\cdot\cdot}{O}}: & \ddot{\underset{\cdot\cdot}{S}} & :\ddot{\underset{\cdot\cdot}{O}}: \\ & :\ddot{\underset{\cdot\cdot}{O}}: & & :\ddot{\underset{\cdot\cdot}{O}}: & \end{array}\right]^{2-}$$

in which two sulfate groups are bridged by two peroxide oxygens. Tetrathionate lacks the powerful oxidizing properties of peroxydisulfate.

cause the quantitative formation of the tetrathionate molecule is not known.[19] Most oxidizing agents convert thiosulfate at least partially to sulfate. In fact, some sulfate will be formed in the reaction with iodine if the solution is appreciably alkaline, apparently because iodine disproportionates (see page 452),

$$I_2 + OH^- = HIO + I^-$$

and the hypoiodous acid thus formed oxidizes thiosulfate to sulfate. Because of this reaction of iodine and the subsequent disproportionation of hypoiodite to iodate and iodide,

$$3IO^- = 2I^- + IO_3^-$$

iodometric titrations are usually not attempted in solutions in which the hydrogen-ion concentration is less than approximately 10^{-9} *M*.

The uniqueness of the reaction of Equation 10 would appear to restrict the usefulness of thiosulfate solutions to the determination of iodine; actually such solutions are used conveniently for the determination of almost any oxidizing agent capable of oxidizing iodide stoichiometrically to iodine. The iodine so produced is titrated with the standard thiosulfate solution:

$$\text{unknown}_{(\text{oxidized})} + I^-_{(\text{excess})} = \text{unknown}_{(\text{reduced})} + I_3^-$$

$$I_3^- + 2S_2O_3^{2-} = 3I^- + S_4O_6^{2-}$$

The preparation and stability of thiosulfate solutions and the conditions under which the iodine-thiosulfate reaction is stoichiometric will be discussed in greater detail in Chapter 21.

A brief historical review of the development of the present general method for the iodometric determination of oxidants may be of interest. The first use of the method for a specific determination appears to have been made by Duflos who in 1845 proposed that Fe(III) be determined by addition of excess iodide and titration of the iodine produced with a standard stannous chloride solution.[20]

Then, in 1853, Bunsen—for whom the Bunsen burner is named—applied a variant of the method to the determination of some eighteen oxidants in

[19] Studies of the mechanism of the reaction between tri-iodide and thiosulfate have shown that an intermediate is first formed, $I_3^- + S_2O_3^{2-} = S_2O_3I^- + 2I^-$, which then reacts with thiosulfate, $S_2O_3I^- + S_2O_3^{2-} = S_4O_6^{2-} + I^-$, to give the over-all reaction. The intermediate can also react with iodide, $2S_2O_3I^- + I^- = S_4O_6^{2-} + I_3^-$, which accounts for the reappearance of iodine sometimes noted near the end point. See A. D. Awtrey and R. E. Connick. *J. Am. Chem. Soc.*, **73**, 1341 (1951); a description of a method for studying a fast reaction and references to earlier studies will be found.

[20] A. F. Duflos. *Chem. Apothekerbuch*, **2**, 101 (1845).

which the oxidant was treated with concentrated hydrochloric acid, the solution heated, the chlorine produced distilled into an iodide solution, and the resulting iodine titrated with a standard sulfurous acid solution.[21] This method was extensively used for many years, even though sulfurous acid solutions are unstable because of air oxidation and the volatility of sulfur dioxide. It is an interesting coincidence that in the same year Schwartz[22] proposed that thiosulfate be employed as a titrant for iodine solutions; its use has gradually replaced the previously proposed reductant solutions.

THE OXYGEN ERROR

By making use of the two following half-cells and their standard potentials

$$2I^- = I_2(s) + 2e^- \qquad E° = -0.54 \text{ v}$$

$$2H_2O(l) = O_2(g) + 4H^+ + 4e^- \qquad E° = -1.23 \text{ v}$$

one can obtain the following cell reaction and potential

$$4I^- + O_2 + 4H^+ = 2I_2(s) + 2H_2O \qquad E°_{cell} = 0.69 \text{ v}$$

from which the equilibrium expression and constant can be obtained[23]

$$\frac{1}{[I^-]^4 P_{O_2} [H^+]^4} = 10^{47}$$

The value of this constant indicates that there is a very strong tendency for oxygen to oxidize iodide; calculations indicate that the equilibrium iodide concentration would be of the order of 10^{-12} M in a solution saturated with air in which the $[H^+]$ was 1 M. The fact that iodometric procedures involving acid solutions can be carried out in air again emphasizes that we cannot predict the rate of a chemical reaction from equilibrium data. However, this tendency is present and is a potential source of what is called the "oxygen error," and, depending upon the other reactions involved and various conditions, can lead to significant and even gross errors. In solutions with hydrogen-ion concentrations about 0.4 M the oxidation of iodide by oxygen of the air becomes appreciable. Moreover, the oxidation reaction is induced by light and is catalyzed by various ions, such as Cu^{2+}. In addition, the reaction may be induced by another oxidation-reduction reaction. For example, in the case of the oxidation of iodide by dichromate, used in Procedure 21-2, the

---

[21] R. W. Bunsen. *Liebigs Ann. Chem.*, **86**, 265 (1853).

[22] H. Schwartz. *Praktische Anleitung zu Maasanalysen.* (2nd ed.) Braunschweig, 1853.

[23] See Chapter 6 regarding the use of half-cell reactions and their potentials to obtain cell reactions and potentials, and the calculation of equilibrium constants from the latter.

oxidation of iodide by oxygen can cause serious errors unless special precautions are taken. Since standard tri-iodide solutions contain appreciable concentrations of iodide ion, and since in titrations of oxidizing substances with thiosulfate an excess of iodide is added (in most cases in an acidic solution) to the oxidant, the *oxygen error* has to be given consideration in all iodometric titrations. Techniques for minimizing this error are described in Procedures 21-2, 21-3, and 21-4.

THE REVERSIBILITY OF THE IODIDE-IODATE REACTION

The disproportionation reactions of iodine and of hypoiodite shown in Equations 5 and 6 on page 452 occur in alkaline solutions. In acid solution the overall reaction is quantitatively reversed so that iodate and iodide react to give iodine. If an excess of iodide is present so that tri-iodide ion forms, the reaction is represented as

$$IO_3^- + 8I^- + 6H^+ = 3I_3^- + 3H_2O \tag{11}$$

Potassium iodate can be purchased in primary standard quality; therefore this reaction is used for the standardization of thiosulfate solution in Procedure 21-3 and is discussed further there. This reaction is quite unusual in that the equilibrium and, to a certain extent, the rate are such that it proceeds quantitatively from left to right if the hydrogen-ion concentration is appreciably greater than 10^{-7} *M*. In fact, this dependency is so critical that a determination of the acid concentration of a solution can be made by adding an excess of iodide and iodate and then titrating the iodine that is formed with thiosulfate.

OUESTIONS AND PROBLEMS

1. What would one predict regarding the stability of tri-iodide solutions from an inspection of the iodide-iodine and water-oxygen half-cell potential? What are the experimental facts? Explain.

2. What factors account for the reversibility of reactions between iodide and many oxygen anions?

3. What can you state regarding the oxidation states of the sulfur atoms in (*a*) thiosulfate, (*b*) tetrathionate, and (*c*) peroxydisulfate?

Calculations Involving Chemical Equilibria

(See Chapter 21, page 485, for stoichiometric calculations involving iodometric methods.)

4. The solubility of iodine, I_2, in water at 25°C is 0.00134 *F*; its solubility in a 0.00500 *F* potassium iodide solution is 0.00388 *F*. Calculate the formal solubility of iodine in a solution 0.100 *F* in potassium iodide. *Ans.* 0.0521 *F*.

5. Calculate the formal concentrations of potassium iodide and iodine in a solution prepared by dissolving 25 g KI and 12.7 g I_2 in water and diluting to 1 liter. Calculate the molar concentration of iodine in such a solution. Calculate the partial pressure of iodine above such a solution. (The vapor pressure of solid iodine is 0.4 mm at 25°C.)

6. In carrying out a chromate determination, 50 ml of a 0.100 *N* chromate solution were pipetted into a 600-ml flask, the flask was filled with CO_2, 3 g of solid KI and the proper amount of acid were added; then the solution was diluted to 100 ml and allowed to stand under CO_2 until the reaction was quantitatively complete and the iodine in the solution and in the space above the solution were in equilibrium. Just before the titration, the CO_2 was inadvertently turned on and substantially all of the iodine in the vapor phase was swept away. Calculate the percentage error caused by the loss of the iodine that was in the vapor phase. (Essential data will be found in Problems 4 and 5.) *Ans.* 0.07%.

7. You wish to analyze solutions containing H_3AsO_3 and are told that this can be done by titration with a standard solution of I_2 dissolved in KI. To test the method you prepare 0.10 *N* solutions of H_3AsO_3 (by dissolving As_2O_3 in water) and of KI_3 and make the following experiments. Explain the observations.

   Experiment 1. 10 ml of the H_3AsO_3 were titrated slowly with the KI_3. At first the color of the KI_3 disappeared rapidly, but after 2 ml–3 ml, the color in the titrated solution persisted for longer and longer periods and became permanent after approximately 5 ml of KI_3 had been added.

   Experiment 2. A 1 g portion of solid $NaHCO_3$ was added to 10 ml of the H_3AsO_3 and the solution titrated with the KI_3. The color of the KI_3 disappeared rapidly until an equivalent volume of the KI_3 had been added.

   Experiment 3. A 1 g portion of solid NaOH was added to 10 ml of the H_3AsO_3 and the solution titrated. The KI_3 color disappeared rapidly even after an equivalent volume of KI_3 had been added.

8. In standardizing an approximately 0.1 *N* iodine solution, 0.2-g samples of arsenious oxide were weighed out and dissolved in 10 ml of 6 *N* NaOH, and the solution was diluted to 150 ml, treated as indicated below, and then titrated to a starch end point. Predict the result of each titration, explaining in detail the basis for your prediction.

   SUGGESTIONS: Consider calculating first an approximate value of the hydrogen-ion concentration before beginning the titration, in order to ascertain if the initial hydrogen-ion concentration is within the limits prescribed; see the Discussion on pages 451–453. Next, assume that the titration reaction proceeds quantitatively and make an approximate calculation of the final hydrogen-ion concentration. Since, because of the uncertainties involved, the *p*H need be known only approximately in order to make the desired prediction, the effect of the ionization of the arsenic acid on the final *p*H may be neglected in reaching a first approximation.

Sample I. The solution was made neutral to litmus with HCl.
Is this solution buffered?

Sample II. The solution was saturated with CO_2 gas.
Write the equations for the reactions taking place when (*a*) a limited quantity of CO_2 is passed into a large excess of NaOH solution, and (*b*) such a solution is saturated with CO_2.

Sample III. The solution was made neutral to litmus with HCl and saturated with CO_2 gas.

Sample IV. To the solution was added 10 ml of 6 F H_3PO_4.

Sample V. The solution was made neutral to litmus, and 10 ml of 6 F Na_3PO_4 were added.

Sample VI. The solution was made neutral to litmus, and 10 ml of 6 F Na_2HPO_4 were added.

Sample VII. The solution was made neutral to litmus, 10 ml of 1.5 F Na_2CO_3 were added, and the solution was then saturated with CO_2.

Sample VIII. To the solution was added 2.8 g of solid P_2O_5.

Sample IX. The solution was neutralized, and 10 ml of 6 F Na_2HAsO_4 were added.

Sample X. The solution was made neutral to litmus, and 10 ml of a solution 6 F in Na_2SO_3 and 0.6 F in $NaHSO_3$ were added.

9. The proposal is made that the arsenic in a sample of Na_3AsO_4 be determined by introducing the sample into an acid solution containing an excess of iodide and then titrating the liberated iodine with 0.1 N thiosulfate.
   (*a*) Formulate the mass-action expression for the quinquepositive arsenic-iodide reaction and calculate the equilibrium constant from the half-cell potentials involved.
   (*b*) Calculate the minimum hydrogen-ion concentration required so that not more than 0.001 % of the arsenic remains in the quinquepositive state. Make what you would consider as reasonable assumptions in regard to size of sample, volume of titrated solution, concentration of iodide, and means of determining the end point.

   State briefly any other assumptions that are involved in the practical use of this value for the minimum $[H^+]$.

   (*c*) Outline briefly a procedure for carrying out this determination.
   (*d*) State and discuss the various factors that should be considered in making an experimental study of the proposed method.

CHAPTER 21

Oxidation-Reduction Reactions. Iodometric Methods (II)

The preceding chapter presented a general discussion of the principles of iodometric methods. As explained there, iodometric methods can be divided into two classes. First, the determination of reductants by titration with an iodine solution. Second, the determination of oxidants by reduction with an excess of iodide and titration of the iodine produced with a thiosulfate solution. Determinations of this second class are of more analytical importance and in this chapter selected procedures of this type will be considered in detail. These include the preparation and storage of a thiosulfate solution, two methods for the standardization of such solutions, and the determination of the copper in an alloy. These procedures illustrate the general considerations involved in the iodometric determination of oxidants.

PROCEDURE **21-1**

Preparation of a Thiosulfate Solution

Outline. Sodium thiosulfate is dissolved in water that has been sterilized by boiling. Sodium carbonate is added to make the solution slightly alkaline to inhibit decomposition by bacterial action.

DISCUSSION

Sodium thiosulfate is uniquely suited for the titration of iodine for two reasons: first, its reaction with iodine is rapid, stoichiometric, and quantitative; and second, its aqueous solutions are quite stable with respect to oxidation by oxygen of the air. As was discussed in Chapter 20, reducing solutions, in general, suffer from instability because of air oxidation.

Anhydrous sodium thiosulfate is too hygroscopic for use as a primary standard; and although the pentahydrate ($Na_2S_2O_3 \cdot 5H_2O$) can be prepared in a very pure state by repeated recrystallizations from water and drying under carefully controlled conditions, this preparation is not justified because of some uncertainty as to the stability of the solution, especially when it is first prepared.

Instability of Thiosulfate Solutions

While thiosulfate solutions are quite stable against air oxidation, changes in the concentrations of the solutions are observed. Numerous explanations have been advanced as to the causes of the instability of thiosulfate solutions. It was originally suggested that, since thiosulfate is decomposed by acid, the effect was due to carbon dioxide present in the water, the reaction being

$$S_2O_3^{2-} + H_2CO_3 = HSO_3^- + HCO_3^- + S(s)$$

and it was assumed that the solution would become stable after the carbon dioxide had been exhausted. However, since sulfite in neutral solution will reduce two equivalents of iodine, such solutions should increase in normality, but this is not generally observed. Furthermore, since the second proton of sulfurous acid is more highly ionized than the first proton of carbonic acid, this effect would be progressive, which it appears not to be. Finally, as a matter of experiment, thiosulfate solutions have been found to be relatively stable in the presence of carbon dioxide.

It has also been assumed that traces of metal ions in the water catalyze the air oxidation of thiosulfate. Thus, although the potential of the reaction

$$2H_2O(l) = O_2 + 4H^+ + 4e^- \qquad E^\circ = -1.23 \text{ v}$$

would indicate that oxygen should oxidize thiosulfate quite completely, the rate of the reaction is such that solutions saturated with oxygen are very little affected.[1] However, if a trace of copper salt is present, it may react with thiosulfate as follows:

$$2Cu^{2+} + 2S_2O_3{}^{2-} = 2Cu^+ + S_4O_6{}^{2-}$$

The cuprous ion is then oxidized by oxygen,

$$2Cu^+ + \tfrac{1}{2}O_2 + H_2O = 2Cu^{2+} + 2OH^-$$

and thus the oxidation of the thiosulfate continues as long as oxygen is present.[2]

Subsequent investigations[3] have shown that the major cause of the instability of thiosulfate solutions is sulfur-consuming bacteria and that sterile solutions are quite stable. Slightly alkaline solutions (H^+ concentrations of from 10^{-8} to 10^{-9}) seem to inhibit this bacterial action. Numerous experiments have confirmed the fact that 0.1 F solutions that are prepared from freshly boiled water to which has been added 100 mg of sodium carbonate per liter and that are protected from bacterial infection are subject to very little change and may be used as soon as prepared.

Reactions between Iodine and Thiosulfate

It is assumed in iodometric methods that the equation

$$I_2 + 2S_2O_3{}^{2-} = 2I^- + S_4O_6{}^{2-}$$

represents quantitatively the reaction between thiosulfate and iodine.

---

[1] Kilpatrick and Kilpatrick. *J. Am. Chem. Soc.*, **45**, 2132 (1923).

[2] This is an example of the previously discussed "homogeneous catalysis," inasmuch as only one phase is present. Heterogeneous catalysis is usually assumed to take place because of an increase in the activity of the reactants at some surface.

[3] Kilpatrick and Kilpatrick, op. cit.; Mayr. *Z. Anal. Chem.*, **68** 274 (1926).

However, under certain conditions sulfate may be formed:

$$4I_2 + S_2O_3^{2-} + 10\,OH^- = 2SO_4^{2-} + 8I^- + 5H_2O$$

and, as is indicated, its formation is favored by hydroxide ion. It is generally assumed that the mechanism of this reaction involves the formation of hypoiodite, according to the reaction

$$I_2 + OH^- = HIO + I^-$$

since hypoiodite is known to oxidize thiosulfate to sulfate as follows:

$$4HIO + S_2O_3^{2-} + 6OH^- = 2SO_4^{2-} + 4I^- + 5H_2O$$

Experimentally, it is found that acid or even neutral solutions, 0.05 F in iodine, may be precisely titrated with thiosulfate or into thiosulfate; the order of the titration not affecting the results. With more dilute iodine solutions, where the disproportionation into hypoiodite is more pronounced, and with solutions that may be slightly basic (containing, for example, an excess of hydrogen carbonate ion, HCO_3^-, and not being saturated with CO_2), the iodine should be titrated into the thiosulfate. With titration in this order, the rate of reduction of the iodine by the thiosulfate is so rapid that disproportionation and subsequent sulfate formation do not take place to an appreciable extent. In general, the hydrogen-ion concentration should not be less than approximately 10^{-9} M during an iodometric titration; moreover, as explained below, under no conditions should the thiosulfate solution be acidified.

It is an easily verified fact that the addition of a thiosulfate solution to even a dilute acid solution will cause decomposition of the thiosulfate with a resultant precipitation of sulfur,

$$S_2O_3^{2-} + 2H^+ = H_2SO_3 + S(s)$$

The reaction is not rapid, and the precipitate of sulfur appears after an interval of time that depends upon the concentration of the acid. Since the H_2SO_3 may either react with iodine or escape from the solution as SO_2, in one case increasing and in the other case decreasing the reducing equivalents in the solution, it is reasonable to question the use of thiosulfate in acid solution. As a matter of experiment, thiosulfate solutions may be titrated *into* iodine solutions that are as concentrated in H^+ as 3 M to 4 M if the solution is *effectively stirred* during the addition of the thiosulfate; the rate of oxidation of the thiosulfate by iodine is so rapid that no appreciable decomposition of the thiosulfate takes place. As stated above, thiosulfate solutions should not be acidified before being titrated. As a matter of practice, in the vast majority of cases, the thiosulfate will be titrated into acid solutions of iodine. Since sodium carbonate is often added to thiosulfate solutions, it is preferable to make neutral iodine solutions slightly acid before they are titrated; otherwise the titrated solution may become unduly alkaline before the completion of the titration.

INSTRUCTIONS (Procedure 21-1: Preparation of a Thiosulfate Solution)

Take 1 liter of distilled water in a large flask, cover it with a watch glass, and heat just to boiling for 5 minutes; then allow the water to cool. Weigh out 25 g of $Na_2S_2O_3{\cdot}5H_2O$ (take 30 g if the solution is to be used for gravimetric titrations) and 0.1 g of Na_2CO_3. Dissolve the weighed reagents in the freshly boiled water, and transfer the solution to a clean bottle with a ground-glass stopper (Note 1).

NOTE

1. If a large volume of solution is prepared, the storage bottle should be fitted with a siphon tube and the inlet tube provided with a soda-lime tube packed with sterile cotton. Since most of the solution may be lost should the rubber or glass stopcock of the siphon tube begin to leak, a convenient modification of this arrangement is described in Chapter 10, page 194.

STARCH INDICATOR SOLUTIONS

The nature of the starch-iodine blue color has been mentioned on page 458.

The indicator solution can be prepared from potato, arrowroot, or rice starch. These starches contain amylose, also designated as β-amylose, which is the so-called "soluble starch," and amylopectin, also designated as α-amylose, which is quite insoluble. When the starch grains are subjected to boiling water, they burst, and upon standing the amylopectin and other insoluble material settles out and can be eliminated by decanting the clearer portion of the solution; amylopectin and iodine form reddish-colored compounds that are slow to decolorize. Cornstarch contains a much higher percentage of amylopectin and should not be used.

If starch solutions are not kept sterile, they decompose because of bacterial action, and are likely to form a reddish color; HgI_2 is sometimes used as a bactericide. Similar products are formed by the hydrolysis of starch in acid solutions, but this effect is not troublesome in titrations unless the acid concentration is greater than 3 to 4 M. Starch is coagulated if it is introduced into concentrated iodine solutions. Always withhold the indicator from the titrated solution until the iodine color becomes uncertain.

Directions for the preparation of starch solutions are given in Appendix 7. A convenient substitute for starch indicator solutions is available from the Fisher Scientific Company. This product, Thyodene, is added as the solid to the solution being titrated, dissolves very rapidly, and gives excellent indicator action. The problems of solution preparation and sterilization can be avoided by its use.

STANDARDIZATION OF THIOSULFATE SOLUTIONS

Various methods are available for the standardization of thiosulfate solutions. These solutions are always titrated against iodine solutions, therefore the fundamental primary standard is iodine. Pure iodine can be readily prepared, but it has such a relatively high vapor pressure that the weighing and transferring of it to the solution for titration require a special technique; therefore, this method is not generally recommended except as a reference method for checking the accuracy of other methods.

Usually recourse is made to the titration of oxidants that can be obtained in primary standard purity and that when treated with an excess of iodide are capable of producing an equivalent amount of iodine in solution—it is seen that these titrations are equivalent to the determination of oxidants and belong to the second class of iodometric methods. Among the primary standards commonly used are potassium dichromate, potassium iodate, potassium hydrogen iodate [$KH(IO_3)_2$], potassium bromate, and potassium ferricyanide. Standard solutions of potassium permanganate are frequently used because of their availability and because of the rapidity and ease with which the standardization can be made. Metallic copper is sometimes used as a primary standard for thiosulfate solutions that are to be used primarily for the analysis of copper ores and alloys. The metal is converted to cupric ion by dissolving it in nitric acid; the reaction between cupric ion and iodide is discussed in Procedure 21-4.

Procedures follow for the standarization of thiosulfate solutions against potassium iodate and potassium dichromate. Potassium dichromate is probably of the most practical use, and the method illustrates not only the effect of hydrogen-ion concentration on both the equilibrium and rate of the reaction of oxygen acids with iodide but also the errors caused by the induction of the oxygen–iodide reaction by another reaction. The potassium iodate method illustrates the use of reactions between compounds of the same element in different oxidation states.

PROCEDURE 21-2

Standardization of Thiosulfate Solutions against Potassium Dichromate

Outline. (*a*) A solution of potassium dichromate is treated with an excess of potassium iodide in hydrochloric acid, and (*b*) the tri-iodide thus produced is titrated with the thiosulfate solution; starch is used as indicator. The reactions involved are

$$Cr_2O_7^{2-} + 9I^- + 14H^+ = 2Cr^{3+} + 3I_3^- + 7H_2O$$

and

$$I_3^- + 2S_2O_3^{2-} = 3I^- + S_4O_6^{2-}$$

DISCUSSION

Potassium dichromate is highly recommended as a primary standard for the standardization of thiosulfate solutions because (1) it can be obtained for that purpose from the Bureau of Standards, (2) it has a relatively high equivalent weight, (3) the solid is nonhygroscopic and exceedingly stable, (4) solutions that are stable indefinitely can be prepared by direct weighing, and (5) the conditions under which the reaction with iodide can be used for very exact standardizations of thiosulfate solutions have been determined by exhaustive experimental studies.[4]

The reaction between iodide and dichromate, to give iodine and tripositive chromium, illustrates several of the factors that have to be considered in developing iodometric methods. In an alkaline solution iodine will oxidize tripositive chromium to chromate, whereas in an acid solution (10^{-4} *M* or greater in hydrogen ion) iodide is quantitatively oxidized by dichromate. However, at this minimum hydrogen-ion concentration, the rate of the oxidation is too low for the reaction to be of practical use. Experimentally, it was found[5] that if the hydrogen-ion concentration is greater than 0.2 *M* and the iodide concentration greater than 0.05 *M*, the oxidation of the iodide is quantitatively complete in 5 minutes.

---

[4] McCroskey. *J. Am. Chem. Soc.*, **40**, 1664 (1918); Vosburgh. *J. Am. Chem. Soc.*, **44**, 2120 (1922); Bray and Miller. *J. Am. Chem. Soc.*, **46**, 2204 (1924); Popoff and Whitman. *J. Am. Chem. Soc.*, **47**, 2259 (1925).

[5] Ibid.

If the hydrogen-ion concentration is much above 0.4 M, the oxygen error becomes appreciable. As was mentioned in the earlier discussion of this error, the oxidation of iodide by oxygen is induced by the dichromate-iodide reaction. Vosburgh,[6] and Bray and Miller[7] found experimentally that this effect varied with the iodide concentration and even with the order of mixing of the iodide and dichromate, being greater when the dichromate was added to the iodide and acid than when the iodide was added to the dichromate and acid; these articles should be consulted for further details.[8] The procedure below is based upon these experimental studies, and the results obtained should be accurate within the limits of the titrimetric measurements involved.

INSTRUCTIONS (Procedure 21-2: Standardization of Thiosulfate Solutions Against Potassium Dichromate)

I. Preliminary Titration

(Optional. Consult your instructor. This titration provides experience in recognizing the disappearance, first, of the iodine color, and, second, the starch–iodine color in the presence of the Cr(III) color discussed in Note 5 below.)

Pipet into a 500-ml flask 25 ml of a 0.100 N $K_2Cr_2O_7$ solution. Add 50 ml of water, and proceed as directed in the second and subsequent paragraphs of the next section. Only one titration is required.

II. Standardization

Accurately weigh out into 500-ml flasks three 0.2-g portions of dry $K_2Cr_2O_7$ (Notes 1 and 2). Dissolve the solid in 50 ml of water.

NOTES

1. Material of primary standard purity can be obtained from the Bureau of Standards or purchased from sources specializing in analytical chemicals.[9] Materials of other than this grade often contain CrO_3, and therefore the salt should be purified by recrystallization from distilled water. The first portion of the crystals to appear should be discarded, and contamination from dust should be carefully avoided. The crystals should be dried to constant weight at 180°C–200°C.

---

[6] Ibid.

[7] Ibid.

[8] A discussion of the intermediate oxidation states of chromium involved in certain induced reactions is given by H. A. Laitinen (*Chemical Analysis*. New York: McGraw-Hill, 1960, pp. 462–471).

[9] Experiments have shown that several of the commercially available reagent grades of $K_2Cr_2O_7$ gave standardization values agreeing within 0.1% with Bureau of Standards material.

Dissolve 3 g of KI in 50 ml of water, add to it 5 ml of 6 *F* HCl (Note 3), and immediately add this to the dichromate solution. Gently swirl the solution for 2 or 3 seconds (Note 4), then close the flask with a watch glass, and allow it to stand in a dark place for 5 minutes.

Dilute the solution with 300 ml of water, washing down the sides of the flask. While slowly swirling the solution, titrate with the thiosulfate solution; use a volumetric or a gravimetric buret (consult your instructor). When the yellow color of the iodine becomes indistinct (Note 5), add 5 ml of starch solution and slowly titrate to the disappearance of the blue color of the starch. Repeat the titration with two other portions of the $K_2Cr_2O_7$ and calculate the formality or the weight formality of the thiosulfate solution (Note 6).

NOTES

2. If a standard solution of the dichromate is desired, proceed as follows:

Weigh out accurately about 2 g of dry $K_2Cr_2O_7$, transfer it to a 500-ml volumetric flask that has been calibrated against a 50-ml pipet (see page 414), dissolve the crystals, and dilute to the mark with water at room temperature. Mix the solution.

Pipet 50 ml of the $K_2Cr_2O_7$ solution into a 500-ml flask and proceed as directed in the second and subsequent paragraphs of the procedure above.

3. When the HCl is added to the iodide, no iodine color should develop; such color indicates the presence of iodate in the KI or, more rarely, of oxidizing agents, usually chlorine, in the acid. If these cannot be avoided, a correction should be made by dissolving another portion of KI in the same volume of water and acid and titrating it with thiosulfate. This is preferable to reducing the iodine with thiosulfate before adding the solution to the dichromate, since a slight excess of thiosulfate may be added, and since this, as well as the tetrathionate, may be oxidized to sulfate by the dichromate. The acidified iodide solution should not be allowed to stand for any prolonged length of time because oxidation by the air will take place.

4. If iodine solutions are violently shaken in conical flasks or allowed to stand in open vessels, a significant loss of iodine vapor will occur. It has been found that when approximately 30 ml of 0.1 *N* iodine in 4% KI solution in a 250-ml flask was swirled gently for 1 minute, about 0.2% of the iodine was lost. Stoppering the flask did not appreciably reduce the loss. When 50 ml of the same solution was left in an open beaker for 15 minutes, a loss of 0.9% was noted.

5. Continue the titration without starch until the yellowish tinge of the iodine can no longer be detected with certainty. This instruction is somewhat difficult to follow because of the pale-green, in some cases purplish, color due to the chromic ion present, but this color is so reduced by the dilution of the solution that there is no excuse for adding the starch until within 0.2 ml to 0.3 ml of the end point. Note also that the final end point will be a change from the starch blue to the chromic ion color; however, the change caused by even half a drop of 0.1 *N* thiosulfate is so distinct that it can be easily detected.

6. Values obtained by this procedure should agree to within 2 parts per 1000.

PROCEDURE 21-3

Standardization of Thiosulfate Solutions against Potassium Iodate

Outline. (*a*) A solution of potassium iodate is treated with an excess of potassium iodide in hydrochloric acid, and (*b*) the tri-iodide that is produced is titrated with the thiosulfate solution. The sequential reactions involved are

$$IO_3^- + 8I^- + 6H^+ = 3I_3^- + 3H_2O$$

and the tri-iodide, thiosulfate reaction.

DISCUSSION

Potassium iodate, potassium hydrogen iodate [$KH(IO_3)_2$], and potassium bromate have all been proposed as primary standards for thiosulfate solutions, and all share the disadvantage of having small equivalent weights. Only 0.14 g of the potassium iodate, which has the largest equivalent weight of the three, is equivalent to 40 ml of 0.1 *N* thiosulfate, so that a weighing error of 0.2 mg will introduce an error of 0.14% in the standardization. This error can be minimized by preparing standard solutions of these substances and taking aliquot portions for the titrations. Potassium hydrogen iodate is an interesting compound in that it also can be used as a primary standard for bases. Its use for extremely accurate work has been questioned; it appears doubtful that potassium acid iodate of primary standard quality can be prepared because of contamination by potassium dihydrogen iodate. Potassium bromate has the disadvantage that the reaction between bromate and iodide ions is somewhat slow, so that a higher acid concentration is required than with iodate, with the accompanying danger of oxygen error. Experiments have shown that the bromate–iodide reaction is catalyzed by small quantities of a molybdate and use has been made of this fact in procedures for the iodometric determination of bromate.

Potassium iodate is used in the procedure below because it can either be purchased in a high state of purity or can be easily purified by recrystallization and because it reacts rapidly and stoichiometrically with iodide under a wide range of conditions. The unusual dependence of the iodate-iodide reaction upon the hydrogen-ion concentration was discussed in Chapter 20.

INSTRUCTIONS (Procedure 21-3: The Standardization of a Thiosulfate Solution Against Potassium Iodate)

Weigh out accurately 1.2–1.5 g of pure dry KIO_3 (Note 1) and transfer it to a 500-ml volumetric flask that has been calibrated against a 50-ml pipet (see page 414). Dissolve the crystals and dilute with water at room temperature to the mark. Mix the solution.

Pipet 50 ml of the KIO_3 solution into a 200-ml flask. Dissolve 3 g of KI in 25 ml of water, add 2 ml of 6 *F* HCl (Note 3, Procedure 21-2), and immediately add this to the iodate solution. Slowly swirl the solution (Note 4, Procedure 21-2) and titrate with the thiosulfate; use a volumetric or a gravimetric buret (consult your instructor). When the iodine color becomes indistinct, add 5 ml of starch indicator solution and titrate to the disappearance of the starch-iodine color. Repeat the titration with two other portions of the iodate solution and calculate the formality or the weight formality of the thiosulfate solution (Note 2).

NOTES

1. Reagent-grade KIO_3 has been found satisfactory for all except the most accurate work. The salt is easily recrystallized and should then be dried at 160°C–180°C. Some samples of reagent have been observed to change to a pale buff color upon being dried as specified here. Results obtained by the use of such samples agreed within 1 part per 1000 with those by nondarkened samples.
2. Values obtained from such titrations should agree within 2 parts per thousand.

Iodometric Determination of Copper in an Alloy

Outline. (*a*) The alloy is treated with nitric acid to dissolve the copper. Sulfuric acid is added and the solution evaporated to remove nitrous acid and the oxides of nitrogen. (*b*) The solution is diluted and buffered to *p*H 2–*p*H 2.5. (*c*) Excess potassium iodide is added and the resultant tri-iodide titrated with thiosulfate to a starch end point. The reactions are

$$2Cu^{2+} + 5I^{-} = 2CuI(s) + I_3^{-}$$

and

$$I_3^{-} + 2S_2O_3^{2-} = 3I^{-} + S_4O_6^{2-}$$

Thiocyanate is added near the end point to make the Cu^{2+}-I^{-} reaction more quantitative and to remove adsorbed iodine from the CuI precipitate.

DISCUSSION

(*a*) Dissolving the Sample

Two preliminary considerations arise in the dissolving of an alloy preparatory to determining its copper content. First, the copper must be brought into solution in the cupric state without the introduction of any interfering constituents. Second, one must ascertain what other metals are present in the alloy and whether they are likely to cause an error in the titration. The method given below is applicable to the determination of copper in such alloys as brasses (which range from 60% to 90% copper and the remainder zinc, with smaller amounts of lead and tin in special alloys), bronzes (which may contain up to 30% tin), and German silvers (which contain up to 20% nickel), and where only traces of iron are likely to be present. These alloys are readily dissolved in nitric acid; this acid is used here even though care must be taken to remove its reduction products from the solution after dissolving the sample. Nitrous acid and the oxides of nitrogen rapidly oxidize iodide, but they can be removed by boiling; the oxidation of iodide by the nitric acid is slower, but most of the nitric acid must be removed before the solution is buffered or the desired *p*H will not be maintained. Therefore, sulfuric acid is added and the solution is heated until sulfuric acid fumes are given off. This treatment expels the more volatile components of the solution and leaves essentially a sulfuric acid solution of the sample.

Behavior of Certain Elements in this Procedure. As stated above these alloys may contain zinc, nickel, tin, lead, and very small amounts of iron in addition to copper. Zinc and nickel cause no significant effects but the behavior of the other metals is of interest. When an alloy containing tin is dissolved in nitric acid, the tin is oxidized to the quadripositive state. In this oxidation state the acidic property of the element becomes so pronounced that hydrolysis takes place, resulting in the formation of a hydrous acidic oxide that is frequently called *metastannic acid.* Although the formula is at times written as $SnO_2 \cdot 4H_2O$, H_2SnO_3, $H_2Sn(OH)_6$, or $SnO_2 \cdot xH_2O$ there is evidence of polymerization during hydrolysis and doubt that the precipitate is of uniform composition. Like hydrous silicon dioxide, $SiO_2 \cdot 2H_2O$, this precipitate is colloidal in nature, and quantitative precipitation is attained only by long heating and partial dehydration. In the procedure below, small quantities of tin may not cause a visible precipitate unless the nitric acid is evaporated almost to dryness; larger quantities of tin will cause a visible precipitate. Because of its acidic nature, this precipitate is not dissolved by even concentrated solutions of non-complex-forming acids, such as nitric and perchloric acids, but is dissolved by the fuming sulfuric acid; the exact formula of the complex formed under these conditions is not known. The stannic oxide adsorbs cupric copper and prevents its reduction, therefore it is desirable that the precipitate be dissolved. When the sulfuric acid is fumed until most of the water is removed (indicated by the solution losing most of its blue color) a precipitate of anhydrous copper sulfate, light gray tinged with green, is likely to separate. When the sulfuric acid solution is subsequently diluted, this precipitate dissolves and the solution is again colored blue by the hydrated cupric ion, $Cu(H_2O)_4^{2+}$.

When a large amount of lead is present, a precipitate of lead sulfate will also form in the fuming acid; small amounts of lead will remain in solution but will be quantitatively precipitated when the sulfuric acid is diluted if the solution is cooled and allowed to stand. The increased solubility of lead sulfate in the fuming acid is caused by the sulfate ion being converted to monohydrogen sulfate ion. It is desirable that the lead be precipitated as quantitatively as possible; otherwise upon addition of iodide a yellowish precipitate of lead iodide is formed which makes the detection of the end point more difficult.

Experiments have shown that ferric iron is reduced by the iodide and causes positive errors in this procedure; with quantities of less than 5 mg of iron these errors can be reduced to less than 0.1 ml of 0.1 *N* thiosulfate if phosphoric acid is added after the sulfuric acid is neutralized. In solutions containing phosphoric acid, ferric iron forms complex compounds, such as $FeHPO_4^+$, that are sufficiently stable to shift the redox equilibrium and to decrease the rate of reduction of the iron by iodide. (See page 421.)

Antimony and arsenic are not likely to be present in considerable quantities in copper alloys; in amounts of less than 50 mg they do not cause significant errors in the procedure below.

(*b*) Adjustment of the Hydrogen-Ion Concentration

Copper is frequently used as a primary standard for thiosulfate solutions, and numerous investigations have been made to determine the factors affecting the accuracy of the method.[10] These studies have shown that the titration proceeds more rapidly and quantitatively if the hydrogen-ion concentration is adequate—greater than approximately 10^{-4}—to repress the partial hydrolysis of cupric ion. If the *p*H is less than approximately 0.5, oxidation of the iodide by the oxygen of the air becomes significant and recurrent end points may be obtained. High concentrations of acetate or other ions that may form complex compounds with cupric ion are undesirable. In the procedure below the solution is buffered with a sulfate—hydrogen sulfate ion buffer; the sulfuric acid solution is first neutralized with ammonia, then sufficient sulfuric acid is added to give the desired *p*H, by means of the sulfate-monohydrogen sulfate (HSO_4^-) buffer system.

(*c*) The Iodide-Cupric-Ion Reaction

If an excess of a soluble iodide is added to a solution containing cupric ions, the copper is reduced to the cuprous state, a precipitate of cuprous iodide is formed, and an equivalent amount of the iodide is oxidized to elementary iodine. By titrating this iodine with standard thiosulfate, one can make a determination of the amount of copper present.

An inspection of the standard potentials of the half-cell reactions

$$Cu^+ = Cu^{2+} + e^- \qquad (E° = -0.17 \text{ v})$$

and

$$3I^- = I_3^- + 2e^- \qquad (E° = -0.54 \text{ v})$$

[10] Gooch and Heath. *Am. J. Sci.*, **24**, 65 (1907); Bray and McCay. *J. Am. Chem. Soc.*, **32**, 1199 (1910); Peters. *J. Am. Chem. Soc.*, **34**, 422 (1912); Popoff. *J. Am. Chem. Soc.*, **51**, 1299 (1929); Crowell. *Ind. Eng. Chem.*, **11**, 159 (1939); Hammock and Swift. *Anal. Chem.*, **21**, 975 (1949).

leads to the disturbing prediction that under equilibrium conditions the reaction

$$2Cu^{2+} + 3I^- = 2Cu^+ + I_3^- \qquad (E^\circ_{cell} = -0.37 \text{ v}) \qquad (1)$$

would tend to proceed from right to left. However, the low solubility of cuprous iodide will tend to shift the equilibrium toward the right by maintaining the concentration of the cuprous ion at a very low value. In fact, from the potential for the half-cell reaction

$$CuI(s) = Cu^{2+} + I^- + e^- \qquad (E^\circ = -0.85 \text{ v})$$

it is seen that the equilibrium for the reaction

$$2Cu^{2+} + 5I^- = 2CuI(s) + I_3^- \qquad (E^\circ_{cell} = +0.31 \text{ v}) \qquad (2)$$

favors the formation of the products on the right. A calculation of the equilibrium constant enables one to predict that the reaction should proceed to an extent well within the usual quantitative limits, especially if an excess of iodide is present in the solution; in addition, the reduction of the tri-iodide by thiosulfate as the titration proceeds will cause the reaction to proceed even more quantitatively toward the right.

The Use of Thiocyanate. There are two common sources of error in the iodometric method for copper. First, if the titration is made in the presence of acetate or other complex-forming anions, which are frequently used for their buffering action, or if the *p*H of the solution is such as to cause partial hydrolysis of the cupric copper, low titration values and recurrent end points are obtained. The second source of error exists in the tendency of the cuprous iodide precipitate to adsorb iodine. This adsorbed iodine causes the end point to be less apparent, since it imparts a buff color to the cuprous iodide precipitate and also may cause premature and transient end points.

The suggestion has been made that in order to minimize these difficulties, a soluble thiocyanate be added to the solution just before the end point is reached.[11] Two effects are obtained by this addition. First, since cuprous thiocyanate is less soluble than cuprous iodide, the equilibrium involving the reduction of cupric ion by iodide (Equation 2 above) is shifted to the right, and the equation for the complete reaction may be written as follows:

$$2Cu^{2+} + 3I^- + 2SCN^- = 2CuSCN(s) + I_3^-$$

Thus the addition of the thiocyanate permits satisfactory titrations to be

[11] Foote and Vance. *J. Am. Chem. Soc.*, **57**, 845 (1935).

made in the presence of complex-forming anions and at *p*H values that would otherwise cause the titration to be incomplete and the end points recurrent.

Second, the cuprous iodide precipitated before the addition of the thiocyanate is at least partially metathesized to cuprous thiocyanate, and there seems to be a preferential adsorption of thiocyanate in place of the iodine; as a result the precipitate is less colored, and less iodine is adsorbed. These effects contribute toward making the end point more readily perceived, especially by one not familiar with the titration, and in bringing it into closer agreement with the equivalence point. The thiocyanate should not be added until most of the iodine has been reduced by the thiosulfate; otherwise, significant reduction of the iodine by the thiocyanate will occur.

INSTRUCTIONS (Procedure 21-4: The Iodometric Determination of Copper in an Alloy)

I. Preliminary Titrations

(Optional. Consult your instructor. These titrations provide experience in recognizing the starch-iodine end point in the presence of the precipitate and show the effect of the thiocyanate.)

Pipet into a 200-ml flask 25 ml of 0.100 *N* (0.100 *F*) $CuSO_4$. Add 10 ml of 9 *F* H_2SO_4. Cool the solution. Treat it by the third and subsequent paragraphs of the following section.

Repeat the titration but omit the addition of the KSCN. Note any differences in the appearance of the end point or in the volume of $Na_2S_2O_3$ required.

II. The Determination of Copper

Weigh samples of the clean dry alloy or metal (Note 1) of such size as to require from 25 ml to 35 ml of the standard volumetric $Na_2S_2O_3$ solution or from 15 g to 20 g of the standard gravimetric solution (Note 2). Transfer the samples to 200-ml flasks.

NOTES

1. Read the section in Chapter 10 (pages 181–182) for general directions for obtaining a representative sample of various materials. If the tool used for milling, drilling, or cutting a metallic sample has been oiled or greased, the material should be washed with alcohol, then with ether, and finally dried.

2. If pure copper is being used for standardization purposes or if the approximate composition of the alloy is known, the weight of sample to be taken can be calculated. With an unknown alloy, one sample should be analyzed before the remainder of the samples are weighed out.

Add 5 ml of 6 F HNO_3, cover the flask with a watch glass, and warm the solution as required to dissolve the alloy without danger of loss by spattering. When solution is essentially complete (Note 3), remove the watch glass, add 10 ml of 9 F H_2SO_4, and evaporate on a sand bath or hot plate (or directly over a burner if the solution is *continuously* kept in swirling motion) until the H_2SO_4 gives off copious dense white fumes (Note 4). Allow the mixture to stand until it cools somewhat (Note 5); then cool the flask and pour slowly into it, 1 ml at a time, 5 ml of water (Note 6). Add 15 ml of water, boil the solution for 1 or 2 minutes (Note 7), and then cool it to room temperature.

While *constantly swirling* the cold H_2SO_4 solution, add slowly from a dropper 15 F NH_3 until the first perceptible blue color (of the cupric ammine complex) is obtained (Note 8). Add 3 F H_2SO_4 *dropwise* until this blue color just disappears; then add just 1 ml in excess (Note 9). Again cool the solution to room temperature.

NOTES

3. With alloys high in tin, a white turbidity or suspension of hydrous stannic oxide may be present at this point. This precipitate is dissolved in the fuming sulfuric acid because of the formation of a complex compound of uncertain composition.

4. The solution is boiled in order to expel the HNO_2 and nitrous oxides formed during the solution of the alloy; these compounds rapidly oxidize iodide. It is essential that the H_2SO_4 be made to fume in order to remove the excess of HNO_3; otherwise the subsequent buffering of the solution would not give the desired pH. The more transparent HNO_3 fumes produced as the solution becomes concentrated should not be mistaken for the dense white fumes of H_2SO_4. The latter can be recognized by the choking sensation that even a very small amount causes. It is to be noted also that the H_2SO_4 will not begin to fume until the solution is reduced to about one-half the volume of 9 F H_2SO_4 added. Do not fume the H_2SO_4 unnecessarily. *Use a hood for this evaporation!*

5. *Allow the flask to cool before using tap water*; otherwise it is likely to crack.

6. Add the water slowly and keep the solution cool; otherwise spattering may result.

In general, concentrated sulfuric acid should be added to an aqueous solution, not in reverse order. In this procedure this order is reversed because transfer of the solution from one vessel to another is thereby avoided.

A precipitate of $PbSO_4$ may form when the solution is diluted and cooled; it can be disregarded (see Discussion above).

7. This boiling is a precautionary measure taken to ensure removal of any remaining oxides of nitrogen.

8. The solution should be cool and the concentrated ammonia added slowly to the acid solution in order to avoid danger of spattering. Concentrated ammonia is used to avoid dilution of the solution.

Dissolve 4 g of KI in 10 ml of water, and add this to the solution of the alloy (Note 10). Swirl the solution gently and continuously and immediately titrate it with standard $Na_2S_2O_3$ solution—use the volumetric buret or the gravimetric buret, depending upon whether titrating with the volumetric or gravimetric standard solution. Add the thiosulfate rapidly until the color of the iodine becomes indistinct (Note 11). Add 3 ml of starch indicator solution and titrate almost to the disappearance of the starch color. Add 2 g of KSCN, swirl the solution, and titrate to the disappearance of the starch color (Note 12).

From the volume or weight of standard thiosulfate used, calculate the percentage of copper in the alloy (or the formality or weight formality of the thiosulfate in case a standardization is being made).

NOTES

The approximate volume of ammonia required to make the solution alkaline with ammonia can be estimated from the sulfuric acid known to be present. Then the ammonia can be added in small portions until the blue color of the copper ammonia complex is first perceived. The intense blue of the complex should be distinguished from the paler greenish blue of the hydrated cupric ion. Avoid splashing the ammonia on the sides of the flask and mix the contents thoroughly by swirling.

9. In case the alloy may contain significant amounts of iron, substitute 2 ml of 15 F H_3PO_4 (85%) for the 1 ml of 3 F H_2SO_4.

10. The volume of the solution should not exceed 50 ml at this point. Larger volumes will require a proportionately large amount of KI; otherwise the change of color at the end point will be less distinct.

11. Anyone accustomed to this titration can carry it to within 1 ml to 2 ml of the end point before adding the starch; however, the presence of the buff-colored precipitate usually confuses a person not familiar with the process. To defer the addition of the starch and approach the final end point more closely, let the mixture in the flask come to rest; then hold the tip of the buret close to the surface of the solution and note if the next drop of thiosulfate causes a local bleaching of the solution.

12. If in doubt as to whether the end point has been passed, note if an additional milliliter of starch darkens the solution. In case the end point is slightly overrun, the mixture can be back-titrated with a standard iodine solution. A large excess of thiosulfate cannot be accurately back-titrated because of the decomposition of the thiosulfate to sulfite in an acid solution.

APPLICATIONS OF IODOMETRIC METHODS INVOLVING THE IODIDE-IODINE HALF-CELL

Methods involving the iodide-iodine half-cell have been so extensively used that it is beyond the scope of this text to list all of them. A few of those

TABLE 21-1

Common Applications of Iodometric Methods

I. Titration of Reductants with Iodine Solutions

Reductant and Half-cell Reaction	Conditions for Titration	Table Notes
$H_3AsO_3 + H_2O = H_3AsO_4 + 2H^+ + 2e^-$	pH 6–8	
$SbOC_4H_4O_6^- + H_2O = SbO_2C_4H_4O_6^- + 2H^+ + 2e^-$	pH 6–8	1
$SnCl_4^{2-} + 2Cl^- = SnCl_6^{2-} + 2e$	HCl solution	
$H_2S(aq) = S(s) + 2H^+ + 2e^-$	acidic solution	2
$HCN + I^- = ICN + H^+ + 2e^-$	CO_3^{2-}-HCO_3 buffer	
$Fe^{2+} + P_2O_7^{4-} = FeP_2O_7^- + e^-$	HCO_3^--H_2CO_3	
$H_2SO_3 + H_2O = H_2SO_4 + 2H^+ + 2e^-$	acidic solution	

II. Determination of Oxidants by Addition of Iodide and Titration with Thiosulfate

Half-cell Reaction and Oxidant	Conditions for Titration	Notes
$H_3AsO_3 + H_2O = H_3AsO_4 + 2H^+ + 2e^-$	$[H^+]$ 3–4	3
$2Cr^{3+} + 7H_2O = Cr_2O_7^{2-} + 14H^+ + 6e^-$	$[H^+]$ 0.2–0.4	4
$CuI(s) = Cu^{2+} + I^- + e^-$	pH 4–2	
$Fe^{2+} = Fe^{3+} + e^-$	pH < 3	
$Mn^{2+} + 4H_2O = MnO_4^- + 8H^+ + 5e^-$	pH 4–2	
$Mn^{2+} + 2H_2O = MnO_2(s) + 4H^+ + 2e^-$	HCl or H_3PO_4	5
$2Cl^- = Cl_2(aq) + 2e^-$	pH 7–3	6
$Cl^- + H_2O = HClO + H^+ + 2e^-$	pH 7–3	
$I_2 + 6H_2O = 2IO_3^- + 12H^+ + 10e^-$	pH 7–3	7
$IO_3^- + H_2O = IO_4^- + 2H^+ + 2e^-$	pH < 7	
$I_2 + 8H_2O = 2IO_4^- + 16H^+ + 14e^-$	acidic solution	

1. As would be predicted from their positions in the periodic table, arsenic and antimony are quite similar in their properties. As a result the principles involved in the titration of Sb(III) are very similar to those applying to the standardization of an iodine solution by means of arsenious oxide (see pages 451–453). [For a study and very complete bibliography of various methods for the determination of antimony, see McNabb and Wagner. *Ind. Eng. Chem.*, anal. ed., **2**, 251 (1930).] Antimony is below arsenic in the periodic table and is more basic in character. Therefore if a hydrochloric acid solution of tripositive antimony is diluted largely or is neutralized, precipitation of antimonyl chloride, SbOCl, or of the hydrous oxide results, and in the subsequent titration the iodine would react only slowly with the precipitate. This precipitation is prevented by the addition of tartaric acid, which forms a soluble ion, $SbOC_4H_4O_6^-$; antimonic oxide is also insoluble in water and in solutions of non-complex-forming acids, but it dissolves in tartrate solutions, forming the ion $SbO_2C_4H_4O_6^-$.

2. Metal ions, such as zinc, cadmium, and lead, can be precipitated as sulfides, the precipitates treated with an appropriate acid and the sulfide oxidized with standard iodine solution.

3. Sb(V) can be reduced by iodide under essentially the same conditions used for As(V).

4. Metal ions, such as barium, strontium, and lead, can be precipitated as chromates, the precipitates treated in an acid solution with excess iodide and the iodine titrated with thiosulfate.

5. This procedure can be applied to PbO_2, Pb_2O_3, and Co_2O_3.

6. This procedure can be used for Br_2.

7. Bromate and chlorate can be determined similarly. They tend to react slowly with iodide; therefore molybdenum compounds are frequently used as catalysts.

more frequently employed are mentioned below and summarized in Table 21-1. Methods involving other oxidation states of iodine are discussed in Chapter 22.

These methods can be conveniently classified into three general types, as follows:

1. Methods involving the direct titration of a reducing constituent with iodine. Certain elements that are thus determined, their change in oxidation state, and the nature of the solution in which the titrations are made are shown in Table 21-1.
2. Methods involving the indirect determination of oxidizing constituents by adding an excess of iodide to the oxidant and titrating the resulting iodine with thiosulfate. These titrations are almost invariably carried out in acid solutions and some of the most frequently used are shown in Table 21-1.
3. Indirect methods involving the precipitation of certain cations with either (*a*) a reducing or (*b*) an oxidizing anion followed by determination of the anion by method 1 or 2. An example of type (*a*) is the determination of zinc or cadmium by precipitation as sulfide, treatment of the precipitate with excess iodine, and titration of excess iodine with thiosulfate. The reaction in acid solution is

$$ZnS(s) + I_3^- = Zn^{2+} + S + 3I^-$$

An example of type (*b*) is the determination of lead by precipitation as chromate, treatment of the precipitate in acid solution with excess iodide, and titration of the resulting iodine with thiosulfate.

QUESTIONS AND PROBLEMS

Stoichiometric Calculations

1. What is the normality (as a standard oxidizing agent) of a 0.1 *F* KI_3 solution?

2. An iodine solution was standardized against 0.2100 g of As_2O_3 (pages 450–453). The titration required 40.20 ml of the iodine solution. Calculate the normality of the iodine solution. *Ans.* 0.1056 *N*.

3. A standard solution of potassium dichromate was prepared by weighing out and dissolving 4.500 g of $K_2Cr_2O_7$, then diluting the solution to exactly 1 liter. A 50.00-ml portion treated by Procedure 21-2 required 42.37 ml of a thiosulfate solution. Calculate the normality of the thiosulfate solution.

4. In the standardization of a thiosulfate solution 0.1250 g of KIO_3 was weighed and dissolved in 50 ml of water; 2 g of KI dissolved in 10 ml of water were added, followed by 2 ml of 6 *F* HCl. The solution was titrated with the thiosulfate, 41.02 ml

being required. Calculate the normality of the thiosulfate. Calculate the percentage error caused by 0.2 mg weighing error. Ans. 0.16%

How could the procedure be modified to minimize this weighing error?

5. A 0.2042-g sample of a red brass containing copper and zinc was analyzed by Procedure 21-4, and 27.02 ml of 0.1019 *N* thiosulfate were used. Calculate the percentage composition of the alloy. *Ans.* 85.7% Cu.

6. Stibnite ores (crude Sb_2S_3) can be analyzed iodometrically by the following procedure: the sample is dissolved in HCl [the H_2S produced reduces any Sb(V) to Sb(III)], tartaric acid is added to form soluble tartrate complexes with the antimony, the solution is buffered with a hydrogen-carbonate-ion–carbonic acid buffer, and standard I_3^- solution is added to the starch iodine end point. When a 0.6612-g sample of a stibnite ore was treated in this way, 40.10 ml of a 0.1014 *N* iodine solution were required for the titration. Calculate the percentage of antimony trisulfide in the ore. What would be the effect on this determination of the presence of arsenic in the ore?

7. A determination of the hydrazine (N_2H_4) in a solution was made by adding to the acid solution an excess of an iodate solution. The iodate oxidizes the hydrazine to nitrogen. After the completion of the reaction between the iodate and the hydrazine, an excess of potassium iodide was added and the solution was titrated with thiosulfate. The data obtained were as follows: iodate used, 60.00 ml of 0.02000 *F* KIO_3; iodide used, 3.0 g of solid KI; thiosulfate used, 32.00 ml of 0.1000 *F* $Na_2S_2O_3$. Calculate the weight of hydrazine in the solution. *Ans.* 32.05 mg.

NOTE: In solving this and similar problems, the principle of the conservation of equivalents should be used, since for the over-all processes the total equivalents of oxidizing agent must be equal to the total equivalents of reducing agent, and the equivalents of either provided by a reactant will be determined by its *initial* and *final* oxidation states. In this case the oxidizing agent is the iodate, and the oxidizing equivalents added are determined by the quantity of iodate added and by the initial and final oxidation states of the iodate iodine—the intermediate reactions are of no consequence. Therefore the reducing equivalents of hydrazine and of thiosulfate must equal the equivalents of iodate. Since the initial and final oxidation states of the potassium iodide are the same, the quantity present does not enter the calculation.

8. An analyst substituted standard solutions of $NaHSO_3$ and of Na_2SO_3 for the $Na_2S_2O_3$ used in Procedure 21-4 for the determination of copper.

   (*a*) When involved in that procedure what is the normality of 0.1 *F* solutions of each of the following:

   (*i*) $CuSO_4$ (*iii*) Na_2SO_3

   (*ii*) $Cu(NO_3)_2$ (*iv*) $NaHSO_3$

   (*b*) The titration of 10.0 ml of a KI_3 solution required 2.50 ml of a 0.100 *F* Na_2SO_3 solution. Calculate the formality and normality of the KI_3 solution.

   (*c*) The titration of 20.0 mg of copper(II) in 12 ml of an H_2SO_4 solution required 5.00 ml of a $NaHSO_3$ solution to react with the liberated iodine. Calculate (*i*) the formality of the sodium hydrogen sulfite solution and (*ii*) the number of mg of $NaHSO_3$ per 10.0 ml of the solution.

(*d*) Calculate the number of mg of copper(II) present in 8.0 ml of a solution if 1.60 ml of 0.50 *F* Na_2SO_3 were required for a titration.

9. A permanganate solution was standardized against sodium oxalate and then used to standardize a thiosulfate solution. A sample of brass (containing Cu, Pb, Sn, and Zn) was treated with nitric acid, the mixture evaporated almost to dryness, dilute nitric acid was added, and the nitric acid precipitate was filtered out, ignited to 1000°C–1200°C, and weighed. An excess of sulfuric acid was added to the filtrate, the mixture was evaporated to fuming, diluted with water, and the precipitate was filtered, dried at 500°C–600°C, and weighed. The filtrate was evaporated to 25 ml, made just alkaline with ammonia, then just acid with sulfuric acid; then an excess of 2 ml of 3 *F* H_2SO_4 was added, followed by 4 g of KI, and the resulting solution was titrated with the thiosulfate solution.

The data obtained were as follows:

Weight of sodium oxalate	0.2009	g
Volume of permanganate used by oxalate	31.50	ml
Volume of permanganate used for thiosulfate standardization	25.00	ml
Volume of thiosulfate used by permanganate	23.75	ml
Weight of brass sample	0.3979	g
Weight of "nitric acid precipitate"	0.01262	g
Weight of "sulfuric acid precipitate"	0.0495	g
Volume of thiosulfate used for copper titration	40.00	ml

Calculate:

(*a*) The formal concentration of the permanganate solution.
(*b*) The normal concentration of the permanganate in the oxalate reaction.
(*c*) The normal concentration of the permanganate solution if used to titrate a neutral solution of manganous sulfate. *Ans.* 0.05709 *N*.
(*d*) The formal concentration of the thiosulfate solution in the standardization procedure.
(*e*) The normal concentration of the thiosulfate when used in the copper titration.
(*f*) The normal concentration of the thiosulfate solution when used in the following reaction:

$$4Br_2 + 5H_2O + S_2O_3{}^{2-} = 2HSO_4{}^- + 8H^+ + 8Br^-$$

(*g*) The complete percentage composition of the alloy assuming that only four constituents are present.

State possible effects of carrying out the following:

(*1*) The permanganate-oxalate titration in a neutral solution.
(*2*) The permanganate-iodide reaction in a neutral solution.
(*3*) The iodine-thiosulfate titration in a neutral solution.
(*4*) The cupric copper-iodide reaction in a neutral solution.
(*5*) The iodine-thiosulfate titration in an alkaline solution.

Calculations Involving Chemical Equilibria

10. Why does solid I_2 not precipitate in Procedure 21-4 when large quantities of copper are present?

11. Calculate and insert the proper value in the blank space of the table below:

Formal Solubilities of Iodine (25°C)

Solution	Water	0.00500 *F* KI	0.100 *F* KI
I_2 (*formality*)	0.00134	0.00388	—

12. The vapor pressure of solid iodine at 25°C is 0.40 mm of mercury. What is the partial pressure of iodine above a saturated solution at 25°C? Calculate the partial pressures of iodine above the KI solutions of Problem 11 above.

13. It was desired to determine the amount of arsenite and also of arsenate in a sample of crude sodium arsenite. A 0.5000-g sample of the material was taken and dissolved in 100 ml of 2 *F* HCl. One ml of a 0.0025 *F* ICl solution was added and the mixture was titrated with 0.0300 *F* $KMnO_4$ solution to an end point. *o*-Phenanthroline ferrous complex was used as the indicator, and 25.00 ml of the permanganate were required. Then 25 ml of 12 *N* HCl and 3 g of KI were added, and the solution was allowed to stand for 5 minutes. The resulting iodine was titrated with 0.1000 *F* $Na_2S_2O_3$ to a starch end point: 50.00 ml of the thiosulfate were required.

    (*a*) Write equations for the reactions involved.
    (*b*) What is the mechanism involved in the action of ICl? Write equations.
    (*c*) Discuss the possibility of using KIO_3 instead of ICl. What are the advantages and disadvantages of allowing the solution to stand after the addition of KI?
    (*d*) Calculate the percentage of arsenite as As_2O_3 and arsenate as As_2O_5.
    (*e*) What would be the minimum hydrogen-ion concentration possible for all but 0.001% of the arsenate to be reduced in a titration similar to that involved in the second part of this determination? Assume that the initial amount of Na_3AsO_4 was 2.5 mmoles, that 3 g of KI were used, that the I_3^- concentration was 5×10^{-6} *M* at the end point, and that the final volume was 250 ml.

CHAPTER **22**

Oxidation-Reduction Reactions. Elements Frequently Involved in Titrimetric Methods

This chapter presents a general discussion of certain of those elements and their compounds involved in redox reactions of importance in titrimetric methods of analysis. As is shown in Figure 22-1 these elements appear in the periodic table as several groups of contiguous members and there appears to be some advantage in making use of this grouping in the following discussion.

The first of these groups is shown in Figure 22-1 and begins with titanium of column IVb and includes the next four transition elements, namely, vanadium, chromium, manganese, and iron. It is of interest that the elements of this group together with cerium and iodine are constituents of most of the standard solutions used in redox titrations.

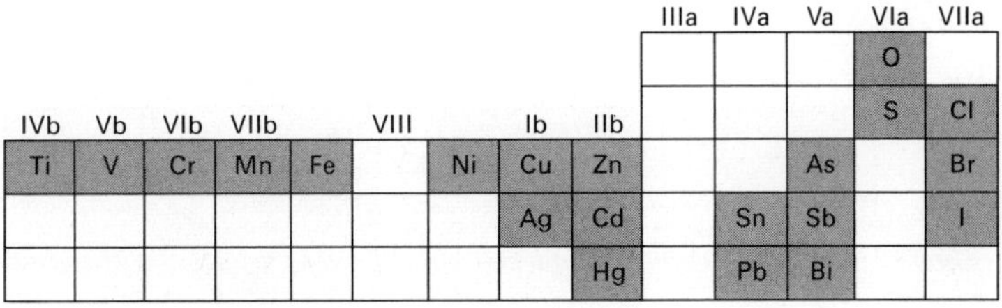

IVb	Vb	VIb	VIIb		VIII		Ib	IIb	IIIa	IVa	Va	VIa	VIIa
												O	
												S	Cl
Ti	V	Cr	Mn	Fe		Ni	Cu	Zn			As		Br
							Ag	Cd		Sn	Sb		I
								Hg		Pb	Bi		

FIGURE 22-1
Periodic positions of elements of redox analytical importance.

Proceeding from the left to right across the table there are three elements—nickel, copper, and zinc—with silver, cadmium, and mercury below them, which constitute a group of six elements whose common analytical characteristic is that they are all used, to varying extents, in the metallic state as pretitration reductants—the reduction of other elements to definite oxidation states prior to a titration with an oxidant. Although they are toward the right side of the periodic table these elements are classified as metals, have only stable positive oxidation states, and, with the exception of zinc, the oxides of their lowest oxidation states show no amphoteric properties.

Further to the left is a group comprising tin, lead, arsenic, antimony, and bismuth. These elements are in the transition region between the typically metallic and nonmetallic elements and the last three are classified as metalloids. Their properties vary and their uses range from metallic reductants to powerful oxidants.

The two elements oxygen and sulfur are of analytical interest because water is such a universally used solvent; the O_2^{2-} and S_2^{2-} ions are useful oxidants; hydrogen sulfide is not only a reductant but the most extensively used precipitant in qualitative systems and quantitative separations.

The last group comprises three elements of column VIIa, the halogens; these elements are typically nonmetallic and their oxides are acid-forming. These groups are discussed in more detail below.

TRANSITION ELEMENTS OF ANALYTICAL IMPORTANCE

This group, the transition elements, comprise the five elements of the first long row or period of the periodic table, beginning with titanium, atomic number 22, and extending through iron, number 26. It is an interesting fact that these five elements are involved in a surprising majority of oxidation-reduction methods of titrimetric analysis. The position of these transition elements in the periodic table, their atomic numbers, electronic configurations, and more analytically important oxidation states are shown in Figure 22-2.

IVb	Vb	VIb	VIIb	VIII
22 Ti +3 +4 [Ar] $3d^2 4s^2$	23 V +4 +5 [Ar] $3d^3 4s^2$	24 Cr +2 +3 +6 [Ar] $3d^5 4s^1$	25 Mn +2 +3 +4 +6 +7 [Ar] $3d^5 4s^2$	26 Fe +2 +3 [Ar] $3d^6 4s^2$

FIGURE 22-2

Transition elements of analytical importance.

TABLE 22-1
Analytically Important Half-Cell Reactions and Potentials of Certain Transition Elements

Half-cell reaction	Potential
$\mathbf{Cr^{2+}} = Cr^{3+} + e^-$	+0.408
$V^{2+} = V^{3+} + e^-$	+0.26
$H_2(g) = 2H^+ + 2e^-$	+0.000
$\mathbf{Ti^{3+}} + H_2O = TiO^{2+} + 2H^+ + e^-$	−0.099
$V^{3+} + H_2O = VO^{2+} + 2H^+ + e^-$	−0.34
$MnO_4^{2-} = \mathbf{MnO_4^-} + e^-$	−0.564
$\mathbf{Fe^{2+}} = Fe^{3+} + e^-$	−0.782
$VO^{2+} + 3H_2O = \mathbf{V(OH)_4^+} + 2H^+ + e^-$	−1.00
$2H_2O = O_2(g) + 4H^+ + 4e^-$	−1.229
$Mn^{2+} + 2H_2O = MnO_2(s) + 4H^+ + 2e^-$	−1.24
$2Cr^{3+} + 7H_2O = \mathbf{Cr_2O_7^{2-}} + 14H^+ + 6e^-$	−1.33
$Mn^{2+} + 4H_2O = \mathbf{MnO_4^-} + 8H^+ + 5e^-$	−1.45
$MnO_2(s) + 2H_2O = \mathbf{MnO_4^-} + 4H^+ + 3e^-$	−1.59

NOTE: Ions appearing in standard solutions are in boldface type.

The most obvious explanation for the analytical use of these elements is that all have more than one oxidation state and each can exist in aqueous solutions in one or more of these states for considerable periods. This makes possible their use as standard solutions.

Table 22-1 shows the more analytically important half-cell reactions of these elements; where an ion is used as a constituent of a standard solution it appears in **boldface**. The hydrogen and oxygen half-cells are given as reference values. (A much more comprehensive table of half-cell potentials is given in Appendix 4.)

Standard Solutions of Oxidants

The most important use of these elements is as standard solutions of oxidants and the three elements so used in order of their importance are manganese (VII) as permanganate (discussed in Chapter 18), chromium (VI) as chromate or more frequently dichromate, and vanadium (V) as vanadate. Cerium, a member of the lanthanide series, is used extensively in the quadripositive state as a standard oxidant; such methods are given in Chapter 19.

Permanganate Solutions. Because of the strong oxidizing tendency of MnO_4^- and of the numerous lower oxidation states of manganese, standard solutions of permanganate have been used in a wider range of methods than any other single oxidant. This diversity of methods and oxidation states is illustrated in Table 22-2 below.

By far the most frequent use of permanganate solutions involves titrations in strongly acid solutions in which the reduction product is manganous ion; examples of such methods are discussed in detail in Chapter 18.

TABLE 22-2

Types of Permanganate Titrations

Titration Reaction	Titration Solution	Comments
$MnO_4^- \rightarrow Mn^{2+}$	Acid; $[H^+] > 0.5$	Most extensively used titration
$MnO_4^- \rightarrow$ Mn(III) complex	pH 6–pH 7	Complexing agents used: F^- and $H_2P_2O_7^{2-}$
$MnO_4^- \rightarrow MnO_2(s)$	Neutral or slightly acid	
$MnO_4^- \rightarrow MnO_4^{2-}$	Alkaline; $[OH^-] > 1$	Ba^{2+} present to precipitate $BaMnO_4$

The reduction product can be changed from the dipositive to the tripositive state if the titration is made in an approximately neutral solution, and if certain ligands, such as fluoride or pyrophosphate, are present that form stable complex ions with Mn(III); the simple tripositive ion, Mn^{3+} is unstable because of the disproportionation reaction

$$2Mn^{3+} + 2H_2O = Mn^{2+} + MnO_2(s) + 4H^+$$

The use of the dihydrogen pyrophosphate ion, $H_2P_2O_7^{2-}$, has been studied by Lingane and Karplus[1] for the determination of manganese by such a titration, and Watters and Kolthoff[2] have studied the Mn(II), Mn(III), pyrophosphate system.

In approximately neutral solutions and in the absence of a ligand forming a stable complex, the reduction product of a permanganate titration will be $MnO_2(s)$. Initially, strong reductants will give Mn^{2+}, but because of the reaction

$$3Mn^{2+} + 2MnO_4^- + 2H_2O = 5MnO_2 + 4H^+$$

a stable visual MnO_4^- end point will not be obtained until the Mn(II) is oxidized to MnO_2. The Volhard method for determining manganese is based on the above reaction and has been discussed in Chapter 18.

It has been shown that in alkaline solutions the reduction of permanganate to MnO_2 by certain reductants involves manganate, MnO_4^{2-}, as an intermediate step. Accordingly methods have been proposed that are carried out

---

[1] J. J. Lingane and R. Karplus. *Ind. Eng. Chem.*, anal. ed., **18**, 191 (1946).

[2] J. I. Watters and I. M. Kolthoff. *J. Am. Chem. Soc.*, **70**, 2455 (1948). They found a value of −1.01 v for the formal potential of the half-cell reaction

$$Mn(H_2P_2O_7)_2^{2-} + H_4P_2O_7 = Mn(H_2P_2O_7)_3^{3-} + 2H^+ + e^-$$

in a solution 0.4 *F* in pyrophosphate and having a *p*H of 2.

in strongly alkaline solution and in which the manganate stage is stabilized by the addition of barium ion to cause precipitation of $BaMnO_4$ (K_{sp} = 2.5×10^{-10}). Such methods have not been widely used. A discussion and review of these methods is given in the monograph, *Newer Methods of Volumetric Chemical Analysis*, translated by Oesper (New York: Van Nostrand, 1938).

Dichromate Solutions. Like permanganate, chromate solutions were first used for the determination of ferrous iron and Frederick Penny[3] appears to have been the first to propose the titration with the use of ferricyanide as an external indictor.[4] He first used potassium chromate, but soon recognized the advantages of the dichromate.

As seen in Table 22-1, the standard potential of the $Cr^{3+}Cr_2O_7^{2-}H^+$ half-cell is -1.3 v, but various studies have shown that the formal potentials in 1 *F* hydrochloric, sulfuric, and perchloric acids are nearer -1.0 v. As would be expected from the hydrogen-ion dependence, the potential rapidly becomes less oxidizing with decrease of acid concentration.

Standard solutions of potassium dichromate offer certain advantages; they can be prepared directly by weighing primary standard material and diluting to volume or weight, they are extremely stable, and titrations can be made in hydrochloric acid solutions. For this latter reason, dichromate solutions formerly were extensively used for the titration of ferrous salts in hydrochloric acid solutions where the titration with permanganate is troublesome. Dichromate solutions have the disadvantage that the color of the chromate is not sufficiently intense for it to be used to obtain the end point of the titration (especially in the presence of the green chromic ion that is the reduction product); therefore, some form of an indicator has to be used, or the end point must be obtained potentiometrically; diphenylamine sulfonate is most frequently used as an internal indicator. Ceric sulfate solutions have replaced dichromate solutions for many purposes because of the

---

[3] Frederick Penny. *Advan. Sci.* **18**, [2], 58 (1850).

[4] This method became known as the Penny method for iron. The end point was obtained by removing and mixing drops of the titrated solution with drops of a potassium ferricyanide solution until no further formation of a blue precipitate or color was observed.

The extremely insoluble blue compounds formed by Fe(II) and $Fe(CN)_6^{3-}$ (hexacyanoferrate(III), or ferricyanide), Turnbull's Blue, and by Fe(III) and $Fe(CN)_6^{4-}$ (hexacyanoferrate(II) or ferrocyanide), Prussian Blue, are interesting because of the uncertainty as to the oxidation states of the iron atoms. Evidence indicates that because of the equilibrium

$$Fe^{3+} + Fe(CN)_6^{4-} = Fe^{2+} + Fe(CN)_6^{3-}$$

the precipitates obtained in the presence of potassium salts are the same substance, have the formula $KFeFe(CN)_6$, and the oxidation states are $KFe(III)[Fe(II)(CN)_6]$. See discussion by L. D. Hansen, W. M. Litchman, and G. H. Daub. *J. Chem. Ed.*, **46**, 46 (1969).

less oxidizing formal potential of the dichromate, and because of troublesome rate effects and side reactions in certain cases.[5]

Vanadate Solutions. Vanadium can exist as V(V), V(IV), V(III), and V(II), but the only half-cell reaction that has had any significant use involves V(V) as an oxidant and its reduction to V(IV); as seen in Table 22-1, the standard half-cell potential is −1.00 v. Thus, vanadate solutions are comparable to dichromate solutions as an oxidant. Such solutions have been suggested as replacements for dichromate and ceric solutions for various determinations, but have not found extensive use.

Standard Solutions of Reductants

Standard solutions of reductants are used to a limited extent for two reasons. First, as is seen from Table 22-1, any reductant with a half-cell potential more positive than that for oxygen would be thermodynamically unstable if exposed to air. Therefore, in order to keep their concentrations constant, elaborate means for protecting their solutions from exposure to the air are necessary; frequently they are kept and delivered under an inert atmosphere such as hydrogen or nitrogen. Alternatively, they can be standardized before each period of use. Second, a reductant with a potential approximately equal to or more positive than that of the hydrogen half-cell would also be unstable with respect to oxidation by hydrogen ion, and, unfortunately, all three of the standard reductants shown in Table 22-1, Cr(II), Ti(III), and Fe(II), require an acid solution to prevent hydrolysis effects. Fortunately, the rates at which oxygen and hydrogen ion react with the above reductants are not extremely rapid.

Chromium(II) Solutions. As seen in Table 22-1, chromium(II) is the strongest reductant listed and is the strongest used as a standard solution. This use is possible only because of the slowness of its reaction with hydrogen ions; if properly prepared and stored, solutions have been found to be stable for months.[6] A method for the determination of nitrate by reduction to ammonia has been devised by Lingane and Pecsok that provides an interesting example of the use of a catalyst.[7] The rate of the reaction between nitrate ion and chromous ion in acid solutions was found to be inconveniently slow even when an excess of the Cr(II) was added. However, it was known that the reduction of nitrate by titanous ion to nitric oxide, NO, was relatively

[5] Laitinen (*Chemical Analysis.* New York: McGraw-Hill, 1960) gives a discussion of certain of these rate effects.

[6] Lingane and Pecsok. *Anal. Chem.*, **20**, 425 (1948); H. W. Stone. *Anal. Chem.*, **20**, 747 (1948).

[7] Lingane and Pecsok. *Anal. Chem.*, **21**, 622 (1949).

rapid and that the reduction of titanic ion by Cr(II) was also rapid. Therefore, Lingane and Pecsok reasoned that a titanium salt, either Ti(IV) or Ti(III), should act as a catalyst for the nitrate-chromium(II) reactions. Their reasoning proved to be correct.

Titanium(III) Solutions. As is apparent from Table 22-1, titanous ion is not as powerful a reductant as chromous ion. Solutions can be prepared from commercially available titanous chloride or by reduction of Ti(IV) compounds with amalgamated zinc. Standard solutions should be kept under an inert atmosphere, but they are not as rapidly oxidized by oxygen as are Cr(II) solutions.

Iron(II) Solutions. At one time the compound known as Mohr's salt, $FeSO_4 \cdot (NH_4)_2SO_4 \cdot 6H_2O$, was extensively used as a primary standard for the standardization of permanganate solutions and for the preparation of standard ferrous solutions. The compound is difficult to obtain in a high state of purity and is now used infrequently as a primary standard. Ferrous sulfate solutions are frequently used in conjunction with permanganate solutions; the titration of ferrous iron with permanganate is discussed in detail in Chapter 18.

ELEMENTS USED AS METALS FOR PRETITRATION REDUCTIONS

The metals of columns VIII, Ib, and IIb shown in Figure 22-1 have been used in metallic form as agents for reducing various elements to a specific oxidation state prior to titrations with solutions of standard oxidants. Their positions in the periodic table, electronic configurations, and stable oxidation states are shown in Figure 22-3.

Zinc

As is evident from the half-cell potentials shown in Table 22-3, zinc is the most reducing of these metals and it is the one most frequently used. It is usually lightly amalgamated; the mercury surface has a high hydrogen *overvoltage* (see pages 621–624) that minimizes oxidation of the zinc by hydrogen ion. Most frequently, the zinc is placed in a vertical tube and the solution to be reduced poured through the tube; such an apparatus is known as a Jones reductor.

The most common reductions for which zinc is used are Fe(III) to Fe(II), Ti(IV) to Ti(III), V(V) to V(II), Mo(VI) to Mo(III), and U(VI) to U(III) and U(IV); before the titration the U(III) is oxidized to U(IV) by exposure to air.

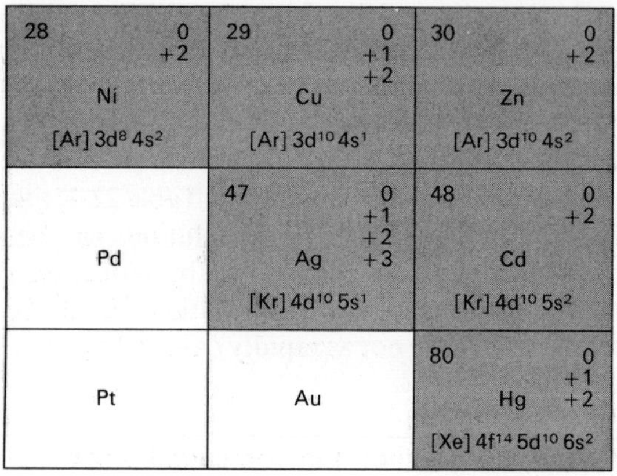

28 Ni (0, +2) [Ar] $3d^8 4s^2$	29 Cu (0, +1, +2) [Ar] $3d^{10} 4s^1$	30 Zn (0, +2) [Ar] $3d^{10} 4s^2$
Pd	47 Ag (0, +1, +2, +3) [Kr] $4d^{10} 5s^1$	48 Cd (0, +2) [Kr] $4d^{10} 5s^2$
Pt	Au	80 Hg (0, +1, +2) [Xe] $4f^{14} 5d^{10} 6s^2$

FIGURE 22-3
Elements used as pretitration metal reductants.

The zinc potential is sufficiently powerful to reduce these elements to the metallic state, but the reduction is carried out in acid solutions where the hydrogen ion tends to act as a so-called potential mediator and prevents reductions requiring potentials more positive than that of the hydrogen–hydrogen-ion half-cell. The only stable positive oxidation state of zinc is Zn(II).

The remainder of the metals of this group have not been used extensively as general reductants but have been used as selective reductants, alone or in conjunction with precipitating anions, complex ions, or in the form of liquid amalgams.

TABLE 22-3
Half-Cell Potentials of Elements Used as Metallic Reductants

Half-reaction	Potential
$Zn(s) = Zn^{2+} + 2e^-$	+0.76
$Cd(s) = Cd^{2+} + 2e^-$	+0.41
$Ni(s) = Ni^{2+} + 2e^-$	+0.25
$Cu(s) + Cl^- = CuCl(s) + e^-$	−0.14
$Cu^+ = Cu^{2+} + e^-$	−0.17
$Ag(s) + Cl^- = AgCl(s) + e^-$	−0.22
$2Hg(l) + 2Cl^- = Hg_2Cl_2(s) + 2e^-$	−0.27
$Cu(s) = Cu^{2+} + 2e^-$	−0.34
$2Hg(l) = Hg_2{}^{2+} + 2e^-$	−0.79
$Ag(s) = Ag^+ + e^-$	−0.80
$Ag^+ = Ag^{2+} + e^-(1\ F\ HNO_3)$	−1.91

TABLE 22-4
Reducing Action of Liquid Amalgam Reductors

Element and Initial Oxidation State	Oxidation State after Reduction by Amalgam[1]				
	Zn	Cd	Bi	Pb	Sn
Ti(IV)	Ti(III)	Ti(III)	Ti(III)	Ti(III)	Ti(III)
V(V)	V(II)	V(II)	V(IV)	V(II)	V(II)
Cr(V), (III)	Cr(II)	Cr(III)	Cr(III)	Cr(III)	Cr(III)
Fe(III)	Fe(II)	Fe(II)	Fe(II)	Fe(II)	Fe(II)
Mo(VI)	Mo(III)	Mo(III)	Mo(III), (V)	Mo(III)	Mo(III)
U(VI)	U(III, (IV)	U(IV)	U(IV)	U(IV)	U(IV)

[1] With certain metals the final oxidation state will depend upon the nature and concentration of acid present.

Amalgams

In about 1922 an extensive series of studies was begun of the use of various liquid amalgams for the pretitration reduction of both cations and anions as well as certain organic compounds.[8] The reduction is carried out in specially designed separatory funnels and the procedure has several advantages: (1) by selection of the proper metal selective reductions can be achieved; (2) similar to the Jones reductor, the mercury surface decreases hydrogen evolution with the more active metals; (3) in most cases the reduction can be carried out in a much shorter time than with a solid amalgam as in the Jones reductor; (4) removal of the excess reductant is easily done by use of the special separatory funnel; and (5) the amalgam can be easily prepared, renewed, and stored.

Table 22-4 shows some of the reduction products of various amalgams.

Cadmium Metal as Reductant

Cadmium metal has been used as a general pretitration reductant and procedures involving the amalgam have been described; they have not been used extensively. Cadmium, like zinc, has only one stable positive oxidation state, Cd(II).

Mercury and Silver

Mercury has been used as a reductant, usually in hydrochloric acid solutions, where the product is Hg_2Cl_2. Investigations have shown that if oxygen is

[8] An extensive review of the use of such amalgams is given by W. I. Stephen (*Ind. Chemist*, **29**, 31, 79, 128, 169, 1953). Briefer reviews are given by Kolthoff and Belcher (*Volumetric Analysis*. Vol. 3. New York: Wiley-Interscience, 1957, p. 18) and by Laitinen (op. cit., p. 353).

present hydrogen peroxide is likely to be formed;[9] therefore, oxygen should be excluded from the solutions.

Silver has also been used as a reductant, and in the presence of hydrochloric acid has a reducing action similar to that of mercury. It is used in a tube similar to but shorter than a Jones reductor because less bulk of metal is required. Contrary to the action of mercury in hydrochloric acid solutions, it has been shown[10] that no peroxide is formed in the presence of air.

Silver can exist as Ag(II) and Ag(III). Silver(I) can be oxidized in nitric acid solutions by ozone to Ag(II) and Ag(III) and the solutions have been used as oxidants; silver(II) oxide is commercially available and has been used as an oxidant.[11]

Nickel and Copper

Nickel as foil has been proposed for reducing Sn(IV) to Sn(II)[12] in hydrochloric acid solutions. The time required for the reduction is less if an antimony salt is added; apparently the metallic antimony deposited on the nickel reacts more rapidly than does the nickel. Copper powder has been used for reducing Fe(III) to Fe(II) in sulfuric acid solutions.[13] Otherwise these two metals have not been commonly used as metallic reductants.

Copper in the +1 state tends to disproportionate unless stabilized by formation of insoluble compounds such as CuCl or CuI or complex ions such as $CuCl_3^{2-}$ or $Cu(CN)_2^-$. The reaction of Cu(II) with iodide is discussed in Procedure 21-4.

ELEMENTS OF GROUPS IVa AND Va OF ANALYTICAL INTEREST

The electronic structures and common oxidation states of these elements are shown in Figure 22-4, and half-cell potentials of analytical importance are shown in Table 22-5.

Arsenic

In redox titrimetric analysis arsenic is primarily known because of the reaction

$$H_3AsO_3 + I_3^- + H_2O = H_3AsO_4 + 3I^- + 2H^+$$

[9] Furman and Murray. *J. Am. Chem. Soc.*, **58**, 429, 1689 (1936).
[10] Walden, Hammett, and Edmonds. *J. Am. Chem. Soc.*, **56** 350 (1954).
[11] Lingane and Davis. *Anal. Chim. Acta*, **15**, 201 (1956).
[12] Hallett. *J. Soc. Chem. Ind.*, **35**, 1087 (1916).
[13] Percival. *Ind. Eng. Chem.*, anal. ed., **13**, 71 (1941).

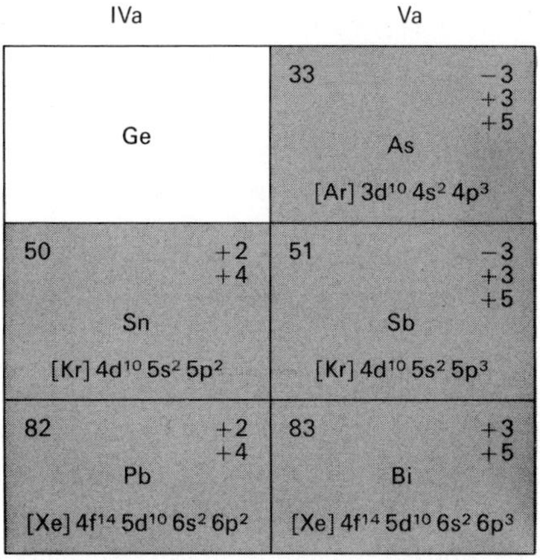

IVa	Va
Ge	33 As −3, +3, +5 [Ar] $3d^{10} 4s^2 4p^3$
50 Sn +2, +4 [Kr] $4d^{10} 5s^2 5p^2$	51 Sb −3, +3, +5 [Kr] $4d^{10} 5s^2 5p^3$
82 Pb +2, +4 [Xe] $4f^{14} 5d^{10} 6s^2 6p^2$	83 Bi +3, +5 [Xe] $4f^{14} 5d^{10} 6s^2 6p^3$

FIGURE 22-4

A discussion of this reaction is given on pages 450–456. Arsenic can be determined by oxidation of As(III) to As(V) in acid solutions by such strong oxidants as bromine or bromate, permanganate, iodate (IO_3^- to ICl), and ceric sulfate; electrolytically generated bromine is used in Procedure 29-1.

Antimony

Like arsenic, antimony is determined by titration in a neutral solution with iodine. However, antimony is below arsenic in column Va, and, as would be expected, its oxides are less acidic and are quite insoluble in neutral solutions. For this reason an organic complexing agent, tartaric acid, is added to form

TABLE 22-5

Half-Cell Potentials of Analytical Importance of Certain Elements of Columns IVa and Va

Half-reaction		
$AsH_3(g) = As(s) + 3H^+$	$+3e^-$	+0.6
$SbH_3(g) = Sb(s) + 3H^+$	$+3e^-$	+0.5
$Sn(s) = Sn^{2+}$	$+2e^-$	+0.14
$Pb(s) = Pb^{2+}$	$+2e^-$	+0.12
$Sn^{2+} = Sn^{4+}(1\ F\ HCl)$	$+2e^-$	−0.14
$Sb(s) + H_2O = SbO^+ + 2H^+$	$+3e^-$	−0.21
$As(s) + 3H_2O = H_3AsO_3 + 3H^+$	$+3e^-$	−0.24
$Bi(s) + H_2O = BiO^+ + 2H^+$	$+3e^-$	−0.31
$H_3AsO_3 + H_2O = H_3AsO_4 + 2H^+$	$+2e^-$	−0.58
$Sb_2O_3(s) + 2H_2O = Sb_2O_5(s) + 4H^+$	$+4e^-$	−0.7
$Pb^{2+} + 2H_2O = PbO_2(s) + 4H^+$	$+2e^-$	−1.47
$2Bi^{3+} + 5H_2O = Bi_2O_5 + 10H^+$	$+4e^-$	−1.7

the soluble ions that are assumed to be $SbOC_4H_4O_6^-$ and $SbO_2C_4H_4O_6^-$. The titration reaction is carried out under the same conditions as is the titration of H_3AsO_3.

Finely powdered antimony metal was once used as a means of reducing Sn(IV) to Sn(II) in hot hydrochloric acid solutions,[14] but has been superceded by other metals.

Bismuth

The metal has been used, commonly as an amalgam, to a limited extent as a selective reductant. Bismuth is much more commonly used as a powerful oxidant both in quantitative and qualitative methods in the form of the so-called sodium bismuthate. This material will oxidize Mn(II) to MnO_4^- and Ce(III) to Ce(IV). The excess can be removed easily by filtration, which adds to the convenience of the preoxidation.

The behavior of bismuth and the other elements of column Va of the periodic system are illustrative of the change in acid-base properties as one goes down a column of the periodic table. At the top of column Va we find nitrogen, which is predominantly an acid-forming element even in its tripositive state. As we pass down this column, from phosphorus to arsenic, to antimony, to bismuth, we find the tripositive state progressively losing its acidic character, and becoming more basic; bismuth trioxide is very insoluble even in concentrated hydroxide solutions. However, when oxidized to the quinquepositive state, bismuth still retains enough acidic character to form salts such as $NaBiO_3$, sodium metabismuthate. This compound (as well as the acid, $HBiO_3$, and the oxide, Bi_2O_5) is unstable; the commerical product, sold as sodium bismuthate, is usually a mixture of $NaBiO_3$ and an oxide having the empirical formula BiO_2; this latter compound can be considered to be either $BiBiO_4$, bismuth orthobismuthate, or $BiO{\cdot}BiO_3$, bismuthyl metabismuthate. The oxidation reaction can be formulated in terms of the oxide as follows:

$$2Bi^{3+} + 5H_2O = Bi_2O_5 + 10H^+ + 4e^- \qquad E^\circ = -1.7$$

The $HBiO_3$ that is assumed to be formed when $NaBiO_3$ is added to an acid solution is such a powerful oxidizing agent that it decomposes, with formation of O_2 and Bi(III), as follows:

$$2HBiO_3 + 6H^+ = 2Bi^{3+} + O_2 + 4H_2O$$

This reaction is slow at 20°C but becomes very rapid in hot solutions;

[14] Ibbotson and Brearly. *Chem. News*, **84**, 167 (1901).

therefore oxidations should be made in cold solutions; the excess reagent can be removed by filtration or by heating the solution if the oxidation product is stable.

Tin and Lead

These elements are both used as metallic reductants although not extensively. Tin is used most frequently in the bipositive state for the pretitration reduction of ferric iron in hydrochloric acid solutions. This process is used and discussed in Procedure 18-3. Lead dioxide is a powerful oxidizing agent in concentrated nitric acid solution and at one time was used for the oxidation of Mn(II) to permanganate; it has been replaced by peroxydisulfate catalyzed by silver, by periodate, and by sodium bismuthate.

ELEMENTS OF GROUP VIa OF ANALYTICAL IMPORTANCE

Oxygen and sulfur are the only elements of group VIa of significant analytical use. Their electronic structures and common oxidation states are shown in Figure 22-5.

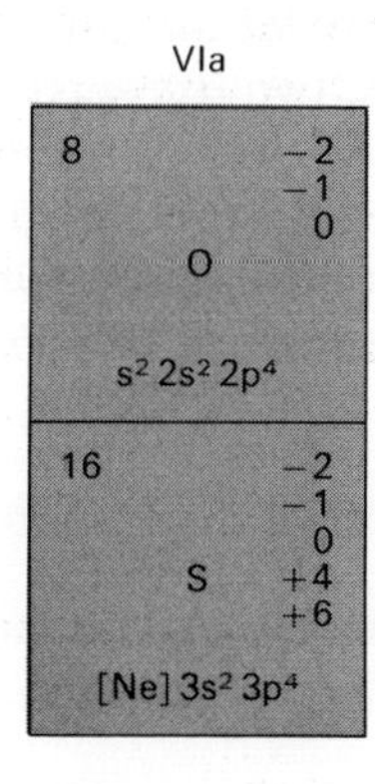

FIGURE 22-5

Oxygen

Ordinary oxygen, O_2, is infrequently used analytically, but atmospheric oxygen is frequently a nuisance in processes involving reducing agents.

Ozone, O_3, however, has been used as an oxidant and were ozone generators more readily available would warrant much more extensive use. As the half-cell potential shown in Table 22-6 indicates it is one of the most powerful oxidants, and, for example, is capable of oxidizing silver(I) in nitric acid

TABLE 22-6
Analytically Important Half-Cell Reactions and Potentials of Oxygen and Sulfur

Half-cell reaction	E°
$S^{2-} = S + 2e^-$	+0.51
$2S^{2-} = S_2^{2-} + 2e^-$	+0.48
$SO_2(g) + 2H_2O = HSO_4^- + 3H^+ + 2e^-$	−0.14
$H_2S = S(s) + 2H^+ + 2e^-$	−0.17
$2S_2O_3^{2-} = S_4O_6^{2-} + 2e^-$	−0.17
$H_2O_2 = O_2 + 2H^+ + 2e^-$	−0.69
$2H_2O = O_2(g) + 4H^+ + 4e^-$	−1.23
$2H_2O = H_2O_2 + 2H^+ + 2e^-$	−1.77
$2HSO_4^- = S_2O_8^{2-} + 2H^+ + 2e^-$	−2.05
$O_2 + H_2O = O_3 + 2H^+ + 2e^-$	−2.07

solutions to a mixture of Ag(II) and Ag(III). It will also oxidize manganese(II) to permanganate; this oxidation is slow but is catalyzed by the presence of a silver salt.

A very convenient feature of ozone as a pretitration oxidant is that one can remove the excess by sweeping an inert gas through the solution. The reduction of ozone is unusual in that only one of the three oxygens is reduced to the binegative state and the other two form O_2.

Hydrogen peroxide, likewise the peroxide ion, are interesting compounds structurally and chemically in that the two oxygens are held together by a single electron-pair bond and therefore the average oxidation state of the oxygen atoms is -1. As a result, hydrogen peroxide can act either as a reducing agent or as an oxidizing agent, but by different half-cell reactions. When it acts as a reducing agent, the oxygen is oxidized to the elementary state according to the half-cell reaction

$$H_2O_2 = O_2(g) + 2H^+ + 2e^- \qquad E^\circ = -0.69 \text{ v}$$

When it acts as an oxidizing agent, the oxygen is reduced to the binegative state as follows

$$2H_2O = H_2O_2 + 2H^+ + 2e^- \qquad E^\circ = -1.77 \text{ v}$$

If these two half-cell reactions are combined, one obtains the following cell reaction and cell potential

$$2H_2O_2 = O_2(g) + 2H_2O \qquad E^\circ_{\text{cell}} = 1.08 \text{ v}$$

The value of the cell potential indicates that H_2O_2 is thermodynamically unstable and has a strong tendency to decompose to water and oxygen. Such is the case, but fortunately the rate of the decomposition reaction is so slow that pure H_2O_2 and its solutions are common reagents. As will be shown below, manganese salts, as well as many other compounds, will catalyze the decomposition and great care has to be taken to exclude them from the commercial preparations. Also, whether peroxide reacts with a given substance as an oxidizing or as a reducing agent depends upon various factors, such as the other half-cell potential involved and the *p*H of the solution.

Hydrogen peroxide is a very weak acid, the constant for the ionization of the first proton being approximately 1.5×10^{-12}. Because of this fact, the salts of hydrogen peroxide (such as sodium peroxide, Na_2O_2, and the peroxides of other metals of the first and second columns of the periodic table) are converted to hydrogen peroxide in acid solutions and in general can be determined by titration with an oxidant such as permanganate; the permanganate is reduced to manganese(II). Hydrogen peroxide forms a series of complex peroxyacids with a considerable number of the elements found in the third to sixth groups of the periodic table. In general, these complex peroxyacids can be considered as being formed by the replacement of an oxide oxygen by the peroxide radical, O_2^{2-}. For example, sulfuric acid can form either peroxymonosulfuric acid, H_2SO_5, or peroxydisulfuric acid, $H_2S_2O_8$. The striking colors formed when hydrogen peroxide is added to acid solutions of Cr(VI), V(V), or Ti(IV), are caused by the formation of peroxyacids; these colors are used as qualitative tests for these elements and in the case of V and Ti for colorimetric determinations.

Reactions of Peroxide with Various Oxidation States of Manganese. As stated above, permanganate is reduced to manganous ion by peroxide in an acid solution. If permanganate is added to a neutral or slightly alkaline solution of peroxide, the permanganate is *reduced*, but only to manganese dioxide, the reaction being

$$3H_2O_2 + 2MnO_4^- = 3O_2 + 2MnO_2 + 2H_2O + 2OH^-$$

If the resultant mixture containing the manganese dioxide is acidified and an excess of peroxide is present or hydrogen peroxide is added, the manganese dioxide is *reduced* by the peroxide to manganous ion as follows:

$$MnO_2 + H_2O_2 + 2H^+ = O_2 + Mn^{2+} + 2H_2O$$

However, if peroxide is added to an alkaline solution containing either manganous ion or manganous hydroxide, the manganese is oxidized by the peroxide to manganese dioxide, as follows:

$$Mn^{2+} + 2OH^-[\text{or } Mn(OH)_2] + H_2O_2 = MnO_2 + 2H_2O$$

This reaction is utilized in systems of qualitative analysis, and in quantitative procedures, for the precipitation of manganese and its separation from the amphoteric elements aluminum, chromium, and zinc.

Thus, it may be observed that (*a*) in an acid solution, manganese dioxide *oxidizes* peroxide and is reduced to dipositive manganese and (*b*) in an alkaline solution, dipositive manganese *reduces* peroxide with formation of manganese dioxide. Therefore the possibility exists that there may be *p*H conditions at which both of these reactions may take place in the same solution, with the net result being the catalytic decomposition of peroxide. This effect is seen if the two equations above are combined (after the second equation is rewritten in terms of H^+) as follows:

$$MnO_2 + H_2O_2 + 2H^+ = O_2 + Mn^{2+} + 2H_2O$$

$$Mn^{2+} + H_2O_2 = MnO_2 + 2H^+$$

$$\overline{\qquad\qquad\qquad\qquad\qquad\qquad\qquad\qquad}$$

$$2H_2O_2 = O_2 + 2H_2O$$

One can demonstrate this catalytic effect by repeatedly adding a large excess of H_2O_2 to a suspension of MnO_2 or $Mn(OH)_2$ in a solution buffered to a *p*H value of approximately 8 and noting the vigorous and continued evolution of oxygen.[15]

Peroxide is an especially convenient pretitration reagent, since it is usually possible, because of its instability, to decompose the excess by boiling the solution. Thus, chromium compounds can be oxidized to chromate, and the excess peroxide can be removed by boiling the solution prior to its determination with a standard reductant.

EXPERIMENT: Reactions of Peroxide With Various Oxidation States of Manganese

(*a*) *Permanganate-Peroxide Reaction in an Acid Solution.* Pipet 1.00 ml of 1 *F* (3%) H_2O_2 into a 400 ml beaker containing 75 ml of water and 5 ml of 3 *F* H_2SO_4. Add one drop of standard 0.02 *F* $KMnO_4$. Note the result. (With very pure chemicals the first drop may fade slowly; if so, repeat the observation with 1 or 2 more drops. The autocatalytic effect is similar to that observed with the permanganate-oxalate reaction of the experiment on page 400.) Titrate the solution with the standard 0.02 *F* $KMnO_4$ until an easily visible color is obtained. Reserve the solution for use in (*b*) below.

1. What is the average oxidation number of the oxygen atoms in H_2O_2?
2. What is the product when H_2O_2 is oxidized?
3. Formulate the half-cell reaction. What is the value of the standard potential?

[15] For a discussion and experimental study of this catalytic decomposition, see D. B. Broughton and R. L. Wentworth. *J. Am. Chem. Soc.*, **69** 741, 744 (1947).

4. What is the equivalent weight of H_2O_2 as a reducing agent?
5. Formulate the equation for the oxidation of H_2O_2 by permanganate in an acid solution.

(*b*) *The Reaction of Permanganate with Manganous Ion in a Neutral or Slightly Alkaline Solution.*

To the titrated solution from (*a*) add 0.1 ml (3 drops) of the 0.02 *F* $KMnO_4$. Then add 6 *F* NaOH, one drop at a time, until the permanganate color fades and a brownish precipitate appears.

6. Write an equation for the reaction taking place.
7. Could this reaction be used for the determination of manganese?
8. What would be the product of the reduction of permanganate by H_2O_2 in a neutral or slightly alkaline solution?

(*c*) *The Manganese(II)—Oxygen Reaction.*

To the mixture from (*b*) above add 10 ml of 6 *F* NaOH.

9. Write an equation for the reaction taking place. (Remember that oxygen is the product of the oxidation of peroxide in (*a*) above. This reaction is the basis of a method for determining the oxygen content of aqueous solutions.)

(*d*) *The Manganese(II)—Peroxide Reaction.*

To the mixture from (*c*) add 0.2 ml of 1 *F* H_2O_2, one drop at a time. Note the effect of each drop.

10. (*a*) What is the product when H_2O_2 acts as an oxidizing agent? (*b*) Formulate the half-cell reaction. (*c*) What is the value of the standard potential?
11. Formulate the reaction for the oxidation by peroxide of dipositive manganese hydroxide to manganese dioxide.
12. What would be the product of the treatment of a neutral solution of a manganous salt with an *excess of sodium peroxide*?

Na_2O_2 is used frequently in qualitative systems of analysis for the separation of iron, manganese, nickel and cobalt, all of which form insoluble hydrous oxides, from aluminum, zinc, and chromium; their hydrous oxides are amphoteric and form hydroxide anions in alkaline solutions. In addition, the chromium(III) is oxidized to chromate.

(*e*) *The Catalytic Decomposition of Hydrogen Peroxide by Oxides of Manganese.*

To the mixture from (*d*) add 2 ml of 1 *F* H_2O_2, then swirl the mixture until effervescence ceases. Repeat this process.

13. Explain the repeated effervescence observed above.
14. Formulate the reaction for the oxidation by manganese dioxide of H_2O_2, with formation of oxygen, in an acid solution.
15. By combining this equation with that from 11 above, show that manganese oxides in alkaline solutions can catalyze the decomposition of H_2O_2 into water and oxygen.

Sulfur

Sulfur is extensively used as the disulfide, S_2^{2-}, in systems of qualitative analysis for the oxidation in alkaline solutions of SnS and Sb_2S_3 to their higher oxidation states and the subsequent formation of the soluble thio salts, SnS_3^{2-} and SbS_4^{3-}.

Hydrogen sulfide is used to a limited extent as a pretitration reducing agent; any excess can be removed by an inert gas or by boiling the solution. It is also a fundamental reagent in systems of qualitative analysis for the precipitation of certain groups of metal sufides as well as of individual elements.

The most important redox analytical uses of oxygen and sulfur are found in their compounds with each other. The first of these is thiosulfate, $S_2O_3^{2-}$, and its reaction with iodine to give tetrathionate, $S_4O_6^{2-}$. This reaction, which is fundamental to iodometric methods of volumetric analysis is discussed in Chapters 20 and 21.

Sulfur dioxide, and sulfurous acid, its hydration product, are used as a pretitration reductant. Like hydrogen sulfide, any excess can be removed by an inert gas or by boiling.

Peroxydisulfate, $S_2O_8^{2-}$, has two oxygens with a peroxide bond and is a powerful oxidant. It is capable of oxidizing Mn(II) to permanganate, but MnO_2 tends to be formed unless the reaction is catalyzed by the presence of a silver salt. It is used in Procedure 30-1 for the oxidation of any iron carbide and graphite that remains after a steel sample has been treated with nitric acid.

ELEMENTS OF GROUP VIIa OF ANALYTICAL IMPORTANCE

The three halogens, chlorine, bromine, and iodine, are all utilized in volumetric methods and Figure 22-6 shows their periodic relationships, electronic

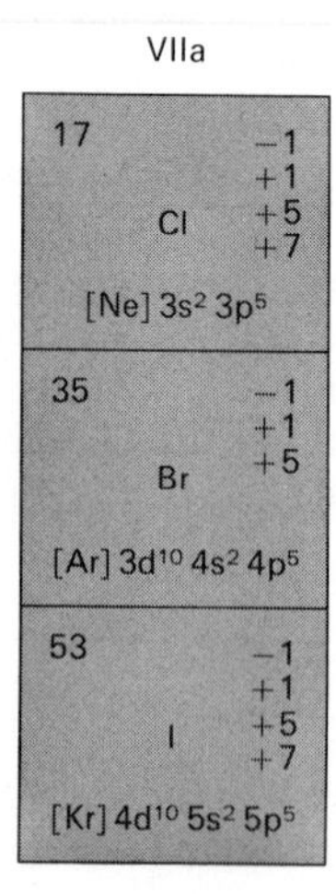

FIGURE 22-6

TABLE 22-7
Half-Cell Potentials of the Halogens of Analytical Importance

Reaction		Potential
$2I^- = I_2(s)$	$+ 2e^-$	-0.5355
$3I^- = \mathbf{I_3^-}$	$+ 2e^-$	-0.536
$I_2 + 4Cl^- = 2ICl_2^-$	$+ 2e^-$	-1.056
$2Br^- = Br_2(l)$	$+ 2e^-$	-1.065
$2Br^- = Br_2(aq)$	$+ 2e^-$	-1.087
$ClO_3^- + H_2O = ClO_4^- + 2H^+$	$+ 2e^-$	-1.19
$I_2 + 6H_2O = \mathbf{2IO_3^-} + 12H^+$	$+ 10e^-$	-1.195
$ICl_2^- + 3H_2O = \mathbf{IO_3^-} + 2Cl^- + 6H^+$	$+ 4e^-$	-1.23
$2Cl^- = Cl_2$	$+ 2e^-$	-1.36
$Cl^- + H_2O = HClO + H^+$	$+ 2e^-$	-1.50
$Br_2 + 6H_2O = 2BrO_3^- + 12H^+$	$+ 10e^-$	-1.52
$IO_3^- + 3H_2O = \mathbf{H_5IO_6} + H^+$	$+ 2e^-$	-1.60

NOTE: Species shown in boldface are used in standard solutions.

configurations, and more analytically important oxidation states. Fluorine, although capable of acting as an extremely energetic and powerful oxidant (the F^--$F_2(g)$ potential is -2.87 v), has not found significant analytical use. The analytically important half-cell reactions and their potentials are shown in Table 22-7; constituents of frequently used standard solutions are in **boldface**.

Iodine

The iodide-iodine half-cell is one that has found extensive use in analytical chemistry and the many interesting features of these methods are discussed in Chapters 20 and 21.

Potassium iodate can be purchased in a high state of purity and standard solutions can be prepared by direct weighing. Such solutions have some interesting usages because of the reaction

$$IO_3^- + 5I^- + 6H^+ = 3I_2 + 3H_2O$$

The equilibrium constant and to a certain extent the rate are such that the reaction proceeds quantitatively from left to right if the hydrogen-ion concentration is significantly greater than 10^{-7} *M*. As one would expect, the reaction is highly dependent on the hydrogen-ion concentration, since it occurs to the sixth power in the equilibrium expression. Because of these facts a standard iodate solution can be substituted for a tri-iodide solution for the titration of reductants in acid solutions; the iodate is reduced to iodide by the reductant and continued addition of the iodate oxidizes the iodide to iodine, which can be detected by its color or by the characteristic reaction with starch. The reaction is also the basis of methods for the determination of iodate and of hydrogen ion.

Standard iodate solutions are also used in an extensive series of methods that involve the reduction of iodate to iodine monochloride. The half-cell reaction is

$$I_2 + 4Cl^- = 2ICl_2^- + 2e^- \qquad E^{\circ\prime} = -1.05 \text{ v}$$

The use of iodine monochloride as a catalyst and potential indicator has been discussed in Procedure 18-2 in connection with the standardization of permanganate solutions against arsenious oxide. However, the iodine monochloride end point finds most extensive application in connection with titrations with a standard solution of potassium iodate.[16] The titration of arsenious acid in solutions 3 *F* to 6 *F* in hydrochloric acid with a standard solution of potassium iodate will serve to illustrate the reactions taking place. Initially the iodate will be reduced to iodine,[17]

$$5H_3AsO_3 + 2IO_3^- + 2H^+ = 5H_3AsO_4 + I_2 + H_2O$$

and iodine will accumulate in the solution. After oxidation of the arsenious acid, further addition of the iodate causes oxidation of the iodine,

$$2I_2 + IO_3^- + 10Cl^- + 6H^+ = 5ICl_2^- + 3H_2O$$

and the end point is taken when the iodine color can no longer be detected in an organic solvent, such as carbon tetrachloride, that is shaken with the aqueous solution. Relatively high chloride and hydrogen-ion concentrations are required to stabilize the ICl_2^- complex and to cause the above reaction to proceed rapidly.

The ease with which standard iodate solutions can be prepared from potassium iodate and the extreme stability of these solutions, together with the sensitivity and permanence of the iodine monochloride end point, have led to the use of this end point in the determination of a large number of compounds.

The iodine monochloride end point has been utilized in a series of procedures in which standard oxidants other than iodate were used. In these procedures a small quantity of an iodine monochloride solution was added

---

[16] Laitinen, op. cit., pp. 426–429.

[17] H_3AsO_4 in an *acid* solution tends to oxidize iodide to iodine; therefore the initial reduction of the iodate will proceed to iodine only. If stannous tin were being titrated, the initial reduction product would be iodide until all of the stannous tin were oxidized; further addition of iodate would then oxidize the iodide to iodine. In this latter case the initial appearance of iodine could be taken as the end point, or the titration could be continued to the iodine monochloride end point.

to the reductant; depending upon the reductant either iodide or iodine was formed. The solution was then titrated to an iodine monochloride end point under conditions similar to those used with iodate but with a standard oxidant capable of oxidizing iodine to iodine monochloride; permanganate, ceric sulfate, and dichromate have been used.[18] Since the iodine monochloride added initially is reoxidized to iodine monochloride when the end point is taken, it is not necessary to know the exact quantity taken; equivalent quantities of iodide, iodine, or iodate could be used, but a correction for the quantity taken would be required.

Periodic acid exists as metaperiodic acid, HIO_4; paraperiodic acid, H_5IO_6; and as dimerized mesodiperiodic acid, $H_4I_2O_9$. The meta acid is highly ionized; the para acid is weak, K_{A_1} being 5×10^{-4}, but is the predominant species in acid solutions. The equilibrium between the two acids shows the effect of the hydrogen ion

$$IO_4^- + H^+ + 2H_2O = H_5IO_6$$

Metal salts precipitating from acid solutions are usually the meta form because of the weakness of the para acid.

It is of interest to note that the large iodine atom is able to coordinate six oxygen atoms in H_5IO_6; the smaller chlorine atom can hold only four, and perchloric acid appears to exist only as $HClO_4$.

As is seen from Table 22-7 periodic acid is a strong oxidant. It is used in Procedure 30-1 for the oxidation of manganese to permanganate prior to the colorimetric determination of that element.

Bromine

Solutions of bromine have been used as standard oxidants, but the volatility of bromine makes such solutions unstable. The addition of bromide to a bromine solution decreases the partial pressure of the bromine because of the formation of a tri-bromide ion, similar to tri-iodide; the equilibrium constant for the formation of the complex $Br_2(aq) + Br^- = Br_3^-$ is 17. Fortunately the bromate-bromide reaction, like the iodate-iodide one, is so dependent upon the hydrogen-ion concentration that neutral solutions containing bromate and bromide are quite stable, but upon being acidified react to give bromine. Therefore for titrations in acid media a bromate-bromide solution can be used as though it were a bromine solution. Although a large number of

[18] E. H. Swift. *J. Am. Chem. Soc.*, **52**, 849 (1930); E. H. Swift and C. Gregory. *J. Am. Chem. Soc.*, **52**, 901 (1930); E. H. Swift and C. S. Garner. *J. Am. Chem. Soc.*, **58**, 113 (1936); E. H. Swift, R. A. Brown, and E. W. Hammock. *Anal. Chem.*, **20**, 1048 (1948); E. H. Swift, D. Beavon, and E. W. Hammock. *Anal. Chem.*, **21**, 970 (1949).

inorganic reductants can be determined by bromine or bromate-bromide solutions, they are more frequently used for the titration of organic compounds. Two types of reactions, addition and substitution, are involved.

Addition titrations are applicable to unsaturated compounds and involve addition of bromine to a double bond; the reaction with ethylene is an example,

$$\mathrm{H{-}\overset{\displaystyle H}{\overset{|}{C}}{=}\overset{\displaystyle H}{\overset{|}{C}}{-}H + Br_2 = H{-}\underset{\displaystyle Br}{\underset{|}{\overset{\displaystyle H}{\overset{|}{C}}}}{-}\underset{\displaystyle Br}{\underset{|}{\overset{\displaystyle H}{\overset{|}{C}}}}{-}H}$$

In substitution reactions a hydrogen atom is replaced by bromine. The reaction with 8-hydroxyquinoline is an example

$$\begin{array}{cccccc} & & & \mathrm{OH} & & \\ & & & | & & \\ & \mathrm{N} & & \mathrm{C} & & \\ \mathrm{H{-}C} & & \mathrm{C} & & \mathrm{C{-}H} \\ | & & \| & & | \\ \mathrm{H{-}C} & & \mathrm{C} & & \mathrm{C{-}H} \\ & \mathrm{C} & & \mathrm{C} & & \\ & | & & | & & \\ & \mathrm{H} & & \mathrm{H} & & \end{array} + 2\mathrm{Br_2} = \begin{array}{cccccc} & & & \mathrm{OH} & & \\ & & & | & & \\ & \mathrm{N} & & \mathrm{C} & & \\ \mathrm{H{-}C} & & \mathrm{C} & & \mathrm{C{-}Br} \\ | & & \| & & | \\ \mathrm{H{-}C} & & \mathrm{C} & & \mathrm{C{-}H} \\ & \mathrm{C} & & \mathrm{C} & & \\ & | & & | & & \\ & \mathrm{H} & & \mathrm{Br} & & \end{array} + 2\mathrm{H^+} + 2\mathrm{Br^-}$$

This particular reaction is of analytical importance because many cations can be precipitated as the metal hydroxyquinolate, then dissolved in hydrochloric acid and the cation determined by titration of the quinoline.

Chlorine

Chlorates are extensively used in qualitative analysis for the precipitation from concentrated nitric acid solutions of manganese as hydrous manganese dioxide and its separation from the elements that are found with it in the group precipitated by ammonium sulfide in an ammonia solution (see Chapter 9).

The perchlorate ion is a curiosity because of the tremendous difference in its oxidizing properties in dilute and concentrated acid solutions.

At one time, perchloric acid was not used extensively in analytical processes for two reasons. First, the cost was prohibitive, and second, there was a general apprehension as to the possibility of explosions. The cost of the acid is now more reasonable, and extensive experience has shown that, if properly handled, perchloric acid can be used with relative safety. Aqueous perchloric acid is an extremely stable reagent when dilute, reduction being impossible except by a few very powerful reducing agents such as chromous and titanous salts, and even these react slowly. When the acid is heated to fuming, it becomes a powerful oxidizing agent, probably because of the high hydrogen-ion activity under these conditions, and chromic and manganous ions are oxidized to chromate and manganese dioxide, respectively. If the acid is heated to fuming with organic substances, especially alcohols, or with certain reducing agents, such as antimonous salts, *violent and dangerous explosions will occur.* However, it has been shown by Noyes and Bray (*Qualitative Analysis for the Rare Elements*, New York: Macmillan, 1927) that if nitric acid is added to such mixtures, they can be fumed, or even evaporated to dryness, and in no case was it possible to cause the mixture to explode. Others have reported explosions with nitric and perchloric acid mixtures (*Ind. Eng. Chem.*, news ed., **15**, 214, 332, 1937); therefore, care should be taken to carry out such procedures so that, should an explosion occur, personal injury will not result.

QUESTIONS AND PROBLEMS

Redox Reactions of Manganese

See Chapter 18 concerning the use of permanganate as a standard oxidant in acid solutions.

1. *The Titration of Peroxide with Permanganate.* In the analysis of a "3% H_2O_2" solution, a 1.824 g sample was transferred to a beaker containing 200 ml of 0.3 *F* H_2SO_4, and the solution titrated with 0.02000 *F* $KMnO_4$ to a permanganate end point; 30.00 ml were required.
   (*a*) Calculate the percentage by weight of H_2O_2 in the solution.
   (*b*) Calculate the formality of the solution (assume a density of one). *Ans.* (*a*) 1.481%.

2. *The Oxidation of Manganese(II) with Sodium Peroxide.* A chemist desires to oxidize the manganese in 100 ml of 0.010 *F* $MnSO_4$ to MnO_2 by means of sodium peroxide.
   (*a*) Calculate the stoichiometric weight of the peroxide.
   (*b*) Why should a large excess be added?

3. *Titrations Involving the Formation of* MnO_4^{2-}. Calculate the MnO_4^{2-}-MnO_4^- potential in a solution in which the OH^- was 1 *M*, the MnO_4^- was 1×10^{-5} *M* (approximately that necessary to perceive the purple color) and the MnO_4^{2-} was 0.02 *M* (produced by titrating a reductant). Repeat the calculation for a solution that was saturated with $Ba(OH)_2$($K_{sp} = 5 \times 10^{-3}$), in which the OH^- was 1 *M* and the MnO_4^- was 1×10^{-5} *M*.

   Discuss your results with reference to the observation that permanganate oxidizes iodide to iodate in alkaline solution in the absence of $Ba(OH)_2$ but to periodate in its presence.

4. *The Titration of Mn(II) to Mn(III).* The stoichiometry of the Lingane and Karplus method (see page 492) for titrating Mn(II) to Mn(III) with permanganate was studied

as follows:

A 100 ml portion of an 0.02 *F* $KMnO_4$ solution was pipetted into a flask, 20 ml of 2 *F* H_2SO_4 added, then 1 *F* Na_2SO_3 added dropwise until the solution was decolorized. The solution was then boiled for 5 minutes.

(*a*) Write an equation for the reduction reaction.
(*b*) Why was the solution boiled?

The solution was cooled, 22 g of $Na_4P_2O_7 \cdot 10H_2O$ were added, and the volume made 250 ml.

(*c*) Calculate the approximate *p*H of the solution.
(*d*) Discuss the uncertainties involved in the calculation.

The solution was then titrated with the original $KMnO_4$ solution to a potentiometric end point.

(*e*) Calculate the volume that should have been required. *Ans.* 25.0 ml.

5. *The Use of Dichromate as a Standard Oxidant.* Chromite is a mineral having the empirical formula $FeO \cdot Cr_2O_3$. Ores containing it are an important source of chromium; they also may contain varying amounts of Ca, Mg, Al, Ni, and Mn as oxides, silicates, or carbonates. In analyzing such an ore the 0.2000 g sample was fused with sodium carbonate and excess peroxide in a nickel crucible, the crucible and contents boiled with 100 ml of water until the fusion mass was completely disintegrated, the mixture filtered, the filtrate acidified with sulfuric acid, 50.00 ml of 0.0950 *F* $FeSO_4$ added, and the solution then titrated with 12.20 ml of 0.01700 *F* $K_2Cr_2O_7$; diphenylamine sulfonate was used as indicator. (See pages 409–413.)

   (*a*) What compounds would be present in the filtered residue?
   (*b*) What ionic species would be present in the filtrate?
   (*c*) Calculate the percentage by weight of Cr_2O_3 in the ore. *Ans.* 44.42%.

6. *Vanadate as a Standard Oxidant.* The suggestion has been made that ammonium metavanadate, NH_4VO_3, be used for the direct preparation of a standard vanadate solution (Bishop and Crawford, *Analyst*, **75**, 273, 1950). Such a solution was prepared by weighing 12.45 g of the salt, dissolving it in 50 ml of 9 *F* H_2SO_4 and diluting to exactly 1 liter.

   (*a*) What is the formality of the solution?
   (*b*) What is the normality when used to titrate ferrous iron?

   Vanadate solutions have the advantage over permanganate and dichromate solutions that they can be used to titrate ferrous iron in hydrochloric acid solutions as concentrated as 4 *F*.

   A 25.00 ml portion of a solution of ferric chloride in 3 *F* HCl was placed in a separatory funnel, the air displaced by carbon dioxide, mercury added and the mixture shaken until after the solution was colorless. The solution was separated from the mercury, phosphoric acid was added, and it was titrated with the standard vanadate, with diphenylamine sulfonate as indicator; 28.40 ml were required.

(*c*) Why was the air displaced? Furman and Murray (*J. Am. Chem. Soc.*, **58**, 429, 1936) have shown that mercury in the presence of hydrochloric acid first reduces oxygen to peroxide, $2Hg(l) + 2H^+ + 2Cl^- + O_2 = Hg_2Cl_2(s) + H_2O_2$; prolonged treatment reduces the peroxide, $2Hg(l) + 2H^+ + 2Cl^- + H_2O_2 = Hg_2Cl_2(s) + 2H_2O$.
(*d*) Write an equation for the reaction taking place when the iron solution was shaken with the mercury.
(*e*) Predict the result of shaking a ferric sulfate solution with mercury.
(*f*) Why was phosphoric acid added?
(*g*) Write an equation for the titration reaction.
(*h*) Calculate the formality of the ferric chloride solution. *Ans.* 0.1209 *F*.

7. *Use of the Iodine Monochloride End Point.* See pages 508–509 for a discussion of this end point and its application to the estimation of tripositive arsenic.

(*a*) What weight of KIO_3 should be taken to prepare a liter of solution that would be exactly 0.1 *N* when used in titrations with the iodine monochloride end point? What advantages do iodate solutions possess as standard oxidizing agents?
(*b*) Standardization of an iodate solution against arsenious oxide. An iodate solution of unknown concentration was standardized as follows: 0.2014 g of As_2O_3 was weighed out, dissolved in 20 ml of cold 12 *N* HCl, and transferred to a flask with a ground-glass stopper, 5 ml of CCl_4 were added, and the solution was titrated with the iodate solution until the iodine color that developed at the beginning of the titration was just removed; 41.07 ml of the iodate were required. Write the equations for the successive reactions taking place and calculate the normality and formality of the iodate solution.
(*c*) A 0.5146-g sample of a stibnite material (crude Sb_2S_3) was dissolved in HCl, the solution then transferred to a flask with a ground-glass stopper and titrated to an iodine monochloride end point with standard KIO_3 solution. The titrated solution was kept between 2.5 *F* and 3.5 *F* in hydrochloric acid [see Hammock, Brown, and Swift, *Anal Chem.*, **20**, 1048 (1948)]. A 0.02487 *F* KIO_3 solution was used for the titration, and 35.46 ml were required. Calculate the percentage of Sb_2S_3 in the stibnite. *Ans.* 58.22%.
(*d*) A solution of sodium vanadate, $Na_3VO_4 \cdot 16H_2O$, was analyzed as follows: A 25.00 ml portion was pipetted into a flask, and 5 ml of CCl_4 and 5 ml of 6 *F* HCl were added, and then 25.00 ml of 0.05414 *F* KI. The solution was then titrated with 0.02500 *F* KIO_3 (concentrated HCl being added so that the final solution was approximately 6 *F* in HCl) until the CCl_4 showed no iodine color; 13.46 ml were required. Under the end-point conditions the vanadium was in the quadripositive oxidation state. Calculate (1) the normality of the vanadate solution and (2) the grams of $Na_3VO_4 \cdot 16H_2O$ per liter.
(*e*) To 10.00 ml of a solution of As_2O_3 in 12 *F* HCl were added 5 ml of CCl_4 and 5 ml of 0.1 *F* ICl in 4 *F* HCl. A color was produced in the CCl_4. The solution was then titrated with 0.04000 *F* $KMnO_4$ until this color disappeared, 20.00 ml of the $KMnO_4$ solution being required. Write the equations for the successive reactions taking place. Calculate the weight of As_2O_3 in the original solution. *Ans.* 0.1979 g.

(*f*) A 0.5000-g sample of a mixture of $KClO_4$ and $KMnO_4$ was added to 25.00 ml of 0.2000 *F* KI and 25 ml of 12 *F* HCl, and an iodine color resulted. This solution was then titrated with a 0.2000 *N* $KMnO_4$ solution until the iodine color disappeared, 25.00 ml being used. Calculate the weight of $KMnO_4$ in the original sample.

NOTE: Perchlorate is not reduced under the above conditions.

(*g*) To 25.00 ml of a solution containing SnI_2 and KI dissolved in oxygen-free 12 *F* HCl were added 5 ml of CCl_4, and the solution was titrated to the first appearance of an iodine color with a 0.0500 *F* KIO_3 solution; 8.33 ml of the iodate were required. Then 5 ml of 12 *F* HCl were added and the titration was continued until the iodine color was no longer perceptible; an additional 54.17 ml of the iodate were used. Write the equations for the successive reactions taking place. Calculate the formal concentration of the SnI_2 and of the KI in the solution. *Ans.* 0.0500 *F* SnI_2, 0.1000 *F* KI.

(*h*) Write equations for the reactions taking place if CCl_4 solutions containing (*a*) iodine, (*b*) bromine, and (*c*) chlorine were shaken with NaOH solutions.

CHAPTER **23**

Ionization Reactions. Acid-Base Equilibria

The preceding four chapters have presented the use of oxidation-reduction reactions in titrimetric methods. In Chapter 7, a general introduction to ionization reactions was given. In this chapter and the three that follow we will consider the use of ionization reactions as the basis of acid-base titrimetric methods. In Chapter 27 methods based upon ionization reactions involving complex compounds will be presented.

CALCULATION OF EQUIVALENCE-POINT CONDITIONS

Expressions were derived in Chapter 7 for the hydrogen-ion concentrations at the equivalence points of various titrations. Those expressions are very useful, but must be used with care since the derivations used there do not show clearly all the approximations that were used. We will give now a more general and detailed treatment of each of the titrations considered in Chapter 7.

Titrations Involving a Strong Acid and a Strong Base

As was shown in Chapter 7 the reaction between a strong acid and a strong base can be considered to be analytically complete. The hydrogen and hydroxide ions at the equivalence point will be determined by the ionization of water and at 25°C will be

$$[H^+] = [OH^-] = \sqrt{K_w} = 10^{-7}$$

During the course of the titration these concentrations will be determined by the concentration of excess acid or base in the solution.

Titrations Involving a Strong Acid and a Weak Base

At the equivalence point the species in solution will be the result of the reaction

$$H^+A^- + BOH = H_2O + B^+A^- \tag{1}$$

and will be identical with the species in a solution that consists of the equivalent amount of the salt B^+A^- dissolved in the appropriate volume of water. Let us examine this solution. If Equation 1 is written in the reverse direction it shows the hydrolysis of BA:

$$B^+A^- + H_2O = BOH + H^+ + A^-$$

This hydrolysis can be represented in terms of the following equilibria:

$$BOH = B^+ + OH^- \qquad K_B = \frac{[B^+][OH^-]}{[BOH]} \tag{2}$$

$$HOH = H^+ + OH^- \qquad K_W = [H^+][OH^-] \tag{3}$$

There are two other independent relationships that can be used to provide the additional equations necessary to solve for the four unknown concentrations: $[H^+]$, $[OH^-]$, $[B^+]$, and [BOH]. These relationships are the *charge balance* and the *species balance*, commonly called the *mass balance*. It is observed that solutions are electrically neutral; therefore the sums of positive and of negative charges must be equal.

$$[A^-] + [OH^-] = [B^+] + [H^+] \tag{4}$$

Moreover, since the only source of A^- and B^+ in the solution is the BA added, the concentration of A^- must be equal to the sum of the concentrations of the species that contain B:

$$[A^-] = [B^+] + [BOH] = c \tag{5}$$

In this expression c is the initial or formal concentration of the salt, BA.

Combination of Equations 4 and 5 gives

$$[OH^-] = [H^+] - [BOH]$$

and if substitution is made from the equilibrium expressions, Equations 2 and 3, the equation becomes

$$[H^+] = \frac{K_W}{[H^+]} + \frac{[B^+][OH^-]}{K_B} = \frac{K_W}{[H^+]} + \frac{K_W[B^+]}{K_B[H^+]}$$

or

$$[H^+]^2 = K_W + \frac{K_W[B^+]}{K_B}$$

For those cases where $c \gg$ BOH (that is, the concentration of the unhydrolyzed salt is large compared to that of the base formed by hydrolysis), the expression can be written

$$[H^+]^2 \cong K_W + \frac{K_W}{K_B}c$$

and

$$[H^+] \cong \sqrt{K_W + \frac{K_W}{K_B}c} = \sqrt{K_W\left(1 + \frac{c}{K_B}\right)}$$

If c/K_B is much greater than 1 (that is, $c \gg K_B$), then

$$1 + \frac{c}{K_B} \cong \frac{c}{K_B}$$

and

$$[H^+] \cong \sqrt{\frac{K_W c}{K_B}}$$

This is identical to the approximate expression obtained in Chapter 7 for the hydrogen-ion concentration at the equivalence point in this titration.

Titrations Involving a Strong Base and a Weak Acid

The titration reaction is

$$HA + B^+OH^- = H_2O + B^+A^-$$

and at the equivalence point the solution is identical with one prepared from the salt BA and water:

$$B^+A^- + H_2O = HA + B^+ + OH^-$$

The equilibria that are involved are

$$HOH = H^+ + OH^- \qquad [H^+][OH^-] = K_W \tag{6}$$

and

$$HA = H^+ + A^- \qquad \frac{[H^+][A^-]}{[HA]} = K_A \tag{7}$$

A third expression results from the electrical neutrality of the solution,

$$[B^+] + [H^+] = [A^-] + [OH^-] \tag{8}$$

and a fourth equation comes from the species balance:

$$[B^+] = [A^-] + [HA] = c \tag{9}$$

Here again, c is the initial or formal concentration of the salt, BA.
Combination of Equations 8 and 9 gives

$$[H^+] = [OH^-] - [HA]$$

substitution for $[OH^-]$ and $[HA]$ leads to

$$[H^+] = \frac{K_W}{[H^+]} - \frac{[H^+][A^-]}{K_A}$$

and therefore

$$[H^+] = \sqrt{\frac{K_W K_A}{[A^-] + K_A}}$$

Unless the acid is very weak $[A^-]$ will be large compared with $[HA]$; therefore, c, the formal concentration of the salt, can be substituted for $[A^-]$, the molar concentration:

$$[H^+] \cong \sqrt{\frac{K_W K_A}{c + K_A}}$$

For those cases where $c \gg K_A$, the expression becomes

$$[H^+] \cong \sqrt{\frac{K_W K_A}{c}}$$

This same expression is obtained by substituting $K_W/[H^+]$ for $[OH^-]$ in the approximation,

$$[OH^-] \cong \sqrt{\frac{K_W}{K_A}c}$$

which was obtained in Chapter 7 for the hydroxide-ion concentration.

Titrations Involving a Weak Acid and a Weak Base

The hydrogen-ion concentration at the equivalence point of such a titration can readily be calculated from the equilibria involved in the hydrolysis of the salt. This is true because the equivalence-point solution is identical to the solution of the salt of the weak acid and the weak base at the same concentration.

Although the over-all reaction can be written

$$B^+A^- + H_2O = BOH + HA$$

three reactions are involved

$$B^+ + H_2O = BOH + H^+$$

$$A^- + H_2O = HA + OH^-$$

and

$$H^+ + OH^- = H_2O$$

The ionization equilibria and constants are

$$H_2O = H^+ + OH^- \qquad [H^+][OH^-] = K_W \tag{10}$$

$$HA = H^+ + A^- \qquad \frac{[H^+][A^-]}{[HA]} = K_A \tag{11}$$

and

$$BOH = B^+ + OH^- \qquad \frac{[B^+][OH^-]}{[BOH]} = K_B \tag{12}$$

The statements of electrical neutrality,

$$[H^+] + [B^+] = [OH^-] + [A^-] \tag{13}$$

and species balance,

$$[B^+] + [BOH] = [A^-] + [HA] = c \tag{14}$$

give the necessary additional relationships to permit the solving of the problem.

Equations 13 and 14 can be combined to give

$$[\mathrm{H}^+] = [\mathrm{OH}^-] + [\mathrm{BOH}] - [\mathrm{HA}]$$

Substitution for the three terms on the right gives

$$\begin{aligned}[\mathrm{H}^+] &= \frac{K_\mathrm{W}}{[\mathrm{H}^+]} + \frac{[\mathrm{B}^+][\mathrm{OH}^-]}{K_\mathrm{B}} - \frac{[\mathrm{H}^+][\mathrm{A}^-]}{K_\mathrm{A}} \\ &= \frac{K_\mathrm{W}}{[\mathrm{H}^+]} + \frac{[\mathrm{B}^+]K_\mathrm{W}}{K_\mathrm{B}[\mathrm{H}^+]} - \frac{[\mathrm{H}^+][\mathrm{A}^-]}{K_\mathrm{A}}\end{aligned}$$

From this we obtain

$$[\mathrm{H}^+] = \sqrt{\frac{K_\mathrm{A}K_\mathrm{W}K_\mathrm{B} + K_\mathrm{A}K_\mathrm{W}[\mathrm{B}^+]}{K_\mathrm{A}K_\mathrm{B} + K_\mathrm{B}[A^-]}}$$

For cases where

$$[\mathrm{B}^+] \gg K_\mathrm{B}, [\mathrm{H}^+] \cong \sqrt{\frac{K_\mathrm{A}K_\mathrm{W}[\mathrm{B}^+]}{K_\mathrm{A}K_\mathrm{B} + K_\mathrm{B}[\mathrm{A}^-]}}$$

and for cases where

$$[\mathrm{A}^-] \gg K_\mathrm{A}, [\mathrm{H}^+] \cong \sqrt{\frac{K_\mathrm{A}K_\mathrm{W}K_\mathrm{B} + K_\mathrm{A}K_\mathrm{W}[\mathrm{B}^+]}{K_\mathrm{B}[\mathrm{A}^-]}}$$

If both $[\mathrm{B}^+] \gg K_\mathrm{B}$ and $[\mathrm{A}^-] \gg K_\mathrm{A}$, then

$$[\mathrm{H}^+] \cong \sqrt{\frac{K_\mathrm{A}K_\mathrm{W}[\mathrm{B}^+]}{K_\mathrm{B}[\mathrm{A}^-]}}$$

When $[\mathrm{B}^+]$ and $[\mathrm{A}^-]$ are large in comparison with [BOH] and [HA], then $[\mathrm{B}^+] \approx [\mathrm{A}^-] \approx c$, and

$$[\mathrm{H}^+] \approx \sqrt{\frac{K_\mathrm{A}K_\mathrm{W}}{K_\mathrm{B}}}$$

It is interesting to observe that when the explicit approximations that were made in this derivation are acceptable, then the hydrogen-ion concentration at the equivalence point is independent of the formal concentration, c.

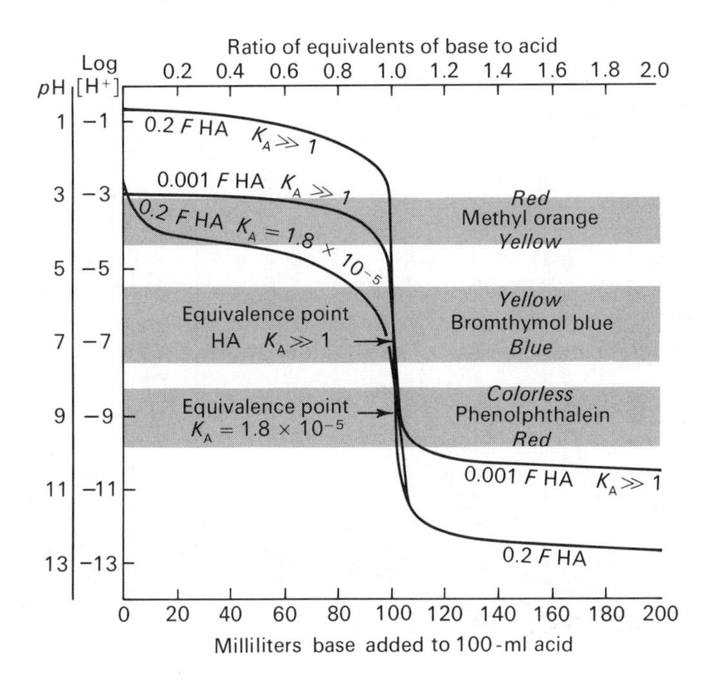

FIGURE 23-1

*p*H and $[H^+]$ changes during titrations of acids with a strong base of equivalent concentration. Transition ranges of selected indicators.

CHANGES IN HYDROGEN AND HYDROXIDE-ION CONCENTRATIONS DURING NEUTRALIZATION TITRATIONS. TITRATION CURVES

Titrations of Strong Acids with Strong Bases

The factor of greatest analytical importance during the course of a neutralization titration is the change in the hydrogen-ion concentration of the solution, because it is this change that permits the end point to be determined. The hydrogen-ion concentration existing at any point during a titration involving a strong acid, $K_A \gg 1$, and a strong base, $K_B \gg 1$, can be calculated directly from the excess of either the acid or base present, since it is assumed that they are completely ionized. At the equivalence point

$$[H^+] = [OH^-] = 10^{-7}\,M \text{ (at 25°C)}$$

and the solution is said to be neutral. In these strong-acid—strong-base titrations the hydrogen-ion concentration changes very rapidly with the

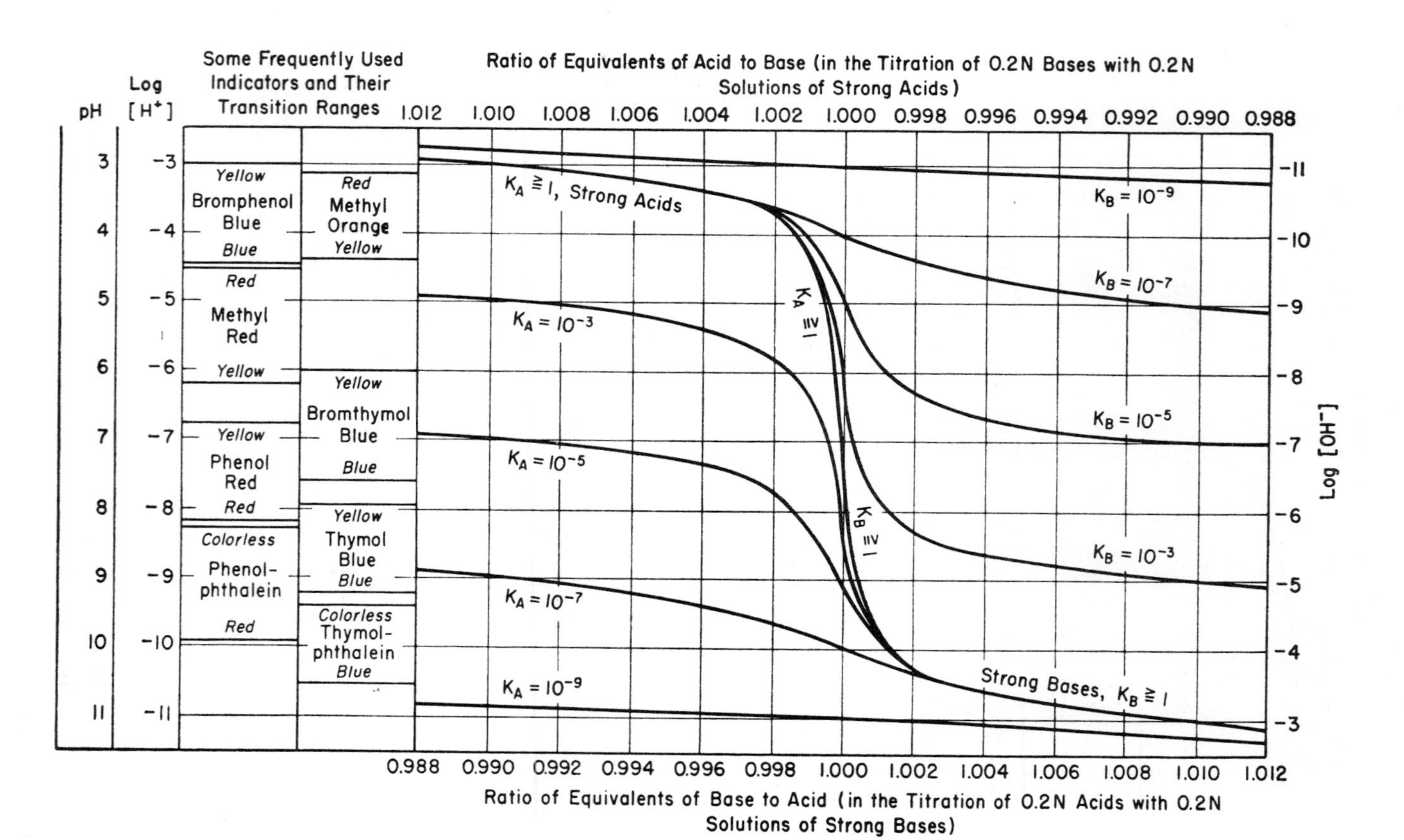

FIGURE 23-2

Hydrogen- and hydroxide-ion concentration changes during neutralization (and displacement) titrations. Indicator transition ranges. The changes shown are from 1% before to 1% after the equivalence point.

addition of acid or base, especially near the equivalence point. This concentration change during the titration of a strong monoprotic acid with a strong monohydroxy base (such as NaOH) is shown in Figure 23-1; such strong acids are HCl, HNO_3, and $HClO_4$. The hydrogen-ion concentrations have been plotted both logarithmically and as *p*H values (the use of the latter is explained on pages 93–94). This figure shows the changes during the course of titrations of 100 ml of such acids, 0.2 *F* and 0.001 *F*, with solutions of a strong base of equivalent concentration; the plot is extended beyond the equivalence point to cover the addition of an equal excess of the base. The curve for the titration of the 0.001 *F* acid shows the effect of the concentration of the titrated acid on the hydrogen-ion concentration during the titration and on the rate of change near the equivalence point. Thus it is seen that with the 0.2 *F* acid any one of the indicators shown would show a change from acidic to basic colors quite close to the equivalence point; with the 0.001 *F* acid, the methyl orange would begin to show a change when the ratio of base to acid was about 0.6 or shortly after half the equivalent quantity of base had been added. Note the similarity of these curves to the titration curves for precipitation and redox titration on pages 67 and 394.

In Figure 23-2 similar curves have been plotted for the range from 1% before to 1% beyond the equivalence point; expansion of the curves in this range enables one to predict more exactly the accuracy of titrations with various indicators. In addition, curves are shown for the titration of monoprotic acids of various strengths with strong monohydroxy bases, and of monohydroxy bases of various strengths with strong monoprotic acids. Also shown are the color transition ranges for a group of representative indicators. This figure enables one to select an appropriate indicator for substantially all titrations involving monoprotic acids and monohydroxy bases, and to predict the probable end-point errors of the titrations. As will be explained later the curves are also applicable to titrations of salts of weak monoprotic acids with strong acids and of salts of weak monohydroxy bases with strong bases.

Titrations of Weak Acids with a Strong Base

The calculation of the hydrogen-ion concentrations at various points during the titration of a weak acid with a strong base is not so simple as is that for a strong-acid–strong-base titration. As an example, consider the titration of acetic acid ($K_A = 1.8 \times 10^{-5}$ at 25°C) with sodium hydroxide. The equation for the reaction is

$$CH_3COOH + OH^- = CH_3COO^- + H_2O$$

and the equilibrium expression is

$$\frac{[CH_3COO^-]}{[CH_3COOH][OH^-]} = \frac{1.8 \times 10^{-5}}{10^{-14}} = 1.8 \times 10^9 \qquad (15)$$

We will consider three points in the titration and calculate the hydrogen-ion concentration at each of them. One is the equivalence point and the others occur shortly before and after the equivalence point.

Before the Equivalence Point. If 100 ml of 0.200 F CH_3COOH are taken and 99 ml of 0.200 F NaOH are added to it, then

$$100 \text{ ml} \times 0.200 \frac{\text{mmole}}{\text{ml}} = 20.0 \text{ mmole of } CH_3COOH$$

and

$$99 \text{ ml} \times 0.200 \frac{\text{mmole}}{\text{ml}} = 19.8 \text{ mmole of NaOH}$$

are mixed. In the solution there is the excess of 0.2 mmole of CH_3COOH plus that present because of the incompleteness of the titration reaction. The molar concentration of acetic acid is, therefore, $0.2/199 + x$ and that of acetate ion is $19.8/199 - x$, where x is the molar concentration of hydroxide ion. Substitution into the equilibrium expression for the titration gives

$$\frac{\dfrac{19.8}{199} - x}{\left(\dfrac{0.2}{199} + x\right)x} = 1.8 \times 10^9$$

If the additive term, x, is assumed to be small, the expression is readily solved to give $[OH^-] = 5.5 \times 10^{-8}$ and $[H^+] = 1.8 \times 10^{-7}$.

A much simpler approximate solution is obtained by noting that as long as an appreciable excess of CH_3COOH is present (as long as $[CH_3COOH]$ in the above equation is large in comparison to x), the $[H^+]$ can be calculated from the familiar relation

$$[H^+] = K_A \frac{[HA]}{[A^-]}$$

Also, since there is a ratio of the concentrations of HA and A^-, the $[H^+]$ is approximately independent of the volume or of dilution.

At the Equivalence Point. Here it is more convenient to consider the problem as one of hydrolysis, since the solution is the same as one containing an equivalent amount of salt dissolved in water. There have been added 20.0 mmoles of NaOH to the 20.0 mmoles of CH_3COOH. The molar concentrations of CH_3COOH and OH^- can be assumed equal,[1] and if x represents the concentration of either of these species then $20.0/200 - x$ is the molar concentration of CH_3COO^-. Substitution of these values into the equilibrium expression, Equation 15, gives $[OH^-] = 7.5 \times 10^{-6}$ and $[H^+] = 1.3 \times 10^{-9}$. This value shows that at the *equivalence point* of this titration the solution has already become distinctly *alkaline.* (By contrast, the equivalence-point solution in the titration of a strong acid with a strong base will be neutral, i.e., pH = 7.)

After the Equivalence Point. As soon as the excess $[OH^-]$ is large, relative to the $[OH^-]$ from the hydrolysis ($7.5 \times 10^{-6}\ M$), the calculation is the same as that for the strong-acid–strong-base titration. From the excess of NaOH added and the volume of the solution one calculates the $[OH^-]$ and hence the $[H^+]$. For example, if 101 ml of 0.200 F NaOH have been added to the 100 ml of 0.200 F CH_3COOH, the OH^- concentration is

$$[OH^-] \approx \frac{0.2 \text{ mmole}}{201 \text{ ml}} = 1 \times 10^{-3}\ M$$

The approximation is good, for the maximum $[OH^-]$ from the hydrolysis of the CH_3COO^- is $7.5 \times 10^{-6}\ M$ and therefore is insignificant compared with the above value.

Summarizing, it is seen that the hydrogen-ion concentration can be calculated for frequent intervals during a titration and that a titration curve similar to those for other types of reactions can be constructed from these data. This has been done and is shown in Figure 23-2, where the $[H^+]$ and $[OH^-]$ at any point during the titration of monoprotic acids having various ionization constants with strong bases, and of monohydroxy bases having various ionization constants with strong acids, can be approximated by means of the curves.[2] These curves are calculated for titrations with 0.2 N solutions; in general the titration errors will be less with more concentrated solutions and greater with more dilute solutions. The case for the titration of

[1] It is well to remember that while CH_3COOH in the solution results only from the incompleteness of the reaction between OH^- and CH_3COOH, the OH^- is present both because of that incomplete reaction and also because of the slight dissociation of water. The latter does not make a significant contribution in the example being considered.

[2] As will be shown in Chapter 25, these curves cannot be used for titrations involving the partial neutralization of polyprotic acids such as the titrations of H_3PO_4 with NaOH to $H_2PO_4^-$ or HPO_4^{2-}.

weak acids with weak bases will not be considered; it is relatively unimportant in analytical work because of the incompleteness of these reactions and the consequent slow rate of change of the $[H^+]$ near the equivalence point.

Displacement Titrations

The use of the titration curves in Figure 23-2 can be extended to displacement reactions, permitting prediction of the changes in the hydrogen-ion concentration during the titration of a salt of a weak acid with a strong acid. Upon hydrolysis the salt of a weak acid and a strong base produces hydroxide ion and can therefore be considered to be a weak base:[3]

$$A^- + H_2O = HA + OH^- \tag{16}$$

The equilibrium expression for this hydrolysis reaction is

$$\frac{[HA][OH^-]}{[A^-]} = K_H \tag{17}$$

where K_H, called the hydrolysis constant, is the reciprocal of the neutralization constant, K_N;

$$K_H = \frac{1}{K_N} = \frac{K_W}{K_A}$$

Note the parallel between the following equations and Equations 16 and 17:

$$BOH = B^+ + OH^-$$

$$\frac{[B^+][OH^-]}{[BOH]} = K_B$$

Hence, when acid is added to a solution of A^-, the $[OH^-]$ and the $[H^+]$ will change in the same manner as they would in a solution of a base with an ionization constant of the same value as the hydrolysis constant of the salt. Therefore, to use Figure 23-2 for a displacement titration of the salt of a weak acid, calculate the hydrolysis constant of the salt and trace the curve for a base with an ionization constant of that value. If the titration is of the salt of a weak base and a strong acid, follow the curve for the corresponding acid. The use of Figure 23-2 is thereby extended to displacement titrations.

---

[3] Note that under the Brønsted definition, A^- is a base, since it is a proton acceptor.

DETERMINATION OF HYDROGEN-ION CONCENTRATIONS AS A MEANS OF OBTAINING THE END POINTS OF ACID-BASE TITRATIONS

The Hydrogen Electrode

So far only the changes in the hydrogen-ion concentration during the course of neutralization and displacement reactions have been considered without reference to possible means of determining the end point of titrations involving reactions of these types. It was seen in the discussion of oxidation-reduction titrations that the end point could be determined potentiometrically by inserting into the solution an inert conducting electrode and measuring the change in potential during the course of the titration. In the absence of strongly oxidizing or reducing systems it is possible to do the same thing in a neutralization titration by saturating the solution with hydrogen gas at constant pressure and measuring the changes in the potential of the half-cell reaction

$$H_2(g) = 2H^+ + 2e^-$$

during the course of the titration. Also, since the standard potential of this reaction is known, it is possible to make determinations of the hydrogen-ion concentration (more properly, the activity) in the solution if the pressure of the hydrogen gas is maintained at some known value; this is the principle and procedure used in the determination of hydrogen-ion concentrations by means of the so-called "hydrogen electrode."[4]

Although the hydrogen electrode is employed in making precise measurements of hydrogen-ion activities, it has certain limitations for practical use. Among these are the requirements of a source of extremely pure hydrogen gas, of means for accurately adjusting or determining the hydrogen pressure above the solution, and of special treatment of the surface of the platinum electrode in order to attain half-cell equilibrium. Moreover, there is interference by oxidizing agents, and inactivating or "poisoning" of the electrodes by various substances. For these reasons other electrode systems have generally replaced the hydrogen electrode for routine use. The *quinhydrone electrode* was used extensively for hydrogen-ion concentration measurement for many years until it was supplanted by the *glass electrode* for most applications.

---

[4] The electrical equipment used is essentially the same as that outlined in the method for making potentiometric titrations. Instead of the platinum wire used there as the inert electrode, it is necessary to use a metallic electrode, usually platinum that has been coated with an amorphous deposit of that metal; such a surface more quickly reaches an equilibrium with the hydrogen gas. Some device must also be provided for keeping the electrode and the solution immediately surrounding it saturated with the hydrogen gas at a constant known pressure.

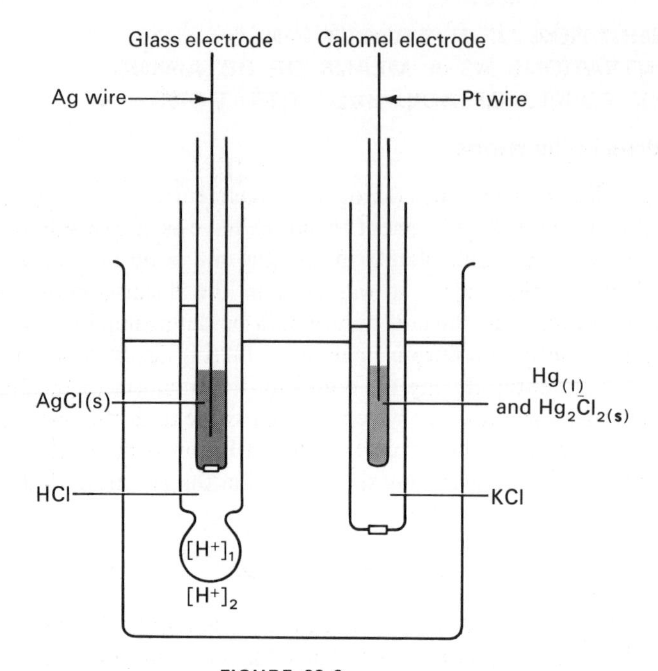

FIGURE 23-3
A glass-electrode system.

The Glass Electrode

If a hydrogen electrode is placed in a solution of constant hydrogen-ion concentration contained in a thin-walled glass bulb,[5] if the bulb is then immersed in a surrounding solution, and if the potential of the electrode within the bulb is monitored (by means of a salt bridge and a standard cell, as is usual in such potential measurements), it will be found that the potential of the electrode will change with changes in the hydrogen-ion concentration of the *surrounding* solution as though the hydrogen-ion concentration of the inner solution had changed a corresponding amount. This phenomenon is the basis for the use of the so-called "glass electrode" for determining hydrogen-ion concentrations and is employed in modern *p*H meters. A diagram of a glass electrode, and of a standard "calomel cell," which is frequently used in conjunction with it, is shown in Figure 23-3. In effect, the

[5] Usually a silver-silver chloride electrode is used in commercial glass electrodes. The purpose of the inner electrode appears to be merely to make electrical connection to the inner wall of the glass bulb. M. T. Thompson (*J. Res. Natl. Bur. St.*, **9**, 853, 1932) asserted that the entire inner electrode could as well be replaced by a metal film deposited on the inner wall of the glass bulb.

glass acts as a hydrogen-ion-permeable membrane that reflects the potential difference between the hydrogen ions on the two sides of the membrane. It is found that the EMF across the glass membrane, E_g, changes with the concentrations (again, more exactly the activities) of the hydrogen ions on the two sides, $[H^+]_1$ and $[H^+]_2$, according to the expression

$$E_g = E' - 0.059 \log [H^+]_2/[H^+]_1$$

where E' is a potential characteristic of the electrode being used. Consequently, if the hydrogen-ion concentration within the bulb is fixed, the potential across the membrane will vary with changes of the hydrogen-ion concentration in the solution surrounding the bulb as shown in the above expression.

A unique and important feature of the glass electrode is that, unlike the hydrogen electrode or the quinhydrone electrode, its potential is not affected by the presence of oxidizing or reducing agents in the outer solution. Originally, measurement of potential changes of the glass electrode required exceedingly sensitive galvanometers (because of the high resistance of the glass bulb), but this obstacle has been overcome in modern *p*H meters by electronic amplification to such an extent that *p*H values can be read directly on calibrated meters. Glass-electrode *p*H meters have largely replaced other methods for the routine determination of hydrogen-ion concentrations.[6] Such a meter will be used in Procedure 25-3.

Neutralization Indicators

The potentiometric method for determining the end point of neutralization titrations, involving the use of the various electrodes described above, is often avoided, since even with the glass electrode it requires considerable time.[7] The end point of such titrations is most frequently obtained by the use of organic compounds called *neutralization, acid-base,* or *pH indicators.* For

[6] For discussions of the glass electrode see Dole. *The Glass Electrode.* New York: Wiley, 1941; Kolthoff and Laitinen. *pH and Electro Titrations.* (2nd ed.) New York: Wiley, 1941; and G. Eisenman. *Glass Electrodes for Hydrogen and other Cations.* New York: Marcel Dekker, 1967. This book contains a comprehensive treatment not only of the hydrogen-ion-specific glass electrode but of more recently developed membrane electrodes that are specific for other cations. Much useful information is available in the descriptive literature that can be obtained from the manufacturers of glass-electrode *p*H meters. Articles by Perley (*Anal. Chem.*, **21**, 394, 559, 1949) show the effect of the composition of the glass membrane on the response to *p*H of the electrode. Recent developments in ion-specific electrodes and their applications are reviewed in *Chem. Eng. News*, p. 41, June 29, 1970.

[7] If a large number of titrations is to be made on very similar samples an automatic titrator, which adds the titrant until a pre-set potential is reached, can be used to advantage. For a discussion of such titrators, see J. J. Lingane. *Electroanalytical Chemistry.* (2nd ed.) New York: Interscience, 1958, Chap. 8.

the present these compounds can be considered as weak acids, usually organic, possessing the distinguishing characteristic that the acid exhibits a different color from that of the basic ion; thus these compounds show a color change upon being neutralized. If these indicators are assumed to be simple acids, their ionization can be represented as

$$HIn = H^+ + In^-$$

with the equilibrium expression,

$$\frac{[H^+][In^-]}{[HIn]} = K_{In}$$

Here K_{In} is the ionization constant of the indicator acid, or, as it is more commonly termed, the indicator constant. Rearranging this expression in the form

$$\frac{[In^+]}{[HIn]} = \frac{K_{In}}{[H^+]}$$

we see that the fraction of the indicator existing in either form will be determined by the value of K_{In} and by the hydrogen-ion concentration of the solution. Also, we see that the color that the indicator exhibits will be determined by the fraction of the indicator that has been converted to the salt or basic (In^-) form and the fraction remaining in the acidic (HIn) form.

Thus if we have a solution in which the indicator is c formal and we let x be the fraction converted into the basic form, then cx is the molar concentration of In^- and $c(1 - x)$ that of the HIn, and we obtain

$$\frac{[H^+](x)}{(1 - x)} = K_{In}, \qquad \text{or} \qquad \frac{(x)}{(1 - x)} = \frac{K_{In}}{[H^+]}$$

From this it is apparent that the indicator is 50% transformed when the $[H^+]$ is equal to the value of K_{In}. Moreover, it is seen that if the $[H^+]$ of the solution is known (is determined, for example, by use of one of the electrode systems mentioned above) and the fraction of the indicator transformed is determined (for example, by colorimetric measurements), then K_{In} can be calculated; or, that if K_{In} is known and x is determined, the $[H^+]$ of a solution can be determined.

The mechanism of the color change of indicators is not so simple as has been assumed above but usually involves equilibria that can be represented as follows:

$$HIn' = HIn'' \qquad \text{and} \qquad HIn'' = H^+ + In''^-$$

where HIn′ would represent the colorless acid, and this would be in equilibrium in an acid solution with HIn″, the colored form. This latter is present in only small amounts, but being more highly ionized, is converted into its salt (In''^-, colored) upon addition of base. It is seen that combining the equilibrium expressions

$$\frac{[HIn'']}{[HIn']} = K_1 \qquad \text{and} \qquad \frac{[H^+][In''^-]}{[HIn'']} = K_2$$

gives an expression,

$$\frac{[H^+][In''^-]}{[HIn']} = K_1K_2 = K_{In}$$

in which, as long as [HIn″] and $[In'^-]$ are small in comparison to [HIn′] and $[In''^-]$, the total indicator in the acid form can be substituted for the HIn′ and that in the basic form for In''^-; thus, it is justifiable to treat indicators as though they were simple acids.

The structural changes accompanying the color transition are shown below for paranitrophenol, which is one of the less complicated of the frequently used organic indicators.

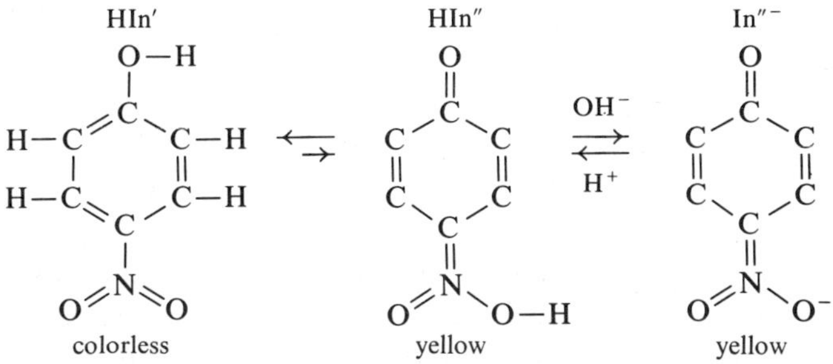

$$\underset{\text{colorless}}{HIn'} \rightleftarrows \underset{\text{yellow}}{HIn''} \underset{H^+}{\overset{OH^-}{\rightleftarrows}} \underset{\text{yellow}}{In''^-}$$

It is also seen that even though the indicator be a base, $InOH = In^+ + OH^-$, the ionization expression

$$\frac{[In^+][OH^-]}{[InOH]} = K_B$$

can be combined with that for water, $[H^+][OH^-] = K_W$, to give

$$\frac{[InOH][H^+]}{[In^+]} = \frac{K_W}{K_B} = K_{In}$$

which again gives the relation between the basic and acidic forms in terms of the $[H^+]$.

From the foregoing it is seen that an indicator changes in color when the $[H^+]$ of the solution changes through such a range that a sufficient fraction of the indicator is converted from one form to the other to result in a color change visible to the eye. The percentage conversion necessary to cause a visual color change depends upon the specific nature of the indicator, and upon whether the change is from colorless to a colored form or from one color to another. In general, the human eye cannot detect the presence of less than 10% of one form of an indicator in the presence of the other, and therefore the entire transition range of most indicators extends over approximately two *p*H units. However, by titrating to either the acid or basic color of an indicator (rather than merely to any part of the color transition range) one can restrict the end point *p*H to about 1 *p*H unit. By matching the titration solution with a reference solution containing the same volume of solution and indicator (and approximately the same amount of salts or of other compounds as are present in the titrated solution at the equivalence point), one can adjust the end point to within a few tenths of a *p*H unit. Figure 23-2 shows not only the change in $[H^+]$ during the titration of various acids and bases but also the so-called "transition ranges" of several of the more common indicators. Therefore, it is possible to select from Figure 23-2 those indicators that will give color changes near the equivalence point of a titration. It is seen that any of the indicators listed can be so used as to give a color change within 0.2% of the equivalence point when strong acids are titrated with strong bases, but that such is not the case when weak acids are titrated with strong bases or weak bases with strong acids. The transition ranges of indicators are shifted somewhat by changes in (1) temperature, (2) the ionic concentration of the solution (this is usually termed the *salt effect*), and (3) the solvent (for example, by the addition of alcohol to the solution). Usually the magnitudes of these effects are not such as to cause serious errors in titrations, but they are of importance in the use of indicators to determine the hydrogen-ion concentration of a solution.[8]

The use of Figure 23-2 in the selection of the proper indicators for specific titrations and in estimating the probable errors involved is illustrated below.

Let us assume that one wishes to titrate a weak monoprotic acid, $K_A = 1 \times 10^{-5}$, with a strong monohydroxy base, and that approximately 0.2 F solutions are used. Which of the indicators shown on Figure 23-2 will provide

[8] For a discussion of these effects and of the indicator method of determining hydrogen-ion concentrations, see Kolthoff. *The Colorimetric and Potentiometric Determination of* pH. New York: Wiley, 1931; Clark. *The Determination of Hydrogen Ions.* (3rd ed.) Baltimore: Williams and Wilkins, 1928; Kolthoff and Rosenblum. *Acid-Base Indicators.* New York: Macmillan, 1937; O. Tomiček. *Chemical Indicators.* trans. A. R. Weir. London: Butterworth, 1951.

an accuracy within about 0.2%? This latter stipulation requires that when the end point is taken, the ratio of equivalents of base added to acid initially present be within 0.998 and 1.002. In the figure the curve labeled $K_A = 10^{-5}$ shows the *p*H changes in such a titration when this ratio changes from 0.990 to 1.010 or from approximately 1% before to 1% beyond the equivalence point. Calculations show that the initial $[H^+]$ of a 0.2 *F* solution of the acid would have been approximately 1.4×10^{-3}, and all of the indicators shown would have been in their acid forms. When the ratio is 0.990 the curve shows the *p*H would be 7 and bromphenol blue, methyl orange, and methyl red would have changed from ther acid to their basic forms; use of any of these would have given errors exceeding 1%. Bromthymol blue would have shown a slow but perceptible change; phenol red would begin to change at this point and continue until the ratio was approximately 0.998—such a slow change is undesirable. Thymol blue would change between ratios of 0.998 and 1.000; phenolphthalein between 0.999 and 1.001; thymolphthalein between 1.000 and 1.002.

QUESTIONS AND PROBLEMS

Problems Involving Chemical Equilibria

1. With the aid Figure 23-2, predict which of the indicators listed will give results accurate to approximately 0.2% in the following titrations, approximately 0.2 *N* solutions being used. Where the error exceeds 0.2%, indicate its sign and probable magnitude.

   (*a*) NH_3 with HCl.
   (*b*) Benzoic acid with KOH.
   (*c*) Aniline with HCl.
   (*d*) KCN with HCl.
   (*e*) Oxalic acid with NaOH (*i*) by titrating one proton—see footnote 2, page 525, (*ii*) by titrating both protons.
   (*f*) Hydrazine with HCl.
   (*g*) Potassium hydrogen phthalate with (*i*) NaOH, (*ii*) HCl.
   (*h*) Borax with HCl.

2. (*a*) You are given 100 ml of a solution that is exactly 1 *F* in CH_3COOH. Calculate the *p*H of this solution.
   (*b*) You wish to add a sufficient volume of exactly 1 *F* sodium hydroxide to give a solution with a hydrogen-ion concentration of 3.6×10^{-5}. Calculate the required volume. *Ans.* 33 ml.
   (*c*) What indicator would you use if you wished to make a quantitative titration of the acetic acid solution with the sodium hydroxide?

3. Calculate to within ±5% the $[H^+]$ existing when to 100 ml of 0.200 *F* H_2SO_4 there are added
   (*a*) 98 ml of 0.400 *F* NaOH.
   (*b*) 100 ml of 0.400 *F* NaOH.
   (*c*) 102 ml of 0.400 *F* NaOH.

4. Having Indicator A ($K_{In} = 8 \times 10^{-10}$) and Indicator B ($K_{In} = 9 \times 10^{-5}$), which would you use in titrating acid A ($K_A = 9 \times 10^{-6}$)? Base B ($K_B = 1 \times 10^{-5}$)? Salt KA ($K_A = 10^{-10}$)?

   Calculate approximately the error that would result from using methyl orange when standardizing sodium hydroxide with potassium acid phthalate (Procedure 24-4).

   Assume the end point to be taken at the midpoint of the methyl orange range, approximately 5×10^{-5}.

5. To 100 ml of 0.200 *F* NH_3 ($K_B = 1.8 \times 10^{-5}$) was added exactly 1 ml of 1.0×10^{-5} *F* HIn ($K_{In} = 1.0 \times 10^{-10}$; HIn colorless, In^- red).

   An end-point comparison solution was made by adding 0.050 ml of HIn to 50 ml of 0.010 *F* NaOH in a beaker similar to that containing the NH_3. The NH_3 solution was titrated with 0.200 *F* $HClO_4$ ($K_A = 10^8$) until the color of the titrated solution matched that of the comparison solution—the two solutions were viewed vertically.

   (*a*) Calculate the volume of the $HClO_4$ added. *Ans.* 77 ml.
   (*b*) Calculate the percentage error in the titration.
   (*c*) Comment on the competence of the titrator.
   (*d*) What would be the effect of viewing the solutions horizontally?

6. (*a*) Calculate the value of the constant K_N for the neutralization of ammonia with hydrochloric acid.
   (*b*) Calculate the hydrogen and hydroxide ion concentrations and the *p*H and *p*OH resulting when to 100.0 ml of exactly 0.2 *N* NH_3 there are added (*i*) 99.8, (*ii*) 100.0, and (*iii*) 100.2 ml of exactly 0.2 *N* HCl.
   (*c*) Which of the indicators shown in Figure 23-2 could be used to give results correct to within approximately 0.2% in the titration of ammonia with a strong acid? *Ans.* (*a*) 1.8×10^9, (*b–i*) *p*H = 6.56.

7. *Transition range of acid-base indicators.* The general statement is made that the human eye cannot detect less than about 10% of one color of a two-color indicator in the presence of the other color. Assuming this to be true show that the transition range of a two-color indicator usually extends over about 2 *p*H units.

8. In the titration of an organic acid ($K_A = 2 \times 10^{-5}$) with a strong base, an indicator with a constant, K_{In}, having the value 5×10^{-6} was used. Assuming that the color change was noted when the indicator was 20% transformed to the basic form, calculate (*a*) the hydrogen-ion concentration at the end point, (*b*) the ratio of organic acid to anion (salt) at that point, and (*c*) the relative percentage error in the titration of the acid.

9. *Displacement titrations.* Calculate the hydrogen-ion concentration at the equivalence point in the titration of a 0.2 *F* KCN solution with a 0.2 *N* solution of a strong acid. *Ans.* 1.4×10^{-5}.

10. *Displacement titrations.* Assume, in the titration of the KCN in Problem 9, that the hydrogen-ion concentration changes as it would in the titration of a 0.2 *N* solution of a base having an ionization constant numerically equal to the hydrolysis constant of KCN. Which of the indicators listed in Figure 23-2 could be used to give results correct to within approximately 0.2%?

11. *Buffer solutions.* Calculate the hydrogen-ion concentration of a solution prepared by mixing 200 ml of 1 *F* CH_3COOH and 200 ml of 1 *F* CH_3COONa. Calculate approximately the hydrogen-ion concentration if (*a*) 100 ml of the solution were diluted to 200 ml, (*b*) 100 ml of 0.1 *F* HCl were added to 100 ml of the solution, and (*c*) 100 ml of 1 *F* HCl were added to 100 ml of the solution. *Ans.* (*c*) 0.25 *M*.

12. *Buffer solutions.* It is desired to prepare 100 ml of a solution having an initial hydrogen-ion concentration of 3.6×10^{-5} *M* and of such composition that, upon the formation of 10 mmoles of hydrogen ion during the course of a reaction, the hydrogen-ion concentration will not exceed 7.2×10^{-5}. Calculate the minimum initial concentrations of acetic acid and of acetate that will provide the desired conditions.

13. *Dilute solutions of weak acids.* Calculate the hydrogen-ion concentration of a solution that is 1.0×10^{-6} *F* in HCN. *Ans.* 1.1×10^{-7}.

NOTE: If the approximate expression $[H^+] = \sqrt{K_A c}$ is used, one obtains a value that indicates an alkaline solution. Why is this value obviously incorrect?

In order to eliminate this fallacy, the ionization of both the weak acid and the water must be considered, thus:

$$HA = H^+ + A^- \qquad \frac{[H^+][A^-]}{[HA]} = K_A \tag{1}$$

$$HOH = H^+ + OH^- \qquad [H^+][OH^-] = K_W \tag{2}$$

Since the solution must be electrically neutral, a third equation can be obtained as follows:

$$[H^+] = [A^-] + [OH^-] \tag{3}$$

Substituting for $[A^-]$ and $[OH^-]$, one obtains

$$[H^+] = K_A \frac{[HA]}{[H^+]} + \frac{K_W}{[H^+]}$$

and

$$[H^+]^2 = K_A[HA] + K_W$$

or

$$[H^+] = \sqrt{K_A[HA] + K_W}$$

If c represents the formal concentration of the acid, c will usually be large compared with $[A^-]$, and therefore

$$[H^+] = \sqrt{K_A c + K_W}$$

When $K_A c \gg K_W$, this expression reduces to the approximate expression given at the beginning of this note.

14. In an experiment to determine the rate of the hydrolysis in acid solution of acetamide, which may be represented as

$$CH_3CONH_2 + H_2O = CH_3COOH + NH_3$$
$$NH_3 + H^+ = NH_4^+$$

it was essential that the *p*H of the reaction solution be maintained between the values 3.69 and 4.00.

(*a*) If a formic acid-formate buffer is to be used, calculate the ratio of concentrations of formic acid to formate ion that will provide the initial *p*H (3.69). *Ans.* Ratio = 1.

(*b*) Calculate the minimum initial concentrations of formic acid and formate ion, keeping the ratio calculated in (*a*), that will maintain the proper *p*H range in the solution, if the hydrolysis reaction produces 0.50 mmoles of NH_3 and the solution volume is 100 ml.

(*c*) An experiment similar to the above was carried out with neither formic acid nor formate ion present. The initial volume was again 100 ml and the initial *p*H was adjusted to 3.69 with HCl and NaOH. Again, the hydrolysis reaction produced 0.50 mmoles of NH_3. Calculate the final $[H^+]$ of the solution. *Ans.* 4.8×10^{-11}.

15. Benzoic acid–ammonium benzoate buffer systems are sometimes used in the separation of tripositive metals (iron, aluminum, chromium) from such bipositive metals as zinc, nickel, cobalt, and manganese. If 500 mg of zinc are to be retained in 100 ml of solution, what is the maximum ratio of benzoate to benzoic acid that can be used? *Ans.* 66.

CHAPTER 24

Ionization Reactions. Titrations with Monoprotic Acids and Monohydroxide Bases

The theory of acid-base reactions in aqueous solutions was discussed in Chapters 7 and 23. In this chapter procedures are given for preparing and standardizing solutions of sodium hydroxide and hydrochloric acid, and for the titration of carbonate-hydroxide mixtures. The standard solutions are used also in Chapters 25 and 26.

Unless standard acids and bases are to be used for some specific purpose, it is desirable that they possess the following general qualifications:

1. They should be highly ionized, since, as has been mentioned, if the titration of a weak base with a weak acid is attempted, the rate of change of the hydrogen-ion concentration near the equivalence point is likely to be so slow as to prevent a precise determination of the end point. (That is, the titration curve will not be sufficiently steep in the equivalence point region.)
2. They should be so soluble that solutions as concentrated as 0.5 *N* and in some cases even 1 *N* can be prepared.
3. They should form soluble salts, since the formation of a precipitate during the titration may obscure the end point, or the precipitate may adsorb the indicators.
4. They should be stable compounds. Oxidizing or reducing agents can be affected by the presence of extraneous material (such as dust, organic matter, or even the oxygen of the air), and are also likely to

react with the indicators. Many acid-base indicators are unstable with respect to oxidation or reduction, being decomposed or converted into colorless compounds. Volatile compounds such as ammonia are difficult to preserve in storage without elaborate precautions. Substances that rapidly attack glass containers also require special apparatus in storage.

No one acid or base meets all four of these qualifications. Sulfuric acid, which is nonvolatile, forms insoluble salts with alkaline-earth hydroxides. Nitric acid, although relatively stable when cold and dilute, may act as an oxidizing agent in hot solutions or may contain traces of nitrous acid, which reacts with certain indicators, notably methyl orange and methyl red. Perchloric acid is a strong acid, is nonvolatile, and is stable toward reduction in dilute solutions, but the potassium and ammonium salts are only moderately soluble. Although hydrogen chloride is a gas, it is so highly ionized in aqueous solutions that its partial pressure from even 0.5 N solutions is so small that they can be boiled for considerable periods of time without appreciable loss if the solution is not allowed to concentrate by evaporation. Since most chlorides are soluble and since hydrochloric acid is relatively inert toward oxidation or reduction, it is extensively used as a standard acid.

Standard bases present more difficulties in storage; they tend to attack glass containers and to absorb various gases, especially carbon dioxide. Barium hydroxide is sometimes used as a standard base but is sparingly soluble and forms insoluble salts. Aqueous ammonia solutions lose ammonia. Potassium hydroxide has no distinct advantage over sodium hydroxide and is more expensive; therefore sodium hydroxide is the more generally used reagent.

Preparation of Carbonate-free Solutions of Sodium Hydroxide

Outline. The sodium hydroxide is first prepared as a concentrated solution in which sodium carbonate is only slightly soluble. The sodium carbonate (present as an impurity in the solid NaOH) is removed and the concentrated and carbonate-free solution is then diluted with freshly boiled water.

DISCUSSION

Sodium hydroxide cannot be used for the direct preparation of a standard solution because (*a*) it is not available in a sufficiently pure state, (*b*) it is too hygroscopic, and (*c*) it is too reactive with the carbon dioxide of the air. Reagent grade sodium hydroxide contains about 3% water along with significant but variable amounts of sodium carbonate. A pellet of NaOH rapidly becomes moist when exposed to the air; with continued exposure the pellet is converted to sodium carbonate. Carbonate is undesirable in standard alkali solutions because, when it is titrated with acid and a strongly acidic indicator (for example, methyl orange) is used, the carbonate is converted to carbonic acid, so that two equivalents of the acid are used for each mole of carbonate present. When it is titrated with acid and a strongly basic indicator (for example, phenolphthalein) is used, the carbonate is converted to hydrogen carbonate (if the solution is cold) and thus neutralizes only one equivalent of the acid; furthermore, this latter end point is not obtained precisely except under very strictly controlled conditions. Hence, a sodium hydroxide solution that contains carbonate will have different normalities depending upon what indicator is used. For these reasons it is an advantage to have standard alkali solutions that are relatively free of carbonate. It is very difficult to obtain carbonate-free sodium or potassium hydroxide, but fortunately sodium carbonate is relatively insoluble in a 50% sodium hydroxide solution. Therefore, by preparing such a solution and allowing it to stand until the sodium carbonate settles out, or by filtering or centrifuging out the precipitate, one can prepare a carbonate-free solution by dilution with water that is free of carbon dioxide.[1]

[1] For studies of this method of preparing carbonate-free sodium hydroxide solutions, see Allen and Low. *Ind. Eng. Chem.*, anal. ed., **5**, 192 (1933); also Han and Chao. *Ind. Eng. Chem.*, anal. ed., **4**, 229 (1932); and Han. *Ind. Eng. Chem.*, anal. ed., **6**, 209 (1934); **9**, 140 (1937).

INSTRUCTIONS (Procedure 24-1: Preparation of Carbonate-free Solutions of Sodium Hydroxide)

Weigh out 16 g of sodium hydroxide, dissolve it in 15 ml of water, transfer the solution to a 50-ml test tube (Note 1), stopper the mixture with a rubber stopper, and then allow it to stand in a vertical position until the precipitate settles (Notes 2, 3). Quickly pipet out 10 ml (or 12 ml for a gravimetric solution) of the clear solution without stirring up the residue, and transfer it to a plastic (polyethylene or equivalent) bottle that has been fitted with a two-hole rubber stopper carrying a siphon tube with a short rubber tube and pinch clamp at the bottom in one hole and a short inlet tube to which a soda-lime tube is attached in the other hole (Note 4). Immediately dilute the solution to a liter with distilled water (Note 5), swirl until mixed, and fit the stopper into place.

NOTES

1. This tube should be of resistance glass or, preferably, a plastic, such as polyethylene, because of the rapidity with which ordinary "soft" glass would be attacked by the concentrated alkali.

2. The settling of the sodium carbonate may require considerable time, usually several days. Therefore, if a centrifuge is available, the process can be expedited by centrifuging the mixture until the precipitate is thrown out, or the mixture can be filtered with the aid of suction through an asbestos or sintered-glass filter and collected directly in the container in which it is to be stored.

3. Since it requires some time for the sodium carbonate precipitate to settle and since there is considerable loss of sodium hydroxide in preparing the 50% solution for only 1 liter of the diluted hydroxide, it is suggested that a stock supply of the concentrated solution be prepared in a suitable container and kept available in the laboratory.

4. If a glass bottle is used it is desirable that the inside surface be protected by a layer of paraffin wax. To apply this layer, clean and dry the bottle, warm it in an oven or immerse it in hot water, and pour into it sufficient melted paraffin wax to cover the inner surface. Roll the bottle until the sides are completely covered, tilt it until the bottom is uniformly covered, and allow it to cool in an upright position.

5. According to Kolthoff (*Biochem. A.*, **176**, 101, 1926) water in equilibrium with air of the usual carbon dioxide content is only 1.5×10^{-5} F in carbon dioxide. This amount is negligible in alkali solutions of the concentrations ordinarily used. However, concentrations 10 to 20 times greater than this equilibrium value are often found in distilled water; such excess concentrations can be removed by boiling for a few minutes.

The higher concentration of carbon dioxide sometimes found in distilled water causes a difference of only about 0.1% in the apparent concentrations of 0.2 N solutions of NaOH when they are titrated to the methyl orange and to the phenolphthalein end points. If accuracy greater than $\pm 0.1\%$ is required, all water used in the preparation of the standard solutions and in the titrations should be boiled prior to its use.

Preparation of Hydrochloric Acid Solutions

Outline. Commercial concentrated hydrochloric acid is diluted with distilled water to the approximate concentration desired.

DISCUSSION

Standard solutions of hydrochloric acid can be obtained by preparing constant-boiling acid[2], which is approximately 6 *N*, and then diluting weighed amounts of this to the desired volume. This procedure has much to recommend it, since standard solutions accurate to 0.1% can be readily obtained and since stock solutions of the constant-boiling acid are quite stable and can easily be stored; its principal drawback is the time required to set up the apparatus and to carry out the distillation. More frequently hydrochloric acid solutions are prepared by simply diluting concentrated acid to approximately the desired normality and then standardizing this solution against an appropriate primary standard.

INSTRUCTIONS (Procedure 24-2: Preparation of Hydrochloric Acid Solutions)

Calculate the volume of the reagent grade concentrated hydrochloric acid (approximately 12 *F*) required to prepare 1 liter of 0.2 *N* volumetric solution or 1 kg of 0.24 *WN* gravimetric solution. Measure out this volume of the acid and dilute it to 1 liter. (Since the solution is to be standardized an approximate volume adjustment is adequate for either solution.) Transfer the solution to a bottle with a ground-glass stopper, mix it thoroughly, and allow it to cool to room temperature.

[2] Hulett and Bonner. *J. Am. Chem. Soc.*, **31**, 390 (1909); Morey. *J. Am. Chem. Soc.*, **34**, 1032 (1912); Foulk and Hollingsworth. *J. Am. Chem. Soc.*, **45**, 1220 (1923); Shaw. *Ind. Eng. Chem.*, **18**, 1065 (1926); Titus and Smith. *J. Am. Chem. Soc.*, **63**, 3266 (1941).

PROCEDURES **24-3 (A and B)**

Reaction of Strong Acids and Strong Bases. Determination of Titrimetric Ratios

Outline. In Procedure 24-3A the sodium hydroxide solution from Procedure 24-1 is titrated with the hydrochloric acid solution from Procedure 24-2 with methyl orange as indicator.

In Procedure 24-3B the hydrochloric acid solution is titrated with the sodium hydroxide with phenolphthalein as indicator.

The titrimetric ratio of the solutions is calculated from data obtained.

The titrated solutions are used in experiments that show the effects obtained with these indicators in titrations involving weak acids and weak bases.

DISCUSSION

Once the titrimetric ratio of the acid and base solutions is determined, if either solution is subsequently standardized, the normality of the other can be calculated by means of this ratio. In general, in titrimetric work, such a pair of solutions is advantageous since it permits one to back titrate in case a titration is overrun.

The titrations of this procedure involve examples of the reaction of a strong acid with a strong base. As has been pointed out in the discussion of such reactions, the hydrogen-ion concentration at the end point will be 10^{-7} (in the absence of carbon dioxide), and reference to the titration curves of Figure 23-2 indicates that the rate of change of the pH with change in ratio of acid to base is such that most of the indicators listed can be used with an accuracy of better than 0.2% when solutions 0.2 N (or more concentrated) are used. In the procedure below, titrations are made using, first, methyl orange, and then phenolphthalein, in order that experience can be acquired in their use.

In strong acid-strong base titrations, when indicators are used with transition ranges having pH values less than about 4, or more than about 10, consideration has to be taken of the volume of the standard solution required to lower or to raise the pH of the titrated solution from that of the equivalence point to that required to cause the indicator transition. The effect of using indicators having transition ranges significantly different from the equivalence point *p*H is illustrated with methyl orange and phenolphthalein in the titrations below. It should be realized that the magnitude of the errors caused by neglecting this effect will increase proportionally as more dilute standard solutions are used.

Usually the volume of the standard solution required to change the indicator from one form to the other is negligible since the quantity of the indicator required is normally very small (because of the intensity of the indicator color).

After completing the titrations given below with each indicator, the titrated solutions are used for experiments that show the nature of the end points obtained and the errors involved in using the indicators for the titration of a weak base with a strong acid and the titration of a weak acid with a strong base.

INSTRUCTIONS (Procedure 24-3: Reaction of Strong Acids and Strong Bases. Determination of the Titrimetric Ratios. A. The Use of Methyl Orange as Indicator)

I. Volumetric Titration

Fill a buret with the 0.2 *N* hydrochloric acid. Prepare a comparison solution by taking 100 ml of water in a 200-ml conical flask and adding 2 drops of methyl orange indicator. Pipet a 25-ml portion of the 0.2 *N* sodium hydroxide into a 200-ml conical flask. Add 50 ml of water and 2 drops of methyl orange indicator. Titrate the base with the acid until the first distinct change in color of the indicator from yellow toward reddish yellow or pink is obtained (Note 1). Reserve this solution for use in Experiment I below.

Titrate the comparison solution with the acid, using fractional drops, until a color is obtained that matches that of the titrated solution (Note 2). Reserve

NOTES

1. If the end point is overrun, add 1.00 ml of the solution being titrated by means of a 1-ml pipet and again titrate to the transition color.

2. Methyl orange will show a perceptible transition from yellow towards pink when it is from 5% to 20% transformed to the acid form, or at about an average *p*H of 4. Therefore, it is seen that 0.05 ml of 0.2 *N* acid should be required to obtain this *p*H value in 100 ml of water.

Theoretically, the uncorrected end point would more clearly coincide with the equivalence point if the titration were performed with base to the absence of the methyl orange pink color. In general practice, the appearance of a color can be perceived more reproducibly than can its disappearance.

this solution for Experiment II below. Subtract the volume of acid required for the comparison solution from that used in the titration and calculate the volumetric ratio of the acid and base solutions.

Repeat the titration and calculation with a second portion of base (Note 3).

NOTES

3. Duplicate values of the ratio should agree to within 2 parts per 1000.

II. Gravimetric Titration

Weigh 10- to 15-g portions of the 0.24 *WN* sodium hydroxide into 200-ml conical flasks. Proceed as in the volumetric titration above except titrate with the gravimetric buret.

From these two titrations calculate the gravimetric ratios.

EXPERIMENT 24-3A: I. Titration of a Weak Base with a Strong Acid—Methyl Orange as Indicator

Add to the first reserved solution from above 5 ml of 1 *F* NH_4Cl (or about 0.25 g of the solid).

(*a*) Note and explain any observable change in the solution.

(*b*) This solution simulates the end point of an acid-base titration. To what type of acid-base titration does it belong?

Add, dropwise, standard 0.2 *F* HCl or NaOH until the color matches that of the comparison solution.

(*c*) Predict from the volume of standard solution required the sign and approximate magnitude of any error when aqueous ammonia is titrated with a strong acid with methyl orange as indicator. Check this prediction by use of Figure 23-2.

EXPERIMENT 24-3A: II. Titration of a Weak Acid with a Strong Base—Methyl Orange as Indicator

Add to the reserved comparison solution from above 5 ml of 1 *F* CH_3COONa (or 0.7 g of reagent-grade $CH_3COONa{\cdot}3H_2O$).

(*a*) Note and explain any observable change in the solution.

(*b*) What type of acid-base titration does this solution simulate?

Titrate this solution with standard 0.2 *F* HCl or NaOH until the original color is restored—the titration may be done rapidly.

(*c*) Predict from this titration the sign and approximate magnitude of any error if acetic acid is titrated with a strong base with methyl orange as indicator.

(*d*) Assume that the indicator changes at a *p*H of approximately 4 and calculate the ratio of base added to acetic acid initially present required to produce this *p*H.

(*e*) What does this result indicate regarding the sign and magnitude of any error in such a titration?

(*f*) Predict the sign and magnitude of the titration error from the titration curves of Figure 23-2. Compare this result with that from (*e*).

INSTRUCTIONS (Procedure 24-3: Reaction of Strong Acids and Strong Bases. Determination of the Titrimetric Ratios of the Acid and Base Solutions. B. The Use of Phenolphthalein as Indicator)

I. Volumetric Titration

Pipet 25 ml of the 0.2 *N* hydrochloric acid into a 200-ml conical flask. Add 50 ml of water (Note 5, Procedure 24-1) and 2 drops of phenolphthalein indicator solution. Prepare a comparison solution by adding to another 200-ml flask 100 ml of water and 2 drops of phenolphthalein indicator solution.

Fill a buret with 0.2 *N* sodium hydroxide (Note 4). Stopper the buret except when withdrawing solution (Note 5). Titrate the acid with the base until the first perceptible pink color can be detected in the solution (Note 1). Stopper the flask (Note 6). Reserve it for use in Experiment I below. Then titrate the comparison solution with the base until a color is obtained that matches that of the titrated solution (Note 7). Stopper the flask and reserve it for use in Experiment II below. Record the volume of NaOH (Note 8).

NOTES

4. Do not allow the sodium hydroxide to stand in the buret for longer than the time required for the titrations. Upon completion of the titrations, drain the base from the buret; then wash it with 1 *F* HCl and finally with water.

5. The buret is stoppered in order to avoid absorption of carbon dioxide by the base. When titrating, loosen the stopper sufficiently to permit withdrawal of the solution. Before beginning a titration, withdraw and discard 0.3 ml–0.5 ml of the base. The solution at the tip absorbs carbon dioxide rapidly.

When phenolphthalein or any other indicator of similar transition range is used it is preferable to add the base from the buret; there is less chance for absorption of carbon dioxide in the buret than when the base is exposed to the air in a beaker or flask throughout the titration.

6. The pink of a phenolphthalein end point will gradually fade upon exposure to the air, the more rapidly if the solution is vigorously shaken, because of absorption of CO_2.

7. The *p*H at which the color will be perceived with phenolphthalein will vary from 9 to 10, depending upon the quantity of indicator added and the individual observer.

Subtract the volume of base required for the indicator correction from that used in the titration and calculate the volumetric ratio of the acid and base solutions. Repeat the titration and calculation with a second portion of the acid (Note 3).

NOTES

With a two-color indicator the perception of the end point is dependent primarily upon conversion of a certain *fraction* of the indicator from one form to the other, and therefore the end point is not critically dependent upon the concentration of the indicator.

With a one-color indicator, such as phenolphthalein, the perception of the end point is dependent upon conversion of a certain *concentration* of the indicator to the colored form, and therefore the *p*H at which perception of the end point occurs is dependent upon the total concentration of the indicator.

This effect will be readily observed if additional phenolphthalein is added to either the blank or the titrated solution above.

8. If the sodium hydroxide contains carbonate, the results obtained with phenolphthalein will be subject to error. Methyl orange will change color only after the carbonate has been changed to H_2CO_3 (or CO_2 and water), and thus absorption of CO_2 by the base will not change its titer when it is titrated with HCl. However, phenolphthalein will show an acid reaction when carbonate has been changed to HCO_3^-, and therefore absorption of CO_2 by the base will change its apparent titer. If the sodium hydroxide is suspected of containing a significant amount of carbonate, proceed with the titrated solution as follows:

> *Test and Correction for Carbonate.* Remove the stopper from the titrated solution, boil the solution vigorously for 30 seconds, stopper the flask loosely, and then cool the solution. Compare the color with that of the "blank." If the color of the titrated solution is substantially more intense, add standard acid from a 1-ml *measuring* pipet until the solution is colorless, note the volume added, and then add a slightly larger additional volume. Boil and cool the solution as before. Again titrate with the NaOH to a match with the color of the "blank" solution. From the total volume of NaOH solution (corrected for the indicator correction) and the total volume of acid, calculate the volumetric ratio of the acid and base solutions.

If the standard base contains a significant quantity of carbonate, the carbonic acid formed during the titration will prevent the phenolphthalein from turning pink until it has been converted to HCO_3^-. When the solution is boiled, the reaction $2HCO_3^- = CO_2 + H_2O + CO_3^{2-}$ will cause the solution to become more alkaline, and the pink color of the phenolphthalein will become more intense. The comparison of colors should not be made until the solution is cooled, since phenolphthalein is less sensitive to hydroxide ion at higher temperatures.

II. Gravimetric Titration

Weigh 10- to 15-g portions of the 0.24 *WN* hydrochloric acid into 200-ml conical flasks. Proceed as in the volumetric titration above except titrate with the weight buret.

EXPERIMENT 24-3B: I. Titration of a Weak Acid with a Strong Base—Phenolphthalein as Indicator

Add to the first reserved flask from above 5 ml of 1 *F* CH_3COONa (or 0.7 g of reagent-grade $CH_3COONa{\cdot}3H_2O$.

(*a*) Note and explain any observable change in the solution.

(*b*) This solution simulates the end point of an acid-base titration. To what type of acid-base titration does it belong?

Add, dropwise, standard 0.2 *F* HCl or NaOH until the color matches that in the comparison solution.

(*c*) Predict from the volume of standard solution required the sign and magnitude of any error when acetic acid is titrated with a strong base with phenolphthalein as indicator. Check this prediction by use of Figure 23-2.

EXPERIMENT 24-3B: II. Titration of a Weak Base with a Strong Acid—Phenolphthalein as Indicator

Add to the second reserved solution from above 10 ml of 1 *F* NH_4Cl (or about 0.5 g of the solid).

(*a*) Note and explain any observable change in the solution.

(*b*) What type of acid-base titration does this solution simulate?

Titrate with standard 0.2 *F* HCl or NaOH until the original color is restored—the titration may be done rapidly.

(*c*) Predict from the titration the sign and approximate magnitude of any error in the titration of aqueous ammonia with a strong acid with phenolphthalein as indicator.

(*d*) Assume that the indicator change was observed at a *p*H of approximately 10 and calculate the ratio of acid added to ammonia initially present to produce this *p*H.

(*e*) What does this ratio indicate regarding the sign and magnitude of any error in such a titration?

(*f*) Predict the titration error by use of the titration curves of Figure 23-2.

NOTES

Upon addition of acid, the carbonate remaining is converted to HCO_3^- and the solution again becomes colorless. Upon introduction of the additional acid, the HCO_3^- is converted to H_2CO_3, which is expelled as CO_2 when the solution is subsequently boiled.

STANDARDIZATION OF HYDROXIDE SOLUTIONS

General Discussion. Primary Standards for Hydroxide Solutions

Among the primary standards against which solutions of highly ionized bases can be precisely standardized are constant-boiling hydrochloric acid,[3] benzoic acid (C_6H_5COOH), potassium hydrogen phthalate ($KHC_8H_4O_4$), commonly called potassium acid phthalate, and sulfamic acid (HSO_3NH_2).[4] Oxalic acid is sometimes used, but the hydrated compound ($H_2C_2O_4 \cdot 2H_2O$) is somewhat difficult to prepare with an exactly known water content, and the anhydrous acid is too hygroscopic for practical use. Potassium hydrogen iodate, [$KH(IO_3)_2$], has been recommended and is an interesting compound since it can be used as both an acid-base and oxidation-reduction primary standard. However, a question has been raised as to its purity; it is not extensively used except for special purposes. Constant-boiling hydrochloric acid is an excellent standard if it is available; however, for a single standardization its preparation is a lengthy process. Benzoic acid can be obtained from the Bureau of Standards, and its only disadvantages are that the resublimed acid is so voluminous as to be somewhat inconvenient to weigh unless it is melted and that the acid is so slightly soluble in water that it must be dissolved in alcohol for the titration. Potassium hydrogen phthalate, which is also supplied by the Bureau of Standards, has neither of the difficulties mentioned in connection with benzoic acid and in addition has a high equivalent weight; it is probably more generally used than any of the others.

Titration of Weak Acids with Strong Bases. When a solution of potassium hydrogen phthalate is titrated with an equivalent amount of sodium hydroxide, the change in hydrogen-ion concentration, and therefore the titration curve (see Figure 23-2) in the region of the equivalence point, is essentially the same as would have been obtained by a similar titration of a monoprotic acid having an ionization constant of the same value as the second ionization constant for phthalic acid. As has been mentioned before, the curves of Figure 23-2 can be applied to the titration of polyprotic acids *only when the last proton is being neutralized.*

The ionization constant for the second hydrogen of phthalic acid is 3.1×10^{-6} and it can be calculated (with the assumption that the titration is made with 0.2 N solutions and that therefore the final concentration of the salt is 0.1 F) that the $[H^+]$ at the equivalence point will be 5.6×10^{-10}.

---

[3] See Footnote 2.

[4] See Butler, Smith, and Audrieth (*Ind. Eng. Chem.*, anal. ed., **10**, 690, 1938) for a discussion and review of the literature concerning sulfamic acid.

This value is obtained approximately, and more readily, by referring to Figure 23-2 and tracing to the equivalence point an imaginary curve for an acid whose constant is 3×10^{-6}. It is seen that this equivalence-point pH value lies within the transition range for phenolphthalein and that, furthermore, when the ratio of the equivalents of base to acid is 0.998, the phenophthalein is largely in its acid form (the indicator constant, K_{In}, of phenolphthalein has the value 2×10^{-10}) and that when this ratio is 1.002, the indicator is largely in its basic form. Thus it can be predicted that with 0.2 N solutions and with phenolphthalein as the indicator, the second proton of phthalic acid can be titrated with sodium hydroxide (or any strong base) to an accuracy within 0.2%.

One should realize, however, that the change of $[H^+]$ in the equivalence-point region with a given change in ratio of base to acid is much less when titrating a weak acid than a strong one; otherwise stated, the slope of the titration curve is much less steep. Therefore, in order to obtain the same accuracy in titrating a weak acid as a strong acid, one must take the end point within much closer pH limits. This is accomplished, first, by selection of an indicator of the proper pH range; and second, when higher accuracy is required, by use of a comparison solution having the equivalence-point pH and containing the indicator in the same concentration as is used in the titration.

In most cases, a solution having the equivalence-point pH is easily prepared by adding to the comparison solution an approximately equivalent quantity of the *salt that is formed by the titration.* For example, in the procedure below, the student should add to the comparison solution an amount of potassium or sodium phthalate approximately equivalent to the monohydrogen phthalate being titrated; if the desired salt is not readily available, a substitute buffer system is employed. By means of such a reference solution, one should be able to adjust the pH of the titrated solution to within 0.2–0.4 of a pH unit of that of the comparison solution, which should represent the equivalence-point pH. When the equivalence-point pH is such that the solution tends to absorb carbon dioxide, it may be more expedient to carry out the titration to the end point, then titrate the comparison solution to a similar color and make any necessary correction.

Titration of Weak Bases with Strong Acids. It is obvious that methyl orange could not be used as the indicator for the above titration of a weak acid. However, for the titration of a weak base such as of aqueous ammonia ($K_B = 1.8 \times 10^{-5}$) with a strong acid, it is seen that the titration curve would indicate that the $[H^+]$ at the end point would be approximately 7×10^{-6}, close to the transition range of methyl orange; that when the ratio of acid to base was 0.998, the indicator would be very largely in the basic form; and that when the ratio was 1.002, it would be sufficiently converted to the acid form for a color change to be observed, so that an accuracy within 0.2% could be attained. As with the titration of the phthalic acid, the slope of the titration curve is such that the accuracy of the titration is improved by the use of a comparison solution. If the aqueous ammonia were being titrated with hydrochloric acid, the comparison solution would be prepared from an approximately equivalent amount of pure ammonium chloride.

From these examples it is seen that with the aid of Figure 23-2 the accuracy with which any monoprotic acid (or the last proton of a polyprotic acid) can be titrated with a strong base, or with which any monohydroxy base can be titrated with a strong acid, can be predicted if the transition range of the indicator is known. Since either methyl orange or phenolphthalein can be used for the titration of strong acids with strong bases, and since weak acids can be titrated by the use of phenolphthalein and weak bases by the use of methyl orange, these two indicators can be used for nearly all of the neutralization titrations usually made. However, some of the indicators whose transition ranges are shown in Figure 23-2 have specific advantages for certain titrations; these will be discussed later.

PROCEDURE 24-4

Standardization of a Sodium Hydroxide Solution Against Potassium Hydrogen Phthalate

Outline. Weighed, dry samples of potassium hydrogen phthalate are dissolved in water and titrated with the sodium hydroxide solution. Phenolphthalein is used as the indicator. The titration reaction is

$$HC_8H_4O_4^- + OH^- = C_8H_4O_4^{2-} + H_2O$$

DISCUSSION

It was mentioned above that potassium hydrogen phthalate is probably the most generally used primary standard for hydroxide solutions. Phthalic acid is a dicarboxylic acid having the structural formula

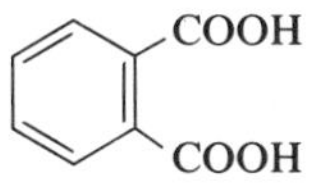

COOH

COOH

The potassium salt is used because it can be readily obtained in a very pure state by recrystallization, because of its moderate solubility, and because its solubility has a high thermal coefficient. The compound conforms to the other requirements of a primary standard and is commercially available in primary standard quality. Conversely, the difficulty of obtaining potassium or sodium phthalate of high purity from commercial sources forces one to consider means of preparing comparison solutions other than the dissolving of an equivalent quantity of the salt of the titrated acid. Two alternatives are provided in the procedure below.

The first is to use the salt of an acid that is of approximately the same strength as the acid being titrated. The second constant for succinic acid is 2.8×10^{-6}; that for phthalic acid is 3.1×10^{-6}. Therefore the substitution of sodium succinate for phthalate provides a comparison solution having very nearly the same hydrogen-ion concentration as the equivalence point.

The second alternative method provided here for the preparation of the comparison solution makes use of a buffer solution having the desired *p*H. An equimolar solution of ammonia and ammonium ion is readily prepared and has the appropriate hydrogen-ion concentration.

INSTRUCTIONS (Procedure 24-4: Standardization of a Sodium Hydroxide Solution Against Potassium Hydrogen Phthalate)

Dry about 6 g of $KHC_8H_4O_4$ at 120°C for 1 hour and allow the crystals to cool in a desiccator (Note 1). Weigh out accurately about 1.5 g of the phthalate and dissolve it in 50 ml of water in a 300-ml flask (Note 2). Add 1 drop of phenolphthalein indicator to the solution and titrate it with the carbonate-free 0.2 *N* sodium hydroxide solution (use the 0.24 *WN* NaOH and a gravimetric buret if a gravimetric titration is being made) to be standardized (Note 3) until the first perceptible pink color is obtained (Note 4). Stopper the flask (Note 5).

Add to a similar flask a volume of water equal to that of the titrated solution, a weight of sodium or potassium phthalate approximately equivalent to the hydrogen phthalate taken (Note 6), and the same amount of

NOTES

1. $KHC_8H_4O_4$ is not hygroscopic. If properly prepared and dried, and thereafter protected from excessive moisture, it should not contain over 0.05% water; consequently, the preliminary drying may often be dispensed with. It has also been shown[5] that standard solutions of the hydrogen phthalate are stable for long periods; it is therefore possible to prepare such a solution and use it as needed.

2. Carbon dioxide in this water, or carbonate in the sodium hydroxide, would cause errors in this standardization if present in considerable amounts, because they would be converted to hydrogen carbonate (HCO_3^-) at the phenolphthalein end point. The extent of the error from CO_2 in the water is usually small unless the water is supersaturated with CO_2. See Note 5, Procedure 24-1.

3. Do not allow alkaline solutions to stand in burets with glass stopcocks for any longer time than is absolutely necessary. Rinse the buret, first with dilute acid and then with water, as soon as the alkali is removed. Burets with tips connected by short pieces of rubber tubing (Mohr type) can be used for all work in which a high degree of precision is not required.

4. Be careful to use the same amount of indicator for each titration and for the blanks. As has been stated, the *p*H range at which the color change is observed will depend on the concentration of the indicator, especially a one-color indicator.

5. A solution that is alkaline to phenolphthalein will slowly absorb carbon dioxide from the air and become colorless.

6. If pure solid sodium or potassium phthalate is not available, an equivalent weight of sodium succinate, $Na_2C_4H_4O_4 \cdot 6H_2O$, can be substituted. The second ionization constant of succinic acid is 2.8×10^{-6}. A suitable buffer solution can also be prepared by adding 10 ml of 1 *F* NH_4Cl (or 0.53 g) and 25 ml of 0.2 *F* NaOH to 65 ml of water.

---

[5] Hoffman. *J. Res. Natl. Bur. of Std.*, **15**, 503 (1935).

indicator solution. Carefully titrate with the sodium hydroxide solution until a color matching that in the first flask is obtained. Subtract the amount required for the blank from the first titration.

Weigh and similarly titrate one or more samples of the phthalate, and from the data calculate the normality of the sodium hydroxide solution.

STANDARDIZATION OF ACID SOLUTIONS

The compounds that are most extensively used as primary standards for the standardization of solutions of strong acids are sodium carbonate, sodium tetraborate decahydrate (borax), and potassium iodate.

Titrations involving the first two of these substances are examples of the reaction of the salt of a weak acid with a strong acid, and the general principles of such titrations have been discussed on pages 98 and 526. Sodium carbonate is used in Procedure 24-5 for the standardization of the hydrochloric acid solution from Procedure 24-2, and the details of the titration are discussed there. The use of borax and of potassium iodate for the standardization of acids presents some interesting features that are discussed below.

Borax as an Acidimetric Standard

The titration of borax, $Na_2B_4O_7 \cdot 10H_2O$, at first appears complicated by the equilibria involved in borate-boric acid solutions. When borax is dissolved in water, a reaction with an equilibrium constant of about 10^{-3} takes place:

$$B_4O_7^{2-} + 5H_2O = 2H_3BO_3 + 2H_2BO_3^-$$

When such a solution is titrated, the pH change near the equivalence point is essentially that which would be obtained by titrating an equivalent quantity of a salt of the formula NaH_2BO_3. Therefore, since boric acid can be treated as a monoprotic acid having a constant equal to 6.4×10^{-10}, the hydrolysis constant for such a salt is calculated as follows:

$$K_H = \frac{K_W}{K_A} = \frac{10^{-14}}{6.4 \times 10^{-10}} = 1.6 \times 10^{-5}$$

Tracing a curve for a base whose constant is 1.6×10^{-5}, we see that at the equivalence point the $[H^+]$ is approximately 8×10^{-6} and that methyl red (or methyl orange) would be a very satisfactory indicator for the titration. Such is the case, and various investigators have confirmed that pure borax can be used accurately and conveniently as a standard for strong acids.[6]

---

[6] Kolthoff and Sandell. *Ind. Eng. Chem.*, anal. ed., **3**, 115 (1931); Vandaveer, *J. Assoc. Offic. Agr. Chemists*, **22**, 563 (1939).

Potassium Iodate as an Acidimetric Standard

The use of potassium iodate presents an interesting application of an oxidation-reduction reaction to an acidimetric titration. Mention was made on pages 463 and 507 that the reaction

$$IO_3^- + 5I^- + 6H^+ = 3I_2 + 3H_2O$$

is highly dependent upon the hydrogen-ion concentration and that it will proceed from left to right so long as the *p*H is significantly less than 7. Therefore, if one adds to a known amount of iodate in a neutral solution an excess of iodide, no appreciable reaction takes place. If, by means of a strong acid, hydrogen ion is added to the solution, iodine is formed and the hydrogen ion is converted to water so long as an appreciable concentration of iodate is present. However, when addition of acid is continued to the point where an amount equivalent to the iodate has been added, the hydrogen-ion concentration will increase rapidly; a plot of the $[H^+]$ against the ratio of equivalents of acid to equivalents of iodate will give a curve having the typical characteristics of a titration curve.

The use of an indicator to obtain an end point for this reaction would be difficult because of the iodine color. However, by the addition of an excess of thiosulfate the iodine color is completely eliminated because of the reaction

$$I_2 + 2S_2O_3^{2-} = 2I^- + S_4O_6^{2-}$$

and an indicator such as methyl orange can be used to give a color change in the region of the equivalence point. The removal of the iodine has the additional advantage that the equilibrium of the iodate-iodide-hydrogen ion reaction is shifted still more toward the right. Both thiosulfuric and tetrathionic acids are highly ionized; therefore thiosulfate and tetrathionate solutions are essentially neutral and do not interfere with the titration.

The reactions involved in the use of potassium iodate as an acidimetric standard are illustrated in the following experiment.

EXPERIMENT: Use of KIO_3 as an Acidimetric Standard

Weigh out 50 mg of KIO_3 and dissolve it in 25 ml of water in a 125-ml conical flask. Add 0.5 g of KI. Note any effect.

Add from a buret 2 ml–3 ml of 0.2 *F* HCl. Note any effect. Write an equation for any observable reaction.

Dissolve 0.5 g $Na_2S_2O_3{\cdot}5H_2O$ in 10 ml of water and add the solution to the flask. Note any effect. Write an equation for any reaction.

Add 1 drop of methyl orange indicator solution. Titrate the solution with the 0.2 *N* HCl until the first permanent change of the indicator is observed.

If one took exactly 0.05 g of KIO_3 and used exactly 7 ml of HCl for the titration, what would be the formality of the HCl?

Since none of the three substances mentioned above can be obtained from the Bureau of Standards as primary standards, for very exact work sodium oxalate from the Bureau is frequently used and is converted to carbonate by heat. Fortunately, primary standard quality sodium carbonate (assaying 99.95%–100.05% Na_2CO_3) is now commercially available and it is used in Procedure 24-5.[7]

---

[7] For an experimental comparison of sodium carbonate, borax, and potassium iodate as primary standards for acid solutions, see Kolthoff. *J. Am. Chem. Soc.*, **48**, 1447 (1926).

Standardization of a Hydrochloric Acid Solution Against Sodium Carbonate

Outline. Primary standard Na_2CO_3 is dried; a weighed sample is dissolved in water and titrated with the HCl solution to be standardized. Methyl orange is used as indicator.

The titration reaction can be written as

$$CO_3^{2-} + 2H^+ = H_2CO_3$$

DISCUSSION. DISPLACEMENT REACTIONS

The titration of sodium carbonate, the salt of a weak acid and a strong base, with a strong acid is not a neutralization reaction according to the classical definition, since the formation of water is not primarily involved, but is a displacement reaction. The titration reaction is

$$CO_3^{2-} + 2H^+ = H_2CO_3$$

and the equilibrium expression is

$$\frac{[H_2CO_3]}{[CO_3^{2-}][H^+]^2} = K_D$$

where K_D is obviously the reciprocal of the total ionization constant of carbonic acid.[8] In the discussion of *displacement reactions* (pages 98 and

[8] It should be noted that the above reaction proceeds in two steps, as follows:

$$CO_3^{2-} + H^+ = HCO_3^- \quad (1)$$

and

$$HCO_3^- + H^+ = H_2CO_3 \quad (2)$$

The end point of the first reaction can be obtained by the proper use of phenolphthalein (or other indicators having approximately the same *p*H range). However, the *p*H change near the equivalence point is so much less pronounced that for the second reaction that it should not be used for highly precise titration. Simpson (*Ind. Eng. Chem.*, **16**, 709, 1924) has recommended a mixed indicator of 6 parts thymol blue and 1 part cresol red when it is desired to use this end point for the analysis of carbonate–hydrogen-carbonate mixtures. The indicator changes from a violet-purple color in carbonate solution to a rose at the end point and becomes orange-yellow with excess acid.

526), it was pointed out (1) that the solution of the salt of a weak acid and strong base would be alkaline by hydrolysis, (2) that at the equivalence point there would be essentially a solution of the weak acid, and (3) that upon addition of excess strong acid, a rapid rise in $[H^+]$ would result. Therefore it would seem that a satisfactory titration could be made through the selection of an indicator that has a transition range close to or slightly above that given by a dilute solution of the weak acid. For example, consider the titration of sodium carbonate. It is to be observed that a solution saturated with carbon dioxide is 3.4×10^{-2} *M* in H_2CO_3;[9] and since the ionization constant for the first hydrogen of carbonic acid is 3×10^{-7}, one can calculate that the $[H^+]$ of the solution at the equivalence point is approximately 10^{-4} *M*. The transition range for methyl orange extends from about 4.5×10^{-5} to 10^{-3}; therefore, although methyl orange is perceptibly transformed to the acid form in such a solution, the further addition of acid causes a pronounced additional change toward the pink color. This end point is frequently used for routine titrations but is not so sensitive as would be desired. Because of this, for exact titrations such as are required for a standardization procedure, it is advisable to titrate with the acid to the first perceptible change, to heat the solution sufficiently to expel the carbon dioxide, then to cool it (since the methyl orange transition range is shifted in hot solutions), and to finish the titration with the cold solution. By this means a very satisfactory end point is obtained.

It was also pointed out in the discussion of displacement reactions that the curves of Figure 23-2 could be extended to such titrations in that the titration with a strong acid of the salt of a weak acid and strong base was equivalent to the similar titration of a base with an ionization constant of the same value as the hydrolysis constant of the salt.

In the application of this principle to the titration of a carbonate, only the first ionization constant of carbonic acid is to be considered, since near the equivalence point of the titration the second proton would contribute very little to the hydrogen-ion concentration of the solution. The value of the first ionization constant for carbonic acid is 3×10^{-7}; therefore,

$$K_H = \frac{K_W}{K_A} = \frac{1 \times 10^{-14}}{3 \times 10^{-7}} = 3.3 \times 10^{-8}$$

From Figure 23-2, it will be observed that *if no carbon dioxide escaped* from the solution, the hydrogen-ion concentration at the equivalence point would be considerably greater than 10^{-4}. Furthermore, the rate of change of

[9] This value is actually $[CO_2(aq)] + [H_2CO_3]$. See page 101 for a discussion of the hydration of CO_2 and other anhydrides and the conventions used in treating equilibria involving such species.

*p*H, and therefore the rate of transition of the indicator, would be so slow that the end point would be exceedingly sluggish. The advantage of heating the solution in order to expel the carbon dioxide is again evident.

Use of Mixed Indicators

As has been explained, in titrations involving displacement reactions the rate of change of the *p*H near the equivalence point is much less pronounced than that obtained in the titration of strong acids with strong bases; the same is true of the titration of weak acids or bases. It is therefore desirable to employ indicators that have very narrow transition bands at some definite *p*H value. In order to obtain this value, mixtures of two indicators or of an indicator and an inert dyestuff have been employed. The use of such an indicator is of advantage in this procedure. (See Note 1 below.) For a discussion and an extensive list of such indicators, see Kolthoff and Stenger. *Volumetric Analysis*. Vol. 2. New York: Wiley-Interscience, 1947, p. 56.

INSTRUCTIONS (Procedure 24-5: Standardization of a Hydrochloric Acid Solution against Sodium Carbonate)

Transfer 2 g–2.5 g of primary standard grade Na_2CO_3 to a weighing bottle and dry for an hour at 150°C. Allow the material to cool in a desiccator, then stopper the bottle and keep it closed except when removing material (Na_2CO_3 is appreciably hygroscopic).

Weigh out accurately 0.4 g–0.5 g of the prepared Na_2CO_3 into a 200-ml flask, dissolve it in 50 ml of water, add 2 drops of methyl orange indicator solution (Note 1), and titrate with the HCl solution, adding the acid slowly

NOTE

1. The titration can be made more rapidly by adding 1 drop of phenolphthalein and rapidly titrating until the red color disappears; this is the end point of the reaction

$$CO_3^{2-} + H^+ = HCO_3^-$$

The methyl orange can then be added and slightly less than the same amount of acid added without overrunning of the end point. Take care to avoid loss of solution by the evolution of CO_2.

A mixed indicator, made by dissolving 0.1 g of methyl orange and 0.25 g of indigo carmine in 100 ml of water, gives a much sharper end point, especially if the work is done under artificial light. The indicator is greenish in alkaline solutions, and turns very sharply to gray at the end point (*p*H = 4) and to violet with an excess of acid. If it is available, the use of this indicator is advised; 2 or 3 drops per 100 ml of solution are most satisfactory.

while inclining the flask, until the first perceptible change in color of the methyl orange (from a clear yellow toward pink) is obtained. Use the same volume of water and 2 drops of methyl orange contained in a similar flask as a reference solution. Heat the titrated solution to boiling for 2 or 3 minutes, frequently swirling it, cool it to room temperature by running tap water over the outside of the flask while swirling the contents, and again titrate to the first perceptible change of the indicator. Titrate the comparison solution with the acid until it matches the color of the titrated solution.

Repeat this process with one or more samples of the Na_2CO_3 and calculate the normality (or the weight normality) of the HCl.

Determination of the Total Alkalinity of Carbonate-Hydroxide Materials

Outline. This procedure is used for the analysis of soluble carbonates and hydroxides. The titration involves the reaction of a strong acid with a strong base and with the salt of a weak acid. The ionic reactions are

$$H^+ + OH^- = H_2O \tag{3}$$

$$2H^+ + CO_3{}^{2-} = H_2CO_3 \tag{4}$$

The second reaction was involved in Procedure 24-5 above and determines the indicator to use. The discussion of that procedure should be reviewed.

DISCUSSION

This procedure is applicable to the determination of the total alkalinity of water-soluble materials containing the hydroxides and carbonates of the alkali metals. Typical commercial compounds of this type are soda ash (anhydrous Na_2CO_3), pearl ash (anhydrous K_2CO_3), washing soda (hydrated Na_2CO_3), and lye or caustic soda (NaOH). Since these and similar crude compounds are frequently used as commercial neutralizing agents, the results of the analysis are frequently reported as the percentage of sodium oxide (Na_2O), which indicates their total alkalinity.

The materials are frequently nonhomogeneous. Therefore, recourse is often made to the expedient of weighing out a rather large sample, diluting to a known volume, and taking aliquot portions for analysis. In general, the purposes for which the analyses are made do not demand the highest accuracy; therefore emphasis is placed upon rapidity of execution.

Since all these compounds are composed of carbonates, or are contaminated with them, the principles involved in obtaining the end point are the same as those discussed in Procedure 24-5. In the procedure below, the end point is taken in the presence of carbon dioxide in order to conserve time, and use is made of a comparison solution buffered to the *p*H of a solution saturated with carbon dioxide.

INSTRUCTIONS (Procedure 24-6: Determination of the Total Alkalinity of Carbonate-Hydroxide Materials)

Weigh out accurately a 5-g–6-g sample, transfer it to a 250-ml beaker, and dissolve it in 100 ml of water (Note 1). Transfer the solution (cooling it to room temperature if necessary) to a 250-ml volumetric flask and dilute to the calibration mark.

Pipet 25 ml of the solution into a 200-ml conical flask; then add 25 ml of water and 2 drops of methyl orange indicator solution. To a similar flask add 75 ml of 0.01 *F* $KHC_8H_4O_4$ and exactly the same volume of indicator solution (Note 2).

Titrate with the standard volumetric or gravimetric HCl until a color match between the titrated and the comparison solution is obtained (Note 3). In case the end point is overrun, pipet 1 ml of the solution of the sample (or of 0.2 *N* base) into the flask and again titrate to the end point.

Similarly titrate a duplicate portion of the solution (Note 4). Unless otherwise instructed, calculate the *total available alkalinity* expressed as percentage Na_2O by weight.

NOTES

1. The above weight of sample is suitable for a soda ash material. If other materials are being analyzed, the sample size should be adjusted accordingly.

2. A 0.01 *F* potassium hydrogen phthalate solution has a *p*H of approximately 4 and contains 2 g of $KHC_8H_4O_4$ per liter.

In principle, it would be more accurate to use for comparison a solution saturated with carbon dioxide, and this can be done. However, such solutions are not stable because of loss of CO_2 and therefore are not so convenient as the phthalate solution. The methyl orange should show a definite transition color in the phthalate solution.

The mixed indicator mentioned in Note 1, Procedure 24-5, is also applicable to this titration.

The color change of bromphenol blue from blue to yellow is more readily perceived by some persons, and this indicator can be substituted for the methyl orange. It should show a transition greenish-blue color in the phthalate solution.

Since the sensitivity of perception of various colors differs widely with observers, each individual should experiment with the optional indicators available for a specific titration in order to determine which is most satisfactory for him.

3. The color of the methyl orange change is particularly difficult to obtain by incandescent light; daylight or a "daylight" fluorescent lamp should be used if possible.

4. Duplicate titrations should not vary by over 0.1 ml of 0.2 *N* solution.

QUESTIONS AND PROBLEMS

Stoichiometric Calculations

1. In the preparation of a supply of constant-boiling hydrochloric acid, it was found that the density of the distillate was 1.09620 at 760 mm and 25°C. A gravimetric analysis of a 1.5694-g (air-weight) sample of the distillate gave 1.2486 g of silver chloride. Calculate the normality of constant-boiling hydrochloric acid under the above conditions.

2. Foulk and Hollingsworth (*J. Am. Chem. Soc.*, **45**, 1220, 1923) give the following table:

Composition of Constant-boiling Hydrochloric Acid

Atmospheric Pressure During Distillation (mm of Hg)	Percentage by Weight of HCl in Distillate (vacuum weight)	Air Weight of Distillate Containing 1 mole of HCl
770	20.197	180.407
760	20.221	180.193
750	20.245	179.979
740	20.269	179.766
730	20.293	179.555

Calculate what weight of the acid (distilled at 740 mm) should be taken in order that just 40 ml of an exactly 0.1 *N* NaOH solution may be used in a titration. *Ans.* 0.7196 g.

3. When standardizing approximately 0.2 *N* sodium hydroxide solutions against potassium acid phthalate by Procedure 24-4, calculate the maximum and minimum weights of phthalate that can be taken without more than 45.0 ml or less than 35.0 ml of the hydroxide being required for the titration.

4. A 0.6273-g sample of benzoic acid was used in the standardization of a carbonate-free sodium hydroxide solution; 27.58 ml of the hydroxide were used. The titration of the "blank solution" required 0.04 ml of hydroxide.

   Which of the indicators shown in Figure 23-2 could have been used to give results correct to within 0.2% (assuming the sodium hydroxide solution to be carbonate-free)?

   Calculate the normality of the sodium hydroxide solution. What would be the effect if the solid sodium hydroxide from which the solution just standardized had been prepared were to contain 1% by weight of Na_2CO_3?

   How could the titration have been modified to give correctly the total alkalinity (carbonate and hydroxide) of the solution?

5. A 0.4250-g sample of Na_2CO_3 was titrated with a hydrochloric acid solution as directed in Procedure 24-5; the titration required 32.46 ml of the acid.

   Calculate the normality of the acid solution. *Ans.* 0.2470 *N*.

   If thymol blue had been added to the sodium carbonate solution before the titration, what color would have been observed? After addition of approximately what volume of the acid would a color change have occurred? What would have been observed if the solution had been boiled after this color change?

6. In the standardization of a hydrochloric acid solution, a 0.6824-g sample of sodium oxalate was weighed into a clean crucible and "ignited," that is, heated to give Na_2CO_3, then treated by Procedure 24-5; 35.00 ml of the HCl were used in the titration. Calculate the normality of the HCl. How could the titration have been carried out if phenolphthalein were the only indicator available?

   The heating of another sample of $Na_2C_2O_4$ was carried out in such manner that 1 % of the oxalate was converted to Na_2O. What error was caused in the standardization of the HCl?

7. A supply of borax was prepared, and a 1.2540-g sample was weighed out and titrated with an HCl solution (see page 553); 28.43 ml of the solution were used. Calculate the normality of the HCl solution. *Ans.* 0.2313 *N*.

8. Pure KIO_3 crystals weighing 0.2542 g were dissolved and titrated with an HCl solution by the process given in the experiment on page 554. The titration required 35.21 ml. The titration of a "comparison solution" required 0.04 ml of the acid. Calculate the normality of the acid. *Ans.* 0.2026 *N*.

9. *Analysis of Carbonate-Hydrogen Carbonate Mixtures.* A sample of technical sodium carbonate was analyzed for carbonate and hydrogen carbonate (HCO_3^-) as follows: A 2.100-g sample was dissolved and diluted to the mark in a 250-ml flask. A 50.00-ml portion was pipetted out and titrated as directed in Procedure 24-5, 37.80 ml of 0.2016 *N* HCl being required. A second 50.00-ml portion was taken and 10.00 ml of 0.2021 *N* NaOH were added. A 2-g portion of solid $BaCl_2{\cdot}2H_2O$ was then dissolved in water and added, and the mixture was titrated at once with the standard HCl, phenolphthalein being used as indicator, until the pink color disappeared; 8.62 ml of the acid were required. Calculate the percentages of Na_2CO_3 and $NaHCO_3$ in the material.

   NOTE: More representative samples are obtained by weighing out a large sample, diluting to volume, and taking an aliquot portion. The above method is also applicable to the analysis of mixtures of carbonate and hydroxide. The total alkalinity is determined as in the first step, the carbonate is then precipitated with barium chloride, and the hydroxide is titrated as in the second titration above.

10. *Analysis of Soda Ash.* In the analysis of a commercial soda ash, a 4.541-g sample was dissolved and diluted to exactly 250 ml in a volumetric flask. A portion of exactly 25 ml was then pipetted out and titrated with 0.1996 *N* HCl, the titration being carried out as directed in Procedure 24-6; 40.46 ml of the acid were required. Calculate the percentage of Na_2O in the soda ash.

NOTE: As soda ash is frequently a nonhomogeneous material, it is advantageous to dissolve a large sample and take aliquot portions of the solution.

How could the titration be performed if phenolphthalein were the only indicator available?

11. *The Kjeldahl Method for Organic Nitrogen.* This is an extensively used process applicable to organic compounds in which the nitrogen is in the trinegative state. The material is first heated with concentrated sulfuric acid until decomposition is essentially complete. In this first step of the process, which is termed the *digestion*, the organic material is dehydrated and the carbonaceous material oxidized by the concentrated sulfuric acid, sulfur dioxide being the principal product from the reduction of the sulfuric acid; K_2SO_4 is usually added to raise the temperature. The next step of the process, termed the *distillation*, comprises diluting the concentrated acid, adding an excess of sodium hydroxide, and distilling the ammonia. The ammonia is usually absorbed in an excess of a standard acid. The final step consists in the titration of the excess acid.

    A 0.6734-g sample of nitrogenous organic material was added to a solution of concentrated H_2SO_4 and K_2SO_4 and the mixture "digested" until a colorless solution was obtained. The solution was allowed to cool, then was diluted, and sodium hydroxide was added. The ammonia formed during the digestion was then distilled and collected in exactly 50 ml of 0.1002 *N* HCl. The excess acid was back-titrated with 25.42 ml of 0.0998 *N* NaOH, methyl red being used as the indicator.

    (*a*) Calculate the percentage of nitrogen in the original sample.
    (*b*) Would phenolphthalein be a suitable indicator? Calculate the approximate NH_3/NH_4^+ ratio at a phenolphthalein end point.

12. *Analysis of Concentrated Acids.* In the analysis of a glacial acetic acid, an approximately 5-g sample was transferred by means of a dry dropper to a previously weighed weighing bottle and the closed bottle was again weighed. The acid was then diluted and washed into a 500-ml volumetric flask, with a large funnel in the mouth of the flask to prevent loss, and the solution was diluted to the calibration mark and thoroughly mixed. A portion was transferred into a flask with a 25-ml pipet and titrated with sodium hydroxide, with phenolphthalein being used as indicator. The data obtained were as follows:

Weight of weighing bottle	20.367 g
Weight of weighing bottle and acid	25.199 g
Volume of 25-ml pipet	25.02 ml
Volume of 500-ml flask	500.6 ml
Volume of NaOH used	19.92 ml
Normality of the NaOH	0.2012 *N*

    Calculate the percentage concentration and the formal concentration of the acid, assuming the density of the acid to be 1.045. *Ans.* 99.7%, 17.35 *F*.

    Which of the indicators shown in Figure 23-2 could be used in the titration to give results correct to within 0.2%.

13. *The Analysis of Acetates.* A sample of crude calcium acetate was added to a phosphoric acid solution and the acetic acid was distilled, quantitatively condensed, and absorbed in an excess of standard sodium hydroxide; the sodium hydroxide solution was then titrated with standard hydrochloric acid. The following data were obtained:

Weight of sample	0.7000 g
Volume of NaOH (0.2000 *N*)	50.00 ml
Volume of HCl (0.2000 *N*)	12.40 ml
Indicator	Phenolphthalein

Calculate the percentage by weight of calcium acetate, $Ca(CH_3COO)_2 \cdot H_2O$, in the original product.

What would be the effect upon the analysis of absorption of carbon dioxide by the NaOH during the distillation? How could this effect be eliminated?

14. A 0.2000-g sample of "oleum" (fuming sulfuric acid, which may be considered to be a solution of sulfur trioxide in anhydrous sulfuric acid) was weighed out and introduced into 100 ml of water and titrated to an end point with 20.50 ml of 0.2150 *N* NaOH.

Which of the indicators shown in Figure 23-2 could be used with an accuracy of within 0.2%?

Calculate the percentage by weight of (*a*) sulfur trioxide and (b) sulfuric acid in the oleum.

15. In 125 ml of distilled H_2O there were dissolved 2.000 g of KIO_3 and 3.000 g of KI; then 25.00 ml of an HCl solution were pipetted into this solution. From another pipet, 25.00 ml of 0.1000 *N* $Na_2S_2O_3$ were added, and then 26.00 ml more of the standard thiosulfate were added from a buret. Next, 3 ml of starch solution were added. Then 0.85 more ml of the thiosulfate was required for the starch-iodine end point. What was the normality of the acid? *Ans.* 0.2074 *N*.

(See Chapter 23 for problems involving chemical equilibria.)

CHAPTER **25**

Ionization Reactions. Titrations of Polyprotic Acids

In Chapter 24 methods involving monoprotic acid and displacement titrations were considered. While phthalic acid is diprotic, the titration of the potassium hydrogen phthalate (Procedure 24-4) involved only the second proton. Similarly carbonic acid is a diprotic acid, but in the displacement titration of carbonate (Procedure 24-5) only the addition of the second proton had to be considered in evaluating the *p*H changes near the equivalence point. It has been mentioned that the curves of Figure 23-2 are applicable to the titration of polyprotic acids *only* when the last proton is being removed. Evaluating the *p*H changes when one or more untitrated protons remain necessitates quite different considerations. The titration of the first and second protons of phosphoric acid will serve to illustrate this problem.

The first and second ionization constants of phosphoric acid are $K_1 = 7.5 \times 10^{-3}$ and $K_2 = 2 \times 10^{-7}$. Since the first proton has the larger constant and is the stronger acid, it might be reasoned by analogy with the curves of monoprotic acids shown in Figure 23-2 that the slope of the titration curve in the region of the equivalence point for the titration of the first proton would be steeper than that in the region of the equivalence point for the titration of the second proton. An inspection of Figure 25-1 shows that this is not the case. The reason is indicated by the generalized equation

$$H_2A = OH^- = HA^- + HOH$$

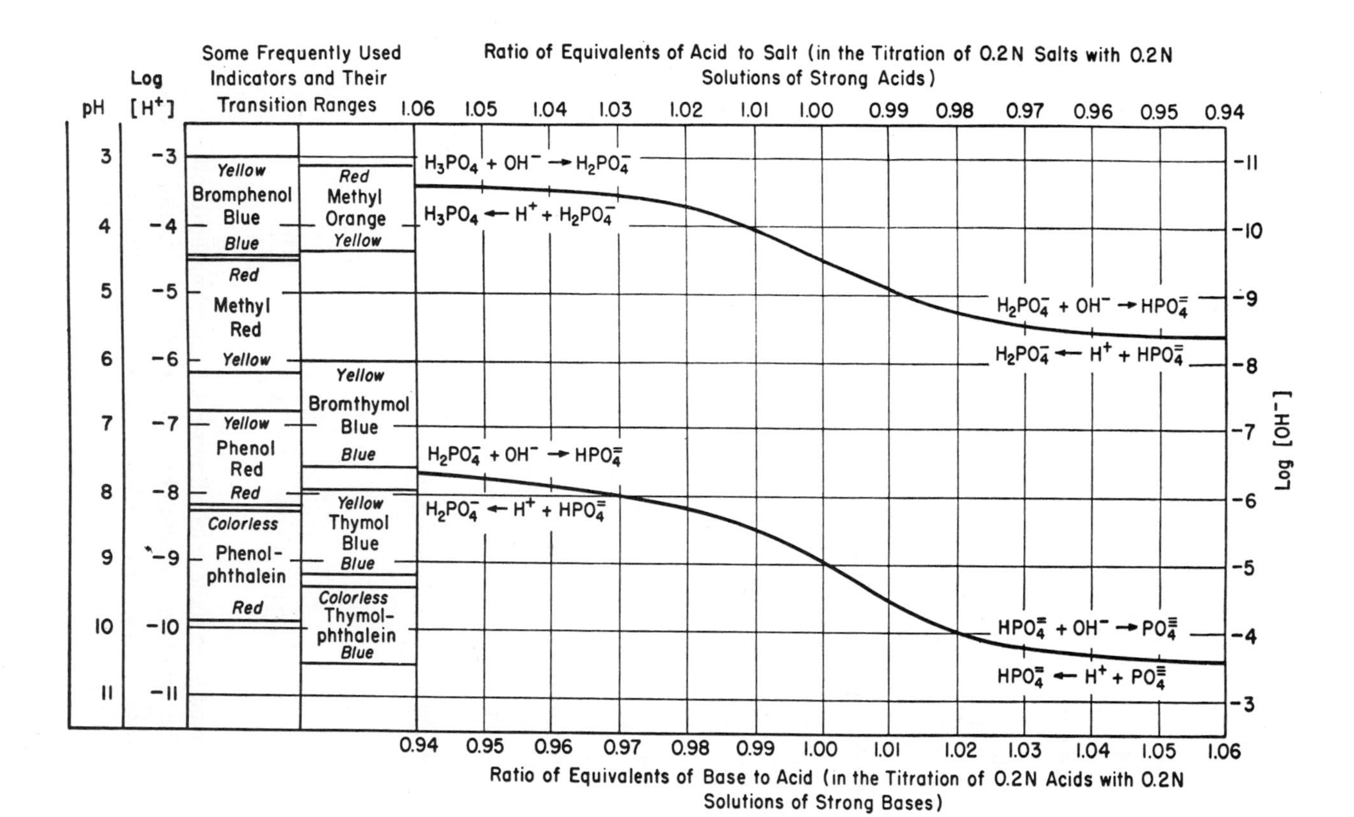

FIGURE 25-1

Hydrogen and hydroxide-ion concentration changes during neutralization titration of H_3PO_4 and during titration of PO_4^{3-} and HPO_4^{2-}.

which shows that at the equivalence point of the titration, the reaction product is not a salt but another acid. The ionization of this acid causes the $[H^+]$ of the solution to be higher than that of a salt of a monoprotic acid having the same constant; in addition, this acid buffers the solution as more base is added. In fact, a strong analogy can be drawn between the titration of the first proton of a diprotic acid and the titration of one acid in a solution of two monoprotic acids having differing ionization constants. In such a case the slope of the curve at the equivalence point upon titration of the stronger acid would be determined not only by the value of the constant of that acid but also by the *difference between the ionization constants* of the two acids. It is obvious that if the constants for the two acids have the same value, there would be no break in the titration curve when base equivalent to one of the acids had been added.

THE *p*H AT THE EQUIVALENCE POINT

The calculation of the $[H^+]$ at the equivalence point of the titration of one proton of a diprotic acid (or of either one or two protons of a triprotic acid) is more complicated than the analogous case of the titration of a weak monoprotic acid with a strong base. This is true because not only the hydrolysis of the acid salt formed but also the acid ionization of that salt has to be taken into account.

Consider the titration of the first proton of H_2A with NaOH. The equivalence-point solution is identical with a solution of Na^+HA^- of the same formal concentration.

The HA^- tends to react in two ways, as an acid and as a base. The equilibria which show these two kinds of behavior are

$$HA^- = H^+ + A^{2-} \qquad K_{A_2} = \frac{[H^+][A^{2-}]}{[HA^-]} \tag{1}$$

and

$$H_2A = H^+ + HA^- \qquad K_{A_1} = \frac{[H^+][HA^-]}{[H_2A]} \tag{2}$$

Equation 2 read from right to left shows HA^- accepting a proton—behaving as a base. The equation is written here in the reverse direction because this is the conventional way to show the equilibrium for the dissociation of an acid. A third equilibrium expression pertinent to this Na^+HA^- solution is the water constant:

$$H_2O = H^+ + OH^- \qquad K_W = [H^+][OH^-] \tag{3}$$

There are five unknown concentrations, $[H^+]$, $[OH^-]$, $[H_2A]$, $[HA^-]$, $[A^{2-}]$, so two more independent equations are needed in order that their values can be calculated. One of these is the charge balance equation,

$$[Na^+] + [H^+] = [OH^-] + [HA^-] + 2[A^{2-}] \qquad (4)$$

and another the species balance.[1]

$$[Na^+] = [H_2A] + [HA^-] + [A^{2-}] \qquad (5)$$

These five equations can be solved to determine the $[H^+]$ at the first equivalence point. If Equation 5 is subtracted from 4, the result is

$$[H^+] = [OH^-] + [A^{2-}] - [H_2A] \qquad (6)$$

Equations 1, 2, and 3 can be substituted into this to give

$$[H^+] = \frac{K_W}{[H^+]} + \frac{K_{A_2}[HA^-]}{[H^+]} - \frac{[H^+][HA^-]}{K_{A_1}}$$

which is rearranged to give

$$[H^+] = \sqrt{\frac{K_{A_1}K_W + K_{A_1}K_{A_2}[HA^-]}{K_{A_1} + [HA^-]}} \qquad (7)$$

Equation 7 is a rigorous expression[2] for $[H^+]$ in terms of acid and water constants and the concentration of $[HA^-]$. We will now show approximations that are valid under certain conditions. For those cases where the fraction of the original acid salt, HA^-, converted to A^{2-} and H_2A is not large, or where $[HA^-] \gg [A^{2-}] + [H_2A]$, the total or formal concentration of the

---

[1] Two observations here may be helpful. First, Equations 2 and 3 can be combined to represent the hydrolysis of the HA^-. One way of describing the solution at the first equivalence point is to say that it is a solution of HA^- that tends to dissociate and that also tends to hydrolyze. Second, the factor 2 in Equation 4 may appear misplaced at first glance. Remember, however, that Equations 4 and 5 are not chemical equations but are mathematical expressions concerning equalities of sums of concentrations. Consider, for example, the soluble ionic substance $BaCl_2$. A statement of its dissolving is the chemical equation

$$BaCl_2 = Ba^{2+} + 2Cl^-$$

while a statement of the relationship of concentrations is

$$[Cl^-] = 2[Ba^{2+}]$$

This latter equation states that the molar chloride-ion concentration is two times that of the barium ion. The preceding equation states that for each barium ion produced there are two chloride ions. The chemical equation and the mathematical equation make rather similar statements but in quite different form.

[2] As explained previously, in this text concentrations are used in place of activities; to be completely rigorous Equations 1, 2, and 3 would require the use of activities.

acid salt, C, can be substituted for $[HA^-]$; thus,

$$[H^+] = \sqrt{\frac{K_{A_1}K_{A_2}C + K_{A_1}K_W}{K_{A_1} + C}}$$

Where $K_{A_2}C \gg K_W$, this expression can be reduced to

$$[H^+] = \sqrt{\frac{K_{A_1}K_{A_2}C}{K_{A_1} + C}}$$

and where $C \gg K_{A_1}$, we obtain

$$[H^+] = \sqrt{K_{A_1}K_{A_2}}$$

When the second proton of a triprotic acid, such as H_3PO_4, is titrated, the expression becomes

$$[H^+] = \sqrt{K_{A_2}K_{A_3}}$$

These expressions give very convenient means for approximate calculations of the *p*H at the equivalence point of such titrations.[3]

The *slope* of the titration curve at the equivalence point of such titrations will be determined largely not by the ionization constant of the proton being removed but by the difference between that constant and the constant for the next proton, or more correctly the ratio of the two constants.[4] In general the titration is not feasible unless this ratio is greater than 10^4, preferably 10^5. The slope of the titration curve will be much less steep than when a monoprotic acid is being titrated with a constant of the same value, since the *p*H change with addition of a given quantity of base is much less but the $[H^+]$ at the equivalence point will be greater. These effects are shown by a comparison of the titration curves of Figure 23-2 and Figure 25-1.

---

[3] As an aid in deciding when use of the simple formula is justified, it can be shown that so long as

$$K_{A_2}C > 3K_W \quad \text{and} \quad C > 3K_{A_1}$$

the approximations involved will not cause errors in the calculated *p*H of greater than about one-tenth of a *p*H unit. It will be seen that for a 0.1 *F* Na_2HPO_4 solution, $K_{A_2}C > 3K_W$, but that this is not true for a 0.01 *F* solution.

[4] Roller (*J. Am. Chem. Soc.*, **50**, 1, 1928; **54**, 3485, 1932) has shown that at the equivalence point the rate of change of *p*H with change of volume of standard solution is proportional to the square root of the ratio of the two constants. Bates (*J. Am. Chem. Soc.*, **70**, 1579, 1948) gives a treatment of the general problem involved in the calculation from experimental data of the ionization constants of polyprotic acids. Treatments of the titration of polyprotic acids will be found in Kolthoff and Stenger. *Volumetric Analysis.* New York: Wiley-Interscience, Vol. 1. 1942, pp. 26–31, 56–58, and 151–153; Vol. 2. 1947, pp. 130–147.

In Figure 25-1 titration curves are plotted over the range from 6% before to 6% after the equivalence point for the titration of the first proton of H_3PO_4,

$$K_{A_1} = 7.5 \times 10^{-3}$$

and the second proton of H_3PO_4

$$K_{A_2} = 2 \times 10^{-7}$$

It is evident from the curves of Figure 25-1 that in order to carry out these titrations with an accuracy of better than $\pm 0.5\%$, the *p*H at the end point must be controlled within approximately ± 0.1 *p*H unit. This control can be achieved only by resorting to a potentiometric titration or by very careful matching of the indicator color with comparison solutions. This latter method is employed in Procedure 25-1 and 25-2 below, and a glass-electrode *p*H meter is made use of in Procedure 25-3.

PROCEDURE **25-1**

Analysis of Phosphoric Acid Solutions

Outline. An approximate determination of the concentration of a phosphoric acid solution is made by a preliminary titration to the methyl orange end point. Comparison solutions are prepared based on this approximate determination, and titrations are made to the first and to the second equivalence points. The titration reactions are

$$H_3PO_4 + OH^- = H_2PO_4^- + H_2O$$

and

$$H_3PO_4 + 2OH^- = HPO_4^{2-} + 2H_2O$$

DISCUSSION

The titration of a solution of phosphoric acid with a standard solution of a strong base illustrates the principles discussed above. The titration can be made by neutralizing either the first, or the first and second protons. An approximate formula whereby the *p*H at the equivalence point of either of these titrations can be calculated has been given above, and the titration curves in the region of the equivalence points are shown in Figure 25-1. In both cases the change in *p*H is so gradual that, when indicators are used, a very careful matching of the titrated solution against a prepared comparison solution is necessary in order to avoid errors of even several percent. Since the ratio

$$K_{A_2}/K_{A_3}$$

is greater than the ratio

$$K_{A_1}/K_{A_2}$$

the slope of the titration curve at the second equivalence point is somewhat steeper than that at the first, but experimentally this difference is hardly significant. Since twice the volume of standard solution is used in neutralizing the additional proton, the same deviation of end point from the equivalence point would cause only half the relative error when both protons are titrated.

The third ionization constant of phosphoric acid is so small,

$$K_{A_3} = 5 \times 10^{-13}$$

that titration of the third proton is not practical; this is evident from the curves shown in Figure 23-2. Such a titration can be made,

$$HPO_4^{2-} + OH^- = PO_4^{3-} + HOH$$

if the phosphate-ion concentration is decreased by precipitation of an insoluble phosphate. Calcium has been used for this purpose, but the method is subject to errors because of the formation of salts such as $CaHPO_4$.

Indicators that have been used for the titration of the first proton are methyl orange, bromcresol green (transition *p*H, 4.0 to 5.6), and methyl red. In all cases the color of the solution being titrated must be matched closely against one containing the same volume, and the same concentrations of indicator and of KH_2PO_4. Phenolphthalein and thymolphthalein have been used for the titration of the second proton. When phenolphthalein is used, the first color is observed before the equivalence point; therefore, a minimum amount of indicator should be added and the color should be matched against a prepared solution of Na_2HPO_4. With thymolphthalein, sufficient indicator should be added to the comparison solution to give a faint but distinct blue color, and then the same amount should be used in the titration.

INSTRUCTIONS (Procedure 25-1: Analysis of Phosphoric Acid Solutions)

A. Volumetric Titrations

Obtain a supply of the phosphoric acid to be analyzed. Consult your instructor regarding the volume required, also the number and type of titrations to be made. If given no information as to the concentration, proceed as directed in Section I below.

I. Approximate Determination of the Acid Concentration. Pipet 5 ml of the acid into a 125-ml flask, add 25 ml of water and 1 drop of methyl orange indicator. Titrate the solution with standard 0.2 *N* NaOH until the color of the solution changes from orange to yellow. Calculate the formal concentration of the acid.

II. Titration of the First Proton. Pipet into a 200-ml flask that volume of the H_3PO_4 that will require 25 ml–45 ml of the 0.2 *N* NaOH to neutralize *one proton*. Dilute the solution to about 50 ml.

Weigh out an amount of KH_2PO_4 (Note 1) equivalent to the sample of phosphoric acid taken, dissolve it in 100 ml of water in a 200-ml flask, and add just 2 drops of methyl orange indicator solution (Note 2).

Add just two drops of the indicator solution to the solution of the sample and titrate with standard 0.2 *N* NaOH until the color of the titrated solution matches that of the comparison solution (Note 3). From the volume of standard base used, calculate the formal concentration of the H_3PO_4 solution.

III. Titration of the First and Second Protons. Pipet into a 200-ml flask that volume of the H_3PO_4 that will require 25 ml–45 ml of the 0.2 *N* NaOH to neutralize the *first two protons.*

Weigh out Na_2HPO_4 (Note 4) equivalent to the H_3PO_4 sample and dissolve it in 100 ml of water in a 200-ml flask. Add 1 drop of phenolphthalein indicator (Note 5). Stopper the flask (Note 6).

Add 1 drop of indicator to the H_3PO_4 sample. Titrate with standard 0.2 *N* NaOH to a color matching the comparison solution (Notes 3 and 6). From the volume of NaOH used, calculate the formality of the H_3PO_4.

B. Gravimetric Titrations

Proceed as directed above in the volumetric titrations except use the gravimetric standard NaOH solution and a weight buret. Use samples of such size that 10 g–15 g of the standard solution will be required in the titrations. Calculate the weight formal concentration of the phosphoric acid.

NOTES

1. Reagent-grade KH_2PO_4 should be used. Material specially prepared for use as a buffer agent is to be preferred and is available commercially. The comparison and the titrated solutions should both be at room temperature.

2. Bromcresol green or methyl red can be used instead of methyl orange. Exactly the same quantity of indicator should be added to the comparison and titrated solutions; otherwise a color match at the same *p*H is more difficult.

3. The approach to the end point is facilitated if three comparison solutions are prepared and 0.1 ml of standard acid is added to the first solution and 0.1 ml of standard base to the third.

4. Reagent-grade anhydrous Na_2HPO_4 specially prepared for buffer solutions can be obtained commercially and should be used if available.

5. Only 1 drop of phenolphthalein is used, since the indicator is largely transformed at the *p*H of the Na_2HPO_4 solution. With phenolphthalein, a one-color indicator, the greater the concentration, the sooner the perception of the color.

Thymolphthalein can be used instead of phenolphthalein. In this case a larger

The Analysis of Strong Acid–Phosphate-Hydroxide Solutions

Outline. The *p*H of the unknown solution is measured. This information is used to determine what species can be present. Titrations are to be planned and performed to determine which species are present and their concentrations.

DISCUSSION

The species that can be present in such solutions range from a strong monoprotic acid, such as HCl, through H_3PO_4, $H_2PO_4^-$, HPO_4^{2-}, and PO_4^{3-} to a strong base, such as NaOH. Therefore the analysis of such solutions, which involves the identification of the species and their estimation, at first appears to be very complicated. However, the problem is simplified by certain considerations. The first of these is the fact that not more than two of the phosphate species can exist in the same solution in significant concentrations. Thus if H_3PO_4 is present, the only other phosphate species that could be present in analytically significant concentration is $H_2PO_4^-$; H_3PO_4, and HPO_4^{2-} would react to give $H_2PO_4^-$, and the equilibrium constant for the reaction

$$H_3PO_4 + HPO_4^{2-} = 2H_2PO_4^-$$

is approximately 4×10^4. Similarly if PO_4^{3-} is present only HPO_4^{2-} can be present in significant concentration since PO_4^{3-} and $H_2PO_4^-$ will react to give HPO_4^{2-}, and the constant for that reaction is 4×10^5. Therefore by titrating with base or acid, depending upon the initial *p*H of the solution,

NOTES

quantity of indicator should be used (3 drops–5 drops), since so small a fraction is transformed at the equivalence point that the color may not be readily matched,

6. Solutions of this *p*H tend to absorb CO_2 from the air; therefore the comparison solution should be kept stoppered and should be prepared only when it is needed. Excessive shaking of the solution should be avoided.

to both the $H_2PO_4^-$ and HPO_4^{2-} end points, the one can determine the two constituents.

The situation when a strong acid and H_3PO_4 (or a strong base and PO_4^{3-}) may be present involves the possibility that three species may exist in equilibrium in significant concentrations. Thus the first proton of H_3PO_4 is so extensively ionized,

$$K_{A_1} = 7 \times 10^{-3}$$

that significant quantities of a strong acid, of H_3PO_4, and of $H_2PO_4^-$ are possible. Likewise the hydrolysis of PO_4^{3-} is so extensive, the constant for the reaction

$$PO_4^{3-} + HOH = HPO_4^{2-} + OH^-$$

being 2×10^{-2}, that PO_4^{3-} is equivalent to a moderately strong base and significant quantities of HPO_4^{2-}, PO_4^{3-}, and OH^- can be in equilibrium. However, by noting the initial pH of the solution and then titrating with either standard base to both the $H_2PO_4^-$ and HPO_4^{2-} end points, or with standard acid to both the HPO_4^{2-} and $H_2PO_4^-$ end points, one can determine not only whether a strong acid or a strong base is present alone or with H_3PO_4 or PO_4^{3-}, respectively, but also the formal concentrations of these species.[5]

Two examples of the interpretation of analyses of strong acid–phosphate-hydroxide solutions follow:

A solution gave an acid reaction to methyl orange, required a certain volume of standard base to titrate the first proton, then required an equal volume of the base to titrate the second proton—H_3PO_4 is the essential constituent.

Another solution gave an alkaline reaction to both methyl orange and phenolphthalein, required a certain volume of standard acid to reach the phenolphthalein end point, then required a larger volume of the acid to titrate from that end point to a methyl orange end point, PO_4^{3-} and HPO_4^{2-} are the constituents.

The discussion above has considered solutions in which only orthophosphoric acid and orthophosphate are present. The analysis of solutions that may contain the salts of pyrophosphoric acid, $H_4P_2O_7$, metaphosphoric acid, HPO_3, and the polyphosphoric acids, such as hexametaphosphoric acid, $(HPO_3)_6$, is beyond the scope of this text.[6]

---

[5] The analytical information will enable one to reproduce the equilibria in the solution, but not necessarily the method by which the solution was prepared. Thus identical solutions can be prepared by beginning with 1 mole of NaOH and 1 mole Na_3PO_4 or with 2 moles NaOH and 1 mole Na_2HPO_4.

[6] For a discussion of such analyses see Kolthoff and Stenger, op. cit., Vol. 2, pp. 135–144; Jones. *Ind. Eng. Chem.*, anal. ed., **14**, 536 (1942); Bell. *Anal. Chem.*, **19**, 97 (1947).

INSTRUCTIONS (Procedure 25-2: Analysis of Strong Acid–Phosphate-Hydroxide Solutions)

Obtain a supply of the solution to be analyzed. (The possible constituents are HCl, H_3PO_4, NaH_2PO_4, Na_2HPO_4, Na_3PO_4, and NaOH).

Use a small portion to determine the *p*H by means of wide-range indicator papers or a glass-electrode *p*H meter (see the first three paragraphs of Procedure 25-3).

Plan and carry out titrations (refer to Procedure 25-1) that will enable you to reproduce the solution being analyzed. (Either acid-base indicators or a *p*H meter can be used. Consult your instructor.)

Use of the Glass-electrode *p*H Meter. Determination of Titration Curves and Indicator Transition Ranges During Titrations of Polyprotic Acids

Outline. A *p*H meter is used to obtain data for the plotting of the titration curve for a polyprotic acid and for determining the transition range of an indicator.

DISCUSSION

The general features of the glass-electrode *p*H meter have been discussed previously (page 528) and should be reviewed. Since various types of *p*H meters are on the market, the instructions given below are of a very general nature. The catalogues, manuals, and instruction sheets of various manufacturers should be consulted for the specific details of operation for the instrument of a particular manufacturer.

Also, the procedure given below is given in general terms rather than with specific reference to any particular acid or indicator.

INSTRUCTIONS (Procedure 25-3: Use of the Glass Electrode *p*H Meter. Determination of Titration Curves and Indicator Transition Ranges During Titrations of Polyprotic Acids)[7]

Obtain a *p*H meter and read carefully the manufacturer's instructions. Consult the instructor for any additional instructions in regard to its use.

[7] *Suggestions to Instructors.* The titration of a particular acid can be made a class problem, and individual students or pairs of students can be assigned or allowed to choose specific phases for investigation. As an example, the various titrations of phosphoric acid could be taken as the class problem. When time is limited, stock solutions of 0.2 *F* H_3PO_4 and 0.2 *F* NaOH can be provided, together with as many indicators as are desired. Titrations of the first proton can be made by taking 25 ml of standard acid and adding 50 ml of water; for titrations of the second proton, 25 ml of acid and 25 ml of water; for the third proton (in the presence of $CaCl_2$), 25 ml of acid and no water. In this way the volume at the end point will be the same in all titrations. Suggested indicators for the titration of the first proton are methyl orange, methyl red, bromphenol blue, bromcresol green, and a mixed indicator of 3 parts bromcresol green and 1 part methyl red. Possible indicators for the second proton titration are thymolphthalein, phenolphthalein, thymol blue, cresol red, and a mixed indicator of 1 part cresol red and 3 parts thymol blue. The effect on the titration curve and the indicator of making the titrated solution 3 *F* or saturated in sodium chloride or sulfate may be investigated.

Check the instrument to ascertain that it is functioning properly (Note 1). If measurements are to be made at *p*H values greater than 9, obtain an electrode designed for such measurements (if it is available). Also procure extended electrodes that can be immersed directly in the solution to be titrated, if these are available. Turn on the meter switch and allow the instrument to "warm up" for the period specified by the manufacturer.

Obtain a standard buffer solution and calibrate the meter. See that the buffer solution is at the temperature for which the *p*H is stated and that the temperature adjustment of the instrument is properly set. Repeat the measurement after 30 seconds to check the constancy of the value. If it is available, use a buffer solution having a *p*H close to that of the equivalence point of the titration (Note 2). Rinse the electrodes thoroughly and immerse them in distilled water.

Prepare or obtain a standard 0.2 *F* solution of the acid and a standard 0.2 *F* NaOH (Note 3).

Pipet a suitable quantity of the acid to be titrated, adjust the solution to the desired volume, add the indicator specified, and check the temperature of the solution (Note 4). If possible, obtain and use a mechanical stirrer; stop the stirrer while making measurements (Note 5). Measure the initial *p*H of the

NOTES

1. Glass electrodes can be easily broken and are relatively expensive (their cost ranges from $25 to $45 for general-purpose electrodes). Before using glass electrodes, rinse them with water, then immerse them in water for some time, preferably for an hour. When stored for short periods of time, they should be kept immersed in water, and not dried.

2. If a wide range of *p*H values is to be covered, at least two buffer solutions having *p*H values near the extremes of the range should be used. Make a check on the calibration of the meter at the conclusion of each titration; malfunctioning of batteries or electrodes may cause drifting of readings or erratic values during a titration.

Do not waste the buffer solutions; use small beakers or weighing bottles in order to minimize the volume necessary to obtain readings.

3. The ratio of base added to acid present can be calculated from standardization values of the acid and base or by taking the inflection point of the titration curve as the end point of the titration.

4. Set the temperature adjustment of the instrument to agree with that of the solution. Compensate properly for temperature changes during the titration.

5. If the titration is to be extended to *p*H values of greater than 9, the titration vessel should be covered in order to exclude CO_2. A suitably perforated piece of cardboard can be used.

A stream of air free from carbon dioxide can be used as an alternative means of stirring the solution and will minimize absorption of carbon dioxide.

solution (Note 6); wait for 15 sec–30 sec and repeat the measurement. Add a predetermined volume of the standard NaOH, wait for 15 sec–30 sec while the solution is stirred, and then measure the *p*H again, repeat this process until a value constant to within 0.02 *p*H unit is obtained (Note 7). Continue this process until the titration has been carried over the desired base/acid ratio.

Observe and record the first perceptible change in color of the indicator, the point or region of maximum change, and the point after which there is no further perceptible change (Note 8).

Plot the data in the form shown in Figure 25-1. If a considerable range of the titration has been covered, construct one plot to show the complete range, the other to show on a larger scale the region within 2%–5% of the equivalence point. Construct a plot of $\Delta p\text{H}/\Delta v$ against v, the volume of base added, for a region within 1%–2% of the equivalence point.

Draw what conclusions you can as to the accuracy of the titration when made (*a*) potentiometrically and (*b*) with the indicator that was used. Discuss the agreement of the *p*H data with those calculated from the available ionization constants. Estimate how closely the end-point *p*H values must be controlled in order to obtain an accuracy of $\pm 0.2\%$; of $\pm 0.5\%$.

NOTES

6. If the curve is to cover only the region of the equivalence point, no *p*H measurements need be taken until a volume of base has been added that gives the desired initial ratio of equivalents of base to acid.

7. The volume of base to be added between making measurements will be determined by the region of the curve to be plotted. Until the ratio of base to acid has reached 0.80, the *p*H change is usually so gradual and the curve so linear that increments of several ml can be added; this volume should be decreased in the immediate region of the equivalence point so that the change in *p*H between increments remains approximately 0.1 unit–0.3 unit. (The instructor should be consulted in this regard.)

8. The first and last perceptible changes of the indicator can be detected more easily if comparison solutions are used in which the indicator is completely converted by means of a slight excess of acid or base.

PROCEDURE 25-4

The Titration of Amino Acids

Outline. The molecular weight and acid constants of an amino acid are determined by titration with standard base and acid. A *p*H meter is used for plotting the titration curve.

DISCUSSION

Amino acids are organic compounds containing both an amino group, $-NH_2$, and a carboxyl group,

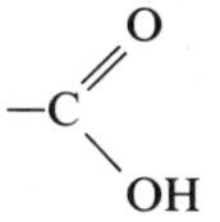

```
    O
   //
 -C
   \
    OH
```

The simplest acid, glycine, can be considered as derived from acetic acid

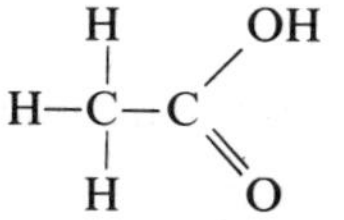

```
   H   OH
   |  /
H--C--C
   |  \\
   H   O
```

be replacing a methyl hydrogen with an amino group, thus producing the compound

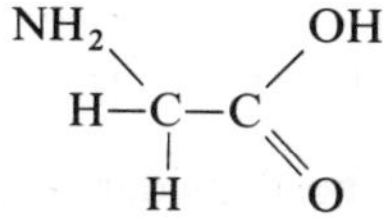

```
NH2     OH
   \   /
 H--C--C
    |  \\
    H   O
```

Glycine is a monoaminomonocarboxylic acid; amino acids can contain more than one amino group and more than one carboxyl group. Amino acids are of fundamental biochemical importance, since they are the structural units of proteins; twenty three of these acids have been identified as protein constituents and nine of these are essential to human nutrition.[8]

[8] These so-called essential amino acids are isoleucine, leucine, methionine, phenylalinine, threonine, and valine—all monoaminomonocarboxylic acids—together with lysine, a diaminomonocarboxylic acid, and histidine and tryptophan, which are more complicated and have heterocyclic rings, that is, groups containing rings in which both carbon and nitrogen atoms are present. The structural, chemical, and biological properties of the amino acids are discussed in most general chemistry texts.

The presence of both an amino group and a carboxyl group makes an amino acid an amphiprotic compound—that is, one that can accept a proton and act as a base and can donate a proton and act as an acid. When an amino acid is dissolved in water, there is evidence from conductivity and ion migration measurements that the molecule becomes polar in nature and has a high dipole moment. This phenomenon is attributed to an internal ionization resulting in a dipolar compound, or as it is commonly called, a "zwitterion" (from the German word meaning *hybrid*).

The structure of the glycine zwitterion can be represented as

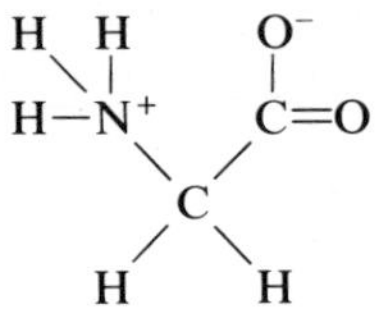

$$\begin{array}{ccccc} H\ \ H & & & & O^- \\ \diagdown | & & & & | \\ H-N^+ & & & & C=O \\ & \diagdown & & \diagup & \\ & & C & & \\ & \diagup & & \diagdown & \\ & H & & H & \end{array}$$

If an acid is added to a solution containing the zwitterion, we obtain the glycinium cation (analogous to the ammonium ion)

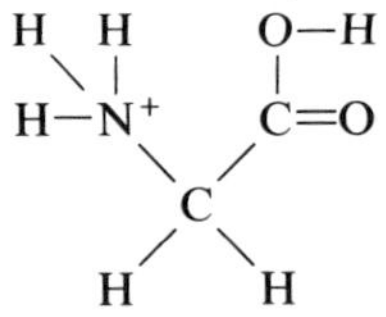

$$\begin{array}{ccccc} H\ \ H & & & & O-H \\ \diagdown | & & & & | \\ H-N^+ & & & & C=O \\ & \diagdown & & \diagup & \\ & & C & & \\ & \diagup & & \diagdown & \\ & H & & H & \end{array}$$

If a base is added we obtain the glycinate anion

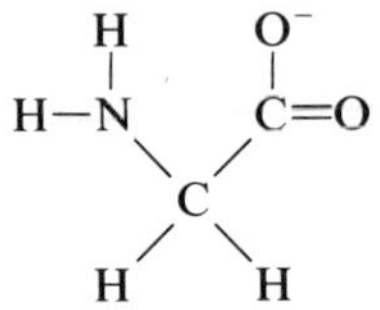

$$\begin{array}{ccccc} H & & & & O^- \\ | & & & & | \\ H-N & & & & C=O \\ & \diagdown & & \diagup & \\ & & C & & \\ & \diagup & & \diagdown & \\ & H & & H & \end{array}$$

The glycinium cation can be considered to be a diprotic acid, H_2G^+. The dissociation equilibria in which this is involved are

$$H_2G^+ = H^+ + HG \qquad K_{A_1} = 4.5 \times 10^{-3} \tag{8}$$

and

$$HG = H^+ + G^- \qquad K_{A_2} = 1.7 \times 10^{-10} \tag{9}$$

where HG represents the amino acid, H_2G^+ is the protonated form, and G^- the deprotonated form.

When glycine (or any monoaminomonocarboxylic amino acid) is added to water several equilibria are established. The glycine tends to react in two ways, as an acid and as a base. The equilibria that show these two kinds of behavior are given in Equations 8 and 9 above. In addition there is the water equilibrium:

$$[H^+][OH^-] = 1 \times 10^{-14} \qquad (10)$$

An expression that will enable one to calculate the $[H^+]$ of the glycine solution can be derived by making use of Equations 8, 9, and 10 together with the equations for species balance

$$[H_2G^+] + [HG] + [G^-] = C \qquad (11)$$

and charge balance

$$[H^+] + [H_2G^+] = [OH^-] + [G^-] \qquad (12)$$

In Equation 11, C represents the formal concentration of the amino acid.

Appropriate substitutions lead to

$$[H^+] + \frac{[H^+][HG]}{K_{A_1}} = \frac{K_w}{[H^+]} + \frac{[HG]K_{A_2}}{[H^+]}$$

and this can be rearranged to

$$[H^+] = \sqrt{\frac{K_{A_1}K_w + K_{A_1}K_{A_2}[HG]}{K_{A_1} + [HG]}}$$

Note that this expression is essentially the same as that given on page 569, Equation 7, for the calculation of the $[H^+]$ at the equivalence point when titrating the first proton of a diprotic acid with a strong base or, in other words, that of the acid salt HA^-. Note also that when the fraction of the glycine (G) converted to H_2G^+ and to G^- is small, or where $[G] \gg [H_2G^+] + [G^-]$, as is the case with amino acids, the total or formal concentration of G, C can be substituted for [G], therefore

$$[H^+] = \sqrt{\frac{K_{A_1}K_w + K_{A_1}K_{A_2}C}{K_{A_1} + C}}$$

By making use of this equation one calculates that the $[H^+]$ of a 0.01 F glycine solution is 7.3×10^{-7}.

The Isoelectric Point

The fact that the above glycine solution is acid indicates that the acid ionization has taken place to a greater extent than has the hydrolysis reaction; as a result, $[G^-]$ is greater than $[H_2G^+]$. If an acid is added to this glycine solution, a point will be reached when $[G^-] = [H_2G^+]$; this point is known as the *isoelectric point*, and the pH at this point is a characteristic property of an amino acid. The solubility of an amino acid is at the minimum at the isoelectric point and there will be no net migration of the compound upon passage of an electric current.

An expression for calculating the *p*H of a monoaminomonocarboxylic amino acid at the isoelectric point can be derived by first multiplying the expressions for K_{A_1} and K_{A_2} to give

$$[H^+]^2 = K_{A_1}K_{A_2}\frac{[H_2G^+]}{[G^-]}$$

Then, since by definition at the isoelectric point $[H_2G^+] = [G^-]$, one obtains

$$[H^+] = \sqrt{K_{A_1}K_{A_2}}$$

Also, it is seen that the $[H^+]$ at the isoelectric point is independent of the concentration of the amino acid.

The $[H^+]$ of the 0.01 *F* glycine solution considered above is 8.8×10^{-7} at the isoelectric point. The more concentrated the amino acid solution, the less the difference between the $[H^+]$ of the pure solution and that at the isoelectric point.

The Analysis of Amino Acids. Separations

It is apparent from the above discussion that whether an amino acid is present in a solution primarily as an anion, a neutral molecule, or a cation will depend upon the *p*H of that solution. Since the acid constants of various amino acids vary significantly, one can so control the *p*H of a solution containing two amino acids having different constants that, as an example, a larger fraction of one than of the other will exist as a cation. Under these conditions, if the solution is subjected to an electric field, one amino acid will migrate to the cathode more rapidly than the other. These considerations are the basis of separations of amino acids, and proteins by electrophoresis, that is, methods depending upon the movement of charged species, particularly colloidal particles and macromolecular ions, under the influence of an electric field. In addition, amino acid separations can be made by ion exchange and chromatographic methods; again separations are expedited by proper *p*H adjustment. Table 25-1 shows the structures and acid ionization constants of most of the amino acids occurring in proteins.

Titrations

Amino acids can be titrated with either strong acids or strong bases. Figure 25-2 shows the *p*H changes taking place when a 0.01 *F* solution of a glycinium

TABLE 25-1
Some Representative Amino Acids of Proteins. Structures and Acid Ionization Constants

Amino Acids of Proteins		K_{A_1}	K_{A_2}	K_{A_3}
Monoaminomonocarboxylic Acids				
Glycine	$H-C(NH_2)(H)-COOH$	4.5×10^{-3}	1.7×10^{-10}	—
*Threonine	$HO-C(CH_3)(H)-C(NH_2)(H)-COOH$	2.8×10^{-3}	2.5×10^{-10}	—
*Methionine	$H-C(H)(H)-S-C(H)(H)-C(H)(H)-C(NH_2)(H)-COOH$	5.3×10^{-3}	6.2×10^{-10}	—
*Leucine	$H-C(CH_3)(CH_3)-C(H)(H)-C(NH_2)(H)-COOH$	4.4×10^{-3}	2.5×10^{-10}	—
*Isoleucine	$CH_3-C(H)(H)-C(H)(CH_3)-C(NH_2)(H)-COOH$	4.4×10^{-3}	2.1×10^{-10}	—
*Phenylalanine	$C_6H_5-C(H)(H)-C(NH_2)(H)-COOH$	1.5×10^{-2}	7.4×10^{-10}	—
Cysteine	$HS-C(H)(H)-C(NH_2)(H)-COOH$	1.1×10^{-2}	4.4×10^{-9}	5.3×10^{-11}
Monoaminodicarboxylic Acids				
Aspartic Acid	$HOOC-C(H)(H)-C(NH_2)(H)-COOH$	8.3×10^{-3}	1.2×10^{-4}	1.1×10^{-10}
Hydroxyglutamic Acid	$COOH-C(H)(H)-C(H)(HO)-C(NH_2)(H)-COOH$	4×10^{-3}	5.8×10^{-5}	7.8×10^{-10}

TABLE 25-1—contd.

	Diaminomonocarboxylic Acids			
Arginine	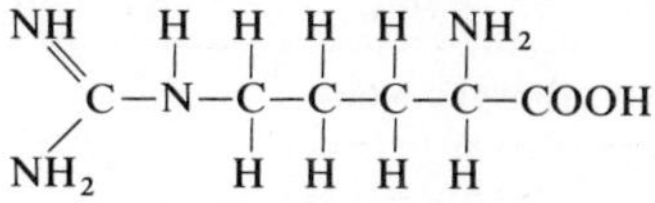 NH H H H H NH_2 C—N—C—C—C—C—COOH NH_2 H H H H	6.8×10^{-3}	9.1×10^{-10}	3.3×10^{-13}
*Lysine	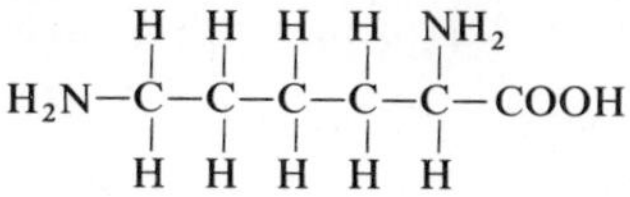 H H H H NH_2 H_2N—C—C—C—C—C—COOH H H H H H	6.6×10^{-3}	8.9×10^{-9}	3.0×10^{-11}
	Diaminodicarboxylic Acids			
Cystine	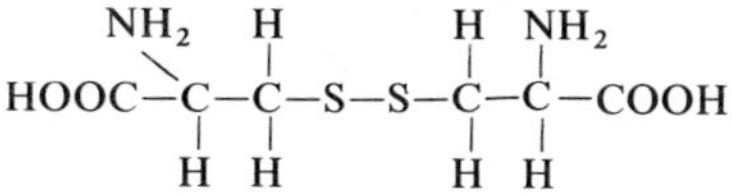 NH_2 H H NH_2 HOOC—C—C—S—S—C—C—COOH H H H H	9.1×10^{-2}	8.9×10^{-3}	1.0×10^{-8} K_{A_4} 5.6×10^{-11}
	Heterocyclic Ring Amino Acids			
Histidine	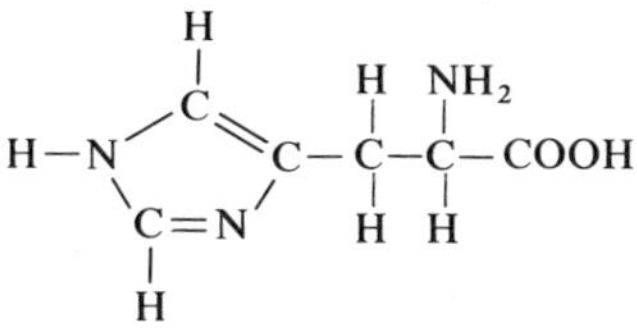 H C H NH_2 H—N C—C—C—COOH C=N H H H	1.5×10^{-2}	1.0×10^{-6}	6.8×10^{-10}
Tryptophan	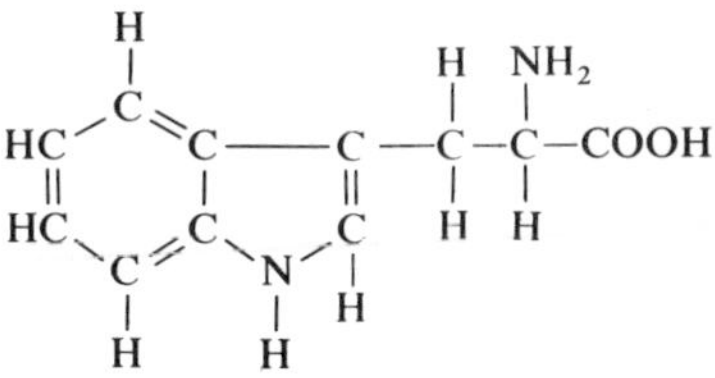 H C H NH_2 HC C——C—C—C—COOH HC C C H H C N H H H	4.2×10^{-3}	4.1×10^{-10}	—
	Amide Group Amino Acids			
Asparagine	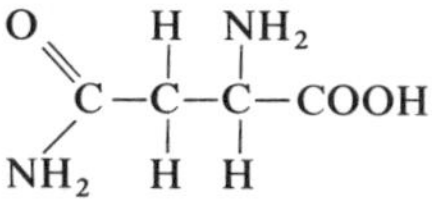 O H NH_2 C—C—C—COOH NH_2 H H	9.6×10^{-3}	1.6×10^{-9}	—
Glutamine	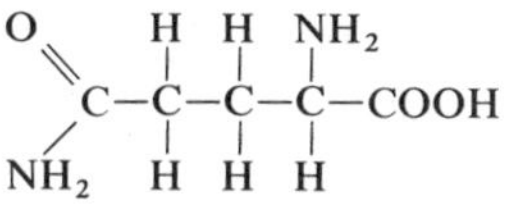 O H H NH_2 C—C—C—C—COOH NH_2 H H H	6.8×10^{-3}	7.4×10^{-10}	—

NOTE: For a more complete list of the principal amino acids of proteins and a discussion of amino acids and proteins, see L. Pauling, *General Chemistry*. (3rd ed.) San Francisco: W. H. Freeman and Company, 1970, pp. 770–781.

* Essential amino acids.

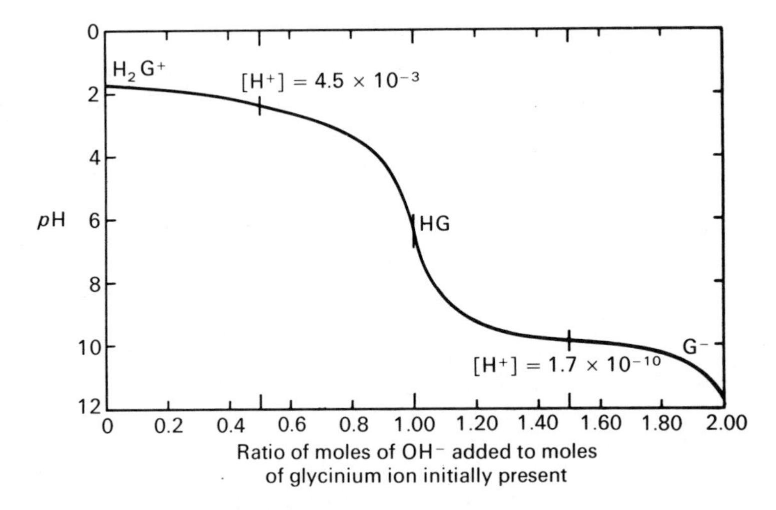

FIGURE 25-2

The titration of 0.1 *F* glycinium ion to glycinate ion with a strong base.

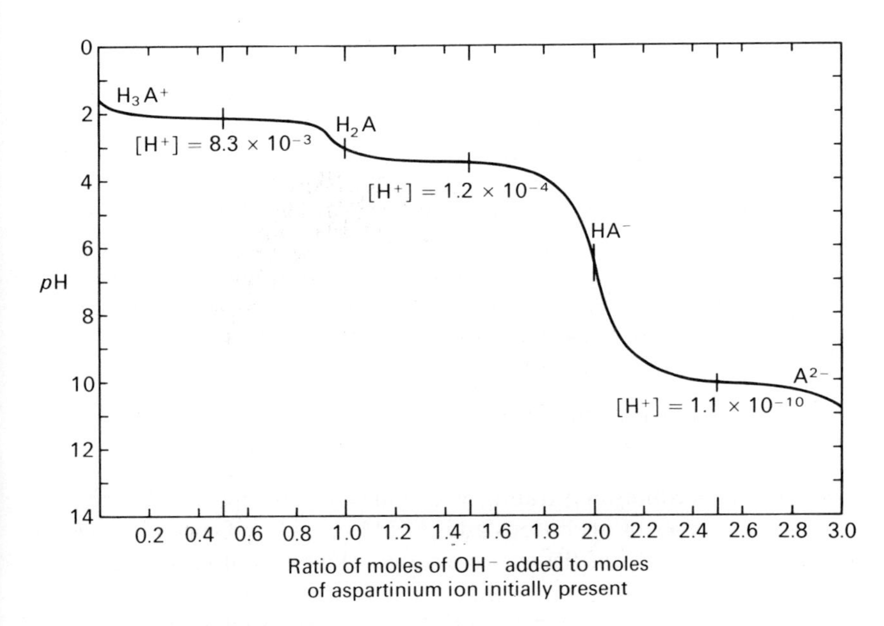

FIGURE 25-3

The titration of 0.1 *F* aspartinium ion to aspartinate ion with a strong base.

salt is titrated with a strong base to the glycinate ion. Figure 25-3 shows the curve obtained when an aspartinium salt is similarly titrated; aspartic acid is a monoaminodicarboxylic acid. By using a *p*H meter to follow such titrations, one can obtain information as to the number of amino and carboxylic groups present and the values for the ionization constants. Instructions for making such measurements are given below.

More accurate stoichiometric analyses can be obtained by making such titrations in nonaqueous solvents. As explained in the discussion of the use of such solvents (pages 104–105), the titration of very weak bases with strong acids can be made much more satisfactorily in anhydrous, "glacial," acetic acid than in water. Perchloric acid is most frequently used with such acetic acid systems. The ionic species present in a solution of perchloric acid in anhydrous acetic acid are $CH_3COOH_2^+$ and ClO_4^-, and the titration of glycine can be written as

$$NH_2CH_2COOH + CH_3COOH_2^+ = {}^+NH_3CH_2COOH + CH_3COOH.$$

The acid strength of $CH_3COOH_2^+$ in anhydrous acetic acid is so much greater than that of H_3O^+ in aqueous solutions that the reaction with glycine —as well as with other amino acids, amines, and salts of organic acids— is much more quantitative and better end points are obtained.

INSTRUCTIONS (Procedure 25-4: The Titration of Amino Acids)

Consult your instructor regarding the procedure to follow. If time is limited, a determination of the values of the successive acid constants of a known amino acid is suggested. A more challenging experiment is the identification of an amino acid by means of acid-base titrations. Experience should be obtained by measurements on a monoaminomonocarboxylic acid before working with a more complex acid.

A. Determination of the Acid Constants of a Known Amino Acid

Outline a procedure for titrating your amino acid. This should require a minimum number of operations and should provide sufficient information to permit calculation of the acid constants. Have this procedure approved by your instructor before proceeding.

Obtain the amino acid from your instructor and accurately weigh out a portion that will require about 20 ml of your 0.2 *F* NaOH for titration of a single proton.

Obtain a pH meter (see Procedure 25-3) and follow your approved procedure. Note that in the regions of the equivalence points and midway between equivalence points, the measurements need to be made frequently and with special care if meaningful values of the acid constants are to be obtained. Stirring must be done very thoroughly after each addition of acid or base. You should obtain a plot of the general form of those in Figures 25-2 and 25-3. From this plot determine values of the acid constants of your amino acid.

B. Identification of an Amino Acid

Outline a procedure for the identification of an amino acid by acid-base titrations and have it approved by your instructor. Obtain a sample of an unknown amino acid, and analyze it by your procedure.

Note from Figures 25-2 and 25-3 that the inflection regions in the titration curves may be easily missed; add the titrant in small portions with thorough stirring between pH readings (Note 1).

NOTE

1. The effective acid strength of an amino acid can be increased by a factor of about 10^3 if the solution contains approximately 16 % of formaldehyde. This effect is attributed to the following generalized reaction

$$H_2C{=}O + H_2N{-}R{-}C(=O)O^- = H_2C{=}N{-}R{-}C(=O)O^- + HOH$$

The hydrolysis of the amino acid anion is decreased by this reaction, therefore the $[OH^-]$ at the end point is decreased and better end points are obtained. This titration of an amino acid in the presence of formaldehyde is called the *formol titration* and is discussed by Kolthoff and Stenger (*Volumetric Analysis.* Vol. 2. New York: Wiley-Interscience, 1947) and by Harris (*Proc. Roy. Soc., London,* **B95**, 500, 1923; **B104**, 412 1939). Instructions for the titration can also be found in most elementary biochemistry texts.

QUESTIONS AND PROBLEMS

Stoichiometric Calculations

1. *Analysis of Carbonate-Hydrogen Carbonate Mixtures.* In the analysis of a mixture of sodium carbonate and hydrogen carbonate, a 5.00-g sample was dissolved and diluted to exactly 250 ml in a volumetric flask. A 50.00-ml portion was pipetted into a large flask and diluted to 200 ml, the solution was cooled, 1 drop of a 1 % phenolphthalein

indicator solution was added, and the solution was titrated slowly until the color was the same as that in a similar flask containing 0.5 g of pure $NaHCO_3$ and the same volume of water and indicator; 25.75 ml of 0.2201 *N* HCl were used for the titration. Exactly 50 ml of the standard HCl were pipetted into the solution, which was then heated to boiling for 10 minutes and titrated with 0.2142 *N* NaOH to the first appearance of the pink color; 2.74 ml were required. Calculate the percentage of Na_2CO_3 and $NaHCO_3$ in the mixture.

NOTE: A reference solution is necessary in obtaining the first end point, or an error of several percent may result. Simpson (*Ind. Eng. Chem.*, **16**, 709, 1924) recommends a mixed indicator of thymol blue and cresol red in the ratio of 6 to 1 for the titration. This titration is not advised for the analysis of mixtures containing high ratios of hydrogen carbonate to carbonate.

2. A solid commercial preparation was found by a qualitative analysis to give an aqueous solution that could contain as possible constituents only the following ions in significant amounts: (*a*) the common alkali metal cations, (*b*) carbonate, (*c*) hydrogen carbonate, and (*d*) hydroxide.

   In carrying out a quantitative analysis, 50.00 ml of the original solution of the preparation, made by dissolving 145.0 g of the solid in exactly 1 liter of water, were titrated in a cold dilute solution with 0.500 *N* HCl, thymol blue being used as indicator; 100.0 ml of acid were required.

   Next, to 25.00 ml of the original solution were added 20.00 ml of 0.5000 *N* NaOH and a large excess of barium chloride. The resulting precipitate was filtered out, washed, and reserved. The filtrate was titrated in a cold dilute solution with 0.1000 *N* HCl, thymol blue being used as indicator; 37.50 ml of the acid were required.

   Finally, the precipitate was added to a solution containing 75.00 ml of 1.000 *N* HCl, and the mixture was boiled for 5 minutes, cooled, and titrated in a cold dilute solution with exactly 0.333 *N* NaOH, thymol blue being used as indicator; 37.50 ml of the base were required.

   State (*a*) two methods by which an original solid material giving a solution of the above properties could be made, (*b*) the *p*H of the original solution above, (*c*) the *p*H of a solution made by diluting the original solution with 3 volumes of water, and (*d*) your critical opinion of the method of analysis and the accuracy of the various data. *Ans.* (*b*) 10.8.

   NOTE: Read the Discussion of Procedure 25-2 as an aid to solving this problem.

3. *Analysis of Orthophosphate Mixtures.* An analysis of an orthophosphate buffer solution was desired. A 25.00-ml portion was pipetted into a flask and titrated with 0.2014 *N* HCl, with methyl orange being used as indicator, until the color matched that in a comparison solution containing 0.75 g of KH_2PO_4 in the same volume of water; 23.25 ml of the acid were used. Another 25.00-ml portion was taken, thymolphthalein indicator was added, and the solution was titrated with 0.1892 *N* NaOH until the color matched that in a solution of the same volume containing 0.75 g of Na_2HPO_4; 26.61 ml of the base were used.

Calculate the formal concentrations of the buffer constituents.

Calculate the *p*H of the buffer solution. *Ans.* 6.7.

With 1.00 *F* H_3PO_4 and 3.00 *F* NaOH solutions available, give data by means of which 100 ml of a similar buffer solution could be prepared.

4. A solution may contain the following possible constituents: H_3PO_4, NaH_2PO_4, Na_2HPO_4, Na_3PO_4, NaOH. Two drops of Indicator *A* ($K_{In} = 1.6 \times 10^{-9}$) were added to a 25.00-ml portion and the solution was titrated with 0.200 *N* H_2SO_4 until 80 % of the indicator was in the basic form; 28.50 ml of the acid were required. Two drops of Indicator *B* ($K_{In} = 10^{-5}$) were added and the solution was again titrated with the H_2SO_4 until 20 % of the indicator was in the basic form; 17.75 ml of the acid were required. Calculate the formal concentrations of the above constituents present in the solution in significant amounts. Show justification for any of the assumptions made.

5. An analyst knew that a solution to be analyzed had been prepared by dissolving NaOH and As_2O_5 in water.

   In preliminary tests both Indicator *A* ($K_{In} = 1.6 \times 10^{-9}$) and Indicator *B* ($K_{In} = 1 \times 10^{-5}$) gave acid reactions in the solution.

   Two drops of Indicator *B* were added to a 25.00-ml portion of the solution, and the solution was titrated with 0.200 *F* KOH until 25 % of the indicator was in the basic form; 15.00 ml were required.

   Two drops of Indicator *A* were added to another 25.00-ml portion, and the solution was titrated with 0.200 *F* KOH until 70 % of the indicator was in the basic form; 55.0 ml were required.

   Calculate what weight (in grams) of NaOH and As_2O_5 would be required to prepare 1 liter of a similar solution. *Ans.* 8.00 g NaOH, 36.8 g As_2O_5.

6. A solution may contain the following constituents: HCl, H_3PO_4, NaH_2PO_4, and Na_2HPO_4.

   A 25.00-ml sample was taken and titrated with 0.2000 *N* NaOH to a bromphenol blue end point (transition point *p*H 4.4), 32.50 ml being required.

   Another 25.00-ml sample was taken and titrated with 0.2000 *N* NaOH to a thymolphthalein end point (transition point *p*H 9.5), 48.00 ml being required.

   Calculate the concentrations of the substances present in significant quantities in the original solution.

Calculations Involving Chemical Equilibria

7. *Buffer Solutions*

   (*a*) To 1 liter of water 10 ml of 10 *F* HCl are gradually added. Calculate approximately the H^+ concentration when 0, 0.10, 1.0, 4, 5, 6, and 10 ml have been added. Plot the *p*H as ordinate against the corresponding number of milliliters of reagent added. (In this and the following parts of this problem, neglect the change in volume with addition of reagents.)

(*b*) Repeat these calculations and plot the *p*H for the case where 10 ml of 10 F NaOH are similarly added to 1 liter of water. On the right side of the plot write in a scale of the values of $-\log(OH^-)$ corresponding to the *p*H scale on the left side.
(*c*) Repeat these calculations and plot the *p*H for the case where 10 ml of 10 F HCl are added to 1 liter of 0.05 F NaOH.
(*d*) Repeat these calculations and plot the *p*H for the two cases where (*1*) 10 ml of 10 F HCl, and (*2*) 10 ml of 10 F NaOH are similarly added to 1-liter portions of a solution 2 F in Na_2HPO_4 and 1 F in NaH_2PO_4.

Compare the change in the hydrogen-ion concentration in (*a*), (*b*), and (*c*) with that in (*d*).

8. A 23.0-g sample of pure As_2O_5 was dissolved in water and the solution was diluted to 1 liter.

(*a*) Calculate the approximate value of $[H^+]$ of the resulting solution.
(*b*) Solid NaOH was then added to the solution in the following stages: 2.0 g, 4.0 g, 6.0 g, 8.0 g, 10.0 g, and 12.0 g. Neglect the volume change and calculate an approximate value for the $[H^+]$ at each stage.

State the color of each of the following indicators in the original solution and at each stage of the addition of NaOH; (*1*) methyl orange, (*2*) methyl red, (*3*) phenolphthalein, (*4*) thymolphthalein.

State which of the stages you would select as the basis for a quantitative titration. Justify your choice! *Ans.* (*a*) $[H^+] = 3.3 \times 10^{-2}$.

9. Make an approximate calculation of the $[PO_4^{3-}]$ in solutions which are (*a*) 1.0×10^{-2} F in H_3PO_4 and 0.50 F in HNO_3, and (*b*) 1.0×10^{-2} F in H_3PO_4 and 5×10^{-5} F in HNO_3. *Ans.* $[PO_4^{3-}] = 2 \times 10^{-17}$.

CHAPTER 26

Ionization Reactions. Total Cations by an Ion-Exchange Process

A very brief introduction to the ion-exchange process as a means of effecting separations was given in Chapter 9 (page 154). In this chapter an application is made of an ion-exchange method to the determination of the total cation content of a solution.

The ion-exchange properties of certain clays and minerals have been known for over a century[1] but widespread use of these properties for analytical purposes began only after the development of organic ion-exchange resins in 1935.[2] These resins are polymers of highly cross-linked hydrocarbons that have had either acidic or basic functional groups introduced into their structures. Those resins that contain acidic functional groups are *cation-exchange resins* while those that contain basic functional groups are *anion-exchange resins*.

The functional groups which characterize cation-exchange resins are commonly phenolic, R—OH (R indicates the resin molecule), carboxylic acid, R—COOH, or sulfonic acid, $R—SO_2OH$. The anion-exchange resins frequently contain an amine and the active group can be considered as $R—NH_3OH$.

Ion-exchange resins swell and shrink with addition and removal of water, which in the case of some resins is absorbed to the extent of about one gram

[1] H. S. Thompson. *J. Roy. Agr. Soc. Engl.*, **11**, 68 (1855); J. T. Way. *J. Roy. Agr. Soc. Engl.*, **11**, 313 (1850).

[2] B. A. Adams and E. L. Holmes. *J. Soc. Chem. Ind.*, **54**, 1 (1935).

per gram of dry resin. As a result of this water absorption, the ionic group on the resin has characteristics very similar to those it would have in an aqueous solution. Although the polymeric resin has very low solubility in aqueous solutions, it is found that water that has been de-ionized by means of ion-exchange is contaminated with organic matter as the result of the slight solubility of the resin.

The exchange reaction for a cation-exchange resin that has sulfonic acid functional groups can be expressed in simplified form as

$$R{-}SO_2O^-A^+ + B^+ = R{-}SO_2O^-B^+ + A^+$$

where R is the resin, $-SO_2O^-$ is the sulfonate group, and A^+ and B^+ are the two ions involved in the exchange. The equilibrium constant expressed correctly in terms of activities is

$$\frac{a_{R-SO_2O^-B^+}a_{A^+}}{a_{R-SO_2O^-A^+}a_{B^+}} = K$$

If activities and concentrations are related through activity coefficients, γ:

$$a_x = [x]\gamma_x$$

then

$$\frac{[R{-}SO_2O^-B^+]}{[R{-}SO_2O^-A^+]} = K\left(\frac{\gamma_{R-SO_2O^-A^+}\gamma_{B^+}}{\gamma_{R-SO_2O^-B^+}\gamma_{A^+}}\right)\frac{[B^+]}{[A^+]} = K'\frac{[B^+]}{[A^+]}$$

where K' is an "apparent equilibrium constant" called the *exchange constant*. It is looked upon as the relative "affinity" coefficient of the resin for the two ions; affinity as used here indicates the attractive force between the resin and an ion. If $K' > 1$ then it is seen that B^+ has a greater attraction for the resin than has A^+. Measurements of the exchange constant permit one to place ions in the order of their attraction for a resin. For example, Kressman and Kitchener[3] have shown that the attractive forces for singly charged ions are in the order

$$Li^+ < H^+ < Na^+ < K^+ = NH_4{}^+ < Rb^+ < Cs^+ < Ag^+$$

and those for doubly charged ions in the order

$$Be^{2+} < Mn^{2+} < Mg^{2+} = Zn^{2+} < Cu^{2+} = Ni^{2+} < Co^{2+} < Ca^{2+} < Sr^{2+} < Pb^{2+} < Ba^{2+}$$

[3] T. R. C. Kressman and J. A. Kitchener. *J. Chem. Soc.*, **1949** 1190, 1201, 1208, 1211.

The exchange constant becomes a much less precise indication of the relative attractions of two ions of different charge since different powers of the species appear in the expression. In general, however, the attraction increases with the charge upon the ion. Thus, doubly charged ions have smaller attractive forces than triply charged, but greater than singly charged ions.

The equilibrium that obtains between the aqueous and the resin phases permits an ion to be removed from one solution by the resin and subsequently exchanged to another aqueous phase. For example, Ag^+ is seen to have a greater attraction for the resin than has H^+. If a resin that has been charged with H^+ (that is, is in the form $R{-}SO_2O^-H^+$) is equilibrated with a neutral solution of $AgNO_3$, the exchange

$$R{-}SO_2O^-H^+ + Ag^+ = R{-}SO_2O^-Ag^+ + H^+$$

occurs; if the exchange constant has an appropriate value, the silver ion can be quantitatively removed from solution and an equivalent quantity of H^+ is released from the resin. The silver ion can be recovered from the resin and replaced by H^+ if the resin is now repeatedly equilibrated with a more concentrated solution of a strong acid. While the attraction of H^+ for the resin is less than that of Ag^+, the treatment of the resin with a solution that has a high hydrogen-ion concentration and that initially contains no silver ion results in release of silver ion from the resin. It should be apparent that, because of the unfavorable exchange constant, several equilibrations with portions of the acid will be required to get quantitative removal of the Ag^+.

It is possible to force an exchange reaction to occur quantitatively in a single equilibration by means of a reaction other than the exchange reaction. For example, in the exchange

$$R{-}SO_2O^-H^+ + Na^+OH^- = R{-}SO_2O^-Na^+ + H_2O$$

the equilibrium constant for the formation of water contributes greatly to the exchange of Na^+ for H^+. Similarly, if a sodium chloride solution is equilibrated with a cation-exchange resin charged with silver ion, the formation of silver chloride forces the exchange of Na^+ for Ag^+.

BATCH VERSUS COLUMN TECHNIQUES

We have seen that multiple equilibrations may be necessary to get quantitative exchange in a resin. A resin supported in a long narrow column permits one to gain the benefits of repeated equilibrations without the tedious transferrings and washings associated with multiple-batch operations. Consider a column of resin that has been charged with H^+. A neutral $AgNO_3$ solution is poured into the top of the column. The silver ions, having greater attraction for the resin than have the hydrogen ions, replace the latter at the top of the column. The solution flowing down the column contains fewer silver ions, but comes into contact with new resin that contains no silver ions. The equilibration at this lower level of the column leaves the silver-ion concentration in solution further depleted. Since these equilibrations occur continuously as the solution descends the column, the result is a very complete removal of silver ions.

In the reverse process, when silver ions are being replaced on the resin and washed from the column by hydrogen ions, the high concentration of the latter causes replacement of Ag^+ at the top of the column. This solution now contains Ag^+, but as it flows downward, new acid solution comes in contact with the top portion of resin and more Ag^+ is released in the exchange process. Soon the top portions of resin contain virtually no Ag^+ and eventually the column is back to the state of being charged with H^+.

APPLICATIONS OF ION-EXCHANGE

A. Removal of Interfering Ions

Frequently in a quantitative determination there are species present that must be removed before a satisfactory determination can be made. For example, in the determination of the sulfur in iron pyrites (crude FeS_2) by a gravimetric method, the sulfur is oxidized to sulfate, precipitated as $BaSO_4$, and weighed. In the process the iron is oxidized to ferric ion, which coprecipitates with the $BaSO_4$ to such an extent that serious errors result. Various approaches are used to decrease this coprecipitation. The iron may be precipitated and removed as the hydroxide, or it may be reduced to the ferrous state in which its coprecipitation is much less serious. As another alternative, the iron may be removed by ion exchange and replaced with hydrogen ion. The tripositive iron has a much greater attraction than hydrogen ion for the resin, so complete removal of Fe^{3+} is achieved readily. The equilibrium expression for the exchange is represented as

$$3R{-}SO_2O^-H^+ + Fe^{3+} = (R{-}SO_2O^-)_3Fe^{3+} + 3H^+$$

After use the resin column is regenerated with 6 *F* HCl, which replaces the Fe^{3+} with H^+.

B. Water De-ionization

By the use of two exchangers, one a cation-exchange resin charged with H^+ and the other an anion-exchange resin charged with OH^-, one can de-ionize water by replacing positive ions with H^+ and negative ions with OH^-. The two resins can be in separate columns or mixed in a single bed. The latter system is capable of achieving exceedingly low ion concentrations with a single column. With the resins separated, several sequential passes through alternating cation- and anion-exchange resins are required to give water of correspondingly low ionic concentration. The mixed bed imposes, of course, a regeneration problem. This is met either by the use of resins of sufficiently different resin particle size that they can be separated by screening or of sufficiently different densities that backwashing with water will separate them.

This ion-exchange de-ionization process can be used to produce water of such low ionic content that for most purposes it is a satisfactory replacement for distilled water.

C. Determination of Total Cation Concentration

A cation-exchange resin is used in Procedure 26-1 below to determine the total cation concentration in a solution. The cations are replaced in the solutions by H^+ from the resin, and the hydrogen-ion concentration is determined by titration with standard base. This method is particularly useful for determinations of such difficult-to-determine substances as the alkali metals and for monitoring the salt content of potable waters and industrial brines.

D. Separations

Chromatographic methods have been discussed in Chapter 9, pages 151–154, and ion-exchange chromatography is the name given to separations based on the differential attractions of ions for an ion-exchange resin. The solution containing the constituents to be separated is added to the top of the resin column (the stationary phase) and is followed by an appropriate solution that causes migration of the exchange species down the column. The migration occurs at different rates for different ions, so separations result. An outstanding example of the possibilities of this method is the separation of the rare earths by use of a citrate solution for elution of the sample.[4] In such a case the separation depends both upon the small differences in attractions of the metal ions for the resin and upon the differences in the formation constants for the metal-ion–citrate complexes. The extraordinary similarities of the chemical properties of the rare-earth elements makes separations by classical chemical methods virtually impossible. Prior to the ion-exchange method of Ketelle and Boyd the best separations were achieved by an

[4] B. H. Ketelle and G. E. Boyd. *J. Am. Chem. Soc.*, **73** 1862 (1951).

extended and time-consuming process of repeated fractional crystallizations. A clear indication of the state of the separation problem of these very similar elements by earlier methods is the statement, "only Ce, Eu, and Yb can be recovered approximately quantitatively in a high degree of purity."[5] Now, in large part because of ion-exchange methods, all the rare earths are available commercially as the 99.9% pure metal, with the exception of promethium, which does not occur naturally.

Other very difficult separations have also been accomplished by means of ion-exchange chromatography. The alkali metals and alkaline earth can be separated as can complex mixtures of amino acids.[6]

---

[5] D. M. Yost, H. Russell, Jr., and C. S. Garner. *The Rare-Earth Elements and Their Compounds.* New York: Wiley, 1947.

[6] For more detailed discussions of the applications of ion-exchange methods see O. Samuelson. *Ion-Exchange Separations in Analytical Chemistry.* New York: Wiley, 1963; J. Inczedy. *Analytical Applications of Ion Exchangers.* New York: Academic Press, 1966. The biennial Analytical Reviews section of *Analytical Chemistry* reports on current work. For example, see *Anal. Chem.*, **40**, 51R, 136R (1968), and each preceding two years.

Determination of Total Cations by Ion-Exchange

Outline. An unknown solution containing a mixture of cations is passed through a cation-exchange resin that has been charged with H^+. The effluent solution is titrated with standard NaOH solution; the quantity of base required is a measure of the total equivalents of cations present.

DISCUSSION

The method used here permits the determination of the total *equivalents* of cations present in the unknown sample but not the total *moles*, for a singly charged cation replaces one hydrogen when the exchange reaction occurs, a doubly charged cation replaces two, and so forth. Such information is often valuable because it permits otherwise difficult determinations to be made. If a sample is known to contain just two cations, of which only one can be determined readily to the desired accuracy, the determination of total cations present by ion-exchange makes possible the calculation of the quantity of the difficult-to-determine cation by difference. Determination of the alkali metal ions to accuracies of a few parts per thousand is not readily accomplished by conventional analytical methods but becomes possible by the technique used here.

In the procedure below, the total equivalent concentration of a simple mixture of cations is determined by titration of the H^+ released when a measured volume of the unknown solution is passed through an acid-charged cation-exchange resin.

INSTRUCTIONS (Procedure 26-1: Determination of Total Cations by Ion-Exchange)

(1) Preparation of the Exchange Column

Assemble the apparatus as shown in Figure 26-1. Weigh 20 g–22 g of a high-capacity cation-exchange resin (Notes 1 and 2) into a 100-ml beaker and

NOTES

1. The resin Dowex 50 has a high capacity, a wide *p*H range, and stability at elevated temperatures. Other similar resins are Amberlite IR-120 and Nalcite

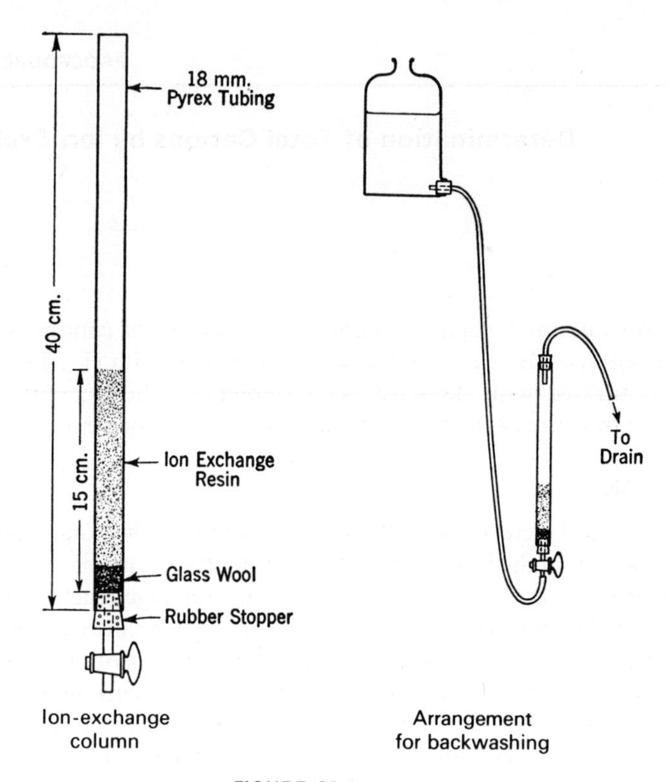

FIGURE 26-1
Ion-exchange apparatus.

add 30 ml–40 ml of water. Stir the mixture vigorously and quickly pour it into the column (Note 3).

NOTES

HCR; these should be equally satisfactory. If suitable analytical-grade (usually designated as AG) resins of high capacity are available, they should be used. They are essentially free of soluble impurities and of "fines," or very fine particles.

2. The apparatus described in this procedure illustrates the essential steps in the operation of an ion-exchange column. In case many determinations are to be made, it is recommended that a larger column made of 2′ of 25-mm tubing and containing 75 g–80 g of a high-capacity resin be used; such a column will remove the cations from at least 10 samples before backwashing and regeneration are necessary; however, larger volumes of water and acid will be needed for each step of the procedure. Such a column will shorten the analysis time and reduce the total volume of regenerating acid required.

3. If necessary, use additional water to transfer the resin to the column. Such a column of resin is termed a *resin bed*.

Connect rubber tubing from a distilled-water reservoir to the outlet of the column. Fit a one-hole rubber stopper provided with a short piece of glass tubing into the top of the column. Connect the rubber tubing to the glass tubing and lead it to a drain. Pass a stream of distilled water into the bottom of the column; regulate the flow with the stopcock so that the resin bed rises slowly (Notes 4 and 5). When particles are in motion throughout the entire length of the column of resin, adjust the rate of flow to 30 ml–35 ml per minute (Note 6). Pass at least 200 ml of water through the column, then drain most of the water out of the column, but leave *at least 1 cm–2 cm of water above the top of the resin* (Note 7).

Regenerate the resin by passing 25 ml of 6 *F* HCl (Note 8) through the column at the rate of 15 ml–20 ml per minute (Note 9). At the same rate of flow rinse the column with distilled water until the *p*H of the effluent solution matches that of the distilled water (Note 10).

NOTES

4. This process of reversing the flow of water through the column in order to clean and redistribute the resin is termed *backwashing*. If this operation is omitted, channels form throughout the resin column and the effectiveness of the column material is reduced.

5. A water reservoir located 2′–3′ above the column is convenient for backwashing the resin bed.

6. Control the rate of flow so that the resin particles can be seen to move in all parts of the bed, but avoid loss of resin in the wash water.

7. After backwashing the column, *always* maintain the liquid level *above* the top of the resin; otherwise, air will be trapped between particles and the effectiveness of the resin may be seriously impaired.

When draining the column, determine the approximate maximum rate of flow. If the maximum flow rate is less than 40 ml per minute, remove the resin to a 150-ml beaker. Add 125 ml of water and stir vigorously. Allow a few seconds for the bulk of the resin to settle; then decant 50 ml–75 ml of the suspension of the fine particles. Return the remainder of the resin to the column as directed in the first paragraph of the procedure.

8. If the resin is new, it is probably in the form of the sodium resin, and 100 ml–150 ml of acid should be used to convert it to the acid resin; otherwise sodium will be introduced into the solution. Inasmuch as the resin bed entrains 15 ml–20 ml of solution, a representative sample of the influent solution cannot be obtained until about 25 ml of effluent has passed through the stopcock.

With new resins the effluent may be colored because of removal of soluble colored organic material; they should be washed until colorless. Some sulfate may also be washed from new resins of the sulfonic acid type. Therefore, test the last portion of the wash water used after the conversion to the acid resin by adding 1 ml of 0.5 *F* $BaCl_2$ and 3 ml of alcohol to 5 ml of the wash solution collected in a 13 × 100-mm test tube; no turbidity should be visible after 5 minutes on looking lengthwise through the test tube.

(2) Removal of the Cations from the Sample

Pipet a 25-ml portion of the unknown solution into the column so that the stream from the pipet flows onto the wall of the column just above the liquid level; if this is done carefully the resin bed will not be appreciably disturbed. Pass the solution through the column at the rate of 25 ml–35 ml per minute (Note 11), and collect the effluent in a 600-ml beaker. Observe carefully whether the effluent solution contains any resin particles; if so, pass the solution through the column a second time. A total of at least 200 ml of water should be used to wash the column. For the last 50 ml of wash water, reduce the rate of flow to 15 ml per minute. Continue the washing until the *p*H of the effluent is the same as that of the distilled water.

(3) Titration of the Acid

Add 2 drops of phenolphthalein indicator (Note 12) to the combined effluent solution, and titrate with standard 0.2 *N* NaOH solution. See Procedure 24-3B for instructions regarding the titration and the preparation of end-point comparison solutions. Calculate the normal concentration (equivalents per liter) of total cations in the sample.

Duplicate samples may be put through the column without regeneration.

(4) Regeneration of the Resin Column

Immediately after using the column, pass 150 ml of 6 *F* HCl through the column; the first 50 ml may go through at the rate of 30 ml per minute, but the rate should be reduced to 10 ml per minute for the remainder of the acid. Wash with water at a rate of 15 ml–20 ml per minute until the *p*H of the effluent is at least 4. Leave 3 cm–5 cm of water above the resin.

NOTES

9. Marks made at 5-ml intervals on the upper part of the column are convenient for estimating the flow rate.

10. Wide-range indicator paper is convenient for testing the effluent solution. Use a minimum of 100 ml of water to wash the column.

11. Removal of the cations may be incomplete at faster flow rates; greatly reduced rates of flow will prolong the procedure unduly.

12. If it is known that only salts of strong acids are present, methyl orange can be used as indicator.

QUESTIONS AND PROBLEMS

1. As mentioned on page 595, an exchange of one ion for another can be greatly enhanced by combining a favorable chemical reaction with the exchange reaction.

An example given there made use of the formation of water to improve the removal of Na^+ from solution.

Suggest five other exchange reactions that could be forced to occur more completely by a concurrent chemical reaction. In each case write both of the reactions.

2. Ion-exchange resins have been used for the concentration of very dilute metal-ion solutions. Uranium, for example, has been concentrated as much as 1000-fold by this method. Outline a technique by which a cation could be concentrated on an ion-exchange resin.

3. Slightly soluble salts can be dissolved by an ion-exchange method in which an aqueous mixture containing the solid is shaken with an excess of an appropriate resin; one ion of the solid is exchanged for an ion that forms a soluble salt.

   Outline a procedure for dissolving $BaSO_4$. Indicate the species with which the resin is charged and the reactions that occur in dissolving the $BaSO_4$.

CHAPTER 27

Ionization Reactions. Titrations Involving Complex Compounds

Some complexes of metal ions with inorganic ligands and with certain organic polydentate ligands were discussed in Chapter 7. In this chapter ethylenediaminetetraacetic acid (EDTA), the most widely used polydentate titrant, is discussed in more detail and a procedure for its use for the titrimetric determination of Ni(II) is given.

STRUCTURE OF EDTA

EDTA is derived from ethylenediamine, $H_2N{\cdot}CH_2{\cdot}CH_2{\cdot}NH_2$, and has the following structure:

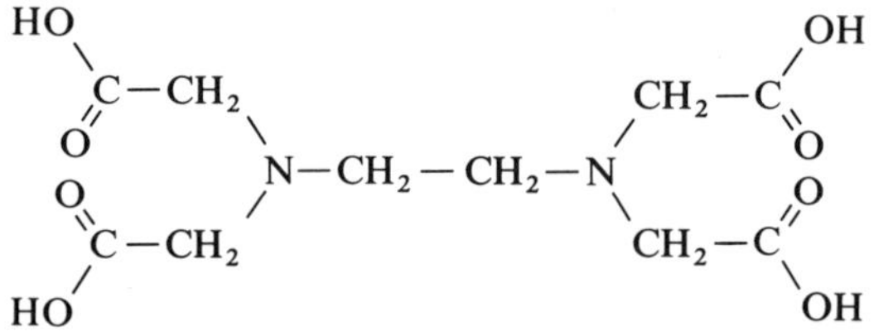

```
HO                                        OH
  \                                      /
   C—CH2                        CH2—C
  //      \                      /      \\
 O         \                    /         O
            N—CH2—CH2—N
 O         /                    \         O
  \\      /                      \      //
   C—CH2                        CH2—C
  /                                      \
HO                                        OH
```

Because of the available electron pairs on the two nitrogens and on one of the oxygens of each acetate group it is capable of forming as many as six bonds with a metal ion and is therefore called a hexadentate ligand.

EDTA has been the most extensively studied of the polydentate compounds, and procedures for the titration of most cations except the alkali metal ions have been developed. The individual bonds that are formed between EDTA and a cation are not fundamentally different from those that ammonia forms with the same cation; the differences in the complexes result from the unusual structural characteristics of the EDTA molecule.

COMPLEXES OF EDTA AND METAL IONS

We will consider the metal ion, M^{n+}, that forms complexes with both NH_3 and EDTA. If to the solution of a metal ion, M^{n+}, an aqueous solution of NH_3 is added gradually, one might expect the concentration of M^{n+} to change drastically when one equivalent of NH_3 has been added, just as $[Cl^-]$ changes drastically in the titration of Cl^- with Ag^+ (see page 61). The observed effect, however, is quite different, because the ammine complexes that form do so in a stepwise manner. An example is seen in Table 7-2, page 111, where the dissociation constants for the Cd(II) ammonia complexes are given. The total range of constants is only from 1.2×10^{-1} to 2.2×10^{-3}. The over-all dissociation constant for the reaction

$$Cd(NH_3)_4{}^{2+} = Cd^{2+} + 4NH_3$$

is $K_1K_2K_3K_4 = 7.5 \times 10^{-8}$. This small value appears to indicate that a titration would be successful, but such is not the fact because the complexes of NH_3 and Cd(II) form gradually and not by distinct steps. If the negative logarithm of $[Cd^{2+}]$ is plotted against the ratio,

$$\frac{\text{moles } NH_3 \text{ added}}{\text{moles Cd(II) present}}$$

the plot has no sharp inflection as has the typical titration curve.

The problem is closely related to that encountered when a mixture of two weak monoprotic acids having identical acid constants is titrated with a strong base. In such a case there is no inflection in the titration curve indicating when one of the acids has been titrated, for both acids are neutralized simultaneously throughout the addition of base. Moreover, if the acids are very weak, the neutralization constant is so small that there is no significant inflection in the titration curve when a quantity of base equivalent to the sum of the acids has been added.

On the other hand, when EDTA is added to a solution of M^{n+} a complex forms in which a single EDTA molecule supplies all the complexing groups that M^{n+} can accommodate. In the case of the ammonia complex, each NH_3 formed a single independent bond with the metal ion; but with the EDTA

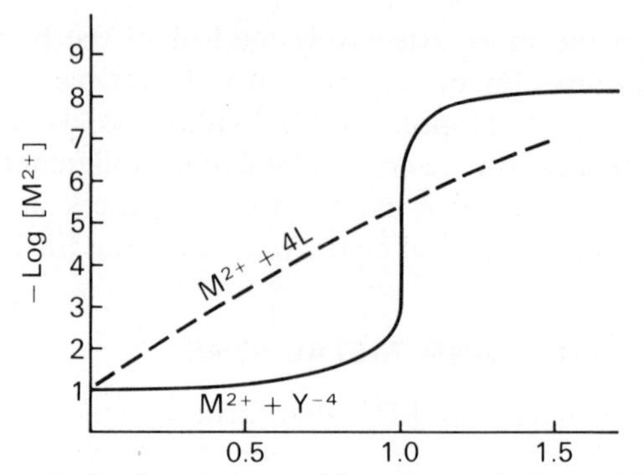

FIGURE 27-1

Titration curves for the single step formation of MY^{2-} and the stepwise formation of $ML_4{}^{2+}$, both with over-all formation constant 10^8.

complex, once a nitrogen or oxygen atom has attached itself to the metal ion, the other five atoms capable of bonding are held by the EDTA molecule very near to the positions that they must occupy in order to form bonds. The result is that the bonding occurs as a single step to form a complex of exceptional stability.

Figure 27-1 illustrates the difference in stepwise and single step formation of complexes. The two curves are for the titrations (*a*) of a metal ion, M^{2+}, with a neutral unidentate ligand (such as NH_3), which forms $ML_4{}^{2+}$, and (*b*) of the metal ion, M^{2+}, with a polydentate ligand (such as EDTA), which forms MY^{2-}. The over-all formation constant is assumed to be the same, 10^8, in each case. Thus, for the unidentate ligand, L,

$$M^{2+} + L = ML^{2+} \qquad K_1 = \frac{[ML^{2+}]}{[M^{2+}][L]}$$

$$ML^{2+} + L = ML_2{}^{2+} \qquad K_2 = \frac{[ML_2{}^{2+}]}{[ML^{2+}][L]}$$

$$ML_2{}^{2+} + L = ML_3{}^{2+} \qquad K_3 = \frac{[ML_3{}^{2+}]}{[ML_2{}^{2+}][L]}$$

$$ML_3{}^{2+} + L = ML_4{}^{2+} \qquad K_4 = \frac{[ML_4{}^{2+}]}{[ML_3{}^{2+}][L]}$$

and

$$K_1K_2K_3K_4 = K' = \frac{[ML_4{}^{2+}]}{[M^{2+}][L]^4} = 10^8$$

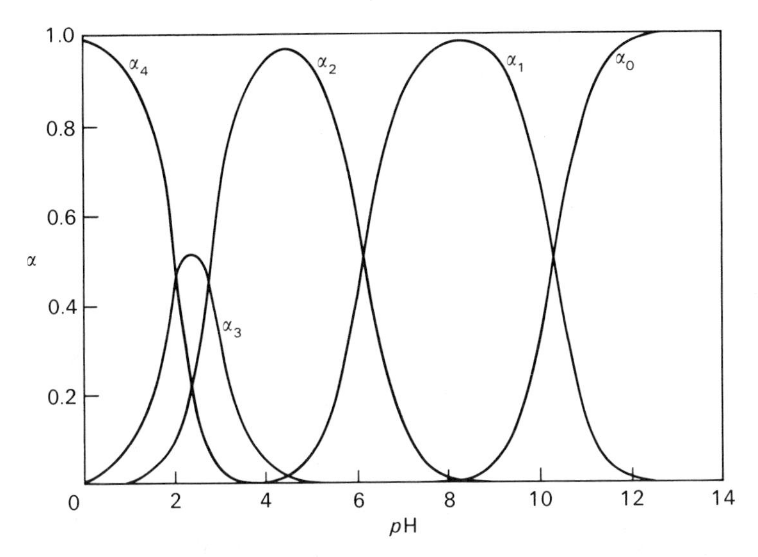

FIGURE 27-2
Fraction of EDTA present in various ionic forms: α_4, as H_4Y; α_3, as H_3Y^-; α_2, as H_2Y^-; α_1, as HY^{3-}; α_0, as Y^{4-}.

For the polydentate ligand, Y^{4-},

$$M^{2+} + Y^{4-} = MY^{2-} \qquad K = \frac{[MY^{2-}]}{[M^{2+}][Y^{4-}]} = 10^8$$

The effect of the successive complexes of M^{2+} and L is to smooth out the titration curve until no inflection is apparent; the titration would be impossible. The titration of M^{2+} with Y^{4-} appears to be practicable, for the complex is formed in a single step giving a strong inflection in the titration curve.

EDTA is a polyprotic acid having four replaceable hydrogens; the successive ionization constants are as follows:

$$K_1 = 1 \times 10^{-2},$$
$$K_2 = 2 \times 10^{-3},$$
$$K_3 = 7 \times 10^{-7},$$
$$K_4 = 5.5 \times 10^{-11}.$$

Figure 27-2 shows the fraction of the total EDTA present that exists as each of the species H_4Y, H_3Y^-, H_2Y^{2-}, HY^{3-}, and Y^{4-} as a function of the *p*H. At *p*H 6, for example, H_2Y^{2-} and HY^{3-} are the predominant species and are

TABLE 27-1
Dissociation Constants for EDTA Complexes

NaY^{3-}	2×10^{-2}	CoY^{2-}	8×10^{-17}
AgY^{3-}	5×10^{-8}	CdY^{2-}	3×10^{-17}
		PbY^{2-}	6×10^{-19}
BaY^{2-}	2×10^{-8}	NiY^{2-}	4×10^{-19}
SrY^{2-}	2×10^{-9}	CuY^{2-}	1×10^{-19}
MgY^{2-}	2×10^{-9}	HgY^{2-}	7×10^{-23}
CaY^{2-}	3×10^{-11}		
		AlY^{-}	7×10^{-17}
		CeY^{-}	4×10^{-16}
MnY^{2-}	3×10^{-14}	FeY^{-}	8×10^{-26}
FeY^{2-}	4×10^{-15}		
ZnY^{2-}	3×10^{-17}	ThY	6×10^{-24}

present in approximately equal concentrations. At *p*H 8 HY^{3-} is predominant and very little of any other species is present. At *p*H 10 the concentrations of HY^{3-} and Y^{4-} are about equal. In a solution in which H_2Y^{2-} is the predominant species the reaction with metal cations can be summarized by the equation

$$M^{n+} + H_2Y^{2-} = MY^{(4-n)-} + 2H^+ \tag{1}$$

As is apparent from this equation, the formation of the complex is influenced by the hydrogen-ion concentration of the solution.

This control of the concentration of the complexing species, Y^{4-}, over a wide range of values is one of the two factors that contribute to the ability of EDTA to serve for selective titrations of one metal ion in the presence of others. The other factor is the great range in the magnitudes of the dissociation constants of the metal-ion–EDTA complexes, as shown in Table 27-1. The constants are given for the expression $[M^{n+}][Y^{4-}]/[MY^{(4-n)-}] = K$. Figure 27-3 gives the minimum *p*H at which satisfactory titrations of various metal ions can be made.

The species most commonly used for the preparation of standard EDTA solutions is the disodium salt, Na_2H_2Y; the acid, H_4Y, has limited solubility in water, and the basic salt, Y^{4-}, hydrolyzes so extensively that its solutions are highly alkaline. The reaction shown in Equation 1 contributes, therefore, two moles of hydrogen ion to the solution for each mole of metal ion titrated. Because of the dependence of the EDTA–metal-ion titration upon the hydrogen-ion concentration, it is essential for most such titrations that the solution be buffered; otherwise, in many cases the increase in $[H^+]$ during the titration would so shift the equilibrium that the reaction would be incomplete. Ammonia–ammonium-ion buffers are often used to control the *p*H,

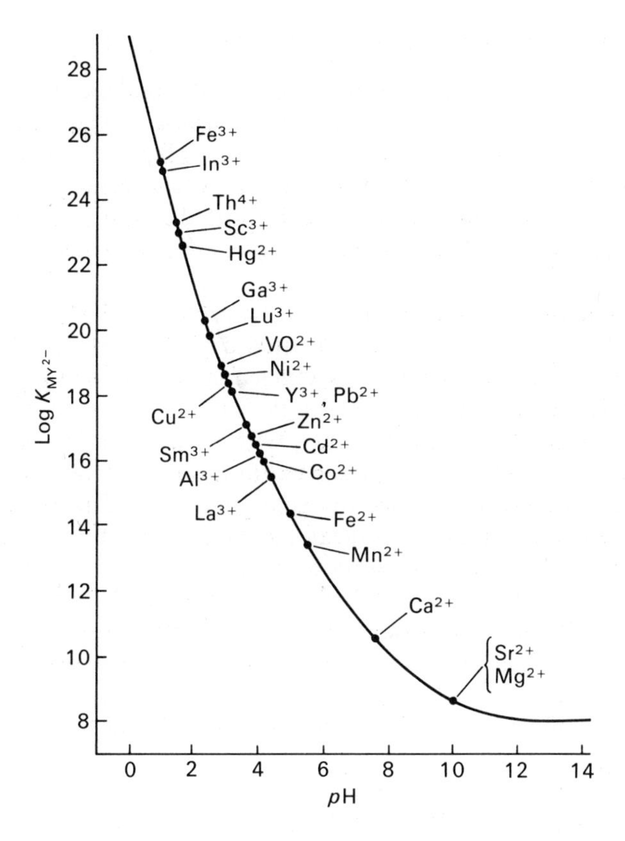

FIGURE 27-3

Minimum *p*H for effective titration of various metal ions with EDTA. Reprinted with permission from C. N. Reilley and R. W. Schmid. *Anal. Chem.*, **30**, 947 (1958).

and these have the added advantage of preventing precipitation of the oxides or basic salts of such metal ions as Cd(II), Cu(II), Ni(II), and Zn(II). The formation constant for the EDTA complex with the metal ion being titrated is so much larger than that for the ammine complexes that the titration is still well within quantitative limits.

Complexing agents may be added to permit or to enhance the titration of one substance in the presence of another. Thus, while Ni^{2+} and Pb^{2+} form EDTA complexes of very similar stability (Table 27-1), if cyanide ion is added to a mixture of the two, Pb^{2+} can be selectively titrated because of the formation of the very stable $Ni(CN)_4{}^{2-}$. This use is frequently called *masking* and the added ligand a *masking agent*.

INDICATORS FOR EDTA–METAL-ION TITRATIONS

The extensive use of EDTA as a titrant for metal ions depends in large part upon the availability of what are called *metallochromic indicators*, which are themselves chelates having the property of forming highly colored complexes with metal ions. In general these indicators are also acid-base indicators; that is, they are weak acids and change color with the addition or removal of a proton. This fact is another reason for close control of the *p*H of the titrated solution.

Eriochrome Black T, for example, is a tri-protic acid having second and third ionization constants of approximately 5×10^{-7} and 3×10^{-12}. The first proton is essentially completely ionized in aqueous solutions. If Eriocrome Black T is represented by H_3In, the colors of the species are H_2In^- red, HIn^{2-} blue, and In^{3-} yellow orange. With several metal ions, Eriochrome Black T forms one-to-one complexes that are wine-red colored. If a small quantity of this indicator is added to a solution of a metal ion buffered to *p*H 7–*p*H 11, the red complex of the indicator and metal ion is formed. During the titration, the EDTA forms complexes with the metal ion until, in the region of the equivalence point, the metal-ion concentration drops very rapidly; correspondingly, the Eriochrome Black T concentration increases rapidly and its color, blue in the *p*H region 6.5–11.5 where HIn^{2-} predominates, becomes apparent.

Murexide is the ammonium salt of purpuric acid and has the structural formula

```
              NH4+
         H   O-            O  H
         |   |             ‖  |
         N—C               C—N
        /     \\          /     \
   O=C          C—N=C          C=O
        \      /          \     /
         N—C               C—N
         |   ‖             ‖  |
         H   O             O  H
```

It is used as an indicator in titrations of Ni^{2+} with EDTA and can be represented as H_4In^-. The first proton is highly ionized, the others only weakly; pK_{A_2} is 9.2 and pK_{A_3} 10.5. The approximate dependence upon *p*H of the species and hence the color is as follows:

*p*H	8	9	10	11	12
Species	H_4In^-		H_3In^{2-}		H_2In^{3-}
Color	red-violet		violet		blue

The color of the nickel-murexide complex is orange-yellow; the color change at the end point is from orange-yellow to red-violet or violet since the titration is ordinarily done in an ammonia–ammonium-ion buffer.

PROCEDURE **27-1**

Preparation and Standardization of a Solution of EDTA

Outline. (*a*) A solution of the disodium salt of EDTA is prepared, and (*b*) this is standardized against zinc metal. Eriochrome Black T is used as the indicator.

DISCUSSION

(*a*) Preparing the Solution

The un-ionized acid, EDTA, is seldom used in the preparation of standard solutions because of its low solubility. The disodium ethylenediaminetetra-acetate dihydrate is most commonly used. Until recently this salt was not commercially available in high enough purity to be used for the direct preparation of standard solutions, but it is available now as a special product assaying 99.98 + %.[1] Direct preparation of the EDTA solution from this highly purified reagent can be used as a substitute for the standardization against zinc.

The EDTA solution used here is 0.01 F; because of the very small dissociation constants for metal-EDTA complexes accurate titrations can be made with such dilute solutions.

(*b*) Standardization Again Zinc

Analytical-reagent-grade metallic zinc is used as the primary standard. The zinc is available in granular and stick form. The latter form has much lower surface area so is less subject to surface oxidation. The granular form is more convenient to use; standardizations against this form agree within 0.1% with those against the stick zinc.

The weight of zinc required for titration with the dilute EDTA solution is so small that weighing of individual samples is impracticable; therefore a standard solution of the zinc is prepared and aliquots are used for the titrations.

[1] J. T. Baker Chemical Company, Phillipsburg, New Jersey.

INSTRUCTIONS (Procedure 27-1: Preparation and Standardization of a Solution of EDTA)

A. Volumetric Solution

Preparation of the Solution. Weigh approximately 1.9 g of disodium ethylenediaminetetraacetate dihydrate and dissolve this in 500 ml of water. Swirl until the solid is completely dissolved and the solution thoroughly mixed (Note 1). Transfer the solution to a storage bottle (Note 2).

Standardization of the Solution. Weigh out 0.33 g of analytical reagent grade zinc (Note 3). Dissolve the metal in 5 ml of 6 *F* HCl. When the metal is completely dissolved, add 50 ml of water, check that the solution is at room temperature, then quantitatively transfer it to a 500-ml volumetric flask, dilute to the mark, and mix thoroughly.

Pipet a 25-ml portion of the zinc solution into each of three 200-ml conical flasks. Add to each 10 ml of an ammonia–ammonium-chloride buffer solution, 0.2 *F* in NH_4Cl and 1.2 *F* in NH_3 (or, alternatively, 0.1 g NH_4Cl and 2 ml 6 *F* NH_3). Dilute to 50 ml and add 1 drop of Eriochrome Black T indicator solution (freshly prepared, see Appendix 7).

Prepare a comparison solution by adding to a 200-ml conical flask 10 ml of buffer solution, 1 ml of the EDTA solution, 1 drop of the indicator and diluting to the same volume.

Titrate the zinc solution with the EDTA until the first change from purplish-red toward blue is observed—thereafter titrate slowly until a blue matching the comparison solution is obtained.

Repeat the titration on the other samples and calculate the concentration of the EDTA solution.

B. Gravimetric Solution

Proceed as for the volumetric solution, except weigh out 2.3 g of EDTA for preparing the solution, and use the gravimetric buret in the titrations.

NOTES

1. EDTA dissolves slowly; be certain that no particles remain undissolved.
2. A polyethylene bottle is preferred for this solution because metal ions are slowly leached out of glass by the EDTA. The effect of this is seen as a gradual decrease in the effective concentration of the solution. Goetz, Loomis, and Diehl (*Anal. Chem.* **22**, 798, 1950) report a decrease in concentration of about 1% after 1 month's storage in glass.
3. The zinc is not dried because of the surface oxidation of the metal that occurs at elevated temperatures.

Determination of Nickel by Titration with EDTA

Outline. Ni(II) in a solution is determined by titration with EDTA; murexide serves as the indicator.

DISCUSSION

The complex formed between Ni^{2+} and Eriochrome Black T is so stable that the indicator is not displaced by EDTA from the Ni^{2+}. Therefore a different indicator is required; murexide (page 610) is used. The titration is made at *p*H 10, the same as used for the titration of zinc in Procedure 27-1, and the same ammonia–ammonium-chloride buffer system is used.

INSTRUCTIONS (Procedure 27-2: Determination of Nickel by Titration with EDTA)

A. Volumetric Titration

With a 25-ml pipet, transfer a portion of the unknown Ni(II) solution to each of three 200-ml conical flasks. Add 10 ml of NH_3-$NH_4{}^+$ buffer, 0.2 F in NH_4Cl and 1.2 F in NH_3 (see Procedure 27-1, Instructions). Dilute to 50 ml, add 0.2 g of solid murexide indicator (Note 1), and titrate with the standard EDTA solution until a permanent color change from yellow to the purple of the free indicator is obtained.

Titrate the remaining samples in the same manner. From the volume of EDTA solution used calculate the concentration of the nickel solution.

B. Gravimetric Titration

If a gravimetric standard solution of EDTA was prepared, proceed as in the volumetric titration, except use the gravimetric buret.

NOTES

1. Solutions of murexide are unstable; the solid indicator is an intimate mixture of murexide and sodium chloride. (See Appendix 7 for its preparation.)

QUESTIONS AND PROBLEMS

1. Explain two purposes of the ammonia in Procedure 27-1.

2. Discuss the effect of titrating Zn^{2+} in an unbuffered solution initially at *p*H 10 with EDTA.

3. Chromium is present as chromate in a sample. Indicate a method that involves the use of a standard EDTA solution by which the chromium could be quantitatively determined.

4. How can the extraordinary stability of EDTA–metal-ion complexes relative to NH_3–metal-ion complexes be accounted for?

5. EDTA solutions are frequently standardized against reagent-grade $CaCO_3$; the latter is dissolved in acid solution that is boiled to expel CO_2, and the solution is buffered at *p*H 10 and titrated with EDTA. In such a standardization, 0.7840 g of $CaCO_3$ was treated by the above precedure. The Ca^{2+} solution was diluted to 1000 ml and a 25.00 ml aliquot portion was taken for the titration; 21.08 ml of EDTA solution were required in the titration. Calculate the formal concentration of the EDTA. *Ans.* 0.00929 *F*.

6. Nickel can be determined by EDTA titration with Eriochrome Black T as indicator in a back-titration procedure. An excess of the standard EDTA solution is added to the Ni(II) solution; Eriochrome Black T is added and the excess EDTA is then titrated with a standard Mg(II) solution.

   In such a determination 5.00 ml of a Ni(II) solution was buffered to *p*H 10 and 25.00 ml of 0.01050 *F* EDTA solution were added. In the back-titration 7.90 ml of 0.01248 *M* Mg^{2+} solution were used. Calculate the formal concentration of the Ni(II) solution. *Ans.* 0.0328 *F*.

7. Construct the titration curve for the standardization titration of Procedure 27-1 with the assumption that the initial volume is 100 ml and that sufficiently concentrated EDTA solution is used that the volume change is negligible.

8. On page 609 the selective titration of Pb^{2+} in the presence of Ni^{2+} was mentioned. Cadmium(II) can similarly be masked to permit the EDTA titration of Pb(II) in its presence. A 25.00-ml portion of a solution containing Cd^{2+} and Pb^{2+} was titrated in a NH_4^+-NH_3 buffer and required 35.80 ml of 0.0100 *F* EDTA. A second 25.00-ml aliquot was titrated in a CN^--HCN buffer at the same *p*H and required only 6.94 ml of the same EDTA. Calculate the formal concentrations of Cd(II) and Pb(II) in the solution. *Ans.* 0.00277 *F* Pb(II).

SECTION V

ELECTRICAL AND OPTICAL METHODS

CHAPTER 28

Electrolytic Methods. Gravimetric

GENERAL PRINCIPLES

The gravimetric procedures of Chapter 14 have involved the formation and separation from the solution by filtration of precipitates that were then heated until of definite composition, and finally weighed. These precipitates have been the result of metathetical reactions, for example,

$$Ag^+ + Cl^- = AgCl(s)$$

and

$$Ba^{2+} + SO_4{}^{2-} = BaSO_4(s)$$

As discussed in Chapter 9, there are also gravimetric procedures in which the formation of the precipitate is dependent upon oxidation-reduction reactions. Thus manganese is frequently precipitated from nitric acid solutions as the result of its oxidation to the quadripositive state by means of chlorate or other oxidizing agents. The half-cell reaction and standard potential involved in this precipitation of manganese are as follows:

$$Mn^{2+} + 2H_2O = MnO_2(s) + 4H^+ + 2e^- \qquad E° = -1.24 \text{ v}$$

In principle, any half-cell reaction with a potential sufficiently negative to the value shown above could be used to cause the precipitation of the manganese dioxide. When a gravimetric determination is made, the manganese dioxide is converted by heat to Mn_3O_4, in which form it is weighed.

Similarly, mercury can be reduced from the mercuric state to the metal by a reducing agent, such as hypophosphorous acid, which is capable of causing the following half-cell reaction to proceed quantitatively from right to left:

$$Hg = Hg^{2+} + 2e^- \qquad E° = -0.85 \text{ v}$$

If instead of using chemical agents to cause these half-cell reactions to proceed in the desired direction, one causes them to take place at suitable electrodes, by means of the application of the required potential from an external source, the process is termed an *electrolytic deposition.* If, in addition, the precipitate is caused to adhere to the electrode and its weight is determined by weighing the electrode before and after the deposition, the process becomes a gravimetric *electrolytic determination.*

Thus it is seen that electrolytic determinations are gravimetric methods in which the desired precipitation half-cell reaction is caused to take place at an electrode by application of a suitable potential rather than by the addition of a chemical agent. In addition, the separation of the precipitate from the solution is usually effected, not by filtration, but by causing it to be deposited in an adherent form on a suitable inert electrode.

The Equilibrium Decomposition Potential

Let us consider the results obtained by placing two electrodes of a relatively inert metal, such as platinum, in a solution containing, for example, cupric sulfate and sulfuric acid, and then applying a constantly increasing potential difference between the two electrodes. Current flow through such a solution requires that at the cathode electrons pass into the solution as the result of a half-cell reaction, and also that at the anode electrons pass to the electrode as the result of a half-cell reaction. Therefore, in order to obtain a significant current, the potential difference between the electrodes has to be raised to a value that will cause these two half-cell reactions to take place.

For the constituents of the above solution, the two half-cell reactions that would seem most likely to occur at the cathode are

$$H_2 = 2H^+ + 2e^- \qquad E° = 0.00 \text{ v}$$

$$Cu = Cu^{2+} + 2e^- \qquad E° = -0.34 \text{ v}$$

Since the standard potential of the latter indicates that it has a much greater tendency, under standard conditions, to proceed from right to left than has the first, one would predict that, with cupric- and hydrogen-ion concentrations of the same order of magnitude, the first cathode reaction would be the deposition of copper.

At the anode, possible half-cell reactions that should be considered are

$$2H_2O = H_2O_2 + 2H^+ + 2e^- \qquad E° = -1.78 \text{ v}$$

$$2HSO_4^- = S_2O_8^{2-} + 2H^+ + 2e^- \qquad E° = -1.5 \text{ v}$$

$$2H_2O = O_2 + 4H^+ + 4e^- \qquad E° = -1.23 \text{ v}$$

Since the last half-cell reaction requires the lowest potential to cause it to proceed from left to right, it will be the one that, under standard conditions, will tend to take place first if equilibrium is attained.

The equilibrium potential of the resulting cell will be the difference between these two half-cell potentials, that is,

$$E_{\text{cell}} = E_{\text{cathode}} - E_{\text{anode}}$$

and these two half-cell reactions can be combined to give the cell reaction and potential; thus,

$$\begin{array}{rr} 2Cu = 2Cu^{2+} + 4e^- & E° = -0.34 \text{ v} \\ 2H_2O = O_2 + 4H^+ + 4e^- & E° = -1.23 \text{ v} \\ \hline 2Cu + O_2 + 4H^+ = 2Cu^{2+} + 2H_2O & E^0_{\text{cell}} = +0.89 \text{ v} \end{array}$$

Under the conventions regarding sign that have been adopted, the positive value for the cell potential, +0.89 v, indicates that there is a tendency for the cell reaction to proceed from left to right. Stated differently, if a cell were constructed, as indicated in Figure 6-1, but with one half-cell a platinum electrode immersed in a sulfuric acid solution saturated with oxygen and the other a copper electrode (or a platinum electrode plated with copper) immersed in a copper sulfate solution, there would be a tendency for oxygen to be reduced at the platinum electrode and for copper to be oxidized and pass into solution from the copper electrode, with the net result being the cell reaction shown above proceeding from left to right. If standard conditions prevailed, that is, if the copper and hydrogen ions were at unit activity and the oxygen gas were at 1 atmosphere, this cell would have an equilibrium or reversible potential of 0.89 v. An opposing potential of this value would be necessary in order to prevent current flow in the direction indicated, and this value would have to be exceeded if one desired to cause the current to flow in the opposite direction and thus to deposit copper from the solution. This potential value, called the *equilibrium decomposition potential* of the cell, is of fundamental importance in the consideration of electrolytic methods, since it is obvious that in order to cause electrolysis, an applied potential *exceeding* this value must be used. The value of 0.89 v was obtained for the above cell under standard conditions; if the concentrations of the cupric or hydrogen ions are at other than unit activity or if the pressure of the oxygen gas is at other than 1 atmosphere, the equilibrium decomposition potential will have a different value, which can be calculated by means of the Nernst equation.

Concentration Polarization

Since the potentials of the electrodes are dependent upon the ions in equilibrium with them, the potential of the copper–cupric-ion half-cell will become more positive as the concentration of the cupric ion is decreased. Therefore, if during an electrolytic deposition of copper the solution were not effectively stirred, the concentration of the cupric ion around the electrode would have a lower value than that of the solution as a whole, and the potential of the half-cell would become more positive. This effect, termed *concentration polarization*, in general is undesirable, since it requires the application of a greater potential difference across the electrodes, and this in turn may cause other half-cell reactions to take place with a resultant lowering of current efficiency. In most electrolytic processes this concentration polarization can be minimized by stirring the solution. The more one tries to shorten the time required for an electrolytic determination by increasing the rate of deposition, the more effective must be the means provided for stirring the solution in order to prevent concentration polarization effects.

It should be understood that concentration polarization can be made very small in a given electrode reaction if the current is low and if the stirring is very effective; however, the fact that diffusion of ions requires finite time means that if an electrode reaction is taking place there will be a finite concentration polarization. Since some species is being produced or consumed at the electrode surface when current flows, the concentration of this species will be different at the surface than in the bulk solution. The concentration (more precisely, the activity) at the surface controls the potential, but controls it at a potential different from that which would result if the bulk solution with its different concentration were in equilibrium with the electrode.

Hydrogen Formation During Electrolysis

In an electrolytic determination involving the reduction of a metal and its deposition on the cathode, the simultaneous formation of hydrogen gas may cause certain undesirable results. These are (1) a tendency for the deposit to be porous and nonadherent, (2) a low current efficiency, and (3) a variable hydrogen-ion concentration. In the case of the copper-sulfate–sulfuric-acid solution considered above, it was obvious that copper would be the first reduction product at the cathode. However, as the cupric-ion concentration becomes smaller, the potential at the cathode becomes more positive and, if the process is continued, eventually will reach a value that will cause reduction of hydrogen ion. If the hydrogen-ion concentration is 1 M and the hydrogen gas is at 1 atmosphere pressure, there will be no reduction of hydrogen ion (assuming that concentration polarization is prevented) until the potential at the cathode reaches a value of zero (by definition). At this potential and under equilibrium conditions, one calculates that the cupric-ion concentration will have been reduced to approximately 10^{-12} M. Therefore, these calculations indicate that the deposition of copper from even strongly acid solutions, or, if desired, the separation of copper ion from hydrogen ion by electrolytic reduction, is a practical procedure; and such is the case.

In the problem of the electrolytic reduction of zinc, similar considerations lead to very different conclusions. The standard potential of the half-cell reaction

$$Zn = Zn^{2+} + 2e^-$$

has a value of approximately +0.76 v. Let us assume that we desire the zinc-ion concentration at the *end* of the deposition to be 10^{-4} M. The cathode potential required would be

$$E_c = +0.76 - \frac{0.059}{2} \log (10^{-4})$$

or approximately +0.87 v. At 1 atmosphere pressure of hydrogen gas, the hydrogen-ion concentration in equilibrium with this cathode potential would be approximately 10^{-15} M; this means a solution 10 M in hydroxide ion. Since in any such solution the zinc-ion concentration would be drastically lowered by the formation of zincate ion, the electrolytic deposition of zinc would not seem to be a practical process. However, the electrolytic deposition of zinc from even *acid* solutions is a process of considerable commercial and analytical importance; this anomaly is due to an effect that is termed *overvoltage*, which is discussed below.

Overvoltage

The equilibrium decomposition potential is, as its name implies, the value of the cell potential under equilibrium or completely reversible conditions, and measurements of its value should be made with zero current flowing through the cell. It is found, however, that if electrode potential measurements are made when a finite current is flowing, the potentials of the electrodes change in such a way as to decrease the current flow, and this effect will be obtained in spite of all efforts to prevent concentration polarization.[1] This change in potential value is known as the *overvoltage*. The overvoltage of the cell will be the sum of the overvoltages at the two electrodes, and the magnitude of the individual electrode or half-cell overvoltages will be highly dependent upon the half-cell reaction involved and the conditions prevailing at the time.

The following general statements can be made regarding overvoltage:

1. The magnitude of the effect will be highly dependent upon the electrode reaction. It will tend to be large when a gas is being produced and small for the deposition of a metal.
2. When a gas is being produced, the magnitude of the overvoltage will depend upon the chemical and physical nature of the electrode. The overvoltage when a gas is produced at a metal electrode will vary from metal to metal, and with the same metal will depend upon the purity of the metal and the nature of the exposed surface.
3. The overvoltage becomes greater as the current density is increased. (Current density is defined as the current divided by the area of the electrode; it is usually expressed in amperes per square centimeter.)
4. In general, the higher the temperature, the less the overvoltage will be.
5. Other conditions that have been thought to affect overvoltage values are the length of time during which the electrolysis has been in progress and the composition of the electrolyte. Thus, certain colloids appear to cause an increase in some overvoltages.

For a brief introduction to the subject of overvoltage see E. H. Lyons, Jr., *Introduction to Electrochemistry* (Boston: Heath, 1967) and for a very detailed discussion see K. J. Vetter, *Electrochemical Kinetics* (New York: Academic Press, 1967); the latter gives a more exhaustive discussion of the various sources of overvoltage.

The nature and magnitude of certain of the above effects are illustrated by the data shown in Table 28-1.

Regardless of the cause, the effect of overvoltage on the deposition of zinc can be seen by repeating the calculations made above and assuming that the current density has a value of 0.01 ampere per square centimeter of electrode

[1] This distinction between concentration polarization, when the electrode is assumed to be in reversible equilibrium with the solution in its immediate vicinity, and overvoltage, when the electrode is not in a reversible state, is amplified by E. H. Lyons, Jr. (*Introduction to Electrochemistry*. Boston: Heath, 1967, p. 52 ff) and by J. J. Lingane (*Electroanalytical Chemistry*. New York: Wiley-Interscience, 1958). Recent discussions of the kinetics of electrode processes have been given by B. E. Conway and M. J. Salomon. *J. Chem. Ed.*, **44**, 554 (1967); R. J. Parsons. *J. Chem. Ed.*, **45**, 390 (1968); K. J. Laidler. *J. Chem. Ed.*, **47**, 600 (1970).

TABLE 28-1

Hydrogen and Oxygen Overvoltage on Metals (25°C)*

Current Density (amp/ cm^2)	Pt		Au	Hg	Fe	Zn	Cu	Sn
	Smooth	Platinized						
				Hydrogen				
0.0001	—	0.003	0.122	0.6	0.218	—	0.351	0.340
0.001	0.024	0.015	0.241	0.9	0.404	0.716	0.479	0.856
0.010	0.068	0.030	0.390	1.04	0.557	0.746	0.584	1.077
0.100	0.288	0.041	0.588	1.07	0.818	1.064	0.801	1.223
1.000	0.676	0.048	0.798	1.11	1.292	1.229	1.254	1.231
				Oxygen				
0.001	0.721	0.398	0.673	—	—	—	0.422	—
0.010	0.85	0.521	0.963	—	—	—	0.580	—
0.100	1.28	0.638	1.244	—	—	—	0.660	—
1.000	1.49	0.766	1.63	—	—	—	0.793	—

* These data are from *International Critical Tables*, Vol. VI, pages 339–340. The following statements preface the more complete data given there:

"Overvoltage is the potential necessary to discharge the ion in question in excess of that necessary for reversible discharge, the reversible and irreversible electrodes being under the same conditions of temperature, pressure, electrolyte, etc.

"The actual magnitude of any overvoltage cannot be specified precisely because of uncontrollable variability. The usual reproducibility is not better than 0.05 volt.

"Electrolyte for hydrogen overvoltages is 2 *N* H_2SO_4; for oxygen overvoltages 1 *N* KOH."

surface. If a layer of zinc has been formed, in effect we have a zinc electrode and a hydrogen overvoltage of approximately 0.75 v. Under these conditions the potential, E, of the hydrogen electrode under standard pressure and concentration would be the standard potential, $E°$, plus the overvoltage, η, thus:

$$E = E° + \eta = 0 + 0.75 = +0.75 \text{ v}$$

Calculations made above indicated that a cathode potential of $+0.87$ would have to be applied in order to reduce the zinc-ion concentration to 10^{-4} *M* and that under *equilibrium conditions* the hydrogen ion would have to be below 10^{-15} *M* in order not to cause hydrogen gas formation. Repeating this calculation for a current density of 0.01 and a zinc electrode, we have

$$+0.87 = 0 + 0.75 - \frac{0.059}{2} \log \frac{[H^+]^2}{P_{H_2}}$$

or a hydrogen-ion concentration of approximately 10^{-2} *M* instead of 10^{-15} *M*. The uncertainties connected with the determination of overvoltage values and in controlling the many conditions affecting them are to be strongly emphasized; as a result, such calculations as the above are not to be considered as the basis for quantitative predictions but as indications of the magnitude of overvoltage effects, and as explanatory of certain otherwise anomalous effects observed in the analytical and industrial application of electrolytic methods.

The Decomposition Potential During Electrolysis

The equilibrium decomposition potential discussed above is applicable only to equilibrium conditions or to measurements made with zero current flow. In order to calculate the decomposition potential of a cell with a finite current, we must take into account not only the equilibrium cathode and anode potentials, E_C and E_A, but also the cathode and anode overvoltage values, η_C and η_A, and the internal resistance of the solution, that is, the voltage needed because of the Ohm's law relation

$$E = IR$$

where E is the potential (volts), I the current (amperes), and R the resistance (ohms).

The decomposition potential during an electrolysis will therefore be given by the expression

$$E_D = (E_C - E_A) + (\eta_C + \eta_A) + IR$$

We have calculated above that the equilibrium decomposition potential of the cell reaction

$$2Cu + O_2 + 4H^+ = 2Cu^{2+} + 2H_2O$$

was +0.89 v under standard conditions in regard to concentrations and pressures. Now let us calculate the applied potential difference required to pass a current of 1 amp through the cell under the same conditions. The first term of the above expression, $(E_C - E_A)$, will have the same value as formerly, that is, +0.89 v. The overvoltage of cupric ion at a copper cathode is so small that it can be neglected. The overvoltage of the oxygen at the anode will depend upon the current density. If the area of the cathode is 100 sq cm, the current density will be 0.01 amp/cm^2, and the overvoltage shown in Table 28-1 is 0.85 v. Therefore the second term of the above expression, $(\eta_C + \eta_A)$, will be $(0 + 0.85) = 0.85$ v.[2] If the internal resistance of the cell is

[2] Overvoltage always increases the required applied potential difference, and the sum of the anode and cathode overvoltage values can be added to the equilibrium cell potential. Cathodic overvoltage causes the cathode potential to be more positive, and anodic overvoltage causes the anode potential to be more negative.

0.20 ohm, the potential required to maintain a current of 1 ampere through this resistance will be

$$E = 1 \times 0.20 = 0.20 \text{ v}$$

Therefore the total external potential required to maintain a current of 1 ampere through the cell with all concentrations at unit activity and gases at 1 atmosphere pressure will be

$$E = [-0.34 - (-1.23)] + (0 + 0.85) + 0.20 = 1.94 \text{ v}$$

Obviously, if we continue the electrolysis with a current of 1 ampere for a finite time, the cupric-ion concentration will decrease and the hydrogen-ion concentration will increase. As a result the cathode potential will become more positive and the anode potential will become more negative; and the total cell potential, and therefore the required applied potential difference, will become greater.

SEPARATIONS BY ELECTROLYTIC DEPOSITION

Methods of effecting electrolytic separations are of two general types and are called *constant current* and *controlled potential* methods. In the first of these, the current is kept essentially constant during the electrolysis; when the solution becomes depleted of the element being deposited, the current is maintained by a constituent of another half-cell provided for that purpose. In the second type the potential at an electrode is so controlled that only the element of interest is deposited. These methods, as well as certain other means for effecting electrolytic separations, are discussed below.

I. Constant-Current Methods

Most of the commercially available instruments for constant-current electrolytic determinations are provided with only an ammeter for measuring the current and a voltmeter for measuring the applied potential difference. To carry out an electrolytic determination, the current is adjusted to a predetermined value by means of a variable resistance in the circuit and is maintained by subsequent adjustment of the external resistance, which in effect changes the potential difference applied across the electrodes. If the current is maintained constant throughout the electrolysis, it is obvious that as the concentration of the metal being deposited at the cathode becomes very small, the potential at that electrode must be increased until some other half-cell reaction takes place and carries the current. In most cases this second reaction is the reduction of hydrogen ion to hydrogen gas. Therefore this type of procedure is generally restricted to the deposition of those metals, such as copper, that can be precipitated quantitatively in acid solutions, and to the separation of such metals from those whose half-cell potentials are positive to the potential of the hydrogen–hydrogen-ion half-cell under the conditions of the electrolysis—that is, with the overvoltage of hydrogen under the conditions prevailing taken into account. This method is useful when a rapid deposition is required and where a relatively high potential difference can be maintained in order to increase the current. The hydrogen–hydrogen-ion half-cell is therefore used in a manner somewhat analogous to a buffer system, in that so long as an excess of hydrogen ion is present in the solution, the cathode potentials will not change to such an extent that metals with more positive half-cell potentials will be reduced. The separation of copper from zinc with the deposition being made from nitric or sulfuric acid solutions is an example of such a process. It is obvious that this method cannot be used for the separation of two metals whose half-cell potentials are quite close together; in such cases one of the methods discussed below has to be used.

II. Controlled Potential Methods

In considering the reduction of cupric, zinc, and hydrogen ions, we have seen that as we decrease the concentration of a given positive ion, we must apply an increased potential at the cathode in order to continue the electrolysis and that eventually, if other positive ions are present, they will begin to be reduced. An inspection of the Nernst equation shows that for each tenfold decrease in the concentration of a unipositive ion, the cathode potential becomes more positive by approximately 0.06 v; the change will be 0.03 v for a dipositive ion. This observation provides an approximate method of predicting the extent to which two metals can be separated by controlling the applied cathode potential. In making a quantitative deposition of a metal, let us assume that not over 0.001% is to be left in the solution. This means that the concentration of the metal will be reduced by a factor of 10^{-5} and that the cathode potential will become more positive by $5 \times 0.06 = 0.3$ v if the metal ion being reduced is unipositive. This value indicates that, neglecting overvoltage effects, we need a half-cell potential difference of 0.3 v between two metals, if the metal being reduced is unipositive and we desire to be able quantitatively to reduce this unipositive metal without danger of simultaneous reduction of the other metal; with a dipositive ion being reduced, the required potential difference would be 0.15 v.

The constant-current method described above cannot be used where the applied potential must be controlled within such narrow limits as even 0.3 v. As stated above, in that method only the total potential difference is measured, and it is varied in order to maintain the current approximately constant. Even if the total applied potential difference were maintained constant, as is sometimes done, the potential at the two electrodes would vary widely, since the ions involved in the cathode and anode reactions will change in concentrations at varying rates, and since the IR drop will also change, thus changing the fraction of the total potential drop that takes place at the electrodes.

Because of these factors, for effective use of the potential control principle it is not sufficient to maintain a constant total potential difference between anode and cathode. It is essential that the potential of the "working electrode," that is, the one on which the deposition is taking place, be measured and controlled independently. An auxiliary reference electrode is used for the measurement; potentiostatic power supplies are available commercially that maintain the potential between the working electrode and the reference electrode at a constant value by varying the electrolysis current. Figure 28-1 shows how a manually controlled cathode potential deposition might be made. The desired value of the cathode potential can be adjusted on the potentiometer, E_x, and the external resistance R varied until the galvanometer G shows no deflection. Since the cathode potential will vary during the course of the electrolysis, as explained above, it is necessary to make frequent adjustment of the external resistance R in order to maintain constancy of the cathode potential. Experimentally this would be tedious at best. Manual control can be replaced to advantage by automatic potentiostats. They range from a device similar to that of Figure 28-1, in which the potentiometer has a motor-driven self-balancing system, to electronic devices. They give much more precise control, since they do not depend upon imbalance and mechanical adjustment.

III. Use of Complex-Ion Formation

Since the potential at which a metal can be reduced from a solution is dependent upon the concentration of the metal ion, it follows that the addition to the electrolyte of any compound that will change the metal-ion concentration will change this reduction, or, for electrolytic work deposition, potential. This effect is illustrated in Table 28-2. Column 2 shows the standard half-cell potential for the metal against its cation; column 3, the calculated potential for the metal against solutions in which the cation is 0.1 M; columns 4 and 5, the potentials in solutions in which the metal is 0.1 F (but existing largely as the complex) and the concentration of cyanide ion is 0.1 M and 1 M, respectively. In the cyanide solutions the potentials of all four metals become more positive as the result of complex formation. However, the effect is much more pronounced in the case of the copper, and it is possible experimentally

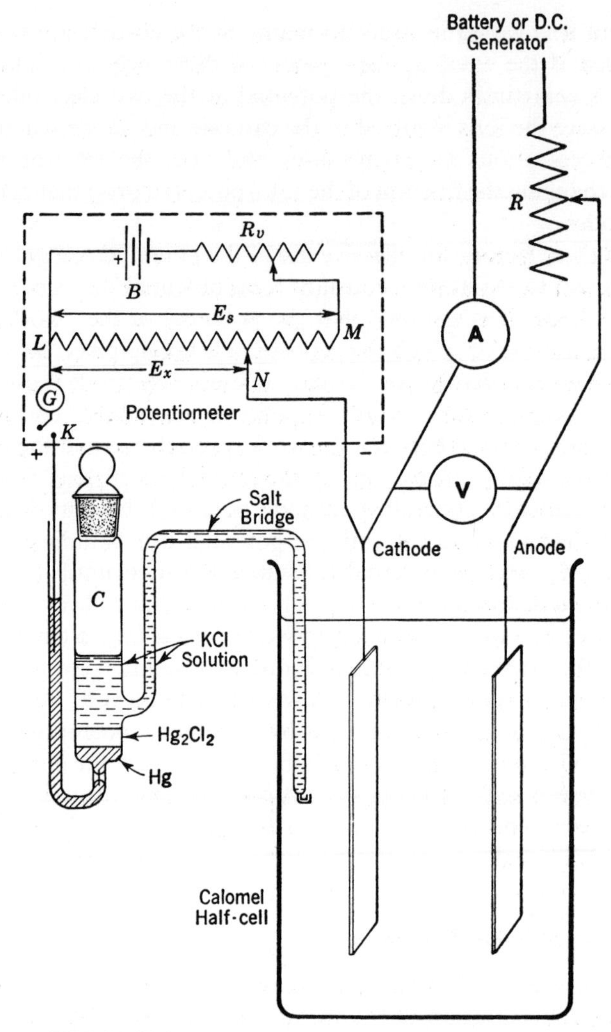

FIGURE 28-1

Apparatus for electrolysis with manual cathode-voltage control. (See Figure 18-4 for explanation of the potentiometer.)

to cause zinc to be selectively plated in the presence of copper from a concentrated cyanide solution. The commercial electroplating of brass, that is, the simultaneous deposition of copper and zinc, can be accomplished by proper adjustment of the cyanide concentration.

In addition to cyanides, other complex-forming compounds such as ammonia, oxalates, fluorides, and chlorides are used in electrolytic proce-

TABLE 28-2

Effect of Complex Formation on the Deposition Potential of Certain Metals

Metal	$E°$ ($M = M^{n+} + ne^-$)	Equilibrium Deposition Potential		
		$[M^{n+}] = 0.1$	$[CN^-] = 0.1$	$[CN^-] = 1$
Zn	+0.76	+0.79	+1.16	+1.28
Cd	+0.40	+0.43	+0.81	+0.93
Cu	−0.34*	−0.31	+0.99	+1.15
Ag	−0.80	−0.74	+0.38	+0.50

* This value is for the half-cell reaction $Cu = Cu^{2+} + 2e^-$; for $Cu = Cu^+ + e^-$, $E° = -0.51$. Cupric copper is reduced by cyanide in alkaline solutions, with formation of the cuprous cyanide complex. Values for the dissociation constants of the various cyanide complexes are given in Appendix 2.

dures, not only to achieve certain separations but also to attain more adherent deposits. This latter effect is discussed below.

IV. Anodic Deposition of Acidic Oxides

It has been mentioned previously that there is a general tendency for the so-called metallic elements to become more acidic in character as their oxidation state increases. As a result certain elements, such as lead and manganese, are predominantly basic in their lower oxidation states; but when they are oxidized to the quadripositive state, they lose their basic properties and form acidic oxides that are insoluble even in concentrated acids, provided they do not form complex ions.

In the case of manganese the oxidation to the quadripositive state, with the precipitation of manganese dioxide, is most commonly effected by the use of a chemical oxidizing agent such as chlorate in a nitric acid solution. This procedure is used extensively in systems of qualitative analysis for the separation of manganese from the other common elements precipitated by ammonium sulfide and ammonia.

The quantitative oxidation of lead to the quadripositive state, with its precipitation as the dioxide, is difficult of attainment in an acid medium by means of chemical agents. However, as mentioned in Chapter 9, Pb^{2+} can be oxidized to PbO_2 by anodic oxidation with resultant deposition on the electrode; this method is frequently used for the separation of lead from various other elements. A quantitative determination of the lead thus separated can be made by weighing the deposit; in some cases the precipitate is treated with an excess of potassium iodide in an acid solution and the iodine thus formed is titrated with thiosulfate. For the separation of cobalt from nickel, an electrolytic method has been proposed[3] that depends upon the anodic oxidation of cobalt and its deposition as Co_2O_3 from a solution buffered to a pH of approximately 5.

Oxide deposits are not in general as adherent as are metal deposits and only limited quantities can be deposited. Electrodes designed for such determinations are usually of large-area platinum gauze and are roughened by sandblasting. Other elements that may be oxidized and at least partially deposited on the anode under various conditions are silver, cobalt, manganese, and bismuth.

V. Anodic Deposition of Certain Anions

Certain anions may be deposited at an anode of a metal that as a cation forms insoluble precipitates with these anions. Thus, under the proper conditions chloride, bromide, and iodide can be precipitated as silver halide at a silver anode, and by proper control of the applied potential separations of certain of these anions can be made.

FACTORS AFFECTING THE DEPOSITION

Physical Characteristics of the Deposit

It was stated in Chapter 12 that the physical nature of a precipitate will be largely determined by the conditions under which precipitation takes place. The same general principles apply to metallic or oxide precipitates that are formed by an electrode reaction. There are certain differences in requirements, however. Thus, in gravimetric methods where the precipitate is separated from the solution by filtration, the usual objective is a precipitate of large crystal size, which can be readily filtered and washed. In electrolytic methods the objective is a precipitate that will adhere to the electrode so that there will be no loss on removal from the solution or on washing; within certain limits, the smaller the crystals of a metallic deposit, the brighter, smoother, stronger, and harder will be the deposit.[4] The factors that affect the nature of the deposit are discussed below.

Current Density. We have seen in Chapter 12 that, other conditions being equal, the less the supersaturation of the solution and the slower the rate of precipitation, the larger the crystal size of the precipitate. Similarly, a very

[3] Torrance. *Analyst*, **64**, 109 (1939).

[4] Discussion of factors affecting the nature of the deposit can be found in W. Blum and G. Hogaboom, *Principles of Electroplating and Electroforming* (3rd ed., New York: McGraw-Hill, 1949), in Glasstone, *Introduction to Electrochemistry* (New York: Van Nostrand, 1942, Chap. 14), in Lyons (op. cit.), and in Lingane (op. cit.).

low current density and slow rate of deposition tend to favor the growth of larger crystals from existing ones rather than the formation of new crystals, and the deposit will tend to be rough and dull, or even nonadherent. An increase in current density, *within certain limits*, results in a deposit of smaller crystals, which give a smoother, denser, and more adherent deposit. However, with very high deposition rates there is a tendency toward irregular deposition, which results in "treelike" growths extending into the solution. Also, with an excessively high current density the solution in the vicinity of the cathode may become depleted in the metal ion that is being deposited (that is, concentration polarization occurs), and hydrogen may be formed. When this occurs the deposit is likely to be porous, spongy, and nonadherent.

Rate of Stirring. Effective stirring of the electrolyte decreases concentration polarization and minimizes the effects resulting from this cause. For this reason all modern electrolytic apparatus provide some means for effective stirring. Among the most common expedients are stirring by means of a stream of air or by rotation of either the anode or the cathode.

Deposition of up to as much as 1 g of metal, which would require many hours when made from unstirred solutions, can be effected within 15 to 20 minutes when rapid stirring is provided.

Type of Ion. It has been found experimentally that many metals that tend to give nonadherent deposits from solutions of their simple positive ions can be plated satisfactorily from solutions in which they exist primarily as complex ions. Thus, the deposit of silver obtained from a nitric or perchloric acid solution tends to be granular and loosely adherent, whereas from ammoniacal or cyanide solutions a bright, smooth, adherent deposit is obtained.

Other metals that are usually deposited from solutions in which they exist largely as complex ions are nickel, cobalt, and cadmium. It is not clear in many cases whether the effect of complex-ion formation is to cause a more regular and crystalline growth of the deposit or whether the use of an alkaline solution minimizes hydrogen formation.

Temperature. Increasing the temperature at which a deposition is made causes various effects. First, the internal resistance between the electrodes is decreased, since an increase in temperature increases the mobility of the ions carrying the current and decreases the viscosity of the solution. Second, concentration polarization is decreased, since a higher temperature increases thermal diffusion and the effectiveness of the stirring. Third, overvoltages are lower at higher temperatures. In cases where the overvoltage is responsible for minimizing hydrogen formation, increasing the temperature may result in a greater tendency toward a spongy deposit. Fourth, solvent effects on the deposit by the electrolyte may be increased. Thus, there is a greater tendency for nitric acid to oxidize copper at elevated temperatures. This oxidation will increase the time and difficulty of attaining complete deposition, but may also result in a smoother, denser deposit, since any metal deposited in a strained or in a finely divided condition will dissolve more readily. Finally, in some cases, deposits of metallic oxides will be less hydrous and more dense and adherent when deposited from hot solutions.

Completeness of the Deposition

In very few cases are electrolytic depositions carried out with even approximately 100% current efficiency; therefore it is not practical to calculate the time required for complete deposition of a given quantity of metal at a given current value. This statement is obviously true of "constant-current" procedures in which hydrogen formation is utilized to prevent the deposition of some metal other than the one being deposited. The time required for the deposition of an element, as well as the amount remaining in solution after removal of the electrode, will be determined by various factors and conditions, some of which will now be discussed.

Applied Potential. One can calculate the equilibrium concentration of a given ion that will result from the application of a given potential. However, so many of the considerations mentioned below enter into the question of attainment of equilibrium that this method of predicting completeness of deposition is of little experimental value.

Hydrogen-Ion Concentration. If the hydrogen-ion concentration of the solution is known, it is possible to calculate the concentration of a given metal at which the deposit will be in equilibrium with the acid and below which reduction of hydrogen ion rather than metal should occur. Again, for practical work such effects as overvoltage and concentration polarization make the results of such calculations of dubious value.

Solvent Effects. As soon as a deposit begins to form, there is the possibility that it may be oxidized if it is a metal or reduced if it is a higher oxide, and thus dissolved by some constituents either initially in the solution or formed at one of the electrodes. An example of the first case arises when a metal is deposited from a nitric acid solution, since most metals are oxidized at a significant rate by nitrate, depending upon the temperature and concentrations. Solubility effects caused by products formed at the electrodes result from the fact that soluble oxidizing agents are likely to be formed at the anode and carried by diffusion and stirring to the cathode, where they may oxidize and dissolve the metal being plated. Similarly, a soluble reducing agent formed at the cathode may pass to the anode and reduce a higher oxide being formed there. The presence of iron in an electrolyte can inhibit the complete deposition of copper and other metals, since the ferric iron that is continuously formed at the anode oxidizes most metallic deposits; the deposition of lead dioxide is likewise inhibited, since the ferrous iron produced at the cathode reduces the lead dioxide. Such effects may be minimized by adding an excess of a suitable reducing agent when depositing a metal, or of an oxidizing agent when depositing an oxide.

Such an effect also can occur if the element being plated can exist in two positive oxidation states. Thus, when copper is deposited from a chloride solution, the first state of reduction is likely to be the formation of the cuprous complex, which is then reoxidized at the anode. The result is a serious loss in current efficiency, and in some cases the cupric copper may cause complete resolution of the previously deposited copper metal. Means employed to eliminate this effect include the following: (1) addition of a reducing agent, such as stannous chloride, capable of reducing the cupric copper to the cuprous state. This agent, if capable of being oxidized at the anode, may also reduce the anode potential to a value below that required to oxidize cuprous copper to the cupric state. (2) Use of a cathode potential that will tend to cause initial reduction of the copper to the metal. (3) Isolation of the anode by means of a salt bridge or porous membrane.

Dissolving of the Deposit after the Current is Broken. Once the current is broken, the effect of the solvent actions mentioned above is much greater, since action of the current to redeposit any dissolved material is no longer present. Moreover, most metals tend to be oxidized by oxygen dissolved in an acid solution. For this reason, steps must be taken to minimize this effect at the conclusion of a deposition. Two alternative procedures are most frequently used. The first one consists in gradual removal of the electrode containing the deposit from the electrolyte, with simultaneous continuous rinsing of the electrode with distilled water. The current is broken only when the electrode is completely removed from the electrolyte. The second method involves siphoning the electrolyte while continuing passage of the current, and at the same time adding water to the containing vessel so that the deposit is never uncovered and exposed to the air. This process is continued until the original electrolyte is so completely replaced by water that no further current passes through the cell.

Apparatus

The elements of the electrical circuit needed for carrying out electrolytic determinations are shown in Figure 28-1 and can be readily assembled. (The calomel half-cell and potentiometric equipment are required only for certain cathode-voltage-control methods.) The method used for stirring the solution will depend upon whether a rotating electrode is used or not. However, various commercial instruments offer many advantages in their flexibility as regards the number of determinations that can be made and the control of such factors as current density, applied voltage, temperature of the electrolyte, and rate of stirring. Catalogues describing these instruments and manuals for their use for a large number of determinations should be consulted.

Electrodes. The electrodes used for quantitative determinations must meet certain requirements that serve to limit the metals that can be used. First, the electrodes must be so inert to attack by the electrolytes used or by the oxygen of the air that they will remain constant in weight during the course of an electrolysis. Second, the anode metal should be one that will not be oxidized during the electrolysis. In general, this requirement means that the anode potential necessary to oxidize the electrode metal in any electrolyte that may be used should be greater than the anode potential required to evolve oxygen at the current density being used, since in most determinations the final anode reaction will involve oxygen formation.

Because of these qualifications, electrodes of platinum are almost universally used for quantitative electrolytic determinations. Even when used as the anode, this metal is not seriously dissolved, except in the presence of ions, such as chloride, that form stable platinum complex ions. Even then, this action can be largely eliminated if an excess of a reducing agent is present in the electrolyte. Thus, lead, tin, and antimony are often deposited from hydrochloric acid solutions containing such reducing agents as hydroxylamine, hydrazine, or formaldehyde.

Since there is a reducing action at the cathode, it is possible to use a less noble metal for that electrode. Cathodes of tantalum have been used to some extent but tend to become brittle if subjected to a vigorous evolution of hydrogen. Copper cathodes have been used as an economy measure for the deposition of copper but are not satisfactory for the most accurate work.

Cathodes of mercury are of interest because the high hydrogen overvoltage makes possible the reduction of even the alkali metals, and the formation of amalgams permits the deposition of certain active metals such as those of the alkali-group and alkaline-earth-group elements, and of certain others that do not form adherent deposits on platinum electrodes.

The procedure given below provides for the determination of copper in solutions or alloys containing no interfering elements. Procedures for the complete analysis of copper alloys containing interfering constituents such as tin and lead, and for the anodic deposition of lead as dioxide can be found in more comprehensive analytical texts and reference books.

PROCEDURE **28-1**

Electrolytic Determination of Copper

Outline. The copper in a solution containing sulfuric and nitric acids is deposited on a platinum cathode by means of controlled current electrolysis. From the weight of the deposit the quantity of copper in the solution is determined.

DISCUSSION

This procedure is designed primarily for the determination of copper in solutions containing copper as the sulfate or nitrate. Similar salts of cadmium, cobalt, nickel, and zinc can be present, but not of lead, tin, or iron. No halides should be present. A supplementary procedure is given in Note 1 below for adapting this procedure to the determining of the copper in alloys containing only the first four elements mentioned.

The deposition of the copper is made from a solution that is approximately 0.5 *F* in sulfuric acid and 0.2 *F* in nitric acid. Sulfuric acid is used for the major amount of the acid, since it has less solvent action on copper; it has been found, however, that the presence of some nitric acid improves the nature of the deposit. This latter effect has been attributed to the cathodic reduction of nitrate minimizing hydrogen gas formation.

The nitric acid at times inhibits the deposition of the last traces of copper; therefore for extremely accurate determinations of copper an electrolyte of sulfuric acid with only a small amount of nitric acid is commonly used; frequently, reducing agents such as urea or hydrazine are added to eliminate the solvent action of any nitrous acid that may be formed.

INSTRUCTIONS (Procedure 28-1: Electrolytic Determination of Copper)

Preparation of the Solution for Electrolysis

Pipet into a conical flask that volume of the solution (Note 1) that is judged to contain 0.4 g–0.6 g of copper. (The solution should be essentially neutral

NOTES

1. In case one has an alloy containing only zinc, cadmium, cobalt, or nickel in addition to copper, the following procedure can be used:

and free of halides.) Add 5 ml of 9 F H_2SO_4 and 3 ml of 6 F HNO_3; then, if necessary, evaporate or dilute to about 50 ml.

Transfer the solution to the electrolysis vessel; use enough water to bring the volume to 100 ml.

Preparation of the Electrodes

Immerse the platinum electrodes in hot 6 F HNO_3 to which has been added 1 ml–2 ml of 3 F KNO_2 (Note 2) until they are free of any previous deposit; then wash with distilled water (Note 3). Rinse repeatedly with small portions of ethyl alcohol (95 %) or with acetone; then dry in an oven at 110°C for 2 or 3 minutes. Place the electrodes in a desiccator until cool; then weigh the cathode.

The Electrolysis

Check that the electrolysis current switch is turned off. Attach the cathode electrode to the negative terminal and the anode electrode to the positive terminal, and adjust them as directed by the instructor (Note 4). If a rotating

NOTES

Weigh out into a 200-ml conical flask a sample of the clean, dry alloy that is judged to contain 0.4 g–0.6 g of copper. Add 10 ml of 6 F HNO_3 and carefully heat until the sample is dissolved. Cool the solution, add slowly and carefully 4 ml of 9 F H_2SO_4, and then evaporate until the sulfuric acid begins to fume. (See Procedure 21-4 for directions concerning this evaporation and the subsequent dilution.) Allow the flask to cool, slowly add 50 ml of water (*as directed in Procedure 21-4*), and then add 3 ml of 6 F HNO_3.

Transfer the solution to the electrolysis vessel using enough water to bring the volume to 100 ml. The solution should be approximately 0.5 F in H_2SO_4 and 0.2 F in HNO_3. If the electrolysis vessel requires a volume significantly different from 100 ml, the concentrations of H_2SO_4 and HNO_3 should be adjusted accordingly. Proceed as directed in this procedure beginning with "Preparation of the Electrodes" (the anode need not be weighed either before or after the electrolysis).

2. The nitrite serves a dual purpose in that it (*a*) reacts much more rapidly with metallic deposits than does nitrate and (*b*) reduces and dissolves lead and manganese dioxide deposits.

3. It is essential that the electrode surface be free of grease films; for this reason do not touch the deposition surface with the hands. In case the wash water does not drain uniformly but collects in streaks or droplets, heat the electrode to dull redness in the oxidizing flame of a Meker burner. Take care to remove all deposits before heating the electrodes.

4. Commercial instruments differ so markedly in construction and operation that no specific directions can be given as to the adjustment of the electrodes and the electrolysis vessel. Consult the instructor for specific instructions.

electrode is to be used, start the rotating mechanism and ascertain that the electrode rotates at the proper speed and without wobbling or touching the other electrode. Adjust the electrodes in the solution so that the cathode is not quite completely covered. Cover the electrolysis vessel with split watch glasses. Start the stirring mechanism.

Electrolyze the solution with a current of 1.8 amperes for 40 minutes. Record the current, the potential difference, and the time. Add 2 ml of 9 F H_2SO_4 to the electrolyte (Note 5). Increase the current to 2.0 amperes and electrolyze for 20 minutes. Raise the level of the electrolyte slightly by addition of water and continue the electrolysis for 10-minute periods until no copper deposit is visible on the newly covered surface of the cathode.

Remove 3 ml of the electrolyte to a small test tube (Note 6) and add a slight excess of 6 F NH_3.

If a *blue color is visible* upon comparison of this solution with water in a similar tube, acidify the solution with 6 F HNO_3, return it to the electrolyte solution, and continue the electrolysis for 20 minutes.

If *no blue color is visible*, make the ammoniacal solution acid with 6 F CH_3COOH; then add 100 mg of Na_2SO_3 and 1 drop of 0.25 F $K_4Fe(CN)_6$. If a *perceptible red-brown precipitate* or *coloration* is obtained, discard the test mixture, continue the electrolysis for 15 minutes, and repeat the test. When only a turbidity or precipitate is obtained, the deposition may be considered complete (Note 7).

NOTES

5. This acid is added to replace that lost by reduction of nitrate to ammonia. Sulfuric acid, rather than nitric, is used since it is less likely to oxidize the copper deposit.

6. Do not at any time expose the copper deposit to the air by lowering the level of the electrolyte; if necessary add more water before removing test portions of electrolyte.

7. Successively more sensitive tests are applied for completeness of deposition of the copper. At least 3 ml of the electrolyte should be taken in order to obtain adequate sensitivity in these tests. If the ammonia test is negative, the quantity of copper removed for the test can be neglected. The test solution should not be returned to the electrolysis vessel after addition of the sulfite and ferrocyanide. If zinc is present a white precipitate of potassium zinc ferrocyanide will be formed upon addition of the ferrocyanide and should not be confused with the colored copper compound. A comparison test made with a solution containing a few milligrams of zinc is an aid in determining the presence of copper.

Experiments have shown that upon addition of the excess of ammonia, 0.1 mg–0.2 mg of copper *in the test portion* will cause a blue color that is distinguishable from a similar blank solution. In *the absence of zinc*, 1 μg of copper will cause a perceptible brownish coloration upon addition of the ferrocyanide; in the presence

When the deposition is considered complete, begin siphoning off the electrolyte into a 500-ml conical flask and at the same time replace the electrolyte with water so that the deposit *is covered at all times*; continue this process until no appreciable current flows. Quickly remove the cathode, rinse it with water, then with alcohol or acetone. Place it in an oven at 110°C for 2 or 3 minutes, allow it to cool in a desiccator, then weigh it.

From the weight obtained calculate (*a*) the formal concentration of cupric copper in the solution or (*b*) the percentage of copper in the alloy.

NOTES

of 5 mg of zinc, the precipitation of zinc ferrocyanide so diminishes the sensitivity of the test that 0.2 mg of copper is required for detection of the brown color.

The sodium sulfite is added to prevent oxidation of ferrocyanide to the yellow ferricyanide by any oxides of nitrogen that may be present.

QUESTIONS AND PROBLEMS

1. A solution 0.02 *F* in $Pb(NO_3)_2$ and 1 *F* in HNO_3 was electrolyzed between polished platinum electrodes with a current of 1 amp. The anode had an area of 100 cm^2, the cathode an area of 20 cm^2, and the ohmic resistance of the cell was 0.5 ohm. Calculate the Pb^{2+} concentration at which oxygen will be formed at 1 atmosphere pressure. *Ans.* 10^{-21} *M*.

2. *Effect of Oxygen Overvoltage on the Deposition of* PbO_2. If the hydrogen-ion concentration of the electrolyte were maintained at 1 *M* and the oxygen partial pressure at 1 atmosphere, (*a*) calculate *under equilibrium conditions* to what value the lead ion (Pb^{2+}) concentration of the solution could be lowered without causing the evolution of oxygen at the anode; (*b*) repeat the calculation for a current of 0.01 ampere/cm^2, assuming that the overvoltage of the lead-ion–lead-dioxide anode is negligible and that the overvoltage for oxygen on the lead dioxide anode has the same value as that for oxygen on a smooth platinum anode. *Ans.* (*a*) 10^8 *M*.

3. Calculate the *equilibrium decomposition potential* of a solution 0.0010 *F* in Ag_2SO_4, 0.010 *F* in NH_3, and 0.10 *F* in $(NH_4)_2SO_4$ (assume that the solution is saturated with oxygen at 1 atmosphere pressure, and that anodic oxidation of ammonia and cathodic reduction of oxygen are negligible). State simplifying approximations.

4. A sample of brass (Cu, Sn, Pb, Zn) was dissolved in 16 *F* HNO_3, the solution was evaporated almost to dryness, dilute nitric acid was added, and the residue was removed by filtration.

   (*a*) Write equations for the reactions taking place.
   The solution was then electrolyzed according to Procedure 28-1.

   (*b*) Write equations for the electrode reactions taking place.
   The anode with its deposit was immersed in a solution containing an excess of H_2SO_4 and KI; then the resulting solution was titrated with thiosulfate.

(*c*) Write equations for the reactions taking place.

The cathode deposit was (*1*) dissolved in HNO_3, (*2*) excess sulfuric acid was added and the solution was fumed, (*3*) the solution was diluted, then neutralized with ammonia, (*4*) the solution was again made acid with sulfuric acid, (*5*) excess KI was added and (*6*) titrated with thiosulfate to a preliminary end point, and (*7*) KSCN and starch were added and titrated to a starch end point.

(*d*) Write equations for the reactions taking place at each of the above steps.

Data obtained: Weight sample 1.000 g.	
Volume of thiosulfate used in titrating:	
Anode product	4.83 ml of 0.1000 *F*
Cathode product	44.15 ml of 0.2500 *F*
Sulfuric acid remaining after fuming	5 ml of 36 *N*
Excess H_2SO_4 added (after neutralization with ammonia)	5 ml of 6 *N*
Concentrations at end point of copper titration:	
(*a*) Tri-iodide ion	2.5×10^{-6} *M*
(*b*) Thiocyanate ion	0.22 *M*
(*c*) Iodide ion	0.12 *M*

Calculate:

(1) The percentages of copper and of lead in the alloy. *Ans.* 5.00% Pb, 70.15% Cu.
(2) The hydrogen-ion concentration of the solution before adding the KI.
(3) The cupric-ion concentration at the end point of the copper titration.

(*e*) State the factors fixing the maximum and minimum hydrogen-ion concentrations between which the copper titration can be made accurately.

CHAPTER 29

Electrolytic Methods. Coulometric

The electrolytic gravimetric methods discussed in Chapter 28 are restricted to those cases in which a solid deposit is obtained in such purity that it can be weighed to the necessary accuracy. In this chapter, we consider coulometric electrolytic methods. Such methods include those *electrolytic processes* in which a determination of the quantity of a constituent is made, not by a measurement of the weight deposited, but by a *measurement of the quantity of electricity* (that is, the coulombs and hence the title) required to react, *directly* or *indirectly*, with that constituent. The three italicized terms should be noted: First, the process is electrolytic in nature. Second, a measurement of the quantity of electricity involved must be made. Third, the constituent being determined can be involved directly or indirectly in an electrode reaction. As will be shown later, this third factor is of importance in the classification of coulometric processes.

A brief historical review of the early developments in coulometry may be of interest and shows the origin of certain terms. If the quantity of electricity involved is to be used for the quantitative determination of a constituent, then the electrode reaction of interest must be capable of yielding "100% current efficiency;" one can use as many significant figures as is required in writing the 100%. The work that was begun at the end of the last century with various types of *coulometers* (or *voltameters* as they were then called) demonstrated that there were electrode reactions that could be made to proceed with 100% current efficiency. Although this work had as its primary objective the determination of the value of the faraday, the principles underlying coulometry were clearly established. For this reason, it is surprising that the analytical implications of this work were not exploited to a significant extent until about forty years of this century had passed. Then, in 1942, Hickling[1] devised an electronic instrument that he called a *potentiostat*, whereby the potential applied to the working electrode could be controlled. By use of this device, he demonstrated that 100% current efficiency could be obtained in the reduction of cupric copper to the metal at a platinum cathode and in the oxidation of iodide to iodine at a platinum anode. In 1945 Lingane[2] made use of a mercury cathode for the determination of copper and other metals, and he later used a silver anode at which the various halides were precipitated as the corresponding silver salts.[3]

There are certain characteristic features of these procedures. First, in all cases the substance to be determined is involved in the half-cell reaction taking place at one of the electrodes; that is, the cupric copper is reduced at the cathode and the halide is either oxidized or precipitated at the anode. Consequently such procedures were called *primary or direct* coulometric methods. Second, in order to obtain 100% current efficiency in such primary processes the potential applied to the working electrode must be controlled and, accordingly, such procedures were designated as *controlled potential* processes; this latter term is more commonly used. The limits within which the potential control must be maintained will depend, first, upon the accuracy of the determination being made, and, second, upon the presence of other electrode-reactive substances in the solution.

In 1938, methods differing in certain characteristics from those described above were proposed by two Hungarian chemists,[4] and published under the title "Coulometric Analysis as a Precision Method."

This title appears to have been the first use of the term *coulometric analysis.* A method developed by these two chemists can be used to illustrate the principles involved. Interested in the determination of thiocyanate, they knew that its anodic oxidation to cyanide and sulfate with 100% current efficiency is not possible. Therefore they added a relatively high concentration of a soluble bromide to an acid solution containing the thiocyanate, anodically produced bromine with 100% current efficiency, and allowed this bromine to quantitatively oxidize the thiocyanate. They used a chemical indicator, measured the quantity of electricity involved by means of a chemical coulometer, and demonstrated that an accuracy of within 1 part per 1000 could be attained. Such a process has subsequently been termed a *coulometric titration.*

---

[1] A. Hickling, *Trans. Faraday Soc.*, **38**, 27 (1942).
[2] J. J. Lingane, *J. Am. Chem. Soc.*, **67**, 1916 (1945).
[3] J. J. Lingane and L. A. Small. *Anal. Chem.*, **21**, 1119 (1949).
[4] L. Szebelledy and Z. Somogyi. *Z. Anal. Chem.*, **112**, 313, 323, 332, 385, 391, 395, 400 (1938).

The next development occurred in this country and was the result of a wartime need to determine small vapor-phase quantities of mustard gas. In the method then used this compound, bis(2-chloroethylsulfide), was collected in an aqueous solution where it hydrolyzed to thiodiglycol

$$(ClCH_2CH_2)_2S + 2H_2O = (HOCH_2CH_2)_2S + H^+ + Cl^-$$

The solution was then titrated with a dilute bromine solution and the thiodiglycol oxidized to thiodiglycol sulfoxide,

$$(HOCH_2CH_2)_2S + Br_2 + H_2O = (HOCH_2CH_2)_2SO + 2Br^- + 2H^+$$

Dilute aqueous bromine solutions are unstable; the frequent standardizations were time-consuming. The titrations of the thiodiglycol with methyl orange as a redox indicator were also time-consuming and the end points uncertain. Because of these difficulties constant-current electrolytic generation of the bromine was substituted for the titration with standard bromine and the end point was obtained by an amperometric method. It was shown that 40 μg to 1400 μg quantities of thiodiglycol could be titrated to about ± 0.5 μg.[5] Subsequently an instrument based upon the same principles was devised for continuously determining the vapor-phase concentration of mustard gas.[6]

Two fundamental differences exist between these last processes and the controlled-potential method described previously. First, in the controlled-potential method the substance being determined was directly involved in the electrode reaction; in coulometric titrations a half-cell reaction was used to generate an active *intermediate* compound, which then reacted with the substance being determined. For this reason such processes are called *secondary* or *indirect* coulometric methods. Second, for reasons that are discussed later, such processes can be caused to take place with the current at a constant value and in such cases are called *constant current* methods. Each of these two types of processes has certain advantages and limitations and these are discussed below.

CONTROLLED POTENTIAL METHODS

These methods can be used to attain a high degree of specificity as to the half-cell reaction taking place. The reduction of one metal in the presence of another and the selective determination of various mixtures of the halides has been demonstrated by Lingane.[7] The use of this fundamental property

[5] J. W. Sease, C. Niemann, and E. H. Swift. *Anal. Chem.*, **19**, 197 (1947).
[6] P. A. Shaffer, A. Breglio, and J. A. Brockman. *Anal. Chem.*, **20**, 1008 (1948).
[7] See Footnotes 2 and 3.

of the controlled potential method minimizes preliminary separations and is one of its valuable features.

In a controlled potential process, the current decreases exponentially and approaches zero asymptotically. For this reason the time required for the process is fixed by the accuracy desired—essentially the ratio of the initial to the final concentration of the constituent being determined—and such a process is not adaptable to rapid determinations. In addition, since the current varys continuously, an accurate measurement of the quantity of electricity involved requires the use of accurate electronic coulometers or chemical coulometers; the latter do not lend themselves to ready use for micro scale operations.

The most fundamental limitation to the development of primary coulometric methods arises from the fact that many substances are not capable of being oxidized or reduced at conventional electrodes with 100% current efficiency. Upon systematically examining a comprehensive list of half-cell reactions one is likely to be surprised to note how many of those reactions exhibit some degree of irreversibility.

CONSTANT CURRENT METHODS

It is apparent that by taking advantage of those intermediate half-cell reactions that can be made to proceed with 100% current efficiency, the coulometric principle can be applied to many of those substances that are not reversibly oxidized or reduced at conventional electrodes. In addition to 100% current efficiency an intermediate half-cell must meet certain requirements. The generated titrant, like other titrants, must react rapidly, quantitatively, and stoichiometrically with the constituent to be determined. As a result much effort has been directed toward studies of such half-cell intermediates. Among the oxidants that have been investigated are bromine, iodine, chlorine, hypobromite, ceric cerium, ferric iron, argentic silver, and manganic manganese. A partial list of the reductants includes cuprous copper in chloride and bromide solutions, ferrous iron, and tripositive titanium. Precipitants have included unipositive silver and mercury; in addition both acids and bases have been coulometrically generated.

The secondary process also allows one to maintain a constant current throughout the process and this permits the attainment of several desirable features. The first of these is rapidity; titrations of 100 seconds or less are conventional. Second, by control of the constancy of the current to within the desired accuracy of the titration, one can substitute for the chemical coulometer the constant-current—time method of measuring the quantity of electricity involved in the titration. Finally, by reducing the current value, one can extend the time required for the titration of micro quantities of constituents to intervals that can be accurately measured. As a result the titration of microgram quantities is readily made with accuracies as good as or better than those attained by the best volumetric methods with milligram quantities. For example, Myers and Swift[8] reported errors consistently less than 0.1 % on 4.5 μmole samples of arsenic.

Finally, coulometry offers obvious advantages in the ever-developing field of automation because an electric current is so much more amenable to automatic control than is a stopcock or other mechanical device. Methods and instruments for automatic control of titrations have been developed, as have methods for obtaining a continuous record of the concentration of a constituent in solutions or in a gas phase. (A comprehensive review of coulometric methods with references to the original literature is given by J. J. Lingane, *Electroanalytical Chemistry*. (2nd ed.) New York: Wiley-Interscience, 1958.)

FARADAY'S LAWS

Faraday's laws, named for their discoverer, Michael Faraday (1791–1867), are the basis for coulometric analysis. These laws state, first, that the total quantity of chemical change produced at an electrode by an electric current is directly proportional to the quantity of electricity that is passed through the electrode, and second, that the weights of different substances produced by the quantity of electricity are proportional to the equivalent weights of the substances.

In more explicit terms, the charge carried by one mole of electrons is called a *faraday* and this quantity of electricity will deposit, for example, one mole of silver by the reaction

$$Ag^+ + e^- = Ag$$

or one-half mole of copper by

$$Cu^{2+} + 2e^- = Cu$$

One *equivalent* of chemical change occurs when one faraday of electricity is passed. The numerical value of the faraday is 96,493 coulombs;[9] a coulomb is the charge that passes a point in one second in a circuit through which there is a current of one ampere. That is, 1 coulomb = 1 ampere-second.

[8] R. J. Myers and E. H. Swift. *J. Am. Chem. Soc.*, **70**, 1047 (1948).

[9] The faraday expressed in other useful units is 23,062 calories per volt equivalent. This value is used on page 82 where the relationship of the standard cell potential and the equilibrium constant is discussed.

METHODS FOR CONTROL OF CURRENT AND POTENTIAL

It is evident that if coulometric methods involve the measurement of the number of coulombs, we must seek techniques for making this measurement. We used Ohm's law in Chapter 28; in its simplest form it is

$$E = IR \tag{1}$$

where E, I, and R are the potential, current, and resistance, respectively. If a constant potential is applied to a system that has constant resistance, the current is seen to be constant also. In such a system the quantity of electricity that passes in a given period of time is easily calculated, for the product of the constant current in amperes (which is coulombs/sec.) and the time in seconds is the quantity of electricity in coulombs. The current in the system can be measured very accurately if a resistance of known value is placed in the circuit and the potential drop across this resistor measured with a potentiometer (page 397). From Ohm's law, Equation 1, the current is calculated. Hence, in a constant-current experiment, if the student measures this current and the time it flows, he can calculate the quantity of electricity (the number of coulombs) that passed through the system.

A simple means of maintaining a constant current is to provide a large potential in the circuit with a very large resistance in series with the electrochemical cell (see Figure 29-1) so that any changes in the cell resistance are relatively very small. For the circuit of Figure 29-1, the current is given by

$$I = \frac{E}{R_1 + R_{cell}}$$

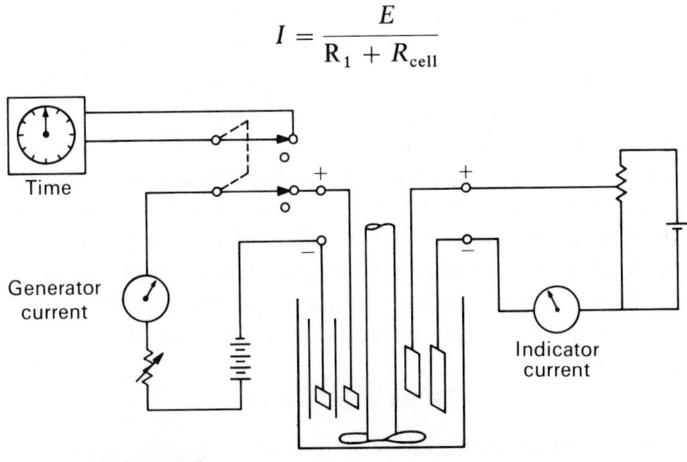

FIGURE 29-1

Basic circuits used for constant current coulometric titrations with amperometric end points. A 45 v battery provided the generator current; a 1.5 v battery the indicator current. (For details see Footnote 8.)

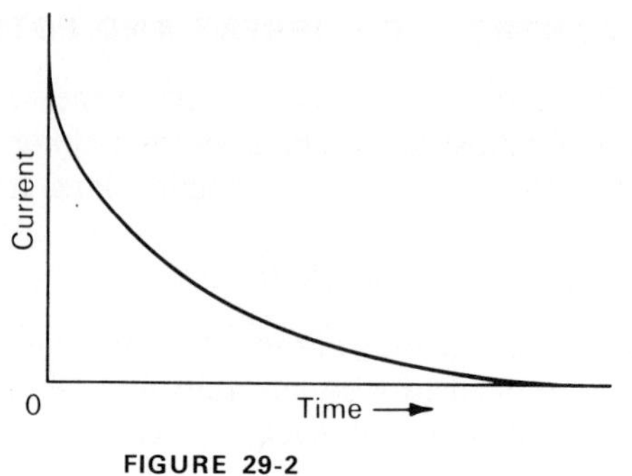

FIGURE 29-2
Typical current-time plot for constant potential coulometry.

Consider a case in which E is 45 v, R_1 is 45,000 ohms, R_{cell} is initially 1000 ohms and changes to 1100 ohms by the conclusion of the experiment because of the electrode reaction. The change in R_{cell} is 10%, but the change in the total resistance is only 1 part in 400 or about 0.22%. Electronic constant-current power supplies of high precision are available commercially or can be readily constructed.

Methods of maintaining a constant potential are discussed on page 628. In a constant-potential experiment, the current does not remain constant with time, so the calculation of the total charge that passes through an electrode is more difficult than in the constant-current case. Figure 29-2 shows the typical shape of a current-time plot during a controlled potential experiment; the area under the curve represents the total electricity. There are several ways by which this total electricity can be measured. Both mechanical and electronic integrators are available that will measure the area under the current-time curve. Another approach is to place a chemical coulometer in series with the electrodes so that the current passing through the cell must also pass through the coulometer. This latter instrument is an electrochemical cell so devised that the effect of the charge passed is readily measured. A silver coulometer, for example, deposits metallic silver on the cathode, which is then weighed to determine the quantity of charge. A hydrogen-oxygen coulometer collects the gases from the electrolytic decomposition of water:

$$2H_2O = 2H_2 + O_2$$

The volume of gases collected is measured at known temperature and pressure, and from this volume the number of coulombs of charge that passed through the system is calculated. Lingane (Footnote 2) has shown that a relatively simple hydrogen–oxygen coulometer is capable of an accuracy of ± 0.1%.

END-POINT METHODS

Visual Method

Visual indicators can be used in coulometric titrations as in volumetric or gravimetric titrations. Such use tends to limit the method to relatively large samples because the sensitivity of the detection of color change is too low to make full use of the inherent accuracy of the time and current measurements.

Potentiometric Method

This method, discussed on page 395, is widely used in coulometric analysis. An auxiliary "sensing" electrode and a reference electrode are placed in the solution and the potential changes are monitered as the constant current flows through the "generating" electrode, for example, the anode at which bromide is oxidized. The titration is stopped when the measured potential corresponds to that of the equivalence point, or when one obtains an inflection point in the titration curve.

Amperometric Method

The basic circuitry is shown in Figure 29-1. The following application illustrates the method. When a potential of about 200 mv is impressed across two platinum-foil electrodes in a stirred solution of 0.2 *F* potassium bromide, there is no appreciable current unless bromine is present; at very low concentrations of bromine the current is found to be linearly related to the concentration. In Figure 29-3 are typical plots of the current versus the concentration of Br_2. The reason that no appreciable current flows if Br_2 is absent is that the applied potential is too small to cause oxidation of bromide ion at the anode and reduction of water at the cathode. However, when there is also a small concentration of bromine, current flows, because now *equal and opposite* reactions can occur at the two electrodes:

$$\text{Anode} \qquad 2Br^- = Br_2 + 2e^-$$

$$\text{Cathode} \qquad Br_2 + 2e^- = 2Br^-$$

With the activities of species the same at one electrode as at the other, the potentials are also equal and opposite. Therefore a potential that is sufficiently large to overcome the overvoltages and the resistance of the system will cause the two half-cell reactions to occur. The magnitude of the current is controlled by the rate at which Br_2 diffuses to the cathode; the high Br^- concentration makes this species always available at the anode to be oxidized.

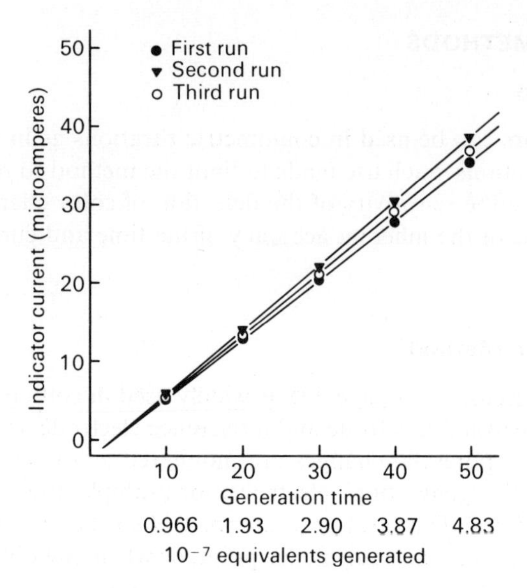

FIGURE 29-3

Indicator current versus time of generation of bromine. Reproducibility of end-point correction measurements are shown. Solutions prepared from 5 ml of 1 *F* sulfuric acid and 5 ml of 2 *F* sodium bromide diluted to 50 ml; titrated at 0.932 milliampere. (For details, see Footnote 8.)

At the low applied potential no Br^- is oxidized except when Br_2 is at the cathode to be reduced simultaneously. Figure 29-3 shows that the indicator current is a linear function of the bromine concentration. Because of this fact, titrations do not have to be stopped at an exact end point; generation can be continued until a current increase is observed, and a correction can be applied from a previously prepared graph or, alternatively, a graph can be prepared by generation of successive increments of bromine.

Other End-Point Methods

Conductimetric and spectrophotometric end-point methods have been used, though much less commonly than potentiometric and amperometric methods. The conductivity of a solution changes during the course of a titration; hence its measurement can be used as an end-point method. A spectrophotometric method is essentially an extension of a visual end-point method by means of the increased sensitivity of a spectrophotometer to changes in wavelength and intensity.

Titration of Arsenic with Electrolytically Generated Bromine

Outline. (*a*) Bromine is generated electrolytically in a sulfuric acid solution at a platinum electrode by a known constant current. (*b*) Arsenic(III) is oxidized to arsenic(V) by the bromine. (*c*) Excess bromine oxidizes methyl orange irreversibly to a colorless form; this bleaching of the solution serves as the end point. (*d*) The quantity of arsenic(III) initially present is calculated from the current, the time, and Faraday's laws.

DISCUSSION

(*a*) Electrolytic Generation of Bromine

Bromine has been suggested as the titrant in redox reactions but has found relatively little use in titrimetric methods because of the volatility of bromine from its aqueous solutions. Unlike tri-iodide, the analogous tribromide ion is not sufficiently stable to make the use of such solutions practical. Bromine is one of the stronger oxidants, the standard potential of the half-cell reaction

$$2Br^- = Br_2(aq) + 2e^-$$

being -1.087 v; in addition the reactions of Br_2 as an oxidant are rapid; therefore, if solutions of bromine were stable they would be of value as a standard titrant.

The problems associated with the loss of bromine by volatilization from standard solutions can be avoided if the bromine is produced in the titration vessel itself. This has been accomplished chemically by means of the reaction

$$BrO_3^- + 5Br^- + 6H^+ = 3Br_2 + 3H_2O$$

A standard solution of bromate that also contains bromide is used to produce an equivalent quantity of bromine when titrated into an acidic solution and this bromine reacts with the substance to be determined. However, this indirect process is less satisfactory than are direct titrations; the end points are less distinct, the oxidations are reported to proceed less smoothly, and the

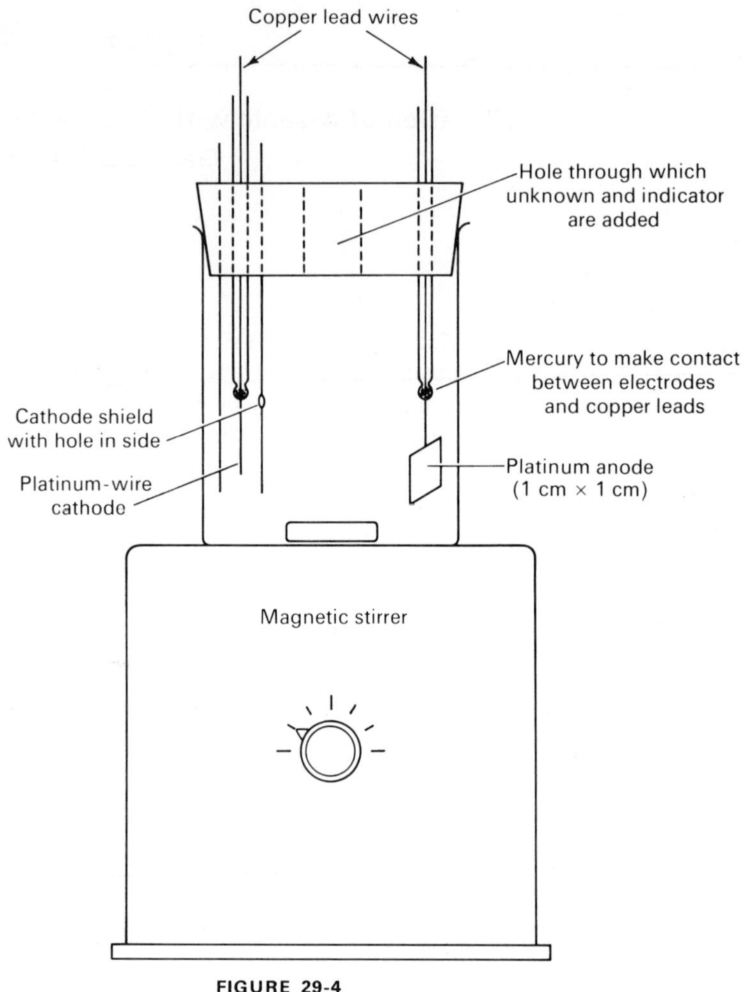

FIGURE 29-4
Coulometric titration cell.

reaction to produce Br_2 occurs only in acidic solutions. An alternative approach is to produce the bromine electrolytically by means of the reaction

$$2Br^- = Br_2(aq) + 2e^-$$

If this is done in the presence of the reducing agent that is to be determined, the bromine concentration does not increase above very low values until substantially all the reductant has been oxidized. Therefore loss of bromine to the gas phase is not significant.

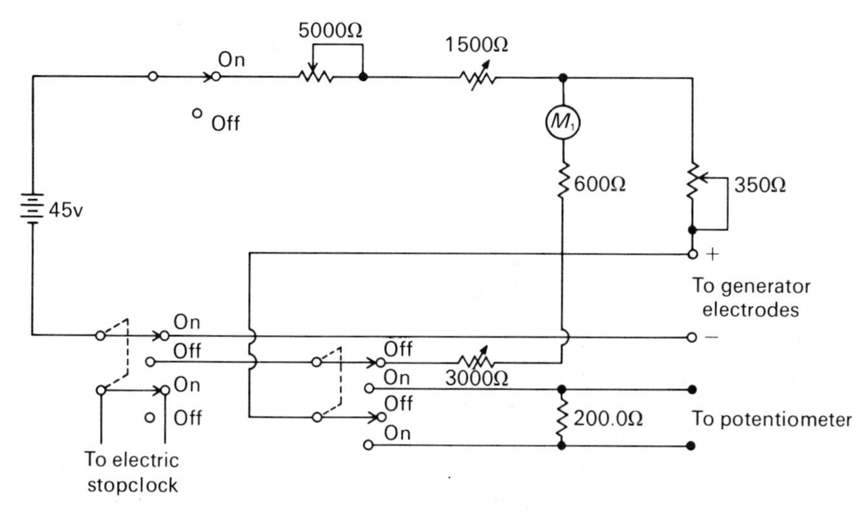

FIGURE 29-5
Simple constant-current source.

Another advantage of the electrolytic generation of the oxidant is the high accuracy to which the quantity produced can be measured, because both time and the constant current can be measured very accurately. This accuracy can be extended to the titration of very small quantities because the current can be maintained at such low values that the titration time can be very accurately measured.

The titration cell consists of a 200-ml beaker without spout fitted with a large rubber stopper (Figure 29-4) with holes suitable for positioning the electrodes. The anode is a platinum foil approximately 1 cm × 2 cm and the cathode is a platinum wire. The latter is in a shield consisting of 10- or 20-mm glass tubing open at the top and bottom; this prevents the hydrogen that is produced at the cathode from being circulated to the anode where some oxidation to H^+ can occur. A hole in the side of the shielding tube is an aid to circulation of the solution.

Commercial constant-current power supplies are available from many suppliers. In addition, constant-current devices of various degrees of elegance are readily constructed. In the original paper describing the coulometric titration of arsenic with bromine a 45-v battery in conjunction with a large resistor was used to give a constant current (Footnote 8). Figure 29-5 shows the constant-current source used for a 10-ma current. The potential drop across a 200.0 ohm precision resistor was measured with a potentiometer to determine the current. The switches are ganged as indicated; the electric stopclock is started and stopped with the current. Simple, easily constructed solid-state power sources for student use are described by Vincent and Ward and by Stock.[10]

The nature of the method is such that three determinations of a sample can be made conveniently in 15 minutes; therefore, provided all preliminary work has been done, a single instrument can accommodate as many as twelve students in a 3-hour laboratory period.

(*b*) Oxidation of Arsenic by Bromine

The conditions under which iodine can be used to oxidize arsenic(III) to arsenic(V) have been discussed in detail on pages 450–456. There it was found that iodine will not quantitatively oxidize arsenious acid if the $[H^+]$ is much above 10^{-5}. Because of the much greater oxidizing tendency of the bromine-bromide half-cell the oxidation takes place rapidly and quantitatively in acid solutions; the titration in the procedure below is made in 0.1 F H_2SO_4.

(*c*) Methyl Orange as a Redox Indicator

Reversible redox indicators were used in the standardization reactions for permanganate and ceric solutions and for the determination of iron with ceric sulfate. Methyl orange is a reversible acid-base indicator (page 543) but is an irreversible potential indicator, that is, it is oxidized to a colorless form but the resulting species is not reduced to the colored form if a reducing agent is added. Thus, a local excess of the oxidant will cause local oxidation of the methyl orange with consequent decrease in the color intensity of the solution. However, in the coulometric method discussed below the extent of oxidation of methyl orange prior to the end point is so small that no color intensity decrease is ordinarily observed until the arsenic is substantially all oxidized.

Instrumental methods of end-point detection are more commonly used in coulometric titrations both because they can be used with greater precision than visual methods and because they lend themselves readily to automatic control. The required instrumentation is simple; an excellent student project is to prepare and use the amperometric indicator system outlined in Figure 29-1 and described in Footnote 8 or the potentiometric end-point system used by Lee and Adams.[11]

---

[10] C. A. Vincent and J. G. Ward, *Anal. Chem.*, **46**, 613 (1969); J. T. Stock, *Anal. Chem.*, **46**, 858 (1969).

[11] J. K. Lee and R. N. Adams, *Anal. Chem.*, **30**, 240 (1958).

INSTRUCTIONS (Procedure 29-1: The Titration of Arsenic with Electrolytically Generated Bromine)

Preparation of the Apparatus

(Consult your instructor.) Check the cell and electrodes (Figure 29-4). Be sure that the stirring bar rotates freely without striking the electrodes. Check that provision has been made for the convenient addition of indicator and unknown solution. By means of a precision resistor and a potentiometer, determine the value of the constant current put out by the instrument being used (Note 1).

Preparation of the Solution

Add to the titration vessel 10 ml of 1 *F* H_2SO_4 and 10 ml of 2 *F* NaBr. Then pipet in (*use a bulb!*) 5.00 ml of the unknown As(III) solution. (This solution should be approximately 0.002 *F* in H_3AsO_3.) Dilute the solution to 100 ml, add 2 drops of methyl orange solution, and adjust the magnetic stirrer so that good steady stirring is accomplished without spattering (Note 2). Prepare a comparison solution consisting of 10 ml 1 *F* H_2SO_4, 90 ml H_2O, and 2 drops of methyl orange solution in a 200-ml beaker.

Titration of the Sample

Close the ganged switch, which simultaneously sends current through the cell and starts the timer. Watch carefully for the first fading of the titrated solution; thereupon, immediately open the switch and wait for 5 seconds (Note 3). Then close the switch for 1 second, open it, wait for 5 seconds, and

NOTES

1. A current of approximately 20 ma is convenient for a 20 μeq sample. An alternative approach is to titrate a known arsenic(III) solution with the instrument and from this titration calculate the current output. If this procedure is to be followed, proceed according to the section "Titration of the Sample," substituting standard 0.002 *F* As(III) solution for the unknown. It is more convenient in such a case to calculate the output of the instrument in micro equivalents per second (μeq/sec) than in milliamperes (ma).
2. A sheet of white paper between the stirrer and the beaker helps in detecting changes in the color intensity.
3. This is to ensure that thorough mixing of the solution has occurred.

check the color of the solution; continue these 1-second periods of bromine generation until the solution is just colorless (Note 4).

Record the time.

Pipet a new 5.00-ml portion of the unknown into the titration vessel (it is not necessary to replace the solution). Add 2 drops of methyl orange, and titrate as before.

End-point Correction

Add 2 drops of methyl orange indicator to the titrated solution. Generate bromine in 1-second periods with 5-second waiting periods between until the solution is colorless. Subtract the generation time required for oxidizing the indicator from the time required for the sample. (Note that 4 drops of the indicator solution were used in the case that a sample was overrun.)

From the measured current and the time required for the titrations calculate the formal concentration of As(III) in the unknown.

NOTES

4. If the initial bromine generation is stopped only when fading of the solution has begun, two or three of the 1-second generation periods should be sufficient. If the titration is overrun, add 1.00 ml (pipet) of the unknown, 2 drops of indicator, and titrate to the end point.

QUESTIONS AND PROBLEMS

1. Explain why bromine can be used much more conveniently in coulometric titrations than in conventional titrimetric methods.

2. Discuss the practicality of standard solutions of chlorine. Chlorine is used as a coulometric titrant. What advantages would you expect it to have over bromine? Disadvantages?

3. The polarity of a direct current power supply is easily determined by touching the leads from the supply to separate points on a filter paper moistened with a potassium iodide solution containing starch. A blue color develops at one of the points of contact.
   (*a*) Write equation(s) for the reaction at this electrode.
   (*b*) Is this the anode or the cathode?
   (*c*) Predict the reaction at the other electrode.
   (*d*) A certain direct current power supply has a maximum voltage output of 0.35 v. Can its polarity be determined by this technique? Explain.

4. A hydrogen–oxygen coulometer is used in series with a constant potential coulometric cell. The gases are collected and measured together; the measured volume was 325.0 ml at 20°C and 700 mm pressure.

   (*a*) How many coulombs of electricity passed through the cells?
   (*b*) If the reaction in the coulometric cell was the reduction of cerium(IV) to cerium(III), how many mg of cerium(IV) were initially present? *Ans.* (*a*) 1.61×10^3 coulombs, (*b*) 2320 mg.

CHAPTER 30

Optical Methods

In all the preceding chapters we have considered methods that have depended upon chemical or electrochemical reactions. The quantity of a reactant was measured or the product was dried and weighed, and from these measurements the quantity of the species being determined could be calculated. In this chapter we consider a method that depends upon the optical properties of the substance of interest. No chemical reaction is involved in the final optical measurement, although frequently a preliminary chemical process is required in order to convert the constituent of interest into a species having suitable optical properties.

ELECTROMAGNETIC RADIATION

The velocity of light in a vacuum, c, is constant and has the value 3×10^{10} cm/sec. The wavelength of the light is represented by the symbol λ and the frequency by ν. The relationship between them is

$$c = \lambda\nu \tag{1}$$

Therefore, yellow light of wavelength 6×10^3 Ångströms (see Table 30-1 for units of length) has a frequency of 5×10^{14} sec^{-1}. The energy of a photon

TABLE 30-1
Units of Length used in Spectrophotometry

Ångström	Å	10^{-8} cm, 10^{-1} mμ
Millimicron	mμ	10 Å, 10^{-6} mm, 10^{-3} μ
Micron	μ	10^{-6} m, 10^{-3} mm, 10^{4} Å

of frequency ν is given by

$$E = h\nu \tag{2}$$

where h, Planck's constant, has the value 6.63×10^{-27} erg-sec. The energy, therefore, increases directly with the frequency and inversely with the wavelength. This latter is clearly seen if Equations 1 and 2 are combined:

$$E = \frac{hc}{\lambda} \tag{3}$$

Thus, short wavelength radiation has high energy and radiation of longer wavelength has lower energy.

Visible light represents only a very small fraction of the electromagnetic spectrum (Figure 30-1). It is beyond the scope of this book to consider in detail the many kinds of transitions that occur when an atom, ion, or molecule is exposed to radiation of different energies. It should be apparent, however, that an x-ray, having high energy, will have a very different effect than will ultraviolet radiation; and that the latter having in turn more energy than infrared radiation will affect the exposed species differently.

The interactions of matter and electromagnetic radiation fall into two broad classes: (1) those in which radiation is absorbed and (2) those in which radiation is emitted.

Absorption methods. In these methods the sample is exposed to radiation, and transmitted and incident radiation are compared, and from the changes that are observed conclusions are drawn concerning the composition of the sample.

Emission methods. In these methods the sample is excited by a flame, high-voltage arc, or high-energy radiation; then the radiation that is emitted by the sample is measured. A commonly observed example of emission is the brilliant yellow color that results when NaCl is introduced into the flame of a burner; this is a characteristic emission of sodium and is used in flame-emission techniques for the detection and estimation of that element.

We are concerned in this chapter with absorption phenomena, those that result from the absorption of radiation. A permanganate solution is purple

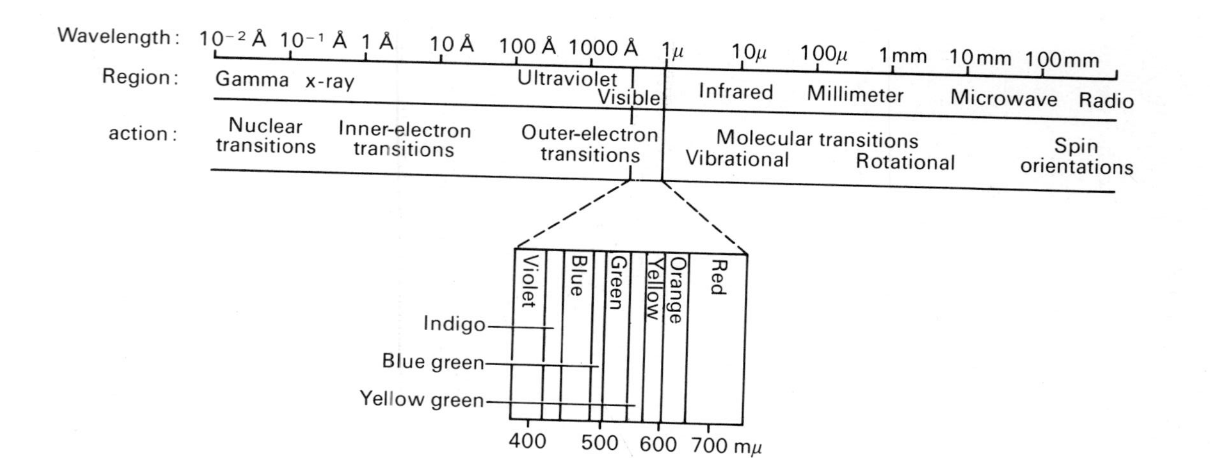

FIGURE 30-1

A portion of the electromagnetic spectrum.

because it absorbs green light and transmits the purple. Ultraviolet and visible light in general causes transitions of valence electrons.[1]

Infrared radiation has lower energy, which in general is not sufficient to cause electronic transitions in atoms, but does cause transitions in the vibrational and rotational energy of molecules. Certain transitions are characteristic of specific groups in organic molecules, and studies by infrared techniques are often effective in identifying these groups in an unknown sample.

Other parts of the spectrum have also proved useful in studying elements and compounds. Figure 30-1 shows some of the general kinds of information derived from studies in various ranges of the spectrum.

MEANS OF ISOLATING PORTIONS OF THE SPECTRUM

A permanganate solution is purple when viewed in white light (light that contains all portions of the visible spectrum) because the absorption of green light is much greater than that of other colors. Therefore, if the intensity of green light is compared before and after passing through a permanganate solution the intensity decrease will be found to be much greater than if white light had been used.

Filters. A simple means of restricting the light to some desired region of the spectrum is to use filters to remove unwanted wavelengths. This technique is used in colorimeters; the result is a wide band of wavelengths in which some regions are transmitted much more strongly than others. Figure 30-2 shows the transmissions of some commercially available filters. A narrower band can often be achieved by a combination of two or more filters with overlapping transmission bands.

Monochromators. A much narrower segment of the spectrum can be selected by means of a monochromator consisting of a prism or grating in combination with a narrow slit. White light is dispersed by the prism or grating and a small segment of the spectrum is allowed to pass through the slit (Figure 30-3). The light that is passed is called *monochromatic*, but this must not be taken to mean that light of only one wavelength is present, for

---

[1] As evidence of this, consider manganese in the +2, +6, and +7 oxidation states. The difference in the manganese in these three states is a difference in the outermost electrons, the 4*s* and 3*d* electrons. If the transitions that result from absorption of visible light involved only electrons of the first and second energy level, these transitions should be unaffected by the oxidation number of the manganese. However, the +2 ion is pale pink, the +6 (manganate) is deep green, and the +7 (permanganate) is purple; this suggests that transitions of valence electrons result from the absorption of visible light.

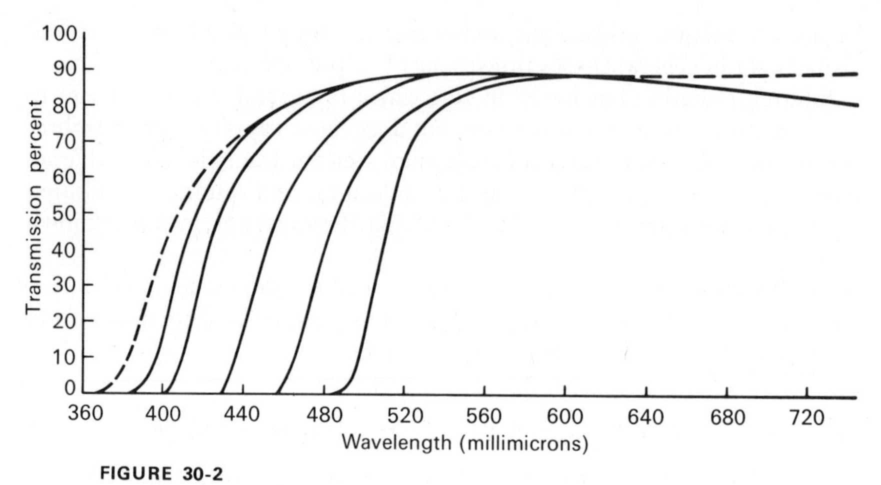

FIGURE 30-2

Transmittances of some commercial filters. (Data from Corning Glass Works.)

the white light is a continuum of wavelengths. Therefore, any finite slit allows a range of wavelengths to pass.

LAWS OF LIGHT ABSORPTION

Lambert's Law

In one form this law states the relationship between the thickness of an absorbing medium (that is, the length of the light path through it) and the ratio of the intensity of the light that enters the medium to that which is transmitted through it. It is a common observation that a colored solution

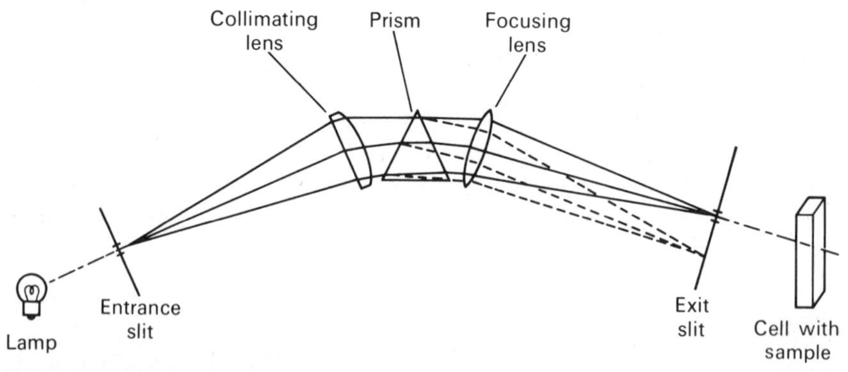

FIGURE 30-3

Monochromator. (After Pecsok, R. L., and Shields, L. D. 1968. *Modern Methods of Chemical Analysis*, New York: Wiley, p. 166.)

appears to be more intensely colored when a long segment is looked through than when the view is through a short segment. This observation is made quantitative by Lambert's Law: *When monochromatic light is passed through a very thin layer of an absorbing medium the decrease in intensity is proportional to both the intensity of the light and the thickness of the medium.*

If $-\Delta I$ is the decrease in intensity (the incident intensity minus the transmitted intensity) that results when radiation passes through a thin layer of an absorbing medium of length Δx, the relationship between these is

$$-\Delta I = kI\Delta x \tag{4}$$

where I is the total intensity falling upon the layer. If Equation 4 is written in the more precise form of the calculus it becomes:[2]

$$-\frac{dI}{dx} = kI \tag{5}$$

This states that the rate of decrease in intensity with distance into the medium is proportional to the intensity. The intensity is lower the farther the light proceeds into the medium, so the differential form of the law (Equation 5) is required to express the relationship correctly.

Equation 5 can be solved to express the Lambert Law in other useful forms. If it is rearranged it becomes

$$-\frac{dI}{I} = k\,dx \tag{6}$$

When $x = 0$ (Figure 30-4) the intensity is I_0, the incident intensity; when $x = b$ the intensity has decreased to I, the transmitted intensity. The definite integral form of Equation 6 is

$$-\int_{I_0}^{I} \frac{dI}{I} = k \int_0^b dx$$

and this when integrated is

$$\ln \frac{I_0}{I} = kb \tag{7}$$

In this form Lambert's law shows the logarithmic relationship that exists between the ratio of intensities of the incident and the transmitted light and

---

[2] The differential form makes the "very thin layer" become a layer of infinitesimal thickness. This is the reason for the greater precision of Equation 5.

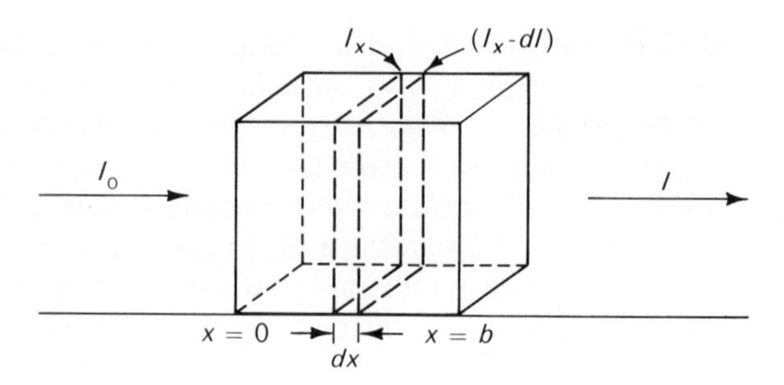

FIGURE 30-4
Absorption by a cell of length b.

the thickness, b, of the medium. Equation 7 can be restated in terms of the common logarithm and solved for I:

$$\log \frac{I_0}{I} = k_1 b$$

therefore,

$$I = I_0 10^{-k_1 b} \tag{8}$$

To obtain a feeling for this relationship, assume that the constant, k_1, is unity and observe how I changes as the length is doubled and tripled:

$$\begin{array}{llll} \text{If} & b = 1 & \text{then} & I = 0.1\, I_0 \\ & b = 2 & & I = 0.01\, I_0 \\ & b = 3 & & I = 0.001\, I_0 \end{array}$$

If $b = 0$ then $I = I_0$, of course, for at $b = 0$ the incident radiation is observed.

Lambert's law is observed by all absorbing materials; there are no known exceptions.

Beer's Law

This law gives a relationship between the concentration of the absorbing species and the ratio of the intensity of the light that enters the medium to that which is transmitted through it. Here, too, there is the common observation that a colored species in solution appears more intensely colored as its concentration increases. Beer's law makes a quantitative statement of this observation: *When monochromatic light is passed through an absorbing medium the decrease in intensity is proportional to both the intensity of the light and the concentration of the absorbing species, if only small changes in concentration are considered.* If $-\Delta I$ is the difference in intensity, I, of the radiation that is passed through two solutions whose concentrations differ by Δc, the relationship between these is

$$-\Delta I = k_2 I \Delta c \tag{9}$$

In the more precise differential form this is

$$-\frac{dI}{dc} = k_2 I \tag{10}$$

which states that the rate of decrease of intensity with concentration is proportional to the intensity. Equation 10 can be solved if it is rearranged and integrated between limits. When the concentration is zero, the intensity is I_0, and when the concentration is c, the intensity is I:

$$-\int_{I_0}^{I} \frac{dI}{I} = k_2 \int_0^c dc$$

Therefore,

$$\ln \frac{I_0}{I} = k_2 c \tag{11}$$

Equation 11 is rearranged to give

$$I = I_0 10^{-k_3 c} \tag{12}$$

Equations 9, 10, 11, and 12 are completely parallel to Equations 4, 5, 7, and 8.

Of greater usefulness than either Equation 8 or 12 is a combined equation that accounts for both path length and solution concentration. The term k, of Equation 8 was constant only because the concentration was kept constant. If the concentration is varied, then $k_2 = \varepsilon c$, where c is the concentration and ε is called the molar absorptivity; to the extent that Beer's law is obeyed by a particular substance ε is a constant for that substance. The molar absorptivity has dimensions liter/mole cm, c is expressed in moles/liter, and b is in centimeters. The equation that expresses the Beer–Lambert law, then, is

$$I = I_0 10^{-\varepsilon c b} \tag{13}$$

The ratio I/I_0 is called the *transmittance* (T) and is a frequently used relationship in spectrophotometry. Equation (13) can be rearranged to give

$$T = \frac{I}{I_0} = 10^{-\varepsilon c b}$$

TABLE 30-2
Terms and Symbols Used in Spectrophotometry

Symbol	Name	Definition
c	concentration	
b	path length	
T	transmittance	I/I_0
A	absorbance	$-\log T$
a	absorptivity	A/b
ε	molar absorptivity	a/c or A/bc

or

$$\log T = -\varepsilon cb$$

Another useful term is the absorbance (A) defined by the equation

$$A = -\log T \tag{14}$$

Hence

$$A = \varepsilon cb$$

The *absorptivity*, a, is

$$a = \frac{A}{b} \tag{15}$$

and is seen to be the absorbance per cm of path length.

Table 30-2 gives definitions of terms used in spectrophotometry. It can be seen from the tabulation that if concentration, path length, and the intensities of incident and transmitted radiation are known, all the other terms can be calculated.

Deviations from Beer's law. Unlike Lambert's law, Beer's law is subject to large deviations. There are often *apparent* deviations that are caused by equilibrium effects when the concentration is changed. For example, dilution of an indicator solution may change the pH sufficiently to cause a significant shift in the $[HIn]/[In^-]$ ratio. In such a case there would be an apparent deviation from Beer's law even if the absorbing species obeyed that law precisely.

Another common cause for apparent deviations from Beer's law is the fact mentioned earlier that the light that passes through the medium is not truly monochromatic but consists of a band of wavelengths. Therefore, the instrument indicates an average absorbance in the band. Figure 30-5 shows the effect of a concentration change upon an absorbance measurement that is made on the side of an absorbance peak. This figure shows two reasons that

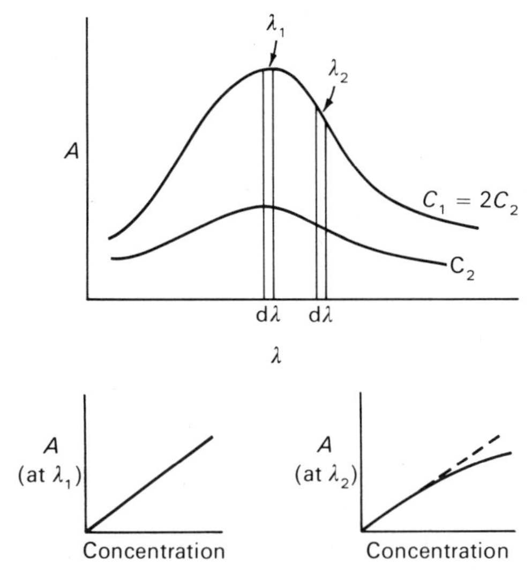

FIGURE 30-5

Effect of bandwidth upon measurements made at absorbance peak (λ_1) and at side of peak (λ_2).

chemists attempt to make measurements at an absorbance peak. First, the changes in absorbance with wavelength are lower than on the slopes; hence the measurement is not so critical upon the wavelength or band width. Second, the increase in absorbance with concentration is greatest at the maximum in the curve, so greater sensitivity is obtained there.

True deviations from Beer's law are less frequently encountered, but do occur in high concentrations where the index of refraction of the absorbed radiation may be changed. For this reason, methods that depend upon Beer's law are ordinarily restricted to concentrations below about 10^{-2} *M*.

The fact that a system shows serious deviation from Beer's law does not rule out the use of spectrophotometry for quantitative determination of the substance. A calibration curve is constructed by the measurement of the absorbance of the substance as a function of its concentration (Figure 30-6). The dashed lines in the figure show how the absorbance of the unknown solution is then used to determine its concentration.

INSTRUMENTS

Instruments for determining the concentration of a substance by making use of the light absorbing properties of the substance range from very simple colorimeters in which the human eye serves as the detector to highly refined instruments.

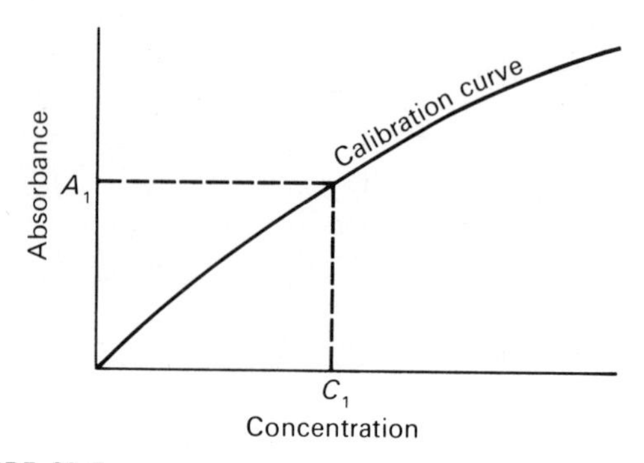

FIGURE 30-6

Use of calibration curves to determine concentration of unknown.

Colorimeters

The most simple method, in principle, to determine the concentration of a colored substance in solution is to prepare a series of solutions having known concentrations and to compare the unknown solution with these to find a match. If all the cells in which the solutions are compared are of the same dimensions, then the unknown has the same concentration as the matching

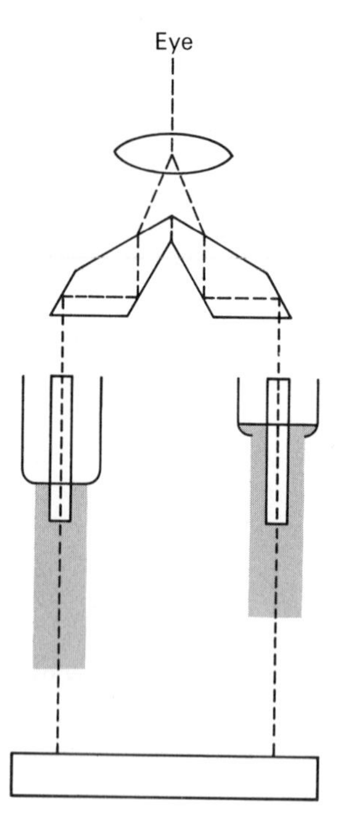

FIGURE 30-7

Duboscq colorimeter, an early form of visual comparitor.

known solution. The method is time-consuming but is not dependent upon Beer's law; it is thus unaffected by deviations from that law.

Visual comparitors (Figure 30-7) permit one to match the colors of a known and an unknown solution by changing the path lengths of the light. It follows from Equation 13 and the fact that I_0, I and ε are the same for both solutions that the products $c_1 b_1$ and $c_2 b_2$ must be equal.

More refined instruments use a photocell to measure the intensity of light and filters to limit the light passing through the sample to the region of wavelengths of maximum absorption.

Spectrophotometers

There is no sharp division between colorimeters and spectrophotometers, but the former term is usually restricted to instruments that are effective only in the visible spectrum and have rather simple instrumentation. The term *spectrophotometer* applies to devices that use a narrow band of wavelengths, obtained by a prism or grating monochromator, and that use a photocell or similar device for the detector. The term is applied both to instruments that function in the visible portion of the electromagnetic spectrum and to those that use other portions of the spectrum.

Figure 30-8 shows a block diagram and an optical diagram of a simple spectrophotometer. In the use of such an instrument the intensity of the radiation that passes through the sample is measured and compared with

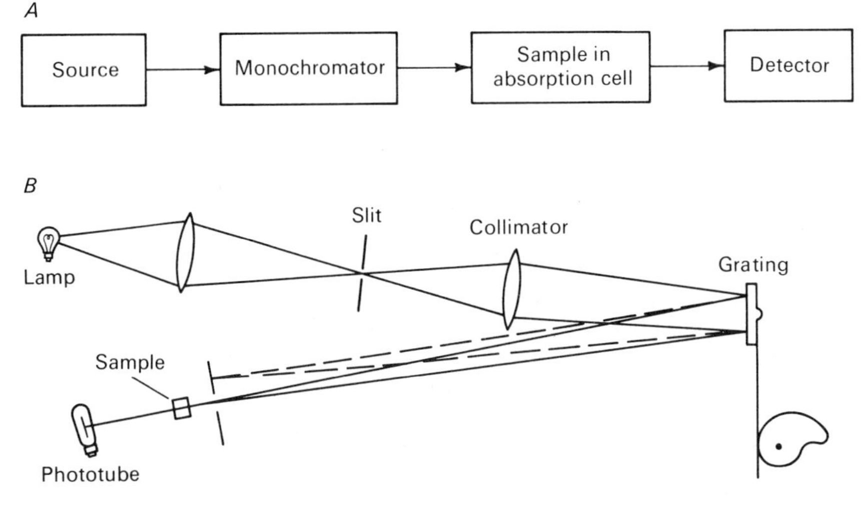

FIGURE 30-8

(*A*) Block diagram and (*B*) optical diagram of a simple spectrophotometer. (After Waser, J. 1964. *Quantitative Chemistry*, New York: Benjamin, p. 239.)

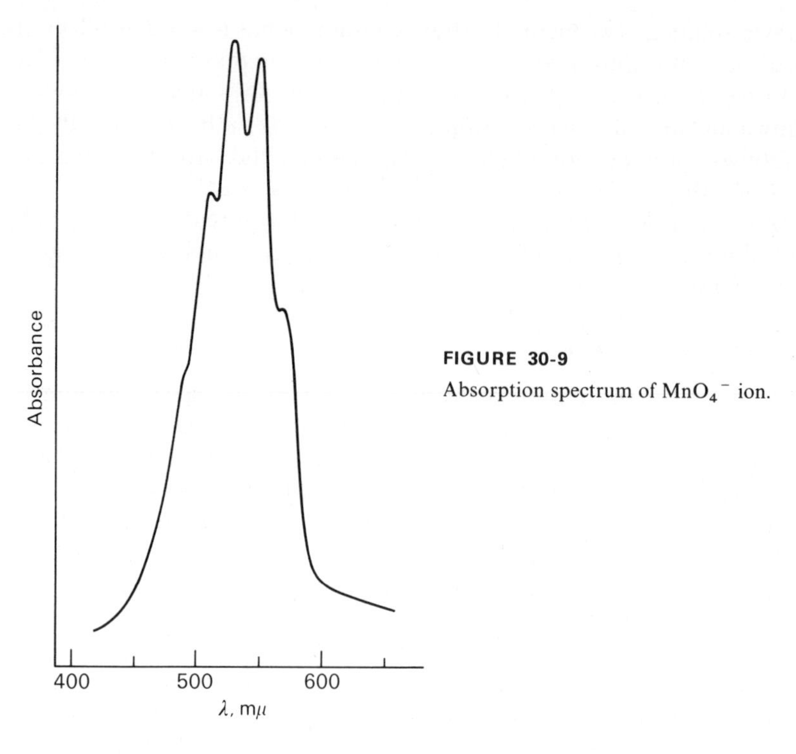

FIGURE 30-9
Absorption spectrum of MnO_4^- ion.

the intensity of the radiation that passes through a *blank*, the sample holder filled with solvent. Double-beam instruments may have either two matched receptors or a chopping system that causes pulses of radiation to pass alternately through the sample and blank. The double-beam instrument has the advantages of giving internal compensation for fluctuations in the intensity of the source and of permitting direct recording of the transmittance of the sample. Such instruments are often equipped with a wavelength sweep mechanism and a recorder so that transmittance is plotted as a function of wavelength. The visible absorption spectrum of permanganate is shown in Figure 30-9. To obtain such a spectrum with a single-beam instrument would require a large number of individual measurements of the transmittance at some small wavelength interval across the visible region. Unless the interval was chosen to be very small, much of the fine detail would be missed.

The Inherent Error in Photometric Measurements

A spectrophotometer measures transmittance, but the concentration of a solution is proportional to the absorbance. Equation 14 shows that transmittance and absorbance are related logarithmically; this relationship results in an error effect that must be considered in all determinations made with a spectrophotometer.

A 0.5% error in the transmittance does not result in a 0.5% error in the absorbance, and indeed causes a relative error of different magnitudes depending upon the absolute magnitude of the transmittance. This is demonstrated below.

Commercial instruments are usually calibrated to read in percentage of transmittance and have absolute precisions of from about ±0.2% to ±1%. Assume that an instrument is used that has an absolute precision of ±0.5% transmittance and that readings are taken at 2%, 50%, and 98% transmittance. To simplify the consideration we will assume that in each case the true percentage of transmittance was 0.5% higher than the value read from the instrument. Thus the true values were 2.5%, 50.5%, and 98.5%. For each value of percentage of transmittance, true and measured, there is a corresponding value for the absorbance, and for each reading the error in the absorbance can be calculated:

Transmittance (%)	Absorbance ($-\log T$)	Relative Error (%)
2.0	1.70	6%
2.5	1.60	
50.0	0.301	2%
50.5	0.297	
98.0	0.009	30%
98.5	0.006	

These figures show that the relative error in absorbance, and hence in concentration, is highly dependent upon the percentage of transmittance.

A much more precise evaluation of the way the transmittance affects the error in absorbance or concentration is obtained by a mathematical approach.

$$A = -\log T = -k \ln T \tag{16}$$

This is differentiated to give

$$dA = -k\frac{1}{T}\,dT \tag{17}$$

If dT is the error in T and dA is the error in A, then $100(dA/A)$ is the percentage

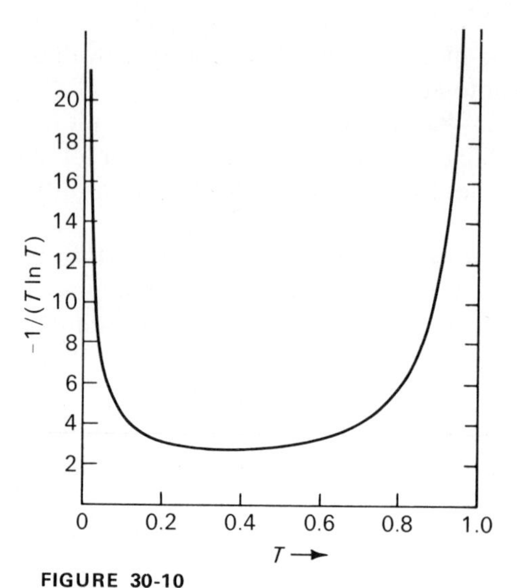

FIGURE 30-10

Plot of $-1/T \ln T$ versus T. (From Waser, J. 1964. *Quantitative Chemistry*, New York: Benjamin.)

of relative error in A. If Equation 17 is divided by Equation 16 the result is

$$\frac{dA}{A} = \frac{dT}{T \ln T}$$

and if this is multiplied by 100,

$$\text{percentage error in } A = \frac{dA}{A}100 = \frac{dT}{T \ln T}100 \qquad (18)$$

A plot of $1/(T \ln T)$ versus T, therefore, will show how the relative error in A depends upon T, if a fixed uncertainty, dT, is inherent in the instrument. Figure 30-10 is such a plot and shows that in the range of transmittance from 0.2 to 0.7 the error in A is small, but that outside this range there is a rapid increase in error.

APPLICATIONS OF OPTICAL METHODS

Optical methods are used extensively in the analysis of organic and biological samples where very complex materials are encountered frequently in trace quantities. These materials can often be detected by their characteristic spectra and can be estimated to the order of plus or minus a few percent. Optical methods similarly find wide application in the detection and estima-

TABLE 30-3
Species Used for Spectrophotometric Determinations of Some Representative Metals

Element	Form in which Analyzed	Color
Ba	indirectly as the chromate	yellow-orange
Cr	$Cr_2O_7^{2-}$	yellow-orange
Co	complex with SCN^- in alcohol-water	blue
Fe	complex of Fe(III) with SCN^-	red
	or complex of Fe(II) with o-phenanthroline	orange-red
Mg	Complex with Eriochrome Black T	red
Ni	complex with dimethyl glyoxime	red
Ti	peroxy complex	yellow
V	peroxy complex	red-brown

tion of traces of metals. For example, the usual titrimetric or gravimetric method is useless in the analysis of a 10-ml sample of water that is 2×10^{-6} *F* in Fe(III), but a colorimetric method is applicable.

Quantitative determinations made by means of absorption methods have typical relative accuracies within 1% or 2%. By special techniques such as differential spectrophotometry (see the general references at the end of this section), determinations of concentrations have in some cases been made to accuracies within 0.1%. The great usefulness of optical methods, however, lies in the rapid analysis of trace substances to much lower relative accuracy. Often there is less need for prior chemical separation and other preparation of the sample than is the case for a conventional analysis. It is the extreme sensitivity of the methods that contributes greatly to their usefulness.

In Table 30-3 are listed only a few representative metals and a method for each by which its colorimetric determination can be made. An indication of the breadth of application of optical methods is given by the fact that Sandell[3] discusses colorimetric determinations for 44 metals plus the rare earth metals as a group; usually several methods for each metal are described. In addition, Boltz[4] discusses colorimetric methods for 11 nonmetals and gives determinations for various oxidation states of nitrogen, chlorine, and sulfur.

Instructions for the determination of manganese in steel are given in Procedure 30-1. This method has been selected as an example for various reasons. (1) The method of dissolving the sample is illustrative of the general principles applicable to dissolving "insoluble" materials. (2) The reactions

[3] E. B. Sandell. *Colorimetric Determination of Traces of Metals.* (3rd ed.) New York: Wiley-Interscience, 1959.
[4] D. F. Boltz. *Colorimetric Determination of Nonmetals.* New York: Wiley-Interscience, 1958.

involved in the dissolving process have certain interesting features. (3) The selection of the oxidant for quantitatively producing permanganate provides a review of the effectiveness of certain powerful oxidants. (4) The spectrophotometric measurements illustrate the precautions required to minimize errors from the presence of other constituents. (5) Finally, and from a practical point of view, the method has been thoroughly tested and extensively used and, furthermore, analyzed samples are readily available.

Many other colorimetric determinations make excellent special problems for qualified students wishing to undertake them. The general reference books given below provide initial source material that can be supplemented by the original literature.

GENERAL REFERENCES

Bauman, R. P. *Absorption Spectroscopy.* New York: Wiley, 1962.

Boltz, D. F. *Colorimetric Determination of Nonmetals.* New York: Wiley-Interscience, 1958.

Ewing, G. W. *Instrumental Methods of Chemical Analysis.* (2nd ed.) New York: McGraw-Hill, 1960.

Pecsok, R. L., and Shields, L. D. *Modern Methods of Chemical Analysis.* New York: Wiley, 1968.

Sandell, E. B. *Colorimetric Determination of Traces of Metals.* (3rd ed.) New York: Wiley-Interscience, 1959.

PROCEDURE **30-1**

Spectrophotometric Determination of Manganese in Steel

Outline. (*a*) A steel sample is dissolved in nitric acid to which ammonium peroxydisulfate is added. (*b*) The solution is diluted and the Mn(II) in an aliquot is oxidized to MnO_4^- with periodate. (*c*) An absorbance curve is made from measurements on a standard permanganate solution. (*d*) The absorbance of the permanganate in the aliquot of the steel solution is measured and compared with the absorbance curve.

DISCUSSION

(*a*) Dissolving the Sample

Most analytical methods are carried out in aqueous solutions; therefore a fundamental problem in analyzing insoluble samples is the preparation of a solution of the material. Dissolving such a material requires that some agent be used that will convert the material into water soluble components. Accordingly, one considers the chemical nature of the material and selects agents that tend to react with compounds of that type. If the material is basic in nature, for example a basic oxide or hydroxide, an acidic agent such as a strong acid would be used; if it is an acidic oxide, an alkaline solution or fusion with a low-melting basic compound (called a flux), such as sodium hydroxide or sodium carbonate, could be employed. If the substance is a reductant, for example a metal or alloy, an oxidizing agent would be added to the solvent solution. Strong acids dissolve reactive metals because of the oxidizing action of hydrogen ions—with more resistant metals, more powerful oxidants are added to the acid solution.

Nitric and hydrochloric acids, alone or in combination, or with added agents are the most frequently used acidic solvents. When treating insoluble salts of volatile acids, "fuming" with an acid having a higher boiling point can be effective. Table 30-4 shows the various types of solvent action obtained by certain commonly used solvents.

In this procedure the sample is first treated with nitric acid and the first effect obtained is hydrogen-ion oxidation of the iron to Fe(II); this is followed by nitrate oxidation to Fe(III). In addition to iron, steels contain from 0.1%

TABLE 30-4

Types of Solvent Action Involved in Preparing a Solution of the Sample

Solvents Used	Type of Solvent Action	Type of Substances Dissolved (Examples)
(*a*) H_2O	Solution (ionization, solvation)	Water-soluble compounds
(*b*) HNO_3 (dilute)	I. Hydrogen ion effects	I.
	1. Neutralization	1. Hydroxides, basic oxides, basic salts
	2. Displacement	2. Salts of weak acids
	3. Oxidation ($2H^+ + 2e^- = H_2$)	3. Certain metals (Zn, Al)
	II. Oxidation ($NO_3^- + 4H^+ + 3e^- = NO + 2H_2O$)	II. Certain vigorous reducing agents (ferrous and stannous salts)
(*c*) HNO_3 (conc. hot)	I, II. Oxidation ($NO_3^- + 2H^+ + e^- = NO_2 + H_2O$)	Reducing compounds (sulfides, alloys, metals, etc.)
(*a*) HCl (conc.)	I, III. Reduction ($2Cl^- = Cl_2 + 2e^-$)	III. Oxidizing compounds—higher oxides (MnO_2), oxidizing salts ($PbCrO_4$)
	IV. Complex ion formation	IV. Compounds of cations forming complex ions ($HgCl_4^{2-}$, $SnCl_6^{2-}$)
(*b*) HCl (excess) and HNO_3 (conc.)	I, II, III, IV.	Those above; also noble metals (Pt, Au) requiring both oxidation and complex formation
(*a*) $HClO_4$, fuming (or H_2SO_4)	I, II (only when fuming)	
	V. Displacement by volatilization (of lower-boiling acids)	V. Salts of volatile acids (sulfides, halides, fluorides, etc.)
(*b*) $HClO_4$, and HF (excess)	IV.	Primarily silicates (formation of H_2SiF_6 and SiF_4)
Na_2CO_3 or NaOH (solutions)	VI. Metathesis (carbonate and hydroxide)	VI. Compounds of cations forming insoluble carbonates and hydroxides
	VII. Hydroxide ion effects	VII.
	1. Neutralization	1. Acidic oxides
	2. Displacement	2. Salts of weak bases
	3. Oxidation	3. Certain reactive metals (Al, Zn)
	4. Hydroxide and oxide complex formation	4. Compounds of elements forming acidic or amphoteric oxides.

NOTE: For the selection and use of fusion agents (fluxes) see Kolthoff, Sandell, Meehan, and Bruckenstein. *Quantitative Chemical Analysis.* (4th ed.) New York: Macmillan, 1969; Hillebrand and Lundell. *Applied Inorganic Analysis.* (2nd ed.) New York: Wiley, 1953.

to 2.0 % carbon; in general the higher the carbon content the harder the steel. Steels usually contain only trace quantities of phosphorus, sulfur, and silicon, but may have been alloyed with other metals to obtain various properties. Among these is manganese for hardness, nickel, cobalt, chromium, copper, and vanadium for elasticity, toughness, and corrosion resistance, and molybdenum and tungsten for use as cutting tools. The solvent action of nitric acid with most of these metals is similar to that with iron and the resulting solution will contain Ni^{2+}, Co^{2+}, Cr^{3+}, Cu^{2+}, $V(OH)_4{}^+$, $Mo(OH)_4{}^{2+}$; tungsten will precipitate as hydrous WO_3.

The carbon will remain primarily as insoluble iron carbides, principally Fe_3C, and colloidal graphite; the latter causes a dark coloration that interferes with spectrophotometric measurements. For this reason an oxidant capable of oxidizing the carbonaceous matter to carbon dioxide is used and experience has shown that peroxydisulfate, $S_2O_8{}^{2-}$, is an effective agent for this purpose. The reaction with carbon can be written as

$$C(s) + 2S_2O_8{}^{2-} + 2H_2O = CO_2 + 4HSO_4{}^-$$

Peroxydisulfate is capable of quantitatively oxidizing manganous ion to permanganate, but the reaction is slow and incomplete under the conditions of this procedure and the principal product is MnO_2. This compound is exceedingly insoluble and resistant to further oxidation; therefore if the solution has a permanganate color or if dark MnO_2 is apparent, a reductant, Na_2SO_3, is added to reduce all manganese compounds to Mn^{2+}; this facilitates the subsequent oxidation to $MnO_4{}^-$.

(*b*) Oxidation of Mn(II) to Permanganate

Various agents have been used in the past for the oxidation of Mn^{2+} to $MnO_4{}^-$. At one time, lead dioxide and sodium bismuthate were extensively used for this purpose, but they have been largely replaced by soluble oxidants. These compounds have been discussed in Chapter 22, pages 500–501.

The two soluble oxidants most frequently used are peroxydisulfate and periodate; their general properties have been discussed in Chapter 22. They have replaced the solid oxidants because both are soluble and colorless; therefore, a filtration to remove the excess oxidant is eliminated.

Periodate is used in this procedure because the oxidation reaction proceeds smoothly without the need of a catalyst and because the stability of periodate in acid solutions makes it possible to keep the permanganate solution for a longer time without loss of color intensity. As explained in Chapter 22, H_5IO_6 is the predominant species in acid solutions of periodate and the reaction with Mn^{2+} can be written as

$$2Mn^{2+} + 5H_5IO_6 = 2MnO_4{}^- + 5IO_3{}^- + 11H^+ + 7H_2O$$

Studies have shown that the reaction is autocatalytic[5]—that is, the rate increases after an initial induction period because a product is formed that acts as a catalyst. The study cited indicates that the MnO_4^- first formed reacts with Mn(II) to produce intermediate oxidation states of manganese that react rapidly with the periodate.

The oxidation is found to proceed more smoothly in the presence of phosphoric acid; without it, solid substances believed to be Mn(III) iodates or periodate are often formed. It is probable that soluble phosphate complexes with Mn(III) and Mn(IV) prevent the precipitation. The phosphoric acid also forms colorless complexes with Fe(III); Fe^{3+} absorbs appreciably at 525 mμ, even in nitric acid solution.

(c) The Spectrophotometric Determination

The maximum in the MnO_4^- absorption spectrum at 525 mμ is used for this determination. A comparison blank consisting of the unoxidized solution of the steel sample is used to eliminate errors caused by absorption by other species at 525 mμ. The effect of absorption by nickel, cobalt, and copper is corrected for by this procedure. Chromium, however, is in the +3 state after the sulfite treatment and is partially oxidized by periodate to $Cr_2O_7^{2-}$, which causes some error by absorbing at 525 mμ. This error can be decreased by using the MnO_4^- peak at 545 mμ.

INSTRUCTIONS (Procedure 30-1: Spectrophotometric Determination of Manganese in Steel)

Dissolving the Sample

Accurately weigh 0.5 g of the steel sample (0.2%–1.5% Mn) into a 100-ml beaker (Note 1), add 25 ml of 6 *F* HNO_3, and warm as necessary to dissolve the sample. When solution appears complete boil for 2–3 minutes to expel the oxides of nitrogen. Allow the solution to partly cool, then cautiously add, in small portions, 1 g of $(NH_4)_2S_2O_8$; allow rapid bubbling to cease before

NOTES

1. A sample adequate for four samples is taken and the resulting solution diluted and aliquots taken. This reduces weighing error and provides reserve solution in case of emergency or of a manganese content either too high or low for optimum measurements.

[5] G. R. Waterbury, A. M. Hayes, and D. S. Martin, *J. Am. Chem. Soc.*, **74**, 15 (1952); references to previous studies are given.

adding a new portion (Note 2). Finally boil gently for 10 minutes to 15 minutes. If MnO_2 (dark) has precipitated or if the solution has the color of permanganate add powdered Na_2SO_3 in 5 mg–10-mg portions until the MnO_4^- and MnO_2 are reduced. Boil the solution 5 minutes to expel the SO_2, then cool it.

Transfer the solution quantitatively to a 100-ml volumetric flask, dilute it to the mark, and mix thoroughly (Note 3). Transfer two 25-ml aliquots of this solution into 100-ml beakers. To each add 5 ml of 15 *F* H_3PO_4.

Transfer the remainder of the solution to a storage container—it may be needed.

Quantitatively transfer one of the 25-ml portions to a 100-ml volumetric flask. Dilute it to the mark, and mix it thoroughly. Transfer this solution to a storage container, and reserve it for subsequent use as a comparison solution in the spectrophotometric measurement.

Oxidation of the Manganese

Add 0.5 g KIO_4 in 25-mg–50-mg portions to the second aliquot solution and boil for 5 minutes. Cool the solution, transfer it to a 100-ml volumetric flask, dilute to the mark, and mix thoroughly.

Preparation of Standard Permanganate Solutions

(In order to save time during this procedure the standard permanganate solution can be prepared in larger quantity and portions issued to students; solutions containing excess IO_4^- are stable. Alternatively, a volume of standard $KMnO_4$ containing approximately 0.1 g of manganese can be transferred to a 100-ml beaker, 15 mg–20 mg of KNO_2 added and the solution treated as directed in the second and third paragraphs below.)

Weigh 0.1 g of 99.8+ % manganese to 0.1 mg into a 100-ml beaker and dissolve the metal in a mixture of 10 ml 16 *F* HNO_3 and 10 ml H_2O.

Boil the solution a few minutes to remove the oxides of nitrogen, cool it, transfer it to a 1-liter flask, and dilute it to the mark with water.

NOTES

2. Both CO_2 and O_2 are evolved. In hot solutions $S_2O_8^{2-}$ oxidizes water, $2S_2O_8^{2-} + 2H_2O = O_2 + 4HSO_4^-$, and excess reagent can be eliminated by boiling the solution.

3. This procedure requires that a 100-ml volumetric flask be used several times. With careful planning before the experiment is started one can conveniently perform the operations with only one flask. Conical flasks or bottles fitted with glass stoppers should be used to store the prepared solutions. *Label these containers.*

Pipet 5-ml, 10-ml, and 15-ml portions of the standard permanganate solution into separate 100-ml beakers, add 5 ml of 15 *F* H_3PO_4, 10 ml 6 *F* HNO_3, and water as needed to make the volume 30 ml. Add 0.5 g KIO_4 in small portions to each and boil the solution for 5 minutes. Cool the solution, transfer it to a 100-ml volumetric flask, dilute to the mark, and mix thoroughly.

Preparation of the Standard Curve

(Detailed instructions for the operation of the spectrophotometer will be provided by your instructor. Instruments are so varied in their operation and so many are available that only very general instructions will be given. In all cases be certain that the absorption cell used is free from fingerprints, and that it is always placed in the instrument with the same orientation. Rinse the cell several times with small portions of the solution whose absorbance is to be measured, fill it, and check carefully for suspended material or air bubbles.)

Prepare a blank solution that has the same concentrations of HNO_3, H_3PO_4, and KIO_4 as have the standard solutions. Measure the absorbance at 525 mμ (Notes 4 and 5) of each of the standard solutions against the blank solution. Plot the absorbances against concentrations for these known solutions (Note 6).

Determination of the Sample Concentration

Measure the absorbance of the steel solution at the same wavelength used for the standard curve; use the reserved unoxidized comparison solution as the blank. Use the standard curve to estimate the percentage of Mn in the steel (Note 7).

NOTES

4. Check the location of the absorbance maximum for the instrument you are using; the wavelength calibration may be inaccurate.
5. If your sample contains chromium the secondary absorption maximum of permanganate at 545 mμ should be used (consult your instructor).
6. This Beer's law plot should be a straight line that passes through the origin.
7. If the measured absorbance of the unknown is greater than 0.7 or less than 0.2 of the original, prepare a new solution of appropriate concentration from the reserved original solution of the sample; it was reserved for such an eventuality.

QUESTIONS AND PROBLEMS

1. V. W. Meloche and R. L. Martin (*Anal. Chem.*, **28**, 1671, 1956) determined rhenium spectrophotometrically as the hexachlororhenate(IV) ion, using the absorption maximum at 281.5 mμ. They reduced the perrhenate ion with chromium(II) chloride in concentrated HCl solution. They obtained the following data, using a 1-cm cell:

*Rhenium (mg)	Absorbance at 281.5 mμ
0.0721	0.100
0.2165	0.296
0.5412	0.742
0.9742	1.333
1.1906	1.626
1.4071	1.929

* This was the weight of rhenium in 50.00 ml of solution.

   (*a*) Calculate the molar absorptivity at each concentration.
   (*b*) Is Beer's law followed?
   (*c*) An unknown solution had absorbance 0.946 at 281.5 mμ. Calculate the rhenium concentration. *Ans.* 7.44×10^{-5} *M*.

2. A method was proposed for the simultaneous determination of Ni(II) and Co(III). (R. D. Whealy and S. O. Colgate. *Anal. Chem.* **28**, 1897, 1956.) Both species form colored complexes with diethylenetriamine; the cobalt complex has maximum absorption at 460 mμ and does not absorb at 850 mμ where the nickel complex has maximum absorption. The nickel complex absorbs sufficiently at 460 mμ that correction for it must be made in calculating the concentration of cobalt in the mixture. Both species follow Beer's law. The absorbance of a 0.010 *M* solution of the Ni complex (all measurements were made in a 1-cm cell) is 0.125 at 850 mμ and 0.008 at 460 mμ. The absorbance of a 0.006 *M* solution of the Co complex is 0.600 at 460 mμ.

   An unknown solution of Ni and Co has absorbances of 0.200 at 460 mμ and 0.750 at 850 mμ when the species are complexed with diethylenetriamine. Calculate the concentrations of nickel and cobalt in the unknown.

3. K. Rowley, R. W. Stoenner, and L. Gordon (*Anal. Chem.*, **28**, 136, 1956) used a spectrophotometric end-point method in the titration of 0.1 mg–5 mg samples of barium with EDTA, Eriochrome Black T as indicator. The absorbances at 650 mμ as a function of ml of 0.00457 *F* EDTA added are as follows:

EDTA Added (ml)	Absorbance
0	0.54
1.00	0.21
2.00	0.19
3.00	0.19
4.00	0.20
5.00	0.22

EDTA Added (ml)	Absorbance
6.00	0.24
7.00	0.30
7.20	0.33
7.40	0.36
7.60	0.40
7.65	0.44
7.70	0.48
7.75	0.50
7.80	0.50
8.00	0.50
9.00	0.51

(a) Plot the absorbance versus the volume of EDTA added. Explain the shape of the curve.

(b) Calculate the weight of barium in the sample titrated.

4. N. H. Furman and A. J. Fenton (*Anal. Chem.*, **28**, 515, 1956) used a spectrophotometric end-point method in the coulometric titration of arsenic(III) solutions with electrolytically generated cerium(IV). A typical titration graph of absorbance versus time of generation is shown below. Explain the difference in the shape of this curve and that obtained in the titration of Problem 3, which also involved a spectrophotometric endpoint.

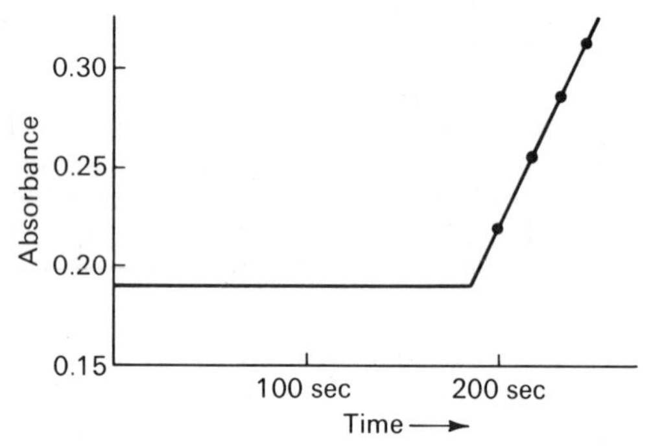

5. Crouthamel (*Anal. Chem.*, **29**, 1756, 1957) reported a molar absorptivity of 52,200 for a complex of technetium and thiocyanate at 513 mμ. Calculate the transmittance in a 1-cm cell of a solution that has a technetium concentration of 0.4 γ (microgram) per ml.

6. O. A. Nietzel and M. A. DeSesa (*Anal. Chem.*, **29**, 756, 1957) report that at 375 mμ with a 1-cm cell, using a thiocyanate complex of uranium in a mixed solvent, they obtained an absorbance of 0.371 with the following solution: 1.00 g of U_3O_8 was dissolved in perchloric acid and diluted to 1 liter. A complex extraction procedure is then carried out upon a portion of this uranium solution; in effect, 0.500 ml of the solution is finally diluted to 35.0 ml in the mixed solvent that contains thiocyanate. Calculate T, a, and ε for the uranium-thiocyanate complex.

APPENDIXES

Ionization Constants of Selected Acids and Bases (at 25°C)

The acids and bases shown below have been selected because of their analytical and general interest. The constants are applicable to aqueous solutions only.

See page 92 for rules for predicting the approximate ionization constants of oxygen acids, and page 101 regarding the partial hydration of certain acid anhydrides and the resulting pseudo ionization constants.

Substance	Formula	Ionization Constant
	Acids	
Acetic	CH_3COOH	1.8×10^{-5}
Arsenic	H_3AsO_4	$K_1 = 5.6 \times 10^{-3}$ $K_2 = 1.7 \times 10^{-7}$ $K_3 = 3 \times 10^{-12}$
Arsenious	H_3AsO_3	$K_1 = 6 \times 10^{-10}$ $K_2 = 3 \times 10^{-14}$
Benzoic	$HC_7H_5O_2$	6×10^{-5}
Boric	H_3BO_3	$K_1 = 6.4 \times 10^{-10}$
Carbonic	H_2CO_3	$K_1 = 3 \times 10^{-7}$ $K_2 = 6 \times 10^{-11}$
Chloracetic	$ClCH_2COOH$	1.5×10^{-3}
Chromic	H_2CrO_4	$K_1 = 1.8 \times 10^{-1}$ $K_2 = 3.2 \times 10^{-7}$
Ferrocyanic	$H_4Fe(CN)_6$	$K_4 = 6.8 \times 10^{-5}$
Formic	$HCHO_2$	2×10^{-4}
Hydrocyanic	HCN	2×10^{-9}
Hydrofluoric	HF	7.4×10^{-4}
Hydrogen sulfide	H_2S	$K_1 = 9.1 \times 10^{-8}$ $K_2 = 1.2 \times 10^{-15}$
Hypochlorous	$HClO$	1.1×10^{-8}

Substance	Formula	Ionization Constant
	Acids	
Hydrazoic	HN_3	2×10^{-5}
Iodic	HIO_3	2×10^{-1}
Nitrous	HNO_2	4.5×10^{-4}
Oxalic	$H_2C_2O_4$	$K_1 = 5.9 \times 10^{-2}$ $K_2 = 6 \times 10^{-5}$
Phosphoric	H_3PO_4	$K_1 = 7.5 \times 10^{-3}$ $K_2 = 2 \times 10^{-7}$ $K_3 = 5 \times 10^{-13}$
Phosphorous	H_3PO_3	$K_1 = 5 \times 10^{-2}$
Phthalic	$H_2C_8H_4O_4$	$K_1 = 1.2 \times 10^{-3}$ $K_2 = 3.1 \times 10^{-6}$
Sulfuric	H_2SO_4	$K_2 = 1.2 \times 10^{-2}$
Sulfurous	H_2SO_3	$K_1 = 1.3 \times 10^{-2}$ $K_2 = 5 \times 10^{-6}$
Tartaric	$H_2C_4H_4O_6$	$K_1 = 1 \times 10^{-3}$ $K_2 = 3 \times 10^{-5}$
	Bases	
Ammonia	$NH_3 + HOH = NH_4^+ + OH^-$	1.8×10^{-5}
Aniline	$C_6H_5NH_2$	4×10^{-10}
Barium hydroxide ($K_{sp} = 5 \times 10^{-3}$)	$Ba(OH)_2$	$K_2 = 2 \times 10^{-1}$
Calcium hydroxide ($K_{sp} = 5 \times 10^{-6}$)	$Ca(OH)_2$	$K_2 = 5 \times 10^{-2}$
Hydrazine	H_2NNH_2	3×10^{-6}
Hydroxylamine	$HONH_2$	1×10^{-8}
Lithium hydroxide	$LiOH$	7×10^{-1}

APPENDIX 2

Dissociation Constants of Complex Ions

In general, these values should be used only to predict the approximate magnitude of the indicated effect. In many cases, in order to conserve space, only the over-all reaction is considered, whereas usually the system is complicated by the stepwise formation of successive complexes. The following sources present more comprehensive data regarding these intermediate complexes: L. G. Sillen, and A. E. Martel, *Stability Constants of Metal-Ion Complexes.* London: The Chemical Society, 1964; K. B. Yatsimirskii, and V. P. Vasil'ev. *Instability Constants of Complex Compounds.* New York: Pergamon, 1960.

The final over-all complex is often formed only at such high ligand concentrations that activity values are quite uncertain.

Hydration of the cations has been neglected; see Chapter 7 regarding the hydration of cations and the nature of the reactions with complexing ligands.

See Table 27-1, page 608 for the dissociation constants of metal-EDTA complexes, also Table 7-1, page 108 and 7-2, page 111 for additional constants of Hg(II) and Cd(II) complexes.

Compound	Dissociation Constant
$Al(OH)_4^- = Al(OH)_3(s) + OH^-$	2.5×10^{-2}
$Sb(OH)_4^- = Sb(OH)_3(s) + OH^-$	10^{-3}
$BiCl_4^- = B^{3+} + 4Cl^-$	3×10^{-6}
$BiI_6^{3-} = Bi^{3+} + 6I^-$	10^{-12}
$Cd(NH_3)_4^{2+} = Cd^{2+} + 4NH_3$	2.5×10^{-7}
$Cd(CN)_4^{2-} = Cd^{2+} + 4CN^-$	10^{-19}
$CdI_2 = Cd^{2+} + 2I^-$	10^{-4}
$CdI_4^{2-} = Cd^{2+} + 4I^-$	5×10^{-7}
$Co(NH_3)_6^{2+} = Co^{2+} + 6NH_3$	10^{-5}
$Co(NH_3)_6^{3+} = Co^{2+} + 6NH_3$	2×10^{-34}
$Cr(OH)_4^- = Cr(OH)_3(s) + OH^-$	10^2
$Cu(CN)_3^{2-} = Cu^+ + 3CN^-$	5×10^{-28}
$CuI_2^- = Cu^+ + 2I^-$	10^{-9}
$Cu(CNS)_2^- = Cu^+ + 2SCN^-$	10^{-12}

Compound	Dissociation Constant
$Cu(NH_3)_3^{2+} = Cu^{2+} + 3NH_3$	3×10^{-11}
$Cu(NH_3)_4^{2+} = Cu^{2+} + 4NH_3$	5×10^{-14}
$CuBr^+ = Cu^{2+} + Br^-$	0.5
$CuCl_2 = CuCl^+ + Cl^-$	4
$CuCl^+ = Cu^{2+} + Cl^-$	0.8
$FeBr^{2+} = Fe^{3+} + Br^-$	2.5×10^{-1}
$FeCl^{2+} = Fe^{3+} + Cl^-$	5.0×10^{-2}
$FeCl_3 = Fe^{3+} + 3Cl^-$	7×10^{-2}
$FeF_3 = Fe^{3+} + 3F^-$	10^{-12}
$FeOH^{2+} = Fe^{3+} + OH^-$	10^{-12}
$Fe(OH)_2^+ = Fe(OH)^{2+} + OH^-$	10^{-11}
$FeHPO_4^+ = Fe^{3+} + HPO_4^{2-}$	10^{-10}
$FeSCN^{2+} = Fe^{3+} + SCN^-$	7.3×10^{-3}
$Fe(C_2O_4)_3^{3-} = Fe^{3+} + 3C_2O_4^{2-}$	about 10^{-10}
$Fe(SO_4)^+ = Fe^{3+} + SO_4^{2-}$	7×10^{-5}
$Pb(CH_3COO)_2 = Pb^{2+} + 2CH_3COO^-$	10^{-3}
$Pb(OH)_3^- = Pb(OH)_2(s) + OH^-$	50
$PbI_3^- = Pb^{2+} + 3I^-$	10^{-5}
$HgCl_2 = Hg^{2+} + 2Cl^-$	2×10^{-15}
$HgCl_4^{2-} = HgCl_2 + 2Cl^-$	1×10^{-2}
$HgI_4^{2-} = Hg^{2+} + 4I^-$	5×10^{-31}
$Hg(CN)_4^{2-} = Hg^{2+} + 4CN^-$	4×10^{-42}
$HgS_2^{2-} = Hg^{2+} + 2S^{2-}$	2×10^{-55}
$Hg(SCN)_2 = Hg^{2+} + 2SCN^-$	3×10^{-18}
$Ni(NH_3)_4^{2+} = Ni^{2+} + 4NH_3$	5×10^{-8}
$Ni(CN)_4^{2-} = Ni^{2+} + 4CN^-$	1×10^{-22}
$SiF_6^{2-} = SiF_4(g) + 2F^-$	2×10^{-15}
$Ag(NH_3)^+ = Ag^+ + NH_3$	4×10^{-4}
$Ag(NH_3)_2^+ = Ag^+ + 2NH_3$	6×10^{-8}
$AgBr_2^- = Ag^+ + 2Br^-$	7×10^{-8}
$AgCl_2^- = Ag^+ + 2Cl^-$	1×10^{-5}
$AgCl_3^{2-} = Ag^+ + 3Cl^-$	5×10^{-6}
$Ag(CN)_2^- = Ag^+ + 2CN^-$	1×10^{-21}
$AgI_3^{2-} = Ag^+ + 3I^-$	10^{-14}
$Ag(SCN)_3^{2-} = Ag^+ + 3SCN^-$	8×10^{-10}
$Sn(OH)_3^- = Sn(OH)_2(s) + OH^-$	2×10^3
$Sn(OH)_6^{2-} = Sn(OH)_4(s) + 2OH^-$	5×10^3
$Zn(NH_3)_4^{2+} = Zn^{2+} + 4NH_3$	3×10^{-10}
$ZnCl_2 = Zn^{2+} + 2Cl^-$	4
$Zn(CN)_4^{2-} = Zn^{2+} + 4CN^-$	10^{-17}
$Zn(OH)_4^{2-} = Zn(OH)_2(s) + 2OH^-$	10

The Solubility-Product Values of Certain Slightly Soluble Compounds (at Room Temperature)

See Table 9-1 for the precipitation *p*H values for the hydrous oxides; solubility product values for certain hydroxides are given below.

The values below have been taken from various sources in the literature and, except for approximate calculations, should be used with caution, as in many cases they appear to have been calculated from formal solubility values without adequate corrections for activity and hydrolysis effects, or for complex-ion formation; a treatment of these factors is given by J. N. Butler (*Ionic Equilibria.* Palo Alto; Addison-Wesley, 1964). Considerable experimental difficulty attends the determination of the solubility of very slightly soluble compounds and widely divergent values are often found in the literature.

Many of the values for the sulfides have been taken from the calculations of I. M. Kolthoff (*J. Phys. Chem.*, **35**, 2711, 1931). For treatments of the problem of the solubility products of the metallic sulfides, refer to this article and to that of Van Rysselberghe and Gropp (*J. Chem. Ed.*, **21**, 96, 1944).

A more comprehensive list of solubility product values is given by L. Meites (*Handbook of Analytical Chemistry.* New York: McGraw-Hill, 1963).

Compound	Solubility Product	Compound	Solubility Product
Barium:		Calcium:	
$BaCO_3$	7×10^{-9}	CaF_2	3×10^{-11}
$BaCrO_4$	3×10^{-10}	$Ca(OH)_2$	5×10^{-6}
BaF_2	3×10^{-6}	$Ca(IO_3)_2 \cdot 6H_2O$	6×10^{-7}
$Ba(IO_3)_2$	1×10^{-9}	$CaC_2O_4 \cdot H_2O$	2×10^{-9}
$BaC_2O_4 \cdot 2H_2O$	2×10^{-8}	$CaSO_4$	2×10^{-4}
$BaSO_4$	1×10^{-10}		
		Cerium:	
Cadmium:		$Ce(OH)_3$	10^{-20}
CdC_2O_4	1×10^{-8}	$Ce(IO_3)_3$	3×10^{-10}
CdS	10^{-28}	$Ce_2(C_2O_4)_3 \cdot 9H_2O$	3×10^{-29}
Calcium:		Cobalt:	
$CaCO_3$	1×10^{-8}	CoS	10^{-27}

Compound	Solubility Product
Copper:	
Cupric:	
$Cu(IO_3)_2$	2×10^{-7}
CuC_2O_4	3×10^{-8}
CuS	10^{-42}
Cuprous:	
$CuBr$	4×10^{-8}
$CuCl$	1×10^{-6}
CuI	4×10^{-12}
$CuSCN$	4×10^{-14}
Iron:	
Ferric:	
$Fe(OH)_3$	10^{-36}
Ferrous:	
$Fe(OH)_2$	10^{-15}
FeC_2O_4	2×10^{-7}
FeS	10^{-22}
Lead:	
$PbBr_2$	7×10^{-5}
$PbCl_2$	2×10^{-4}
$PbClF$	3×10^{-9}
$PbCrO_4$	8×10^{-14}
PbF_2	5×10^{-8}
$Pb(IO_3)_2$	2×10^{-13}
PbI_2	1×10^{-8}
PbC_2O_4	3×10^{-11}
$PbSO_4$	2×10^{-8}
PbS	10^{-28}
Magnesium:	
$MgNH_4PO_4 \cdot 6H_2O$	3×10^{-13}
$MgCO_3 \cdot 3H_2O$	3×10^{-5}
MgF_2	7×10^{-9}
$Mg(OH)_2$	10^{-11}
$MgC_2O_4 \cdot 2H_2O$	1×10^{-8}
Manganous:	
$Mn(OH)_2$	10^{-14}
MnS (pinkish)	10^{-16}
MnS (green)	10^{-22}

Compound	Solubility Product
Mercury:	
HgS	9×10^{-52}
Mercurous:	
Hg_2Br_2	1×10^{-21}
Hg_2Cl_2	2×10^{-18}
Hg_2I_2	7×10^{-29}
Hg_2SO_4	6×10^{-7}
$Hg_2(SCN)_2$	3×10^{-20}
Nickel:	
$Ni(OH)_2$	6×10^{-18}
NiS_γ	10^{-26}
NiS_β	10^{-24}
NiS_α	10^{-19}
Silver:	
CH_3COOAg	4×10^{-3}
$AgBr$	5×10^{-13}
Ag_2CO_3	5×10^{-12}
$AgCl$	1.8×10^{-10}
Ag_2CrO_4	2×10^{-12}
$Ag \cdot Ag(CN)_2$	2×10^{-12}
$AgOH$	10^{-8}
$AgIO_3$	2×10^{-8}
AgI	1×10^{-16}
$Ag_2C_2O_4$	5×10^{-12}
$AgSCN$	1×10^{-12}
Ag_2S	6×10^{-52}
Strontium:	
$SrCO_3$	2×10^{-9}
$SrCrO_4$	3×10^{-5}
SrF_2	3×10^{-9}
$SrC_2O_4 \cdot H_2O$	1×10^{-9}
$SrSO_4$	5×10^{-7}
Zinc:	
$ZnCO_3$	3×10^{-8}
$Zn(OH)_2$	10^{-17}
$ZnC_2O_4 \cdot 2H_2O$	3×10^{-9}
ZnS_α	10^{-25}
ZnS_β	10^{-24}

APPENDIX 4

Selected Standard and Formal Half-Cell Potentials (at 25°C)

Shown below are certain half-cell reactions and potential values selected because of their general interest and because they indicate the redox characteristic of these elements in their various oxidation states.

In Chapter 6 the convention regarding the sign given the potential values is discussed, as is the distinction between *standard* and *formal* potentials.

The half-cell potential values shown below have been collected from various sources. The most frequently used have been W. M. Latimer, *Oxidation Potentials* (2nd ed., New York: Prentice-Hall, 1952), G. Charlot, *Selected Constants: Oxidation-Reduction Potentials* (New York: Pergamon, 1958), and A. J. Bethune and N. A. S. Loud, *Standard Aqueous Electrode Potentials* (Skokie, Illinois: Clifford A. Hampel, 1964). Many values were collected or calculated by Clifford S. Garner from various sources; many of the formal potential values were calculated from measurements made over a number of years by selected students as part of the course in analytical chemistry at the California Institute of Technology.

Reaction	Standard Potential, E° (volts)	Formal Potential, E°′ (volts)			
		HCl (1 *F*)	$HClO_4$ (1 *F*)	H_2SO_4 (1 *F*)	Other Solutions
$K(s) = K^+ + e^-$	+2.922				
$Ca(s) = Ca^{2+} + 2e^-$	+2.88				
$Na(s) = Na^+ + e^-$	+2.712				
$Mg(s) + 2OH^- = Mg(OH)_2(s) + 2e^-$	+2.7				
$Mg(s) = Mg^{2+} + 2e^-$	+2.36				
$Al(s) + 4OH^- = Al(OH)_4{}^- + 3e^-$	+2.35				
$Al(s) = Al^{3+} + 3e^-$	+1.7				
$Zn(s) + 4OH^- = Zn(OH)_4{}^{2-} + 2e^-$	+1.22				
$Mn(s) = Mn^{2+} + 2e^-$	+1.03				
$Sn(OH)_4{}^{2-} + 2OH^- = Sn(OH)_6{}^{2-} + 2e^-$	+0.96				
$Fe(s) + 2OH^- = Fe(OH)_2(s) + 2e^-$	+0.86				
$H_2(g) + 2OH^- = 2H_2O(l) + 2e^-$	+0.829				
$Zn(s) = Zn^{2+} + 2e^-$	+0.758				
$Fe(OH)_2(s) + OH^- = Fe(OH)_3(s) + e^-$	+0.65				
$Cr(s) = Cr^{2+} + 2e^-$	+0.6				
$AsH_3(g) = As(s) + 3H^+ + 3e^-$	+0.6				
$H_3PO_2 + H_2O = H_3PO_3 + 2H^+ + 2e^-$	+0.59				
$Pb(s) + 2OH^- = PbO(s) + H_2O(l) + 2e^-$	+0.58				

$S^{2-} = S(s) + 2e^-$	+0.51				
$SbH_3(g) = Sb(s) + 3H^+ + 3e^-$	+0.5				
$H_2C_2O_4 = 2CO_2(g) + 2H^+ + 2e^-$	+0.49				
$2S^{2-} = S_2{}^{2-} + 2e^-$	+0.48				
$Fe(s) = Fe^{2+} + 2e^-$	+0.44				
$H_2(g) = 2H^+(10^{-7}M) + 2e^-$	+0.414				
$Cr^{2+} = Cr^{3+} + e^-$	+0.4				
$Cd(s) = Cd^{2+} + 2e^-$	+0.41				
$2Cu(s) + 2OH^- = Cu_2O(s) + H_2O(l) + 2e^-$	+0.34				
$Tl(s) = Tl^+ + e^-$	+0.336	+0.551	+0.33	+0.33	
$Ag(s) + 2CN^- = Ag(CN)_2{}^- + e^-$	+0.29				
$Co(s) = Co^{2+} + 2e^-$	+0.28				
$V^{2+} = V^{3+} + e^-$	+0.26		+0.21		
$Ni(s) = Ni^{2+} + 2e^-$	+0.25				
$N_2H_5{}^+ = N_2(g) + 5H^+ + 4e^-$	+0.17				
$Ag(s) + I^- = AgI(s) + e^-$	+0.151				
$Sn(s) = Sn^{2+} + 2e^-$	+0.14		+0.16		
$Pb(s) = Pb^{2+} + 2e^-$	+0.12		+0.14	+0.29	+0.32 (2 F CH_3COONa)
$2Hg(l) + 2I^- = Hg_2I_2(s) + 2e^-$	+0.042				
$2Ag(s) + H_2S(g) = Ag_2S(s) + 2H^+ + 2e^-$	+0.04				
$H_2(g) = 2H^+ + 2e^-$	±0.000	+0.005	+0.005		+0.005 (1 F HNO_3)
$NO_2{}^- + 2OH^- = NO_3{}^- + H_2O(l) + 2e^-$	−0.0				

Reaction	Standard Potential, E° (volts)	Formal Potential, E°′ (volts)			
		HCl (1 *F*)	$HClO_4$ (1 *F*)	H_2SO_4 (1 *F*)	Other Solutions
$Ti^{3+} + H_2O(l) = TiO^{2+} + 2H^+ + e^-$	−0.099			−0.04	−0.12 (2 *F* H_2SO_4)
$Ag(s) + Br^- = AgBr(s) + e^-$	−0.073				
$Ag(s) + SCN^- = AgSCN(s) + e^-$	−0.095				−0.159 (0.1 *F* KSCN)
$Hg(l) + 2OH^- = HgO(s) + H_2O(l) + 2e^-$	−0.098				
$2Hg(1) + 2Br^- = Hg_2Br_2(s) + 2e^-$	−0.139				
$SO_2(g) + 2H_2O(l) = HSO_4^- + 3H^+ + 2e^-$	−0.14			−0.07	
$Cu(s) + Cl^- = CuCl(s) + e^-$	−0.14				
$Sn^{2+} = Sn^{4+} + 2e^-$		−0.14			
$Co(NH_3)_6{}^{2+} = Co(NH_3)_6{}^{3+} + e^-$					−0.16 (3.3 *F* NH_3)
$H_2S(g) = S(s) + 2H^+ + 2e^-$	−0.17				
$Cu^+ = Cu^{2+} + e^-$	−0.17	−0.45			
$2S_2O_3{}^{2-} = S_4O_6{}^{2-} + 2e^-$	−0.17				
$Pt(s) + 4Cl^- = PtCl_4{}^{2-} + 2e^-$	−0.2				
$Sb(s) + H_2O(l) = SbO^+ + 2H^+ + 3e^-$	−0.21		−0.213	−0.213	
$Ag(s) + Cl^- = AgCl(s) + e^-$	−0.224				
$As(s) + 3H_2O(l) = H_3AsO_3 + 3H^+ + 3e^-$	−0.24		−0.249	−0.25	
Saturated calomel half-cell					−0.246
$2Hg(l) + 2Cl^- = Hg_2Cl_2(s) + 2e^-$	−0.269				

Normal calomel half-cell					-0.282
$Bi(s) + H_2O(l) = BiO^+ + 2H^+ + 3e^-$	-0.31				-0.34 (0.02 *F* HCl)
$H_2S(g) + 4H_2O(l) = HSO_4^- + 9H^+ + 8e^-$	-0.316			-0.294	
$V^{3+} + H_2O(l) = VO^{2+} + 2H^+ + e^-$	-0.34			-0.360	
$Cu(s) = Cu^{2+} + 2e^-$	-0.345				
$Fe(CN)_6^{4-} = Fe(CN)_6^{3-} + e^-$	-0.356	-0.71	-0.72	-0.72	-0.56 (0.1 *F* HCl) -0.48 (0.01 *F* HCl) -0.46 (0.01 *F* NaOH)
$4OH^- = O_2(g) + 2H_2O(l) + 4e^-$	-0.40				
$U^{4+} + 2H_2O(l) = UO_2^{2+} + 4H^+ + 2e^-$	-0.4			-0.4	
$S_2O_3^{2-} + 3H_2O = 2H_2SO_3 + 2H^+ + 4e^-$	-0.40				
$PtCl_4^{2-} + 2Cl^- = PtCl_6^{2-} + 2e^-$	-0.44				
$CuCl_2^- = Cu^{2+} + 2Cl^- + e^-$	-0.46				
$MoO^{3+} + 2H_2O(l) = MoO_3(s) + 4H^+ + e^-$	-0.5				
$2I^- = I_2(s) + 2e^-$	-0.536				
$3I^- = I_3^- + 2e^-$	-0.54				
$H_3AsO_3 + H_2O(l) = H_3AsO_4 + 2H^+ + 2e^-$		-0.577	-0.577		
$MnO_4^{2-} = MnO_4^- + e^-$	-0.564				-0.6 (1.5 *F* KOH)
$2Hg(l) + SO_4^{2-} = Hg_2SO_4(s) + 2e^-$	-0.621				
$Hg_2Cl_2(s) + 2Cl^- = 2HgCl_2 + 2e^-$	-0.63				
$H_2O_2 = O_2(g) + 2H^+ + 2e^-$	-0.69				
$C_6H_4(OH)_2 = C_6H_4O_2(quinone) + 2H^+ + 2e^-$	-0.699				
$Sb_2O_3(s) + 2H_2O(l) = Sb_2O_5(s) + 4H^+ + 4e^-$	-0.7				

Reaction	Standard Potential, E° (volts)	Formal Potential, E°′ (volts) HCl (1 *F*)	HClO$_4$ (1 *F*)	H_2SO_4 (1 *F*)	Other Solutions
$SbCl_4^- + 2Cl^- = SbCl_6^- + 2e^-$					−0.75 (3.5 *F* HCl) −0.78 (4.5 *F* HCl) −0.82 (6.0 *F* HCl)
$Fe^{2+} = Fe^{3+} + e^-$	−0.782	−0.700	−0.732	−0.68	−0.61 (1 *F* H_2SO_4 and 0.5 *F* H_3PO_4)
$2Hg(l) = Hg_2^{2+} + 2e^-$	−0.798	−0.274	−0.776	−0.674	
$Ag(s) = Ag^+ + e^-$	−0.799	−0.288	−0.792	−0.77	
$CuI(s) = Cu^{2+} + I^- + e^-$	−0.85				
$I_2(s) + 4Br^- = 2IBr_2^- + 2e^-$	−0.87				
$Hg_2^{2+} = 2Hg^{2+} + 2e^-$			−0.907		
$HNO_2 + H_2O(l) = 3H^+ + NO_3^- + 2e^-$					−0.92 (1 *F* HNO_3)
$Hg(l) + H_2O(l) = HgO(s) + 2H^+ + 2e^-$	−0.93				
$Cl^- + 2OH^- = ClO^- + H_2O(l) + 2e^-$	−0.94				
$NO(g) + 2H_2O(l) = NO_3^- + 4H^+ + 3e^-$	−0.94				
$NO(g) + H_2O(l) = HNO_2 + H^+ + e^-$	−0.98				
$VO^{2+} + H_2O(l) = VO_2^+ + 2H^+ + e^-$	−1.000	−1.02	−1.02	−1.0	
$I_2(s) + 4Cl^- = 2ICl_2^- + 2e^-$		−1.06			−1.05 (4 *F* HCl)
$2Br^- = Br_2(l) + 2e^-$	−1.066				
$\frac{1}{2}I_2(s) + 3H_2O(l) = IO_3^- + 6H^+ + 5e^-$	−1.19				

$H_2SeO_3 + H_2O(l) = H_2SeO_4 + 2H^+ + 2e^-$	−1.2				
$2H_2O(l) = O_2(g) + 4H^+ + 4e^-$	−1.23				
$Mn^{2+} + 2H_2O(l) = MnO_2(s) + 4H^+ + 2e^-$	−1.24		−1.24		
$Tl^+ = Tl^{3+} + 2e^-$	−1.25	−0.77	−1.26	−1.22	−1.23 (1 *F* HNO_3)
$2Cr^{3+} + 7H_2O(l) = Cr_2O_7{}^{2-} + 14H^+ + 6e^-$	−1.33	−1.00		−1.1	
$ICl_2{}^- + 2Cl^- = ICl_4{}^- + 2e^-$					−1.31 (4 *F* HCl)
$2Cl^- = Cl_2(g) + 2e^-$	−1.359				
$Au(s) + H^+ + 4Cl^- = HAuCl_4 + 3e^-$		−1.4			
$Cl^- + 3H_2O(l) = ClO_3{}^- + 6H^+ + 6e^-$	−1.45				
$Mn^{2+} + 4H_2O(l) = MnO_4{}^- + 8H^+ + 5e^-$	−1.45				
$Pb^{2+} + 2H_2O(l) = PbO_2(s) + 4H^+ + 2e^-$	−1.47		−1.47	−1.628	
$Ce^{3+} = Ce^{4+} + e^-$		−1.23	−1.7	−1.44	−1.61 (1 *F* HNO_3)
$Cl^- + H_2O(l) = HClO + H^+ + 2e^-$	−1.50				
$MnO_2(s) + 2H_2O(l) = MnO_4{}^- + 4H^+ + 3e^-$	−1.59				
$2Bi^{3+} + 5H_2O(l) = Bi_2O_5(s) + 10H^+ + 4e^-$	−1.7				
$2H_2O(l) = H_2O_2 + 2H^+ + 2e^-$	−1.77				
$Co^{2+} = Co^{3+} + e^-$				−1.82	−1.83 (1 *F* HNO_3) −1.85 (4 *F* HNO_3)
$Ag^+ = Ag^{2+} + e^-$					−1.91 (1 *F* HNO_3) −2.00 (4 *F* $HClO_4$)
$2HSO_4{}^- = S_2O_8{}^{2-} + 2H^+ + 2e^-$	−2.05				
$O_2(g) + H_2O(l) = O_3(g) + 2H^+ + 2e^-$	−2.07				
$2F^- = F_2(g) + 2e^-$	−2.88				

Species Formed by Certain Elements in Various Solutions

There are shown in the following table the more common elements arranged according to their position in the periodic table. Also shown are the species that they form in their common oxidation states in (*a*) noncomplexing acids, (*b*) neutral solutions, (*c*) hydroxide solutions, (*d*) acid hydrogen sulfide solutions, and (*e*) alkaline sulfide solutions. Complex forming agents are assumed to be absent. The precipitates formed in alkaline solutions are shown as hydroxides; in many cases they are hydrous oxides of uncertain hydration. The colors of precipitates are stated unless they are white.

Periodic Column	Element and Oxidation State	Compounds or Ions Formed				
		1 *F* HNO_3	*p*H = 7–8	1 *F* NaOH	H_2S (Saturated) 0.1 *M* H^+	0.1 *F* Na_2S 1 *F* NaOH
Ia	Li(I)	Li^+	Li^+	Li^+	Li^+	Li^+
	Na(I)	Na^+	Na^+	Na^+	Na^+	Na^+
	K(I)	K^+	K^+	K^+	K^+	K^+
Ib	Cu(I)	*unstable*	*unstable*	Cu_2O(s), *red*	Cu_2S(s), *black*	Cu_2S(s), *black*
	Cu(II)	Cu^{2+}	CuO(s), *black*	CuO(s), *black*	CuS(s), *black*	CuS(s), *black*
	Ag(I)	Ag^+	Ag^+	Ag_2O(s), *black*	Ag_2S(s), *black*	Ag_2S(s), *black*
IIa	Be(II)	Be^{2+}	$Be(OH)_2$(s)	$Be(OH)_3^-$	Be^{2+}	$Be(OH)_3^-$
	Mg(II)	Mg^{2+}	Mg^{2+}	$Mg(OH)_2$(s)	Mg^{2+}	$Mg(OH)_2$(s)
	Ca(II)	Ca^{2+}	Ca^{2+}	Ca^{2+}, $Ca(OH)_2$(s)	Ca^{2+}	Ca^{2+}, $Ca(OH)_2$(s)
	Sr(II)	Sr^{2+}	Sr^{2+}	Sr^{2+}	Sr^{2+}	Sr^{2+}
	Ba(II)	Ba^{2+}	Ba^{2+}	Ba^{2+}	Ba^{2+}	Ba^{2+}
IIb	Zn(II)	Zn^{2+}	$Zn(OH)_2$(s)	$Zn(OH)_3^-$	Zn^{2+}	ZnS(s)
	Cd(II)	Cd^{2+}	$Cd(OH)_2$(s)	$Cd(OH)_2$(s)	CdS(s), *yellow*	CdS(s), *yellow*
	Hg(I)	Hg_2^{2+}	Hg_2O(s), *black*	Hg_2O(s), *black*	HgS(s), *black*; Hg(l)	HgS_2^{2-}; Hg(l)
	Hg(II)	Hg^{2+}	HgO(s), *yellow-red*	HgO(s), *yellow-red*	HgS(s), *black*	HgS_2^{2-}
IIIa	B(III)	H_3BO_3	H_3BO_3, $HB_4O_7^-$	$B_4O_7^{2-}$	H_3BO_3	$B_4O_7^{2-}$
	Al(III)	Al^{3+}	$Al(OH)_3$(s)	$Al(OH)_4^-$	Al^{3+}	$Al(OH)_4^-$
IIIb	Ce(III)	Ce^{3+}	Ce^{3+}	$Ce(OH)_3$(s)	Ce^{3+}	$Ce(OH)_3$(s)
	Ce(IV)	$Ce(NO_3)_6^{2-}$	$Ce(OH)_4$(s), *yellow*	$Ce(OH)_4$(s), *yellow*	*reduced*	*reduced*

Periodic Column	Element and Oxidation State	Compounds or Ions Formed				
		1 *F* HNO_3	*p*H = 7–8	1 *F* NaOH	H_2S (Saturated) 0.1 *M* H^+	0.1 *F* Na_2S 1 *F* NaOH
IVa	C(IV)	H_2CO_3, CO_2	HCO_3^-	CO_3^{2-}	H_2CO_3, CO_2	CO_3^{2-}
	Si(IV)	$SiO_2 \cdot xH_2O(s)$	$SiO_2 \cdot xH_2O(s)$	SiO_3^{2-}	$SiO_2(H_2O)_x(s)$	SiO_3^{2-}
	Sn(II)	Sn^{2+}	$Sn(OH)_2(s)$	$Sn(OH)_3^-$	SnS(s), *brown*	SnS_2^{2-}, SnS(s)
	Sn(IV)	$SnO_2 \cdot xH_2O(s)$	$SnO_2 \cdot xH_2O(s)$	$Sn(OH)_6^{2-}$	$SnS_2(s)$, *yellow*	SnS_3^{2-}
	Pb(II)	Pb^{2+}	$Pb(OH)_2(s)$	$Pb(OH)_4^{2-}$	PbS(s), *black*	PbS(s), *black*
	Pb(IV)	$PbO_2(s)$, *brown*	$PbO_2(s)$, *brown*	PbO_2, $Pb(OH)_6^{2-}$	PbS(s), *black*	PbS(s), *black*
IVb	Ti(III)	Ti^{3+}	$Ti(OH)_3(s)$, *dark*	$Ti(OH)_3(s)$, *dark*	Ti^{3+}	$Ti(OH)_3(s)$, *dark*
	Ti(IV)	TiO^{2+}	$TiO_2(H_2O)_x(s)$	$TiO_2(H_2O)_x(s)$	TiO^{2+}	$TiO_2(H_2O)_x(s)$
Va	N(-III)	NH_4^+	NH_4^+	NH_3	NH_4^+	NH_3
	N(III)	HNO_2	NO_2^-	NO_2^-	*reduced*	*reduced*
	N(V)	NO_3^-	NO_3^-	NO_3^-	NO_3^-	NO_3^-
	P(III)	H_3PO_3	HPO_3^{2-}	HPO_3^{2-}	H_3PO_3	HPO_3^{2-}
	P(V)	H_3PO_4	HPO_4^{2-}	PO_4^{3-}	H_3PO_4	PO_4^{3-}
	As(III)	H_3AsO_3	H_3AsO_3	$HAsO_3^{2-}$	$As_2S_3(s)$, *yellow*	AsS_3^{3-}
	As(V)	H_3AsO_4	$HAsO_4^{2-}$	AsO_4^{3-}	$As_2S_3(s)$ and $As_2S_5(s)$, *yellow*	AsS_4^{3-}
	Sb(III)	$Sb(OH)_2^+$	$Sb(OH)_3(s)$	$Sb(OH)_4^-$	$Sb_2S_3(s)$, *orange*	SbS_3^{3-}
	Sb(V)	$Sb_2O_5(s)$	$Sb_2O_5(s)$	$Sb(OH)_6^-$	$Sb_2S_3(s)$ and $Sb_2S_5(s)$, *orange*	SbS_4^{3-}
	Bi(III)	$Bi(OH)_2^+$	$Bi(OH)_3(s)$	$Bi(OH)_3(s)$	$Bi_2S_3(s)$, *black*	$Bi_2S_3(s)$, *black*

Vb	V(IV)	$V(OH)_2^{2+}$	$V(OH)_4$(s), *dark*	$V_4O_9^{2-}$	$V(OH)_2^{2+}$	VS_3^{2-}
	V(V)	$V(OH)_4^{+}$	V_2O_5(s), *red*, HVO_3	VO_3^{-}	$V(OH)_2^{2+}$	VS_4^{3-}
VIa	S(-II)	H_2S	H_2S, HS^-	HS^-	H_2S	HS^-
	S(IV)	H_2SO_3	SO_3^{2-}	SO_3^{2-}	*reduced to* S	*reduced to* S_2^{2-}
	S(VI)	HSO_4^-	SO_4^{2-}	SO_4^{2-}	HSO_4^-	SO_4^{2-}
	Se(IV)	H_2SeO_3	SeO_3^{2-}	SeO_3^{2-}	Se(s), *red*	Se_2S^{2-}
	Se(VI)	$HSeO_4^-$	SeO_4^{2-}	SeO_4^{2-}	$HSeO_4^-$	SeO_4^{2-}
VIb	Cr(III)	Cr^{3+}	$Cr(OH)_3$(s), *green*	$Cr(OH)_3$(s), *green*	Cr^{3+}	$Cr(OH)_3$(s), *green*
	Cr(VI)	$Cr_2O_7^{2-}$	CrO_4^{2-}, $Cr_2O_7^{2-}$	CrO_4^{2-}	Cr^{3+}	$Cr(OH)_3$(s), *green*
VIIa	F(-I)	HF	F^-	F^-	HF	F^-
	Cl(-I)	Cl^-	Cl^-	Cl^-	Cl^-	Cl^-
	Cl(I)	HClO	HClO	ClO^-	Cl^-	Cl^-
	Cl(V)	ClO_3^-	ClO_3^-	ClO_3^-	Cl^-	Cl^-
	Cl(VII)	ClO_4^-	ClO_4^-	ClO_4^-	ClO_4^-	ClO_4^-
	Br(-I)	Br^-	Br^-	Br^-	Br^-	Br^-
	Br(V)	BrO_3^-	BrO_3^-	BrO_3^-	Br^-	Br^-
	I(-I)	I^-	I^-	I^-	I^-	I^-
	I(V)	IO_3^-	IO_3^-	IO_3^-	I^-	I^-
	I(VII)	H_5IO_6	$H_4IO_6^-$	IO_4^-	I^-	I^-
VIIb	Mn(II)	Mn^{2+}	Mn^{2+}	$Mn(OH)_2$(s)	Mn^{2+}	MnS(s), *pink*
	Mn(IV)	MnO_2(s), *black*	MnO_2(s), *black*	MnO_2(s), *black*	Mn^{2+}	MnS(s), *pink*
	Mn(VII)	MnO_4^-	MnO_4^-	MnO_4^-	Mn^{2+}	MnS(s), *pink*

Periodic Column	Element and Oxidation State	Compounds or Ions Formed				
		1 *F* HNO_3	*p*H = 7–8	1 *F* NaOH	H_2S (Saturated) 0.1 *M* H^+	0.1 *F* Na_2S 1 *F* NaOH
VIII	Fe(II)	Fe^{2+}	Fe^{2+}	$Fe(OH)_2$(s)	Fe^{2+}	FeS(s), *black*
	Fe(III)	Fe^{3+}	$Fe(OH)_3$(s), *brown*	$Fe(OH)_3$(s), *brown*	Fe^{2+}	Fe_2S_3(s), *black*
	Co(II)	Co^{2+}	Co^{2+}	$Co(OH)_2$(s), *pink*	Co^{2+}	CoS(s), *black*
	Ni(II)	Ni^{2+}	Ni^{2+}	$Ni(OH)_2$(s), *greenish*	Ni^{2+}	NiS(s), *black*

Densities of Water and Air at Laboratory Temperatures

Densities of Water (g/ml) at Laboratory Temperatures

Temp. (C°)	Density	Temp. (C°)	Density
15	0.99910	23	0.99754
16	0.99894	24	0.99730
17	0.99877	25	0.99704
18	0.99859	26	0.99678
19	0.99840	27	0.99651
20	0.99820	28	0.99623
21	0.99799	29	0.99594
22	0.99777	30	0.99565

Densities of Dry Air (g/l) at Laboratory Temperatures and Pressures

	Pressure (mm of Hg)			
Temp. (C°)	720	740	750	760
15	1.18	1.19	1.21	1.23
20	1.16	1.17	1.19	1.21
25	1.14	1.15	1.17	1.19
30	1.12	1.13	1.15	1.16

APPENDIX 7

Reagents and Chemicals

The concentrations and method of preparation of the various solutions and reagents required in this text are given below.

The concentrations of all reagents are specifically stated in the procedures, and only a reagent of the specified concentration should be used.

In preparing the reagents indicated below, use only the best available chemicals. The percentage composition and specific gravity of the concentrated acids of various manufacturers may vary slightly from those stated, but not sufficiently to necessitate an adjustment of the quantities given.

SOLUTIONS

Solutions of acids and bases

Acids:

* Acetic, 6 *F*. Dilute 345 ml of 99.5% (glacial) acid to 1 liter.
Hydrochloric, 12 *F*. Use 36% acid of sp. gr. 1.19.
* Hydrochloric 6 *F*. Dilute 500 ml of the 36% acid to 1 liter.
Hydrofluoric, 27 *F*. Use 48% acid.
Hydriodic, 7 *F*. Use 57% acid of sp. gr. 1.70.
Nitric, 16 *F*. Use 69% acid of sp. gr. 1.42.
* Nitric, 6 *F*. Dilute 375 ml of the 69% acid to 1 liter.
Perchloric, 9 *F*. Use 60% acid of sp. gr. 1.54.
Phosphoric, 15 *F*. Use 85% acid of sp. gr. 1.7.
Sulfuric, 18 *F*. Use 95% acid of sp. gr. 1.84.
* Sulfuric, 3 *F*. Dilute 165 ml of the 95% acid to 1 liter.

Bases:

Ammonia, 15 *F*. Use the 29% solution of sp. gr. 0.90.
* Ammonia, 6 *F*. Dilute 400 ml of 15 *F* to 1 liter.
Sodium hydroxide, 6 *F*. Dissolve 255 g of 95% NaOH in water, cool, and dilute to 1 liter.

---

* Suggested desk reagents.

Solutions of Salts

When the same salt is used in more than one concentration, only the preparation of the most concentrated solution is given. The last column shows the weights to be taken of the best available chemicals.

Salt	Concentration	Formula	Formula Weight	Grams Per Liter
Ammonium chloride	1 *F*	NH_4Cl	53.49	53
Barium chloride	0.5 *F*	$BaCl_2 \cdot 2\,H_2O$	244.28	122
Cupric sulfate	0.100 *F*	$CuSO_4 \cdot 5\,H_2O$	249.68	25
Ferric chloride	1 *F*	$FeCl_3 \cdot 6\,H_2O$	270.30	270
Ferric nitrate	0.3 *F*	$Fe(NO_3)_3 \cdot 9\,H_2O$	404.0	121
Manganous sulfate	0.5 *F*	$MnSO_4 \cdot H_2O$	169.01	85
Mercuric chloride	saturated	$HgCl_2$	271.50	100
Mercuric nitrate	0.05 *F*	$Hg(NO_3)_2 \cdot H_2O$	342.62	17
Potassium bromide	2 *F*	KBr	119.01	238
Potassium chromate	0.5 *F*	K_2CrO_4	194.20	97
Potassium dichromate	0.100 *N* (0.0167 *F*)	$K_2Cr_2O_7$	294.19	4.91
Potassium ferrocyanide	0.25 *F*	$K_4Fe(CN)_6 \cdot 3\,H_2O$	422.41	106
Potassium hydrogen phthalate	0.01 *F*	$KHC_8H_4O_4$	204.23	2
Potassium iodide	1 *F*	KI	166.01	166
Potassium nitrite	3 *F*	KNO_2	85.11	250
Potassium permanganate	0.2 *F*	$KMnO_4$	158.04	32
Potassium thiocyanate	0.1 *F*	$KSCN$	97.18	9.7
Silver nitrate	1 *F*	$AgNO_3$	169.87	170
Sodium acetate	1 *F*	$CH_3COONa \cdot 3\,H_2O$	136.08	136
Sodium chloride	1 *F*	$NaCl$	58.44	58
Sodium carbonate	1.5 *F*	Na_2CO_3	105.99	159
Sodium hydrogen carbonate	0.1 *F*	$NaHCO_3$	84.01	8.4
Sodium sulfide	1 *F*	$Na_2S \cdot 9\,H_2O$	240.18	240

Composite Solutions

When the same solution is used in more than one concentration, only the preparation of the most concentrated solution is given. Use the best available chemicals.

Potassium cyanide, 0.2 *F* in KCN and 0.1 *F* in KOH. Dissolve 13 g KCN and 6 g KOH in 1 liter of water.

Arsenious acid, 0.050 *F*. Dissolve 5.0 g As_2O_3 in 10–20 ml 6 *F* NaOH, dilute to about 750 ml, neutralize with HCl using phenolphthalein as indicator (just to colorless), then dilute to 1.0 liter.

Ferrous sulfate, 0.10 *F* in $FeSO_4$ and 0.03 *F* in H_2SO_4. Dissolve 28 g $FeSO_4 \cdot 7\,H_2O$ in water, add 10 ml 3 *F* H_2SO_4 and dilute to 1.0 liter.

Iodine, 0.050 *F* in I_2 and 0.15 *F* in KI. Dissolve 25 g KI in 25 ml H_2O (not more). Add 12.7 g I_2 and stir until completely dissolved. Slowly dilute to 1.0 liter.

Buffer, 0.2 *F* in NH_4Cl, 1.2 *F* in NH_3. Dissolve 11 g NH_4Cl in 100 ml H_2O, add 200 ml 6 *F* NH_3, and dilute to 1 liter.

H_2O_2, 3%. Use the commercial 3% solution.

Stannous chloride, 0.5 *F* in $SnCl_2$ and 3 *F* in HCl. Dissolve 113 g of $SnCl_2 \cdot 2\,H_2O$ (iron-free) in 250 ml of 12 *F* HCl and dilute to 1 liter. Add 10 g–20 g of "mossy" tin to the storage bottle.

Zimmermann–Reinhardt preventive solution, 0.3 *F* in $MnSO_4$, 3 *F* in H_3PO_4, and 2 *F* in H_2SO_4. Dissolve 70 g $MnSO_4 \cdot H_2O$ in 500 ml of water, add slowly with constant stirring 110 ml of 18 *F* (95%) H_2SO_4; then add 200 ml of 15 *F* (85%) H_3PO_4 and dilute to 1 liter.

Iodine monochloride, 0.25 *F*. Dissolve 2.76 g KI in 50 ml of 4 *F* HCl, add 1.78 g KIO_3, and shake in a bottle with a ground-glass stopper with 5 ml of CCl_4. Add dilute (0.05 *F* or less) solutions of KI and KIO_3 as needed until a barely perceptible color persists in the CCl_4 after repeatedly and vigorously shaking the mixture. Dilute to 100 ml with 4 *F* HCl. Prepare 0.0025 *F* ICl by diluting portions of this solution 100-fold with 4 *F* HCl.

Osmium tetroxide, 0.01 *F*. Dissolve 0.25 g OsO_4 crystals in 100 ml of 0.1 *F* H_2SO_4. Prepare as needed and protect from dust and reducing agents.

Suspensions

Starch indicator. Mix 1 g of *soluble starch* and 10 mg to 20 mg of HgI_2 to a thin paste with 20 ml to 30 ml of cold water and pour this into 200 ml of boiling water. Prepare this solution only as needed. Thyodene (see page 471), a starch substitute, can be used.

Asbestos, for quantitative filters. Purchase a washed and ignited long-fiber asbestos (of the amphibole, nonhydrated type). Cut the required amount into lengths of about 0.5 cm and triturate with a little water in a mortar until all large particles are disintegrated. Heat with 6 *N* HCl for several hours, filter on a Büchner funnel, and wash free of chloride. Shake with water and decant and discard the very fine powdery material. Again shake with water and decant so that about one-fifth of the asbestos is poured off. Use this suspension to complete the filtering mat when filtering finely divided precipitates. Add sufficient water to the remainder to make a thin suspension for general use.

Standard Solutions

See the pages indicated below for detailed directions for the preparation and standardization of these solutions.

Constituents	Concentration		Page
	Normal	Formal	
Ceric sulfate	0.1	0.1	437, 440
Hydrochloric acid	0.2	0.2	541, 558
Potassium dichromate	0.1	$\frac{1}{6}$	474
Potassium permanganate	0.1	0.02	403, 413
Potassium thiocyanate	0.1	0.1	372, 380
Silver nitrate	0.1	0.1	352
Sodium hydroxide	0.2	0.2	540, 552
Sodium thiosulfate	0.1	0.1	470, 473

Hydrogen Ion Indicators

Common Name	Molecular Weight	Concentration		Solvent
		Formal ($\times 10^3$)	Grams per Liter	
Bromphenol blue	856	3.5	3	Ethanol
Bromthymol blue	624	1.6	1	Ethanol
Methyl orange	327	3.1	1	Water
Methyl red	269	7.4	2	Ethanol
Phenolphthalein	318	6.3	2	Ethanol
Phenol red	354	2.7	1	Ethanol
Thymol blue	466	2.1	1	Ethanol
Thymolphthalein	430	2.3	1	Ethanol

The alcoholic solutions of the above indicators should not be used for determining the *p*H of unbuffered solutions.

Mixed Indicators

Methyl orange and indigo carmine. Dissolve 1 g of methyl orange and 2.5 g of indigo carmine in 1 liter of water.

Methyl orange and xylene cyanole FF. Dissolve 2 g of methyl orange in 500 ml of water and 2.8 g of xylene cyanole FF in 500 ml of ethanol and mix the two solutions.

Thymol blue and cresol red. Dissolve 3 g of thymol blue and 0.5 g of cresol red in 1 liter of ethanol.

Metallochromic Indicators

Murexide. Grind thoroughly in a mortar 0.2 g of murexide with 100 g NaCl.

Eriochrome Black T. Dissolve 200 mg of solid, reagent grade Eriochrome Black T in a solution of 5 ml of absolute ethanol and 15 ml triethanolamine. Solutions should be freshly prepared every 2 weeks.

Indicator Papers

Litmus

Wide-range indicator test papers.

Potential Indicators

Ortho-phenanthroline ferrous sulfate, 0.025 *F*. Dissolve 7 g of $FeSO_4 \cdot 7\,H_2O$ in water, add 15 g ortho-phenanthroline monohydrate[1] ($C_{12}H_8N_2 \cdot H_2O$). Stir until dissolved and dilute to 1 liter.

Diphenylamine, 0.01 *F*. Dissolve 17 g of diphenylamine, $(C_6H_5)_2NH$, in 1 liter of 18 *F* H_2SO_4.

SOLIDS

Primary Standards

The compounds listed below are useful as primary standards. Those marked with an asterisk may be purchased from the National Bureau of Standards; the others should be of a special grade intended for standardization purposes or should be prepared by the analyst.[2]

---

[1] This compound can be obtained from the G. Frederick Smith Company, Columbus, Ohio; 0.025 *F* solutions of the *o*-phenanthroline ferrous sulfate can be purchased from the same source.

[2] Mallinckrodt Chemical Works, St. Louis 7, Mo., offer the following analytical reagent primary standards with assays from 99.95% to 100.05%; benzoic acid, arsenic trioxide, potassium hydrogen phthalate, potassium dichromate, sodium oxalate, and sodium carbonate.

*Arsenious oxide
*Benzoic acid
*Potassium dichromate
Potassium ferricyanide
*Potassium hydrogen phthalate (acid phthalate)
Potassium iodate
Silver nitrate
Sodium carbonate
*Sodium oxalate
Sodium tetraborate decahydrate (borax)

Reagents

Ammonium ceric sulfate
Ammonium peroxydisulfate
Calcium carbonate
Potassium dihydrogen phosphate
Potassium iodide
Potassium nitrate
Potassium periodate
Potassium thiocyanate
Sodium carbonate
Sodium ethylenediaminetetraacetate
Sodium hydrogen carbonate
Sodium monohydrogen phosphate
Sodium succinate
Sodium thiosulfate
Zinc (granular)

INDEX

Abbreviations, 18
Absolute errors, 27
Accuracy, 27ff
Acetamide, hydrolysis, 536
Acetate, determination of, 565
Acetic acid
  equilibrium calculations, 523ff
  glacial, analysis of, 564
  as a nonaqueous solvent, 105
Acid, 89
  anhydrides of, 101ff, 121
  Arrhenius definition, 87
  Brønsted-Lowry definition, 89
  ionization constants of (table), 683
  Lewis definition, 91
  standardization, 553ff
Acid-base equilibria, 86ff, 515ff
  charge balance, 516
  dilute solutions of weak acids, 535
  mass balance, 516
  pseudo equilibrium constants, 101ff
  species balance, 516
Acid-base indicators, 529ff
  mechanism, 531
Acid-base methods, 537ff
  acetate analysis (problem), 565
  amino acid determination, 581ff
  carbonate-hydrogen carbonate
    analysis, 563
  glacial acetic acid analysis (problem), 564
  polyprotic acid analysis, 572ff
  soda ash analysis, 563
  strong acids, 541ff
  strong bases, 539ff
  titrimetric ratio determination, 543ff
  total cation determination, 593ff
Activity, 39, 49ff
  coefficients
    mean, 50
    table, 52
  Debye-Hückel relationship, 57
  effects in precipitation, 248
  quotient, 77
Adsorption by charcoal (figure), 261
Adsorption chromatography, 152
Air, density of (table), 701
Alkali elements
  determination of, ion-exchange, 599ff
  periodic properties, 117
Alkaline earth elements
  periodic properties, 117
  properties (table), 124
Alkalinity, total, determination of, 560ff
Amalgam reductors (table), 497
Amino acids, 581ff
Ammonia
  complexes
    of cadmium, 605
    of silver, 251
  hydration of, 104

Ammonium ceric nitrate, as a primary standard, 436
Ammonium hydroxide (aqueous NH_3), 104
Amphoteric oxides, definition, 124
Amylose, β, starch indicator, 458
Anhydrides, of acids, 121
Anions, definition, 118
Antimony
  determination of (problems), 445, 486
  tartaric acid complexes of, 499
Arrhenius theory, acids, bases, 87
Arsenic
  colorimetric determination, 680
  coulometric determination, 644, 653ff
  electrogravimetric determination, 649ff
Arsenic acid, iodide oxidation by, 508
Arsenious acid
  reaction with iodate, 407, 508
  reaction with iodine, 450ff
  reaction with iodine monochloride, 407
  solutions, preparation of, 414, 704
Arsenious oxide, as a primary standard, 405, 439ff, 451ff
Asbestos filters, 200ff, 704
Ash content of paper filters, 198
Autocatalytic reactions, definition, 399
Average deviation, 31
Azide determination, using Ce(IV) 432

β-amylose (starch), indicator, 458
Balances, 213ff
  accuracy of weights, 227
  chain-weight, 220
  constant load, 222
  construction, 215
  equal-arm, 215
  micro, 223
  point of rest, 229ff
  pulp, 180
  quartz-fiber, 223
  sensitivity of, 217, 234
  single-pan, 222
  top loading, 180
  torsion, 223
  trip, 180
  triple-beam, 180
Barium manganate, in permanganate methods, 401, 492
Barium sulfate
  coprecipitation
    of chlorides and nitrates (table), 296
    of iron, 293ff
    minimization of, 295
  gravimetric method using, 291ff
  solubility, 47
    in acids, 293
    common-ion effect, 47
    H^+ concentration effect, 48
Base
  Arrhenius definition, 87
  Brønsted-Lowry definition, 89
  ionization constants of (table), 683
  Lewis definition, 91
Basic acetate separation, 155
Benzoic acid
  as a primary standard, 405
  for sodium hydroxide, 548
Benzoic acid-benzoate buffer, 156
Bismuth
  gravimetric determination, 302
  titrimetric determination, 387
  use of, in redox methods, 500
Blank, use of to minimize errors, 29
Borax, as a primary standard, 553
Bromine
  in coulometric methods, 642, 649ff
  standard solutions, 509
  in substitution reactions, 510
Brønsted-Lowry theory, 90, 104
Buffer systems, 99ff
  acetic acid-acetate, 99
  benzoic acid-benzoate, 156
  carbonic acid-hydrogen carbonate, 454
  in iodometric methods, 454ff
  in *p*H controlled separations, 136
  sulfuric acid-hydrogen sulfate, 479
Buoyancy corrections, 225ff
  for gravimetric solutions, 316
Burets
  cleaning, 331
  drainage time (figure), 329
  outflow time (table), 329
  stopcocks, 331
  volumetric, calibration of, 328ff
  weight, use of, 335ff
Burners, 163ff, 280

Cadmium
  ammonia complexes, 605
    table, 111
  EDTA complexes, 112, 605ff
  as a reductant, 497
Calcium
  carbonate
    EDTA standardization, 614
    in Mohr titration, 362
  content, in limestone, 272, 427
Calculations, approximations in, 341
Calibration
  glassware, 319ff
    buret, 328ff
    coefficient of expansion, 320
    flask, 321

Calibration (*continued*)
limits of error (table), 323
transfer pipet, 324ff
weights, 227, 239ff
Calomel electrode, 528
Carbon dioxide
hydration of, 103
solubility, 540
in basic solutions, 539
Carbonate
correction for, 546
mixture of, analysis of, 560
in sodium hydroxide, 539
Carbonic acid
as aqueous CO_2, 103
buffers, in iodometric methods, 454ff
Catalysts
iodine monochloride, 407
manganous ion, 399
osmium tetroxide, 439
in oxidation-reduction reactions, 408
manganese compounds, 504
Cation exchange resin, 593
Cations
definition of, 118
determination by ion-exchange, 593ff
Ceric sulfate solutions
preparation, 436
standardization, 439ff
Cerimetric methods, 432ff
antimony, determination of (problem), 445
azide, determination of, 432
iron, determination of, 442ff
Charge balance, 516
Chelate compounds, 112
Chemical equilibrium, law of, 39
Chlorate
oxidation of iodide, 311
oxidation of manganese, 510
Chloride
complexes
of copper, 106
of iron, 375, 421
of silver, 54, 250, 346
coprecipitation with $BaSO_4$, 296
determination of
gravimetric, 285ff
Mohr, 60ff, 356ff
Volhard, 377ff
oxidation of, by ferrate, 420
Chromate
determination of, iodometric, 464
dichromate equilibrium, 63, 358
titration indicator, Mohr method, 59ff, 349
Chromatography, 151ff
Chromic acid, cleaning solution, 178, 318
Chromous, standard solution, 494
Cleaning solution, chromic acid, 178, 318
Clear point, in precipitations, 288, 347
Cobalt determination, 679
Colorimetry (see Spectrophotometry), 656ff
Common-ion effect, on solubility, 46
Complex ions
dissociation constants (table), 108, 111, 608, 685
in electrogravimetric methods, 627ff
in ionization reactions, 106ff, 604ff
in precipitation, 137, 249ff
Composite solutions, preparation, 704
Concentrated acids
analysis of, 564
dilution of, 188
fuming procedures, 482
weight percent (table), 21
Concentration polarization, 620
Concentration units (table), 17
Confidence limits, 32
Conjugate base, acid, 89
Copper
adsorption on stannic oxide, 478
alloys, 477
ammine complexes, 114
electrogravimetric determination, 635ff
ferrocyanide, use as indicator, 350
halides, solubility of, 345
hydrated ions, 106, 114, 478
iodometric determination of, 477ff
as a primary standard, 471
as a reductant, 498
thiocyanate, and cuprous iodide, 480
Coprecipitation 259ff 259ff
by adsorption, 259ff
of chlorides and nitrates (table), 296
compound formation, 264
of iron, with $BaSO_4$, 293ff
mechanical inclusion, 265
minimization of, 268, 295
mixed crystals, 264
of Mn(II) with MnO_2, 430
variables, 266ff
Coulometric methods, 640ff
arsenic determination, 644, 653ff
bromine, use in, 642, 649ff
constant current, 643
controlled potential, 642
endpoints, 647
indirect methods, 642
instrumentation, 645, 651
Crucibles
filtering, 204ff

Gooch, 200
platinum filtering, 205
porous porcelain, 205
sintered, 204
Cupferron, organic precipitant, 142
Cyanide
complexes (table), 111
Liebig determination, 110, 369
as a masking agent, EDTA methods, 609

Debye-Hückel theory, 49, 57
Decomposition potential, 624
Density of water, air (tables), 701
Desiccators, 175ff
Determinate errors, 28
Dichromate
chromate equilibrium with, 63, 358
in determination of iron, 395
preparation of standard solution, 474, 493
in thiosulfate standardization, 472ff
Dimethylglyoxime, precipitant, 142
Diphenylamine indicator, 412
Displacement reactions, 98, 526, 556ff
Disproportionation, of iodine, 451ff
Dissolving of sample (table), 674
Distillation, 143ff
Distilled water, use of, 207
Distribution
constant, 41ff
liquid-liquid, 143ff
Double salt formation, coprecipitation, 264
Droppers, construction, 167ff

Ebullition tube, 191
EDTA, 112, 604ff
complexes, 604ff
of cadmium, 605
table of, 608
determination of nickel, 613ff
solutions
preparation, 611
standardization, 611ff, 614
storage, 612
titration indicators, 610
Effective concentration (activity), 49
Electrochemical cell, 75
Electrode potentials, 75, 80
Electrode (half-cell) reactions, 74ff
Electrogravimetric methods
apparatus, 633
constant current methods, 643
controlled potential methods, 642
copper determination, 635ff
deposit characteristics, 630ff
equilibrium decomposition potential, 618
Electrolysis, in separations, 150
Electrolytic methods
coulometric (*see* Coulometric), 640ff
gravimetric (*see* Electrogravimetric), 617ff
Electromagnetic radiation, 656ff
Electronegativity, 118ff
En (ethylenediamine), 111
Endpoint
acid-base methods, 599ff
coulometric methods, 643, 647
definition, 60
errors, 64ff, 375ff
iodometric methods, 458
precipitation methods, 59ff, 110, 349
redox methods, 409, 443, 508
Equilibrium constants, 39
displacement, 98
dissociation, for complexes (table), 685
distribution, 41ff
exchange, 594
hydrolysis, 97
ionization, 45
of acids, bases (table), 683
pseudo, 101ff
solubility product, 45ff
table of, 687
from standard potentials, 81
water, 91ff, 97
Equivalence point
in acid-base methods
strong acid-strong base, 92
strong acid-weak base, 94
weak acid-strong base, 95
calculation of, from potentials, 392
definition, 60ff
for polyprotic acid titrations, 566ff
Equivalents
in metathetical reactions, 13
in redox reactions, 14
Eriochrome Black T (EDTA titration), 610
Errors, 27ff
absolute, 27
approximation, 341
calibration (table), 323
determinate, indeterminate, 28
evaluation, 30
instrumental, 29
method, 29
parallax, 330
personal, 29
random, 29
reagent, 29
relative, 27
spectrophotometric, 668
systematic, 28
technique, 29
titration, 64ff, 375

Errors (*continued*)
  weighing, 223ff
Ethylenediamine ferrous sulfate, Ce(IV) standardization, 444
Ethylenediamine tetraacetic acid (*see* EDTA), 112
Evaporation technique, 143, 190
Experimental data, evaluation of, 22ff
Exponential notation, 23

Faraday's laws, 644
Ferrate, oxidation of chloride, 420
Ferric chloride extraction, ether, 146
Ferric thiocyanate complex
  formula for, 349
  Volhard titration endpoint, 108ff, 371ff, 374ff
Ferricyanide, as a primary standard, 471
Ferrocyanide
  determination of zinc, using, 349
  permanganate standardization of, 428
  standard solutions of, 387
Ferrous ammonium sulfate, as a primary standard, 405
Filters
  optical, 659
  platinum, 205
  porous porcelain, 205
  precipitation, 274
  sintered glass, 204
Filters, preparation, use, 274, 282
  asbestos fiber, 200ff
  glass fiber, 204
  paper, 197
Filtration, 188, 196ff, 275ff
Flasks, 192
Fluoride complexes
  of iron, 418
  of the silver group, 345
Formal concentration, 15
Formal half-cell potentials, 79, 434, 689

Gas chromatography, 153
Gay-Lussac titration, for silver, 7, 347
Glacial acetic acid analysis (problem), 564
Glass electrode, 528, 578
Glass expansion, 320
Glass manipulation, 165, 172
Gold, properties of compounds, 344
Gooch crucibles, 200
Gravimetric buret, 315, 335ff
Gravimetric factor, definition, 302
Gravimetric methods (*also see* Electrogravimetric methods)
  definition, 58, 315
  electrolytic, 617
  general principles, 247
  operations, 274
  precipitates used in (table), 271

Half-cell potentials
  analytically important (table), 434, 689ff 496, 499, 502, 507
  convention used, 80
  definition, 76
  effect of concentrations on, 76ff
  formal, 79
  standard, 76
Henry's law, 43
Heterogeneous phase reactions, 39
History of chemistry, 1ff
Homogeneous phase reactions, 39ff
Homogeneous solutions, precipitation from, 257
Hydration
  of ammonia, 104
  of anhydrides, 101ff
  of carbon dioxide, 103
  of cations, 106, 114, 478
  of sulfur dioxide, 101
Hydration, of anhydrides, 101ff
Hydrazine, iodometric determination (problem), 486
Hydrochloric acid
  constant boiling, as a primary standard, 548, 562
  standardization
    against borax, 553
    against iodate, 554
    gravimetric, 301
    against solium carbonate, 556ff
    against sodium oxalate, 563
Hydrogen carbonate buffers, 454
Hydrogen electrode, 76, 527
Hydrogen ion
  concentration, determination of
    indicators, 522, 529ff
    *p*H meters, 529, 578ff
  effect of, on solubility of precipitates, 48, 251
  hydronium ion, 88
Hydrogen peroxide reactions, 502ff
Hydrogen sulfide
  as a precipitant in separations, 138ff
  solubility in water, 56
Hydrolysis
  of Ce(IV), 434
  constant, 97
  of salts in aqueous solutions, 97ff
Hydronium ion, 88
Hydrous oxides, definition, 126
Hydroxide

complexes, 375
precipitations, pH values, (table), 135
Hydroxyquinoline (8-1), precipitant, 143

Indeterminate errors, 28
Indicators
acid-base, 522, 529ff
preparation of, 705
coulometric, 647
external, 349
internal, 348
irreversible, 652
metallochromic, 610, 706
mixed, 558, 706
potential, 409ff, 706
reactions, 59
Induced reactions, 311
oxidation of chloride by MnO_4^-, 419
Instrumental errors, 29
Interionic forces, 46, 49
Iodate
as H_3AsO_3 oxidation catalyst, 407, 508
in hydrazine determination (problem), 486
oxidation of iodide, 507
primary standard for acids, 554
in thiosulfate standardization, 475ff
Iodide
oxidation by arsenic acid, 508
oxidation by chlorate, 311
oxidation by oxygen, 462
reaction with cupric ion, 479ff
reaction with dichromate, 472ff
reaction with iodate, 463, 475ff
silver, Liebig endpoint, 110
Iodine
determination of iron, 457
disproportionation, 451ff
half-cells, 449
oxidation of thiosulfate, 311, 459ff
reaction with H_3AsO_3, 450ff
solid-liquid-gas phases, 42ff
solubility, 447, 448
vapor pressure, 448, 464, 474
Iodine monochloride
endpoint, 508
use as a catalyst, 407ff
Iodometric equilibria, 450ff, 457
Iodometric methods, 466ff
antimony determination (problem), 486
common applications (table), 484
copper determination, 477ff
endpoints, 457
hydrazine determination (problem), 486
hydrochloric acid standardization, 554
iron determination, 457
oxygen error, 448, 462
photochemical effects, 448
thiosulfate, standardization of, 471
Ion exchangers, 154, 593
Ion-exchange methods
alkali metal determination, 599ff
cation determination, 593ff
water deionization, 596

Ionic bonds, definition, 118
Ionic strength, definition, 50
Ionization constants
definition, 45
tables, 108, 111, 608, 683, 685
Ionization reaction, 86ff
Iron
complexes,
chloride, 375, 421
fluoride, 418
hydroxide, 375
phosphoric acid, 421
sulfuric acid, 421
thiocyanate, 375
coprecipitation with $BaSO_4$, 293
determination of, using
Ce(IV), 442
dichromate, 395
iodine, 457
permanganate, 78, 391ff
vanadate, 512
equilibrium with silver, 77ff
Fe(II)-Fe(III) half-cell, 393
ferric reduction by tin(II), 417
ores, dissolution of, 417
orthophenanthroline complex, 409ff
as a primary standard, 405, 439, 445
pyrite, determination of sulfur in, 302
standard solutions of Fe(II), 495
Irreversible reactions, definition, 399
Isoelectric point, amino acids, 583
Isomorphism and coprecipitation, 265

Jones reductor, 495

Kjeldahl method, nitrogen, 564

Lanthanides, oxidation states, 433
Law
chemical equilibrium, 39
Henry's, 43
mass action, 39
phase distribution, 41ff
Le Chatelier's principle, 40
Lead, electrolytic determination, 629
Lead dioxide, as an oxidant, 501
Leveling effect, on acid-base strength

of nonaqueous solvents, 104
of water, 91
Lewis theory, acids, bases, 91
Liebig titration, of cyanide, 110, 369
determination of silver, using, 388
Ligands, in complexes
chelate, 112
inorganic, 106ff
organic, 111ff
polydentate, 111
Limestone, calcium content of (problems), 272, 427
Liquid phase distributions, 41
Liter, definition of, 12

Manganate, and permanganate, 401, 493
Manganese
determination, Williams method, 427
hydrogen peroxide reactions with, 503ff
Volhard determination of, 400
Manganous ion
coprecipitation with MnO_2, 430
as redox catalyst, 399, 420
Mass (vacuum weight) 213, 319
Mass balance, 516
Mass-action law, 39
Mean, definition of, 31
Mean activity coefficient, 50
Measuring pipet, 324
Mechanical inclusion, coprecipitation, 265
Median, definition of, 31
Meniscus, reading of, 170, 330
Mercury
chloride complexes, 107
chloride, in iron determination, 419, 442
halide complexes (table), 108
overvoltage on, 495
as a reductant, 497
Volhard determination of, 107, 384, 385
Metallochromic indicators, 610
Metalloids, definition, 117
Metals, definition, 117
Metastannic acid, 478
Metathesis, 13
Meter, definition of, 11
Method errors, 29
Methyl orange
neutralization indicator, 522, 543
potential indicator, 652
Metric system, 11ff
Mixed crystals, coprecipitation, 264
Mixed indicators, 558
Mohr titration, of chloride, 59ff, 356ff
Molal concentration, 16
Molar concentration, 15
Mole fraction, 42
Monochromators, 659
Murexide (EDTA indicator), 610

Nernst equation, 76ff, 82
Neutralization, 87, 90
definition, 87, 90
hydrogen ion changes in, 516, 567
indicators, 529ff
mechanism, 531
Nickel
determination of
using EDTA, 613ff
using silver, 389
spectrophotometric, 679
as a reductant, 498
Nitrate, coprecipitation with $BaSO_4$, 296
Nitric acid, reaction products of, 383
Nitrogen, Kjeldahl method (problem), 564
Nitrous acid
oxidation of iodide, 482
oxidation of thiocyanate, 375
Nonaqueous solutions, acid-base reactions in, 104
Nonmetals, definition, 117
Normal concentration, 15
Notebook, use of, 162
Numbers, 23ff
exponential notation, 23
percent uncertainty, 26
reliability, 23

Occlusion and coprecipitation, 265
Oleum (fuming sulfuric acid) analysis of (problem), 565
Optical methods (spectrophotometry), 656ff
Organic precipitants, 142
Orthophenanthroline Fe-complex, 409ff
Osmium tetroxide, catalyst, 446
Overvoltage
definition, 621
table, 623
Oxalate
as a primary standard, 405
reaction with permanganate, 399, 400
Oxalic acid as a primary standard
for permanganate, 399
for sodium hydroxide, 548
Oxidation state, 14, 70ff
Oxidation-reduction reactions, 14, 39, 69ff
catalysts in, 407
and half-cell potentials, 69ff
in titrimetric methods, 489ff
Oxides
acidic, 125
amphoteric, 126
basic, 125

hydrous, 126
ionic and covalent, 120
periodic properties, 127
precipitation of (table), 135
separation by precipitation, 134ff
Oxygen acids, strengths of, 92, 122
Oxygen error, 462
Ozone, as an oxidant, 501

Paper chromatography, 153
Paper filers, 197ff
Parallax errors, 330
Paranitrophenol indicator, 531
Perchloric acid, as an oxidant, 510
Periodic acid, as an oxidant, 509
Periodic properties
of elements, 116ff
in aqueous solutions, 697ff
in basic solutions (figure), 128
used in separations, 128ff
oxides, 127
sulfides (figure), 140
Permanganate, 391ff
chloride oxidation, by 419
determination of
ferrocyanide (problem), 428
iron, 417
manganese (problem), 427
peroxide (problem), 511
endpoint
concentration, 392
indicators, 411
stability, 398
half-cells, 392, 397
H^+ concentration, 398
iron determination, 417ff
photochemical effect, 402
solutions
preparation, 402
stability, 402
standardization
against arsenious oxide, 406ff
against ferrous ammonium sulfate, 405
against oxalic acid, 399
against potassium ferrocyanide, 428
against sodium oxalate, 399
titration curves, 394ff
titrations, and barium manganate, 401, 492
types of reactions, 491ff
Peroxide
catalytic decomposition, 504
determination by permanganate, 511
Peroxydisulfuric acid, 503
oxidation of carbon in steel, 675
oxidation of manganese, 506
*p*H
definition and use, 93ff
determination of, 527ff
meters, 529, 578ff
papers, 206
Phase equilibria, 40ff
Phase separations
gas-liquid, 41, 44, 143ff
gas-solid, 143ff
liquid-liquid, 41, 145ff,196
solid-liquid, 44, 133ff, 195
Phenolphthalein, 522, 546
Phosphoric acid
complexes of iron, 421
determination of, by NaOH, 572
Zimmermann-Reinhardt solution, 424
Photochemical effects
iodometric methods, 448
permanganate, 402
silver chloride, 287
Pipet
calibration, 324
tolerances, 326
types, 325
Platinum-ware, use and care of, 280, 283, 300
Point of rest
definition of, 229
determination
long swing method, 231
short swing method, 233
Polarization, concentration, 620
Policeman, 277
Polydentate ligands, 111
Polyprotic acids, 566ff
EDTA, 607
*p*H at equivalence point, 568
titration curves, 567
Potassium bromate, as a primary standard, 471, 475
Potassium dichromate
as a primary standard, 405
standard solution, 474
Potassium ferricyanide, as a primary standard, 471
Potassium ferrocyanide, as a primary standard, 428
Potassium hydrogen iodate, as a primary standard
for sodium hydroxide, 548
for thiosulfate, 471, 475
Potassium hydrogen phthalate, as a primary standard, 551
Potassium iodate, as a primary standard
for strong acids, 554
for thiosulfate, 475

Potential
  decomposition, 624
  effect of concentration on, 76
  formal, 79, 689ff
    table of, 689ff
  indicator, 409, 652
  mediator, 84
  sign convention, 80
  standard, 76, 689ff
Potentiometer, 396
Potentiometric titrations, 395ff
Precipitates, 188
  composition and stability of, 269ff
  drying procedures, 269ff, 279
  table of, 271
  transfer of, 277
  washing, 275ff
  weighing, 280
Precipitation, 58ff, 247ff
  from homogeneous solution, 257
  mechanism of, 254ff
  methods, 58ff, 342ff
  supersaturation and, 253
Precision, 27ff
Primary standards, 310, 404
  ammonium ceric nitrate, 436
  arsenious oxide, 405ff, 439ff
  benzoic acid, 405
  borax, 553
  calcium carbonate (problem), 614
  ferrous ammonium sulfate, 495
  oxalic acid, 399, 548
  potassium bromate, 471, 475
  potassium dichromate, 405, 471, 472
  potassium ferricyanide, 471
  potassium ferrocyanide (problem), 428
  potassium hydrogen iodate, 471, 475, 548
  potassium hydrogen phthalate, 405, 551
  potassium iodate, 471, 475, 507, 554
  sodium carbonate, 405, 555ff
  sodium oxalate, 405
  sulfamic acid, 548
  table of, 706
Pseudo equilibrium constants, 101ff
Purity of reagents, 208
Pycnometer, calibration of (problem), 340
Pyrite, iron ore, determination of sulfur in, 302

Q, activity coefficient, 77
Q values, in error analysis, 34ff
Qualitative analysis, 4, 128ff
Quartz-fiber microbalance, 223

Random errors, 29
Range, definition of, 31
Reagent bottles, 186, 192, 310
Reagents,
  concentrations, 702
  drying of, 206
  preparation, 702
  purity of, 208
Redox reactions, 14, 69ff
Reference half-cell, 76
Relative errors, 27
Replacement, displacement reactions, 98, 526, 556ff
Reprecipitation, 269
Resins, exchange, 593

Safety, in laboratory procedures, 161ff
Salt bridge, 75
Sampling procedures, 181ff
Separations
  gas-liquid, 41, 44, 143ff, 195
  gas-solid, 143ff
  liquid-liquid, 41, 145ff, 196
  multiple operations, variants, 148ff
    chromatography, 151ff
    continuous extraction, 149
    electrolysis, 150
    fractional distillation, 148
    ion-exchange, 154
  qualitative systems, 128ff
  by selective precipitation
    oxides, 128, 134ff
    sulfides, 138ff
  by selective precipitation, oxides, 134ff
  solid-liquid, 44ff, 133ff, 195, 196ff
Significant figures, 24ff
Silver
  ammine complexes, 251
  chloride, 53ff
    complexes, 54, 60, 250, 346
    in gravimetric determination, 285
    in titration, 60ff
    photochemical effects, 287
    solubility, 54, 249
    un-ionized molecules, 60
  chromate, 46, 59
    as an indicator, 59ff, 349
    solubility of, 63
  cyanide, 110
  Gay-Lussac determination, 347
  halides
    complexes, 346
    coprecipitation of, 262
  iodide
    adsorption of silver (I) (figure), 263
    in Liebig endpoint, 110
  in Liebig determination, 110, 388
  nitrate, standard solutions, 351ff

as a reductant, 497
salts, solubility (table), 360
sulfate, solubility (figure), 50
thiocyanate, 78, 371
titrations
curves, 61
equivalence point phenomena, 62
Volhard determination of, 371ff
Soda ash analysis, 563
Sodium carbonate, as a primary standard, 405, 556ff
Sodium hydroxide, standardization
against benzoic acid, 562
against constant boiling HCl, 548
against potassium hydrogen phthalate, 551ff
against sulfamic acid, 548
Sodium oxalate, as a primary standard, 405
for Ce(IV), 439
for permanganate, 399
Sodium succinate, 552
Solid solutions in coprecipitation, 264
Solubility
barium sulfate, 47ff, 292
carbon dioxide, 540
gold halides, 344
hydrogen sulfide, in water, 56
iodine, 447
silver salts (table), 360
silver sulfate (figure), 50
thallous chloride (figure), 53
Solubility effects, 248
common ion, 47
complex ions, 52, 249
coprecipitation, 266
hydrogen ion, 251
particle size, 254
solvent, 252
temperature, 252
time, 253
Solubility product, 45ff
silver halides (table), 346
table of, 687
Solvation, definition, 88
Solvent, action on samples (table), 674
Solvent extraction, 44, 145ff, 147
Species balance, 516
Spectrophotometric methods, 671
arsenic (problem), 680
colbalt (problem), 679
manganese, 673ff
nickel (problem), 679
rhenium (problem), 679
Spectrophotometry, 656ff
absorption, 657ff
Beer's law, 662
Lambert's law, 660
electromagnetic radiation, 656ff
emission, 657
errors, 668
filters, optical, 659
instrumentation, 665
monochromators, 659
sample preparation (table), 674
Standard deviation, 32
Standard half-cell potentials, 76
equilibrium constants, from, 81
Standard hydrogen electrode, 76
Standard solutions, 17ff, 310, 316
arsenious acid, 414
bromine, 509
Ce(IV), 439ff
chromium(II), 494
direct, indirect preparation, 351
hydrochloric acid, 553ff
iron(II), 495
potassium dichromate, 493
potassium permanganate, 402ff, 491ff
silver nitrate, 351ff
sodium hydroxide, 548ff
table of, 705
thiocyanate, 374
thiosulfate, 471
titanium(III), 495
transferring, 326
vanadate, 494
Stannic oxide, 478
Stannous chloride
ferric ion reduction, 417
reaction with mercuric chloride, 419
reaction with permanganate, 312
Starch, iodometric indicator, 458, 470, 704
Starch substitute, 471
Statistical treatment of data
average deviation, 31
confidence limits, 32
mesn, 31
median, 31
normal distribution, 30
probability, 32
Q values, 34ff
range, 31
rejection
Q test, 34
$2.5\bar{d}$ rule, 34
$4\bar{d}$ rule, 34
standard deviation, 32
true value, 27, 32
Stirring rods, 167
Stoichiometric reactions, definition, 311

Sublimation, 143
Substitution method of weighing, 225
Sulfamic acid, as a primary standard, 548
Sulfate
  complexes of Ce(IV), 435
  complexes of iron, 421
  gravimetric determination of, 291ff
Sulfides
  in *p*H controlled separations, 138ff
  in qualitative analysis, 505
  precipitation of (table), 139
Sulfur dioxide
  hydration of, 102ff, 122
  as a reductant, 506
Sulfuric acid
  gravimetric standardization of (problems), 272, 304
  hydrogen sulfate buffer, 479
  method
    of fuming, 482
    of diluting, 482
  from sulfur trioxide, 102
  Zimmermann-Reinhart solution, 424
Sulfurous acid, from sulfur dioxide, 102ff
Systematic errors, 28

Technique errors, 29
Tetrathionate ion, structure, 460
Thallous chloride solubility, 52
Thin-layer chromatography, 152
Thiocyanate
  in copper(II)-iodide reaction, 480
  nitrous acid oxidation of, 375
  solutions
    preparation of, 372
    standardization, 374
  titration of mercury, 385
  titration of silver, 384
Thiosulfate
  oxidation state of sulfur in, 71
  reaction with iodine, 459ff, 468ff
  solutions, 467ff
  standardization
    against bromate, 471
    against copper, 471
    against dichromate, 472ff
    against ferricyanide, 471
    against iodate, 475ff
  structure of, 461
Thyodene, starch substitute, 471
Titanium(III), standard solution, 495
Titration, 59, 309ff
  blank, use of, to minimize errors, 29
  curves
    acid-base monoprotic, 521
    amino acid titrations, 587
    permanganate against iron, 394
    polyprotic acids, 567
    precipitation reactions, 67
    redox reactions, 396
  definition, 309
  error calculations, 64ff, 375ff
  gravimetric, 310
  volumetric, 310
Titrimetric methods (*see* Specific ion or Reagent)
  definition, 309
  general principles, 309
  operations, 317
Transfer pipet, calibration of, 324
Transferring
  precipitates, 277
  solids, 180
  solutions, 184
  standard solutions, 246
Transition range, acid-base indicators, 522
Transposition method of weighing, 225
Tri-iodide ion, 447
True value, definition, 27, 32

Un-ionized compounds (*see* Complex ions), 39, 53, 604ff
  inorganic ligands, 106
  mercuric thiocyanate, 107ff
  organic ligands, 111ff
  silver chloride, 60
Units, 11ff
  chemical, 12
  concentration, 11
  metric, 15

Vanadate
  as an oxidant, 512
  standard solution of, 494
Volatilization, 143ff
Volhard titration
  chloride, 377ff
  manganese, 400, 430
  mercury, 107ff, 384ff
  silver, 371ff
Volumetric glassware
  buret, 328
  calibration of, 319
  coefficient of expansion, 320
  flask, 192
  pipet, 324
Volumetric procedures, definition, 7, 314

Wash bottles, 171
Water
  deionization, 596
  density of (table), 701

dissociation constant of, 91ff, 97
viscosity of, 252
Weighing (*see* Balances), 213ff
buoyancy corrections, 225ff
difference method, 183
direct method, 183
errors, elimination of, 223
principles, 213, 217ff
transposition method, 225
Weighing bottles, 177
Weight buret, use of, 315, 335ff
Weight calibration, 227, 239ff
Weight-formality, 16
Weight-molarity, 16
Weight-normality, 16

Zeolites and ion-exchange resins, 154
Zimmermann-Reinhardt solution
in iron determination, 421, 424
preparation, 704
Zinc
coprecipitation of, 263
in EDTA standardization, 611
potassium ferrocyanide, 349
as a reductant, 495
Zwitter ion, in amino acids, 582